HVACR

Level 2

SIXTH EDITION

Pearson

NCCER National Center for Construction Education and Research

NCCER
President and Chief Executive Officer: Boyd Worsham
Vice President of Innovation and Advancement: Jennifer Wilkerson
Chief Learning Officer: Lisa Strite
Senior Manager, Curriculum Development: Chris Wilson
Production Manager: Graham Hack
HVACR Project Manager: Patrick Bruce
Technical Writing Manager: Gary Ferguson
Technical Writer: Troy Staton
Art Manager: Bree Rodriguez
Production Artist: Judd Ivines, Chris Kersten
Permissions Specialists: Adam Black, Sherry Davis
Managing Editor: Hannah Murray
Project Coordinators: Colleen Duffy, Milagro Maradiaga

Pearson
Commercial Product Manager-Associations: Andy Dunaway
Senior Digital Content Producer: Shannon Stanton
Content Producer: Alexandrina Wolf
Associate Project Manager: Monica Perez-Kim
Designer: Mary Siener
Rights and Permissions: Jenell Forschler
Composition: Integra Software Services
Printer/Binder: Lakeside Book Company
Cover Printer: Lakeside Book Company
Text Typefaces: Palatino LT Pro and Helvetica Neue

Cover Image
Standard Heating and Air discovered a shorted coil by first disconnecting the control wires from the contactor. By measuring the resistance through the contactor coil, they were able to see that the resistance was too low. Replacing the contactor solved the problem.

10 9 8 7 6 5 4 3 2 1

7 2025

Pearson NCCER National Center for Construction Education and Research

ISBN-10: 0-13-821016-0
ISBN-13: 978-0-13-821016-8

PREFACE

To the Trainee

The heating, ventilating, air-conditioning, and refrigeration (HVACR) industry doesn't just greatly impact the comfort of our indoor spaces. It also preserves the quality of our food supply and supports manufacturing at many levels. The spread of COVID-19 in recent years elevated the importance of indoor air quality and filtration. HVACR technicians install and maintain our refrigeration systems, ensuring our medical institutions, restaurants, and food processing facilities are maintained to the required standards. HVACR technicians serve a vital role in keeping many industries operational and contributing to the business sectors they serve.

Competent technicians are required to install, maintain, and troubleshoot HVACR systems. They must be able to read and interpret drawings or other specifications to maintain these systems. Technicians understand air movement, temperature and pressure relationships, and heat transfer applications. They often need basic knowledge in other crafts such as sheet metal, welding, pipefitting, and electrical practices to succeed in their chosen field.

The increasing advancement of technology prompts employers to recognize the importance of continuous education and keeping up with the latest equipment and techniques. Technical training and apprenticeship programs provide employers with skilled craftworkers who are well prepared to lead the way and confront new challenges.

NCCER's HVACR program has been designed by highly qualified subject matter experts. The four levels present an apprentice approach to the HVACR field, including both theoretical and practical skills essential to your success as an HVACR technician.

We wish you the best as you begin an exciting and promising career. This newly revised curriculum will help you enter the workforce with the knowledge and skills needed to perform productively in any facet of the HVACR craft.

New with *HVACR Level 2*

NCCER is proud to present *HVACR Level 2*. This new edition includes updates to the curriculum that will help engage you and provide the best training possible. These updates include video resources that depict relevant tasks that HVACR technicians must routinely complete.

All modules have received updated art and tables. Module 03302 received enhanced coverage of oil distributions systems when multiple compressors are employed. Coverage of screw compressor and centrifugal compressor internal operation and capacity control were improved. The new content also includes ASHRAE refrigerant classifications and information related to EPA 608 requirements. Additional updates include current information about the ENERGYSTAR program from the Department of Energy.

We wish you success as you progress through this training program. If you have any comments on how NCCER might improve upon this textbook, please complete the User Update form using the QR code on this page. NCCER appreciates and welcomes its customers' feedback. You may submit yours by emailing **support@nccer.org**. When doing so, please identify feedback on this title by listing #*HVACRL2* in the subject line.

Our website, **www.nccer.org**, has information on the latest product releases and training.

SCAN ME

NCCER Standardized Curricula

NCCER is a not-for-profit 501(c)(3) education foundation established in 1996 by the world's largest and most progressive construction companies and national construction associations. It was founded to address the severe workforce shortage facing the industry and to develop a standardized training process and curricula. Today, NCCER is supported by hundreds of leading construction and maintenance companies, manufacturers, and national associations.

The NCCER Standardized Curricula was developed by NCCER in partnership with Pearson, the world's largest educational publisher.

Some features of the NCCER Standardized Curricula are as follows:

- An industry-proven record of success
- Curricula developed by the industry, for the industry
- National standardization providing portability of learned job skills and educational credits
- Compliance with the Office of Apprenticeship requirements for related classroom training (*CFR 29:29*)
- Well-illustrated, up-to-date, and practical information

NCCER maintains a secure online database that provides certificates, digital credentials, transcripts, and wallet cards to individuals who successfully complete programs under an NCCER-accredited organization or through one of NCCER's self-paced, online programs. This system also allows individuals and employers to track and verify industry-recognized credentials and certifications in real time.

For information on NCCER's credentials, contact NCCER Customer Service at 1.888.622.3720 or visit **https://www.nccer.org**.

Digital Credentials

Show off your industry-recognized credentials online with NCCER's digital credentials!

NCCER is now providing online credentials. Transform your knowledge, skills, and achievements into digital credentials that you can share across social media platforms, send to your network, and add to your resume. For more information, visit **www.nccer.org**.

STANDARD HEATING & AIR CONDITIONING

Cover Image Provider

Established in 1930, Standard Heating & Air Conditioning stands as a family-owned and operated HVAC company based in the Twin Cities of Minnesota. Celebrating a cumulative experience in the heating business exceeding 470 years, the team remains dedicated to ongoing education and the promotion of trade awareness. This commitment is evident through their engagement with the community, which includes participation in charitable initiatives and serving on boards for local technical schools. Standard Heating recognizes their impact extends beyond the replacement of furnaces and air conditioners and embody their mission to "Make Lives Better" through the practice of their trade.

DESIGN FEATURES

Content is organized and presented in a functional structure that allows trainees to access the information where they need it.

Source: Shutterstock@william.hager

Objectives

Successful completion of this module prepares you to do the following:

1. Identify and describe the use of common maintenance materials and components.
   a. Identify and describe common gaskets and packing materials.
   b. Identify and describe common types of shaft seals.
   c. Identify and describe common types of bearings.
2. Describe the properties of common lubricants and how they are applied.
   a. Describe the properties of common lubricants.
   b. Explain how lubricants are selected and applied.
3. Identify and describe how to install various drive belts.
   a. Identify and describe various types of drive belts.
   b. Identify and describe how to align sheaves and calculate drive speeds.
   c. Explain how to install and apply the proper tension to drive belts.
4. Identify and describe basic inspection and maintenance procedures for selected equipment.
   a. Identify common environmental health hazards associated with HVACR maintenance activities.
   b. Describe common inspection and maintenance procedures for gas heating equipment.
   c. Describe common inspection and maintenance procedures for cooling and heat pump systems.
   d. Describe common inspection and maintenance procedures for various system accessories.

Performance Tasks

Under supervision, you should be able to do the following:

1. Lubricate a bearing using a grease gun.
2. Install, align, and adjust a drive belt.
3. Perform a periodic maintenance inspection on a gas furnace and document the inspection results.
4. Perform a periodic maintenance inspection on a cooling system or heat pump system in the cooling mode and document the inspection results.

Digital Resource for HVACR

Scan this code using the camera on your phone or mobile device to view the digital resources related to this craft.

SCAN ME

Overview

HVACR equipment maintenance requires technicians to make intelligent decisions regarding equipment performance and condition. Technicians must also choose the correct materials to use in each situation. Common maintenance items such as air filters, drive belts, and lubricants must be chosen and applied correctly. This module provides guidance in periodic maintenance procedures and the selection of appropriate maintenance materials.

Trainees can navigate *HVACR Level 2* using color coded tabs on the upper right hand corner.

The Objectives list the skills and knowledge trainees need in order to complete the module successfully.

The Performance Tasks give you an opportunity to apply your knowledge to real-world tasks.

Section Openers provide a visual organizational structure for the information. Objectives and Performance Tasks are broken out for each section.

1.0.0 Fasteners

Performance Tasks

1. Torque threaded hardware to a specific torque value.
2. Select the appropriate drill bit and install an anchor in brick or concrete block.

Objective

Identify the types, uses, and installation methods of fasteners.

a. Identify various types and uses of threaded fasteners.
b. Identify and explain how to use taps, dies, and screw extractors.
c. Explain how to install and torque fasteners to a specific value.
d. Identify and explain how to install various types of toggle and anchor bolts.
e. Identify and describe various types of non-threaded fasteners.

The rotational speed of a motor, measured in revolutions per minute (rpm), is determined by the number of stator windings and the frequency of the applied voltage. The speed at which the magnetic field rotates is referred to as its **synchronous speed**. Synchronous motors, those that rotate at precisely the same speed as the magnetic field, are relatively rare in HVACR equipment. They are most often used in applications where precise rotational speed is required, such as a timer or tape recorder.

Standard induction motors experience some *slippage*, or inefficiency, in the operation of the motor. Therefore, the actual speed of standard induction motors is about 95% to 97% of their synchronous speed.

Synchronous speed: The speed at which the magnetic field in a motor rotates, determined by the number of stator windings and the frequency of the applied voltage.

Trade Terms appear on the page adjacent to the text where they are first presented.

Step-by-step presentations and math equations help make the concepts clear and easy to grasp.

3.2.4 Current Imbalance on Three-Phase Motors

Current imbalance between any two legs of a three-phase motor should not exceed 10%. While a voltage imbalance will always produce some current imbalance, a current imbalance can easily occur without a voltage imbalance present. This can happen when an electrical terminal or contact set becomes loose or corroded, causing a higher resistance in one leg of power. Because current follows the path of least resistance, the current in the other two legs will increase. Calculate the current imbalance in a motor using the following steps:

Step 1 Measure the current to the motor on all three phases.

Step 2 Add the three values together, then divide by three to determine an average current. For example, if the three current readings are 54A, 48A, and 50A, the average would be 50.7A, rounded to one decimal place.

Step 3 Select the one current value of the three phases that deviates the most from the average. The 54A value is the farthest of the three from the average, and therefore is selected for the rest of the calculation.

Step 4 Determine the difference between the selected value and the average. In this example, the difference is 3.3A (54A – 50.7A = 3.3A). Then divide this value by the average current:

33A ı 50.7A average = 0.065A

Step 5 Multiply the result by 100 to determine the percentage of imbalance. This result is within the 10% tolerance for current imbalance:

0.065A · 100 = 6.5% current imbalance

SCAN ME

QR codes link trainees directly to digital resources that highlight current content.

Important information is highlighted, illustrated, and presented to facilitate learning.

Placement of images near the text description and details such as callouts and labels help trainees absorb information.

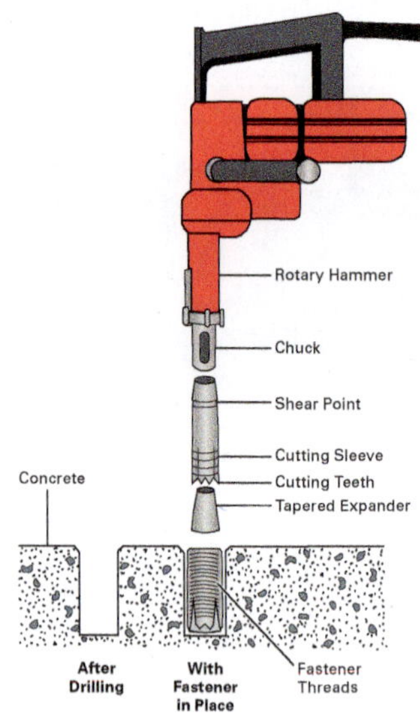

Figure 34 Self-drilling anchor.

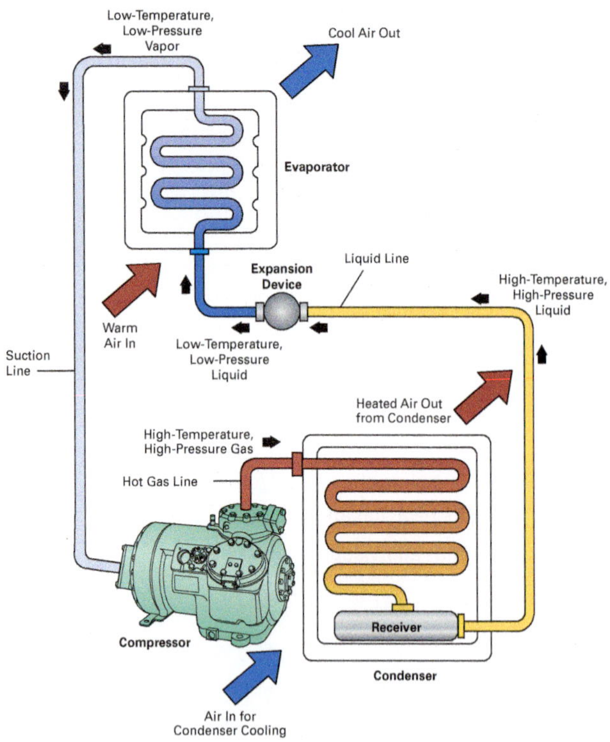

Figure 1 Typical refrigerant circuit.

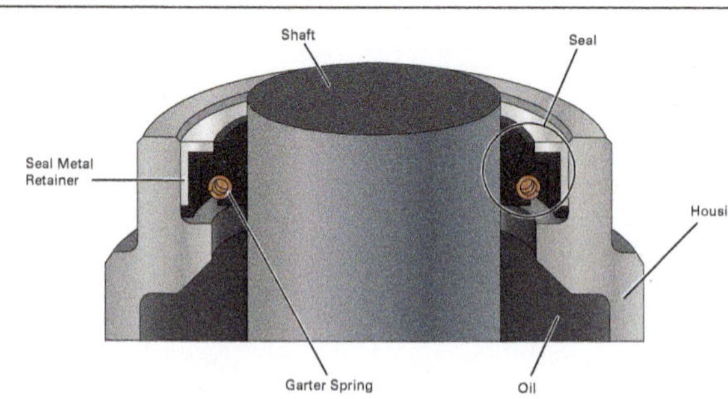

Figure 20 Oil seal, also known as a lip seal.

New boxes highlight safety and other important information for trainees. Warning boxes stress potentially dangerous situations, while Caution boxes alert trainees to dangers that may cause damage to equipment. Notes boxes provide additional information on a topic.

WARNING!

The three-phase transformer hookups shown here are for example only. Three-phase power transformer hookups, as well as their service and maintenance, must be done by properly trained, qualified electricians.

CAUTION

It is very important that capacitors be sized correctly. If you must install or replace them, use the capacitance value specified by the motor manufacturer. However, a capacitor with a higher voltage rating than the original can be used if necessary.

NOTE

Many of today's ductless split systems place the metering device in the outdoor unit, rather than at the evaporator coil inlet. This means that both the suction line and the liquid-vapor line leaving the metering device must be insulated between the indoor and outdoor unit.

Going Green

Semi-Hermetic Compressors

Semi-hermetic compressor housings are very durable and can be reused numerous times. They are either *rebuilt* or *remanufactured*. All the moving parts in a remanufactured compressor are typically replaced. A rebuilt compressor often retains some of its moving parts if the parts meet specifications. Many semi-hermetic compressors replaced each year are rebuilt or remanufactured units that are less expensive than fully new models.

When a semi-hermetic compressor is purchased, the failed compressor should be traded in against the replacement. If not, a *core charge* is applied to the invoice, charging the buyer for the value of the core, or housing. This charge is waived when a trade-in is offered.

Going Green looks at ways to preserve the environment, save energy, and make good choices regarding the health of the planet.

On Site

Suction Line Pitch

As the suction line leaves the evaporator, it should be pitched slightly downward to allow for proper drainage of oil and any liquid refrigerant to move away from the coil and TXV sensing bulb.

Think About It

ERV/HRV Operation

Air enters and exits a home in many ways and at many different times. Bathroom and kitchen exhaust fans, for example, operate intermittently and can exhaust more than 100 cfm of air per fan when in operation. Clothes dryers also send a significant amount of indoor air out of the home.

Now consider that an ERV/HRV consistently exhausts indoor air and replaces it with fresh air from outside. How might the various points of exhaust affect each other? How is the air pressure in the home affected? If you see a problem, how might you solve it in the most energy efficient way?

These boxed features provide additional information that enhances the text.

Did You Know?

The Great San Francisco Earthquake

In 1906, a devastating earthquake struck the city of San Francisco. Many of the city's water mains were broken, resulting in a loss of water pressure. With little or no water pressure available to fight the many fires that started after the earthquake, fires raged out of control, destroying the city.

Review questions at the end of each section and module allow trainees to measure their progress

1.0.0 Section Review

1. Spiral-wound gaskets are primarily used on _____.
   a. low-temperature, low-pressure systems
   b. high-temperature, high-pressure systems
   c. low-temperature, high-pressure systems
   d. high-pressure, low-temperature systems
2. Which of the following statements about oil seals is *correct*?
   a. Oil seals are designed for low-pressure applications.
   b. Oil seals are negative contact, meaning that they do not touch the shaft.
   c. Oil seals are made exclusively from soft metals.
   d. The lip of an oil seal should always face away from the machine.
3. The presence of darkened yet still oily grease on a failed bearing suggests that the *most likely* cause of failure is _____.
   a. oscillatory motion during operation
   b. too little lubricant
   c. high loads or contaminants
   d. excessive external heat

Module 03215 Review Questions

1. The pressure class of pipe flanges, in pounds per square inch, is determined by dividing the tested bursting point by a factor of 3.0.
   a. True
   b. False
2. A lantern ring is found in _____.
   a. compressor cylinders
   b. pump stuffing boxes
   c. mechanical seals
   d. hand valves
3. What percentage of an O-ring's diameter is typically compressed to form a reliable seal?
   a. 10%
   b. 25%
   c. 40%
   d. 60%
4. To create a leak-proof seal around a rotating shaft, mechanical seals usually rely on two perfectly machined faces made from _____.
   a. steel
   b. aluminum
   c. carbon
   d. lithium

NCCERconnect

This interactive online course is a unique web-based supplement in the form of an electronic book that provides a range of visual, auditory, and interactive elements to enhance training. Visit www.nccerconnect.com for more information!

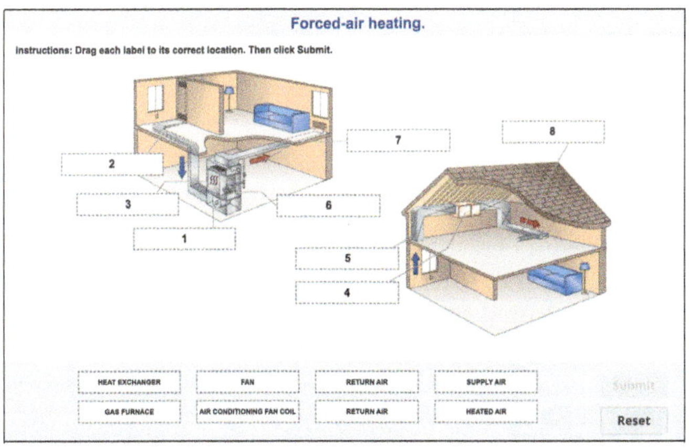

1.0.0 Basic HVACR Principles

can be expensive, electric-resistance heating alone is no longer common in cold climates. It is more likely to be used in warmer climates where heat is seldom required.

Figure 3 Electric furnace.

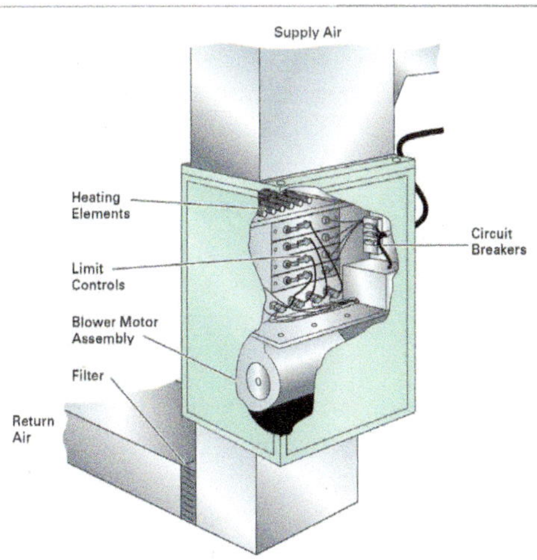

5: Introduction to Cooling (03107)

Refrigeration is a term that describes the transfer of heat from one place, where it is not wanted, to another place where the added heat is insignificant or unimportant. Air conditioning units and refrigerators simply transfer heat out of a space or object and move it elsewhere (*Figure 1*). In most cases, the heat is transferred outdoors. Domestic refrigerators and small commercial reach-in cases reject heat to the space. This heat must be moved outdoors by the building cooling system.

Figure 1 Mechanical refrigeration systems transfer heat from one location to another.

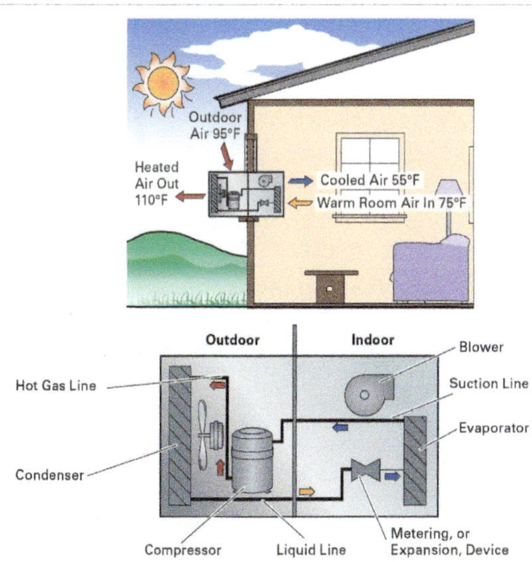

A refrigerant circulates through a refrigerant circuit to enable the process. The refrigerant absorbs

ACKNOWLEDGMENTS

This curriculum was revised as a result of the farsightedness and leadership of the following sponsors:

AGC of Oklahoma Education Foundation
Brown and Root Industrial Services
College of Southern Maryland
Construction Industry Training Council of Washington
Epling Distributing
Haynes Mechanical
International Code Council
LaGuardia Community College
Lennox Industries
Mielke Holdings
Pacific Northwest National Laboratory
Service Experts
Standard Heating and Air Conditioning
Southern Careers Institute

This curriculum would not exist were it not for the dedication and unselfish energy of those volunteers who served on the Authoring Team. A sincere thanks is extended to the following:

Allen Heath
Frank Calderone
James Koehn
Jamie Kono
Jim Cika
Jim Hughes
Joseph Moravek
Kasey Wade
Paul Ringley
Terry Mielke

A final note: This book is the result of a collaborative effort involving the production, editorial, and development staff at Pearson Education, Inc., and NCCER. Thanks to all of the dedicated people involved in the many stages of this project.

NCCER PARTNERS

To see a full list of NCCER Partners, please visit
www.nccer.org/about-us/partners.

SCAN ME

CONTENTS

Module 03316 Customer Relations

Module 03215 Basic Maintenance

Module 03313 Fasteners and Hardware

Module 03206 Alternating Current

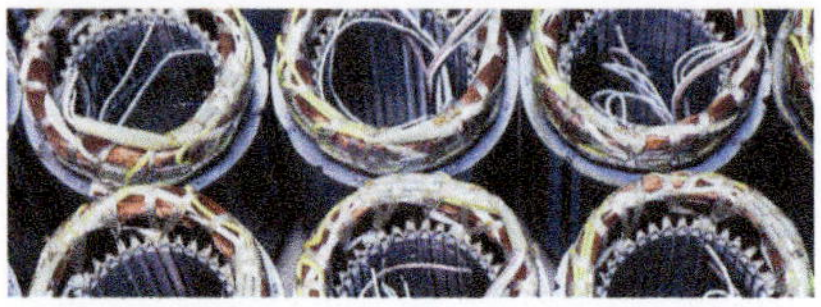

Module 03302 Compressors

Module 03303 Metering Devices

Module 03301 Refrigerants and Oils

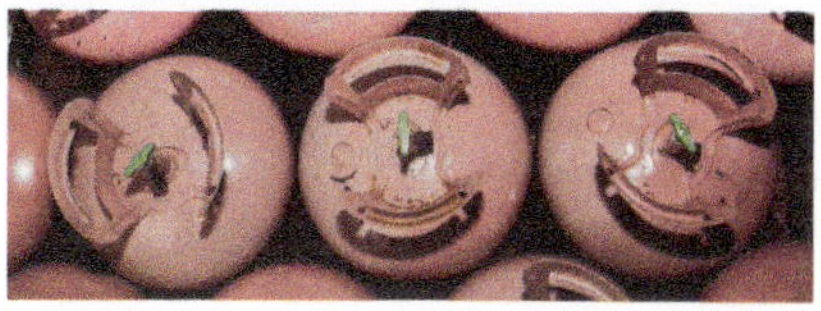

Module 03205 Leak Detection, Evacuation, Recovery, and Charging

Module 03211 Heat Pumps

Module 03403 Indoor Air Quality

Module 03204 Air Quality Equipment

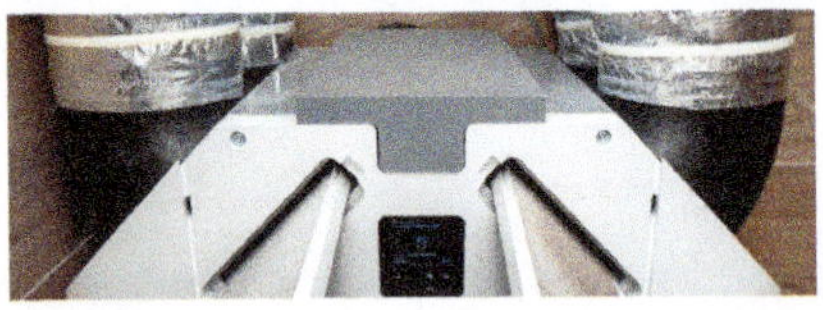

Module 03202 Chimneys, Vents, and Flues

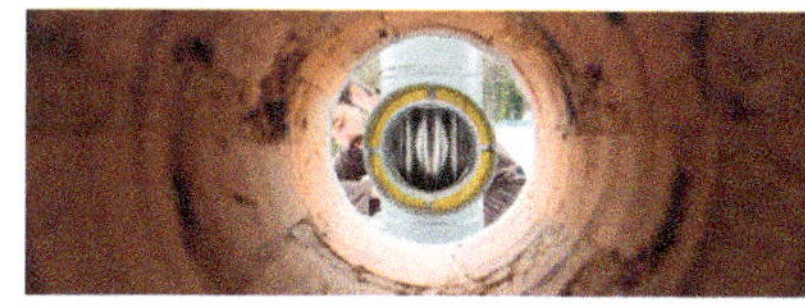

Appendices

MODULE 03316

Customer Relations

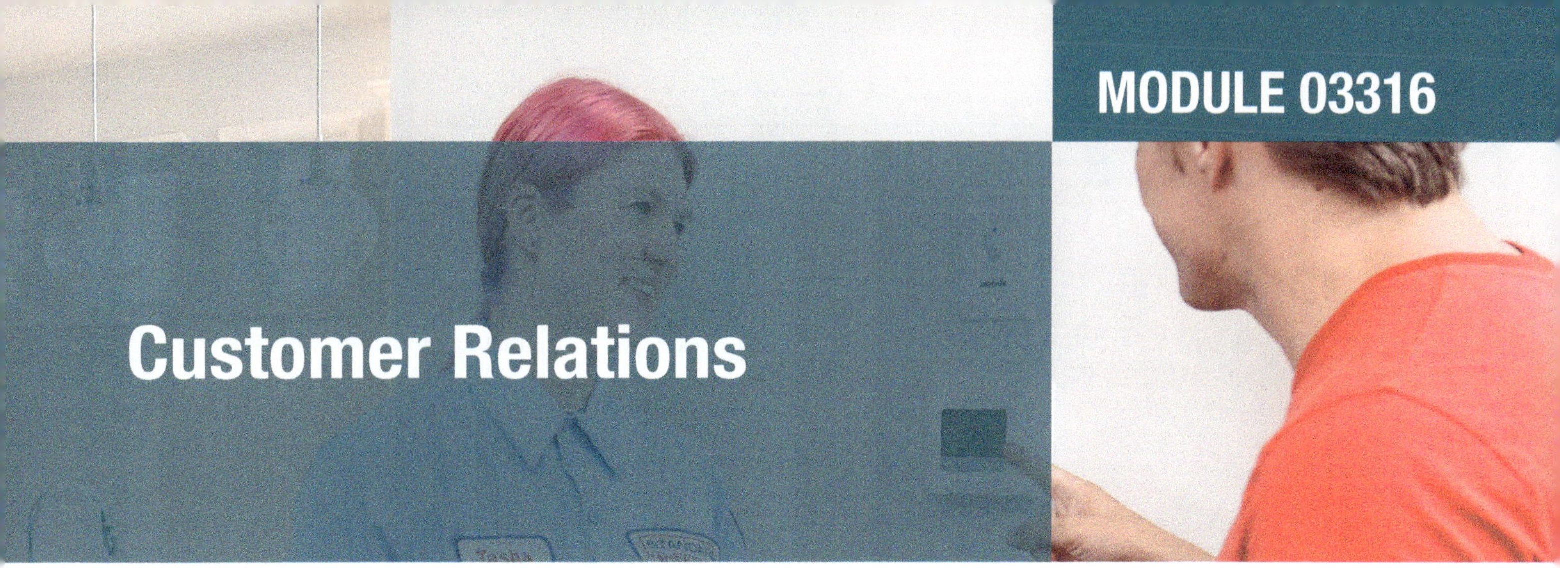

Source: Standard Heating & Air Conditioning

Objectives

Successful completion of this module prepares you to do the following:

1. Explain the service technician's role in customer relations.
   a. Explain how personal habits, behaviors, and attitudes affect customer relations.
   b. Explain how to properly communicate with customers.
2. Describe how professional technicians communicate and conduct themselves throughout a service call.
   a. Describe how to conduct the three phases of a service call.
   b. Describe ways to handle challenging customer situations.

Performance Task

Under supervision, you should be able to do the following:

1. Participate in at least three different role-playing scenarios related to challenging customer service situations.

Overview

To a customer, the HVACR service technician is the face of a company. The customer's attitude toward the company and the likelihood of that customer doing future business with the company are determined largely by the customer's impressions of and interactions with that technician. For those reasons, a service technician must learn how to deal effectively with customers.

NCCER Industry-Recognized Credentials

If you are training through an NCCER-accredited sponsor, you may be eligible for credentials from NCCER's Registry. The ID number for this module is 03316. Note that this module may have been used in other NCCER curricula and may apply to other level completions. Contact NCCER's Registry at 1.888.622.3720 or go to **www.nccer.org** for more information.

You can also show off your industry-recognized credentials online with NCCER's digital credentials. Transform your knowledge, skills, and achievements into credentials that you can share across social media platforms, send to your network, and add to your resume. For more information, visit **www.nccer.org**.

Digital Resources for HVACR

Scan this code using the camera on your phone or mobile device to view the digital resources related to this craft.

SCAN ME

1.0.0 The Technician's Role in Customer Relations

Performance Task

1. Participate in at least three different role-playing scenarios related to challenging customer service situations.

Objective

Explain the service technician's role in customer relations.

a. Explain how personal habits, behaviors, and attitudes affect customer relations.
b. Explain how to properly communicate with customers.

Good customer relations are essential to the success of many businesses, including those in the HVACR industry. Every employee represents the company and influences what customers think about the business. Keeping these impressions positive requires that all employees work toward a satisfactory experience for the customer.

Providing good service is an essential element of this experience. Your role as a technician requires patience, determination, and confidence in your craftsmanship. A jobsite-appropriate appearance, attention to detail, and customer-pleasing attitudes also build and maintain customer goodwill. Customer goodwill results in more business through referrals and loyal customers who consistently depend on your services.

This module provides guidance for dealing with your customers. Although the examples used in the module lean toward the residential market, the basic principles apply to all service environments.

It's All About Perception

Current and potential customers usually base their buying decisions on their personal perceptions. If someone sees you speeding or driving aggressively in your company vehicle, it can affect their willingness to do business with your company in the future. The company name may invoke negative thoughts. If your truck is dirty on the outside or cluttered and disorganized on the inside, it reflects on both you and the organization.

The same applies to your personal appearance, which should reflect that you are prepared and professional. If your hair is untidy, your clothes are dirty or wrinkled, or even if your shirt is untucked, some customers will be put off by it.

Don't be too surprised if your company requires you to cover some or all tattoos while you're on the job, as some customers react negatively to them as well. Always refer to company policies for questions related to your personal appearance and be sure you are prepared to accept them prior to employment.

1.1.0 How Personal Habits, Behaviors, and Attitudes Affect Customer Relations

Positive habits, behaviors, and attitudes help to ensure that a customer's first impression of you and your company is a good one. It also guarantees that the customer's needs are clearly understood and that appropriate customer service is provided to meet these needs. When you provide effective service to your customers, you can increase your company's share of business and position in the market.

Your ability to relate well to customers is critical. As a service technician, you are often the only person from your company that the customer sees. In the customer's eyes, you are the company, so appearances count. First impressions are made only once, and an impression is generally formed in the first 60 seconds. As the old saying goes, "You don't get a second chance to make a first impression."

Making a good first impression will not only make your job easier, but it will also drive customer loyalty and confidence. Positive responses and feedback from customers that reach your office also reinforce your value and enhance your status within the organization.

Because most customers lack the proper training and equipment, they must rely on a professional for help with heating and cooling problems. Although they may not be able to judge a technician's competence, they can and do form an impression based on your appearance and attitude as you approach them. A customer's first glimpse of a technician often establishes their opinion of the worker as well as the company. Think about the impression you make. Consider the following:

- Do you practice good personal hygiene?
- Do you get enough sleep and look alert?
- Do you wear a neat, clean uniform to start the day?
- Do you wear shoe covers or remove your shoes when entering a customer's home or business?
- Is a pencil and pad to take notes part of your daily uniform?
- Do you park on the street, rather than in the customer's driveway? While this is not always practical, it is a best practice. Vehicles often have fluid leaks that stain a driveway and make an otherwise happy customer angry.

Once a customer has formed their first impression of you and your company, your work habits can solidify or change that impression. Here are some questions to ask yourself consistently:

- Were you on time? Keep in mind that customers must often take time away from work or from other activities to be available for your visit. Always respect the time of your clients. If you are unavoidably late, have the courtesy to call and let them know. Apologize immediately for your tardiness. Don't invent excuses. Simply accept responsibility and apologize. Offer a valid explanation when asked to do so.
- Is your tool set complete and organized? Check your tool bag and your truck daily before setting out to make sure you have everything you need in good working order.
- Do you listen carefully while the customer explains their problem? Listening attentively can provide you with useful information for troubleshooting. It also helps you establish **rapport** with the customer.
- Once you understand the problem from the customer's perspective, do you get to work promptly?
- Do you protect the premises and work area? Do you carry rags for cleaning and drop cloths to protect floors and carpets when necessary?
- Do you respect the jobsite by cleaning up after yourself?
- Do you take the time to carefully explain your diagnosis and what the owner can expect after the repairs are completed?
- Do you refrain from smoking, spitting, and similar habits that many people find unacceptable while you are at their home?
- Do you respect company equipment, tools, and vehicles? Trucks and vehicles displaying the company name are rolling advertisements (*Figure 1*). Is your vehicle clean and in good repair? Are you a courteous driver?
- Do you avoid using alcohol and drugs while on the job?
- Do you avoid profanity and horseplay? Customers, especially those with children on the premises, are likely to be offended by inappropriate language. Horseplay leaves the impression that you aren't taking the work seriously.
- Do you avoid wearing political or sports-related articles that might cause a confrontation?

Customer Technical Knowledge and Skills

When talking with customers, never assume they do not possess technical knowledge or skills. Speak in plain terms about your diagnosis and suggested resolution, avoiding jargon and acronyms. Do not use a condescending or "dumbed down" tone or posture.

Your customers will appreciate the clarity and feel respected, regardless of how much technical knowledge they have.

Think About It

Personal Habits, Behaviors, and Attitudes

If you were asked to hire a new service technician for your company, what personal qualities would you consider the most important?

Rapport: A close and agreeable relationship between people that understand and accept each other's thoughts and connect well.

Figure 1 Keep your service vehicle clean.

If you are working at a commercial construction site, you must review and respect the site safety requirements. This includes wearing the proper personal protection equipment (PPE) (*Figure 2*), which generally includes the following:

- Hard hat
- Ear protection
- Safety glasses
- Reflective vest
- Work gloves
- Steel-toed boots

Although your use of PPE is often monitored at construction sites, it is essential to understand that it is equally important to apply it properly regardless of the site. Your customers will likely find it refreshing and a sign of professionalism. More importantly, your safety and well-being demand it.

Figure 2 Common construction site PPE.
Source: Rawf8/Getty Images

Site managers are perfectly within their rights to refuse entry to a worker who is not properly attired. Learn in advance what PPE is required. Don't assume anything since safety requirements can vary from site to site. You will make a poor first impression if the site manager must stop you from entering. However, if you are stopped, your attitude and response will be remembered. Act with humility to leave the right impression.

Every hour of every working day, you are an advertisement for your company. Your appearance and behavior must be consistent with the positive image your company wants you to project.

Off the job, you can increase word-of-mouth advertising (free advertising created by satisfied customers who recommend your company to friends) by having a positive attitude about your employers, coworkers, and customers. Positive statements about your job are free advertising for your company.

Social media has significantly changed how people interact compared to the past (*Figure 3*). A person who is dissatisfied with a company can go online and log a complaint that thousands of potential customers will see. The complaint doesn't even need to be valid to cause damage. Many people these days will not do business with a company without first looking for reviews.

People living in a residential subdivision are often a part of an internet-based social group. They can instantly share a positive or negative opinion of a contractor with hundreds of their neighbors. Both good and bad news travel at lightning speed through social media.

Unfortunately, negative information typically spreads farther and faster than positive information. One reason for this is that the nature of anger and resentment serve as powerful motivators. Happy, pleased clients are less motivated to share information, but they will generally share a positive experience when it is relevant to a conversation.

Figure 3 Social media has connected millions of people who previously had no connection to each other.
Source: Photosbypatrik/Getty Images

1.2.0 Customer Communication

Obviously, it's important to have the necessary knowledge and skill to repair an HVACR system. It is important to be confident in your skills, which comes from hard work and study. However, interacting effectively with customers is an equally important skill. The result of your interaction may determine whether you or a competitor is called the next time.

One of the most important things to remember about communication is that it is a two-way street. Be sure to allow customers a chance to communicate their thoughts and feelings. Part of this process is allowing them to vent their frustrations about the equipment or about the discomfort that the equipment failure caused. For example, a customer who has been without heat on a cold day is likely to be very uncomfortable (*Figure 4*). The customer will often want to share that with you.

NOTE

There are many videos available on YouTube and similar websites that clearly explain and demonstrate active listening. Take some time to research the topic and practice the skill.

Active listening with sincerity and a concern for the situation is required here. Active listening means listening to someone else with obvious interest, while verbally and physically demonstrating your concern and engagement. It is a good idea to practice your active listening skills in your personal life as well, as everyone truly appreciates a listener that provides their undivided attention.

Active listening: Listening attentively while using both verbal and nonverbal indications to demonstrate your interest and engagement.

An important component of active listening is called **paraphrasing**. As you are listening, you are trying to ensure you received the speaker's message clearly. To ensure that you did, repeat the message in your own words. The speaker will either indicate you got it right and possibly add more details or provide more information to clarify their message. Sometimes the paraphrasing only needs to contain a few words; more complicated messages require more effort to paraphrase. Paraphrasing is a valuable communication tool that helps prevent misunderstandings.

Paraphrasing: Expressing what you believe to be the meaning of a speaker's message in different words to ensure you understand the message as intended.

Figure 4 It can be challenging to communicate with unhappy customers.
Source: Cunaplus_M.Faba/Getty Images

Communication by telephone must also be clear and precise. In this situation, only verbal cues can be used to show you are listening actively instead of passively. Here are some tips to follow when making a phone call to a customer:

- Before you call, think about what you want to say and the questions you want to ask. It's always a good idea to jot them down, perhaps leaving room to record some responses.
- When the person answers, identify yourself and your company. Speak clearly and take notes to be sure you are getting the correct information. Repeat it back to make sure.
- Prepare a message in advance in case you need to leave a message. Be ready to give your name and company, the purpose of your call, and a way to contact you. Speak slowly and clearly, especially when you are giving a phone number or email address. Repeat the number or email address for clarity.

1.2.1 Keeping Communications Positive

On every service call, try to keep the communications positive. Here are some simple but effective tips:

- Do the best work that you can—actions speak louder than words.
- Always treat the customer with courtesy, concern, and respect. Treating others as you'd like to be treated is particularly important, especially in unpleasant situations. Remember, it is easy to be courteous when things are going well. A truly mature person is courteous even when things aren't going well.
- Treat each service call as if it were an emergency. To the owner, it is. Service customers do not call for a technician unless they are without heating, cooling, or refrigeration. When comfort or financial loss is at stake, repairs become a top priority.

Empathy: The ability to identify with another person's situation or feelings.

- Show **empathy** for the customer. This means putting yourself in the customer's shoes. Think, for example, about how you would feel on a cold day if your furnace had been off for hours. Be sincere. The customer is likely to detect the fake concern and resent it.
- Demonstrate confidence to help put the customer at ease. Trust yourself to solve the repair issues you are facing. When you need to analyze a situation, it is okay to step away from the equipment and the customer for a few minutes to collect your thoughts. Think about the issue from different angles. Then return to the customer with your renewed confidence showing.

- Leave your personal life at home. Don't let personal issues affect your attitude and your relationships with customers. If you are having personal problems, avoid using customers as sounding boards.
- If you are running late, report this to your dispatcher or directly to the customer. Both the dispatcher and the client must often make other arrangements to accommodate the delay. Don't leave the customer wondering about what is going on. Make sure they know your status and can work with changes. This is also true if you leave the site for parts or lunch and find you can't return as promptly as you thought.

1.2.2 The Positive Approach

The following are some helpful tips to keep things on a positive note:

- *Smile* — Be genuinely interested. You may think you were called to repair a furnace. But you were really called to make the customer's home comfortable and livable again—to provide essential security.
- *Use positive statements with the customer* — Compliment things that are good about the environment. Be an optimist on the job instead of a pessimist.
- *Do not criticize the client or the system* — Avoid negative statements, especially those that simply don't need to be spoken. Be careful of what you say about the brand of equipment, the design of the system, or the quality of the workmanship, because you may offend the customer. Homeowners do not want to hear that you think they own inferior equipment. Even if it is inferior, fix what is possible or explain respectfully to the customer what must be done further. Wait until the problem is clearly solved before you talk about maintenance agreements and their benefits. It's better to bring that up later after you have corrected their problem.
- *Do not criticize your own organization* — Speak well of your employer. Occasionally, you may be called upon to service a system recently repaired by one of your coworkers. Rather than criticizing your colleague's work, apologize for the inconvenience and assure the customer you will solve the problem and report your findings to them directly. Then follow through and take responsibility for any errors on behalf of your employer.

1.2.3 Showing Concern for Customers

Another part of good, positive, on-the-job communications is showing concern for your customers. Try the following approaches:

- *Be a good listener* — Actively listen when the customer explains what went wrong, what was observed, etc. If you yawn, slouch, or let your eyes wander, it is a sign of inattention. Ask questions to gather the information you need to solve the problem. Remember that although the customer may not have the technical vocabulary you do, their words can supply you with important clues. The client's description may seem long and rambling, but it's important to stay patiently engaged. While the situation may be "old hat" to you, it is unique and stressful for the customer.
- *Make the customer's situation the priority* — Always try to look at the problem from the customer's point of view. For example, a customer may be upset that you are late. Perhaps, they are concerned about picking up their kids from school or being on time for an important meeting elsewhere. This adds to the stress of a broken HVACR system, sometimes resulting in a verbal attack. Try turning the conversation around by opening with, "I know that I'm late and I would be surprised if it didn't complicate your day. I sincerely apologize. Please show me to your equipment so we can get you back on schedule quickly." Getting ahead of the anger with the right attitude can be very disarming.
- *Answer questions honestly, but positively* — If a customer asks if their equipment is worn out, you may reply, "It can't be repaired effectively and needs to

be replaced," or "The cost of repairing your unit would be a poor investment due to its age." But avoid negative comments such as, "This pile of junk really needs to go."

- *Respect your customer's opinions* — Never just respond by saying, "You're wrong!" People do not like being told they are wrong in such a direct and personal way. For example, if they say that the thermostat is no good and you suspect there is a simple wiring problem, don't say, "No, you're wrong. It's in the wiring." Instead, say, "I think the wiring is really the source of your problems, but I'll check out the thermostat completely once that's fixed."
- *Don't socialize while on the call* — Service technicians are always on the clock. You really don't want your clients to be billed for social time. If you explain pleasantly that you appreciate the offer of a cup of coffee but that you must decline out of respect for their costs, the customer will not be offended. Also avoid socializing with others. Leave your mobile phone in your pocket unless it's needed for support. Customers that hear a technician constantly in conversation unrelated to their job will not feel good about the quality and speed of their work. Chatting on the phone as you work is also distracting and an unsafe practice.
- *Go beyond "parts changing"* — Make sure that you have completed the appropriate troubleshooting and are confident that the problem has been solved. Spend time looking for additional problems before wrapping up. There may be other problems that cause the system to fail again soon after you leave, resulting in a *callback*. In the HVACR service business, a callback typically means lost revenue. Protect your company and yourself by trying to identify all potential problems and sharing your findings with both the client and your supervisor.
- *Sell the customer only what they need* — Honest service builds trust in your company and will help you bring pride and confidence to your work. There are too many opportunities to succeed in both sales and service without resorting to dishonest tactics. Tell them what you honestly think they need, and don't push products that don't have real benefits.

It has long been said that everyone, regardless of your position, is in sales. All employees have an opportunity to affect sales, either directly or indirectly. Make it easy for your current customers to find the company contact information and share with others by doing one of the following:

- Leave a business card.
- Put a sticker with your company's phone number on the equipment itself or leave one with the customer for placement near the phone.
- Show the customer some simple maintenance techniques that will prolong equipment life, such as changing filters frequently (*Figure 5*). Tell the customer why this maintenance is essential. Suggest that they consider a maintenance agreement with your company if they would rather not handle such things themselves, regardless of the reason.
- Teach the owner how to use their equipment safely and most effectively. For example, show the owner how to program the thermostat (*Figure 6*) for occupied and unoccupied periods to save energy.
- If the opportunity arises, discuss additional or replacement equipment to improve comfort and efficiency. Ask for their permission to allow your sales staff to contact them.
- Perform a personal customer follow-up, if possible, within several days of your repair. It only takes a minute, and the personal touch is appreciated. If you are confident that you will follow through, tell your customer that you will contact them soon to ensure things are working as they should, and establish how they prefer to be contacted.

Figure 5 Changing air filters regularly is an essential step to ensure the reliability of HVAC equipment.
Source: Sucharas wongpeth/Getty Images

Figure 6 Show the customer how to adjust the thermostat for greater comfort and efficiency.
Source: Standard Heating & Air Conditioning

1.2.4 Selling Yourself and Your Company

You may not consider yourself a salesperson, but selling is a key element of your job. It's true for most every job, regardless of where you work. In reality, you are selling at every service call. At a minimum, you are selling yourself as well as the value of your skills.

You are also selling your company, acting as its representative. The impression you make, along with a successful repair, will determine whether the customer is likely to call your company again.

Every service call you make is a selling opportunity. If you handle a service call for a customer without a service agreement professionally, there is an opportunity to sell the advantages of a periodic maintenance program (*Figure 7*).

Figure 7 Discussing the benefits and provisions of a periodic maintenance agreement.
Source: Standard Heating & Air Conditioning

Think About It

Share Your Thoughts

Consider these situations:

1. A customer owns a brand of furnace that is no longer being manufactured. You don't think parts are available to fix the blower assembly. What would you say?
2. A customer called for a technician to fix a furnace. As you enter the jobsite, the customer says, "It's cold in here." What questions would you ask?
3. A customer says a blower and burner are not operating in sequence and the furnace control board is likely at fault. You think the problem is a faulty limit switch. How do you build the customer's confidence in your skills without making them feel as if their input isn't valued?
4. Can you suggest some unique ways of advertising your company's services that are simple and easy to apply?

You also have an opportunity to suggest the replacement of aging equipment. You might be surprised by how practical your customers are. Many of them readily understand the long-term benefit of replacing aging equipment with new, energy-efficient equipment, looking beyond the initial cost.

Lost opportunities can be painful. How would you feel if your company got a $100 service call while a competitor got a $5,000 contract to replace the system for the same client a year later? Note the age of a customer's heating and cooling equipment and pass that information to your company's sales department when there is a valid reason, and the customer is willing to consider the investment.

1.0.0 Section Review

1. True or False: First impressions are usually made in the first 60 seconds of contact.
   a. True
   b. False

2. If you are going to be late for an appointment, the *best* thing to do is _____.
   a. just get there as quickly as you can
   b. call the dispatcher and have them reschedule for another day and time
   c. report it to your dispatcher and/or let the customer know you will be late
   d. leave the first job unfinished to be on time for the next appointment

2.0.0 Handling Service Calls

Performance Task

1. Participate in at least three different role-playing scenarios related to challenging customer service situations.

Objective

Describe how professional technicians communicate and conduct themselves throughout a service call.

a. Describe how to conduct the three phases of a service call.
b. Describe ways to handle challenging customer situations.

There is more to a repair call than simply showing up ready to work. In a typical case, the technician starts by interviewing the customer to gather information and, in a few cases, soothing ruffled feathers. Remember, customers are your most valuable resource. They may not know anything about their equipment, but they do know when something is wrong and how it will affect them. Because you are new on the scene, this interview process is one of the most critical steps in a successful service call.

2.1.0 Conducting the Service Call

The typical service call can be broken down into three major parts:

- *The opening* — Meeting the customer and identifying the problem.
- *Troubleshooting and repair* — Solving the problem.
- *The closing* — Leaving the customer with a positive impression.

Remember that your objective in servicing your customer's equipment goes beyond technical competence. Your objective must include winning the customer's confidence and repeat business.

2.1.1 The Opening

Prepare yourself for that first impression. Your personal appearance and service vehicle should look good. You should also show your respect by appearing promptly and politely at the door. Now what? Try the following:

- *Smile and display confidence and polite respect for the customer* (*Figure 8*). Don't be afraid to make eye contact.
- *Promptly identify yourself.* Many customers waiting for a service worker are naturally cautious. Identify yourself clearly and show an appropriate ID. Many companies provide badges with the worker's photo. Others send out text messages or emails with the technician's photo included so the client knows what the visitor should look like. Also provide a business card with your name on it if you have one.
- *Be respectful.* After the client introduces themselves, you may have a feeling about how they wish to be addressed going forward. If you are unsure, it's acceptable to ask. Above all, concentrate on remembering the customer's name in case you need to get their attention later. At every opportunity, use the words *please, thank you,* and *may I.*
- *Actively listen to your customer to understand the problem.* Ask questions about what the customer has seen, heard, smelled, and felt. Avoid trying to seem unusually smart by using a lot of confusing technical terms. Ask when things occurred and how often they happened. Often, the customer can help speed your diagnosis by providing clues to the equipment problem. When you listen attentively, everyone wins.

  It's important to ask open-ended questions—the kind that result in specific, detailed answers. Asking questions that are easily answered with a

Figure 8 A pleasant greeting is a good way to start a service call.
Source: DragonImages/Getty Images

Figure 9 Disposable shoe covers.
Source: Pongstorn Pix/Getty Images

yes or a no is a painfully slow way to get what you need. You really don't want to be led to the equipment without any information from the customer. Avoid saying anything like, "I'm here to fix your furnace. Where is it?" Instead, say, "I came to repair your furnace. What seems to be the problem?"

- *Do not enter a home until you are invited to do so, either verbally or by gesture.* Perhaps you need to begin outdoors anyway. When you need to come in the first time, ask for permission. Before entering, don some disposable shoe covers (*Figure 9*). Don't put them on until you are at the entrance. Otherwise, they will get soiled as you walk from the truck to the door.

Unhappy Customers

On some service calls, an unhappy and uncomfortable customer greets you. It happens. The customer may have lost heating on a cold day or cooling on a hot day. In most cases, the problem has been going on for hours already and often for more than a day.

At this point, the customer may be looking for someone to blame, as people often do when angry. If it's equipment your company installed, they might be happy to hold you largely responsible, even if you weren't personally involved with the job.

The first order of business is to defuse the situation. Listen attentively and sincerely to the customer's complaints. Ask them to describe how the problem manifested itself. Engage the customer by asking the following questions:

- Is the system operating but not satisfying the thermostat setting?
- Does the amount of airflow feel normal?
- Was there anything else that seemed unusual when it stopped working?
- Were there any unusual noises or odors?

If this is a new customer, ask about previous problems. Engaging the customer in this way not only provides you with valuable information, but it also helps make your customer feel like an ally in the battle to restore normalcy.

2.1.2 Troubleshooting and Repair

The customer has an equipment problem. You are there to solve it. Once you have listened to the customer explain the problem, get right to work to determine the cause (*Figure 10*).

Figure 10 After the greeting and exchange of information, it's time to get to work.
Source: IndyEdge/Getty Images

Practice good work habits. Avoid initiating general conversation while working. Remember, many companies charge customers by the hour for service. The customer must feel that you are using the time wisely, rather than prolonging the visit unnecessarily.

If you make an error, do not try to cover it up by stating that the problem is worse than you feared. Be honest. Accept responsibility for your mistakes and learn from them. A good employer will support you, while also ensuring that you know how to avoid the error in the future. Although constant errors will not be tolerated, a great deal of respect and admiration can be gained by stepping up when you do make an error. Assure your client that any additional repair costs resulting from your error will not be their responsibility.

It is important to arrive prepared, with the proper tools and common replacement parts appropriate to the equipment class. The dispatcher may include the equipment data in the service order if the call is from an existing customer. This makes it easier to be prepared. If you must make a trip out to get something, like a PVC fitting, the customer might think that could result in an unnecessary labor charge. *Flat-rate pricing*, a common approach to service invoicing today, let's the client know what the repair will cost before they are committed to it, regardless of the amount of time expended.

During this part of the service call, you will need your professional skills to do the following:

- Analyze the symptoms.
- Perform necessary testing to isolate the problem.
- Determine the solution.
- Explain the solution to the customer, including an estimate of what it will cost.
- If the customer approves, complete the repair.

While the work is ongoing, maintain a clean, organized work area. Not only does this impress a customer, but it also helps to ensure your work is completed safely. Make sure that all trash is picked up and inventory your tools when the work is done. It's a best practice to ensure that the area is cleaner than it was before by picking up any other trash nearby.

Many successful technicians carry simple cleaning materials to clean the outer cabinets of furnaces and condensing units. Make sure that the equipment serviced looks its best before departing.

Going the Extra Mile

People often neglect to change the air filters for their equipment. Sometimes they are not physically able to do it. If you are on a call for a customer that does not have a service agreement, offer to check the condition of their filters. This takes very little time and could increase the value of your service. Make a note of the filter sizes so you can add that to the records for future visits.

Neglected filters also represent a good opportunity to mention the value of a service agreement. Agreements are easier to sell than you might think, especially if the client likes and believes in you. That might improve your own income as well. But if they don't really like you or respect your work, the chances of selling anything are slim.

2.1.3 The Closing

This very important step helps establish the final image left by you and your company. Do the following once the work is finished:

- Neatly pack your tools.
- Return the premises to its original, or better, condition. (Replace covers, wipe off dirty fingerprints, clean up drop cloths, etc.)
- Explain to the customer that the problem has been solved. Explain what parts you replaced and offer to show the customer the defective parts.

Verify That the System Works

One of the most important things you can do at the end of a service call is to demonstrate to the customer that the system works. Don't leave them wondering about it; show them with supply air temperatures, for example. Also make sure that the thermostat is set as they wish, since you must often adjust the settings during testing.

- Demonstrate that the equipment works. Ask the customer if they would like to perform this task on their own. Being given the choice is comforting and lets them know, at least subconsciously, that you aren't trying to hide anything.
- Wrap up the call by relating your service to the customer's real problem, not the equipment. Rather than say, "I fixed the thermostat; your furnace will work now," try, "I repaired your thermostat so you can be warm and comfortable again." A positive closing builds confidence in you and your company and shows that your mind is in the right place. Stating it this way also reminds them that you aren't just an HVACR service technician, but that you are also a problem solver.

2.2.0 Handling Challenging Customers

When you deal with customers, remember that their wants and needs include safety, comfort, and convenience. If your actions inspire the customer to believe your company can provide those things, they will likely continue to be a client. Sometimes, though, you encounter problems that go beyond failed machinery. For example:

- A customer may be excessively worried about safety.
- Sometimes people are upset about something totally unrelated to the problem, but they take it out on you.
- Customers may be upset about poor equipment performance or the previous bill.

In cases like these, the technician's actions and attitude usually determine whether the visit is a successful one.

The following sections cover several situations that require advanced customer relations skills. Consider how you might handle these clients and compare your ideas with those of your fellow trainees. There may not be a single correct path to avoid a confrontation, but the correct approach and responses can help you through even the most difficult situations.

2.2.1 Fearful Customers

A customer's fears are often based on experience, rumors, or misunderstandings. Try to determine what the source of their fears is and provide information that will help ease their mind. If you listen carefully and take their fears seriously, they will feel reassured and be more willing to listen to your explanation.

Exercise Scenario:

A customer thinks their furnace is leaking because they detected gas due to an extinguished standing pilot light. You have reignited the pilot, tested the output, and are closing the call. What can you say to calm their fears?

Exercise Solution:

You can often discover what is behind their fears by asking some detailed questions. Getting them talking about the details can help them identify what they are truly afraid of. Consider the following questions:

- How often does this happen? When did it start?
- Do you hear anything unusual when it happens?
- What does it smell like to you?
- When it happens, are you doing something different than what you've done in the past?

In response, don't just say, "Yes, when the pilot light goes out there is a light smell. It is fixed." This doesn't address their concerns for the next time it happens. Instead, explain how it works in simple terms to help alleviate their fears: "When a standing pilot flame goes out, a faint gas odor is often detected in this type of furnace. A minute amount of gas escapes but it doesn't collect, as it flows out of the flue piping due to the consistent natural draft through the furnace. The small volume of gas involved isn't hazardous. Your family is not in any danger. But we also want to ensure your comfort as well as your safety. I'll do my best to determine why the pilot flame went out to prevent it from happening again."

This explanation addresses both the technical problem and the fear associated with it.

2.2.2 Opinionated Customers

Opinions are beliefs that may not be entirely based on facts, but also on what seems probable to someone. Everyone has opinions. Everyone also likes to have their opinions validated. When opinions are ridiculed, people feel hurt and offended. The opinions people offer during your visit may or may not be related to the HVACR problem.

It is important that people do not feel their opinions are being discounted or belittled. It's always best to keep the conversation focused on the problem at hand, even if you happen to agree with the opinion.

Exercise Scenario:

The customer has had a bad week, including the failure of their air conditioner. They are trying to draw you into their earlier problems and their opinions about them. You're trying to finish quickly because you have two other service calls before lunch. What do you say?

Exercise Solution:

Taking an obviously negative approach is offensive. Avoid saying, "Yeah, we all have problems. My problem is fixing air conditioners. Where is it?" Instead, say, "It does seem like problems like to come at us all at once! I can help you with one of your problems today, though. How did your cooling problems begin?" This offers sympathy while at the same time diverting to the reason for your visit.

It is never a good idea to initiate or participate in discussions on divisive topics, such as politics or religion. Discussions about any issue that people tend to be passionate about should be avoided. Your position on those matters should not be a factor in whether the client uses your services in the future, and the results of the discussion are unpredictable. The safest approach is to acknowledge the statement with something simple like, "That's an interesting topic for sure!" Then bring the conversation back to the problem you came to solve.

Exercise Scenario:

A client has no airflow from their furnace. You've quickly discovered a broken fan belt. They want to initiate a discussion of local politics while you are working on the repair. They think the mayor is a crook, but assume you disagree since you aren't responding as they hoped. You don't want to be drawn into this discussion. How should you respond?

Exercise Solution:

State a very general fact that shouldn't offend anyone to avoid prolonging or expanding the discussion. Simply agree that politics has always been a concern for those governed, for example. Then turn the conversation back to the problem you're there to solve. You could say, "Politics never fails to be a source of concern for us, does it? But I think getting your system running again and returning comfort to your home might at least make you feel a little better about the situation."

2.2.3 Biased Customers

Some people will bring prejudice into their interactions with others. They may doubt your ability based on age, gender, ethnicity, or any other personal characteristic. In these situations, let your personal integrity and technical skills speak for you.

NOTE

If you find yourself facing racist or sexist harassment, document everything and report the situation to your supervisor.

Exercise Scenario:

Following introductions, a client immediately asks your age and rolls their eyes when you say that you are 21 years old. The customer watches over your shoulder from that point forward, questioning your actions while you work. At the end of the job, the customer insists that you didn't fix the furnace correctly and demands your supervisor's number.

Exercise Solution:

Remain cool and confident throughout the process. Answer the customer's questions calmly and professionally. Demonstrate that the equipment now works and ask if they would like to try it themselves.

When a client asks for your supervisor's number, never hesitate to provide it. When you feel it necessary, document what happened as soon as possible to ensure that you don't forget the details. It is personally challenging to endure prejudice against you—that much is certain. But as was originally said in a Shakespearian play (although stated a little differently), discretion is the better part of valor. In some cases, companies will choose to deny service to clients who refuse to put their prejudice aside, rather than allow their employees to be subjected to it.

2.2.4 Sloppy Customers

As you make service calls, you'll see all kinds of housekeeping habits—or the lack thereof (*Figure 11*). But, of course, you're not there to judge. Keep your thoughts to yourself. Avoid being judgmental in your words as well as your gestures. Since you're a visitor, the owner will often offer an excuse when things are messy.

You can handle such sensitive situations in the following ways:

- *Focus on the task at hand* — Don't look around as if you're judging the site. Avoid comments like "What a mess!" Raised eyebrows and rolled eyes say the same thing nonverbally. Maintain your composure and focus only on factors related to the HVACR issue.
- *Put the owner at ease* — If you feel you must reply, avoid direct agreement with their negative analysis. Say something like, "No need to be concerned about that. It's a part of life for all of us."

Figure 11 Housekeeping and living habits vary widely among people.
Source: Denis Torkhov/Getty Images

- *Be diplomatic* — If belongings are stacked in front of the furnace, bring it to the owner's attention and ask permission to move it. Move only what is necessary and put it back in place when finished unless the material represents a hazard. Bleach and cleaning chemicals, for example, should never be stored close enough to allow vapors to be drawn into the combustion airstream. Let the customer know that this could be a safety hazard or lead to rapid corrosion of the burner components and vent piping. While moving things around, do so with a positive attitude rather than complaining your way through the task.
- *Don't sound like an accuser* — Avoid negative "you" statements. It's always better to say, "The fumes from chemical products can damage your furnace," as opposed to, "You are making a big mistake by placing these things here." Using positive "you" statements, such as, "You did a great job!" is fine, but leave the "you" out of negative statements. This makes it less personal to the listener.

2.2.5 Angry Customers

As mentioned in the previous section, using the word "you" in a sentence is acceptable for positive statements. When you encounter a situation where your client has a reason to be angry, you can agree by saying, "You have every right to be angry."

When their equipment has broken down twice in three days, or when they have had two appointments that ended with a no-show on your part, anger can be justified. Even though you don't likely control your schedule, you will probably be the focal point of their anger. What should you do?

First, let the angry customer vent. Next, acknowledge their feelings and be empathetic. Sometimes, this means saying something like, "I know I would feel the same way." Then assure them that you'll do your best to solve the problem.

Exercise Scenario:

A furnace was repaired just last week. It has broken down again, leaving the family without heat for two days. Their children have colds. The family has already paid a sizable sum for a repair that did not last. As you walk in the door, the customer snarls at you right away. What do you say?

Exercise Solution:

Here is one appropriate answer: "I hear your furnace isn't working again. I'm sure you're very concerned about your family's well-being. I am so sorry that you have experienced a second failure. I understand what a burden it is and sincerely apologize. I'll find out what's wrong and get your system running as quickly as I can. Tell me what's been happening when you try to operate it."

You have shown concern for their situation. You have acknowledged that their concerns are justified. You have also assured them that you will address the problem quickly and to their satisfaction.

2.2.6 Critical Customers

Sometimes you'll encounter someone who can't be satisfied, no matter what you say or do. Often this happens because they are under a lot of stress or have had a bad service experience in the past. When this customer explodes, you may be a convenient target. How do you handle these situations with sensitivity and tact?

You've heard of cases where someone, angry about a bad day at work, comes home and takes it out on the family. This is called *anger displacement* (*Figure 12*). The following are two simple examples of what can generate displaced anger:

- A customer who refused to pay extra for a weekend service call is now dealing with a mess because the pipes in their poorly insulated laundry room froze as a result.
- A customer having trouble at work doesn't really want to be home waiting for a service technician when they should be dealing with their office problems.

Figure 12 We all experience anger, and sometimes our anger is directed at someone that doesn't deserve it.
Source: Antonio_Diaz/Getty Images

Think About It

The Negative Customer

You arrive at the jobsite and the client is clearly unhappy. The builders who just completed the home left several jobs unfinished. The homeowner begins ranting about builders and construction workers in general, including you, even though you've never seen the place before. The rant begins anew while you are actively working on the equipment. How do you handle this difficult customer?

Source: Fokusiert/Getty Images

Think About It

Dealing with a Know-It-All

A homeowner wants to use their existing furnace to heat the home addition they just proudly completed on their own. They have asked you to check out the new ducts and make the connection to the existing duct system; they have already decided the furnace can handle the additional load. They seem to believe they are an expert on everything.

The addition alone is 1,100 ft^2 on the north side, with three outside walls. With some quick math, you determine that the existing furnace is grossly undersized for the additional load. How do you approach this client with the news?

Source: Khosrork/Getty Images

Remember that everyone has bad days. Consider that there are times when you feel displaced anger too. On those days, you can only hope those around you will be tolerant. Give your customer the same understanding you'd like to have.

If the customer explodes, don't take it personally. After all, you may have just been a convenient target. You're not responsible for the problem, but how you handle the situation might provide the bright spot someone needs to adjust their attitude.

If it's manageable, be a good listener and let the customer vent. Empathize with them, acknowledging that you would probably feel the same way under the circumstances. Reassure the customer that you will do your best to fix one particular problem as quickly as you can, hopefully removing one thing from the list of woes.

Then do what you said you would do—solve the problem as quickly and effectively as possible. The ability to do that can be a bigger gift to people than you might realize.

2.2.7 Customers with Unresolvable Problems

Occasionally, you will encounter a problem you can't resolve. For example, the wiring may be too small to support the new unit the customer wants. Maybe a client complains that their system can't keep up with the heat, but it's obvious they added a great deal of living space recently. What should you do?

A poor response would be to say, "I'm sorry. Your electrical service is not adequate to handle this air conditioner. I can't do electrical work, so call us when you get that problem solved."

A better response might be, "I discovered that your electrical service was too small to handle the new air conditioner. We can't solve the problem directly, but here is the name of an electrical contractor we often deal with. Of course, you're free to use any contractor you like. If you prefer, we can subcontract a licensed contractor to do the job, and then we'll complete the rest of the installation as quickly as possible. I do apologize for the inconvenience."

A homeowner often chooses to contact the company that is recommended. Knowing that your company works with them adds a bit of comfort to the choice. Your positive, helpful attitude will leave a favorable impression. You have identified the problem and proposed multiple solutions, then offered to complete the job quickly when the electrical issue has been resolved.

2.2.8 Customers Who Request Help with Odd Jobs

Sometimes people ask a technician to perform a task that is not relevant to HVACR work. In these situations, you'll have to explain why you can't help in the right tone.

Exercise Scenario:

While you are working on a furnace, the customer asks you to move a freezer to another part of the house. What do you say?

Exercise Solution:

A good answer is, "I'm sorry, but company regulations won't let me handle appliances we don't service, due to the liability involved. Also, as you know, you're billed by the hour for my time. I know you're anxious to move that freezer. Perhaps a neighbor can help you. If not, having a moving company assist would likely be less expensive." This response shows that you recognize the inconvenience it causes and are offering alternative thoughts. It also explains why you aren't able to move the freezer yourself.

Liability is a valid answer, rather than just an excuse. If you drop the freezer and it rolls down the steps, who is responsible? Contractors understand that they accept liabilities every day in the HVACR service world. But they aren't interested in accepting liability for something that isn't part of their scope of work.

2.0.0 Section Review

1. True or False: It is a good idea to establish rapport with a customer by initiating conversation about general topics while you are working on their equipment.
   a. True
   b. False

2. If an angry customer starts berating you because equipment your company installed has stopped working, you should _____.
   a. immediately turn around and leave
   b. shift the blame to others in your company
   c. empathize with the customer
   d. tell the customer you don't want to hear it

Module 03316 Review Questions

1. A customer begins forming a first impression about you and your company _____.
   a. after you have fixed their equipment
   b. after you have submitted your bill
   c. within the first minute of your arrival
   d. as soon as you speak

2. You arrive at a service call wearing a NY Yankees ball cap. Unfortunately, the customer is a Boston Red Sox fan who hates the Yankees and immediately starts insulting the Yankees. Looking back on this incident, what do you think you would do in the future?
   a. Turn around and walk away because you won't do work for a Red Sox fan.
   b. Leave the ball cap in the truck to avoid the possibility of confrontation.
   c. Argue the relative merits of the two teams.
   d. Brush past the customer and find the equipment that needs to be repaired.

3. When speaking with a customer, *always* _____.
   a. start by telling them about your training and experience
   b. break the ice by chatting about local politics
   c. tell the customer if you think their equipment is low-quality
   d. show courtesy, concern, and respect

4. True or False: One way to make a good first impression is to use a lot of technical terms during the opening stages of the service call.
   a. True
   b. False

5. A customer is giving a long, boring description of a problem. It is obvious that the customer has no technical knowledge. You should _____.
   a. tune the customer out and start thinking about how you will approach the problem
   b. nod appreciatively and pretend you are interested even if you are thinking about something else
   c. listen actively to establish rapport and learn what you can about the problem
   d. talk about similar problems you have encountered and how you solved them

6. True or False: When gathering information from a customer, it's important to ask open-ended questions.
   a. True
   b. False

7. After telling the customer how long it will take you to complete a repair, you accidentally tear a length of flex duct in the system. Replacing it will lengthen the service call considerably and the customer is paying by the hour. What should you do?
   a. Tell them about your mistake and explain that the cost of that repair will not be included in their bill.
   b. Tell them that you discovered an additional problem with the system that will lengthen the service call and increase their cost.
   c. Tell them the problem was likely caused by another technician who previously serviced the system.
   d. Ignore the damage and let it be discovered during a future service call.

8. The customer is worried because they have read about families dying of carbon monoxide (CO) poisoning from a defective furnace. How should you respond?
   a. Tell them to relax and stop worrying.
   b. Explain that it hardly ever happens but it's good to be afraid.
   c. Calm them and use factual information to ease their mind.
   d. Ask your supervisor to contact the customer.

9. To access a furnace for service, a lot of clutter must be moved from in front of it. The way to deal with this problem is to _____.
   a. refuse to continue working until the customer moves it
   b. move it out of the way and leave it there when you are finished working
   c. call your supervisor and explain the problem
   d. move it out of the way and replace it when the work is finished

10. A client was supposed to have an electrical circuit installed to serve the new humidifier you came to install. The electrical work is not done, so the customer asks you to do it, even though you are not an electrician. What should you do?
    a. Go ahead and do the wiring because it's an easy job.
    b. Explain that the work must be done by a licensed electrician and provide a reference.
    c. Run an extension cord from a nearby electrical outlet so the humidifier can operate until a proper service is installed.
    d. Contact your supervisor and see if it is okay to do the electrical work.

Answers to Odd-Numbered Module Review Questions are found in *Appendix A*.

Answers to Section Review Questions

Answer	Section	Objective
Section One		
1. a	1.1.0	1a
2. c	1.2.1	1b
Section Two		
1. b	2.1.2	2a
2. c	2.2.6	2b

MODULE 03215

Basic Maintenance

Source: William Hager/Shutterstock

Objectives

Successful completion of this module prepares you to do the following:

1. Identify and describe the use of common maintenance materials and components.
   a. Identify and describe common gaskets and packing materials.
   b. Identify and describe common types of shaft seals.
   c. Identify and describe common types of bearings.
2. Describe the properties of common lubricants and how they are applied.
   a. Describe the properties of common lubricants.
   b. Explain how lubricants are selected and applied.
3. Identify and describe how to install various drive belts.
   a. Identify and describe various types of drive belts.
   b. Identify and describe how to align sheaves and calculate drive speeds.
   c. Explain how to install and apply the proper tension to drive belts.
4. Identify and describe basic inspection and maintenance procedures for selected equipment.
   a. Identify common environmental health hazards associated with HVACR maintenance activities.
   b. Describe common inspection and maintenance procedures for gas heating equipment.
   c. Describe common inspection and maintenance procedures for cooling and heat pump systems.
   d. Describe common inspection and maintenance procedures for various system accessories.

Performance Tasks

Under supervision, you should be able to do the following:

1. Lubricate a bearing using a grease gun.
2. Install, align, and adjust a drive belt.
3. Perform a periodic maintenance inspection on a gas furnace and document the inspection results.
4. Perform a periodic maintenance inspection on a cooling system or heat pump system in the cooling mode and document the inspection results.

Overview

HVACR equipment maintenance requires technicians to make intelligent decisions regarding equipment performance and condition. Technicians must also choose the correct materials to use in each situation. Common maintenance items such as air filters, drive belts, and lubricants must be chosen and applied correctly. This module provides guidance in periodic maintenance procedures and the selection of appropriate maintenance materials.

Digital Resource for HVACR

Scan this code using the camera on your phone or mobile device to view the digital resources related to this craft.

SCAN ME

NCCER Industry-Recognized Credentials

If you are training through an NCCER-accredited sponsor, you may be eligible for credentials from NCCER. The ID number for this module is 03215. Note that this module may have been used in other NCCER curricula and may apply to other level completions. Contact NCCER at 1.888.622.3720 or go to **www.nccer.org** for more information.

You can also show off your industry-recognized credentials online with NCCER's digital credentials. Transform your knowledge, skills, and achievements into credentials that you can share across social media platforms, send to your network, and add to your resume. For more information, visit **www.nccer.org**.

1.0.0 Gaskets, Packing, Seals, and Bearings

Performance Task

There are no Performance Tasks in this section.

Objective

Identify and describe the use of common maintenance materials and components.

a. Identify and describe common gaskets and packing materials.
b. Identify and describe common types of shaft seals.
c. Identify and describe common types of bearings.

Gaskets, packing, and shaft seals are used to create a seal between machine parts and prevent leaks. While gaskets are used to seal nonmoving parts, packing and seals are used to seal spaces between moving parts. Bearings support and maintain the position of a rotating shaft, reducing friction and allowing it to rotate freely.

HVACR technicians often need to handle gaskets and packing during routine system installation and maintenance. Their proper installation is critical to the performance and reliability of a system.

Shaft seals may be mechanical or nonmechanical in nature. They are critical to the prevention of leaks around a rotating shaft. Preventing leaks around a spinning shaft despite a difference in pressure is very challenging, as the slightest flaw will reveal itself quickly. Knowing which type of seal to use and how to properly install it is an important skill to learn.

HVACR systems use a variety of bearing types to reduce friction on rotating shafts. All motors, pumps, and compressors have bearings of some type. Some are field accessible while others are not. Today's residential systems have few accessible bearings, but larger commercial/industrial systems typically have more to support larger blowers and motor shafts.

WARNING!

Failure to de-energize and disable HVACR equipment before work begins can result in serious injury or death. Lockout/tagout (LOTO) procedures safeguard workers against unexpected releases from various energy sources. Energy sources can include electrical, mechanical, hydraulic, pneumatic, chemical, and thermal energy. Additional dangers exist if the system contains chemicals or flammable liquids and gases. Exposed blower wheels and fan blades should be blocked to prevent windmilling in windy conditions.

When working on HVACR systems, all equipment that could release energy must be shut down, drained, de-energized, or otherwise rendered harmless. Switches, circuit breakers, valves, and other components must be switched off or closed, and locks or tags must be applied to prevent inadvertent operation.

The appropriate PPE must always be worn when servicing HVACR systems. PPE may include safety glasses, safety shoes, hard hats, gloves, electrical safety gear, fall arrest equipment, dust masks or respirators, and other equipment.

1.1.0 Gaskets and Packing

The basic function of a gasket is to create a seal between parts in a fixed position. Flanged joints in piping systems and compressor cylinder heads are two common examples. Generally, the thinnest gasket that creates an effective and reliable seal is the most efficient and will last the longest.

There are many types of gasket materials (*Figure 1*). The type of gasket used depends on the following:

- The operating temperature range
- The operating pressure range
- The type of connection being made
- The type of fluid to which the gasket is exposed

In the HVACR field, *packing* refers to a rope-like material that is often impregnated with a lubricant and formed into a specific shape. Packing is used to control or prevent leakage around a rotating stem or shaft in valves and pumps. In the case of valves, the valve stem rotates very slowly, so speed and heat production are not factors. Pump shafts, however, rotate continuously and at a high speed. *Figure 2* provides an example of how valve packing is applied.

Figure 1 A collection of flange gaskets.

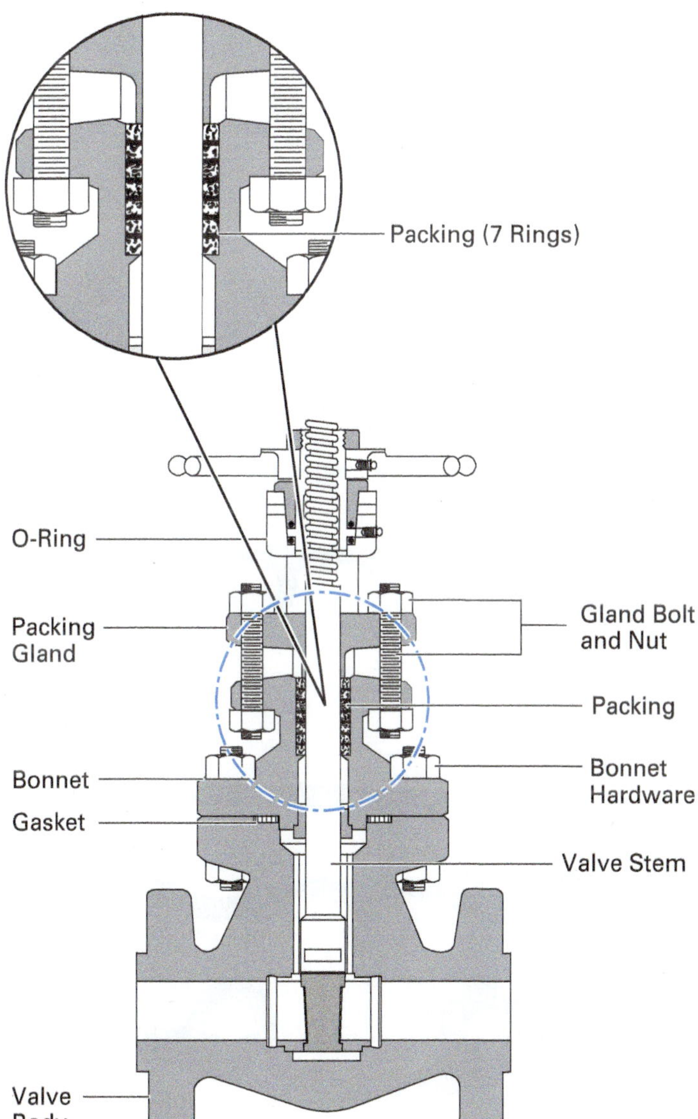

Figure 2 Valve components.

1.1.1 Gaskets

Pre-formed gaskets in standard sizes and patterns are readily available from equipment suppliers and manufacturers. These factory-made gaskets are cut to the required pattern, with holes pre-punched for the specific application.

Gaskets are made from natural or man-made materials, and sometimes both. *Figure 3* shows some common gasket types. Most gaskets you encounter belong to one of the following categories:

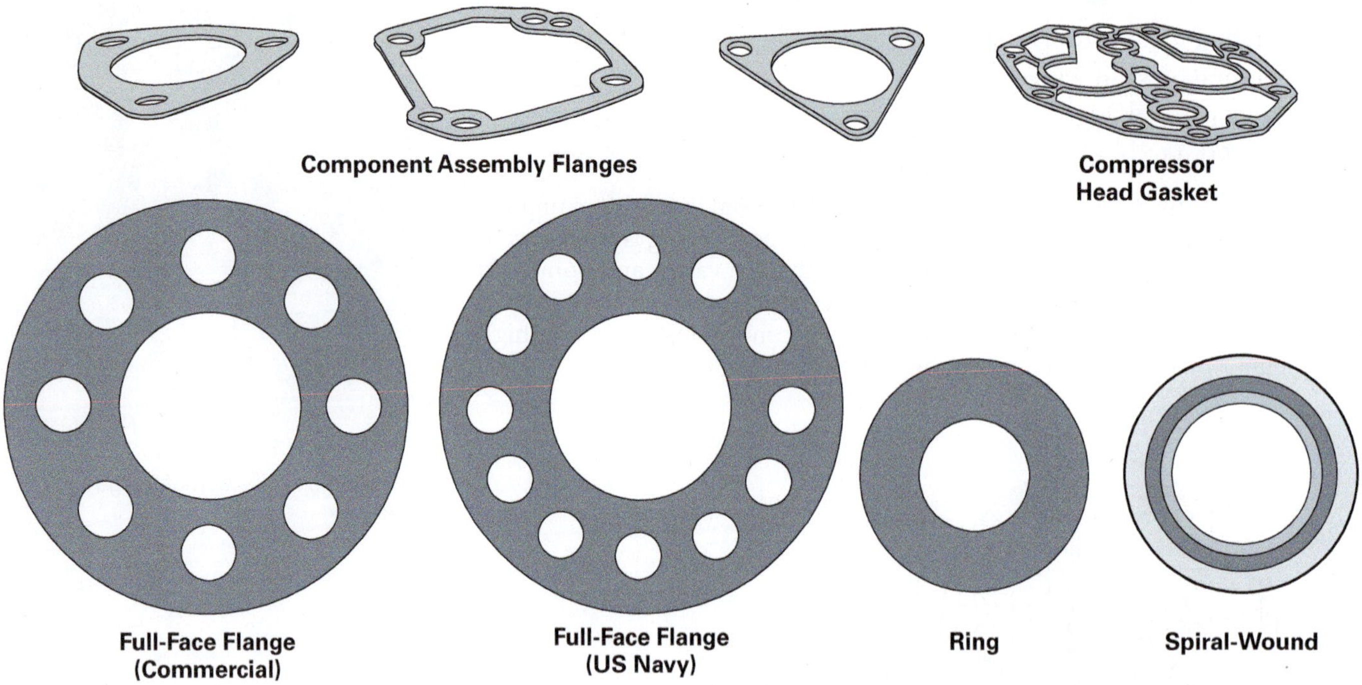

Figure 3 Common gaskets encountered in HVACR work.

- *Component assembly gaskets* — These are made from various materials with many possible uses. They are provided by equipment manufacturers to fit specific components. While many are flat, some do have texture, such as the compressor head gasket shown in *Figure 4*. Some very unusual shapes and sizes fall into this category. Many are virtually impossible to make in the field or shop.

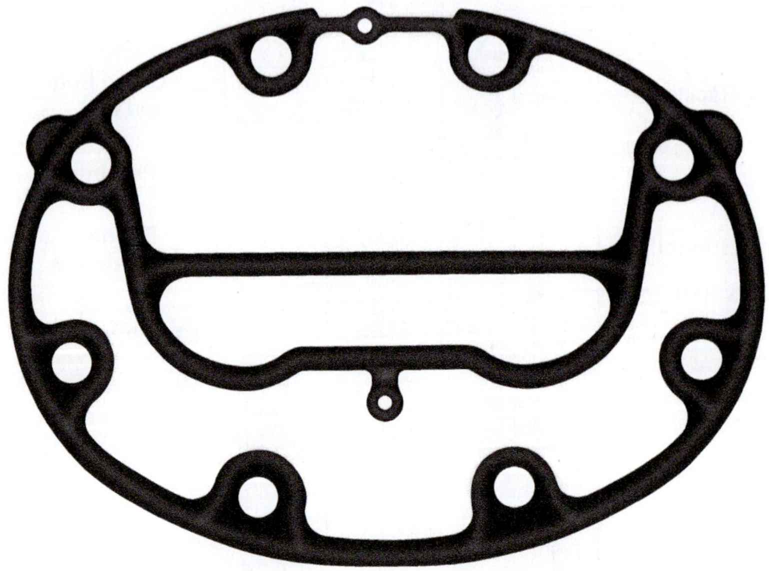

Figure 4 Compressor head gasket for a Carrier semi-hermetic compressor.

- *Full-face flange gaskets* — These are flat gaskets made from various materials to resist the conditions of the application. On a full-face pipe flange, the gasket covers the mating faces completely. As a result, the gasket must have properly spaced and sized holes to accommodate the flange bolts.
- *Ring flange gaskets* — Ring gaskets are also flat gaskets, but they are smaller than a full-face gasket and have no bolt holes. They are made to fit inside the bolt circle of a pipe flange. This type of gasket is used on raised-face pipe flanges.
- *Spiral-wound gaskets* — Spiral-wound gaskets (*Figure 5*) are often made of stainless steel with a graphite insert. They are typically used on high-temperature, high-pressure piping systems. They are crushable gaskets that are very elastic in nature. This allows them to respond to changes in line pressure, thermal shock, and vibration.

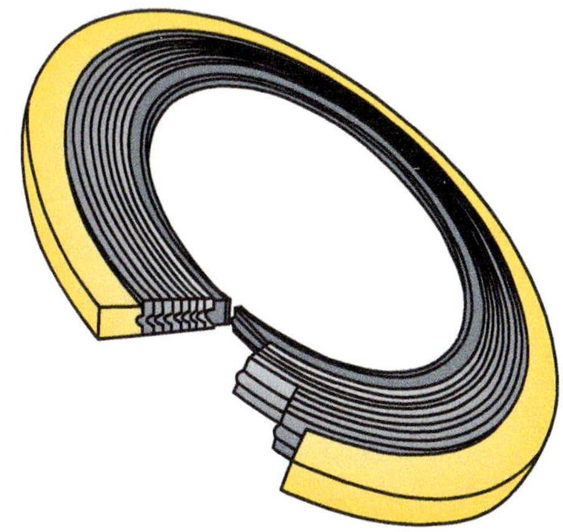

Figure 5 Spiral-wound gasket construction.

Choosing a gasket is not a responsibility to take lightly. The list of materials used to fabricate gaskets is extensive, and each material has its own strengths and weaknesses. For equipment assembly, it is always best to use gaskets provided by the equipment manufacturer. The manufacturer selects the material to ensure its compatibility with the fluids as well as the expected temperature and pressure range. Equipment gaskets are also more likely to be unique in shape or texture.

WARNING!

Always replace a gasket with one provided or specified by the equipment manufacturer. Using an incompatible material may cause an unexpected release of energy or hazardous fluids.

Pipe flanges are divided into pressure classes assigned by the American Society of Mechanical Engineers (ASME) in ASME Standard 16.5, *Pipe Flanges and Flanged Fittings*. Typical pressure classes for forged carbon steel and cast-iron flanges are shown in *Table 1*.

TABLE 1 Flange Pressure Classes

Forged Carbon Steel	Cast Iron
Class 150	Class 25
Class 300	Class 125
Class 400	Class 250
Class 600	Class 800
Class 900	
Class 1,500	
Class 2,500	

The number for each class represents a pressure in pounds per square inch (psi), but it does not necessarily reflect the maximum pressure the flange can withstand under every set of conditions. Within a given pressure class, the maximum pressure rating for a flange is determined by two primary factors: the temperature of the process fluid and the material from which the flange is made. The design pressure rating for flange classes at various temperatures can be found in the tables of ASME Standard 16.5. Ratings are determined based on **hydrostatic testing** to the component bursting point. The burst pressure is then divided by a factor of 3.0 to provide a significant safety margin.

Hydrostatic testing: A method of testing pressure vessels, piping systems, and similar components for leaks by filling them with a liquid (typically water) and then applying pressure using an inert gas to test the component's structural integrity.

Flange gasket selection is, then, affected by the pressure class of the flange set. In addition, the gasket dimensions and bolt patterns vary from class to class for the same pipe size. These differences in physical characteristics help prevent mismatches. To ensure you have the right gaskets, you'll need to know both the pipe size and the pressure class of the flanges.

Figure 6 Gasket scraper.
Source: Courtesy of Stanley Black and Decker, Inc.

Follow these basic steps to install a flange gasket:

Step 1 Clean the surface where the gasket is to be installed. If an old gasket must be removed, use a gasket scraper (*Figure 6*) or a liquid gasket remover for stubborn materials (*Figure 7*). Avoid damaging the surfaces; do not gouge the surface with the scraper. Never use a screwdriver or collapsible knife, and wear the appropriate PPE as described in the scraper instructions, including gloves. Once the gasket has been removed, use an appropriate solvent to clean the mating surfaces as needed.

WARNING!

Wear appropriate PPE as described in the scraper instructions, including gloves. Remember that solvents typically require nonporous gloves, such as nitrile or latex. Follow the solvent instructions carefully and ensure there is adequate ventilation.

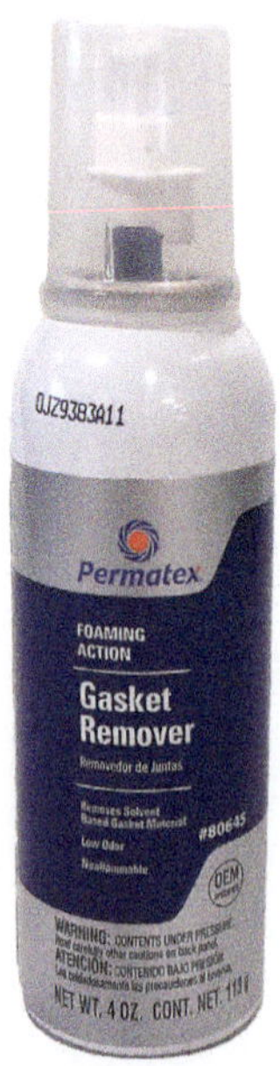

Figure 7 Liquid gasket remover.

Step 2 Inspect the flange surfaces for irregularities or damage. Carefully dress the faces if necessary to remove burrs.

Step 3 If specified, apply an adhesive, sealer, or lubricant to the gasket.

Step 4 Position the gasket before inserting the bolts. Use one or more drift pins to align the bolt holes in the two flanges. However, two pipes should never be forcibly aligned, where stress is constantly applied to one or both pipes.

Step 5 Insert the flange bolts and hand-tighten the nuts. Insert bolts into any holes used for alignment last.

Step 6 Use a calibrated torque wrench to tighten the bolts, following the proper tightening sequence. The sequence varies somewhat with the number of bolts, but the typical method is to torque one bolt and then torque the bolt on the opposite side in a crossover pattern.

- Tighten the bolts in increments of one-third of the final torque, using the crossover pattern consistently. For example, if the torque is specified as 120 foot-pounds (ft-lb), tighten each bolt to 40 ft-lb, then to 80 ft-lb, and finally to 120 ft-lb. In the metric system, the torque would be approximately 163 Nm. The bolts would then be torqued to 54 Nm, 108 Nm, and then to 163 Nm to complete the task.

Step 7 Complete the assembly by tightening each bolt to the specified value in the proper sequence.

CAUTION

Adhesives, sealers, or lubricants should never be used unless specified by the equipment manufacturer. In some cases, small amounts of adhesive can be used to hold the gasket in position during installation. If any such material is used, it must be compatible with the gasket material and the process fluid, and it should be used sparingly.

Around the World

Protection from Asbestos Exposure

The US Environmental Protection Agency (EPA) has prioritized the protection of Americans from asbestos exposure. Many products were banned from the marketplace in 1989, but many others remained on the market. In 2019, a new ruling brought the remaining products, including gaskets and many similar uses, under EPA control.

The new ruling allows the EPA to evaluate the use of asbestos in a product and the level of hazard it represents. Companies are required to notify the EPA at least 90 days before they begin manufacturing, importing, or processing asbestos or asbestos-containing products. The requested use is prohibited until the EPA conducts a thorough review and prepares a response. They now have the power to allow or deny requests on an individual basis.

Thus, there is no guarantee that gaskets and similar materials you'll work with are free of asbestos. Many older asbestos-bearing gaskets also remain in service.

Around the world, many countries have banned all uses of asbestos. However, several countries, including Russia and China, continue to mine asbestos and export the product.

For more information about asbestos and the hazard it represents, visit **https://www.asbestos.com/** and the EPA website at **https://www.epa.gov/**.

1.1.2 Shaft Packing

Shaft packing is a rope-like material that may be square, wedge-shaped, or a unique shape for special applications. The most common form of packing you will likely encounter is braided.

Common packing products include those made from graphite yarn, carbon yarn, flax, graphite foil, aramid, and polytetrafluoroethylene (PTFE). Various types of packing are shown in *Figure 8*.

CAUTION

Do not overtighten flange bolts. This can damage the gasket, elongate the fasteners, or distort the flanges. Determine and use the proper torque for the flanges being assembled.

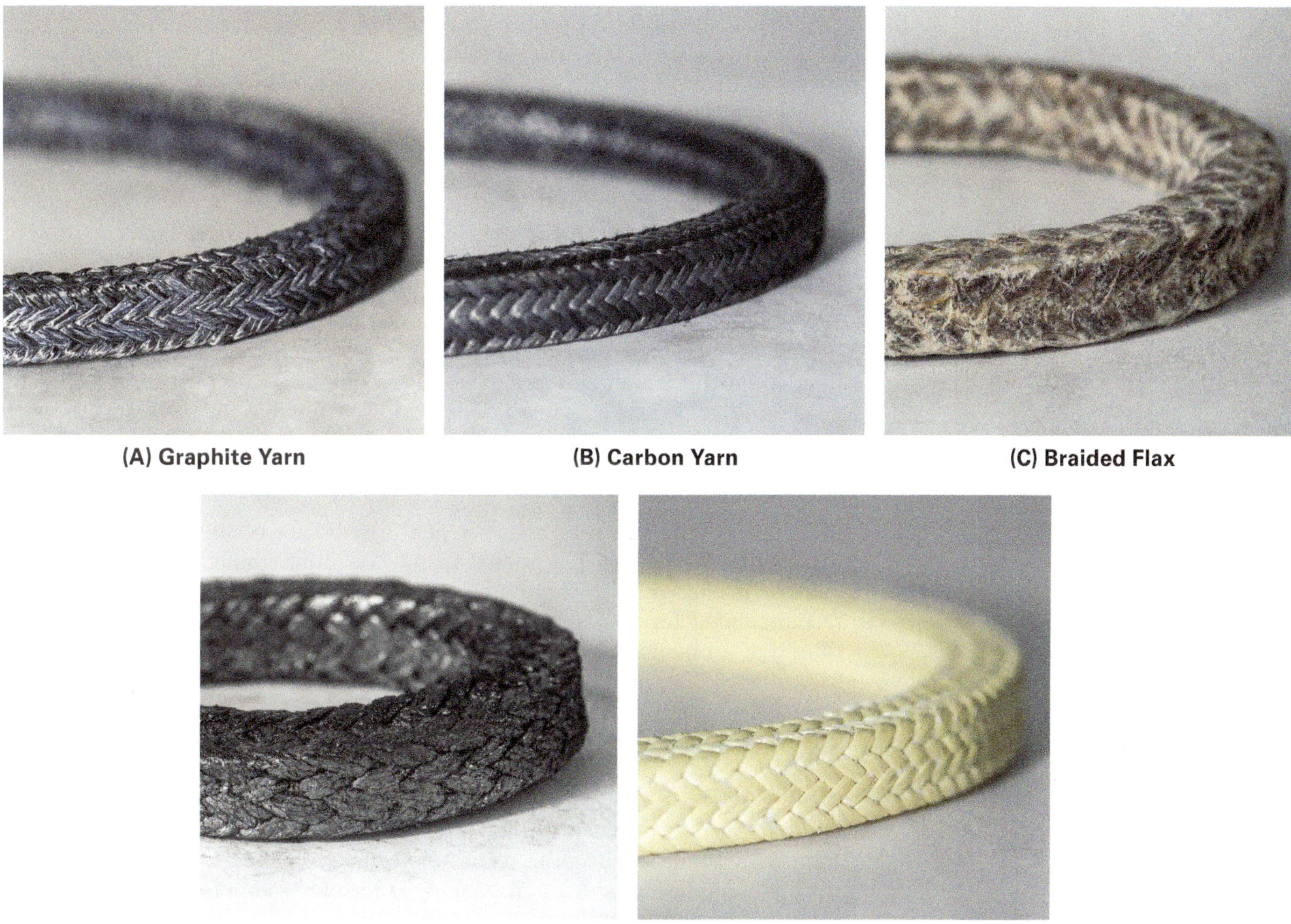

(A) Graphite Yarn **(B) Carbon Yarn** **(C) Braided Flax**

(D) Graphite Foil with Inconel® Wire **(E) Aramid Yarn**

Figure 8 Examples of shaft packing.

Source: Courtesy of SEPCO, www.sepco.com

It is always best to use the packing type as directed or provided by the equipment manufacturer. The following factors are considered by manufacturers when choosing packing:

- Operating pressure range
- Operating temperature range
- Rotational speed range
- Fluid flowing through the pump or piping
- Composition of valve stem or shaft

The following types of packing material are likely to be encountered in HVACR work. Keep in mind that new products are constantly being developed and these materials may be combined into a single product. Refer to *Figure 8* for visual identification.

NOTE

PTFE is often referred to as Teflon®, because the Teflon brand of PTFE is the most recognizable due to its use in cookware and other domestic products.

Caustics: Strong alkaline substances that destroy or damage organic tissue and cause significant metal corrosion.

Graphite Yarn

Graphite is a good heat conductor and dissipates heat from a shaft out to the surrounding structure. This permits higher shaft speeds with less leakage than some other materials. Graphite yarn packing is commonly used on both rotating and reciprocating shafts and valves. It can also be used in applications where strong **caustics**, acids, and chemicals are present. It is also suitable for high-pressure steam applications.

Carbon Yarn

Carbon yarn packing is used on rotating and reciprocating shafts and in high-temperature valves. It is used with many chemicals, except for those that are strong oxidizers. Carbon yarn packing is often treated with graphite particles to improve heat conductivity. It is often combined with PTFE as well.

The rotational speed of a shaft is a factor in packing selection for pumps and similar equipment. In this case, rotational speed is measured in linear units, such as feet per minute (fpm) or meters per second (mps). Carbon yarn packing is generally limited to shaft speeds up to 3,000 fpm (15.2 mps). It can be used for temperatures up to 650°F (343°C).

Vegetable Fibers

Vegetable-fiber packing can be used on rotating and reciprocating shafts. Braided flax fiber is a popular choice. It is recommended for use with brine, cold water, and cold oil. Vegetable-fiber packing is suitable for shaft speeds up to 1,885 fpm (9.6 mps) and temperatures up to 220°F (104°C).

Metals

Metallic packing material consists of thin ribbons of foil manipulated into the shape of packing. The material may be impregnated with oil. Metallic packing is used on centrifugal and reciprocating pumps and valve stems. It is recommended for use with steam, oils, gasoline, air, ammonia, and water. It should never be used on soft metal shafts and rods. Shafts made from brass or bronze, for example, may wear more quickly than the packing.

Metallic packing can withstand temperatures of up to 1,250°F (677°C) and steam pressures up to 5,000 psi (34,474 kPa).

Aramid Yarn

Aramid yarn is a very strong, chemical-resistant packing material. It is commonly braided with other types of yarns, such as PTFE, for use in a wide variety of applications. Aramid packing is recommended for reciprocating and rotary shafts as well as many valve applications.

Aramid yarn is compatible with water, oil, grease, weak acids, and alkaline solutions. It can be used at temperatures up to 500°F (260°C) and rotational speeds up to 1,900 fpm (9.7 mps).

PTFE

Several variations of PTFE are used to manufacture packing. PTFE can be used in the most severe applications related to corrosives and oxidizers. It is used extensively in the food processing industry and anywhere a material approved by the US Food and Drug Administration (USFDA) is required. PTFE is limited to temperatures up to 500°F (260°C). PTFE is also resistant to many corrosive fluids.

1.1.3 Removing and Replacing Packing

The service life of packing depends on the packing material and the application. Packing can last from a few weeks to years. In many cases, packing that begins to leak excessively can be reseated by simply tightening the packing gland. Applying pressure to the packing using the gland is how adjustments are normally made. Eventually, though, the packing becomes too worn or compressed and must be replaced. It is also replaced as a precaution when internal service work is required on a pump or valve.

Before proceeding, understand the following important point about packing. Unlike a gasket, pump packing applications typically have an allowable or specified leak rate. In other words, some amount of leakage is not only expected, but a small amount of leakage is often required to cool and lubricate the packing. In the case of pumps, an apparatus to collect and drain the leakage away is generally provided.

Before new packing can be installed, the old packing must be removed. In the case of a pump, it must be removed from the *stuffing box* (*Figure 9*). Removing packing is a relatively simple procedure, but it is important to avoid damaging the shaft. The procedure for removing packing may vary slightly from one piece of equipment to another. Refer to the equipment manufacturer's documentation for specific procedures.

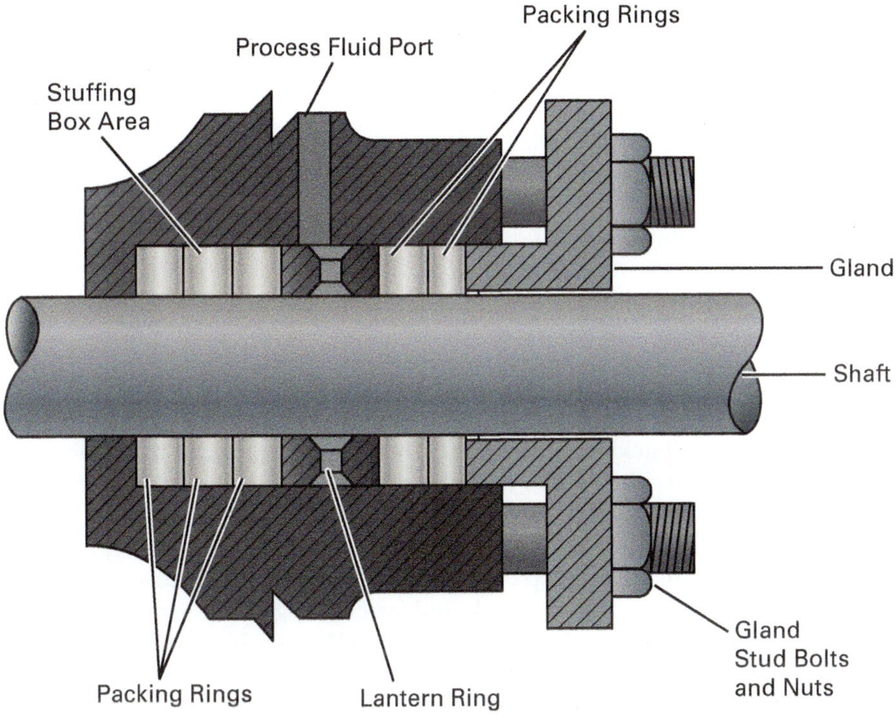

Figure 9 Pump stuffing box assembly.

A packing puller or extractor is typically used to remove the packing from around a shaft (*Figure 10*). Many packing extractors have a flexible shank with either a hook or a spiral at the tip. If the puller is properly screwed into the packing ring, the ring often comes out in one piece. If the ring breaks during removal, all the smaller pieces must be carefully removed. To help prevent breakage, two pullers may be used at the same time to pull the packing from both sides of the shaft.

In *Figure 10*, note the **lantern ring** shown in the space between groups of packing rings. A lantern ring is a spacer that is often placed in a pump stuffing box to support lubrication and cooling. Fluid is routed into and through the ring, then flows into the stuffing box to lubricate and cool both the shaft and the packing. A lantern ring can extend the life of pump packing as well as the shaft. When present, it too must be removed to access the packing rings behind it. However, not all stuffing boxes are equipped with a lantern ring.

Lantern ring: A perforated ring positioned near the center of a pump stuffing box, providing a point of access and a distribution point for fluid entering to cool and lubricate the packing and shaft.

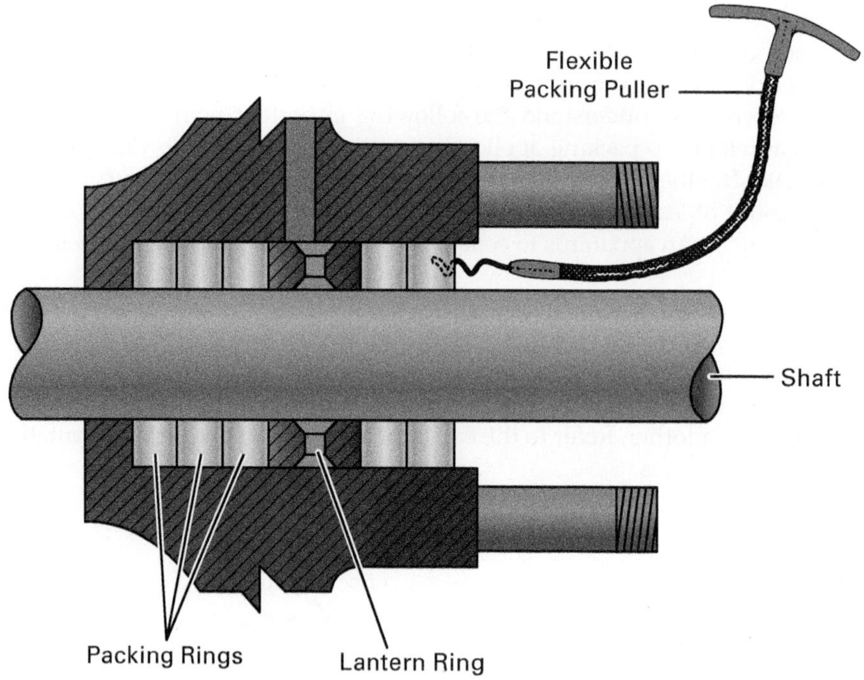

Figure 10 Packing puller screwed into packing.

When using a packing puller, be careful not to scratch the shaft. Keep the puller angled away from the shaft. Even minor shaft damage will cause accelerated packing wear and failure. Any burrs or nicks on the shaft must be removed before the new packing is installed. If there are deep scratches or excessive shaft wear, shaft replacement may be required.

Installing Packing

Installing new packing is not difficult but it does require time to prepare the rings. Unless precut packing is acquired from the equipment manufacturer or similar source, individual rings must be cut from a roll to fit the shaft. It is very important that each ring fits around the shaft tightly, but without being stretched. Because the cut faces of each ring must meet perfectly, the rings must be cut with precision using a sharp enough tool to make a clean cut.

There are two ways to fit and mark and/or cut rings. One way is to wrap the material around the shaft multiple times, taut but without stretching, and then make a straight mark across the packing (*Figure 11*). The number of wraps depends on the number of rings needed. Then remove the packing from the shaft and cut it on a bench with a sharp, sturdy knife.

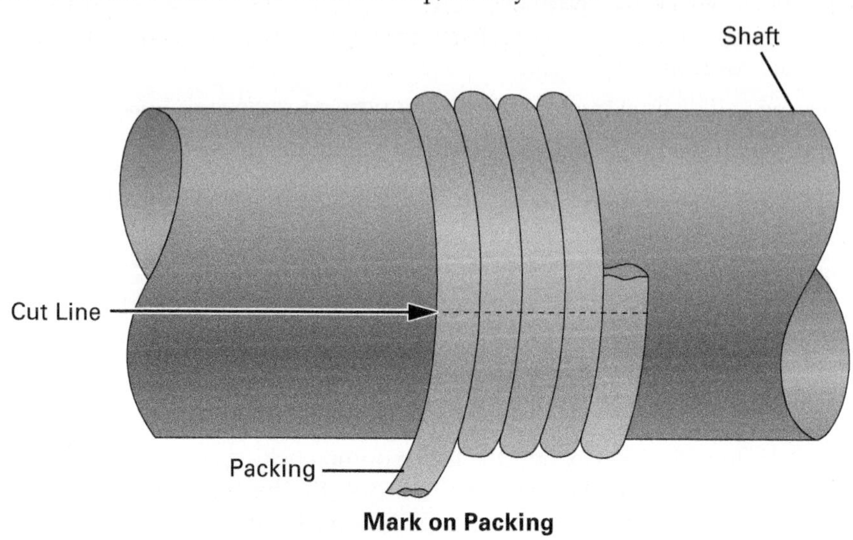

Figure 11 Marking the packing cut lines.

Alternatively, the material can be wrapped around a **mandrel** with the same diameter as the shaft. The packing can then be marked and removed for cutting, or it can be cut while it remains wrapped around the mandrel. It is important that the mandrel be precisely the same size as the shaft, or a poor fit and wasted material will result.

Some packing manufacturers recommend a straight cut; others recommend a cut made at a 45-degree angle (*Figure 12*). Check the manufacturer's documentation or job specifications and follow the recommendations for cutting the packing. When the packing is cut while wrapped around a mandrel, a cut can be made across several wraps of packing at the same time.

Mandrel: A cylindrical or tapered rod or shaft around which other materials can be shaped or forged.

CAUTION

Do not cut packing against the shaft of the equipment being serviced. The knife can damage the shaft.

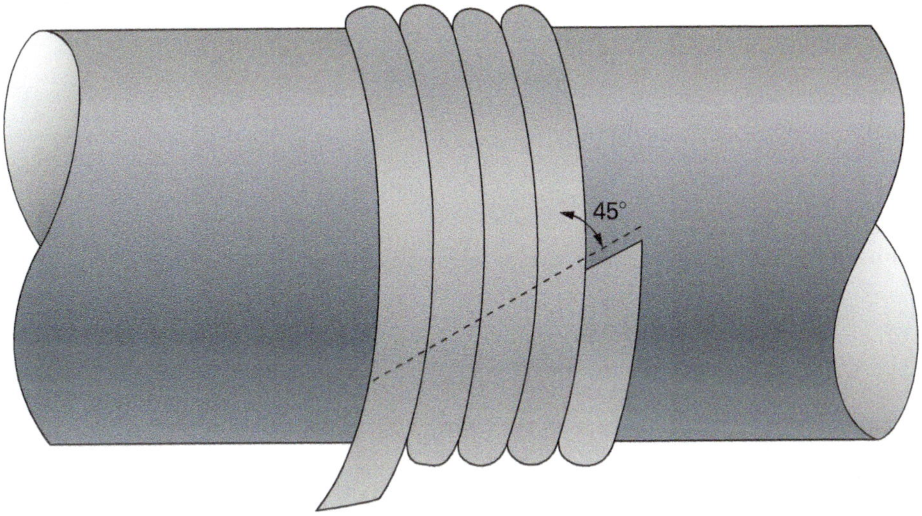

Figure 12 Packing cut at a 45-degree angle.

A clean, straight cut is important. A sharp, sturdy knife can be used, and a small miter box can be used with the knife to ensure a straight cut. Never use a saw with the miter box, as the teeth will damage the material.

Follow these steps to install compression packing:

Step 1 Inspect the stuffing box and shaft to ensure that they are clean and that there are no burrs, nicks, scratches, or signs of excessive wear. Any burrs or nicks on the shaft must be resolved using a fine file or emery cloth. If there are deep scratches or excessive wear is evident, determine if the shaft is serviceable based on the manufacturer's guidance.

Step 2 Acquire the proper type and size of packing for the job. The packing configuration and material are usually determined by a product or mechanical engineer and specified in a work order.

Step 3 Wrap the packing tightly around the shaft or mandrel. The number of revolutions required is equal to the number of rings required for the job.

Step 4 Determine if the packing should be cut straight or at a 45-degree angle. Position the material to minimize waste, starting near the end of the packing. Mark the packing for cutting.

Step 5 Remove the packing from the equipment shaft. If you use a mandrel and plan to cut the packing in place, leave it in position. Make a single cut across all the wraps while the material remains wrapped around the mandrel. Alternatively, place the packing on a hard surface or in a miter box for cutting with a knife.

Step 6 Cut on the marks with a sharp knife, using either a straight cut or an angled cut as directed.

Step 7 If specified by the equipment manufacturer, lubricate the packing rings on the inner and outer surfaces with a commercial packing lubricant. Some products are water-based while others are based on oil or silicone. Ensure the lubricant is compatible with the packing material and the process fluid.

CAUTION

Use an S-shaped twist to slip the rings onto the shaft. Do not bend the ring open in a C shape to slide it over the shaft, because this might crack some types of packing.

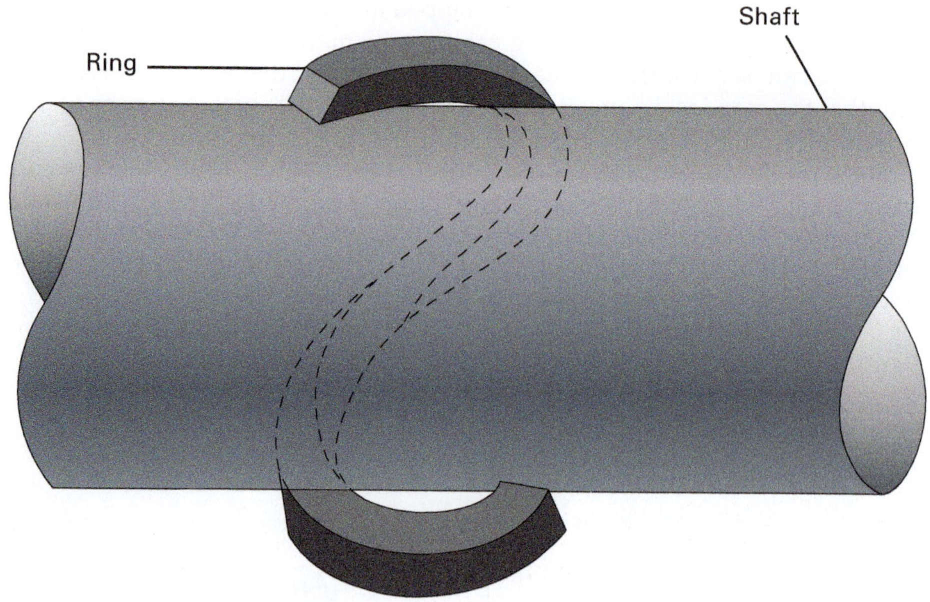

Figure 13 Wrapping a packing ring around a shaft.

CAUTION

Although the stack of packing rings can eventually be pushed down using the gland hardware, this tends to compress the upper rings more than the lower rings. To ensure balanced loading, use a blunt tool to firmly position each ring inside the stuffing box before adding the next ring.

CAUTION

Never position the rings on the shaft with the cuts of adjacent rings aligned.

Step 8 Slip one ring onto the shaft by twisting the material around the shaft in the shape of an S, as shown in *Figure 13*.

Step 9 Slide the ring to the back or bottom of the stuffing box. Use a blunt tool that will not damage the packing, such as a pin punch slightly smaller than the packing width, to firmly position the packing.

Step 10 Slip the second ring onto the shaft and push it into the box, positioning the cut 90 degrees away from the cut in the first ring. The cuts in the rings should be staggered 90 degrees apart as you continue installing the rings. Staggering the location of the cuts helps eliminate leak paths.

Step 11 Install and seat additional rings until the correct number has been installed in front of the lantern ring.

Step 12 Reinstall the lantern ring. The lantern ring must ultimately align with the lubrication inlet port. This will always be present when fluid lubrication is provided to the stuffing box. Ensure the lantern ring is aligned slightly forward (toward you) of the port (*Figure 14*). If the proper size and number of rings have been installed behind the lantern ring, it will align with the port once the packing has been compressed by the gland. Most importantly, the ring should not be positioned beyond the port when pushed against the packing rings in place.

Step 13 Install the remaining packing rings, firmly seating each ring against the one behind it.

Step 14 Install the packing gland. The gland will extend out of the stuffing box more than before because the packing rings have not yet been compressed.

Step 15 Install and tighten the gland nuts enough to apply light pressure to the packing. Consult the pump manufacturer's documentation for any specific instructions related to this step. The gland must not be tightened further until the pump is started; final adjustments are made with the pump in operation. If the packing is compressed too tightly now, the pump motor may not be able to start against the friction.

Step 16 Open the necessary valves and start the pump. With the pump in operation, tighten the nuts until there is slight leakage around the packing. Remember that slight leakage is required for most pumps, to provide lubrication and cooling of the shaft and packing.

Step 17 Check the pump after a few hours of operation to see if additional adjustments are required.

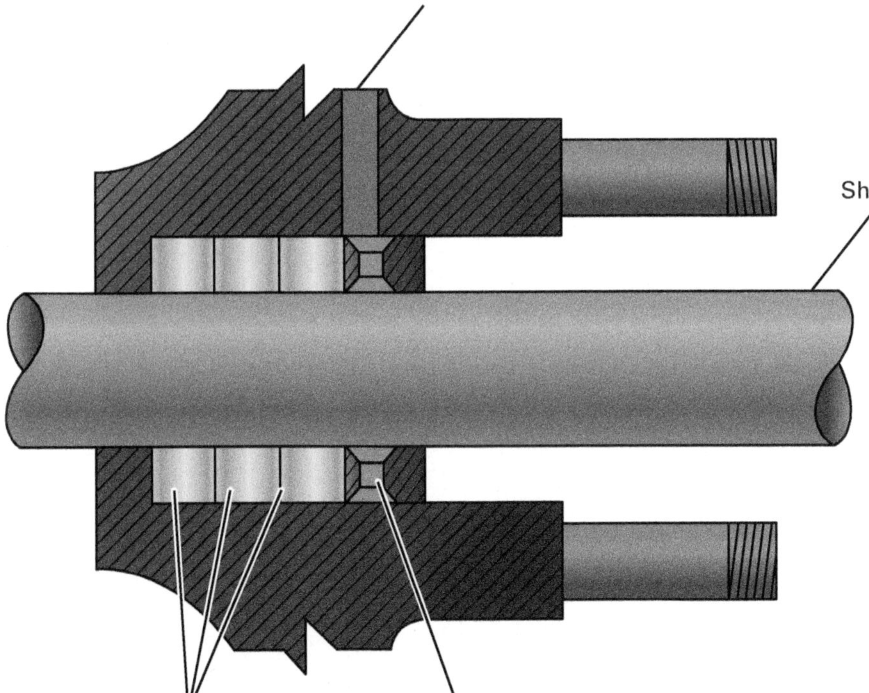

Figure 14 Lantern ring position prior to compression.

1.2.0 Seals

Seals are used to prevent or control leakage around rotating shafts. There are two broad classes of seals: nonmechanical and mechanical.

1.2.1 Nonmechanical Seals

A nonmechanical seal can be used as either a *static* seal or *a dynamic* seal. A static seal is one in which there is no movement between the two joining parts, or between the seal and the mating part. O-rings (*Figure 15*) are examples of seals used as static seals.

A dynamic seal is one in which there is movement between two mating parts, or between one of the parts and the seal. Lip seals, also referred to as *oil seals*, are dynamic seals used for rotating shafts (*Figure 16*).

Figure 15 O-rings.

Source: SergeiRamiltsev/Getty Images

Figure 16 Double-lipped oil seal.

Source: Courtesy of SKF, Lincoln Industrial Products, Alemite LLC

O-Rings

O-rings are generally made from rubber or rubbery synthetic materials. Some of the more common materials are Buna-N (nitrile), ethylene propylene, Viton™, PTFE, silicone, and polyurethane.

One example of a static application is a flange cap (*Figure 17*). When the O-ring meets the surfaces to be sealed, it is slightly distorted by an action called *mechanical squeeze* (*Figure 18*). Mechanical squeeze creates firm contact with the surfaces to be sealed. Engineers generally design the components to allow for about 10 percent of an O-ring's diameter to be compressed. This is very important: if there is too little compression, the O-ring won't provide a reliable seal under pressure. It must be slightly thicker than the groove depth so there will be some compression when the two surfaces come together.

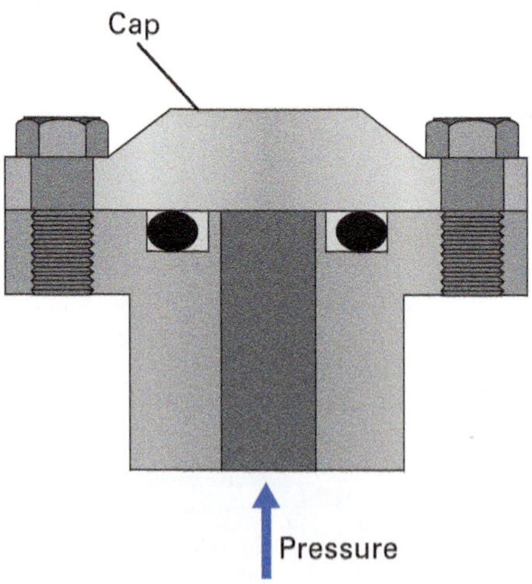

Figure 17 Static O-ring application.

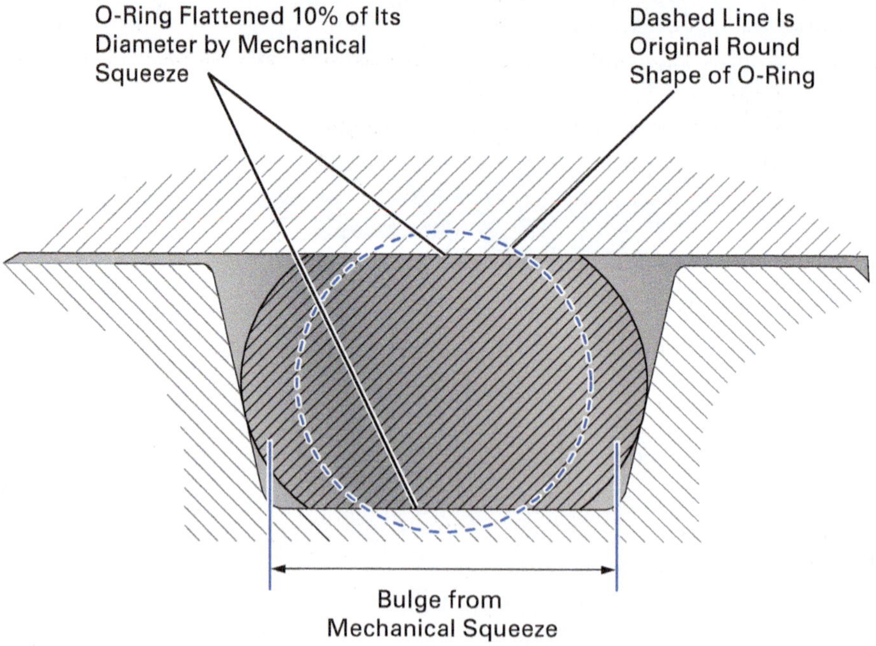

Figure 18 O-ring distortion from mechanical squeeze.

When pressure from the fluid is applied to one side of the O-ring, it tends to slide across the groove, away from the pressure. More pressure, up to the maximum pressure the O-ring can withstand, improves the sealing action.

An O-ring around a sliding shaft is a good example of a dynamic application (*Figure 19*). However, O-rings are only used this way when the shaft or stem turns slowly and intermittently. They are not a good choice for a constantly rotating shaft.

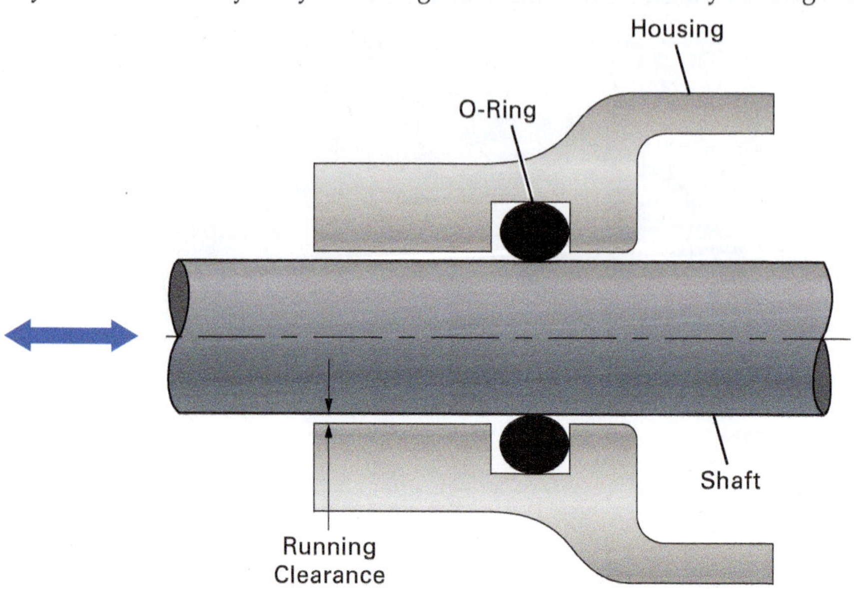

Figure 19 Dynamic application for an O-ring.

O-rings come in a wide variety of sizes and are made from materials that suit the application. When replacing an O-ring, follow the manufacturer's recommendations for size and material to ensure a reliable seal.

Installing an O-ring is a relatively simple procedure. Applying a lubricant is best unless the specifications indicate otherwise. The lubricant aids installation, but it can also be a factor in forming an effective seal. Clear silicone grease is a common O-ring lubricant.

Before lubricating an O-ring, wash your hands and keep dirt and grit out of the grease. One good lubrication technique is to place a small amount of lubricant in a plastic bag along with one or more O-rings. Then massage the O-rings around in the bag to ensure all the surfaces are covered. Leave them in the bag until you are ready to install them, removing one at a time.

To achieve a proper seal and protect the new O-ring from damage, observe the following basic guidelines:

- Use the correct size and material of the O-ring for the application. This information may come from job specifications or the installation, operating, and maintenance (IOM) manuals for the equipment being serviced.
- Make sure the equipment is properly prepared to receive the new O-ring. The groove must be clean and free of sharp edges.
- Lubricate the O-ring and the groove(s) using a lubricant acceptable to the O-ring or equipment manufacturer.
- Do not overstretch an O-ring. If it requires too much effort to fit an O-ring into place, you may have the wrong size. Overstretching can cause small tears that are not visible when it returns to its original shape.
- Do not roll an O-ring down a shaft or other component; slide it on without rolling. This is one of the reasons a lubricant is needed. Rolling causes the O-ring to twist, and it might remain twisted once it reaches the groove. Once the ring is in the groove, move it around and examine it carefully to ensure it is not twisted.
- Follow the same bolt-tightening procedure as you would for flanges and similar assemblies. Do not overtighten the components; use the manufacturer-specified torque value.

Oil Seals

Oil seals are also called *lip seals*. They are designed for low-pressure, positive-contact applications. They are commonly used on rotating and reciprocating shafts to keep fluids in and foreign matter out.

Oil seals (*Figure 20*) are surrounded by a metal retainer that aligns and supports the outside edge of the seal. Most have a metal retainer around the outside to

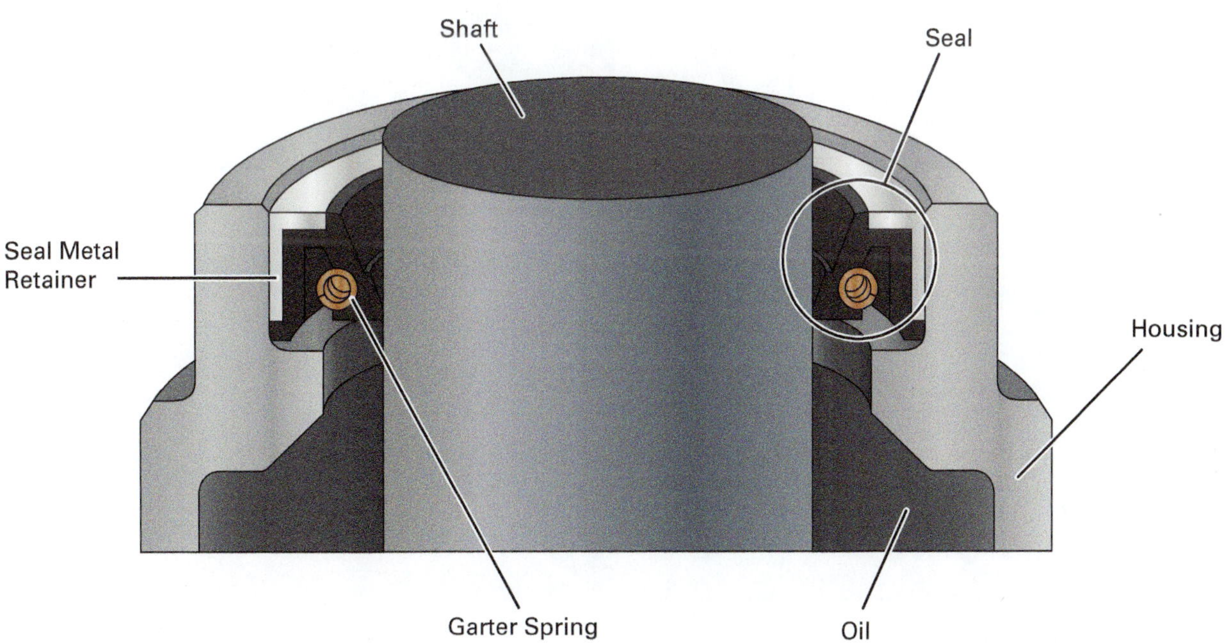

Figure 20 Oil seal, also known as a lip seal.

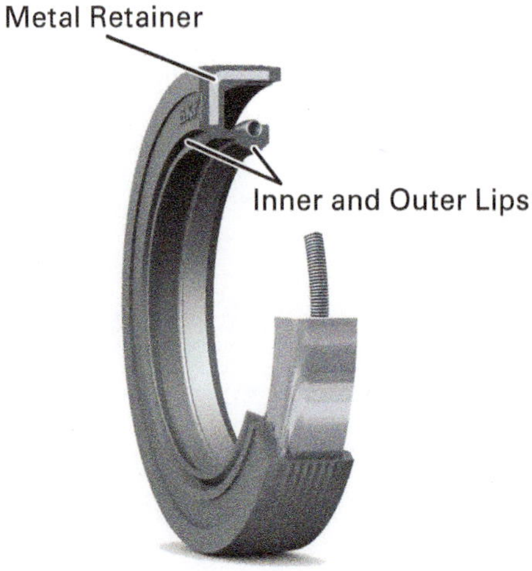

Figure 21 Double-lip seal.

Source: Courtesy of SKF, Lincoln Industrial Products, Alemite LLC

maintain a consistent shape. The flexible lip fits snug around the shaft, held against it by a spring that runs around the inside perimeter of the lip.

Oil seals are usually needed to contain lubricant in a housing, with the lip oriented toward the lubricant or fluid. The seal in *Figure 20* is shown in this position. In this configuration, with the lip facing the lubricant, a lip seal can generally withstand a small positive pressure, such as 10 pounds per square inch (psi).

When dirty conditions exist outside of the housing, the seal may be installed with the lip facing away from the machine to better prevent foreign matter from getting inside. In this position, even a small amount of pressure can force the lip away from the shaft, causing it to leak. For this reason, when the lip of the seal faces out, the housing is usually vented to the atmosphere through what is known as a *breather valve*. This prevents pressure from building up behind the seal.

Many breather valves are simple check valves that open to allow slight pressure to escape but are otherwise closed to keep contaminants from entering. Two separate seals may also be used together in an assembly, with the inner seal facing in and the other facing out. The two seals may be made from different materials. It is very important to properly orient a seal when it is installed.

In some cases, the area being sealed is operating in a vacuum rather than at a positive pressure. Double-lip seals (*Figure 21*) are often used for these applications, and in other situations when keeping a fluid in and dirt out are both priorities. They are also used when there are fluids on each side of the seal that must remain separated.

1.2.2 Mechanical Seals

Mechanical seals provide superior performance when compared to packing and nonmechanical seals. They can provide leak-free service. They are typically used on shafts of open-drive compressors and pumps. *Figure 22* shows a mechanical seal installed in a pump.

Figure 22 Mechanical seal in a pump.

Source: Nordrodon/Shutterstock

Mechanical seals typically seal by mating two carbon surfaces machined perfectly flat. The two sealing faces are referred to as the *rotating face* and the *stationary face* (*Figure 23*). A closing mechanism provides the energy required to keep the seal faces in contact with each other. Some have a single spring like the one shown in *Figure 23*. Others have multiple small springs (*Figure 24*) or metal bellows to push the faces together.

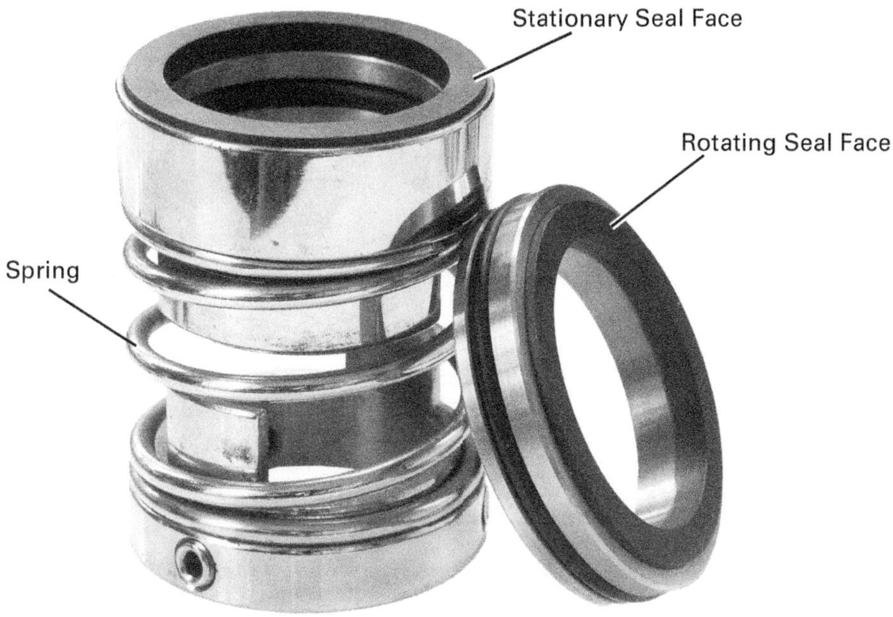

Figure 23 Simple mechanical seal.

Source: nayladen/Shutterstock

Maintaining a reliable seal depends on a thin film of lubricant existing between the sealed faces. The lubricant comes from the equipment sump. The seal parts must be carefully handled and protected to preserve the precise surface finish. A grain of sand between them, for example, can destroy the seal surface in seconds. Mating surfaces should never be placed in contact without some lubrication.

When a seal begins to leak, it must be replaced. Seals should be removed and installed in accordance with the service instructions provided by the equipment manufacturer. Most are not difficult to install, but cleanliness is essential. Remember the following guidelines when working with mechanical shaft seals:

- The seal used must be correct for the application. Replacement seals from the equipment manufacturer are always preferred over generic seals.
- Familiarize yourself with the entire procedure before attempting to install a mechanical seals.
- Lubricate the shaft, O-rings, and seal faces using the lubricant specified by the manufacturer or using the same lubricant as the protected equipment.
- Protect these expensive components. Do not touch the seal faces with your hands. Handle seals only by the edges, with clean hands. Inspect seal faces carefully to ensure that there are no nicks or scratches.
- During assembly, protect any O-rings or other static seals that may be part of the assembly. Sharp edges such as threads and keyways can damage them.
- All parts must fit properly without binding.
- Follow the proper bolt-tightening sequence.

Figure 24 Multi-spring seal.

Source: nayladen/Shutterstock

1.3.0 Bearings

Nearly every piece of equipment with rotating parts has one or more bearings. Bearings support and maintain the position of rotating shafts while reducing friction. Many types of bearings can be replaced in the field, including larger bearings such as *pillow block bearings* and *flange bearings* (*Figure 25*).

The following terms are commonly used to describe bearings and their applications:

- **Axial load** — An external force acting parallel to the shaft.
- **Radial load** — A side, or radial, force applied at right angles to a bearing and shaft, such as that applied by a taut drive belt.
- **Thrust** — A force acting along the axis of a shaft, in either direction.
- **Journal** — The part of a shaft, axle, or spindle that is supported by and revolves within a bearing.

Bearings are divided into two broad classifications: plain and antifriction.

Axial load: An external force acting lengthwise, parallel to the shaft centerline.

Radial load: A side, or radial, force acting perpendicular to a shaft.

Thrust: A force acting along the axis of a shaft, in either direction.

Journal: The part of a shaft, axle, or spindle that is supported by and revolves inside a bearing.

1.3.1 Plain Bearings

Plain bearings (*Figure 26*) are simple in construction, operate efficiently, and can support heavy loads. To some, they don't look much like bearings. They are simple sleeves made from a material softer than the shafts that ride within them. During operation, an oil film develops between the shaft journal and bearing surfaces that overcomes friction.

There are three general classifications of plain bearings. *Radial bearings*, also called sleeve bearings, are used to support radial loads. *Thrust bearings* and *thrust washers* (*Figure 27*) are used to resist axial loads on rotating members. Thrust bearings resist axial loads and/or restrain axial movement of the shaft. Other than having an exceptionally smooth surface, they look much like any other washer. Two or more thrust washers made from low-friction materials are often combined to restrain shaft movement.

Guide bearings are used to guide moving parts in a straight line, such as in machine tools. The most common plain bearings used in HVACR equipment are *sleeve bearings* and *thrust bearings*.

A sleeve bearing provides the bearing surface for shaft rotation and is usually press-fitted into a supporting member. Sleeve bearings operate quietly and have good radial load capacity. A fractional-horsepower blower motor used in a residential furnace is often equipped with sleeve bearings, many of which are permanently lubricated.

Lubrication for a sleeve bearing is important because the bearing must have an oil film between the shaft and bearing surface. Wear occurs quickly when there is no lubrication. During rotation, the shaft essentially floats on the oil film with minimal contact between the shaft and the bearing walls.

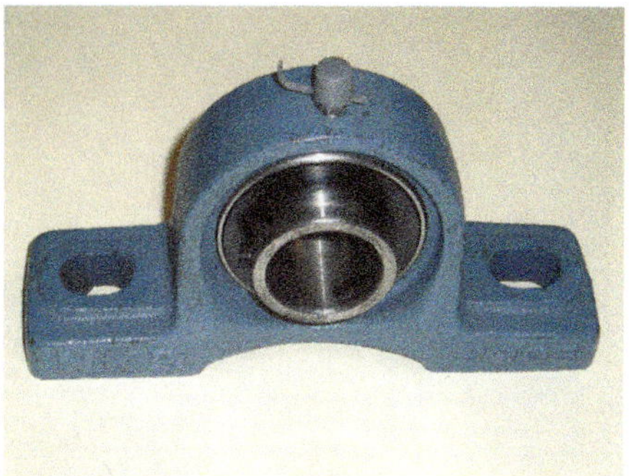

(A) Pillow Block Bearing

(B) Flange Bearing

Figure 25 Pillow block and flange bearings.

(A) Journal Sleeve Bearings

(B) Porous Bronze Sleeve Bearing

Figure 26 Plain bearings.

Source: (A) prabhjits/Getty Images; (B) Courtesy of National Bronze

Figure 27 Ring bearing to be mounted between two thrust washers.

Source: Евгений Харитонов/Getty Images

The simplest and most widely used sleeve bearings are cast bronze and porous bronze, such as the one shown in *Figure 26(B)*. When made from bronze, sleeve bearings offer good resistance to humidity, dirt infiltration, and corrosion.

Cast bronze bearings are normally lubricated with oil. They usually have a reservoir that is filled from the outside through an access port. Cast bronze bearings usually need to be lubricated regularly with the proper lubricant. Porous bronze bearings absorb oil from the reservoir slowly.

Magnetic Bearings

Magnetic bearings are used in some chiller compressors. These bearings support rotating shafts using powerful magnetic fields that enable them to support shafts weighing several tons. Once the bearing is energized, it does not contact the shaft, which eliminates friction. Essentially, the shaft is being levitated, but there is no magic. The powerful magnetic field simply lifts the shaft and centers it in the bearing. One significant advantage to the lack of friction is that the bearing's rotational speed is virtually unlimited.

Although research continues into permanent-magnet bearings that require no power input, electrical power is required to generate the magnetic field. An electric current passed through the bearing windings generates the magnetic field. A backup mechanical bearing may be part of the assembly for critical applications where failure of the main bearing would be catastrophic.

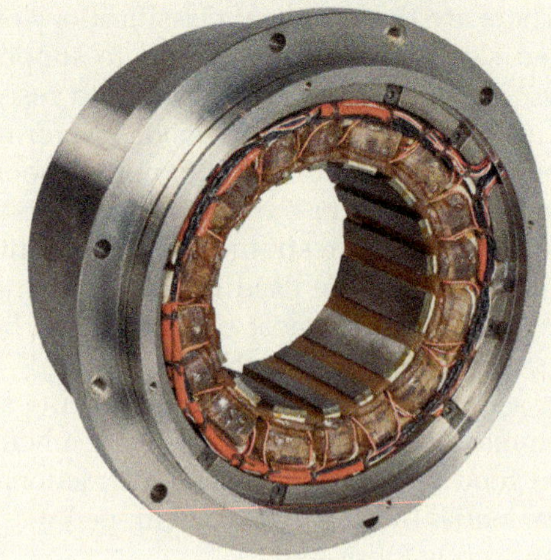

Source: Courtesy of Dover Precision Components and Bearings Plus

1.3.2 Antifriction Bearings

Antifriction bearings are so named because they operate using either balls or rollers between rotating and fixed surfaces to reduce friction. As a result, friction can be reduced to a fraction of that produced by plain bearings. The rolling action of antifriction bearings normally makes them noisier than plain bearings. Because they are made of steel, antifriction bearings can rust when exposed to the elements.

Most antifriction bearings are usually lubricated with grease. The grease helps repel moisture and dissipate heat that builds up in the bearing, but an oil film to reduce friction is not usually necessary for these bearings. Some types are permanently lubricated. The two common types of antifriction bearings are *ball bearings* and *roller bearings.*

Ball bearings (*Figure 28*) consist of outer and inner rings called *races,* with the balls trapped between them. The inner race has a groove around its outside circumference for the balls to roll in. The outer race usually has a similar groove on its inside circumference. A separator or retainer spaces the balls around the tracks.

Ball bearings fall into three classes. *Radial* and *thrust ball bearings* are used in applications with radial loads and axial thrust loads, respectively. *Angular ball bearings* are used in applications that have combined radial and thrust loads, and where precise shaft positioning is needed.

In general, roller bearings (*Figure 29*) have higher load capacities than ball bearings. They are used in heavy-duty, moderate-speed applications. The construction of roller bearings is like that of ball bearings, except that cylindrical rollers are used instead of balls. Also, the outer race is called the *cup,* and the inner race is called the *cone.* The separator or retainer is called the *cage.* Roller bearings generally fall into one of four classes:

- *Cylindrical bearing* — Used for radial loads
- *Thrust bearings* — Used for axial loads (*Figure 30*)
- *Tapered bearing* — Used for radial loads, axial loads, or both, depending on the design
- *Spherical bearing* — Used in applications requiring high load capacity and/or those with a shock load, such as a conveyer (*Figure 31*)

Figure 28 Typical ball bearing.
Source: urfingus/Getty Images

Figure 29 Tapered roller bearing.
Source: surakitsawangchit/Getty Images

Figure 30 Thrust roller bearing.
Source: Valdimir_Timofeev/Getty Images

Figure 31 Spherical roller bearing.
Source: Valdimir_Timofeev/Getty Images

1.3.3 Identifying Bearing Failures

During maintenance, check bearings to confirm that all is well. If you spot problems early, you can fix them before they do serious harm.

When inspecting bearings, pay attention to your senses. Use your eyes to look for anything strange. Examine both plain and antifriction bearings. On antifriction bearings, check the races, cage, seals, and rolling components specifically. Do the parts show unusual wear patterns? Are they discolored? Do you see corrosion or odd colors? Are there wear particles in the lubricant? Is the lubricant discolored or contaminated with water?

Use your sense of touch to find things that your eyes miss. Is the bearing housing hotter than it should be? Does the bearing turn smoothly? If possible, run your finger over the race grooves or the sleeve's inner surface. Do you feel scratches, dents, nicks, roughness, or patterns? Does the lubricant feel gritty?

A technician who is familiar with the normal sounds and behavior of a specific operating component can listen for unusual sounds that may indicate a bearing problem. Use your ears to detect abnormal operating behavior, both while the machine is running and after shutdown. Does the bearing sound normal? Did the machine vibrate excessively or make odd noises? Does it feel "scratchy" when you turn it by hand?

Another method used to detect a bearing problem audibly is to place one end of a steel rod on the bearing housing while the other end is held close to the ear. The rod acts as an amplifier, transmitting noises such as thumping or grinding which would indicate bearing failure.

Use your nose to identify material changes. Does anything smell scorched? Do seals smell like burnt rubber or plastic? Does the lubricant smell abnormal?

Excessive end play (longitudinal movement) in a motor shaft can be an indication of bearing wear. Ball-bearing motors should typically have an end play of about $^{1}/_{32}$" to $^{1}/_{16}$" (0.8 mm–1.6 mm). Sleeve-bearing motors may have an end play of up to $^{1}/_{2}$" (13 mm). Any side-to-side movement of the shaft that can be felt is unacceptable and must be resolved by replacing the motor or having it repaired by a motor shop.

Examining the lubricant on a damaged bearing can help to identify the cause of a bearing failure. *Table 2* summarizes some common problems indicated by lubricant symptoms.

TABLE 2 Lubricant Conditions Related to Common Bearing Failures

Lubricant Type	Symptom	Probable Cause
Oil or grease	Reddish deposit at ball contact	Indicates chafing; look for oscillatory or vibratory motion during operation
Oil or grease	Metallic particles accompanied by indications of wear on bearing components	Lubricant failure or external contamination
Oil	Dry surface with little evidence of residue, or slight brownish haze on ball or raceway surfaces	Insufficient lubricant, either initially or resulting from leakage or migration away from critical areas
Grease	Grease darkened but still oily	High loads or contaminants
Grease	Grease darkened and dry	Extremely high loads or contaminants
Grease	Grease surface dry and hard, soft interior	Extreme external heat

1.3.4 Bearing Removal and Installation

When bearings are damaged or worn out, they must be replaced. Bearings may also be removed to disassemble a piece of equipment for other repairs. Most bearings require special tools or equipment to remove them. *Figure 32* shows a simple bearing puller in use, while *Figure 33* shows a *bearing splitter* in use.

Figure 32 Removing a bearing with a manual puller.

Source: Courtesy of SKF, Lincoln Industrial Products, Alemite LLC

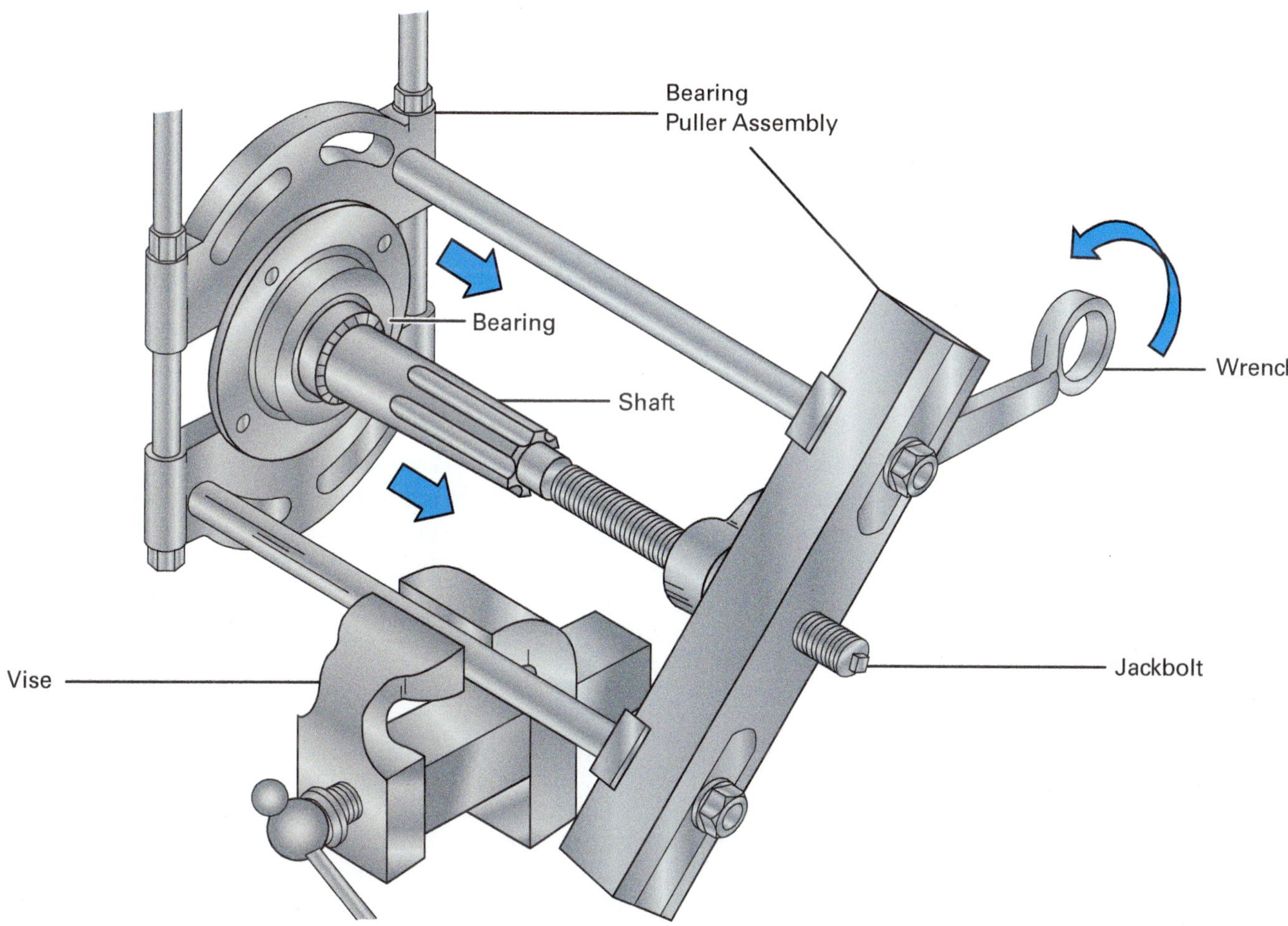

Figure 33 Bearing splitter in use.

Due to the tight fit required, some bearings must be pressed into position (*Figure 34*). Heat may also be required to expand a bearing so that it can slip over a shaft without pressing (*Figure 35*). Once it cools, it fits very tightly to the shaft.

As mentioned previously, pillow block and flange bearings are generally changed in the field. Changing motor bearings is challenging and rarely practical from a labor point of view. Large and unique motors are generally serviced and refurbished in motor shops that have the proper tools, while small motors are rarely worth the cost of the work. They are best collected and sent for recycling.

CAUTION

Unless the manufacturer specifies otherwise, never heat a bearing above 250°F (121°C). Overheating a bearing can change its mechanical properties.

Bearing heaters are generally limited to a maximum setting of 250°F. For this reason, do not use common hot plates or other equipment that can easily exceed this temperature. Heated oil baths for bearings were common many years ago, but due to the safety hazard they represent, they should not be used.

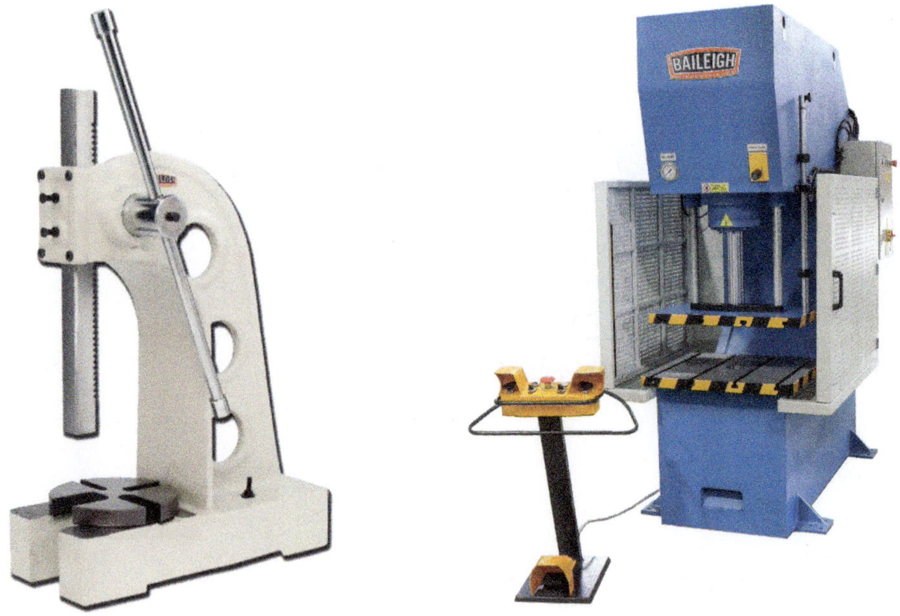

(A) Arbor Press

(B) Hydraulic Press

Figure 34 Bearing presses.

Source: Bailiegh Industrial Holdings LLC

Figure 35 Bearing hot plate.

Source: Courtesy of SKF, Lincoln Industrial Products, Alemite LLC

NOTE

Large pillow blocks may have to be tapped onto the shaft with a hammer. Never strike a bearing directly! Place a wood block on the inner race and tap it with the hammer.

Pillow blocks come with the bearing already in place in the housing. The shaft slides through the slip-fit inner race. A special collar grips the shaft with several setscrews.

Follow these steps to install a preassembled pillow block bearing:

Step 1 Clean the shaft and inspect it for damage.

Step 2 Slide the new bearing over the shaft.

Step 3 Loosely bolt the pillow block to the mounting surface.

Step 4 Align everything properly. Confirm that the shaft slides freely in the bearing.

Step 5 Torque the pillow block bolts to the correct value.

Step 6 Tighten the setscrews to the shaft in the order that the manufacturer specifies, but not too tightly.

Step 7 Rotate the shaft a partial turn and then torque the setscrews to the manufacturer's specified value.

1.0.0 Section Review

1. Spiral-wound gaskets are primarily used on _____.
   a. low-temperature, low-pressure systems
   b. high-temperature, high-pressure systems
   c. low-temperature, high-pressure systems
   d. high-pressure, low-temperature systems

2. Which of the following statements about oil seals is *correct*?
   a. Oil seals are designed for low-pressure applications.
   b. Oil seals are negative contact, meaning that they do not touch the shaft.
   c. Oil seals are made exclusively from soft metals.
   d. The lip of an oil seal should always face away from the machine.

3. The presence of darkened yet still oily grease on a failed bearing suggests that the *most likely* cause of failure is _____.
   a. oscillatory motion during operation
   b. too little lubricant
   c. high loads or contaminants
   d. excessive external heat

2.0.0 Lubrication

Objective

Describe the properties of common lubricants and how they are applied.

a. Describe the properties of common lubricants.

b. Explain how lubricants are selected and applied.

Performance Task

1. Lubricate a bearing using a grease gun.

Any machinery with rotating or moving parts produces *friction*. Friction is the resistance to motion that takes place when two parts move or rub against one another. Lubrication is required whenever friction is a problem and needs to be minimized.

2.1.0 Properties of Lubricants

Lubrication is the application of a substance that reduces friction by creating a slippery film between two surfaces. Lubricants can be made from animal fats, vegetable oils, mineral oils, or synthetic fluids. However, around 90 percent of all lubricants used are made from mineral oils, which come from refined crude oil.

Lubricants can be in the form of a liquid (oil), semisolid (grease), or solid films. Engineers determine the best lubricant for each application based on many factors. One very important factor in the selection of compressor lubricants is the refrigerant in use. Many refrigerants are not compatible with mineral oil.

NOTE

Compressor lubricants are presented in detail in NCCER Module 03301, *Refrigerants and Oils*.

2.1.1 Oils

Oils represent a broad class of liquid lubricants. Lubricating oils are generally made from two types of crude oil: *naphthenic* and *paraffinic*. Naphthenic oils contain very little wax and are good lubricants for many applications. Paraffinic oils are very waxy, so lubricants made from them are used mainly in hydraulic equipment and similar machinery. All lubricants possess certain properties that determine their suitability for a particular use. These properties include the following:

- **Viscosity**
- **Viscosity index (VI)**
- **Pour point**
- **Flash point**
- **Fire point**
- **Oxidation resistance**

Viscosity: The resistance of a fluid to flow, which can vary depending on the temperature of the fluid. For refrigerant oil, the viscosity must be such that the oil will flow and provide proper lubrication over the wide range of temperatures encountered in the refrigerant circuit. See also *pour point*.

Viscosity index (VI): A measure of the rate at which an oil's viscosity changes with temperature.

Pour point: An oil quality related to viscosity. It can be defined as the temperature at which oil first starts to flow.

Flash point: The temperature at which heated lubricant vapors can ignite but cannot sustain combustion.

Fire point: The temperature at which a lubricant will burn if an ignition source is present.

Oxidation resistance: The ability to resist the degradation process that is caused by elements freely combining with oxygen. All petroleum products react with oxygen to some degree, and the process accelerates as temperature increases.

Viscosity

Viscosity is an important property of oil. Viscosity represents the thickness of the oil or its ability to flow at a specific temperature. Low-viscosity oils are light oils that flow freely when poured. High-viscosity oils are heavy oils that flow slowly when poured.

The viscosity rating of oils can be expressed in two ways. Automotive and gear lubricating oils are expressed in Society of Automotive Engineers (SAE) ratings. The other viscosity rating system that applies to industrial lubricants is the Saybolt Universal Seconds (SUS) rating system. This rating system was developed by the American Society for Testing and Materials (ASTM) International.

The metric system has an assigned unit for the relative resistance to flow of materials. This unit, called the **centistoke (cSt)**, represents the relationship between simple viscosity and the density of the lubricant. *Table 3* lists some typical operating viscosity ranges, in centistokes, for common applications. *Figure 36* shows viscosity equivalents among grading systems.

Centistoke (cSt): A metric unit of measure for viscosity, representing centimeters of movement per gram per second.

TABLE 3 Typical Operating Viscosity Ranges

Lubricant	Viscosity Range (cSt)
Clock and instrument oils	5 to 20
Motor oils	5 to 50
Roller bearing oils	10 to 300
Plain bearing oils	20 to 1,500
Medium-speed gear oils	50 to 150

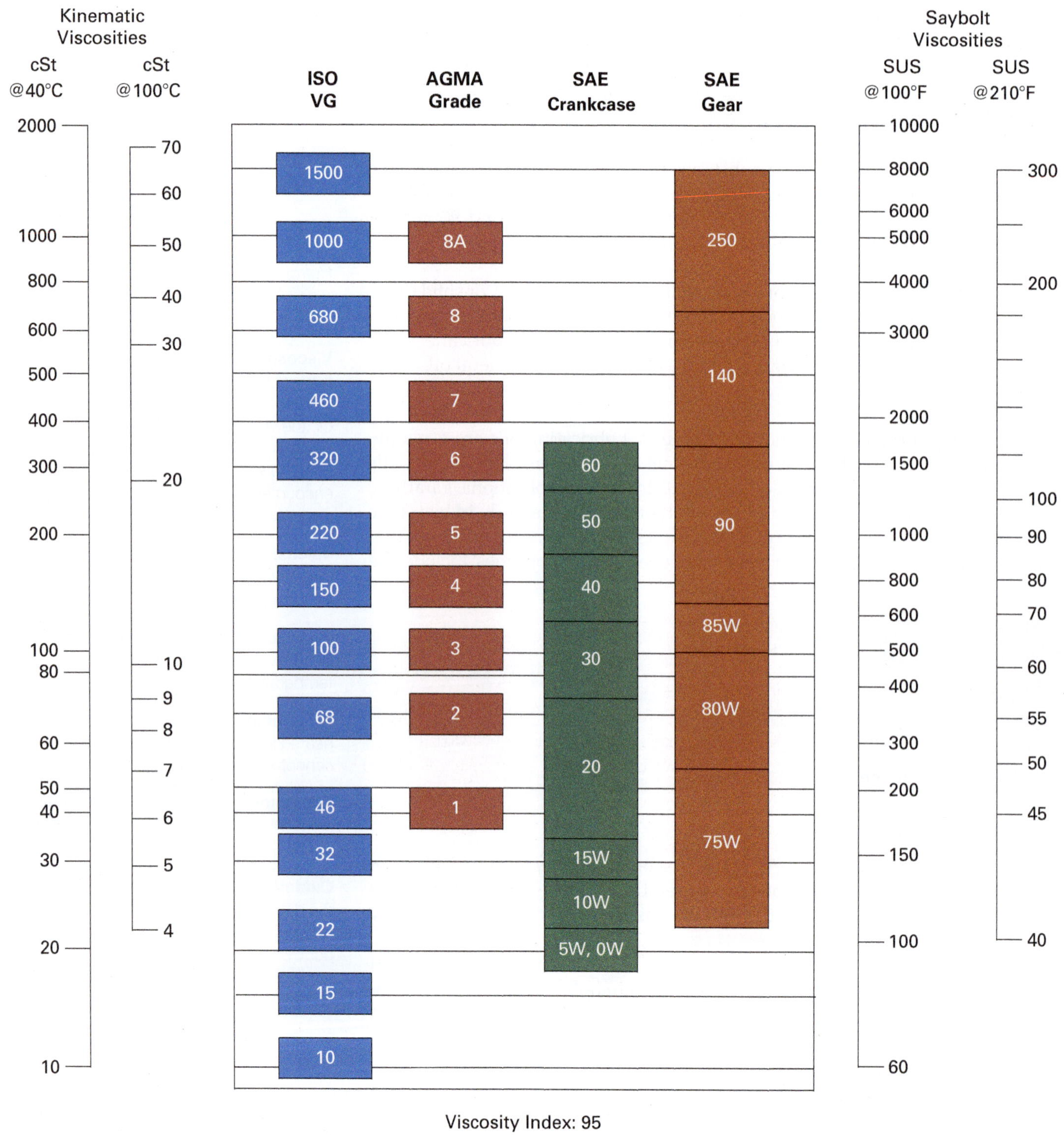

Figure 36 Lubricant viscosity equivalents.

Source: Evolvent Design, https://evolventdesign.com

2.1.2 Grease

Grease is a solid or semisolid lubricant that is formed by adding a thickening agent, usually soap, to oil. Grease typically has at least three ingredients: oil, soap, and at least one additive.

The most important ingredient is the chosen oil, which is usually a mineral oil. Greases for high-temperature, low-speed applications are made with high-viscosity oils; greases for low-temperature, high-speed applications are made with low-viscosity oils.

The most common thickening agent is a soap made from fats, either animal or vegetable, and some type of mineral or metal. Calcium, barium, and lithium are some metals that are routinely used to make the soap base. Grease often contains one or more metals, such as lithium. The function of the soap is to absorb the oil and release it at a slow rate to provide lubrication. The soap itself may have some value as a lubricant, but the oil provides the intended lubrication.

The third element in grease is an additive, although some grease may be free of additives. Additives are used to change one or more characteristics. They help to keep the grease from undergoing chemical changes, give the grease desired properties, or protect metal parts from corrosion.

Grease produces the same lubricating action of oil, but in a different way. Oil reduces friction by forming a liquid film that keeps surfaces apart. Grease, however, is not a liquid and cannot form such a film when it is initially applied. Once the surfaces begin to move, pressure is exerted on the grease and the temperature rises. The grease then releases some of its oil, allowing the lubricating action to begin. When the motion ceases and the pressure and temperature are reduced, grease tends to solidify again.

The following properties are considered when selecting the appropriate grease for an application:

- Consistency
- **Dropping point**
- Pumpability
- Water resistance
- Stability

Dropping point: The temperature at which grease is soft enough for a drop of oil to fall away or flow from the bulk of the grease.

Consistency

A key physical characteristic of grease is its *consistency*, which refers to its hardness or softness. Consistency is determined by measuring the distance, in tenths of a millimeter, that a standard metal cone penetrates the grease under a standard load. This is known as a *penetration test*.

The consistency of grease varies with temperature, and it generally softens as the temperature increases. The National Lubricating Grease Institute (NLGI) has developed a widely used classification system for grease consistency. *Table 4* shows the NLGI grease classification system.

TABLE 4 NLGI Grease Classification System

NLGI Number	Penetration at 25°C (tenths of a mm)	Description of Consistency
000	445 to 475	Fluid
00	400 to 430	Semifluid
0	355 to 385	Very soft
1	310 to 340	Soft
2	265 to 295	Normal
3	220 to 250	Firm
4	175 to 205	Very firm
5	130 to 160	Hard
6	85 to 115	Very hard

NOTE

Some grease products do not list a dropping point because chemical decomposition of the grease begins at a lower temperature than its dropping point. For these products, the grease loses its value before it reaches the dropping point, rendering the information useless.

Dropping Point

When grease is warmed, it eventually reaches a temperature at which a drop of oil will fall away from the mass. This temperature is known as the *dropping point*. The dropping point is normally considered the highest temperature at which the grease can be used, but that is not always the case. Grease may still be a satisfactory lubricant above its dropping point, but it will behave more like an oil than a grease above this temperature.

WD-40®

WD-40® brand products are familiar to most craftworkers as well as homeowners and garage mechanics. The product was developed in 1953 by Norm Larsen, founder of the Rocket Chemical Company. His goal at the time was to develop a dependable water repellant and corrosion inhibitor for nuclear missile system maintenance. He found success on his 40th attempt. Thus, the product name stands for *Water Displacement—40th Attempt*. The Rocket Chemical name was changed to WD-40® Company in 1969, honoring its flagship product.

For more than 65 years, WD-40® Company has provided products and training that support skilled craft professionals, homeowners, and others. WD-40® branded products consist of two product lines. One is the WD-40® Multi-Use Product that is offered in the iconic blue and yellow can and has over 2,000 uses. The other is the WD-40® Specialist® product line that includes specialty lubricants, penetrants, greases, degreasers, and rust management solutions designed for professional craftworkers.

Complete books listing potential uses for the original product have been published, including one titled *WD-40® for the Soul: A Guide to Fixing Everything*. For more information, visit **www.wd40.com**.

Source: Image Courtesy of WD-40 Company, www.wd40.com/techs-and-trades/

2.1.3 Lubricant Additives

An additive is a substance added to adjust the qualities of a lubricant. The following are the most common types of additives in use:

- Pour-point depressants
- Oxidation inhibitors
- Viscosity index improvers
- Antifoam agents
- Corrosion inhibitors
- Extreme-pressure (EP) additives
- Detergents and **dispersants**
- **Emulsifying agents**
- **Demulsifiers**
- Anti-wear agents
- Tackiness agents

Dispersants: An additive used to disperse, or separate, groups of particles to prevent them from coagulating or joining together.

Emulsifying agents: Chemical compounds that support the mixing of two or more normally immiscible liquids, such as oil and water. Emulsifying agents also help prevent the liquids from separating.

Demulsifiers: Chemicals used to separate emulsions, such as oil and water; the opposite of an emulsifying agent.

A few of these additives are described in the sections that follow.

Pour-Point Depressants

Pour-point depressants are additives that lower the temperature at which a lubricant thickens. The pour-point can be lowered by as much as 85°F (30°C) with the addition of 1 percent or less of pour-point depressants.

Viscosity Index Improvers

VI improvers are largely used to thicken a light oil, increasing its viscosity without changing the original viscosity-temperature change coefficient. This enables a lower-viscosity oil to work better at higher temperatures without breaking down. For example, a 10W-base stock oil can be thickened to a 10W-50 product.

Extreme-Pressure Additives

Also called *film-strength additives*, extreme-pressure (EP) additives protect the oil under very heavy loads. At high spots where metal contact and loading are extreme, EP additives form layers of iron chloride and iron sulfide. These layers help prevent destructive welding, excessive metal transfer, and severe surface breakdown. They also bear some of the compressive load and cushion the shock of heavy load engagement.

2.2.0 Selecting and Applying Lubricants

Product and lubrication engineers typically determine the correct lubricant to use in every situation. The two primary factors to consider are load and speed. Other factors that affect selection include operating temperature and other substances that might be present, such as a particular refrigerant.

Load refers to the amount of pressure that is exerted on the lubricant in operation. Higher loads increase the possibility that the molecules in the lubricant will come apart under stress. High-viscosity oils are generally recommended for extreme loads.

At high speeds, surfaces tend to wear faster. This is mostly because the temperature rises. Low-viscosity oils generally provide less friction and better heat transfer at high speeds than high-viscosity oils, unless the temperature is very high. When high loads are coupled with high speed and high temperatures, great care must be taken in the selection process.

Knowing what type of lubricant to use for an application comes with experience and a thorough knowledge of lubricant properties. Considerations of product requirements are often as important as the mechanical requirements of the machinery itself. One example is lubrication in food processing facilities. When food is involved, the lubricant specifications are controlled by United States Department of Agriculture (USDA) standards. The machinery in these facilities requires lubricants that would not endanger the consumer if small quantities made their way into food products. They must also be nonreactive and remain effective in the presence of cleaning products and steam.

Although HVACR technicians should understand the lubricant selection process, the best source of information is always the equipment manufacturer. All manufacturers provide lubricant information for their products. A technician will rarely be expected to choose a lubricant beyond that needed for simple low-pressure, low-speed applications. It is essential to follow the manufacturer's recommendations. In many cases, the use of an unapproved lubricant can void the warranty. If acquiring a particular brand of lubricant is a problem, consult the manufacturer and local lubricant retailers for conversion to an equivalent product.

2.2.1 Grease Lubrication Equipment

The lubrication of most HVACR equipment is done manually. The most common manual lubrication tool used in the field is the lever-type grease gun (*Figure 37*). Most lever guns can develop significant pressure with limited effort. Handheld battery-powered grease guns (*Figure 38*) are also available, which speeds up the work when many points need attention. When using a powered grease gun, following the manufacturer's guidance carefully and wear the appropriate PPE, including gloves.

Figure 37 Lever-operated grease gun.

Source: Prabhjit Singh Kalsi/Getty Images

Figure 38 Battery-powered grease gun.

Source: Courtesy of Stanley Black and Decker, Inc.

Adapters, such as flexible extension hoses and adjustable swivel fittings, are used with grease guns to gain easier access to lubrication fittings in hard-to-reach places. A coupler (*Figure 39*) is required to make a reliable connection to a grease fitting. The coupler on a grease gun is easily changed when necessary. Grease fittings are described later in this section.

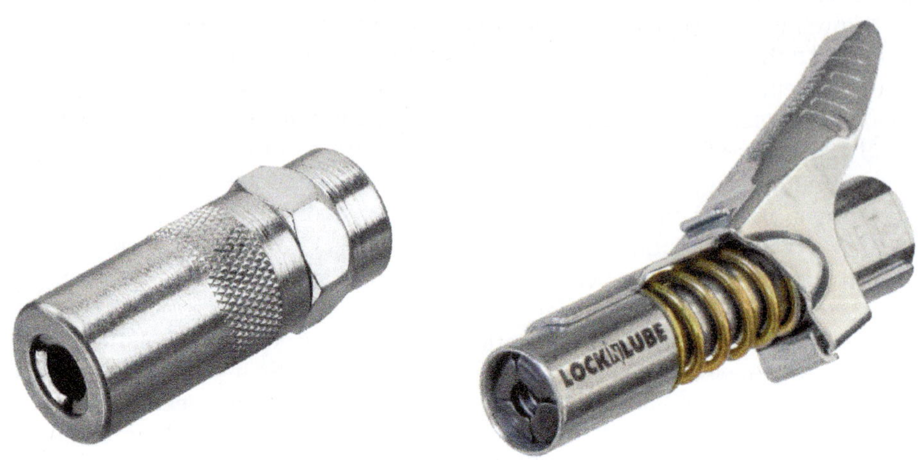

(A) Standard Coupler **(B) Locking Coupler**

Figure 39 Grease fitting couplers.

Source: (A) Courtesy of Stanley Black and Decker, Inc.; (B) Courtesy of LocknLube, locknlube.com

Lever-operated grease guns use preloaded, disposable grease cartridges (*Figure 40*), or they can be reloaded in bulk using a filler pump. HVACR technicians often use disposable cartridges due to their need for mobility.

Figure 40 Grease gun cartridges.

Source: HVACGuru

2.2.2 Using a Grease Gun

Loading and using a grease gun is relatively simple. The common parts of a grease gun are shown in *Figure 41*.

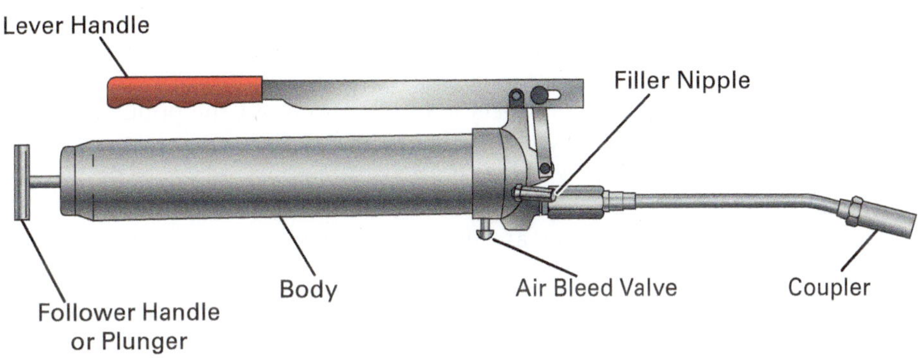

Figure 41 Parts of a lever-operated grease gun.

Before beginning, remember to avoid mixing different greases. Using grease cartridges makes mixing easier to avoid. After inserting a new cartridge, pump out enough grease to push the remains of the old grease out of the hose and fill it with new grease. You'll need to do this anyway, to remove any trapped air. If the gun is being filled from a bulk container with a filler pump, it is best to dedicate the gun to a specified grease. Otherwise, the previous grease product must be thoroughly cleaned from the gun body before changing over. This is a messy but necessary task.

Installing a Grease Cartridge

To replace a grease cartridge, follow the instructions provided by the tool manufacturer. The following is a summary of the basic steps:

Step 1 Pull the follower handle back. The follower is spring-loaded to maintain constant pressure on the grease. As you pull, you are loading the spring and will feel the pressure. Once it is pulled out as far as it will go, latch the rod on the locking mechanism provided.

Step 2 Unscrew the body from the pump assembly at the top. There may be a seal at this joint. If so, make sure it doesn't get thrown out with the old cartridge.

Step 3 Remove the spent cartridge.

Step 4 Remove the dust cap from the new cartridge and insert the open end of the tube so it faces the follower.

Step 5 On the end now facing you, remove the top. For those that have a pull-top, be very careful of the edge to avoid a serious cut.

Step 6 Screw the body back into the pump assembly.

Step 7 Some guns have an air bleed valve, while others don't. But trapped air needs to come out.

- For guns with a bleed valve, the air can be evacuated through the valve. Release the follower, allowing it to put some pressure on the grease. Then open the air vent. You may need to apply a little manual pressure to the follower handle to help it push the grease down and push the air out.
- For guns without a bleed valve, first unscrew the body about one full turn—this will let air leak out around the threads. Then release the follower. Again, you may need to put slight pressure on the follower handle to help force the air out. Then retighten the body.

Step 8 Pump some grease onto a piece of scrap paper or cardboard, or into a suitable waste container. It should flow freely. If it doesn't, or the handle feels spongy or doesn't offer much resistance, there is air still trapped in the body. If grease doesn't come out after a few pumps, try purging the air again following *Step 7*. Pump enough grease out to clear any remaining air bubbles and ensure the previous grease has been cleared from the hose.

Grease Fittings

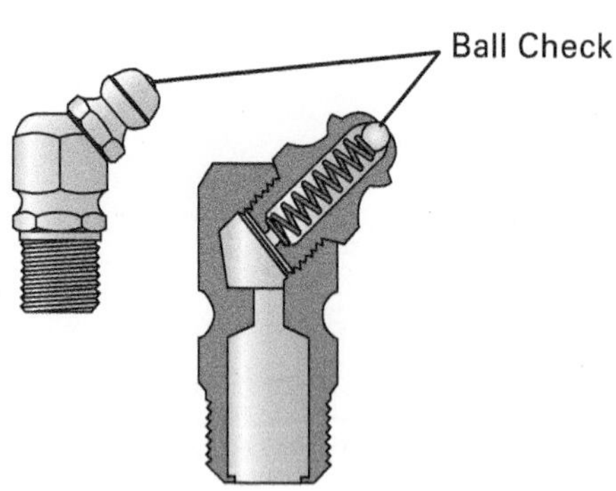

Figure 42 Angled grease fitting (Zerk fitting).

Grease fittings on the equipment allow grease to be injected into the mechanism or bearing (*Figure 42*). A company named Alemite Manufacturing, an active participant in the lubrication products market, introduced the popular quick-connect grease fitting in the early 1900s. The original design was improved upon and patented in 1929 by Oscar Zerk, on behalf of Alemite. Consequently, the fitting is known to many as a *Zerk fitting*.

Two features of these fittings are important. The first is the shape of the head. The head is designed to accommodate a grease gun coupler with spring-loaded jaws that grip the head of the fitting. As the coupler is pushed down on the fitting, the jaws spread apart and then snap closed around the tip. This forms a relatively leak-proof connection. Locking couplers like the one shown in *Figure 39(B)* connect the coupler to the fitting with greater security.

The second feature is internal. Each grease fitting contains a tiny check valve, as shown in *Figure 42*. The spring keeps a ball pressed against the opening, keeping dirt and water out. When a connection is made and the grease gun applies pressure to the ball, the spring is depressed, and grease flows through the fitting. Fittings can be straight or angled, and may have US standard pipe, metric, or machine threads. Some are designed for a simple friction fit in the opening and are tapped into place with a hammer.

Some other variations are available. *Figure 43* shows a grease fitting that has a second check valve inside, for added reliability. Another variation shown is a fitting with a vent slot cut in the sides. This leaves the grease chamber beneath the fitting open to the atmosphere, allowing grease to escape if the cavity is full and the internal pressure rises. There are other variations, but they are primarily found on industrial machinery and equipment.

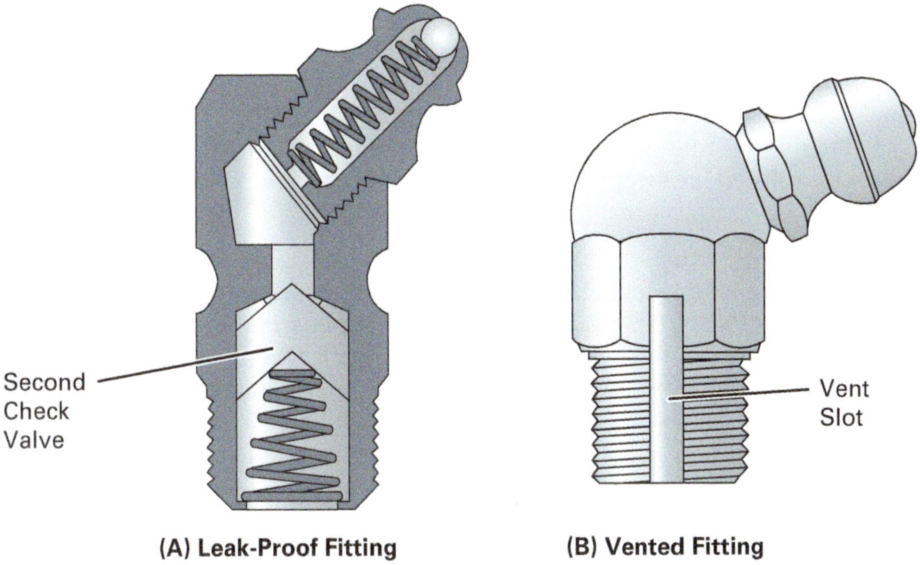

Figure 43 Grease fitting variations.

A relatively high pressure is required to pump grease through a fitting and into the grease cavity around a bearing. Surprisingly, lever guns can develop pressures up to 10,000 psi (~69,000 kPa). Over-pumping the handle can cause excessive grease pressure that could damage grease seals. Avoid over-pressurizing any lubrication point.

CAUTION

Lever-operated grease guns, as well as other types, can develop pressures exceeding 10,000 psi, which can damage bearings and seals. Follow the equipment manufacturer's service instructions carefully to avoid over-pressurizing a lubrication point and damaging the seals. If grease doesn't seem to be flowing, stop and consider the problem rather than continuing to apply pressure.

Greasing a Bearing

Applying grease is simple. Connect the gun coupler to a compatible grease fitting and pump the handle. Knowing when to stop is the key. In every case, it is best to consult the manufacturer's maintenance instructions and follow them carefully. Adding too much grease is usually more destructive than not adding enough.

Assuming the grease gun is prepared with the specified grease, clean the grease fitting on the bearing housing with a rag. While cleaning, check the bearing housing for a relief plug. Small bearings usually don't have a relief port, but many electric motors and pumps do, as shown in *Figure 44*. The relief plug is usually on the opposite side of the housing from the grease fitting. Removing the plug allows the old grease to be pushed out by fresh grease.

Figure 44 Grease relief plug on an electric motor.

Source: Courtesy of Baldor Electric Company

CAUTION

Failure to reinstall the relief plug will result in the grease being forced out of the housing as the equipment warms up and the grease begins to expand and liquify.

Clean the coupler on the grease gun first, then push it firmly on the grease fitting. As you pump, you can watch the old grease come out of the relief port. Stop pumping when the grease appears fresh. Use a rag under the port to capture the grease that pushes out. Clean the old relief port and remember to reinstall the plug!

If there is no relief plug to remove, you must be more careful. When you begin to feel some resistance as you pump, it's time to stop. Here are some additional tips for using a grease gun:

- *Avoid over-greasing* — Over-lubrication is a common source of problems. When there is too much grease in the housing, bearings run hotter. As mentioned previously, it is also easy to damage the bearing seals. Where you see grease squeezing out constantly from the area around a shaft, the seal has likely been damaged. While grease guns can develop thousands of pounds of pressure, most seals can only withstand a few hundred pounds at best.
- *Know the volume of grease needed* — It is always best to know how much grease should be added, and how much the grease gun pumps in each stroke. A flow meter mounted on the grease gun can help. If one is not available, pump about five strokes of grease onto a scale. Weigh it and divide the weight by the number of strokes to get a reasonable idea of the volume per stroke.
- *Pay attention to unsealed bearings* — Not every bearing has a tight seal. For those that don't, stop applying grease as soon as fresh grease appears around the edge of the bearing.
- *Keep fittings clean* — Make sure that both the coupler and the grease fitting are clean, before and after greasing. This prevents pushing dirt and grit into the bearing along with the grease.

2.2.3 Small Motor and Bearing Lubrication

The bearings in many motors are permanently lubricated at the factory and require no field lubrication. The service literature for the equipment being serviced normally states when this is the case. Fractional horsepower motors with sleeve bearings are often permanently lubricated, but a few may require periodic lubrication with oil.

When lubricating motor bearings with oil, it is important to follow the lubrication interval and use the type and quantity of lubricant recommended in the equipment manufacturer's service literature. To lubricate motors and bearings equipped with an oil port, proceed as follows:

- Shut off power to the equipment, then complete lockout and tagout procedures as required.
- Oil the motor using the type and quantity of oil specified by the manufacturer.
- Rotate the shaft while adding the oil to evenly distribute the lubricant.
- Wipe away any excess oil, restore power, and observe the motor operation.

2.0.0 Section Review

1. The viscosity rating that applies to industrial lubricants is known as the ____.
   a. SAE rating
   b. SUS rating
   c. ASTM rating
   d. naphthenic rating

2. After inserting a new grease cartridge, you should pump out enough grease to push the remains of any previous grease product out of the hose, and to ____.
   a. purge any air from the gun and hose
   b. build enough pressure for greasing
   c. test the grease gun
   d. separate the oil from the soap

3.0.0 Drive Belts

Objective

Identify and describe how to install various drive belts.

a. Identify and describe various types of drive belts.
b. Identify and describe how to align sheaves and calculate drive speeds.
c. Explain how to install and apply the proper tension to drive belts.

Performance Task

2. Install, align, and adjust a drive belt.

Belt drives are a quiet, smooth, and economical form of power transmission. They are commonly used in commercial and industrial HVACR equipment. A belt drive consists of a drive **sheave** (or sheaves), a driven sheave, and drive belts to match the size and profile of the sheaves (*Figure 45*). Sheaves are available in many shaft sizes, diameters, and types of construction.

Belt drives can be divided into two basic types: V-belts and **synchronous belts** (*Figure 46*). Nearly all drive belts used in HVACR equipment are V-belts, although synchronous belts may be encountered occasionally. Synchronous belts offer superior grip and maintain the timing between the two rotating shafts because they do not slip. However, these features are not often needed for fans and blowers.

Sheave: A wheel with a groove in which a drive belt rides.

Synchronous belts: Belts with teeth that correspond to grooves in matching sheaves, eliminating slippage and maintaining synchronized rotation of the shafts; also referred to as *timing belts*.

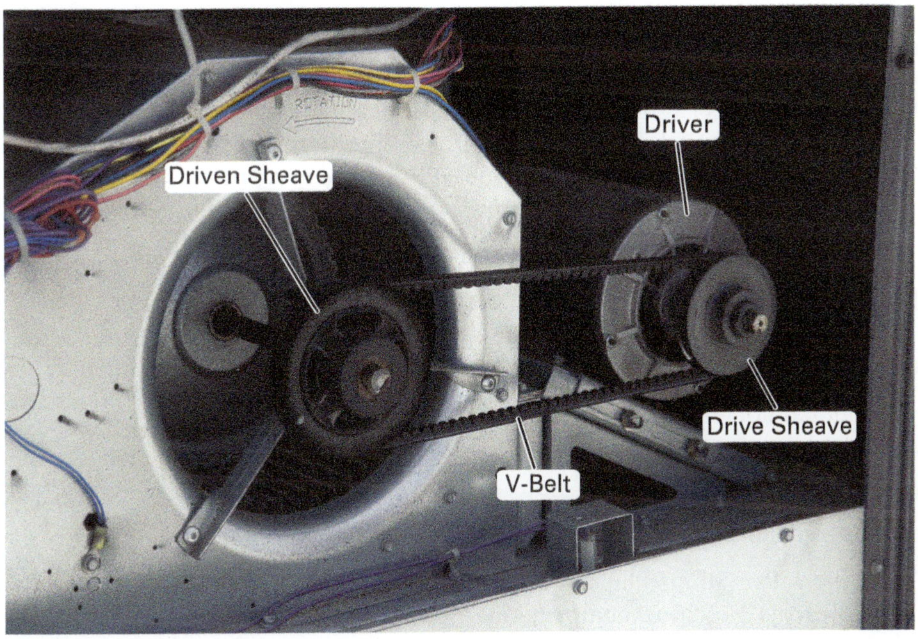

Figure 45 Basic belt drive.

Source: William Hager/Shutterstock

Figure 46 Synchronous belt profile.

Source: Toa55/Shutterstock5

3.1.0 V-Belts

V-belts rely on friction to maintain grip and transfer power. *Figure 47* shows the cross section of a belt where it meets the sheave. The bottom of the belt, at the base of the V, doesn't normally contact the base of the sheave groove. The all-important friction develops only along the tapered, V-shaped sides of the belt. With the belt(s) properly tensioned, friction and grip increase due to a wedging action. When the sides of the belt do not make firm contact with the sheave, excessive slippage results.

V-belts are fabricated with strong cords that run the length of the belt. Their position in the belt structure represents the pitch line of the belt (*Figure 48*). In recent years, the terminology has changed, and it is now often referred to as the *datum line*.

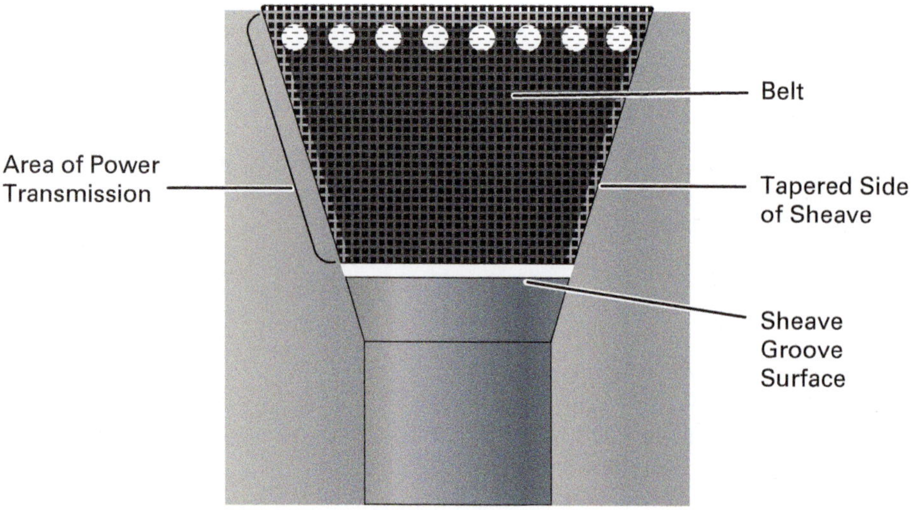

Figure 47 V-belt area of power transmission.

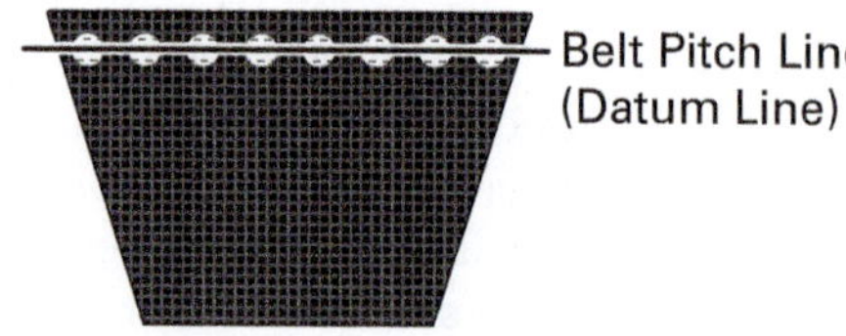

Figure 48 Belt pitch line (datum line).

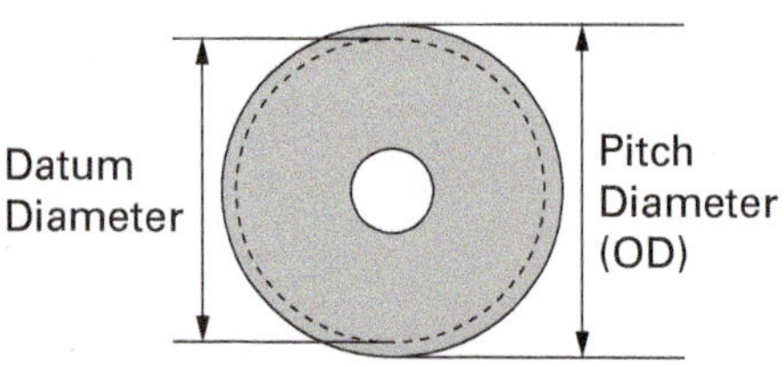

Figure 49 Sheave datum and pitch diameters.

Belt manufacturers generally consider the datum line to be the effective part of the belt because the cords are the foundation of belt construction. Without the cords, drive belts would be relatively weak. Therefore, the length of the belt measured along the datum line (the *datum length*) represents the *effective length* of the belt. The outer circumference of the belt, also provided in many belt specifications, is always slightly longer than the datum length.

For sheaves, the pitch diameter is simply another term for the outer diameter, which is understandably confusing. The *datum diameter* represents the diameter of the circle where the belt's datum line is expected to ride (*Figure 49*).

Understanding these dimensions helps you understand drive belt and sheave specifications and catalog data. Sheaves and belts must be properly matched for the drive to perform as expected.

The types of V-belts that will be introduced in this section include the following:

- Fractional horsepower (FHP) belts
- Classic belts
- Wedge belts
- Banded belts

Notched belts will also be discussed, although they aren't considered a separate category. Notching is an optional feature found in more than one type of belt.

Note that FHP and classic belts can often be mounted on the same sheaves, but not all sheaves are designed to accommodate both types of belts. However, wedge belts require unique sheaves designed to fit their profile. FHP and classic belts should never be mounted on sheaves designed for wedge belts, and vice versa.

3.1.1 Fractional Horsepower Belts

Fractional horsepower (FHP) belts (*Figure 50*) are considered light-duty belts, with only one belt on a drive. As the name implies, this type of belt is usually found on small exhaust hoods, centrifugal fans, and similar applications that are driven by a fractional horsepower motor (less than 1 horsepower). FHP belts may occasionally be found in multiples to accommodate more horsepower and load, but those arrangements are rare.

The size of an FHP belt is indicated by a code marked on the outside of the belt. The first number and letter in the code tell the width of the belt in eighths of an inch. The next three numbers in the code identify the nominal outside length (outer circumference) of the belt, in inches.

Figure 50 FHP V-belts.

For example, a belt marked 4L300 is $\frac{4}{8}$" ($\frac{1}{2}$") wide and 30" long on the outside perimeter. The final 0 in the number is only relevant if the length is not a whole number. Imagine that there is a decimal point between the two zeros. The number 4L305 would indicate that the belt is 30.5" long. However, odd lengths are extremely rare. It is very important to remember that FHP belts are measured on the outside circumference of the belt, not along the inside or datum line.

The depth of the belt depends on the width, with the depth being standard for each given width. FHP belts come in the following standard widths and depths:

- 2L — $\frac{1}{4}$" wide × $\frac{5}{32}$" deep
- 3L — $\frac{3}{8}$" wide × $\frac{7}{32}$" deep
- 4L — $\frac{1}{2}$" wide × $\frac{5}{16}$" deep
- 5L — $\frac{5}{8}$" wide × $\frac{3}{8}$" deep

3.1.2 Classic Belts

Classic belts (*Figure 51*) may also be referred to as *conventional belts.* They are more common than FHP belts in HVACR equipment. They are also commonly used in multiples to increase strength and grip. However, single belt applications are also very common.

Figure 51 Classic V-belts.

Classic belts are used for continuous service and generally outlast FHP belts. The size of classic belts is also indicated by a code marked on the belt, with a letter (A through E) indicating the width accompanied by a number to indicate the length. The number should be considered a nominal length only.

For example, a belt marked A42 is nominally 42" long. However, the 42" length of a classic belt refers to its length along the *inside* of the belt. The outside length is usually 2" to 3" longer. An A42 belt, for example, measures about $44\frac{1}{4}$" on the outside. The effective length of the belt (measured along the datum line) falls somewhere between 42" and $44\frac{1}{4}$".

The sides of a classic belt form an angle of 40 degrees, matching the angle of FHP belts. Because they have the same angle as FHP belts, they can often be mounted on the same sheaves. Classic belts are available in various lengths for each width, and in the following standard widths and heights:

- A — $\frac{1}{2}$" wide × $\frac{5}{16}$" deep
- B — $\frac{21}{32}$" wide × $\frac{13}{32}$" deep
- C — $\frac{7}{8}$" wide × $\frac{17}{32}$" deep
- D — $1\frac{1}{4}$" wide × $\frac{3}{4}$" deep
- E — $1\frac{1}{2}$" wide × $\frac{29}{32}$" deep

3.1.3 Wedge Belts

Wedge belts (*Figure 52*), or *narrow-edge belts,* are typically used in multiples. Multiple wedge belts can be used on drives up to roughly 1,000 horsepower.

Due to their construction, wedge belts can transfer up to three times the horsepower of a classic belt when all other factors are equal. They are often found on applications using smaller-diameter sheaves with shorter center-to-center distances. Wedge belts are not interchangeable with standard multiple belts and should not be mounted on sheaves made for classic or FHP belts.

Like the other types, the size of wedge belts is indicated by a code. The first number in the code identifies the width and cross section of the belt in eighths of an inch. The V following this number identifies it as a wedge belt. An X inserted in the third position indicates it is a notched belt. For example, a 5VX belt is a notched wedge belt.

The next three numbers in the code identify the length of the belt in inches. For example, a belt marked 3V500 is 50" long. Like FHP belts, the final 0 in the number is only relevant if the length is not a whole number.

The length of a wedge belt is measured along the outside perimeter, like FHP belts. Therefore, their datum length will be slightly shorter, but not significantly. The cords are placed very close to the top surface of the belt.

Figure 52 Notched wedge belt.

The sides of a wedge belt form an angle of 40 degrees, but a few variations do exist. Always make sure the sheaves and belts match each other. When changing belts and using another brand, for example, be sure that the belt angle is compatible with the current sheave. When the sheave and belt angles do not match, a significant amount of grip is lost, and the belts wear quickly. The difference in the angle between two belts may be difficult to see.

Common sizes of wedge belts include the following:

- 3V — $\frac{3}{8}$" wide × $\frac{5}{16}$" deep
- 5V — $\frac{5}{8}$" wide × $\frac{17}{32}$" deep
- 8V — 1" wide × $\frac{7}{8}$" deep

Metric Belts

The term *metric belt* refers to belts that are exclusively constructed to metric dimensions and standards. Many belts constructed to US standards of measurement comply with standards from the Rubber Manufacturers Association (RMA). However, some belts constructed to metric standards have the same dimensions as belts constructed to RMA standards. As you review manufacturers' specifications for many of the belts introduced, you will find that they have both metric and US standard dimensions listed.

One popular series of metric belts is the SP/XP series. They are comparable to the US standard classic belt series. The common dimensions of these belts are shown here. The XP series is notched, while the SP series is not. The length may be specified along the datum line, along the outside, and/or along the inside, in metric units.

Conversion charts are available for metric and US standard belts, but most craft professionals agree that it is best not to interchange them. While small differences in length can be dealt with through drive adjustment, the cross sections may vary.

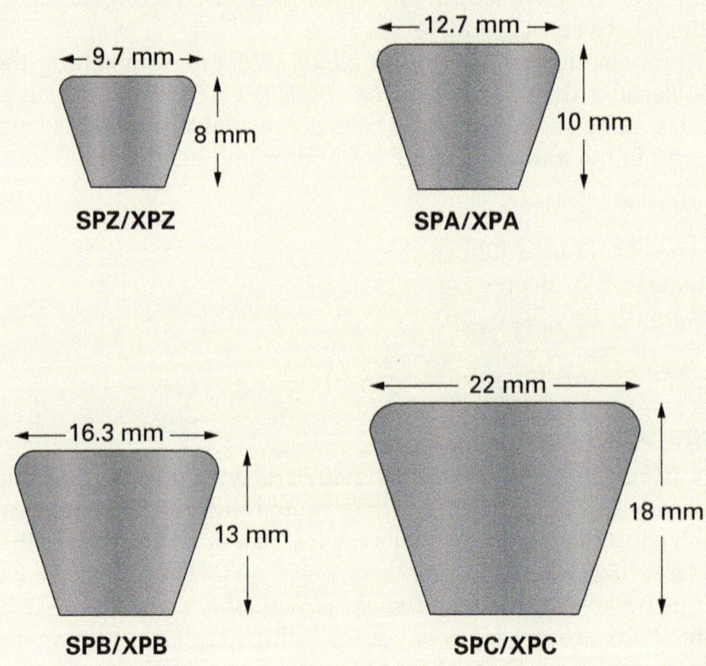

Belt whip: A condition where excess slack in a drive belt causes it to begin riding up and out of the sheave grooves, primarily on the slack side, causing a whipping action. The slack side is the side exiting the drive pulley as it rotates.

Figure 53 Banded belt.

3.1.4 Banded Belts

Banded belts, also known as *polyband belts,* combine multiple drive belts into a single piece (*Figure 53*). Banded belts are often chosen where belt vibration or **belt whip** is a problem.

A set of long, individual belts tend to whip as they rotate. This is especially true if the belts are not perfectly matched, leaving one or more belts slightly longer than the others in a set. These situations are relatively common. Unusual and sudden changes in the driven load while the belt is in motion also cause belt whip. Belt whip can cause singular belts to roll over in the sheave grooves or even fly off completely. Banded belts virtually eliminate belt whip.

Multiple banded belts can also be used together when necessary. For example, two banded belts, each with three belts attached to a single top, can be used together to create a six-band drive.

Both classic and wedge belts are available in a banded configuration. Some manufacturers use a numeral and a slash in front of the belt code to identify it as banded, and to identify the number of belts in the band. For example, a belt code of 6/B80 generally identifies a banded belt composed of six bands built to classic B-belt standards. However, the means of identifying these belts varies by manufacturer.

3.1.5 Notched Belts

Notched belts (*Figure 54*) don't represent a separate class of belts. Both classic and wedge belts are available in notched forms. It is important to recognize that notched belts are not designed for use with sheaves that have a matching profile, like synchronous belts. They are used only with standard sheaves.

Notched belts have several advantages. They are generally more efficient at transmitting power and have a higher load capacity. This is due to the increased edge contact against the sheave along the sides of the belt.

The notched design also makes the belt more flexible. This flexibility makes a notched belt a good choice for small-diameter sheaves because it can wrap around a tighter circle than plain belts. The notches also encourage air movement around the belt as it moves, helping it run cooler.

Notched belts are a bit more sensitive to proper sheave alignment. Slight misalignment may have a limited effect on a plain belt. A notched belt, however, may tend to roll over on the sheave, due to its increased flexibility.

Notched belts are identified with the letter X placed after the letter in the original belt number. For example, an AX42 belt has the same dimensions as an A42 belt but has a notched profile.

Figure 54 Notched V-belt.

3.1.6 Matched Belts

Years ago, when multiple belts were used together on a drive, technicians had to request a matched set of belts. These matched sets were significantly more expensive than common belts, but matched sets were needed to ensure that the belts were precisely the same length. If they were not, one or more belts would carry less tension than the others. The tighter belts would carry far more of the torque load and friction than the looser belts.

Today, quality standards in manufacturing allow the production of belts that are far more consistent. However, it is best to avoid using different belt brands on the same drive. Whenever possible, use belts on a drive from a single manufacturer, and preferably from the same production run. To clarify, belts used in multiples on a drive should be of the same brand, generation, type, and size.

Temporary Belts

There are times when a piece of equipment goes offline due to a damaged or broken drive belt. Although replacement belts may be stocked at most facilities, perhaps the one you need isn't readily available.

HVACR technicians can sometimes turn to temporary belts. Their length is easily adjusted, and products are available to fit FHP and classic belt sheaves.

3.2.0 Sheaves

Figure 55 Examples of sheaves.
Source: Courtesy of MasterDrive, Inc.

Sheaves (*Figure 55*) are the other major components in a belt drive. The terms *sheave* and *pulley* are often used interchangeably, so don't be surprised when you hear both terms. Technically, a pulley represents an assembly of sheaves, ropes or cables, and other components often used for hoisting, managing boat sails, and similar applications. The wheels on which the rope rides through the pulleys are called sheaves. *Sheave* is therefore the correct term for belt drive components, but again, you will often hear the term *pulley* used instead.

Sheaves are often made from cast iron, but they can also be made from cast aluminum or machined from steel. A typical V-belt drive has two sheaves—one mounted on the shaft of the driver and one mounted on the shaft of the driven load.

As shown in *Figure 56*, the drive sheave is usually smaller than the driven sheave. This is because the motor speed, measured in revolutions per minute (rpm), is generally faster than the design speed of the load. For example, one common electric motor speed is 1,750 rpm, but the load may be designed to rotate at 875 rpm—half the speed of the motor shaft.

The rotational speed of the load depends on the speed of the driver and the ratio of the two sheaves. When two sheaves of the same diameter are used, the driven load will rotate at the same speed as the driver. Calculating the speed of a driven load is presented in Section 3.3.2.

Multi-groove sheaves (*Figure 57*) are made for equipment that requires two or more belts on the drive. Multiple belts provide far more grip, which is needed to drive substantial loads. Multi-groove sheaves are also required for banded belts.

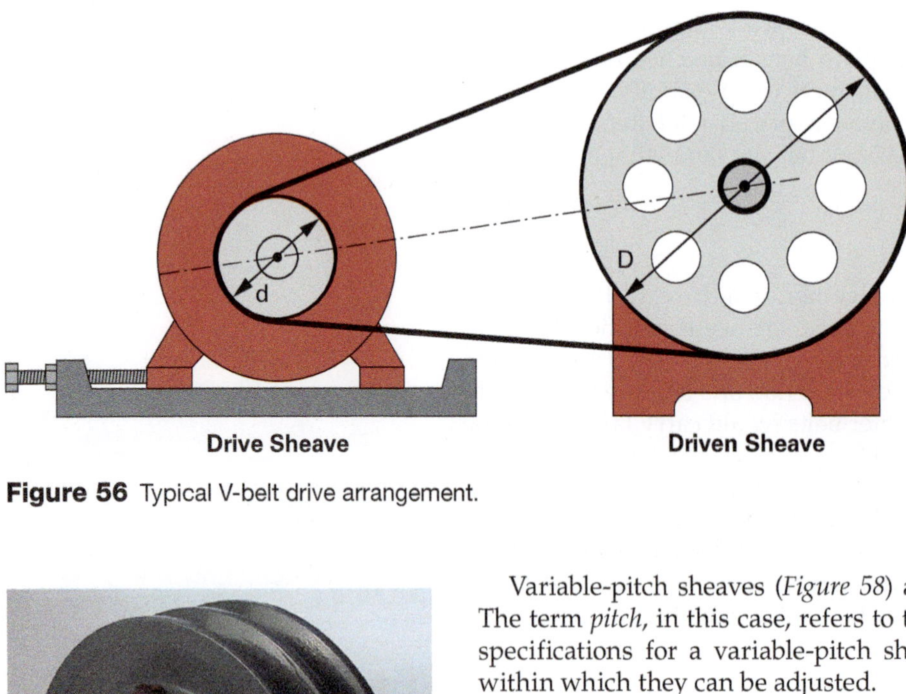

Figure 56 Typical V-belt drive arrangement.

Figure 57 Multi-groove sheave.
Source: Courtesy of MasterDrive, Inc.

Figure 58 Variable-pitch sheave.
Source: Courtesy of MasterDrive, Inc.

Variable-pitch sheaves (*Figure 58*) are very common in HVACR equipment. The term *pitch*, in this case, refers to the position of the belt in the groove. The specifications for a variable-pitch sheave show a range of datum diameters within which they can be adjusted.

Variable-pitch sheaves are made with parts of the assembly threaded onto a hub. Turning the movable portion in or out adjusts the groove width. This causes the belt to ride higher or lower in the sheave groove, changing the speed of the driven load. One or more setscrews lock the movable portions in place once they are positioned. Variable-pitch sheaves for multiple belts are also used, as shown in *Figure 58*.

Variable-pitch sheaves are installed only on the drive shaft, while a standard fixed-pitch sheave is installed on the driven shaft. Ideally, variable-pitch sheaves are used temporarily to adjust the speed of the driven sheave. Once the correct speed is obtained, the variable-pitch sheave is replaced with a standard fixed sheave that matches the adjusted datum diameter. However, in practice, variable-pitch sheaves often remain in service for the life of the equipment.

The *bore* of a sheave refers to the shaft size it fits. A fixed-bore sheave fits only one shaft size. The variable-pitch sheave shown in *Figure 58* is an example of a fixed-bore sheave. To allow a single sheave to be mounted on various shaft sizes, the bore can be cast very large, and the sheave is installed using a bushing that fits a specific shaft size (*Figure 59*). These are referred to as *bushed-bore sheaves.*

Most bushings are tapered and are split along their length, as shown in *Figure 60.* They are often keyed to both the sheave and the shaft. The key for the sheave is visible in *Figure 60,* and the keyway for the shaft can be seen in the bore of the bushing.

(A) Bushed-Bore Sheaves

(B) Sheave with Bushing Installed

Figure 59 Bushed-bore sheaves.

Source: Courtesy of MasterDrive, Inc.

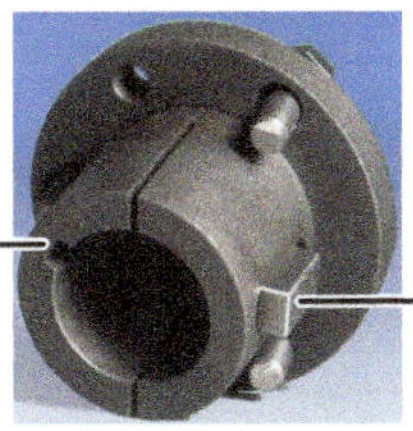

Figure 60 Typical tapered sheave bushing.

Source: Courtesy of MasterDrive, Inc.

Note that the bushing in *Figure 60* has several bolts inserted through the flange. The bolts are used to pull the bushing into the bore of the sheave. The holes in the bushing flange are not threaded. As the bolts are tightened, the taper of the bushing causes it to clamp down tightly on the shaft. For this reason, when installing a bushed-bore sheave on a shaft, you must insert the bushing into the sheave and tighten it while it is on the shaft. You cannot fully install the bushing into the sheave and then slide the assembly over the shaft.

In *Figure 60,* also note the extra hole in the bushing flange with no bolt inserted. There is a second hole on the opposite side that is not visible in the figure. These holes are threaded and are used to push a bushing out of the sheave. Once the bolts securing the bushing are removed, they are threaded into those holes and used as jackbolts to push the sheave out.

It is important to note that, although this style of bushing is very common, there are many other styles. Always consult the manufacturer's literature to determine the correct bushing for a given sheave. It is best to avoid mixing components from different manufacturers.

Many sheaves are designed to accommodate both FHP and classic belts. However, due to small dimensional differences between the two types, they may ride in slightly different positions in the groove. Again, consult the manufacturer's literature to determine what belt types can be used with a given sheave.

Some manufacturers use a sheave numbering system that makes it easier to select belt and sheave combinations. For example, the product number for a sheave designed for 5V wedge belts may also begin with 5V. But it is always best to review the specifications carefully to ensure a match unless you are intimately familiar with a manufacturer's product line.

3.2.1 Sheave Installation and Alignment

There are three types of misalignment to consider when working with sheaves. The following conditions are illustrated in *Figure 61*:

- Parallel misalignment
- Angular misalignment
- Sheave groove (axial) misalignment

Angular misalignment in one plane can be avoided before the sheaves are installed. Begin by ensuring that the two shafts are level. If the shafts are not in

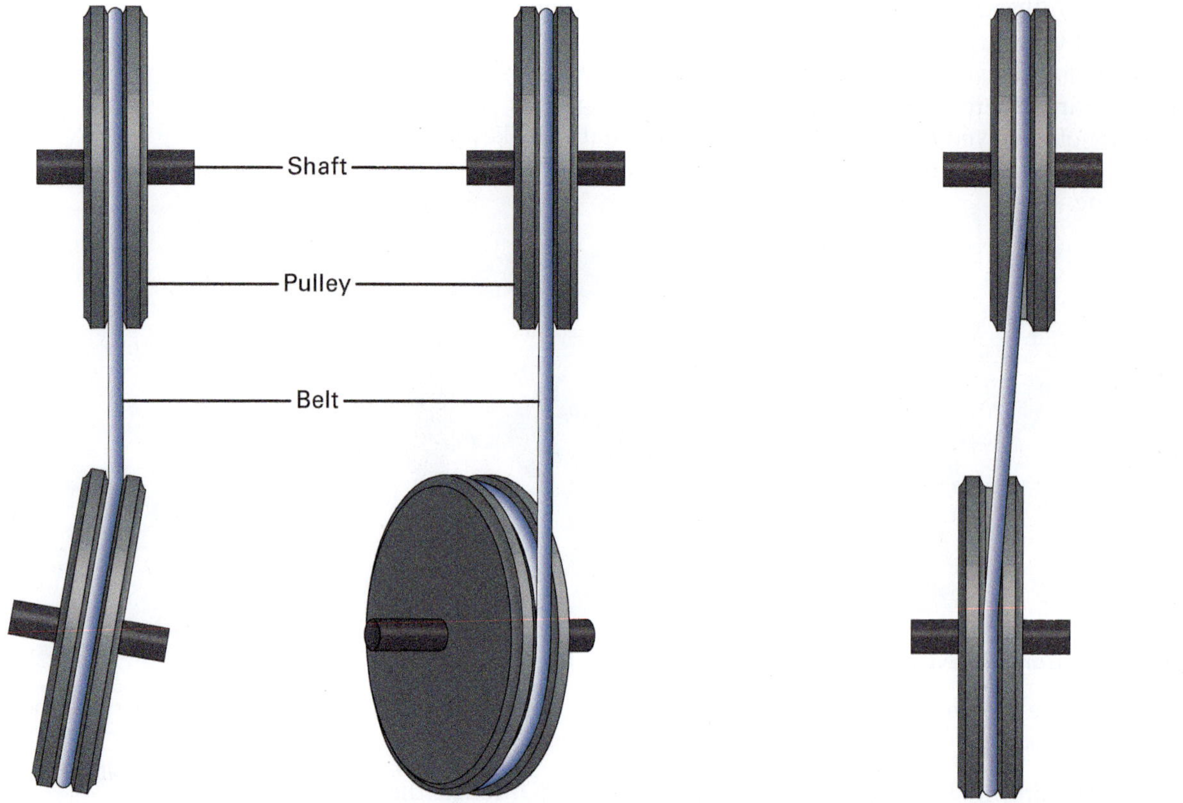

Figure 61 Examples of sheave misalignment.

a horizontal plane by design, then ensure that both shafts have the same inclination angle.

Sheave groove alignment must be checked as the sheaves are installed. The grooves are brought into alignment by moving one of the sheaves along its shaft, either toward or away from the motor.

Aligning sheave grooves raises the question of where the sheaves should be positioned along the shafts. Sheaves should generally be placed close to the driver and driven equipment bearings, but with enough clearance to avoid interference. The farther out on the shafts the sheaves are placed, the more leverage is applied to both bearings when the belt is under tension. This condition is referred to as an *overhung load* (*Figure 62*).

Installing sheaves requires the following basic steps. However, it is always important to review and follow the manufacturer's installation instructions:

Step 1 Inspect the shafts and use a fine file to address any burrs or nicks.

Step 2 Clean both shafts thoroughly, using a solvent and a rag.

WARNING!

Many solvents are toxic and/or flammable. Wear appropriate respiratory protection and gloves when handling these products. Ensure there is sufficient ventilation in the work area. All rags should be disposed of in approved containers.

Step 3 Ensure you have the proper size and length of keystock. When using a bushed sheave, you will often need two keys and perhaps different sizes. Check and clean the keys and address any burrs that may be present by lightly dressing them with a file.

Step 4 Inspect the bore of the sheaves, address any visible flaws, and clean them as necessary. Ensure the setscrews are backed out.

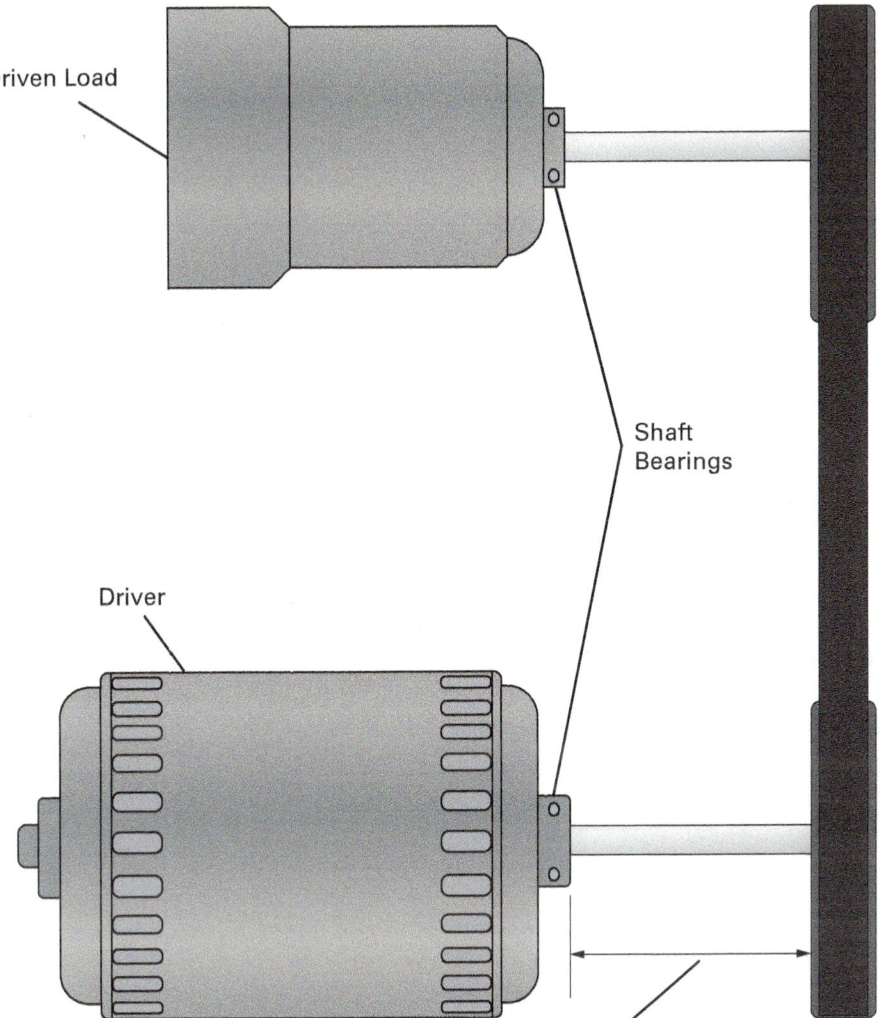

Figure 62 Sheaves placed too far out on the shaft create an overhung load.

Step 5 Slip the driven sheave onto the equipment shaft. If the sheave requires a bushing, loosely mount it to the sheave with the bolts provided. Make sure that keys are properly positioned in their keyways. The sheave is usually positioned close to the motor housing, but it must be far enough away to avoid any possibility of interference with the motor as well as any guards that must be installed.

Step 6 Lock the driven sheave in position. For fixed-bore sheaves, tighten the setscrew(s) while holding the sheave at the desired location. For bushed sheaves, ensure the sheave remains in the desired position as you tighten the bushing mount bolts. Tighten them incrementally, one or two turns at a time, so that the bushing is pulled straight into the bore of the sheave. As you tighten, the bushing will also be clamping down on the shaft. Ensure the sheave remains in the correct position as you engage the bushing.

Step 7 Slip the drive sheave onto the motor shaft, following the same procedures outlined for the driven sheave.

Step 8 Position a straightedge across the faces of both sheaves and move the drive sheave along the shaft until they are perfectly aligned (*Figure 63*). When they are aligned, the straightedge will be flat on the faces of both sheaves. Alternative methods of checking parallel alignment are to pull a wire or string tightly across the faces, or to use laser alignment tools.

Step 9 With the sheave grooves aligned, lock the drive sheave to the motor shaft by tightening the setscrew(s) or drawing the bushing into the sheave bore.

CAUTION

Do not lubricate the bore of the sheave nor the exterior of the tapered bushing before inserting it into the sheave. In some cases, the shaft itself might be lubricated or coated with an anti-seize compound, but never the area where the bushing and sheave bore are in contact.

CAUTION

Sheave bushings typically have specific torque values associated with the mount bolts. Torque the mount bolts to the manufacturer-specified value for the bushing type being installed.

NOTE

Remember that the goal is to align the sheave grooves. Due to differences in construction between the two sheaves, aligning the sides may not result in aligned grooves. If there is a difference in the width of the sheaves, for example, you must factor that in as you position the second sheave. However, aligning the sides will reveal whether the two shafts are parallel. Once parallel alignment is confirmed, you must focus on groove alignment as the second sheave is installed.

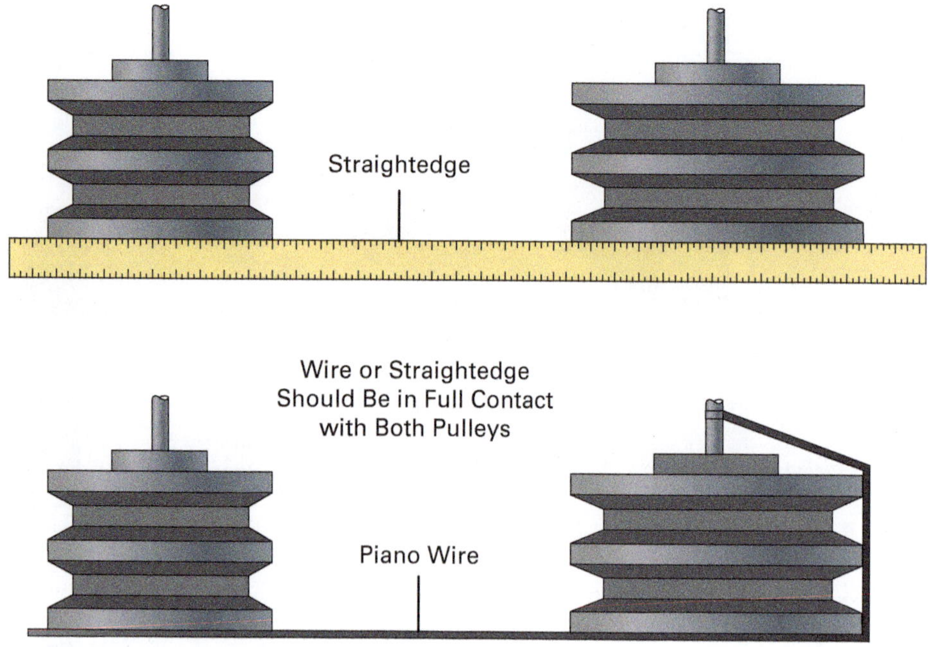

Figure 63 Checking sheave alignment.

Sheave Alignment with Instruments

Sheave alignment with a straightedge, wire, or string may be sufficient for small drives. Larger drives and those where the sheaves are a significant distance apart require a more precise method.

Laser-based alignment tools, also called *sheave lasers*, use laser technology to align sheaves with great precision (*Figure 64*). The alignment components shown in the figure are equipped with powerful magnets, making them easy to attach to the sheave. These tools can detect all three types of misalignment by aligning laser beams with targets.

Figure 64 Laser-based sheave alignment tool, or sheave laser.

Source: Pruftechnik, a Fluke Reliability brand

Note that the tool in *Figure 64* attaches to and aligns the sides of the sheaves. The type shown in *Figure 65* mounts against the grooves, making groove alignment easier. This tool is useful when the driving and driven sheaves have different widths. It is also capable of detecting all three forms of misalignment.

Sheave lasers are accurate across a significant distance, typically 20' to 30' (6 m to 9 m), and can be operated by one person. Adjustments can be made while using the laser as a guide.

Using these tools for sheave alignment is highly recommended. Although belt drives can be forgiving, properly aligned sheaves reduce belt wear and power consumption. The reduction in maintenance costs can save a great deal of money in a large plant. Review the instructions for the chosen tool and follow the manufacturer's guidance carefully.

Figure 65 Laser-based groove alignment tool.

Source: Courtesy of SKF, Lincoln Industrial Products, Alemite LLC

3.2.2 Speed Calculations

As mentioned previously, the speed of the driven load is rarely the same as that of the driver. Most of the time, the load rotates at a lower speed than the driver.

The rotational speed of the driven load is based on the ratio between the diameter of the two sheaves and the rotational speed of the driver. In the unlikely event that you need the load to rotate at the same speed as the motor, you can use two sheaves of precisely the same diameter.

Some speed ratios are very easy to determine. If, for example, you need the load to rotate at exactly half the speed of the driver, then make the driven sheave twice as big. A 3" drive sheave matched with a 6" driven sheave will produce that result.

Unfortunately, the situation is rarely that simple and some calculations are required. The following equation is manipulated to find the information you need:

$$\frac{\textit{Driver rotational speed}}{\textit{Driven rotational speed}} = \frac{\textit{Driven pitch diameter}}{\textit{Driver pitch diameter}}$$

You'll recall that the pitch diameter of a sheave typically represents its outside diameter. For very precise work, you will need to determine the exact datum diameter for the sheaves. However, when working with fixed-pitch sheaves, using the pitch diameters will provide good results. The datum diameter of a variable-pitch sheave can change, so working with the datum diameter is necessary for accuracy when using a variable-pitch sheave.

The four equations used to determine sheave sizes and resulting speeds are as follows:

$$\text{Driven rotational speed} = \frac{P_{driver} \times S_{driver}}{P_{driven}}$$

$$\text{Driven pitch diameter} = \frac{S_{driver} \times P_{driver}}{S_{driven}}$$

$$\text{Driver rotational speed} = \frac{P_{driven} \times S_{driven}}{P_{driver}}$$

$$\text{Driver pitch diameter} = \frac{S_{driven} \times P_{driven}}{S_{driver}}$$

Where:

P_{driver} = Driver pitch diameter
P_{driven} = Driven pitch diameter
S_{driver} = Driver rotational speed
S_{driven} = Driven rotational speed

The following example problem demonstrates the use of these equations. A drive motor rotates at 1,800 rpm and the drive sheave has a pitch diameter of 4". The driven sheave has a pitch diameter of 12". How fast will the driven load rotate?

$$\text{Driven rotational speed} = \frac{P_{driver} \times S_{driver}}{P_{driven}}$$

$$\text{Driven rotational speed} = \frac{4" \times 1{,}800 \text{ rpm}}{12"}$$

$$\text{Driven rotational speed} = \frac{7{,}200}{12"}$$

$$\text{Driven rotational speed} = 600 \text{ rpm}$$

That result makes sense, because the driven pitch diameter happens to be three times that of the drive sheave. Therefore, the speed will be one-third of the driver speed.

Now an engineer asks you to change the rotational speed of the load to 750 rpm—an increase of 150 rpm. You can do this in one of two ways. The drive sheave can be replaced with a larger sheave, or the driven sheave can be replaced with a smaller one. In this example, we will choose to replace the driven sheave.

Use this equation to determine the size of the new driven sheave:

$$\text{Driven pitch diameter} = \frac{S_{driver} \times P_{driver}}{S_{driven}}$$

$$\text{Driven pitch diameter} = \frac{1{,}800 \text{ rpm} \times 4"}{750 \text{ rpm}}$$

$$\text{Driven pitch diameter} = \frac{7{,}200}{750}$$

$$\text{Driven pitch diameter} = 9.6"$$

In summary, replacing the 12" driven sheave with a 9.6" sheave will increase the speed of the driven load from 600 rpm to 750 rpm.

3.3.0 Belt Installation and Tension Adjustment

Belt drives are intended to give many hours of service. Belt failure and excess sheave wear can often be traced to improper belt tension or misalignment, although normal wear and tear does take its toll.

Excess belt tension stretches the belt and places stress on the bearings and shafts. If the belt is too loose, it will slip and wear out quickly. In addition, the blower is unlikely to turn at the desired speed, reducing system airflow. The tension of all drive belts should be adjusted using a tension gauge according to the manufacturer's instructions.

To install the belt(s), adjust the drive motor using the jackbolt on the driver to close the gap between the two pieces of equipment. This will allow you to slip the belts over the sheaves and drape them over the grooves. Do not "roll" belts over the edge of the sheaves under tension.

It is best to approximate the required tension and then run the drive for a short period, allowing the belt to seat in the sheave groove. If a rotational speed for the driven load has been specified, this also provides an opportunity to measure its speed with a tachometer. If a variable-pitch sheave isn't adjusted before the target tension is set, the process must be repeated afterwards. Adjusting a variable-pitch sheave changes where the belt rides in the groove, affecting the tension. If a variable-pitch sheave is not used, speed changes are made by changing one or both sheaves.

If the speed is too low, you can adjust the variable-pitch sheave by bringing the halves closer together, forcing the belt to ride higher in the groove. If the speed is too fast, open the sheave to adjust. Make sure you tighten the setscrew to lock the sheave in position. Then readjust the belt tension.

WARNING!

Never attempt any type of adjustment to a rotating drive. Ensure the drive is de-energized and locked out before working on drive components. Before starting belt-driven equipment, ensure that all belt guards have been properly reinstalled, and all tools and equipment have been removed from the area.

3.3.1 Setting Belt Tension

A popular belt-tensioning tool is shown in *Figure 66*. Its use is based on the premise that the force required to deflect a drive belt crossing a span is directly related to the belt's tension. This tool is also made in assemblies of two and three devices bound together. They are used to measure tension on banded and very large belts, beyond the range of the spring in a single device.

The first step is to determine the required tension for the belt(s). This information can come from a variety of sources. The equipment manufacturer is likely the best source if it provides that information in service manuals. Belt manufacturers are also a reliable source. Perhaps the most accessible source of information, though, is the manufacturer of the tension measurement tool. Most provide tables based on factors such as drive speed and sheave diameter that show the recommended deflection force. *Table 5* is an example of a belt deflection chart.

Figure 66 Mechanical belt tension gauge.

Source: Courtesy of Timken Belts

CAUTION

Table 5 provides an example only and is not recommended for fieldwork. Use current table(s) provided by the equipment, belt, or tensioning tool manufacturer.

TABLE 5 Minimum Belt Deflection Force

Belt Section	Smaller Sheave		Deflection Force[1] (lbf) for a Drive Speed Ratio of:			
	Speed Range (rpm)	Diameter[2] (inches)	1.0	1.5	2.0	4.0+
3VX	1,200–3,600	2.2	2.2	2.5	2.7	3.0
	1,200–3,600	2.5	2.6	2.9	3.1	3.6
	1,200–3,600	3.0	3.1	3.5	3.7	4.2
	1,200–3,600	4.1	3.9	4.3	4.5	5.1
	1,200–3,600	5.3	4.6	4.9	5.1	5.7
	1,200–3,600	6.9	5.0	5.4	5.6	6.2
5VX	1,200–3,600	4.4	6.5	7.5	8.0	9.0
	1,200–3,600	5.2	8.0	9.0	9.5	10.0
	1,200–3,600	6.3	9.5	10.0	11.0	12.0
	1,200–3,600	7.1	10.0	11.0	12.0	13.0
	900–1,800	9.0	12.0	13.0	14.0	15.0
	900–1,800	14.0	14.0	15.0	16.0	17.0
A or AP	1,800–3,600	3.0	2.0	2.3	2.4	2.6
	1,800–3,600	4.0	2.6	2.8	3.0	3.3
	1,800–3,600	5.0	3.0	3.3	3.4	3.7
	1,800–3,600	7.0	3.5	3.7	3.8	4.3
B or BP	1,200–1,800	4.6	3.7	4.3	4.5	5.0
	1,200–1,800	5.0	4.1	4.6	4.8	5.6
	1,200–1,800	6.0	4.8	5.3	5.5	6.3
	1,200–1,800	8.0	5.7	6.2	6.4	7.2
AX	1,800–3,600	3.0	2.5	2.8	3.0	3.3
	1,800–3,600	4.0	3.3	3.6	3.8	4.2
	1,800–3,600	5.0	3.7	4.1	4.3	4.6
	1,800–3,600	7.0	4.3	4.6	4.8	5.3

Notes:
[1]Double the values to tension a new set of belts.
[2]Use closest sheave diameter for sizes not shown.

Tables and charts may also have footnotes, some of which indicate that new belts should be tensioned to a higher value than the value shown on the chart. *Table 5* indicates that new belts should be set to twice the normal tension. This accommodates expected stretching that will occur early in the life of the belt.

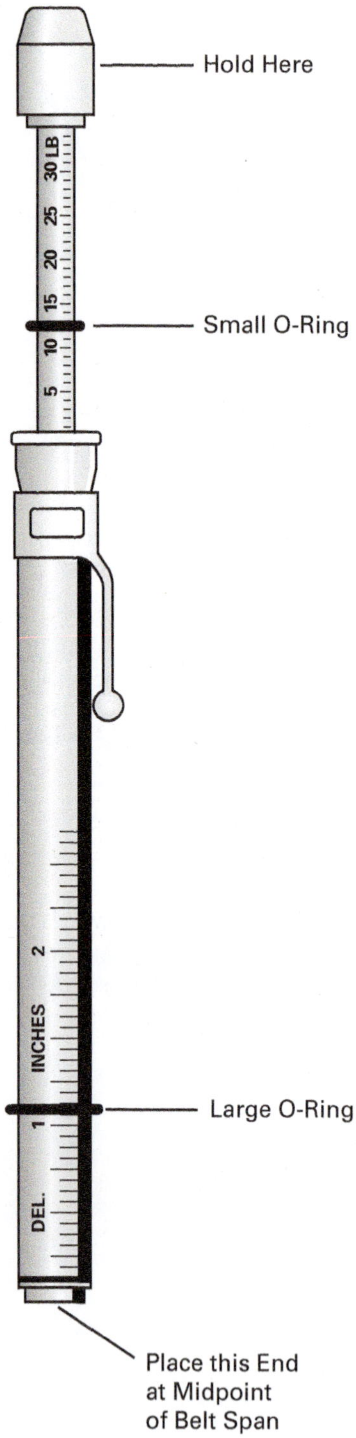

Figure 67 Belt tension gauge settings.

Source: Courtesy of Timken Belts

Always refer to the instructions for the specific type and brand of tool in use. The basic instructions for using the tension gauge shown in *Figure 66* are as follows:

Step 1 De-energize the system and lock out the power source.

Step 2 Measure the distance between the centers of the two sheaves.

Step 3 Refer to *Figure 67*. Set the *large* O-ring to a value that represents the number of inches between sheave centers, multiplied by $\frac{1}{64}$". For example, if the distance is 68", set the large O-ring to $1\frac{4}{64}$" ($1\frac{1}{16}$"), as shown in *Figure 67*.

Step 4 Set the *small* O-ring at zero on the upper scale. Note that it will move as a reading is taken.

Step 5 If there is only one belt on the drive, set a straightedge across the top of the belt between the two sheaves, as shown in *Figure 68*. A string can also be pulled taut between the sheaves if the span is too long for an available straightedge. Note that this step is unnecessary if there is more than one belt.

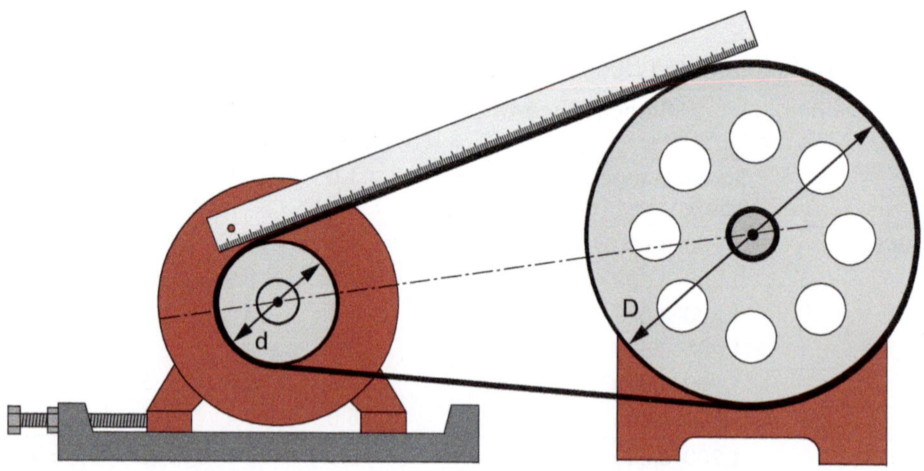

Figure 68 Straightedge placed across the top of the belt.

Step 6 Carefully push the base of the tension gauge down on the belt in the center of the span, with attention to one of these two scenarios (*Figure 69*):

- If there is only one belt, push down until the large O-ring is even with the bottom of the straightedge.
- If there are multiple belts, push down until the large O-ring is even with the top of the adjacent belt. In this case, take a reading from each belt. Remember to reset the small O-ring to zero before taking each measurement.

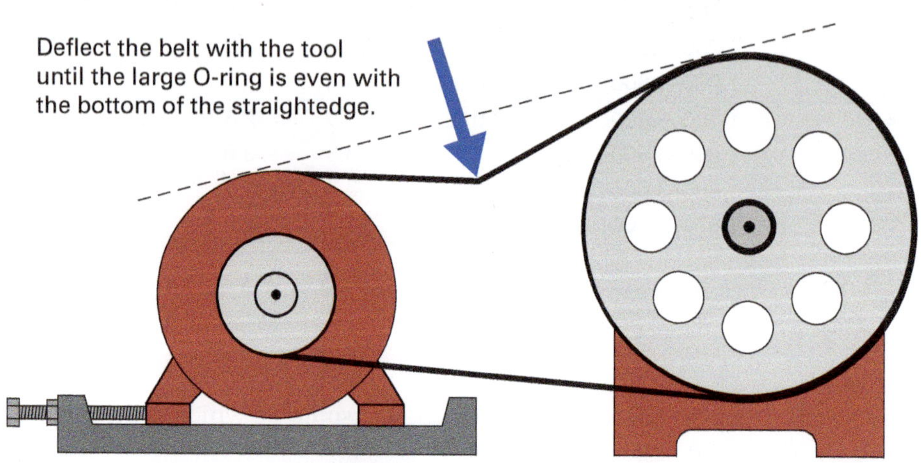

Figure 69 Taking a force reading.

Step 7 Examine the tension gauge and note the new position of the small O-ring. It now indicates the force required to deflect the belt, in pounds. If there are multiple belts on the drive, take a reading on each belt and average the readings.

Step 8 Compare this value to the chart in use. If the force is lower than prescribed, add more tension using the motor jackbolt and measure the force again. If it is higher than prescribed, reduce the tension slightly and measure the force again.

Step 9 With the proper tension established, reinstall any belt guards, remove locks and tags from the power supply, and restore power to the driver. Observe the drive as it operates for a short time to ensure that it is operating as expected.

WARNING!

Do not operate the equipment without the belt guards installed. Serious injury will result if your hand or clothing is caught in the drive. Inspect the blower wheel and immediate area for tools or foreign objects before starting. When initially starting a blower assembly, back away from the immediate drive area to avoid any chance of injury as it starts.

Electronic tension gauges, or *tensiometers*, are becoming increasingly popular. Most of the electronic models base the tension measurement on the frequency of belt vibration. They may also be referred to as *trummeters*.

The model shown in *Figure 70* uses a pulsed beam of light to measure the frequency of a vibrating belt. With the instrument ready and the beam pointed at the belt surface in the correct location, the user merely strums the belt to initiate vibration. Algorithms determine the tension based on the vibration frequency. The digital display can provide readings in units of newtons (N) or pounds of force (lbf).

After installing new sheaves or changing the speed of a blower, the operating current of the motor should always be checked to ensure it remains at or below its ratings. Before checking the current, ensure the equipment is fully assembled and under a normal operating load. An open access door on the return-air side of an air handler, for example, significantly increases airflow and results in higher current readings. After the adjustments are complete, duct velocity measurements may be needed to determine the new air volume.

Figure 70 Electronic tensiometer.
Source: Courtesy Checkline

3.0.0 Section Review

1. What are the nominal dimensions of a 4L460 belt?
   a. $\frac{1}{4}$" wide × 46" long
   b. $\frac{3}{8}$" wide × $5\frac{3}{4}$" long
   c. $\frac{1}{2}$" wide × 46" long
   d. $\frac{1}{2}$" wide × $57\frac{1}{2}$" long

2. A drive motor rotates at 1,750 rpm and the drive sheave has a pitch diameter of 3.5". The driven sheave has a pitch diameter of 10". How fast will the driven load rotate?
   a. 498 rpm
   b. 551.8 rpm
   c. 612.5 rpm
   d. 814 rpm

$$\text{Driven rotational speed} = \frac{P_{driver} \times S_{driver}}{P_{driven}}$$

$$\text{Driven rotational speed} = \frac{3.5" \times 1{,}750 \text{ rpm}}{10"}$$

$$\text{Driven rotational speed} = \frac{6{,}125}{10"}$$

$$\text{Driven rotational speed} = 612.5 \text{ rpm}$$

3. An open blower access door on the return-air side of an air handler will result in a motor current reading that is _____.
   a. abnormally high
   b. abnormally low
   c. the most accurate
   d. inconsistent

4.0.0 Basic Inspection Procedures

Performance Tasks

3. Perform a periodic maintenance inspection on a gas furnace and document the inspection results.
4. Perform a periodic maintenance inspection on a cooling system or heat pump system in the cooling mode and document the inspection results.

Objective

Identify and describe basic inspection and maintenance procedures for selected equipment.

a. Identify common environmental health hazards associated with HVACR maintenance activities.
b. Describe common inspection and maintenance procedures for gas heating equipment.
c. Describe common inspection and maintenance procedures for cooling and heat pump systems.
d. Describe common inspection and maintenance procedures for various system accessories.

The safe and economical operation of HVACR equipment depends on proper periodic service and maintenance. Any deficiencies must be documented, reported to the owner, and resolved with their permission whenever billable work is involved.

An HVACR technician must have the skills to perform inspection and maintenance tasks on a variety of equipment and accessories while observing all necessary safety precautions. Periodic maintenance often represents a significant part of your workday.

One of the most important tasks related to HVACR equipment maintenance is the completion of forms and/or service reports. The well-known saying, "no job is finished until the paperwork is done" applies here. Maintenance documentation in the HVACR industry is recorded and used in many ways. It is important that the documentation be timely, thorough, and understandable.

Forms and reports serve as historical records for the equipment. For this reason, it is important to fill them out completely and accurately. When troubleshooting later, recorded data is useful because it can be compared to current system readings to determine what has changed. Today's available software packages can provide a great deal of data to technicians while they are on site.

Unfortunately, lawsuits are a part of every service industry. This means that the documentation technicians generate may be called into question or handled as evidence in court cases. Forms and reports are often used to prove that specific tasks were performed, or that the installed equipment has met design and customer specifications. The customer's signature is a crucial part of the document, as it typically indicates an acceptance of the services performed. Without a signature, the door remains open for future disagreements.

Three critical items that should appear on all field documents are the make, model, and serial number of the equipment being serviced, which are described as follows:

- The make identifies the name of the manufacturer, such as Carrier, Trane, Goodman, etc.
- The model number identifies the specific style of the unit and many of its features. The number generally identifies its style, capacity, and electrical characteristics at a minimum.
- The serial number is a unique sequential number assigned to the unit by the manufacturer. This number is critical when you need to get replacement parts for the unit. Many units share the same model number, but a different component may have been used in later versions of the same model. The serial numbers help the manufacturer track and maintain records for each unit sold.

Did You Know?

Getting Paid

Many HVACR dispatch and operations management software packages do not allow the processing of payroll until a technician has properly documented their work in the system. This creates some motivation for you to complete your paperwork accurately and in a timely manner!

4.1.0 Environmental Health Hazards

Environmental conditions associated with HVACR maintenance can represent health hazards to technicians. HVAC systems tend to attract hazardous elements for a variety of reasons. These hazards include those associated with animals that tend to dwell in areas where HVACR equipment may be installed.

4.1.1 Rodent Hantavirus Hazard

Most of the hazards encountered by HVACR technicians are electrical or chemical in nature. In some situations, however, biological hazards must also be considered. One such hazard involves rodents nesting in or near HVACR equipment.

According to the National Center for Infectious Diseases (NCID), a component of the Center for Disease Control and Prevention (CDC), hantavirus pulmonary syndrome (HPS) is a potentially fatal respiratory disease. Caused by *hantaviruses* carried by rodents, HPS can be contracted by humans who encounter rodent waste or nesting materials (*Figure 71*). The virus, a distant relative to the more widely recognized Ebola virus, does not spread from person-to-person. While HPS is not common in the United States, HVACR technicians must take precautions to guard against it.

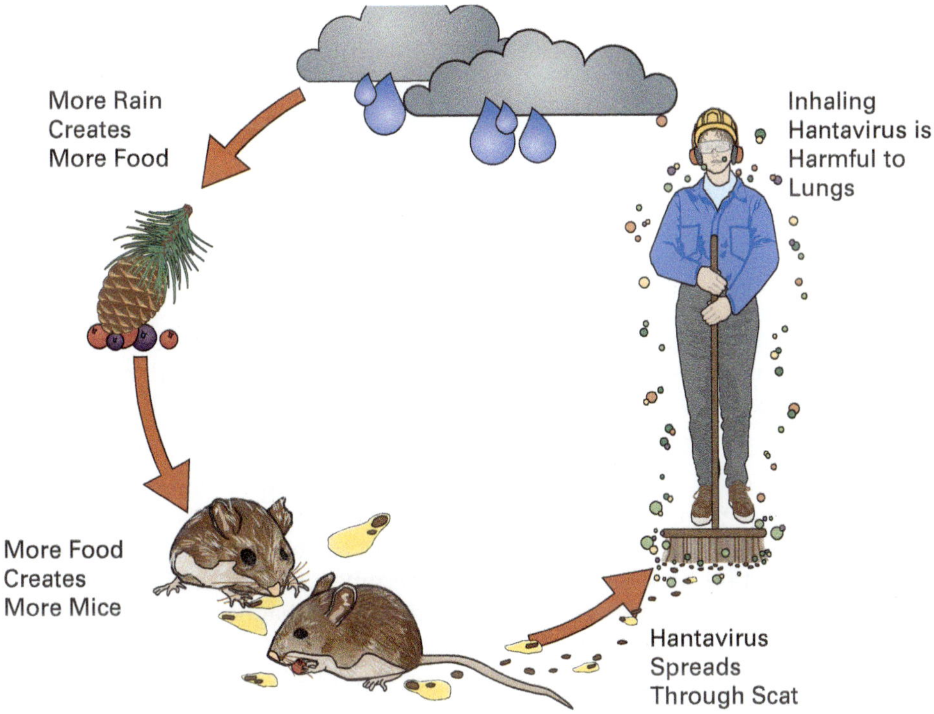

Figure 71 Cycle of hantavirus transmission.

According to the NCID, the initial symptoms of HPS include muscle aches, fever, and fatigue. The symptoms are similar to the flu. An individual may actually begin to feel better after the initial symptoms begin. However, after one or two days, breathing becomes difficult and symptoms worsen. Secondary symptoms include a dry cough, nausea, vomiting, and abdominal pain.

Once in the lungs, hantavirus is very difficult to treat. Early hospitalization and intensive care can increase a victim's chance of survival. According to the CDC, hantavirus has a mortality rate of 38 percent, although advances in medical care are improving the rate.

Another disease caused by hantavirus is *hemorrhagic fever with renal syndrome* (HFRS). Common symptoms at its onset are high fever, lower back and/or abdominal pain, and blood clots that form in the body's smaller blood vessels.

CAUTION

Before applying bleach or other chemicals to clean and sanitize an area, shut down any air systems associated with the area and take steps to prevent undesirable odors and vapors from affecting building occupants.

However, renal (kidney) dysfunction is the primary indicator. Renal failure typically begins within one to two weeks of the early symptoms.

The best way to avoid HPS is to minimize contact with potentially infected materials. The most effective way of preventing hantavirus infection is to eliminate any contact with rodents, but this is not a practical solution for those who must service equipment located in their habitat. Technicians must learn to protect themselves from direct contact and minimize their exposure. While technicians may not see the rodents, their nests and droppings will be visible.

When nests and droppings are apparent, use the following precautions:

- Wear an N95-rated disposable face mask while in the area. Never use a broom or vacuum cleaner to remove the materials unless they have first been sprayed liberally with a 10 percent mixture of bleach to water. Allow the solution to stand for 30 minutes. For maximum protection, wear a full-face respirator with a high-efficiency particulate air (HEPA) filter.
- Wear goggles, disposable coveralls, and heavy latex or rubber gloves when removing the materials. As noted above, treat the debris with a mixture of bleach and water before handling. Use a towel or bedsheet soaked in the same cleaning solution to remove the material.
- Once the materials have been removed, mop or sponge the affected area a second time with the bleach solution. Keep all your safety gear on during this process.
- Double bag the debris along with the cleaning materials. Make sure to disinfect your gloves before removing them, then dispose of them as well. Dispose of the bags properly, following all applicable health codes. This normally involves burning or burying the materials. If neither of these options is feasible, contact the state health department for instructions. Wash up thoroughly with an appropriate disinfectant.

4.1.2 Bird and Bat Excrement Hazards

Bird and bat excrement represent a hazard when servicing some equipment, especially if the units are located beneath areas where birds or bats have roosted. In some cases, the equipment itself may contain bird nests. The following diseases are associated with organisms found in dried bird and bat droppings:

- *Cryptococcosis* — *Cryptococcus neoformans* is a fungus commonly found in the dry dust of pigeon, dove, or bat droppings. Cryptococcosis is a yeast infection contracted by inhaling the mold spores of the organism. The mold cells are 5 to 8 microns in diameter and easily become airborne. An infection may result in *cryptococcal meningitis*, which is difficult to diagnose and can be fatal if not properly and promptly treated.
- *Histoplasmosis* — *Histoplasma capsulatum* is a fungus found in soils throughout the world. It flourishes in soils that have been enriched by bird or bat droppings for three or more years; however, it has also been found in droppings that have not been in contact with soil. The fungus is difficult to eliminate from soil even after the nutrient source is removed. Histoplasmosis is contracted by inhaling the spores of the fungus carried by wind and dust. Most infections produce no symptoms or only a mild flu-like illness. However, untreated chronic or reactivated infections can result in pneumonia, blindness, chronic and progressive lung disease, and even death.
- *Psittacosis* (also called *ornithosis* or *parrot fever*) — *Psittacosis* is caused by inhaling *Chlamydia psittaci* bacteria that are found in feathers and droppings from any infected bird, including pigeons and doves. Because the organism becomes less infectious with time, active roosts are the greatest concern. The disease results in fever, headaches, and muscle pain, with or without obvious respiratory symptoms. Untreated cases can progress to pneumonia or toxemia and can result in death, especially in older people.

- *Rabies* — Rabies is a viral disease and bats are known to be carriers. Contact with the dried blood, urine, or droppings from an infected bat is not a major risk. However, dead bats should be picked up and disposed of using a tool or heavy gloves because the rabies virus can remain infectious in a carcass until decomposition is complete. If even a scratch from a live or dead bat occurs and the bat cannot be retrieved for testing, prompt rabies vaccinations must begin.

Most of the hazards described here attack the human body through the lungs. Breathing infected material, therefore, represents a significant hazard. Technicians should always wear N95-rated breathing masks any time they are working around potentially hazardous materials. It is highly recommended that you wear this same protection any time you are handling air filtration materials or working on the airside of the system. Many common viruses are captured by air filters along with the dust and debris.

If the equipment or the surrounding area is contaminated with bird or animal droppings, the building owner should be notified promptly. It is best for professionals in hazardous material removal to clean and sanitize the area. Once this is done, recommend or take steps to prevent the problem from recurring.

4.2.0 Gas Furnace Inspection and Maintenance

Gas heating is the most common type of heating system in the United States. When properly serviced, gas-fired heating equipment (*Figure 72*) is very effective and reliable.

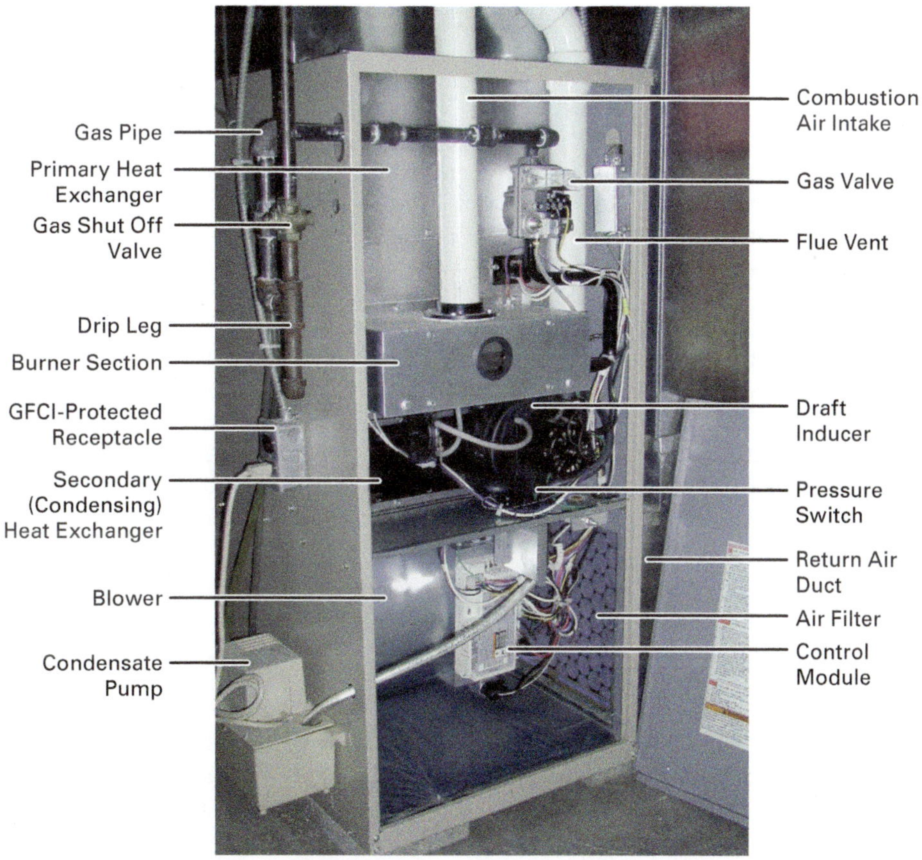

Figure 72 Condensing gas furnace.

To operate safely and reliably, gas-fired furnaces should be serviced at least once a year by a qualified technician. Safety is an especially important consideration because burning fossil fuels can produce hazardous byproducts such as carbon monoxide (CO).

The appearance of the flame and the absence (or presence) of soot deposits on burners and heat exchangers are general indicators of the combustion process. However, it is necessary to use more sophisticated equipment to get an accurate picture of combustion efficiency and CO content. Some CO is often present in the flue gases when a furnace is operating properly. However, poor combustion typically increases the volume.

The careful inspection of fossil-fuel heating equipment is helpful in preventing CO problems. Cracked heat exchangers and inadequate or obstructed vents are factors that can cause excessive CO generation. As part of every service call, check to make sure the vent has a good draft, and the byproducts of combustion are being drawn into the flue.

4.2.1 Gas Heating Safety

Safety is the most important part of your job, for self-protection and for the protection of others. Remember the following general safety guidelines when working with natural gas or propane furnaces:

WARNING!

When performing planned maintenance that requires the dismantling of gas burners or any gas-line components for cleaning or repair, or if a misfire occurs, always shut off the gas supply before beginning work. Allow five minutes to elapse so that any unburned natural gas will dissipate. Leak test the assembly before testing the burner. Never use a flame to check for a gas leak.

WARNING!

Liquid petroleum (LP) vapor is heavier than air and will not rise naturally. The vapor collects in low, calm areas near the floor. If you suspect that a concentration of LP gas is present, do not turn on any lights or electrical appliances. These devices may generate an arc and ignite the gas/air mixture. Open doors and windows and mechanically ventilate the area.

Figure 73 Combustible gas leak detector.
Source: Courtesy of RIDGID Tool Company

Leak reactant solution: Noncorrosive liquid solution that readily foams or forms bubbles in the presence of a leak.

- Conduct a gas leak test after completing any installation, repair, or service. Use a combustible gas leak detector (*Figure 73*) or a **leak reactant solution** applied to fuel tubing connections, pipe joints, and gaskets. Never use a flame to check for a gas leak. Remember that the gas pressure serving furnaces is relatively low; allow sufficient time for a leak to be revealed when using a bubble solution. Tighten the joints or replace gaskets as needed to stop any leaks.
- Disconnect the power supply before connecting or disconnecting any wiring. Always turn off and lock out the power supply unless the service procedure requires the furnace to be powered.

4.2.2 Gas Furnace Inspection and Testing

Before performing any service procedure, check with the customer and listen to any complaints. Make note of any comments that might indicate a problem.

It is always best to follow the manufacturer's guidance for the equipment being serviced. The following sequence is not necessarily appropriate for all units and may need to be adjusted to both suit the specific unit and improve productivity. However, the following sequence provides a good outline for the inspection of a gas furnace.

Begin at the system thermostat and remember the settings so you can leave it as the owner preferred.

NOTE

Appendix B provides a form to serve as a guide during the execution of the related Performance Task.

- Check the thermostat using the following sequence:
  - Move the system switch to the Heat or Auto position. Move the setpoint above room temperature. The heating system should start the ignition sequence, and the indoor blower should run after a short delay.
  - Move the setpoint below room temperature. The burner should shut down. The indoor blower should continue to run for a short time and then stop.
  - To avoid damaging the control transformer, turn off the power to the furnace to de-energize the control circuit. Remove the thermostat to expose the subbase. Ensure the terminal connections are tight.
  - Replace the batteries in digital thermostats annually. Most digital thermostats have a capacitor that powers the thermostat memory and retains the programming for a short period while the batteries are changed. This is also a good time to reset any air-filter change reminders that may be programmed, because the air filter should be changed during this visit.
  - Reinstall the thermostat on the subbase.
- Check indoor blower operation.
  - Check and record the supply voltage at the unit disconnect and ensure it is within the unit nameplate voltage range.
  - With the blower compartment door in place, check and record the current draw of the blower motor. Check it against the nameplate rated load amperage.
  - Turn off the power to the furnace before proceeding.
- Inspect the blower and change the air filter(s).
  - With the system de-energized, don an N95-rated face mask and change the air filter(s). Make sure the filters are properly fitted to the opening so air cannot easily bypass them.
  - If the blower is belt-driven, remove the blower belt and check it for cracks and excess wear. Loosen the motor mount or other adjusting mechanism before trying to remove the belt. Leave the belt off for some of the steps that follow.
  - Check the blower and/or motor bearings for wear. Make sure everything rotates freely and smoothly.
  - Lubricate the blower and/or motor bearings per the manufacturer's instructions, taking care not to over-lubricate. Ensure any sheave setscrews are tight and the sheaves are properly aligned.
  - Clean the blower wheel and compartment with a HEPA vacuum. If the blower wheel is very dirty, remove and clean the wheel with detergent and a brush.
  - Check the blower and/or motor mounts for tightness and make sure any vibration isolators are in place and intact.
  - Check all the electrical wiring related to the blower, looking for sound electrical connections and terminations.
- Inspect the burners, ignition gear, and controls.
  - With the gas and power off, check all the electrical wiring in the burner vestibule, looking for sound electrical connections and terminations.
  - If the furnace has an electronic ignition, inspect the igniter assembly. Use a feeler gauge to measure the gap between the electrode and flame rod or pilot burner (*Figure 74*); a $^{7}/_{64}$" (2.7 mm) gap is common.
  - If the furnace uses a hot surface igniter (HSI) like the one shown in *Figure 75*, examine it during the next furnace startup to ensure it glows hot and there is little hesitation in the ignition sequence. If it seems dim and is slow or fails to ignite the burner, its resistance will need to be tested to see if falls within the required range established by the manufacturer.
  - Clean the flame sensor or rod with fine emery cloth to remove any deposits. Once the furnace is operational, check the flame signal strength with an ammeter. A very small amount of current, usually ranging between

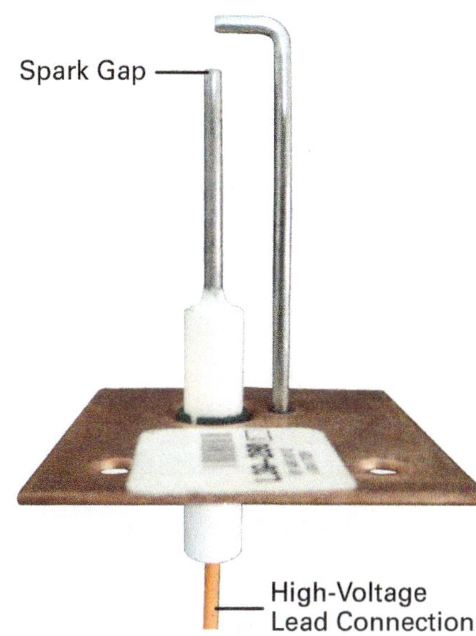

Figure 74 Typical spark ignition electrode and flame rod.

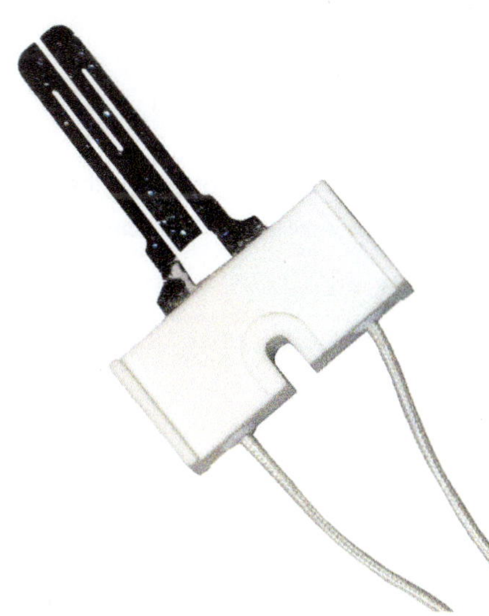

Figure 75 Hot surface igniter (HSI).

CAUTION

When removing burners, use caution to prevent damaging the pilot assembly, HSI, spark ignitor, or flame sensor. These devices may have to be carefully removed before the burners can be removed.

2 microamps (μA) and 6 μA, is expected. Remember that one microamp is one millionth of an amp. The ammeter must be wired in series with the sensor for this measurement.

- Remove and clean the burners inside and out using a soft wire brush and a round brush for the inside.

If there is cause for concern about the condition of the heat exchanger, such as the age of the unit, the heat exchanger must be given some special attention. Irregular flames on the burners after the indoor blower starts is another possible indication of a heat exchanger leak. However, a detailed heat exchanger inspection is not usually needed every year.

WARNING!

The purpose of heat exchanger inspection is to ensure that potentially deadly gases, such as CO, are not being released into the structure. It is an important responsibility. Any furnace with a leaking heat exchanger should be shut down immediately for repair or replacement. Never attempt to repair a hole or crack in a heat exchanger.

Borescope: An optical device that typically uses light-transmitting fiber to display an image through a narrow tube, allowing for the inspection of areas that are otherwise difficult or impossible to see.

Today's heat exchangers can be difficult to inspect due to their narrow, twisting passages. A **borescope** (*Figure 76*) will allow you to see inside many of them, but you may not be able to pass the probe all the way through. However, it is still the best way to get a look inside.

Figure 76 Flexible borescope inspection camera.

Source: Courtesy of RIDGID Tool Company

To gain reasonable access to the heat exchanger cells, you will likely have to remove the burners. Then insert and guide the borescope as needed to inspect the interior. You can also shine a bright light into the heat exchanger cells (top and bottom) and see if any light is visible from the outside. Areas near the burner are more susceptible to cracks due to the expansion and contraction of the metal through repeated heating cycles.

Another method used for a quick test of the heat exchanger is the "standing match" test. Start the system blower but leave the burner off. Light a long match and hold it just inside each heat exchanger cell. If there is a significant leak in one or more cells, the match will waver and may even blow out. Otherwise, the match flame should remain steady.

A combustion analyzer (*Figure 77*) can also be used to test a heat exchanger. Most residential gas furnaces operate with a slightly negative pressure in the heat exchanger. When the indoor blower is running, it's likely that room air will enter the heat exchanger, rather than the byproducts of combustion escaping into the supply airstream.

Monitor the O_2 and CO_2 of the flue gases as the burner ignites and stabilizes. The readings will typically stabilize before the indoor blower starts. Watch the readings carefully as the blower starts. If the heat exchanger is leaking, the O_2 should rise due to the oxygen-rich air entering the heat exchanger downstream of the combustion chamber. The CO_2 reading will most likely decrease at the same time.

If a heat exchanger inspection has been completed, reassemble the furnace and burners as necessary to prepare it for operational testing.

Check the burner manifold pressure during the operational test as follows:

- Ensure power to the furnace is off and the gas supply shutoff valve is closed.
- Remove the pressure tap plug near the outlet of the gas valve (*Figure 78*), or the small pipe plug located on the gas manifold downstream of the gas valve. Either port should provide an accurate pressure reading.

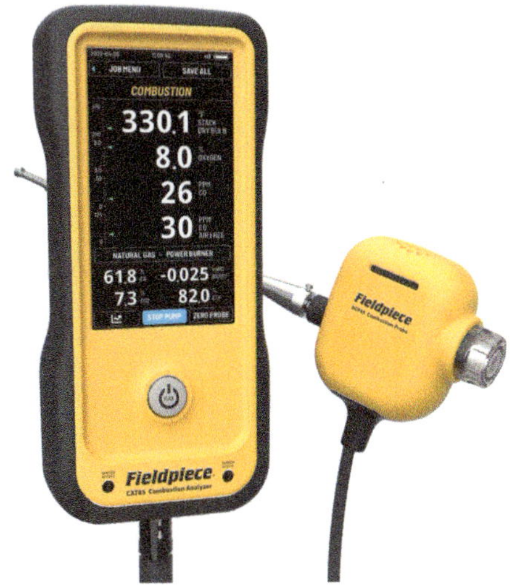

Figure 77 Combustion analyzer.

Source: Courtesy of Fieldpiece Instruments

NOTE

Gas-furnace combustion analysis is presented in NCCER Module 03209, *Troubleshooting Gas Heating*.

Figure 78 Typical furnace gas valve.

- Connect a U-tube manometer, digital manometer, or gas pressure gauge hose to the valve or manifold (*Figure 79*). A $\frac{1}{8}$" MPT × barbed hose fitting is typically required.

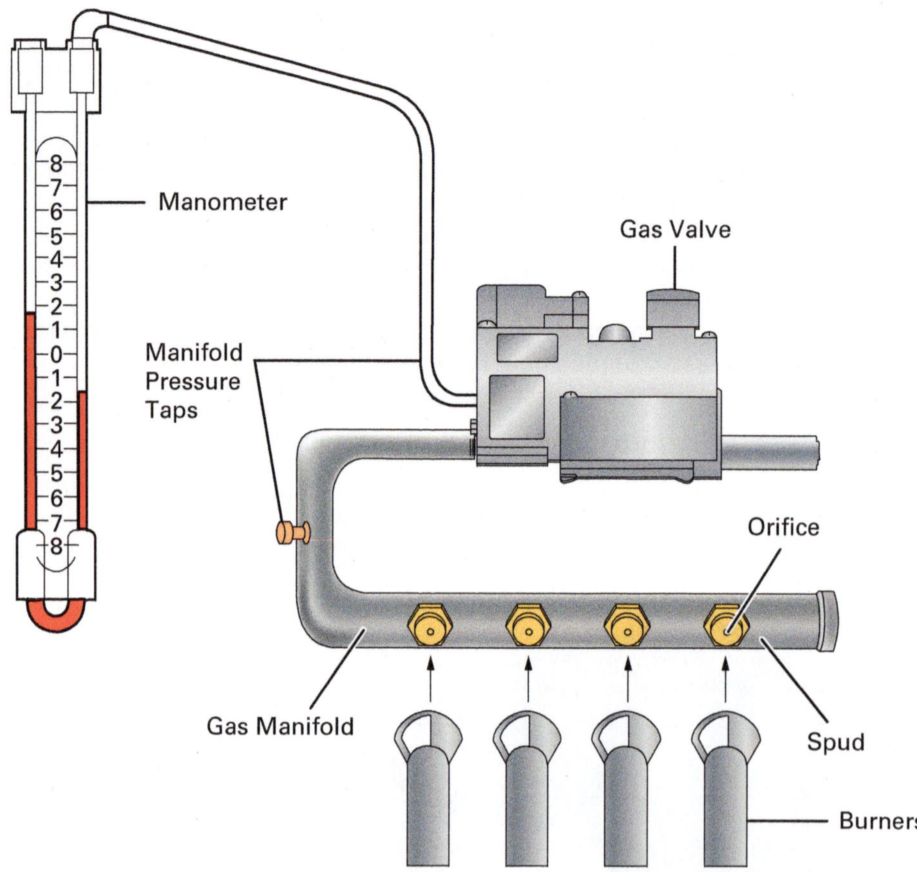

Figure 79 Testing gas manifold pressure.

- With the power to the furnace remaining off, set the thermostat well above the room temperature so it calls for heat when power is restored. This allows you to be at the furnace to observe the operating sequence.
- Open the gas supply valve and then restore power. The heating sequence should begin after a short delay. The draft inducer motor will likely start first to begin the sequence. Check for quiet, even pilot and/or burner ignition.
  - For burners equipped with spark ignition, observe the igniter, and listen for the sound of the ignition arc. Once the pilot flame is ignited and proven, the arcing will stop.
  - For burners equipped with an HSI, observe it as it heats up. Once the HSI proves its integrity, the gas valve opens and the HSI lights the gas directly; there is no pilot flame. A flame sensor, which may be incorporated into the HSI, detects burner ignition and the presence of a flame within a few seconds. The HSI is then de-energized.
- Observe the burner flame. It should be a soft blue, likely with some tiny orange tips. The random orange tips often seen are the result of dust and other particles in the combustion air burning with the gas. If there is yellow in the flame, adjust the primary air supply at the burner if possible. However, most induced draft furnaces do not provide for field adjustment of the combustion air volume. If the flame is erratic, reduce the heating setpoint, then de-energize the furnace after the heat exchanger has cooled and the indoor blower shuts down. Check the burner orifices with a mirror for blockage or damage. Replace any flawed orifices. If deposits need to be cleared, use a torch tip file that matches the size of the orifice (*Figure 80*). Look for anything that may be interfering with the combustion airflow into the combustion chamber. A negative pressure in the building itself can occasionally be a factor.

Gas Pressure Regulators

Most furnace gas valves are not designed to accept a high-pressure gas supply. Most valves are rated for a maximum inlet pressure of only $\frac{1}{2}$ psi. To reduce gas pipe sizes and save installation costs, a home or business gas piping may be designed to provide a 2 psi or 5 psi supply pressure. In these cases, a gas pressure regulator must be installed ahead of the furnace to reduce the pressure to $\frac{1}{2}$ psi or less.

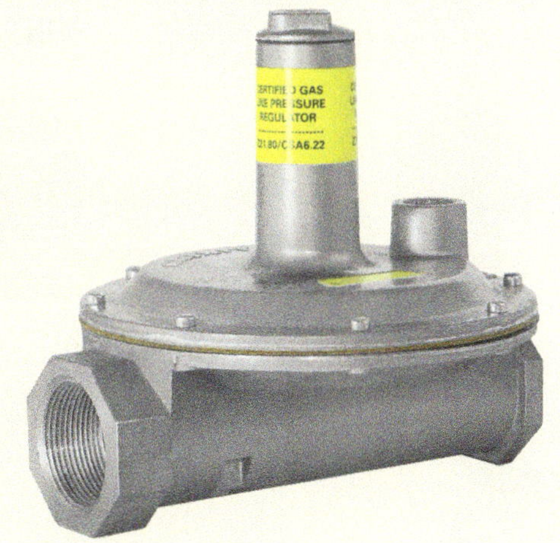

Source: Maxitrol Company

NOTE

On LP burners, the pressure in the tank falls as vapor is extracted (just like a refrigerant cylinder). The burner flame is often affected by the reduced gas pressure. Do not adjust the burner if the fuel level is low. LP burners should not be operated when the tank is at or below 15 percent of its capacity.

CAUTION

The burner orifice is precisely drilled and has a clean, smooth inner surface to provide unimpeded gas flow. Attempting to clean a blockage with a sharp wire or drill bit can scratch the orifice, disrupting the gas flow. Remove and blow out the orifice with compressed air or nitrogen. Clear any deposits with a torch tip file that matches the size of the orifice.

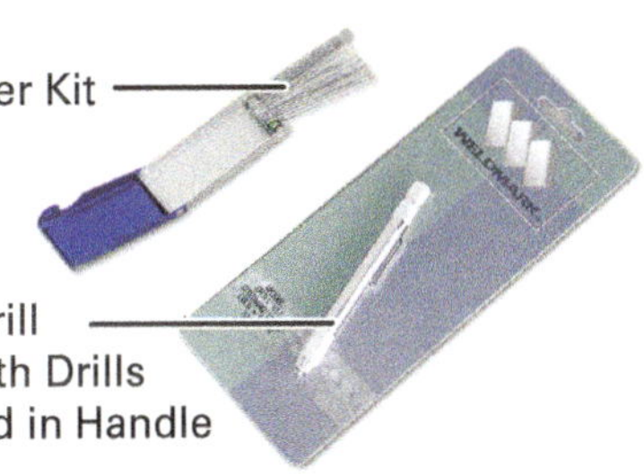

Figure 80 Torch tip cleaning tools.

CAUTION

If a furnace is operated with a temperature rise higher than the specified range, the heat exchanger can degrade prematurely or crack due to excessive expansion and contraction. The high-temperature safety switch may also open and disable the burner.

- Ensure that the blower starts after a short time delay that allows the heat exchanger to warm up.
- Now check and record the gas manifold pressure. Today's more sophisticated burners make it difficult to offer a typical manifold pressure. Traditionally, the common manifold pressure for natural gas has been between 3" and 4" water column (w.c.). For LP gas, the traditional manifold pressure has been 11" w.c. However, it is important to factor in the thermal value of the fuel, as it varies regionally. Different furnace models and brands may also vary in their manifold pressure requirements. What these burners do have in common is a very low pressure in the manifold.
    - If the manifold pressure is over or under the recommended setting for the furnace, adjust the gas pressure regulator in the main gas valve as necessary, following the manufacturer's documentation. Refer to *Figure 78*. Note that the maximum gas supply pressure to the inlet of the main gas valve is typically $\frac{1}{2}$ pound per square inch gauge (psig) (~3.5 kPa). If the manifold pressure is low and will not increase when adjusted, check the inlet pressure to the main gas valve while the burner is in operation to ensure it is adequate (higher than the needed manifold pressure).
- With the manifold pressure correct and furnace operation stable, check the temperature rise across the furnace (*Figure 81*). This value is stated as a range and is found on the furnace rating plate.

Some additional tests require that the furnace be stopped and started several times. These tests are related primarily to safety controls. In many cases, the power to the furnace will have to be cycled off and back on again to reset the controls once certain safety devices have been activated.

- Check the draft pressure switch by disconnecting the hose while the burners are operating. The pressure switch is used to prove that the induced draft fan is working and generating the required pressure. When the hose is disconnected, the burners should shut off immediately. Allow the furnace to cool off, then reconnect the pressure switch and cycle the power switch to reset the controls.
- Check the ignition safety lockout. This locks out the furnace if the main burner fails to light.
    - Close the gas supply shutoff valve while the burner is in operation.
    - The system should lock out after several attempts to reignite the burner. Allow the furnace to cool off, then reopen the valve and cycle the power switch to reset the controls.

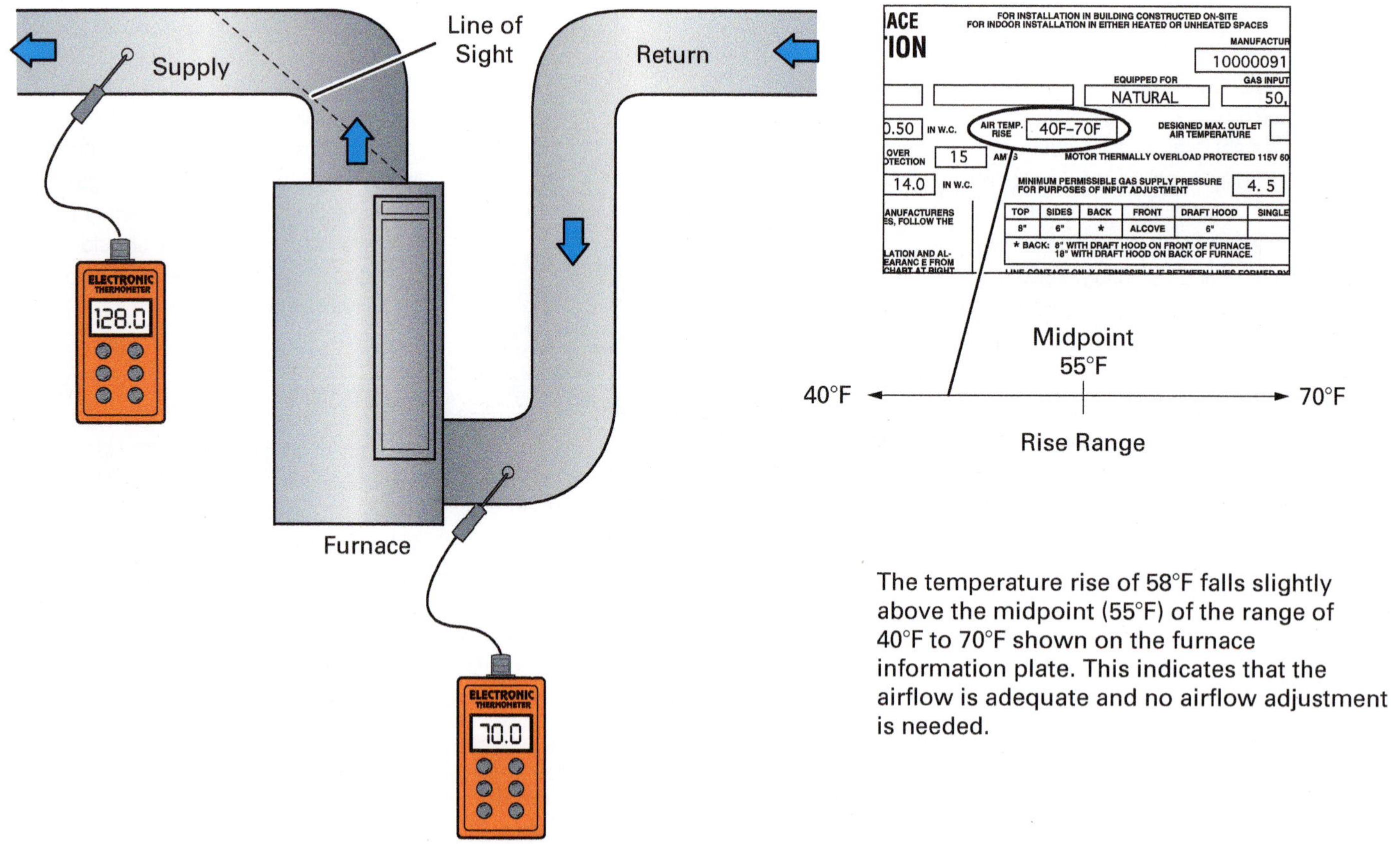

Figure 81 Checking temperature rise.

- Check the high-temperature limit switch. This device shuts down the burners when the heat exchanger temperature is too high, preventing the furnace from overheating. It will usually open rather quickly if the blower does not start.
  - Prevent the blower from operating by disconnecting the control wiring leads.
  - Initiate a call for heat at the thermostat and place a suitable thermometer probe alongside the limit control, if practical.
  - After a few minutes of burner operation, the limit should open and shut down the burner. The temperature and time at which this occurs may be specified in the product literature.
  - If the furnace continues to operate after exceeding the high-limit control setpoint, shut it off immediately and restore blower operation to cool the furnace. Replace the control and check its operation.

Automatic Reset Limit Controls

A limit switch should open rather quickly if the blower motor fails, and it should reset and resume burner operation after a cooling-off period. Many older, spiral-wound, bimetal limit controls were adjustable universal devices that could be used in a wide variety of furnaces. With modern furnaces, it is usually necessary to obtain an exact replacement limit switch to ensure safe and reliable furnace operation. Using a limit switch with the wrong temperature setting can have devastating results.

On condensing furnaces, perform the following additional procedures:

- Check the furnace condensate drain to make sure it is not fouled or leaking. Condensate pumps are often required to collect and pump the condensate away. If present, disassemble and clean the pump and sump.
- If a fin-type secondary heat exchanger is used, clean the coil. To avoid damaging the fins, follow the cleaning procedure in the product literature.
- Check the PVC combustion air and vent piping for damage and ensure the pipes are properly supported as they travel to and through the outside wall.

To wrap up the inspection, disconnect all test instruments and return the furnace to service. Restart the furnace and observe it through at least one complete heating cycle, allowing the thermostat to turn the furnace off when the setpoint is reached. This ensures that all controls are functioning properly.

Many companies leave a copy of the completed service and maintenance checklist with the owner. Discuss any issues that need further attention and inform the owner of any next steps. Then clean up the work area and complete the service call.

4.3.0 Cooling and Heat Pump System Inspection

This section presents a systematic approach to seasonal inspection of common cooling equipment. The information provided earlier for checking thermostats, blower compartments, and blowers—components that are usually a part of a cooling system—will not be repeated here. However, those actions must also be included in a cooling inspection.

4.3.1 Cooling Equipment Safety

In addition to any safety precautions provided in the product literature, the following general safety precautions must be followed when servicing cooling units:

- Be sure all electrical power to the equipment is de-energized. Turn off, lock, and tag disconnect switches unless and until power is needed for operational testing.
- If it is necessary to check voltage or current, be sure to keep one hand outside the unit and avoid touching bare wires and connections.
- Be aware of surfaces that are hot and always remain clear of the outdoor fan blade.

WARNING!

If the compressor terminals are damaged, disturbing them while the system is under pressure could cause them to blow out unexpectedly, causing serious injury. When it is necessary to check the voltage to the compressor, take the measurement at the compressor contactor. It is best to avoid working with the compressor terminals in a pressurized system. When necessary, wear gloves and remain to one side of the terminals, out of the direct path.

Heat Illnesses

A great deal of HVACR maintenance and repair occurs during the summer months. Areas such as attics where HVACR technicians work can become dangerously hot. Exposure to excessive heat can cause weakness, vomiting, and diarrhea, as well as more serious conditions such as *heat stroke.*

Heat exhaustion is generally accompanied by light-headedness, heavy sweating, headache, weakness, thirst, and a rapid heartbeat, in addition to the symptoms noted above. You may also experience heat cramps. A heat stroke is likely at hand when you experience confusion, fainting, seizures, and an elevated body temperature.

If you suffer any of these symptoms, leave the area immediately. Lower your body temperature by resting in a cooler area and applying ice or cold packs. When working in excessive heat, drink plenty of water as well as sports drinks to replace body fluids and restore electrolytes lost to perspiration. Take steps to protect yourself on the job. If you or a coworker experience the symptoms of heat stroke, seek medical attention immediately.

4.3.2 Inspecting a Cooling System

As is the case with heating equipment, always follow the equipment manufacturer's recommendations. The procedures described here may not be appropriate for the system being inspected.

NOTE

Appendix C provides a form to serve as a guide during the execution of the related Performance Task.

Bearing in mind that manufacturer recommendations take precedence over these general procedures, proceed as follows, starting at the indoor unit (a split cooling system is assumed):

- Check all electrical wiring terminations in the indoor unit to ensure they are secure.
- Check the condition of the indoor coil.
  - Remove the coil access panel. Inspect it for accumulated dirt and dust. The coil condition is generally representative of the quality and efficiency of the air filters used.
  - Clean the fins with a vacuum cleaner and brush (*Figure 82*). Be careful not to damage the coil tubing and joints, as the tubing is relatively thin. Some accumulations may be difficult to remove. Remember that the indoor coil is wet when the system operates in the cooling mode. Debris becomes crusty and adheres to the fin surface. On coils that are fouled with dirt or grease, chemical or detergent cleaning may be needed in accordance with the manufacturer's recommendations. Complete removal of the coil from its cabinet or case may also be required. Straighten any bent coil fins with an appropriate fin comb (*Figure 83*).

Figure 82 Dirty indoor coil.
Source: Tanut Nitkumhan/Getty Images

(A) Nylon Combs

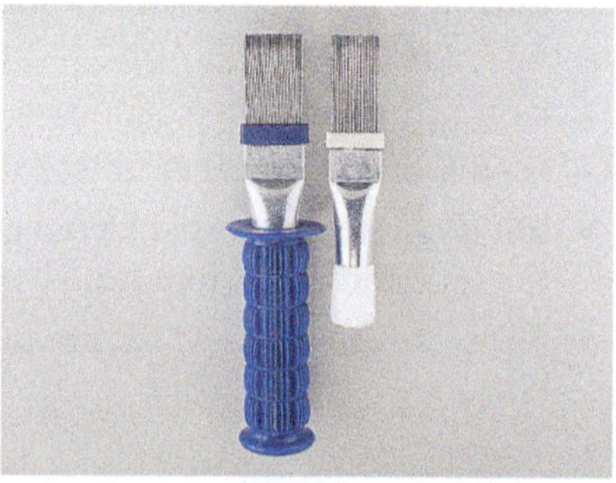

(B) Adjustable Wire Combs

Figure 83 Coil fin combs.
Source: Ritchie Engineering Co., Inc./YELLOW JACKET

  - Check the tubing connections for signs of oil that would indicate refrigerant leakage. If a leak is suspected, follow up with a refrigerant leak detector.
- Inspect and clean the condensate pan and drain.
  - Remove any accumulation of dirt and rust from the condensate drain pan. If local codes permit, tablets or chemicals can be placed in the condensate drain pan to inhibit the growth of slime and algae.
  - Check and clear the drain opening in the pan. Algae may be removed with suitable chemicals; diluted household bleach is also acceptable. Be sure to wear proper gloves when working with any chemical.
  - Pour some water into the condensate pan and check for proper drainage. Clear the drain line if the water does not drain as it should.
- Check the static pressure drop across the indoor coil. Note that this procedure assumes a clean and dry coil. Alternatively, measure and record the external static pressure (ESP) as outlined in NCCER Module 03109, *Air Distribution Systems*.
  - Drill two $\frac{1}{4}$" (6.5 mm) holes in the access panel, one on each side of the coil (*Figure 84*). Be careful not to drill into or damage the evaporator coil.
  - Attach an inclined or digital manometer to static pressure probes and insert them into the holes as shown in *Figure 84*. The higher pressure will be on the entering side of the coil. If necessary, seal around the hose with putty, such as Permagum®.
  - Ensure all access doors are installed. Energize the indoor blower without heating or cooling.

CAUTION

Make sure that any test holes to be drilled in a coil plenum are positioned so that the drill bit will not damage the coil.

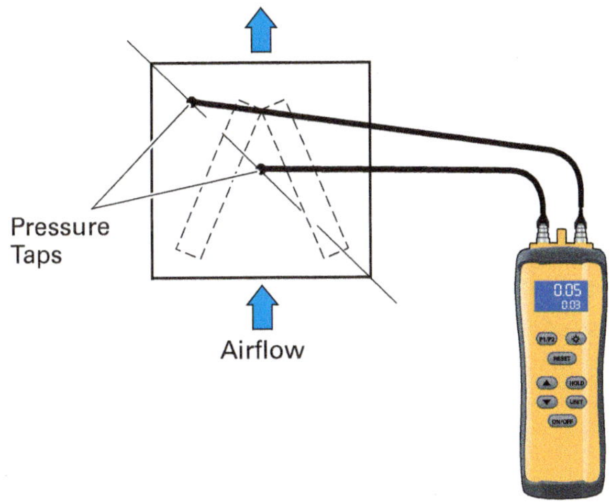

Figure 84 Using a digital manometer to check coil pressure drop.

- The pressure differential between the pressure of the air entering and leaving the coil represents the pressure drop through the coil.
- Refer to the manufacturer's specifications for expected pressure drop and air volume and compare your results.
- If the pressure drop is too high, ensure the coil is clean. As the passages between the fins become blocked, the pressure drop increases. Clean the coil thoroughly if necessary. If the coil is clean, excessive pressure drop likely means that the air volume is higher than needed. If that is the case, reducing the fan speed will lower the pressure drop. A low pressure drop usually indicates a low air volume. Increase the fan speed as necessary to reach the expected pressure drop.
- Remove the static pressure probes and stow the equipment. Fill the test holes with a suitable plug or foil tape.

Coil Cleaning Gear

Effective coil cleaning equipment can save a lot of time and effort. Several systems are available to make the task less of a chore. A technician is shown here using a pump and tank setup that is fully portable. A long-life rechargeable battery provides the power for the pump that develops 125 psig (860 kPa), and the apparatus is filled with water at the source. This eliminates the need to drag a garden hose to the unit. The water and detergent tanks are isolated from each other.

When a garden hose is easily accessible, consider a cleaning gun that mixes water with an inserted detergent tablet. Wands attach to the gun nozzle to reach inside the unit cabinet and wash it from the inside out. Coils should always be cleaned in this manner—forcing debris out the same way it came in.

(A) Portable Pump and Tank System **(B) Cleaning Gun with Detergent Tablet**

Source: Courtesy of SpeedClean

NOTE

If the outdoor coil will be cleaned during the inspection, leave the power off and check the supply voltage after cleaning.

CAUTION

Do not use a pressure washer to clean a coil. The high-pressure stream can easily flatten or damage the aluminum fins. Allow the cleaning solution time to do the work. When washing a coil, prevent water from splashing into motors or electrical components. A simple plastic bag pulled over a motor can provide an effective shield.

Figure 85 Foaming condenser coil cleaner.

Source: Courtesy of SpeedClean

NOTE

If the pressures and temperatures obtained are not as expected per the manufacturer's literature, further analysis is required before adding or removing refrigerant.

Refrigerant charging by both superheat and subcooling methods is presented in detail in NCCER Module 03205, *Leak Detection, Evacuation, Recovery, and Charging*. Troubleshooting and analysis of the refrigerant circuit is presented in NCCER Module 03210, *Troubleshooting Cooling*.

- Inspect and clean the condensing unit.
  - Turn off, lock, and tag the disconnect switch while inspecting the unit. Also, de-energize the power source feeding the control transformer.
  - Check all electrical wiring terminations in the indoor unit to ensure they are secure. Look for discoloration of the terminals and wire insulation.
  - Check the contactor contacts for pitting or slag from arcing.
  - Check the refrigerant piping for evidence of leaks and proper support. Oily patches around fittings often indicate a refrigerant leak.
  - Lubricate the condenser fan motor, if required. Use the lubricant specified by the manufacturer.
  - Restore power to the unit.
  - Check the supply voltage at the contactor and ensure it is within the range specified on the dataplate.
- Clean the outdoor coil (if needed).
  - Ensure the power supplies to the outdoor unit and the control transformer remain off.
  - Remove the condenser fan grille and outdoor fan assembly. In many cases, the entire unit top can be removed with the grille and motor attached. Many condenser coil assemblies consist of two coils back-to-back. These coils may need to be carefully separated for cleaning; a great deal of debris can be trapped between the two coil faces. The top of the unit cabinet must be removed to access the top of the coils.
  - Apply a commercial coil cleaner in the proper mix ratio to the coil from the inside. These cleaners help to loosen dirt and break down grease, then push it out of the coil with a foaming action (*Figure 85*). Always clean and flush the coil from the inside to push the debris out of the entering-air side of the coil. Rinse and thoroughly flush the coil cleaner out.
  - Reassemble the condensing unit to prepare for operational testing.
- Measure and record the electrical operating characteristics and check the refrigerant charge. Consult the equipment manufacturer's service literature for specific information regarding operating pressures and temperatures.
  - Make sure that all system registers, grilles, and diffusers are open to ensure proper airflow.
  - Connect the high-pressure side of the gauge manifold set (*Figure 86*) to the service port on the liquid line. Connect the low-pressure side of the manifold to the suction line port.
  - Restore power to the control transformer; leave the power off to the condensing unit for now.
  - Adjust the thermostat cooling setpoint to well below room temperature.
  - Attach an ammeter to the common lead of the compressor as it leaves the contactor.
  - Restore power to the condensing unit and observe the starting current and behavior of the compressor. There may be a delay of up to 5 minutes before the compressor starts, depending upon the thermostat programming and control arrangement.
  - Allow the system to operate for 5 to 10 minutes until the pressures are stable. This time offers an opportunity to measure the running current of the compressor and fan motor. Record the readings and compare them with the dataplate. Note that a compressor or fan motor may have been replaced, so it is best to check the component nameplate if there is any doubt.
  - Check and record the system refrigerant pressures, superheat, and subcooling. Remember that systems using a fixed restrictor are typically charged by the superheat method; systems using an expansion valve are

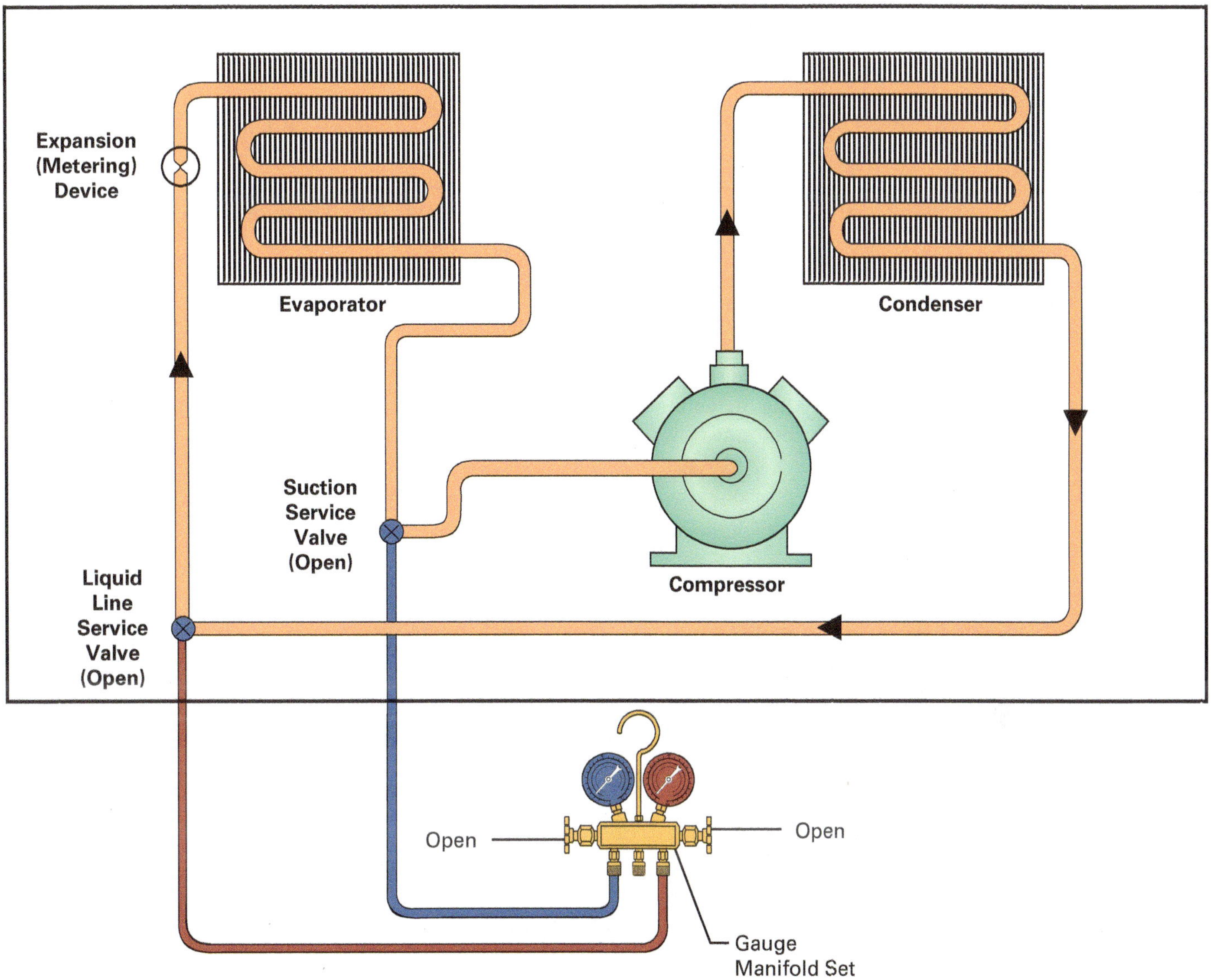

Figure 86 Gauge manifold connections.

typically charged by the subcooling method. However, it is important to check both superheating and subcooling to form a complete picture of performance.

- Check the temperature differential across the indoor coil while the system is still running.
  - Insert thermometers into holes on the entering and leaving sides of the indoor coil. The same holes used to measure the pressure drop across the coil can be used (*Figure 87*).
  - Record the temperature difference (TD or ΔT) across the coil. Generally, it should range between 17°F and 23°F (9°–13°C), but unique operating modes, such as dehumidification, will affect the ΔT. The room temperature and humidity in the conditioned space also affect the ΔT. Use the same thermometer or probe for both measurements to ensure accuracy.
  - Remove the thermometer probes and plug the holes.
  - Make a final walk-around inspection of the complete system. Set the thermostat as the owner prefers. Secure all access doors, clean the work area, complete any remaining documentation, and wrap up the inspection by discussing your findings with the owner.

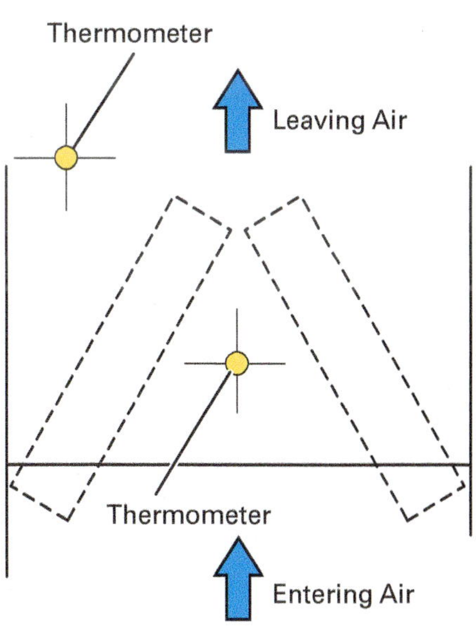

Figure 87 Indoor coil thermometer insertion points.

Some technicians and organizations prefer to check the change in system enthalpy (ΔH) to determine system performance, rather than rely on the ΔT across the coil. To do so, you must first determine the enthalpy, or total heat, contained in the supply and return airstreams.

Start by measuring the dry bulb and wet bulb temperatures in the return duct. Then repeat the process for the supply duct with the system operating and stabilized. For example, suppose the dry bulb temperature of the return air is 78°F and the wet bulb temperature is 67°F. The supply air dry bulb temperature is 59°F and the wet bulb temperature is 58°F. Referring to a psychrometric chart or a psychrometric calculator app, you'll find that the return air enthalpy is 31.55 Btu/lb and the supply air enthalpy is 25.07 Btu/lb.

Next, you'll need to determine the air volume moving through the system. Measure the velocity in the main supply duct and multiply it by the duct square footage to determine the air volume in cubic feet per minute. For this example, the air volume is 850 cfm. Then use the following equation to determine the amount of heat transferred by the system in Btus per hour (Btuh):

$$\text{Total Heat Removed} = (H1 - H2) \times 4.5 \times \text{cfm}$$
$$\text{Total Heat Removed} = (31.55\ \text{Btu/lb} - 25.07\ \text{Btu/lb}) \times 4.5 \times 850\ \text{cfm}$$
$$\text{Total Heat Removed} = 6.48\ \Delta H \times 4.5 \times 850\ \text{cfm}$$
$$\text{Total Heat Removed} = 24{,}786\ \text{Btuh}$$

The final step is the application of a correction factor. Thus far, the equation assumes that the outdoor air temperature is 95°F (35°C). For all other outdoor temperatures, you will need to consult *Table 6*. The correction factor is a multiplier that adjusts the result to accommodate the difference between the actual outdoor temperature and equipment design conditions.

TABLE 6 Capacity Correction Factors

Outdoor Dry Bulb (°F)	Correction Factor
55°F	0.82
60°F	0.84
65°F	0.86
70°F	0.88
75°F	0.90
80°F	0.93
85°F	0.95
90°F	0.98
95°F	1.00
100°F	1.03
105°F	1.06
110°F	1.10
115°F	1.13
120°F	1.18

Note that the air volume in the system is what you might expect from a 2-ton system. The amount of heat transferred out of the airstream is also consistent with a system of that size.

Note that many of today's smart diagnostic tools and instruments can make most or all these calculations by simply inserting the temperature probes into the ducts and requesting the data.

4.3.3 Inspecting a Heat Pump

Many of the inspection procedures for heat pumps are the same as those for cooling units. A few additional steps are needed to check reverse-cycle heating and defrost functions.

CAUTION

A heating performance check should only be made on air-source heat pumps when the outdoor temperature is within the unit's heating operation range per the manufacturer's specifications. Operating a heat pump in warmer temperatures will result in excessive pressures. However, a quick check of the defrost cycle or the reversing valve to ensure it changes position can be safely conducted. Consult the manufacturer's documentation for additional guidance.

Perform the procedures described for cooling equipment and work these additional steps into your process.

- Check basic heat pump operation in the heating mode.
  - Adjust the thermostat to call for heating with the setpoint just slightly above the room temperature. Note that moving the setpoint 2°F to 3°F (1.1°C to 1.7°C) or more above the room temperature will likely energize the supplemental heat. This will not affect most of the inspection, but it is important to be aware that it is likely active. It can be temporarily disabled using the outdoor thermostat (if present) if desired.
  - At startup, the heat pump reversing valve may switch from one position to the other. Most reversing valves, however, are energized in the cooling mode rather than heating. As a result, you won't likely hear the valve change position. Ensure the unit is producing warm air. Operate the system for 5 to 10 minutes to allow the operating pressures and temperatures to stabilize.
- Test the defrost controls. As the approach to defrost control can vary from system to system, refer to the manufacturer's service manual as needed for guidance.
  - Most defrost control boards have a feature that allows you to initiate a defrost cycle manually and run the heat pump through a defrost cycle test mode. Sensor inputs are typically ignored during the test. The test is often initiated by using a jumper to short across two pins (*Figure 88*). Note that the normal defrost cycle is shortened for testing.

Figure 88 Defrost speedup pins.

Source: HVACGuru

- The reversing valve should shift back to the cooling position, routing hot refrigerant to the outdoor coil. The outdoor fan should shut down. If there is frost buildup on the coil, it should begin to melt.
- Ensure the supplemental heat energizes to temper the cold supply air while the unit is in the defrost mode.

- Check the supplemental electric heat package and Emergency Heat operation.
  - Turn off the power supply to the electric heating package as well as that serving the control transformer.

WARNING!

Some systems have more than one power supply. The power supply serving the indoor unit may be separate from that serving the electric heat package. Ensure that the electric heat package is de-energized using a multimeter. Always test the meter on a known energized source before use.

- Check all the electrical wiring for loose connections and burned or discolored insulation. Repair any damaged terminations and replace damaged wiring.
- Identify the conductors powering the individual heating elements and ensure that they are safely accessible for current measurements with a clamp-on ammeter.
- Restore line voltage and control power to the system.
- Adjust the thermostat at least 3°F (1.6°C) above the room temperature. If the system uses an outdoor thermostat to control the availability of electric heat, adjust it to allow all stages to energize. Another option is to set the thermostat in the Emergency Heat position and raise the setpoint. This position should bypass other controls limiting the availability of electric heat. However, note that it also should shut down and lock out the heat pump.
- Check the amperage of each individual heating element. Place the jaws of a clamp-on ammeter around each heating element's power lead and record

the value. If any element is not operating (no current), determine the cause using the appropriate troubleshooting techniques.

- Lower the thermostat setpoint to shut down the heating system.
- Test the system's ability to operate in the Emergency Heat mode if this has not already been done. Switch the wall thermostat to the Emergency Heat mode and raise the setpoint above the room temperature. All the electric heating elements should energize to heat the structure, while the heat pump itself remains off.
- Access the outdoor thermostat (if equipped) and return it to the proper setting. Record the setting in the inspection notes.
- Make a final walk-around inspection of the complete system. Set the thermostat as the owner prefers. If any settings were changed to enable the electric heat for testing, return them to their previous setting. Secure all access doors, clean the work area, complete any remaining documentation, and wrap up the inspection by discussing your findings with the owner.

4.4.0 Inspecting and Maintaining System Accessories

HVACR system accessories must be serviced along with the rest of the system. In addition to the safety precautions specified in the product literature, be sure the electrical power to the accessories is de-energized before accessing them. Open, lock, and tag disconnect switches. If it is necessary to check voltage or current, be sure to keep one hand outside the unit and avoid touching bare wires and connections. Always pay strict attention to any rotating components, heated surfaces, and pinch points.

4.4.1 Humidifiers

Humidifiers require periodic cleaning and maintenance for efficient and safe operation. Any time water is involved, regular maintenance is needed. Humidifier maintenance is best performed at the beginning of each heating season, but some units need additional attention as the season progresses. The water quality has a great deal to do with the frequency of maintenance.

Most humidifiers accumulate calcium and other mineral deposits. As the water evaporates or vaporizes, it leaves behind all the solids it contained. As a result, the operating time as well as the water quality will determine the need for maintenance and cleaning. The humidifier may require more frequent care than the other system components.

Wetted-media humidifiers collect mineral deposits on the media and in their water sump. Biological growth occurs in these areas as well. Thus, the two primary tasks in humidifier maintenance are eliminating mineral deposits and destroying any biological growth. The installation of a water filter capable of capturing minerals and scale before they reach the humidifier can often extend the time between maintenance visits (*Figure 89*).

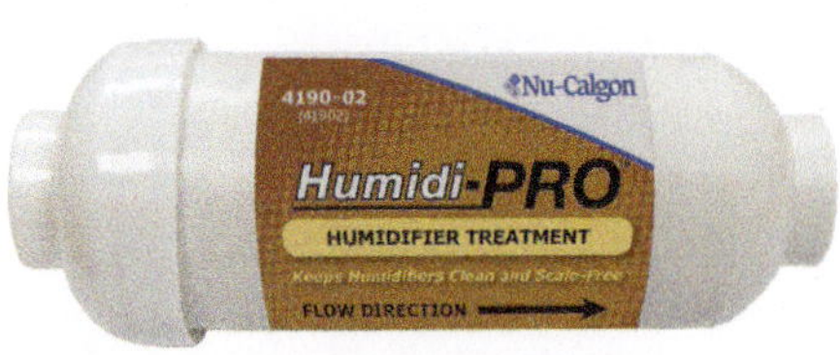

Figure 89 Humidifier water supply filter.

Source: Nu-Calgon

Basic maintenance procedures for a typical wetted-media bypass or fan-powered residential humidifier follow. These steps do not apply to all models. Always follow the manufacturer's recommended service procedures carefully.

- Turn off the power and water supplies to the unit.
- Remove the water panel (*Figure 90*). Drain and clean the interior thoroughly with a product that helps remove mineral scale and neutralizes biological growth. There are many such products available through HVACR distributors.
- Clean the frame for the water panel.
- Make sure the drain spud rests in the opening at the bottom of the housing. Make sure the drain line is intact, properly supported, and free of leaks.
- Reassemble the interior and install a new water panel. Note that some panels have an indicator to show which end of the panel should be placed at the top.

Figure 90 Humidifier water panel.
Source: BanksPhotos/Getty Images

- Service the water valve strainer. The strainer is often located under a cap on the solenoid valve body. It may also be located at the solenoid valve inlet, as shown in *Figure 91*. If so, disconnect the water line to access the strainer. Remove the strainer and flush it with water. Remove any stubborn debris manually, then reinstall the strainer.
- An orifice at the valve outlet meters the water flow rate to the humidifier. Remove and carefully clean the orifice, being careful not to damage or distort the opening. An old toothbrush can safely remove the deposits. Replace the orifice if it is damaged.
- Reinstall the orifice and reconnect the water line to the valve assembly.
- Lubricate the motor (if equipped) per the manufacturer's instructions.

CAUTION

Ensure the water supply is off. Despite the valve being closed, the line may still hold pressure. As you disconnect the line, capture the water in a bucket. It also helps to cycle the solenoid valve open momentarily to relieve the pressure before disconnecting the line.

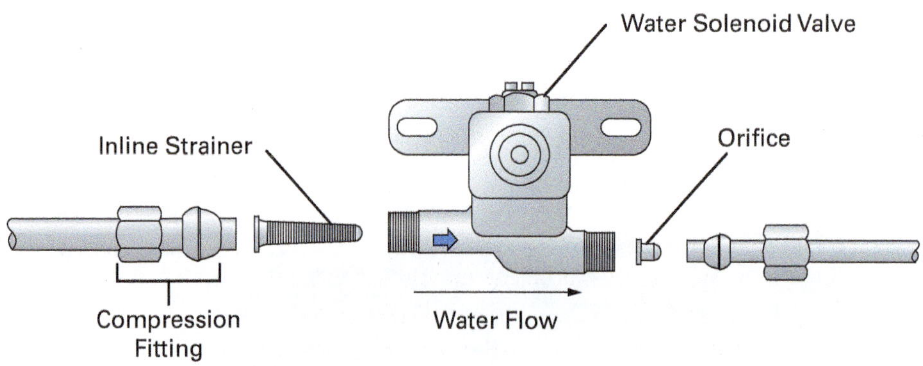

Figure 91 Common strainer location.

- Turn on the power supply and open the water supply valve. Check for leaks at the solenoid valve and other joints that may have been disturbed.
- Operate the humidifier, ensuring there is metered water flow and airflow through the water panel.

Remind the user that the humidifier should be checked regularly for bacterial growth and scale buildup during the heating season.

4.4.2 Electronic Air Cleaners

Basic periodic maintenance helps maintain efficiency and increases the life of an electronic air cleaner (EAC). *Figure 92* provides a view of the typical filter sections inside an EAC. Because EACs often have media-based prefilters, they can also become clogged and reduce airflow.

When servicing any air filtration equipment, wear an N95-rated breathing mask to avoid inhaling airborne particles.

Also remember that the power supply of an EAC can develop as much as 10,000 VDC. Be cautious when working with them and ensure the unit is de-energized while performing maintenance. Do not depend on the EAC power switch alone; disable the power feeding the unit.

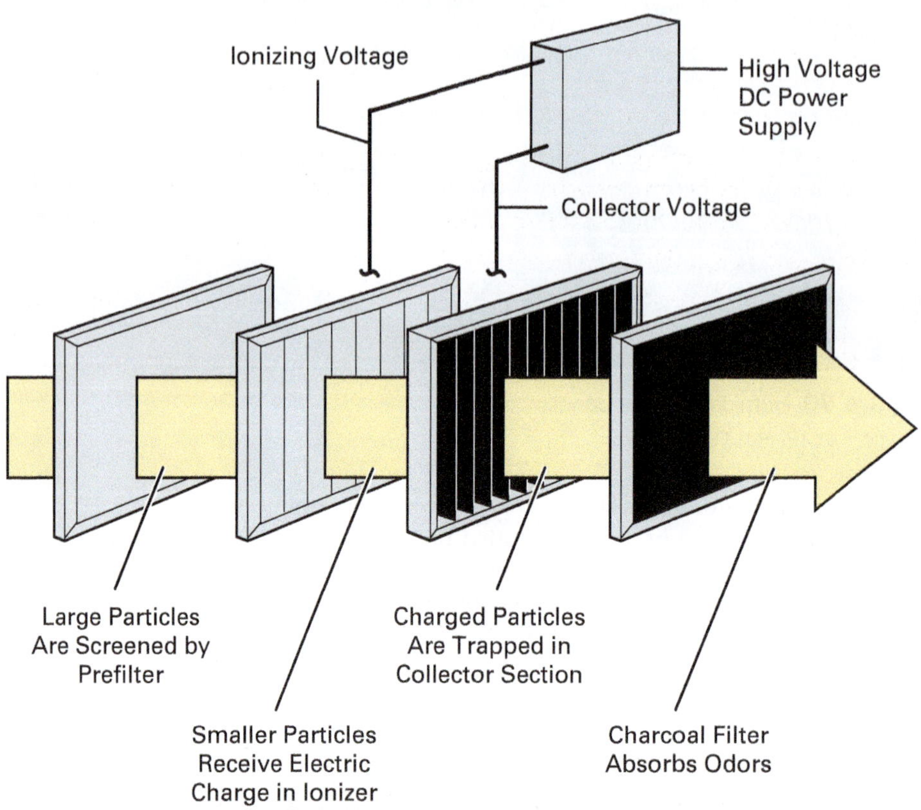

Figure 92 Electronic air cleaner sections.

Did You Know?

Electronic Air Cleaner Maintenance Tip

Unlike standard mechanical filters, electronic air cleaners become progressively less effective as they collect debris. Once dirty, they can also generate annoying arcing sounds. If the unit operates for long periods in this condition, the arcing can create ozone gas (O_3), which can cause respiratory irritation in some people. In addition, excessive ozone contamination can initiate and/or accelerate metal corrosion.

Refer to the manufacturer's service manual for specific service instructions. In general, the following tasks are required:

- De-energize the EAC as well as the furnace or fan coil unit. Most EACs are powered directly from the associated unit. Use a meter to verify the unit is de-energized.
- Remove and clean the collector cell(s).
  - Open the unit cabinet and slide the cell(s) out. Be careful not to damage the cells and avoid cuts from the sharp edges. Gloves are recommended.
  - If the manufacturer allows it, the cells can be placed in a dishwasher. Obtain permission from the customer before using their dishwasher. Place the cells on the bottom rack with the airflow arrows pointing up. Be careful not to damage the **ionizing wires** when handling them. Run the cells through a normal dishwashing cycle with detergent. After washing, inspect the cells for dirt or residue. Rewash if necessary.
  - Alternatively, clean the cells by soaking them in a cleaning solution. Find a container large enough to immerse all the cells. Fill the container with hot water and dissolve an appropriate amount of dishwashing detergent in the water. Allow the cells to soak for 15 to 20 minutes. Then rinse with a fine spray and soak once more in clean hot water. Stand the cells upright to drain. Inspect again for deposits and rewash as necessary.
  - Another alternative is to clean the cells using a commercial cleaning spray. The product shown in *Figure 93* can be diluted with water for soaking the cells or be applied to the cells with a spray bottle. Apply the product generously and to all the surfaces of the cell. Let it stand for 5 to 15 minutes. Use hot water (if available) to thoroughly rinse the cells. Repeat if necessary.
- Inspect the ionizing wires for damage.
  - Check for damage and replace any wires that are damaged or broken. Replace the fine tungsten wire in the cells with parts provided by the manufacturer.
- Clean or replace the prefilter. If it is washable, wash it using the same detergent used for the cells.
- Check the charcoal final filter (if equipped). The charcoal filter reduces odors but is not designed to capture particulates. It should be replaced at least annually, but changing it more often ensures better performance.
- Reassemble the unit after ensuring all the components are dry. They must be completely dry before energizing the power supply. Turn on the system blower and allow it to run for 15 to 30 minutes before energizing the air cleaner.
- Restore power to the EAC.
  - Many of today's advanced furnaces and fan coil units have a solid-state control board with terminals that power the air cleaner only when the blower is powered. Other control possibilities include a sail switch, a blower interlock relay, and a differential-pressure switch. To ensure it works, first energize the furnace blower, then check the EAC power light. If the EAC is not operating and the light is off, troubleshooting will be required.

Ionizing wires: Thin wires in an electronic air cleaner that impart a positive charge to passing particles. The charged particles are then attracted to the negatively charged collector plates.

Figure 93 Special HD CalClean® concentrated detergent.

Source: Nu-Calgon

WARNING!

You could be exposed to a lethal voltage while performing these tests. Avoid contact with exposed wiring and remember that a 10,000 VDC arc can jump a gap ½" (~1.3 cm) wide.

- Check the voltage to the cells.
  - Access and check the incoming voltage to the power pack.
  - Check the voltage output to the collecting section and the ionizing section. The readings should be specified in the manufacturer's literature. Note

that these voltages will likely be 3,000 to 10,000 VDC; use an appropriate meter with the correct range and settings.

- Reinstall the power pack access cover and return the system to service.

4.4.3 Ultraviolet Lamps

Ultraviolet (UV) lamps (*Figure 94*) are used for microbial control and disinfection around evaporator coils, drain pans, blowers, and ducts. Their popularity continues to increase in both the residential and commercial markets. Remember that UV light and the related equipment can be hazardous in several ways. Review the precautions provided by the manufacturer before working with UV light equipment and lamps.

Figure 94 UV-C lights installed in a residential air handling unit.

Source: Fresh-Aire UV

WARNING!

Review the warning labels and cautionary statements provided by manufacturers for UV-C lights before working with them. Make sure you de-energize the lamp(s) before opening the access door. UV light can cause serious eye damage. Allow sufficient time for the lamps to cool before handling them, as the intense heat can cause severe burns. Always wear gloves and safety glasses when handling the lamps.

UV lamps do occasionally fail, but they need to be replaced periodically regardless. Their intensity diminishes over time, making them less effective although they still appear operable. Manufacturers often designate their replacement lamps as one-year or two-year models, referring to the estimated length of time the lamp will produce effective UV-C light.

For that reason, it is important to keep track of the lamp replacement schedule. A small chart mounted beside each unit can be used to keep track of lamp installation dates. Note that UV lamps contain mercury and must be disposed of in accordance with local requirements.

You may also need to wipe the lamps down periodically to remove dust and debris. Remove and carefully clean the lamp with a common glass cleaner and a soft rag. Ensure that the bulb is completely dry before reinstalling it. Also wipe off the lamp base and any reflectors.

NOTE

More information about UV-C systems for HVACR applications can be found in NCCER Module 03204, *Air Quality Equipment*.

CAUTION

UV lamps contain mercury. Do not place old lamps in the trash. Dispose of them properly in accordance with local guidelines, which often require delivery to, or pickup by, an authorized recycler. Fluorescent lamps are typically disposed of in the same manner.

CAUTION

Do not handle UV lamps with your bare hands. Oils from your hands may cause premature failure of some products, depending on the materials used for the glass.

4.4.4 Condensate Pumps

Condensate pumps (*Figure 95*) are used to pump condensate away from evaporator coils when a gravity-based drain is not practical. They may be used to move the water some distance horizontally as well as vertically. Condensing furnaces may also require condensate pumps for operation in the heating mode. As a result, condensate pumps are more common than ever.

Note that condensate from condensing furnaces is an acidic, potentially corrosive solution. Draining or pumping it into sewage or septic systems is typically prohibited unless the condensate is first treated to neutralize the acid. It is best, and often required, to install a condensate neutralizer (*Figure 96*) at the furnace. When installed, these devices will also require periodic service and replacement of the media.

The following basic procedures are generally required periodically to maintain a condensate pump:

- De-energize the pump and the associated equipment. Ensure that the 24 VAC control circuit is also de-energized, because the pump is likely tied into the control circuit. Condensate pumps are often powered by an electrical cord plugged into a receptacle. Simply unplug the power cord.
- Remove the top and pump motor assembly from the sump. The entire assembly is usually designed for easy removal.
- Check and clean the condensate drain from the equipment to the pump. Make sure the condensate drain line is clear of debris and biological

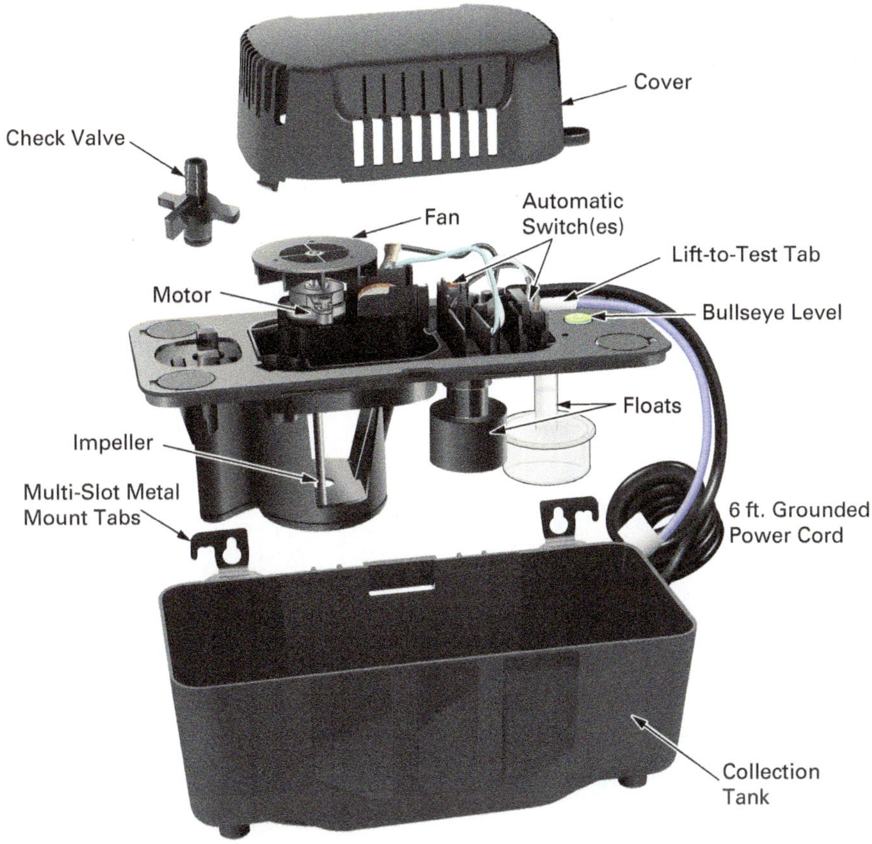

Figure 95 Condensate pump.

Source: Beckett Corporation

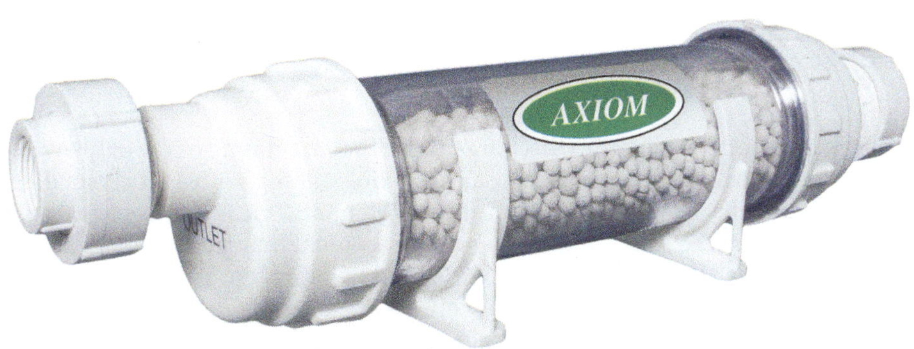

Figure 96 Furnace condensate neutralization kit.

Source: Axiom Industries Limited

Know the Code

Condensate Pumps

Per Section 307 of the 2021 *International Mechanical Code*® (IMC), condensate pumps located in uninhabitable spaces, such as crawl spaces and attics, must be connected to the system in a manner that will prevent the equipment from operating if the pump fails. Note that local codes may require a control interlock regardless of pump and equipment location.

Many condensate pumps are equipped with two floats. One float rises and initiates the pump when the sump is nearly full. The second float is set to a higher level and opens a normally closed switch to disable the system if the pump fails and the water level rises too high in the sump. This second switch is interlocked with the system control circuit.

growth. The drain pan is often treated with biocides to minimize future biological growth in the pan and line (*Figure 97*). These treatments may last 30 to 120 days. Pay careful attention to the instructions and apply them as directed.

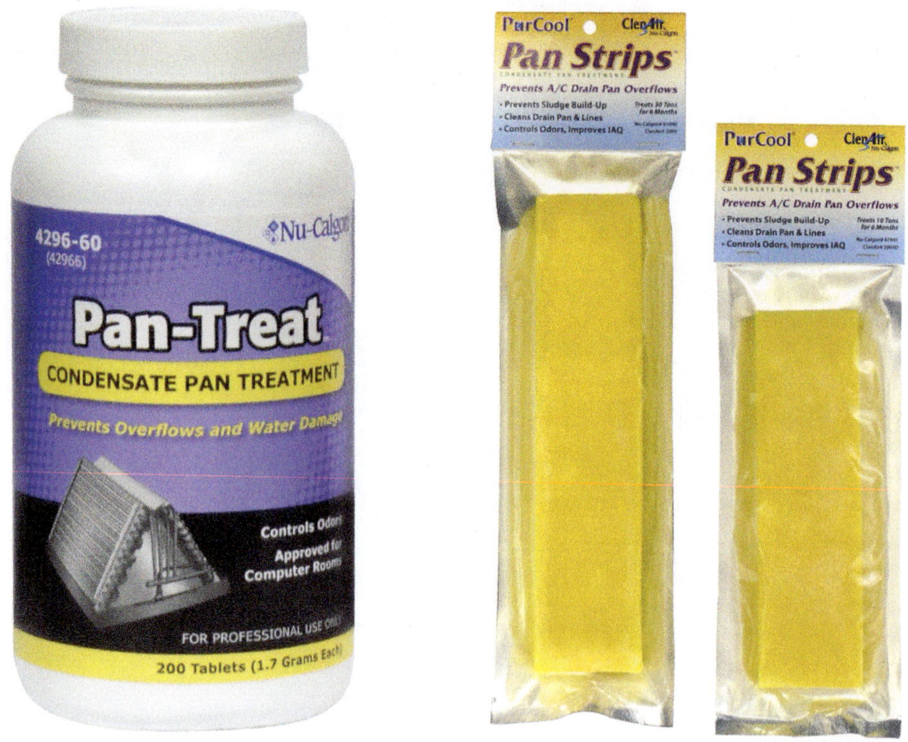

(A) Tablets **(B) Strips**

Figure 97 Examples of condensate pan biocides.

Source: Nu-Calgon

- Check the condensate pump discharge line to ensure it is also free of growth. The pump discharge line is usually clear flexible tubing with a much smaller inside diameter than a gravity drain line. Manually cleaning the tubing is not practical. If the line is badly fouled, replacement may be required. Using biocides such as those shown in *Figure 97* will help prevent biological growth in the pump and discharge line as well. Also check the discharge line and eliminate any kinks or significant sags that form traps.
- Remove and clean the discharge line check valve. The check valve is normally located at the pump outlet, with the discharge tubing connected directly to it.
- Clean the condensate pump sump thoroughly.
- Reassemble the pump and drain lines, then fill the sump with water.
- Restore power to the pump and system control circuit. Check that the float switch activates the pump as expected, and de-energizes the pump when the sump is empty.
- Actuate the float that opens the control circuit and ensure it breaks the control circuit as expected.
- Complete the work by ensuring the control circuit closes when the safety float falls, and the system is operable.

4.0.0 Section Review

1. Bats are known to carry rabies.
   a. True
   b. False

2. A common measurement for the spark gap on an electronic igniter assembly is _____.
   a. $^1/_2$" (12.7 mm)
   b. $^3/_8$" (9.5 mm)
   c. $^9/_{32}$" (7.1 mm)
   d. $^7/_{64}$" (2.7 mm)

3. The two conditions that can result in excessive pressure drop readings across the indoor coil are _____.
   a. low air volume and a clogged coil
   b. excess air volume and a clogged coil
   c. low air volume and a refrigerant overcharge
   d. excess air volume and a low refrigerant charge

4. The water strainer serving a wetted-media humidifier is often found _____.
   a. under a cap on the water solenoid valve body
   b. on the bottom of the media panel
   c. in the water source shut-off valve
   d. on the humidifier drain line

Module 03215 Review Questions

1. The pressure class of pipe flanges, in pounds per square inch, is determined by dividing the tested bursting point by a factor of 3.0.
   a. True
   b. False

2. A lantern ring is found in _____.
   a. compressor cylinders
   b. pump stuffing boxes
   c. mechanical seals
   d. hand valves

3. What percentage of an O-ring's diameter is typically compressed to form a reliable seal?
   a. 10%
   b. 25%
   c. 40%
   d. 60%

4. To create a leak-proof seal around a rotating shaft, mechanical seals usually rely on two perfectly machined faces made from _____.
   a. steel
   b. aluminum
   c. carbon
   d. lithium

5. Which of the following is a type of antifriction bearing?
   a. Sleeve bearing
   b. Roller bearing
   c. Bronze bearing
   d. PTFE bearing

6. Which of the following tools can amplify the sound of a running bearing, helping to determine if a problem exists?
   a. A spring
   b. A borescope
   c. A tensiometer
   d. A steel rod

7. A common motor with acceptable sleeve bearings may have shaft end play of up to _____.
   a. $\frac{1}{16}$" (1.6 mm)
   b. $\frac{1}{8}$" (3.2 mm)
   c. $\frac{1}{4}$" (6.4 mm)
   d. $\frac{1}{2}$" (13 mm)

8. Most lubricants—as much as 90 percent of the market—are made from vegetable oils.
   a. True
   b. False

9. The term *viscosity* refers to _____.
   a. the lowest temperature at which an oil will flow freely
   b. an oil's ability to flow at a specific temperature
   c. the tendency of an oil to release ignitable vapors
   d. the highest temperature at which the oil will flow freely

10. Grease often has some type of mineral or metal added to it, such as _____.
    a. lithium
    b. steel
    c. aluminum
    d. quartz

11. When adding oil to a small motor with a sleeve bearing, it is important to ensure that the shaft does *not* turn while the oil is being added.
    a. True
    b. False

12. A drive belt marked as a 4L250 identifies it as a _____.
    a. classic belt
    b. wedge belt
    c. fractional horsepower (FHP) belt
    d. notched belt

13. In what way are wedge belts different from classic and FHP belts?
    a. Wedge belts cannot be used on small-diameter sheaves.
    b. Wedge belts are deeper (thicker, not wider) than all the other types.
    c. All wedge belts are both banded and notched.
    d. Wedge belts can transfer up to three times more horsepower than the others.

14. Many sheaves are designed to accommodate both FHP and classic belts.
    a. True
    b. False

15. A drive motor is operating with an 1,800-rpm motor that has a 4" sheave. The blower has a 12" sheave, so the blower shaft speed is 600 rpm. What size sheave is needed on the blower shaft to decrease the speed to 540 rpm?
    a. 11.9"
    b. 13.3"
    c. 14.2"
    d. 15.5"

$$\text{Driven pitch diameter} = \frac{S_{driver} \times P_{driver}}{S_{driven}}$$

$$\text{Driven pitch diameter} = \frac{1{,}800 \text{ rpm} \times 4''}{540 \text{ rpm}}$$

$$\text{Driven pitch diameter} = \frac{7{,}200}{540 \text{ rpm}}$$

$$\text{Driven pitch diameter} = 13.3''$$

16. A potentially fatal respiratory condition that comes from rodents but is *not* transmissible between humans is caused by _____.
    a. COVID-19
    b. the Ebola virus
    c. hantavirus
    d. psittacosis

17. The most common manifold gas pressure for LP gas burners is _____.
    a. 0.5 psig
    b. 3.5 psig
    c. 7" w.c.
    d. 11" w.c.

18. A heat pump thermostat will typically attempt to start the electric heat to supplement the heat pump when the setpoint is _____.
    a. 1°F (0.5°C) or more above the room temperature
    b. 1°F to 2°F (0.5°C–1.1°C) or more above the room temperature
    c. 2°F to 3°F (1.1°C–1.7°C) or more above the room temperature
    d. 5°F (2.8°C) or more above the room temperature

19. The solenoid valve of a bypass humidifier controls the water flow rate to the unit using a(n) _____.
    a. orifice
    b. ball valve
    c. pulsing action
    d. water-pressure regulator

20. What part of an electronic air cleaner is used to capture odors from the air passing through?
    a. Ionizing wires
    b. Collector cells
    c. Charcoal filter
    d. Prefilter

Answers to Odd-Numbered Module Review Questions are found in *Appendix A.*

Answers to Section Review Questions

Answer	Section	Objective
Section One		
1. b	1.1.1	1a
2. a	1.2.1	1b
3. c	1.3.3	1c
Section Two		
1. b	2.1.1	2b
2. a	2.2.1	2b
Section Three		
1. c	3.1.1	3a
2. c	3.2.2	3b
3. a	3.3.1	3c
Section Four		
1. a	4.1.2	4a
2. d	4.2.2	4b
3. b	4.3.2	4c
4. a	4.4.1	4d

MODULE 03313

Fasteners and Hardware

Source: Coprid/Getty Images

Objectives

Successful completion of this module prepares you to do the following:

1. Identify the types, uses, and installation methods of fasteners.
   a. Identify various types and uses of threaded fasteners.
   b. Identify and explain how to use taps, dies, and screw extractors.
   c. Explain how to install and torque fasteners to a specific value.
   d. Identify and explain how to install various types of toggle and anchor bolts.
   e. Identify and describe various types of non-threaded fasteners.
2. Identify and describe the installation of various types of vibration isolators.
   a. Identify and describe vibration isolators used to support and suspend equipment and piping.
   b. Identify and describe methods used to restrain equipment during seismic events.

Performance Tasks

Under supervision, you should be able to do the following:

1. Torque threaded hardware to a specific torque value.
2. Select the appropriate drill bit and install an anchor in brick or concrete block.

Overview

During the installation and maintenance of HVACR equipment, technicians work with a variety of fasteners and hardware. These components must be carefully selected for the specific application and installed according to accepted practices. Using the correct components ensures that the assembled system will perform properly and will not fail because of the wrong choice of component or its incorrect installation.

NCCER Industry-Recognized Credentials

If you are training through an NCCER-accredited sponsor, you may be eligible for credentials from NCCER. The ID number for this module is 03313. Note that this module may have been used in other NCCER curricula and may apply to other level completions. Contact NCCER at 1.888.622.3720 or go to **www.nccer.org** for more information.

You can also show off your industry-recognized credentials online with NCCER's digital credentials. Transform your knowledge, skills, and achievements into credentials that you can share across social media platforms, send to your network, and add to your resume. For more information, visit **www.nccer.org**.

Digital Resources for HVACR

Scan this code using the camera on your phone or mobile device to view the digital resources related to this craft.

SCAN ME

1.0.0 Fasteners

Performance Tasks

1. Torque threaded hardware to a specific torque value.
2. Select the appropriate drill bit and install an anchor in brick or concrete block.

Objective

Identify the types, uses, and installation methods of fasteners.

a. Identify various types and uses of threaded fasteners.
b. Identify and explain how to use taps, dies, and screw extractors.
c. Explain how to install and torque fasteners to a specific value.
d. Identify and explain how to install various types of toggle and anchor bolts.
e. Identify and describe various types of non-threaded fasteners.

Fasteners are used to install many parts and many types of equipment. Common fasteners include bolts, screws, pins, and retainers. It is important to select the correct fastener for an application. For example, bolts and screws are graded based on strength. Even if the size of a bolt is correct, it may fail if the wrong grade (strength) is installed.

1.1.0 Threaded Fasteners

Threaded fasteners are one of the most common types of fasteners. Many are assembled with nuts and washers while others are installed in threaded holes.

Types of threaded fasteners that will be covered in this section include the following:

- Machine bolts, machine screws, stud bolts, and cap screws
- Eye bolts
- Nuts
- Inserts
- Flat and lock washers
- Set screws
- Screws

Before reviewing the types and shapes of threaded fasteners, it is important to understand the various threads used. You must also understand the grades of threaded fasteners. The appropriate thread type and grade depends on the application.

1.1.1 Thread Designations

Thread standards: An established set of standards for machining threads.

Fastener threads are made to established **thread standards**. The most common standard is the Unified, or American National, Standard. There are three series defined by the Unified Standard. The following series are based on the number of threads per inch for a fastener of a certain diameter:

- *Unified National Coarse (UNC) threads* — UNC threads are used for bolts, nuts, and threaded rod in all routine applications.
- *Unified National Fine (UNF) threads* — UNF threads are used for bolts and nuts where a finer thread (more threads per inch) and additional holding power are required.
- *Unified National Extra Fine (UNEF) threads* — UNEF threads are used for thin-walled tubing, such as a chrome plumbing drain fitting, and similar materials where the threads cannot be very deep.

Bolt and screw threads are designated by a standard method (*Figure 1*). The standard designations include the following:

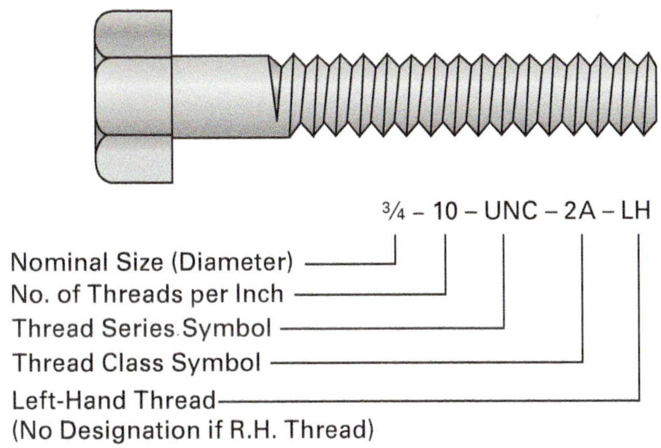

Figure 1 Thread designations.

- **Nominal size** (diameter)
- Number of threads per inch
- **Thread series** symbol
- **Thread class**
- Left-hand thread symbol (Unless otherwise indicated, the threads are right-hand. A symbol is used only for left-hand fasteners.)

The thread class determines how tightly the male (external) and female (internal) threads fit each other. There are three classes for external and internal threads. Classes 1A, 2A, and 3A apply to external threads, such as a bolt. Classes 1B, 2B, and 3B apply to internal threads, such as a nut.

Higher numbers indicate a tighter fit. Therefore, Class 3 threads provide the tightest fit. Class 1 is very rare and is only used when a loose fit is desired. The loose fit provides limited strength and resistance to separation. Class 2 is the most common. Class 3 threads are needed when close tolerances and maximum strength are priorities, but they are more difficult to assemble.

The thread series refers to the coarseness of the threads in terms of threads per inch. Coarse-pitch threads (fewer threads per inch) are by far the most common. Coarse threads are used for all routine applications and offer good resistance to stripping when used with larger diameter fasteners. Coarse threads are easier to form, and nuts start and run down the threads easily unless they are damaged in some way.

Fine-pitch threads have a higher **tensile strength**, and resist loosening when subjected to shock or vibration. Fine threads are useful if a thin nut must be used or where very fine adjustments must be made. However, they are rarely encountered in HVACR work. Extra-fine threads are virtually nonexistent in the craft.

Nominal size: A means of expressing the size of a bolt or screw using its approximate diameter.

Thread series: The combination of diameter and thread pitch, used to identify the coarseness or fineness of a thread.

Thread class: A classification related to the amount of tolerance the threads provide between a bolt and nut, designated by a number and letter combination. Higher numbers indicate a tighter fit.

Tensile strength: The maximum stress a material can endure before being pulled apart.

Metric Threads

Metric threads follow the International Organization for Standardization (ISO) metric thread standards. *Figure 2* shows the designations for metric threaded fasteners. An M or MJ begins the identification, specifying a metric thread.

The nominal size in millimeters comes next. Rather than a "threads-per-centimeter" value, however, metric fastener identifications use thread pitch instead. The thread pitch indicates the distance in millimeters between the crests (tops) of adjacent threads. The final element is the fastener length in millimeters.

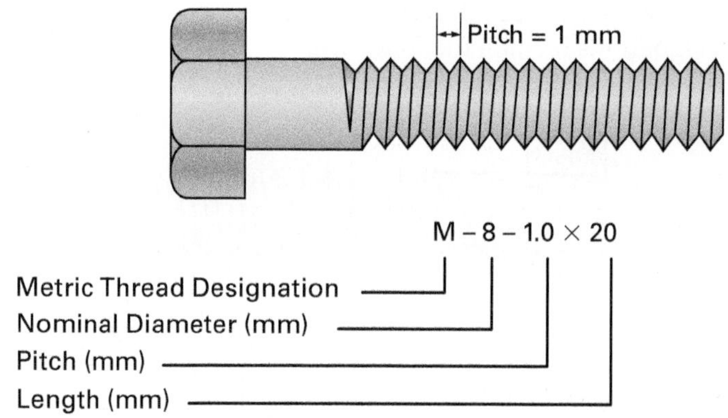

Figure 2 Metric thread designations.

Metric fasteners come in varying degrees of thread fineness too. Unlike US standard fasteners, however, the terms *fine* and *coarse* don't apply to metric fasteners. Instead, the thread pitch value indicates the thread's fineness or coarseness. The smaller the thread pitch value, the finer the threads. Metric fasteners beginning with an M have standard threads. Fasteners beginning with MJ have a modified thread style that is more resistant to fatigue. Some aerospace applications use the MJ style for this reason. An MJ bolt requires an MJ nut (or a hole with MJ internal threads). An M bolt, however, works with both M and MJ nuts (or holes with either type of internal threads). There is little need for MJ-designated hardware in common HVACR applications.

1.1.2 Threaded Fastener Grades

Fasteners may perform differently even if they look identical. Their material of construction and the way that the manufacturer heat-treated it can make a huge difference. A fastener made from low-quality steel is not very strong. It may work fine in a light-duty application. But when put under significant stress, it will fail. A high-quality steel that the manufacturer heat treats is much stronger and performs better under stress.

The strength and quality of a bolt can be determined by special grade markings on the head. The markings represent standards developed by the Society of Automotive Engineers (SAE) and ASTM International. Grade markings are sometimes called *line markings*. *Table 1* shows the SAE and ASTM markings for US standard bolts.

SAE Grade 2, Grade 5, and Grade 8 are the most common grades. Grade 2 is made from low-carbon steel and is used in low-stress applications. For example, a Grade 2 bolt might be adequate for a bracket supporting conduit. Grades 5 and 8 bolts use higher-quality steel that has been quenched and annealed. They are used in applications subjected to higher stress or pressure. For example, the higher-grade bolts are used to secure the cylinder heads of a semi-hermetic compressor. Higher stress or pressure dictates whether a Grade 8 is used rather than a Grade 5.

Higher-grade fasteners can be torqued to higher values. Low-grade fasteners deform more easily and may not hold a torque value as well, relaxing over time. Never substitute a lower grade bolt for a higher grade. For important or demanding applications, always use higher-grade fasteners.

Metric fasteners use their own grading system, as shown in *Table 2*. Instead of markings, numbers on the head identify the grade. Like the US standard bolts, unmarked heads indicate the lowest grade.

TABLE 1 Grade Markings for US Standard Bolts

Grade Marking	Specification	Material
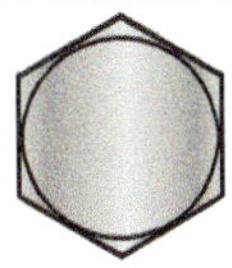	SAE Grade 0	Steel
	SAE Grade 1 ASTM-A307	Low-Carbon Steel
	SAE Grade 2	Low-Carbon Steel
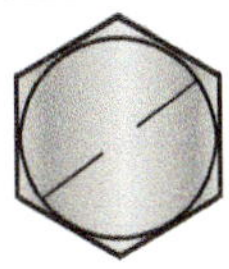	SAE Grade 3	Medium-Carbon Steel (Cold Worked)
	SAE Grade 5 ASTM-A449	Medium-Carbon Steel (Quenched and Tempered)
	ASTM-A3125	Medium-Carbon Steel (Quenched and Tempered)
	ASTM-A354 Grade BB	Low-Alloy Steel (Quenched and Tempered)
	ASTM-A354 Grade BC	Low-Alloy Steel (Quenched and Tempered)
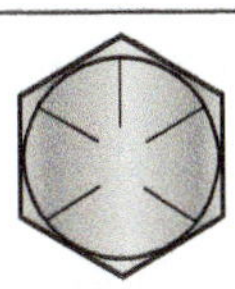	SAE Grade 7	Medium-Carbon Alloy Steel (Quenched and Tempered; Roll Threaded After Heat Treatment)
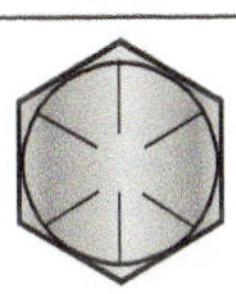	SAE Grade 8	Medium-Carbon Alloy Steel (Quenched and Tempered)
	ASTM-A354 Grade BD	Alloy Steel (Quenched and Tempered)
	ASTM-A3125	Alloy Steel (Quenched and Tempered)

ASTM Specification References:

A307 – *Standard Specification for Carbon Steel Bolts, Studs, and Threaded Rod*

A354 – *Standard Specification for Quenched and Tempered Alloy Steel Bolts, Studs, and Other Externally Threaded Fasteners*

A449 – *Standard Specification for Hex Cap Screws, Bolts and Studs, Steel, Heat Treated*

A3125 – *Standard Specification for High Strength Structural Bolts and Assemblies, Steel and Alloy Steel, Heat Treated*

SAE Specification References:

J429 – *Mechanical and Material Requirements for Externally Threaded Fasteners*

TABLE 2 Grade Markings for Metric Bolts

Grade	Head Markings	Size Range (mm)	Min Proof Strength (10^6 Pa)	Min Tensile Strength (10^6 Pa)	Material
4.6		M5–M36	225	400	Low- or Medium-Carbon Steel
4.8		M1.6–M16	310	420	Low- or Medium-Carbon Steel
5.8		M5–M24	380	520	Low- or Medium-Carbon Steel
8.8		M16–M36	600	830	Medium-Carbon Steel (Quenched and Tempered)
9.8		M1.6–M16	650	900	Medium-Carbon Steel (Quenched and Tempered)
10.9		M5–M36	830	1,040	Low-Carbon Martensite Steel (Quenched and Tempered)
12.9		M1.6–M36	970	1,220	Alloy Steel (Quenched and Tempered)

1.1.3 Machine Bolts, Machine Screws, Stud Bolts, and Cap Screws

Craftworkers use machine bolts (*Figure 3*) to assemble parts and attach equipment. They come in a variety of sizes and grades. Machine bolts are made with diameters ranging from 1/4" to 3" and with lengths from 1/2" to 30". A machine bolt can screw directly into a threaded hole, or it can be secured with a compatible nut.

Machine screws are used for general assembly. They have slotted or recessed heads. Machine screws are available in diameters from #6 through #12 (0.060" to 0.2160"). Fractional diameters range from 1/4" to 1/2" and vary in length from 1/8" to 3". These screws are also made in metric sizes. A machine screw normally mates with an internally threaded hole into which it is tightened or released. Machine screws can also be used with matching nuts.

Cap screws have a more attractive head and are generally used on assemblies where a finished appearance is important. They pass through a clearance hole in one part of the assembly and are screwed into a threaded hole or secured with a nut. They are made to close tolerances with machined or semifinished bearing surfaces under the head. Cap screws come with coarse or fine threads, in diameters from 1/4" to 2", and lengths from 3/8" to 10". They are also made in metric sizes.

Stud bolts are essentially headless bolts. They resemble a piece of threaded rod with an unthreaded section in the middle. One end screws into a threaded hole. A component with a hole is set over the protruding stud. A nut and washer combination clamps the part in place. Some stud bolts have different threads on each end; for example, UNC and UNF.

Threaded rod, also referred to as *all-thread*, is simply metal rod threaded over its entire length. Unlike a stud bolt that usually has threads moving in opposing directions on each end, threaded rod has one continuous thread. Standard lengths are available, such as 6' or 10' lengths, but they can be cut to any length. The rod is usually made of low-carbon steel because the rods are often used in low-load applications. Typical HVACR applications include their use as a pipe hanger or as part of a trapeze assembly for supporting a duct (*Figure 4*).

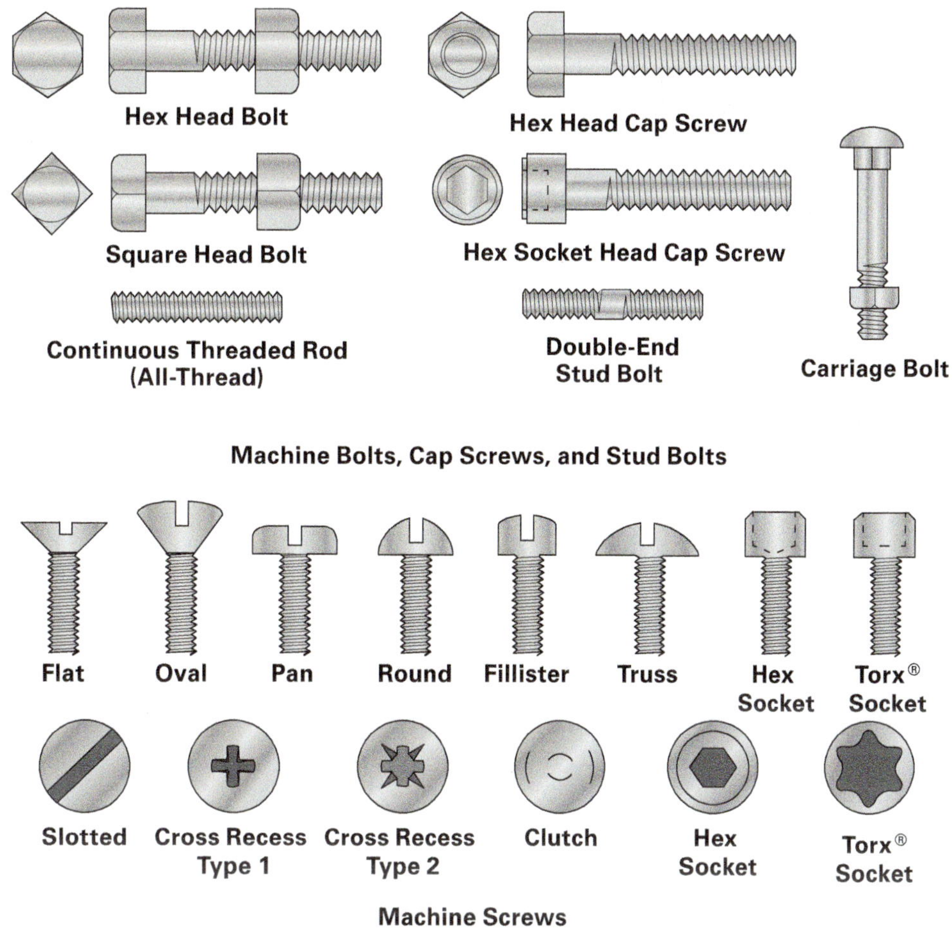

Figure 3 Machine bolts, machine screws, cap screws, and stud bolts.

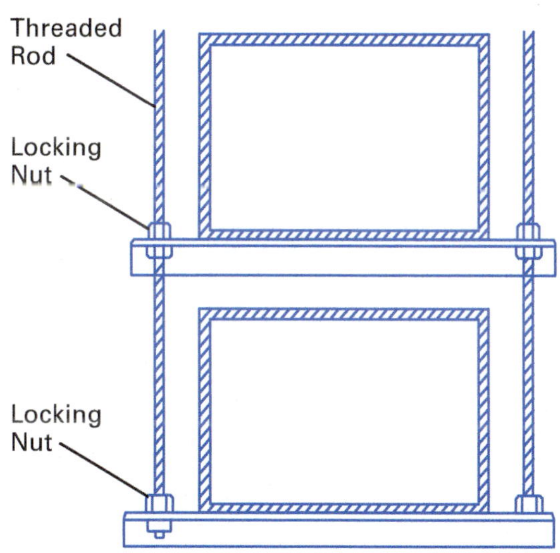

Figure 4 Threaded rods supporting ducts.

1.1.4 Eye Bolts

Eye bolts (*Figure 5*) have a loop, or eye, at one end and a threaded section at the other. Some eyes are formed and welded. Others are forged. Eye bolts never fasten parts together. Instead, they provide attachment points. Some manufacturers insert shoulder-forged eye bolts into equipment to act as lifting points.

WARNING!

Only forged eye bolts can be used for rigging purposes. Common mild-steel bolts are not designed to handle a substantial load.

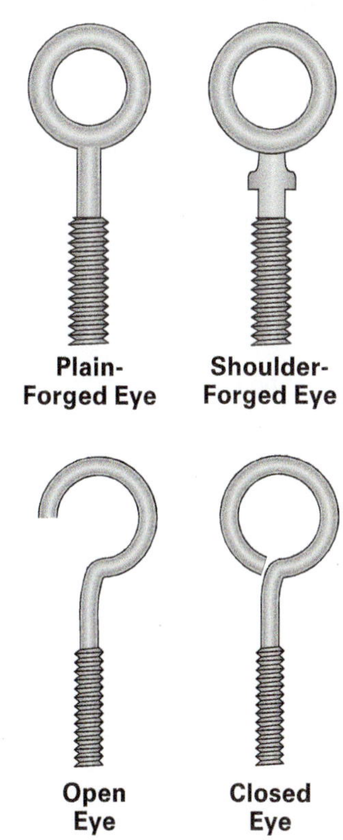

Figure 5 Eye bolts.

1.1.5 Nuts

Nuts used with most threaded fasteners have either hex (hexagonal) or square shapes and are used with bolts having a similar head. *Figure 6* shows a variety of nut designs.

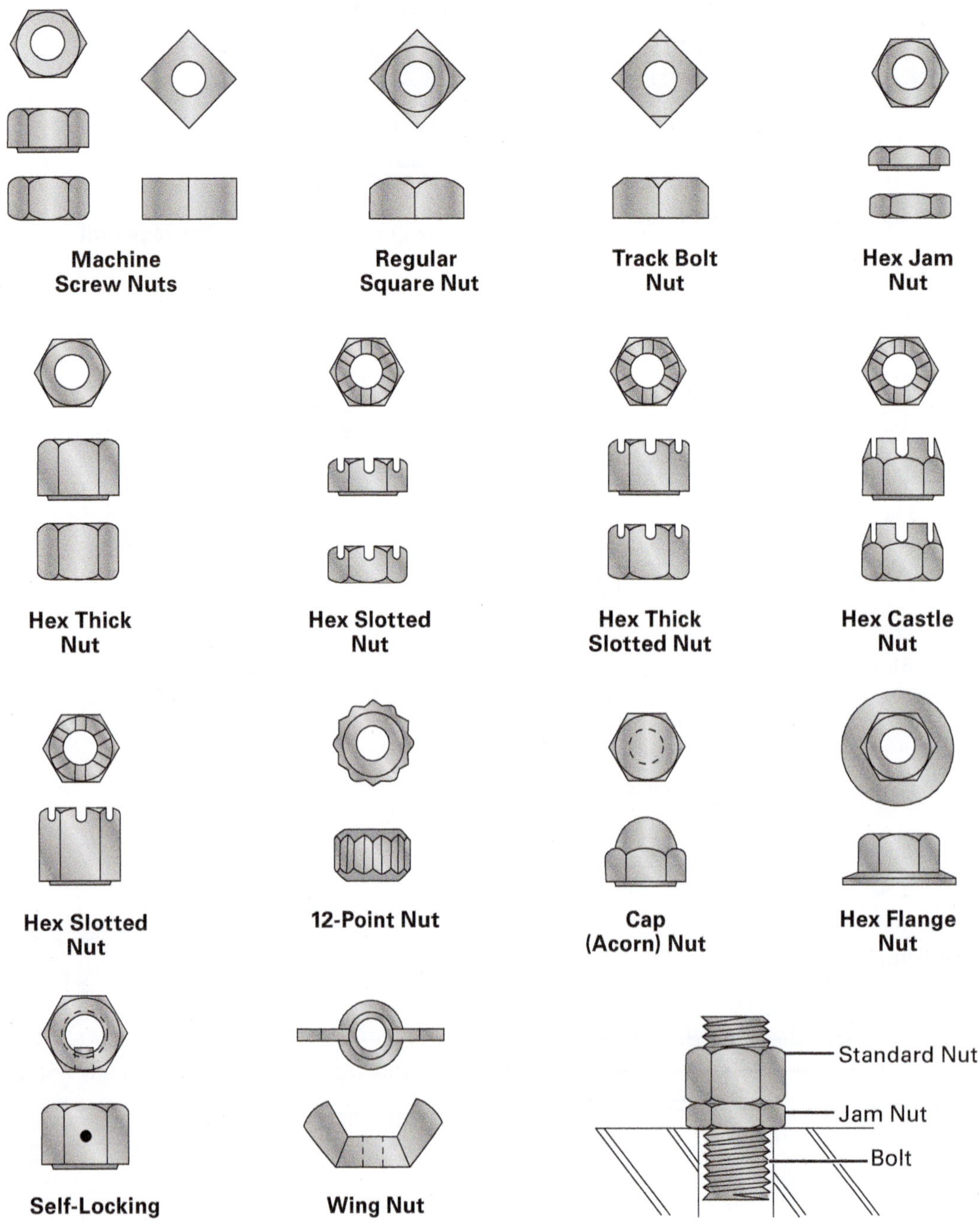

Figure 6 Common nut designs.

NOTE

Cotter pins are discussed further in *Section 1.5.2*.

Some special-purpose nuts include the following:

- *Acorn nut* — Acorn nuts, also known as *cap nuts*, are used when appearance is important or when there are exposed sharp threads on the fastener.
- *Castellated (or castle)* and *slotted nuts* — These nuts provide for an extra locking feature. After the nut is tightened, a *cotter pin* is fitted into one set of slots and through a hole in the bolt. The cotter pin keeps the nut from loosening. These nuts may also be secured with wire.
- *Self-locking nut* — A self-locking nut has a nylon insert or is slightly deformed (intentionally) so it cannot work loose. Once a self-locking nut is used and removed, it must be thrown away and replaced with a new one.
- *Wing nut* — These are used where frequent assembly and disassembly of non-critical parts is required. Wing nuts are designed to be installed and removed with your fingers; no wrench is required.

- *Jam nut* — A jam nut is a thin nut used to lock a standard nut in place. Note that jam nuts are installed on the fastener first, leaving it under the standard nut. The jam nut is often installed on top of the other nut, which is technically incorrect. It is also common to install two adjacent nuts of the same size to provide the same basic locking function, instead of using the thinner jam nut.
- *U-nuts and J-nuts* — U-nuts and J-nuts (*Figure 7*) are used with soft, thin panel materials, or with thin sheet metal. The U-nut and J-nut have a threaded hole or stamped opening that accepts threads. They are slipped onto the edge of thin, relatively soft material to provide a quick method of attachment.
- *Cage nut* — Cage nuts, also shown in *Figure 7*, are also referred to as *nut plates*. They are basically square nuts enclosed in a stamped sheet metal box. The cage has a spring clip at the bottom that can be clipped into a square hole in a rack. The nut is loose enough inside the cage to allow for some small misalignment when mounting equipment.

Figure 7 Cage nuts, J-nuts, and U-nuts.

1.1.6 Helical Inserts

Helical inserts, or Heli-Coils® (*Figure 8*), are a special kind of fastener used to provide high-strength threads in soft metals and plastics. They are also used to replace damaged or stripped threads in a tapped hole. Inserts are made in standard sizes and forms including metric sizes. The insert, which is larger in diameter than the tapped hole, is compressed during installation and allowed to spring back, permanently anchoring the insert in the tapped hole.

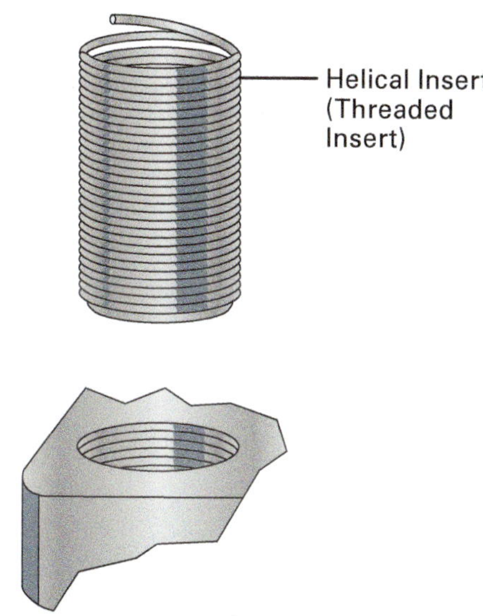

Figure 8 Helical insert.

NOTE

These fasteners are often referred to as *helicoils*, but the term Heli-Coil® is a registered trademark. *Helical insert* and *threaded insert* are the preferred generic names.

1.1.7 Flat and Lock Washers

Flat washers (*Figure 9*) provide an enlarged surface used to distribute the load from bolt heads and nuts. Flat washers are made in light, medium, heavy-duty, and extra heavy-duty series.

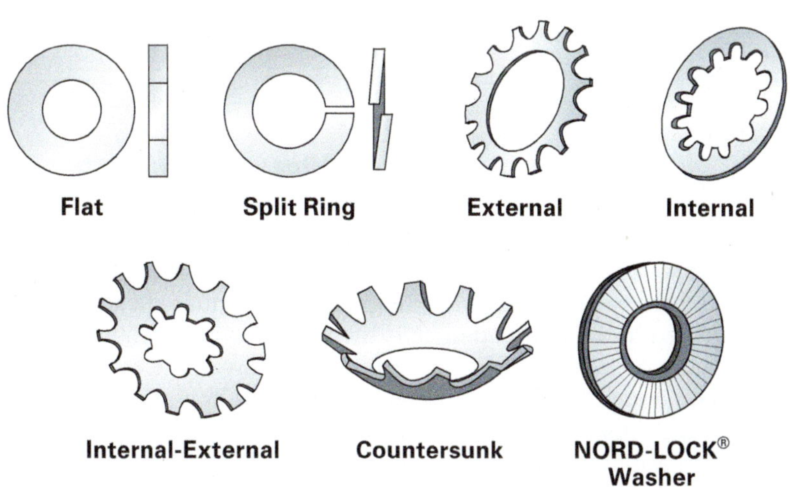

Figure 9 Flat and lock washers.

Fender washers are a type of flat washer, but they have a larger outside diameter to bridge oversized holes, cover a blemish near the hole, or spread the load across a larger surface.

Lock washers are used to keep bolts or nuts from working loose. Some common types of lock washers include the following:

- *Split ring* — Commonly used with bolts and cap screws
- *External* — Provide the greatest resistance to loosening
- *Internal* — Used on small fasteners
- *Internal-external* — Used for oversized mounting hole
- *Countersunk* — Used with flat or oval-head fasteners
- *NORD-LOCK®* — Uses a pair of washers equipped with wedges to hold the fastener in place; unique and rarely found in the HVACR craft

1.1.8 Set Screws

Set screws (*Figure 10*) are usually made of heat-treated steel and are used to secure objects within or against another object. For example, set screws are commonly used to fasten a pulley or gear to a shaft. Set screws are generally classified by head styles and point styles.

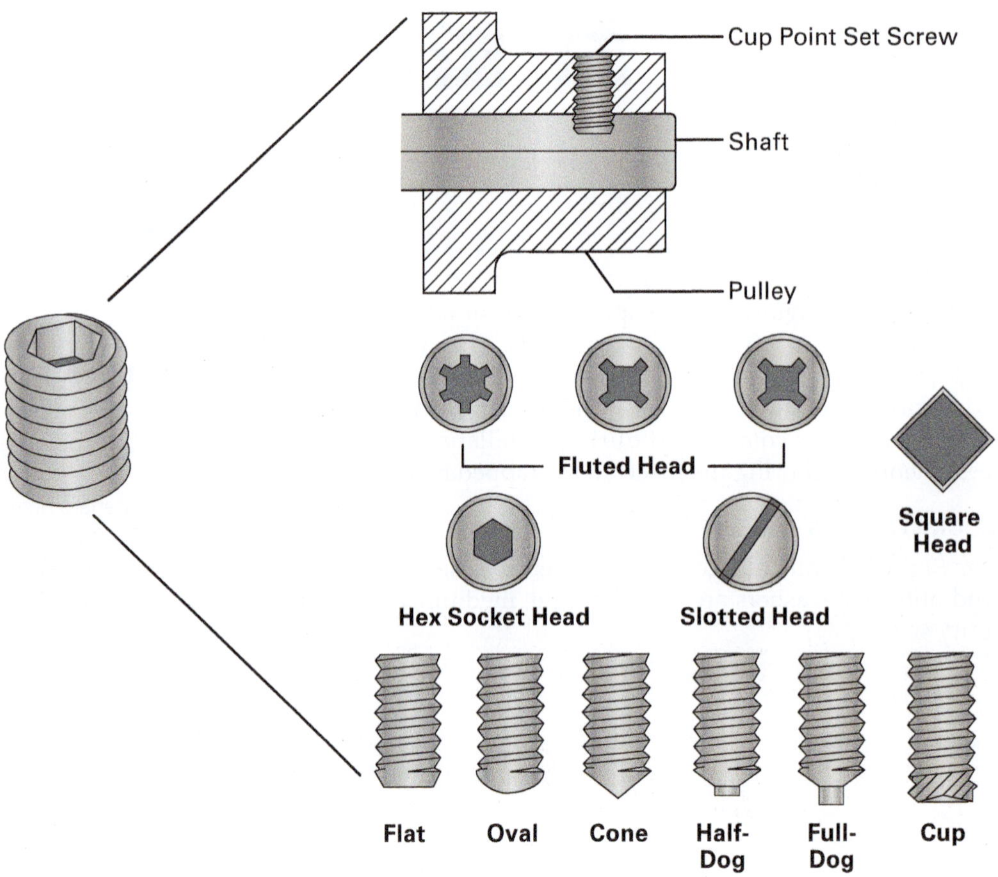

Figure 10 Set screws.

Set screws are usually headless (also known as *blind*), meaning the screw is fully threaded. They can have fluted, hex, slotted, or square heads. Once the set screw is screwed into the threaded hole, the tip protrudes through the hole and usually contacts a shaft, securing a fan blade or blower wheel in place. This is the most common application for a set screw in HVACR work.

The *cup point* set screw is the most common type. It is well suited for use in high vibration applications because it is very durable and shows a high resistance to wear while having strong grip capabilities. The cupped shape provides a locking function that resists loosening.

1.1.9 Screws

Screws resemble bolts but often have tapered bodies, usually with pointed ends. They do not require a threaded hole or a nut to hold them. Many produce their own matching threads when twisted into soft or thin materials. Others grip the edges of holes in thin materials. Screws securely fasten parts together and may be used with parts requiring disassembly.

Screws have many head styles. Some accept ordinary screwdrivers. Others require special driver bits. *Figure 11* shows typical screw head types and drives.

NOTE

Although they are referred to as *screws*, set screws do not have the tapered shape of common screws.

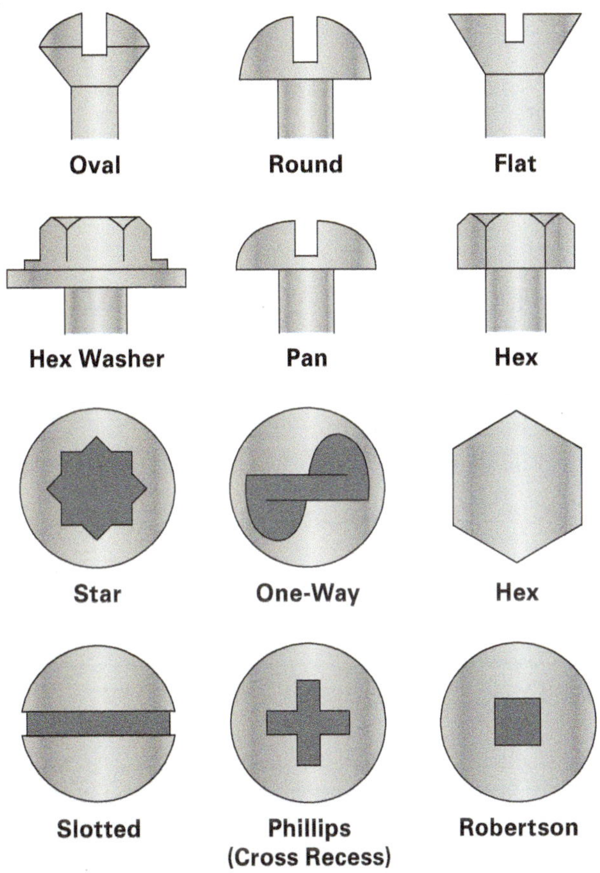

Figure 11 Screw head and drive styles.

Some US standard screws come in numbered sizes, ranging from #2 to #24. Others have fractional diameter designations like $\frac{1}{2}$". Lengths range from $\frac{1}{4}$" to 6". Metric screws follow the same general pattern as bolts, but have their lengths specified in millimeters.

Screws often come with specialized coatings or are plated with certain metals. These features improve corrosion resistance or make the screws more attractive. Screws made from brass or even nylon handle specialized needs.

Lag screws attach heavy loads to wood structures. When used with a lag shield, lag screws can also attach objects to concrete or masonry (*Figure 12*). The shield, often made from lead, fits into a pre-drilled hole. Threading the screw into the shield expands it, causing it to grip the hole. Overtightening can damage the shield, causing it to lose its grip.

Thread-forming screws fasten light-gauge metal parts. They form a thread as they enter the material, eliminating the need for tapping. Because they have a very sharp point, they can drill their own hole in thin metals. For thicker materials, self-drilling thread-forming screws work better. They have a special tip shaped like a drill bit that allows them to drill their own hole.

Thread-cutting (self-tapping) screws are more aggressive, cutting threads into heavier metal as they enter a pilot hole. *Figure 13* shows examples of both screw types.

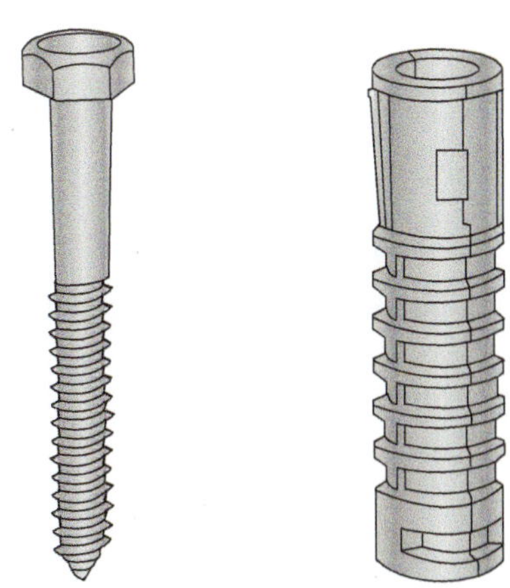

Figure 12 Lag screw and lag shield.

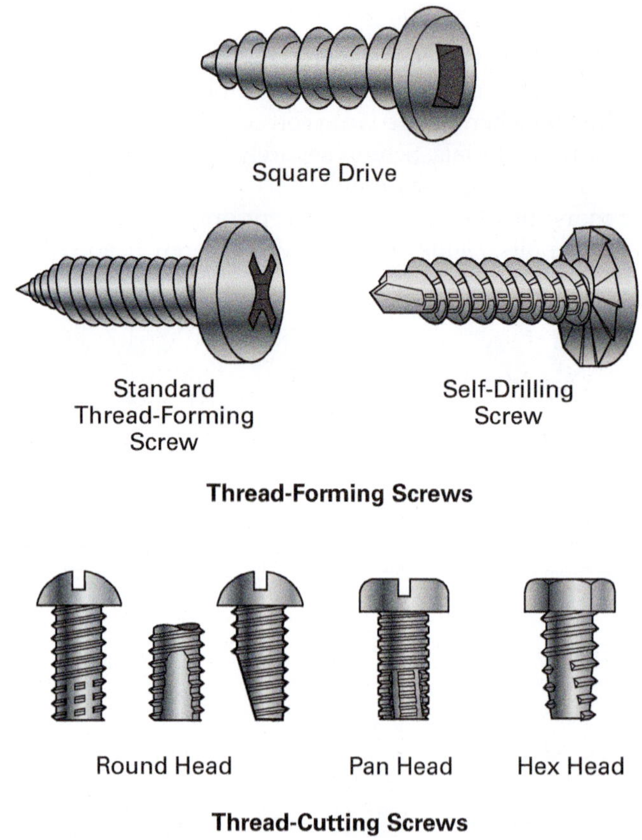

Figure 13 Thread-forming and thread-cutting screws.

1.2.0 Taps, Dies, and Screw Extractors

Taps and dies are used to cut threads in materials such as metal, plastic, and hard rubber. Taps create threaded internal holes for bolts, while dies create external threads on rod. Screw extractors are used to remove fasteners that have broken off flush or below the surface.

1.2.1 Thread Taps

Thread taps (*Figure 14*) are used to cut internal threads in materials; dies are used to cut external threads on bolts and rods. Taps are used to thread holes that go all the way through a piece of material, or holes that end inside the material (blind holes).

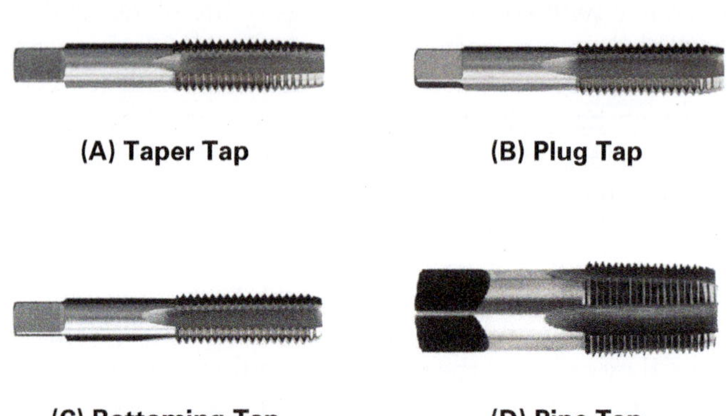

Figure 14 Common thread taps.
Source: Courtesy of Drill America

Note the ends of the taper, plug, and bottoming taps in *Figure 14*. Each has a slight taper at the end, but the tapers differ in length. Common taps for threaded hardware include the following:

- *Taper taps* — Taper taps are used when initially tapping a blind hole or when tapping through holes. They have a chamfer length of 8 to 10 threads.
- *Plug taps* — Plug taps are used following the taper tap in a blind hole. They have a chamfer length of 3 to 5 threads.
- *Bottoming taps* — Bottoming taps are always used for blind holes after the taper and plug taps have been used. They are used for threading the bottom of a blind hole and have a chamfer length of only 1 to $1\frac{1}{2}$ threads.
- *Pipe taps* — As the name implies, pipe taps are used to create pipe threads. The diameter of the tap from end to end of the threaded portion is tapered at the rate of $\frac{3}{4}$" per foot. Although you will rarely need them to fully tap a port to accommodate threaded pipe, they are invaluable for cleaning up damaged threads in a pipe fitting.

1.2.2 Tapping a Hole

The first step in tapping a hole is to drill a hole of the proper size. *Table 3* is one example of a tap drill table. There are many versions available, primarily from tap and drill manufacturers. *Table 3* also provides alternate metric drill sizes as well as US standard drill sizes.

TABLE 3 Common Tap Drill Sizes

Tap Size/Pitch Form	US Tap Drill	Metric Drill	Tap Size/ Pitch Form	US Tap Drill	Metric Drill	Tap Size/ Pitch Form	US Tap Drill	Metric Drill
Unified Taps			**Taper Pipe Taps (NPT)**			M27 - 3.00	$\frac{61}{64}$"	24.00
0-80	$\frac{3}{64}$"	1.25	$\frac{1}{16}$" - 27	D	6.30	M30 - 3.50	$1\frac{3}{64}$"	26.50
1 - 64 UNC	53	1.55	$\frac{1}{8}$" - 27	R	8.70	M33 - 3-50	$1\frac{5}{32}$"	29.50
1 - 72 UNF	53	1.55	$\frac{1}{4}$" - 18	$\frac{7}{16}$"	11.10	M36 - 4.00	$1\frac{1}{4}$"	32.00
2 - 56 UNC	51	1.85	$\frac{3}{8}$" - 18	$\frac{37}{64}$"	14.50	M39 - 4.00	$1\frac{3}{8}$"	35.00
2 - 64 UNF	50	1.90	$\frac{1}{2}$" - 14	$\frac{23}{32}$"	18.00	**Metric ISO (Fine)**		
3 - 48 UNC	47	2.10	$\frac{3}{4}$" - 14	$\frac{59}{64}$"	23.25	M3 - 0.35	37	2.65
3 - 56 UNF	46	2.15	1" - 11.5	$1\frac{5}{32}$"	29.00	M4 - 0.35	27	3.65
4 - 40 UNC	43	2.35	$1\frac{1}{4}$" - 11.5	$1\frac{1}{2}$"	38.00	M4 - 0.50	29	3.50
4 - 48 UNF	42	2.40	$1\frac{1}{2}$" - 11.5	$1\frac{47}{64}$"	44.00	M4.5 - 0.45	21	4.05
5 - 40 UNC	39	2.65	2" - 11.5	$2\frac{7}{32}$"	57.00	M5 - 0.50	16	4.50
5 - 44 UNF	37	2.70	$2\frac{1}{2}$" - 8	$2\frac{5}{8}$"	67.00	M5 - 0.70	18	4.30
6 - 32 UNC	36	2.85	3" - 8	$3\frac{1}{4}$"	82.50	M5 - 0.75	18	4.25
6 - 40 UNF	33	2.95	**Straight Pipe Taps (NPS)**			M5.5 - 0.5	9	5.00
8 - 32 UNC	29	3.50	$\frac{1}{8}$" - 27	S	8.80	M6 - 0.50	$\frac{7}{32}$"	5.50
8 - 36 UNF	29	3.50	$\frac{1}{4}$" - 18	$\frac{29}{64}$"	11.50	M6 - 0.75	5	5.25
10 - 24 UNC	25	3.90	$\frac{3}{8}$" - 18	$\frac{19}{32}$"	15.00	M7 - 0.75	D	6.25
10 - 32 UNF	21	4.10	$\frac{1}{2}$" - 14	$\frac{47}{64}$"	18.50	M8 - 0.50	M	7.50
12 - 24 UNC	17	4.50	$\frac{3}{4}$" - 14	$\frac{15}{16}$"	23.75	M8 - 1.00	J	7.00
12 - 28 UNF	15	4.70	1" - 11.5	$1\frac{3}{16}$"	30.25	M9 - 0.50	Q	8.50
$\frac{1}{4}$" - 20 UNC	7	5.10	$1\frac{1}{4}$" - 11.5	$1\frac{33}{64}$"	38.50	M9 - 1.00	O	8.00
$\frac{1}{4}$" - 28 UNF	3	5.50	$1\frac{1}{2}$" - 11.5	$1\frac{3}{4}$"	44.50	M10 - 0.50	$\frac{3}{8}$"	9.50
$\frac{5}{16}$" - 18 UNC	F	6.60	2" - 11.5	$2\frac{7}{32}$"	57.00	M10 - 0.75	U	9.25

(continued)

TABLE 3 *(continued)*

Tap Size/Pitch Form	US Tap Drill	Metric Drill	Tap Size/ Pitch Form	US Tap Drill	Metric Drill	Tap Size/ Pitch Form	US Tap Drill	Metric Drill
5/16" - 24 UNF	I	6.90	**Metric ISO (Coarse)**			M10 - 1.00	T	9.00
3/8" - 16 UNC	5/16"	8.00	M1.6 - 0.35	3/64"	1.25	M10 - 1.25	11/32"	8.75
3/8" - 24 UNF	Q	8.50	M1.7 - 0.35	55	1.35	M11 - 1.00	X	10.00
7/16" - 14 UNC	U	9.40	M1.8 - 0.35	64	1.45	M12 - 1.00	7/16"	11.00
7/16" - 20 UNF	W	9.90	M2 - 0.40	1/16"	1.60	M12 - 1.25	27/64"	10.75
1/2" - 13 UNC	27/64"	10.80	M2.2 - 0.45	50	1.75	M12 - 1.50	Z	10.50
1/2" - 20 UNF	29/64"	11.50	M2.3 - 0.40	49	1.90	M13 - 1.50	29/64"	11.50
9/16" - 12 UNC	31/64"	12.20	M2.5 - 0.45	46	2.05	M13 - 1.75	7/16"	11.25
9/16" -18 UNF	33/64"	12.90	M2.6 - 0.45	44	2.15	M14 - 1.25	1/2"	12.75
5/8" - 11 UNC	17/32"	13.50	M3 - 0.50	40	2.50	M14 - 1.50	31/64"	12.50
5/8" - 18 UNF	37/64"	14.50	M3.5 - 0.60	33	2.90	M15 - 1.50	17/32"	13.50
3/4" - 10 UNC	21/32"	16.50	M4 - 0.70	30	3.30	M16 - 1.00	19/32"	15.00
3/4" - 16 UNF	11/16"	17.50	M4.5 - 0.75	27	3.70	M16 - 1.25	37/64"	14.75
7/8" - 9 UNC	49/64"	19.50	M5 - 0.80	19	4.20	M16 - 1.50	9/16"	14.50
7/8" - 14 UNF	13/16"	20.40	M5.5 - 0.90	15	4.60	M18 - 1.00	43/64"	17.00
1" - 8 UNC	7/8"	22.25	M6 - 1.00	9	5.00	M18 - 1.25	21/32"	16.75
1" - 12 UNF	59/64"	23.25	M7 - 1.00	15/64"	6.00	M18 - 1.50	41/64"	16.50
1" - 14 UNS	15/16"	23.50	M8 - 1.25	H	6.80	M18 - 2.00	5/8"	16.00
1 1/8" - 7 UNC	63/64"	25.00	M9 - 1.25	5/16"	7.80	M20 - 1.00	3/4"	19.00
1 1/8" - 12 UNF	1 3/64"	26.50	M10 - 1.50	Q	8.50	M20 - 1.50	47/64"	18.50
1 1/4" - 7 UNC	1 7/64"	28.00	M11 - 1.50	3/8"	9.50	M20 - 2.00	45/64"	18.00
1 1/4" - 12 UNF	1 11/64"	29.50	M12 - 1.75	Y	10.20	M22 - 1.00	53/64"	21.00
1 3/8" - 6 UNC	1 7/32"	30.75	M14 - 2.00	15/32"	12.00	M22 - 1.50	13/16"	20.50
1 3/8" - 12 UNF	1 19/64"	32.75	M16 - 2.00	35/64"	14.00	M22 - 2.00	25/32"	20.00
1 1/2" - 6 UNC	1 11/32"	34.00	M18 - 2.50	39/64"	15.50	M24 - 1.00	29/32"	23.00
1 1/2" - 12 UNF	1 27/64"	36.00	M20 - 2.50	11/16"	17.50	M24 - 1.50	7/8"	22.50
1 3/4" - 5 UNC	1 9/16"	39.50	M22 - 2.50	49/64"	19.50	M24 - 2.00	55/64"	22.00
2" - 4.5 UNC	1 25/32"	45.00	M24 - 2.50	53/64"	21.00	M24 - 2.50	27/32"	21.50

Several types of drill bits are represented in the *US Tap Drill* columns. Letter, number, and fractional drill sizes are needed for various tap sizes. Unified taps are based on the US standard, which you'll recall are classified as having either coarse (UNC) or fine (UNF) threads. The pitch of these threads refers to the number of threads per inch (TPI). Threaded hardware begins in machine screw sizes, from #0 to #12, at the beginning of the table. For example, a #6 machine screw has either coarse threads (UNC) with a pitch of 32 TPI, or fine threads (UNF) with a pitch of 40 TPI.

To use the tables, simply look for the machine screw or bolt size you intend to use in the tapped hole. Then read the proper drill size in the column to the right. For example, to tap a hole for a 1/4"-20 UNC bolt, use a #7 drill from a numbered drill index. Alternatively, a metric drill size of 5.1 mm can be used.

The following procedure explains how to tap a blind hole. Through-holes may only require the use of a taper tap, depending on the thickness of the material. To tap a blind hole, you'll need to use all three of the taps. Follow these steps to tap a hole:

Step 1 Determine the size of the hardware and the thread pitch needed. Threading a 1/4"-20 UNC bolt hole in 3/8" (~10 mm) plate is a good place to start. However, thicker material is needed if you intend to drill a blind hole to be tapped.

Step 2 If the hole has not been drilled, use *Table 3* or a similar table to determine the proper drill size, then drill the hole using a drill press. A handheld drill can be used, but the resulting hole will not be as straight and true.

Step 3 Select a taper tap of the proper size and thread pitch. Inspect it to ensure the tap is sharp and clean with no broken or nicked teeth.

Step 4 Attach a suitable tap wrench, place the end of the tap in the hole, and turn it clockwise about one-half to one-and-a-half turns while holding it true and applying light downward pressure. The tapered portion of the teeth will begin to cut.

Step 5 Turn the tap enough so that it can hold itself in place in the hole. After each half-turn to one-and-a-half turns, back the tap up slightly to break the metal chip being created (*Figure 15*). Note that the threads are very fragile at this point; handle the tap with care. Check the tap with a square at two positions to ensure it is perpendicular to the surface of the workpiece. The tap must be square to the surface to ensure the hole is tapped straight.

CAUTION

Do not use common wrenches or similar tools with taps, dies, or extractors. Always use the proper tap or die wrench provided by the manufacturer or an aftermarket source.

NOTE

Although right-hand threads are the most common by far, remember that a left-hand thread tap must be turned counterclockwise when cutting.

Figure 15 Tapping a hole.
Source: Evkaz/Getty Images

Step 6 Lubricate the tap with cutting fluid now, and regularly during the tapping process. Proper lubrication is a must. Use a cutting fluid designed for the material being tapped, such as ferrous metal, nonferrous metal, or plastic. Continue to turn the tap, applying light pressure until the tap feeds itself into the material. Downward pressure is then unnecessary.

Step 7 Continue the tapping process of turning in one-half to one-and-a-half turns at a time, then backing out one-quarter turn until the tap touches the bottom of the hole or you feel it grinding against the chips in the bottom.

Step 8 Back the tap out of the hole and clear the chips.

Step 9 Select and inspect a plug tap for the second phase of the tapping process.

Step 10 Tap the hole again with the plug tap, following the same procedure as with the taper tap. Little or no cutting action will be felt until you approach the bottom of the hole.

Step 11 Once the tap bottoms out, back it out of the hole and clear any chips.

Step 12 Select and inspect a bottoming tap for the final phase of the tapping process.

Step 13 Tap the hole again with the bottoming tap. The tap may need to be removed several times to clear chips until it contacts the bottom of the hole.

Step 14 Back the tap out, clear the chips, and wipe up any remaining lubricant.

Step 15 Clean the taps thoroughly to remove all chips and shavings.

Step 16 Return the tools to their protective case and store them with similar tools that need protection from large, heavy tools.

CAUTION

If the tap binds, carefully back it out to break the chip. Be careful when encountering major resistance; taps are somewhat brittle because of their hardening. Small taps especially can snap unexpectedly and are difficult to remove.

NOTE

Several lubricants, referred to as *tapping fluids*, are specifically designed for thread tapping. Cutting oils used for pipe threading can also be used, but they are messy and do not perform as well for this type of threading.

NOTE

For a through hole, continue tapping until the full, untapered threads of the tap emerge from the opposite side.

Thread Files

Thread files are very handy tools, especially if you often work with and cut threaded rod. Each side of the file is machined to fit a different thread pitch. Most files have a total of eight different faces. Thread files can quickly clean and straighten damaged threads. They are very helpful for removing burrs and realigning the threads of threaded rod at cut end, making it much easier to get a nut started.

Source: Tenbergen at English Wikepedia

1.2.3 Thread Dies

Thread dies are available in several styles and a wide range of both US standard and metric sizes. Like taps, they are typically tapered slightly on one side to help get threading started. Refer to *Figure 16*. The following are common types of dies:

- *Hex rethreading dies* — Hex rethreading dies are also known as *die nuts*. They are used for repairing damaged or rusty threads on bolts or similar hardware. They are hexagon-shaped and can be turned using any type of wrench that will fit. They are not designed to create new threads on round stock. They are used to clean up existing threads.
- *Round threading dies* — Threading dies create new threads but must be used with a diestock (*Figure 17*). They are round, with no flats for a common wrench or tap wrench to engage, preventing the use of the wrong tool. Some are adjustable, having a small slot and an adjusting screw, as shown in *Figure 16*. The die can be adjusted to cut deeper, creating a looser fit. When an HVACR technician needs to thread a piece of rod, no adjustment would normally be needed.
- *Pipe dies* — Like pipe taps, these cut tapered or straight pipe threads.

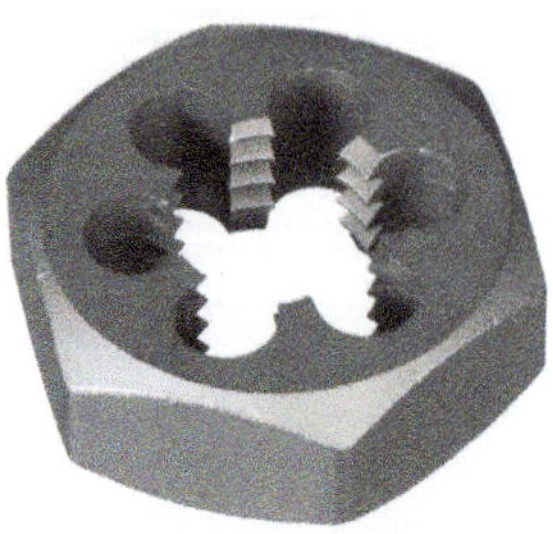

(A) Hex Rethreading Die

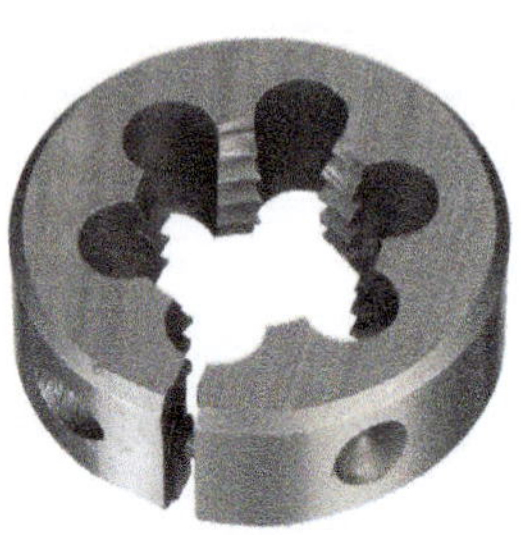

(B) Adjustable Round Thread Die

Figure 16 Common thread dies.
Source: Courtesy of Drill America

Figure 17 Threading die mounted in a diestock.
Source: Courtesy of Drill America

Taps and dies can be obtained in complete sets that include diestocks, tap wrenches, and user guides. *Figure 18* shows a common tap and die set.

Figure 18 Tap and die set.
Source: Courtesy of Drill America

1.2.4 Fastener Extraction

A bolt may become seized if it has been in place for a long time, or perhaps was forced into the bottom of a blind hole. When you try to remove it, the fastener may break. This leaves the bolt or screw locked in place.

To remove a seized fastener, a special tool called a *screw extractor* is used. Screw extractors are commonly referred to as *easy-outs* (*Figure 19*). A screw extractor is a tapered tool with a helical shape traveling in the opposite direction of the fastener threads.

Figure 19 Screw extractor set.
Source: Courtesy of Drill America

Screw extractors are used with a drill of the correct size to drill a hole in the center of the fastener. The taper is inserted into the fastener and turned in the direction that would be used to unscrew the fastener. The spiral grips the inside of the drilled hole, providing grip and encouraging the fastener to turn.

Before attempting to loosen the fastener, saturate the threads with penetrating oil and let the oil soak in. A light tap with a hammer may also help loosen a seized fastener. Applying heat with a torch and/or applying dry ice to the joint may even be necessary.

Fastener Extraction—Art or Science?

Extracting a seized or broken fastener can be a challenge. Special tools such as screw extractors are helpful, but there are many other ways that stubborn fasteners can be removed. Technicians must often get creative.

For example, if the hex head of a bolt is rounded off, a file can be used to flatten two of the sides enough so they can be gripped with a wrench. A hacksaw can be used to cut a slot in the top of a broken bolt so that a screwdriver can be used to remove the stud. Penetrating oil can help break the bond of a rusted connection. The use of heat or cold can cause metal to expand or contract, helping to break the bond holding the bolt. The vibrations caused by light hammer taps can also help break the bond. Seized and broken fasteners can be challenging, but several methods combined with patience usually get the job done.

1.3.0 Torquing Threaded Fasteners

Threaded fasteners must often be tightened by applying a specific amount of force. Otherwise, the amount of force applied is completely random. A physically large worker with a long wrench can apply far more force to a bolt than a small worker with a short wrench.

To ensure some hardware is tightened consistently, the force must be measured. The term used to describe the tightening of a fastener with a specific amount of force is called **torquing**. The specified amount of force to be applied is the *torque*.

Torquing: Applying a specific amount of twisting force to a threaded fastener.

The proper torque to be applied to a fastener should be specified in a manufacturer's assembly instructions or similar sources of information. If the amount of torque is not specified, you'll need to refer to a generic chart. Examples of torquing charts are provided later in this section. To use charts, you must know the characteristics of the hardware, such as its grade or material of construction. This section describes how to properly torque threaded fasteners and tube fittings.

CAUTION

Always refer to the equipment manufacturer's specifications for torque whenever possible. They offer the best and most accurate information. Refer to generic torque tables only when there is no other information available.

1.3.1 Torque Terminology

Technicians should recognize certain torque terms. Equipment, tool, and fastener manufacturers use the following terms regularly:

- **Break-away torque** — This is the torque required to loosen a fastener. Usually, it is higher than the recommended tightening torque value.
- **Set or seizure** — Sometimes during tightening, a fastener will lock up as it approaches the recommended torque. You might hear a popping sound or feel vibration from it when this happens. To free the fastener, loosen it slightly and then re-torque to the proper value. Torque tools will not work accurately on a seized fastener.
- **Run-down resistance** — If the fastener includes extra hardware, such as washers or locknuts, these make tightening the fastener harder. You may have to consider this additional resistance when torquing the fastener and increase the target torque value. Consult the manufacturer for guidance rather than guessing. Improper torquing can have serious consequences.

Break-away torque: The torque required to loosen a fastener.

Set or seizure: An apparent locking effect felt during the last stages of rotation as you approach the final torque value for a fastener, often accompanied by a noticeable popping effect.

Run-down resistance: The torque required to overcome the resistance of associated hardware, such as locknuts and lock washers, when tightening a fastener.

1.3.2 Torque Units

When you use a wrench to torque a bolt, it acts like a lever. You apply a force at the end, and the lever multiplies that force. The longer the lever, the greater the force.

Torque units combine the lever length with the amount of force applied. In the United States, the **foot-pound (ft-lb)** is a common torque unit. Tightening a bolt to 1 ft-lb of torque is like hanging a 1 lb weight on the end of a 1' long wrench. For smaller amounts of torque, the **inch-pound (in-lb)** unit is better. Tightening a bolt to 1 in-lb of torque is like hanging a 1 lb weight on a 1" long wrench (*Figure 20*). For this reason, 1 in-lb equals $^1/_{12}$ ft-lb.

Foot-pound (ft-lb): A torque unit equal to the force of one pound applied to a lever one foot long. One foot-pound equals 12 inch-pounds. Also called the *pound-foot (lb-ft)*.

Inch-pound (in-lb): A small torque unit equal to the force of one pound applied to a lever one inch long. One inch-pound equals $^1/_{12}$ foot-pound. Also called the *pound-inch (lb-in)*.

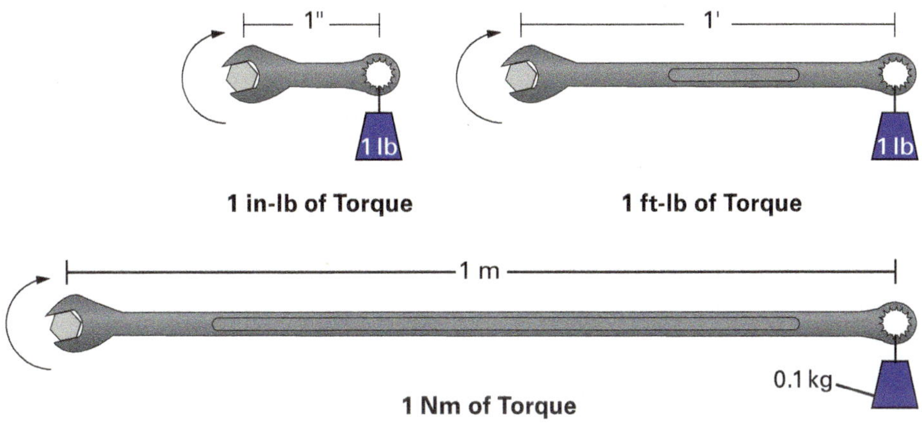

Figure 20 Common torque units.

Countries that follow the metric system use the **Newton-meter (Nm)** as their torque unit. Tightening a bolt to 1 Nm is like hanging a 0.1 kg weight on the end of a 1 m wrench.

Fastener and equipment manufacturers specify tightening torque in one of these three units of force. But how do you know when you have reached the correct torque? Special tools solve this problem.

Newton-meter (Nm): A metric torque unit equal to the force of one newton applied to a lever one meter long.

1.3.3 Torque Tools

Technicians use torque wrenches and torque screwdrivers to tighten bolts and screws to the proper values. These tools indicate when the applied torque reaches the target. They come in several styles and many sizes.

The beam-type torque wrench shown in *Figure 21* has a scale that indicates the amount of torque being applied as it happens. The beam stays in position

Figure 21 Beam-type torque wrench.
Source: Courtesy of Stanley Black & Decker, Inc.

Did You Know?

The Newton-Meter

The newton (N) is the metric unit of force. It is named after Sir Isaac Newton (1643–1727), an early physicist. A 1 kg weight produces 9.8 N of force.

Coincidentally, 1 N of force approximately equals the weight of an apple sitting on the table. It is believed that Newton formulated gravitational theory after watching an apple fall and asking himself why the apple only fell straight down, rather than some other direction.

CAUTION

Click-type torque tools will continue to tighten after the click, leading to over-torquing. Stop pulling as soon as the tool clicks.

as the wrench handle rotates. The scale on the handle moves under the beam, showing the current torque value. The operator simply tightens the fastener until the scale shows the desired torque.

Click-type torque wrenches (*Figure 22*) have an adjustment dial on the handle. After setting the desired torque value, the operator tightens the fastener until the tool noticeably clicks. This indicates the proper torque has been reached.

Electronic torque wrenches offer the most features (*Figure 23*). These include a display and controls for setting the target torque. During tightening, the tool beeps to indicate that it is approaching the correct torque. When it reaches the target, the tool vibrates, emits a steady sound, or flashes a signal. Some tools can store torque values for later transfer to a computer. Factories use these torque records to confirm proper equipment assembly. Remember that they require batteries to operate.

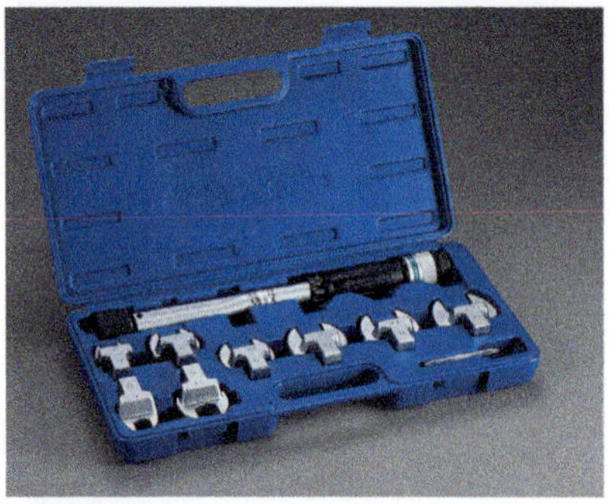

Figure 22 Click-type torque wrench.
Source: Ritchie Engineering Co. Inc./YELLOW JACKET

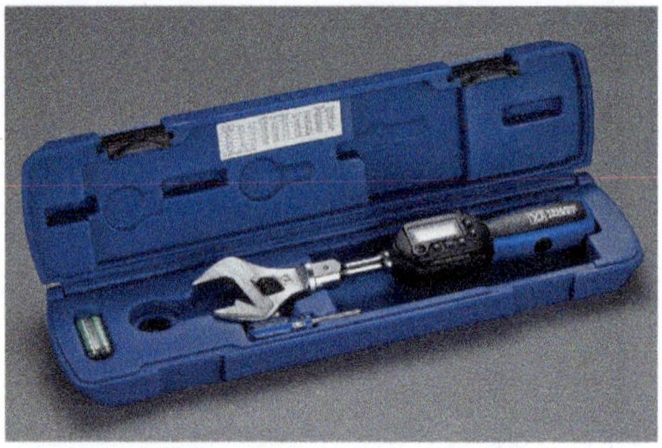

Figure 23 Electronic torque wrench.
Source: Ritchie Engineering Co. Inc./YELLOW JACKET

When torquing a fastener to the specified value, you must choose the correct tool size. Torque wrenches have an established operating range. They tend to be less accurate near the ends of the range. For this reason, choose a tool that places the target torque well inside the range—between 25% and 75% of its range is ideal.

Non-electronic torque tools are rated in inch-pounds, foot-pounds, or newton-meters. If the fastener's torque specification uses a different unit than the tool, convert the torque value to the system you need. *Table 4* shows conversion factors for all three torque units.

TABLE 4 Torque Conversions

To Convert from...	To...	Multiply by
in-lb	ft-lb	0.08333
ft-lb	in-lb	12.00
in-lb	Nm	0.1130
Nm	in-lb	8.851
ft-lb	Nm	1.356
Nm	ft-lb	0.7376

Electronic torque tools can switch torque units at the press of a button. This eliminates manual conversions.

1.3.4 Dry and Wet Torquing

If the equipment manufacturer does not provide fastener torque values, consult the fastener manufacturer. A fastener's recommended torque value depends upon the application. Another factor is whether the fastener is installed "dry" or "wet."

Dry torquing involves assembling the fastener without any kind of lubrication. Unlubricated threads create more friction, making the fastener harder to tighten. This extra effort consumes some of the tightening torque applied by the torque tool.

To compensate for the lost torque, manufacturers increase the recommended value for dry torquing. Torque wrenches measure the total torque applied to a fastener. They have no idea how much of that torque tightens (stretches) the fastener and how much is lost to friction.

Wet torquing involves coating the fastener threads with a lubricant like oil, grease, or graphite. Lubricated fasteners assemble more easily and waste less of the tightening torque. As a result, manufacturers usually specify a target wet torque value that is about 25% to 30% less than the dry torque value. For example, a typical ½"-13 bolt might have a dry torque value of 35 ft-lb and a wet torque value of 23 ft-lb.

Lubricating fasteners also makes them easier to remove in the future. Anti-seize products like Never-Seez® reduce the break-away torque. Some prevent galling, which is the tendency of certain metals to stick to each other after tightening. Many contain both lubricants and powdered metals such as copper.

Always check the fastener manufacturer's guidelines very carefully. Confirm whether they refer to dry or wet torquing, because torque values differ for each. For wet torquing, be sure to use the recommended lubricant. Following the wrong specification or a using different lubricant could result in applying the wrong amount of torque.

1.3.5 Torquing a Fastener or Fitting

Table 5 shows torque specifications for some common ungraded fastener types. *Table 6* provides common torque values for SAE-graded bolts. Always consult the equipment manufacturer's documentation for the torque value before consulting a generic chart. If a chart is used, make sure you have properly identified the grade of the bolt and/or its material of construction.

TABLE 5 Common Torque Values for Ungraded Bolts

Diameter and Threads Per Inch	Torque (Foot-Pounds)		
	Mild Steel	Stainless Steel	Alloy Steel
¼"-20	4	6	8
5⁄16"-18	8	11	16
⅜"-16	12	18	24
7⁄16"-14	20	32	40
½"-13	30	43	60
⅝"-11	60	92	120
¾"-10	100	128	200
⅞"-9	160	180	320
1"-8	245	285	490

TABLE 6 Suggested Torque Values for Graded Bolts

Grade	SAE 1 or 2	SAE 5	SAE 6	SAE 8
Diameter and Threads Per inch	Torque (Foot-Pounds)			
¼"-20	5	7	10	10
5⁄16"-18	9	14	19	22
⅜"-16	15	25	34	37
7⁄16"-14	24	40	55	60
½"-13	37	60	85	92
9⁄16"-12	53	88	120	132
⅝"-11	74	120	169	180
¾"-10	120	200	280	296
⅞"-9	190	302	440	473
1"-8	282	466	660	714

The following steps outline the general procedure for using a torque wrench:

Step 1 Determine the fastener's target torque value. If necessary, convert it to a different torque unit of measure.

Step 2 Select a torque wrench appropriate to the fastener type. Confirm that the fastener's target torque is well inside the tool's operating range.

Step 3 Set the torque on the wrench using the adjustment dial or controls. For a beam-type wrench, you will have to monitor the scale to know when the correct torque is reached.

Step 4 Install the fastener and tighten it by hand. A wrench may be needed if there is excessive resistance.

Step 5 Place the torque wrench on the fastener or fitting, holding it in place in a safe way. Use a backup wrench if needed.

Step 6 Tighten the fastener (*Figure 24*). For click-type or electronic tools, listen for the target torque signal. For a beam-type wrench, watch the indicator for the correct torque. As soon as you reach the target torque, immediately stop applying force.

Step 7 Remove the tool. Set adjustable tools back to zero after each use. Clean and store the tool properly.

CAUTION

When torquing a series of bolts in an assembly, apply only one-third of the torque on each pass. See the next section for more information on tightening sequences.

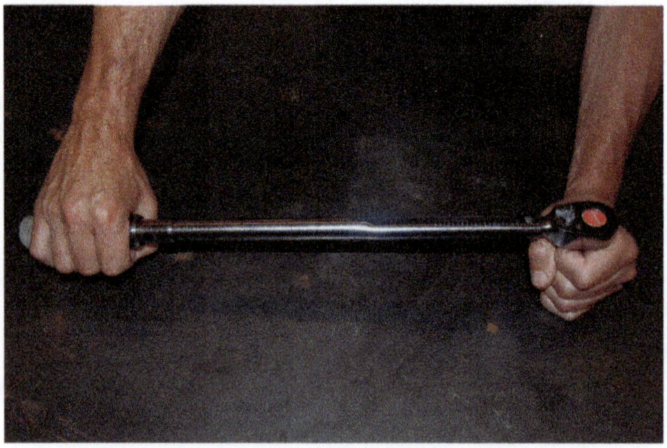

Direction of Rotation Depends on Direction of Threads on Bolt

Figure 24 Using a torque wrench.

WARNING!

Whenever possible, pull a wrench rather than pushing it. This technique reduces the likelihood of finger and knuckle injury if the wrench slips. Wear gloves and always grip the wrench by the handle. Never use extensions or cheater bars on torque wrenches.

Reusing Nuts and Bolts

When new nuts and bolts are torqued, a certain amount of stretching and/or deforming takes place. If the nut/bolt combination is taken apart and reused, additional stretching and deformation takes place, further weakening the hardware. In low-stress applications this may not be a problem. However, in high-stress environments such as the head bolts on a semi-hermetic compressor, it is wise to replace the bolts with new ones. Compressor manufacturers sell replacement cylinder head bolts in complete kits.

1.3.6 Tightening Sequences

When tightening a series of bolts on components like flanges and compressor heads, the bolts must be tightened to the proper torque and in the proper sequence.

Tighten each fastener to about one-third of the target torque in each round. This means that three passes through the sequence are required to finish the job. The numbers in *Figure 25* show the proper tightening sequences for common bolt patterns.

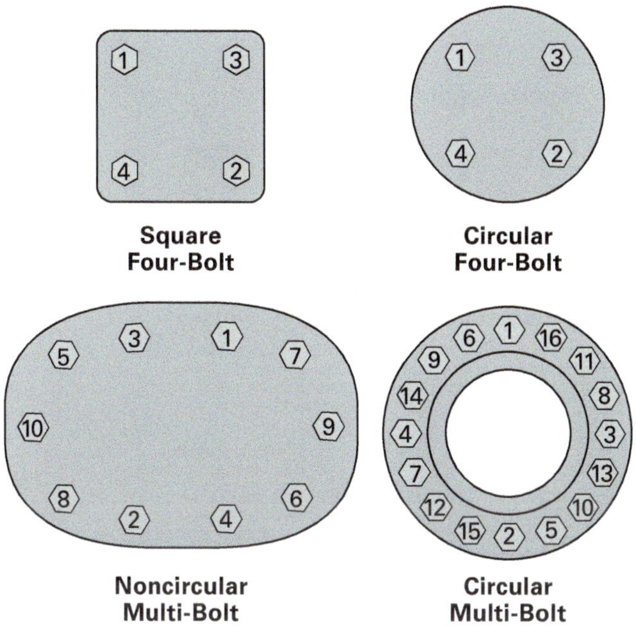

Figure 25 Common tightening sequences.

Follow the proper tightening sequence and tighten the bolts to the final torque on the third pass through the pattern. This prevents warping or damaging the equipment and helps prevent leaks.

1.4.0 Toggle Bolts and Anchor Bolts

Toggle bolts and anchor bolts are specialized fasteners used to secure objects. Toggle bolts are used to fasten a part to a hollow wall or panel. Anchor bolts are used to fasten parts and equipment to concrete and masonry.

1.4.1 Toggle Bolts

Generally, toggle bolts (*Figure 26*) should only be used to support light loads. In many cases, the surface will be sheet rock or a similar wall or siding material. Regardless of the strength of the bolt, the wall material may not be sound enough to reliably support a great deal of weight.

To install a toggle bolt, follow these basic steps:

Step 1 Drill the correct size hole in the wall or panel. The hole must accommodate the wings of the toggle.

Step 2 Remove the screw from the toggle bolt, insert it in the object to be fastened, and then re-insert the screw into the toggle assembly.

Step 3 Compress the wings of the toggle toward the head of the bolt and insert it into the hole. Once this is done, the spring-loaded wings will deploy and engage the back side of the wall.

Step 4 Use a screwdriver to draw the wings against the backside of the wall and secure the object (*Figure 27*). You will likely need to pull the toggle bolt out toward you, allowing the toggles to hit the wall and stop. This allows you to tighten the bolt until the head meets the object being mounted.

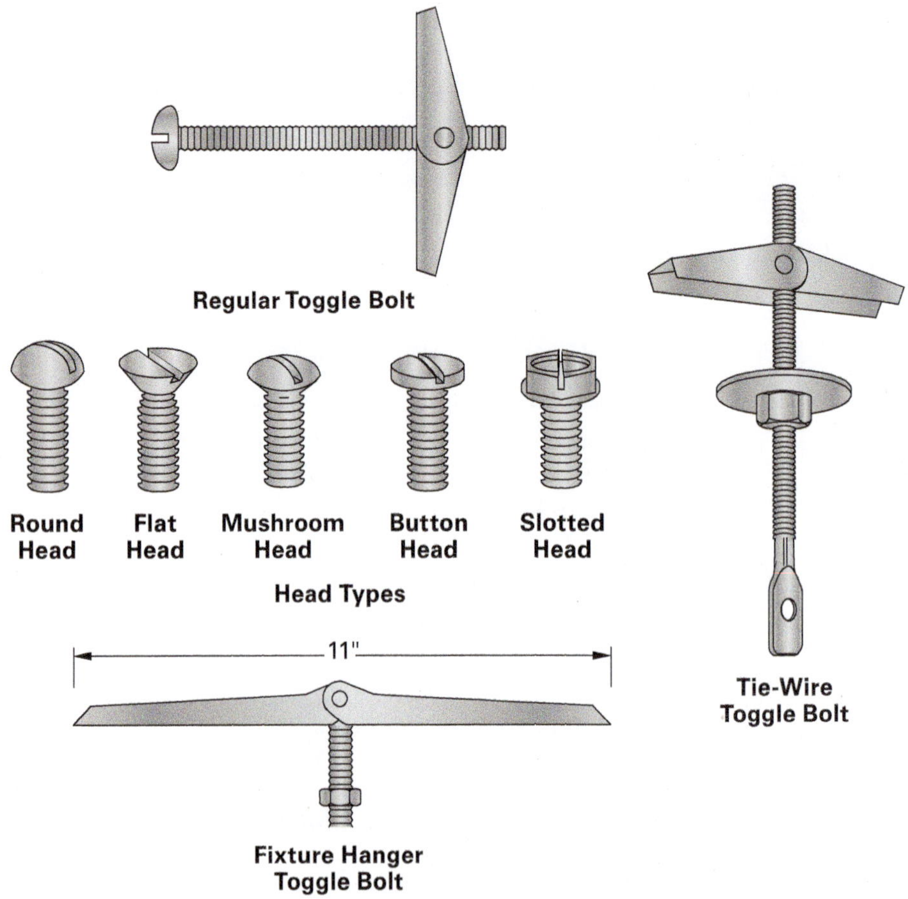

Figure 26 Toggle bolts.

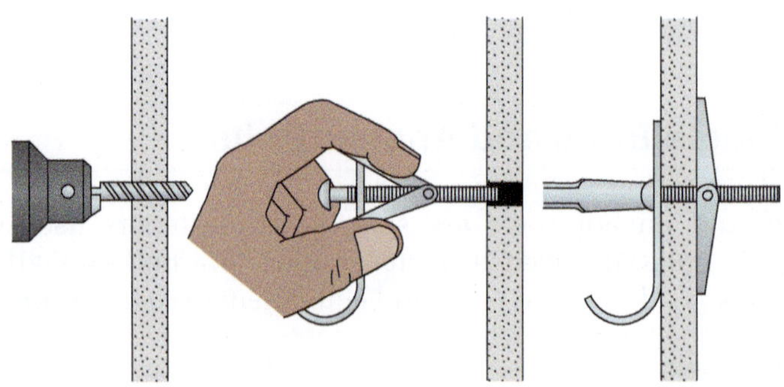

Figure 27 Toggle bolt installation.

1.4.2 Anchor Bolts

Types of concrete and masonry anchors include the following:

- Concrete inserts
- Expanding concrete anchors
- Non-expanding concrete anchors
- Self-drilling anchors
- Tapcon® screws

Concrete inserts (*Figure 28*) are embedded in wet concrete and the hanger extension rod is screwed into the anchor after the concrete has set. The nut may be permanently secured to the insert, but some inserts allow for some adjustment in the position of the nut (*Figure 29*).

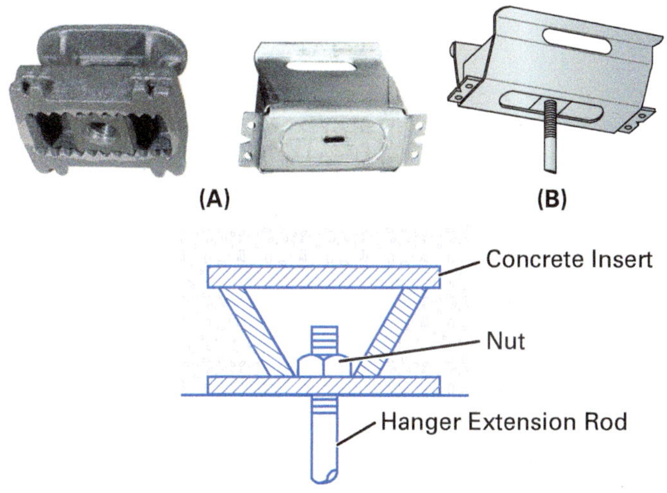

Figure 28 Concrete inserts.
Source: (A) Anvil International & Smith-Cooper International

Figure 29 Concrete insert with a movable nut.
Source: Anvil International & Smith-Cooper International

When installing anchors in hardened concrete, make sure the area where the equipment is to be fastened is smooth so that the equipment will have solid footing. Uneven footing might cause the equipment to twist or vibrate during operation.

Carefully inspect the drill and masonry bit to make sure they are in good condition. If they are not, you could be injured by parts that chip and fly off during operation.

WARNING!

Drilling in concrete generates noise, silica dust, and possible flying objects. Always wear safety glasses, ear protection, and work gloves. Per OSHA regulations, respiratory protection may be required. Ideally, water is used both as a lubricant and to minimize airborne silica dust. Ensure that others in the area also wear the appropriate PPE.

Consult OSHA Standard *29 CFR 1926.1153, Respirable Crystalline Silica*, more commonly known as the Silica Standard, for additional information. Silica dust inhalation is a significant jobsite hazard.

The anchor bolt must extend out of the hole far enough for the threads to pass through the part to be anchored. Before selecting the bolt, consider how long it needs to be. For example, if a vibration-isolating pad is placed on the floor under the equipment foot, this will increase the required bolt length.

Expanding Anchors

Expanding concrete anchors (*Figure 30*) are placed in snug-fitting holes drilled in hardened concrete. As the nut is tightened, it causes the wedge at the base of the anchor to be drawn up into the sleeve (*Figure 31*). The sleeve expands in response, wedging it tightly against the walls of the hole.

When drilling the hole for an expanding concrete anchor, the diameter of the hole should be only slightly larger than the anchor diameter. The anchor manufacturer will specify the correct diameter needed for each size anchor. This is the size that should be used; deviations in either direction will cause problems. If the hole is too large or out-of-round, for example, the anchor may not be able to wedge itself tightly in the hole. The correct drill bit is often provided in a box of anchors.

NOTE

Always consult the anchor manufacturer's installation instructions, which are often found on the box, before beginning. Important measurements and guidance are readily available.

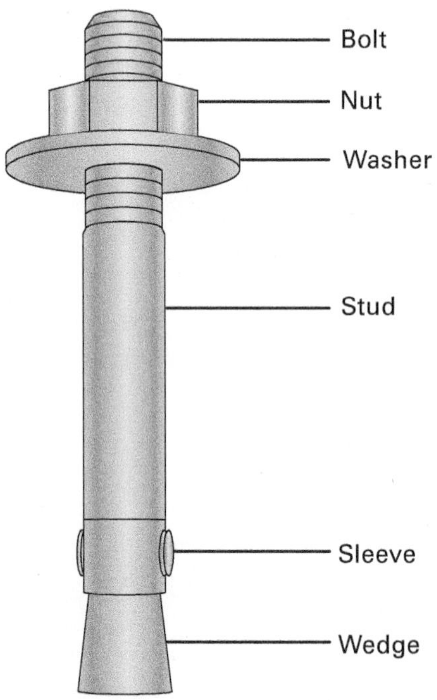

Figure 30 Expanding concrete anchor.

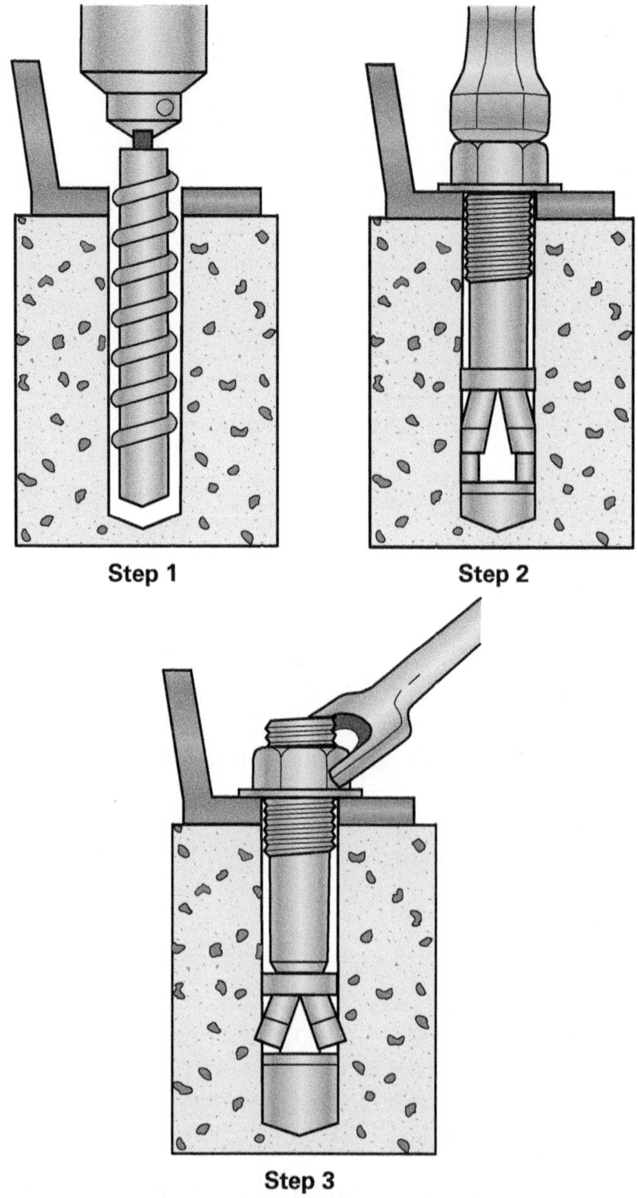

Figure 31 Installing an expanding concrete anchor.

The hole must be drilled to the correct depth. Once you know the appropriate depth, mark the depth on the drill bit shank, then wrap a piece of tape around it at the mark. You'll then be able to tell when you have reached the proper depth.

When drilling, let the drill do the work; avoid pushing down too hard. For the largest holes, start with a small bit and then change to the larger one. Concrete dust may accumulate in the bottom of the hole, preventing complete insertion of the anchor. Vacuum out any accumulated dust using a HEPA vacuum approved for use with silica dust. Then follow the manufacturer's instructions for installing the fastener.

Non-Expanding Anchors

Non-expanding anchors are secured in concrete using an epoxy adhesive or grout (*Figure 32*). When installing this type of anchor, the drill bit must be slightly larger in diameter than the head of the fastener and washer used. The flat washer must be able to fit down inside the hole, just below the surface level.

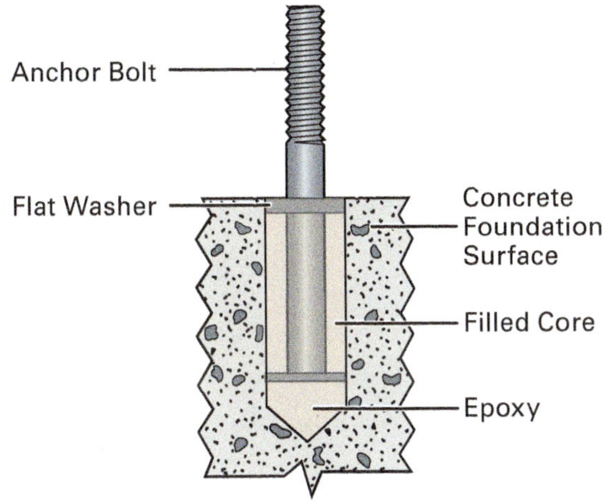

Figure 32 Fastener anchored in epoxy.

Figure 33 Two-part anchor epoxy.
Source: Photo courtesy of Simpson Strong-Tie Company Inc.

The epoxy adhesive is available in small, pre-packaged glass capsules of epoxy and hardener. The glass capsules are dropped directly into the hole, unbroken. The bolt is inserted to break the capsules. Moving the fastener around mixes the two components. The bolt is then left in place until the epoxy hardens. Use care when working with the capsules, as loose glass fragments can cause cuts. Never place your fingers in the hole after the capsule is inserted.

When installing a number of anchors, special packages that dispense the epoxy and hardener in the correct proportion may be a better and less expensive choice (*Figure 33*).

Once the bolt is installed and the hole is filled with grout or epoxy, make sure the bolt is left straight after working the anchor around in the hole to distribute the anchoring material. The washer helps center the bolt. If the grout hardens and the bolt is not straight, it will have to be excavated and redone. Allow the grout or epoxy to fully cure before mounting anything to the anchor bolt.

Self-Drilling Anchors

Self-drilling anchors require a hammer drill to install (*Figure 34*). The fastener itself acts as the drill bit. After drilling the hole, remove the anchor and clean the hole. Insert the expander plug into the sleeve and drive it into the hole with the drill. This action expands the anchor, holding it in place. The top part then snaps off. Finally, a bolt threads into the anchor.

Tapcon® Screws

Tapcon® screws (*Figure 35*) can be used to fasten various components to concrete and masonry in light- and medium-duty applications. Introduced in 1976, these screws are driven directly into concrete after a pilot hole is drilled. Various diameters and lengths are available.

Tapcon® screws are driven into masonry or concrete in much the same way wood screws are driven into wood (*Figure 36*). However, drilling a pilot hole is essential. Failure to use the correct pilot drill results in one of the following two problems:

- If the pilot hole is oversized, the threads of the screw cannot properly engage the masonry or concrete. Under stress, the screw will pull out.
- If the pilot hole is undersized, it is difficult to drive the screw. Over-torquing the screw when it won't turn results in the head breaking off.

The manufacturer will specify the correct size masonry drill bit to use. The correct drill is often packaged with the screws.

Hammer Drills

Hammer drills are commonly used to drill holes in masonry and concrete. The combination of the twisting of the drill bit accompanied by hammer-like blows makes drilling holes in masonry and concrete much easier than is possible with an ordinary drill.

Source: Courtesy of Stanley Black & Decker, Inc.

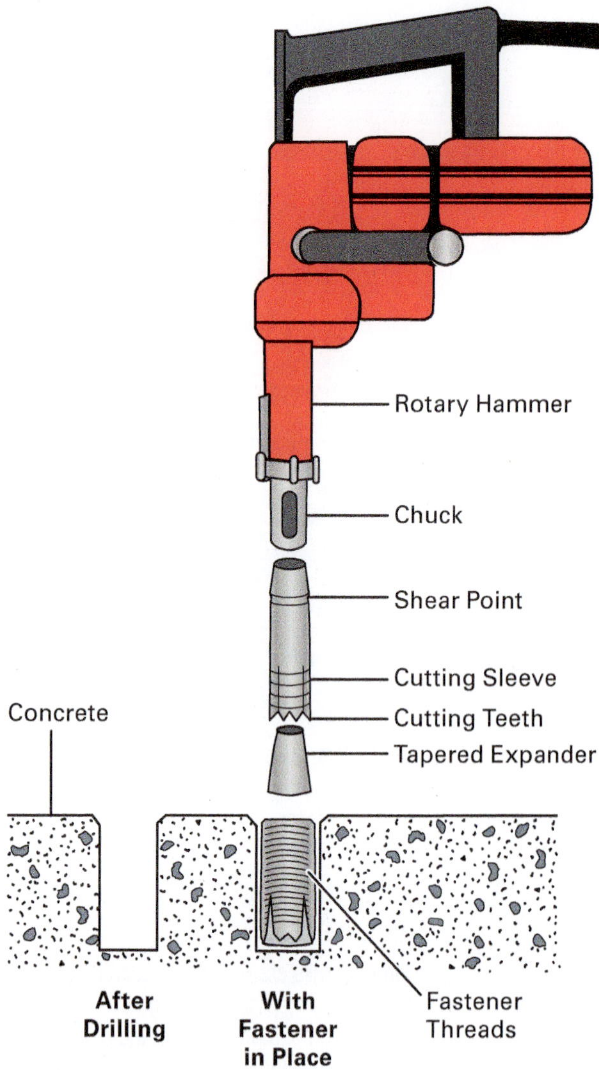

Figure 34 Self-drilling anchor.

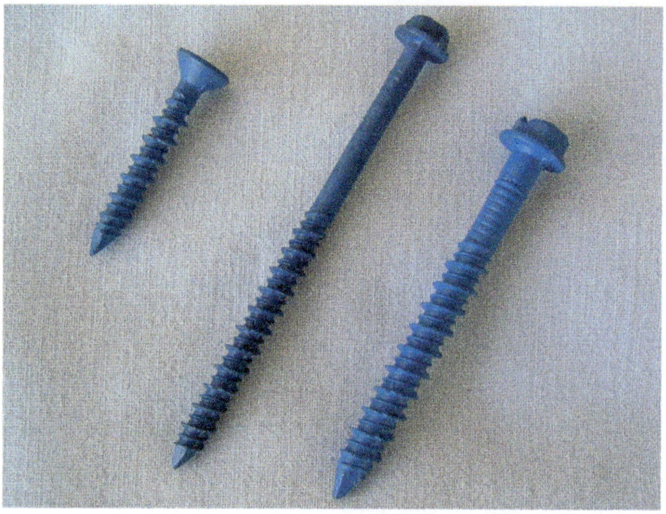

Figure 35 Tapcon® screws.

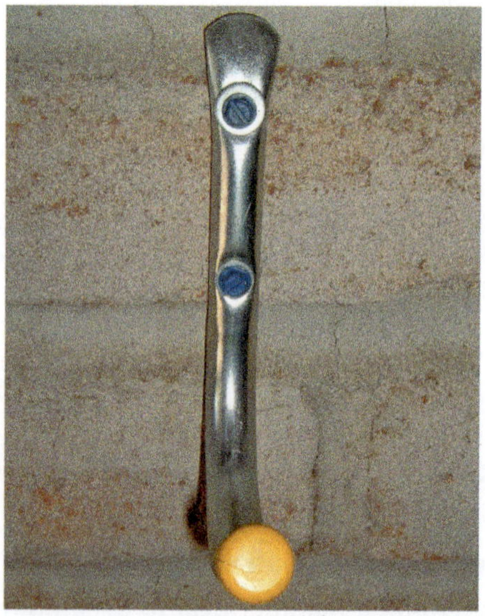

Figure 36 Hanger installed with Tapcon® screws.

1.5.0 Non-Threaded Fasteners

Non-threaded fasteners have many uses. This section introduces the following types of non-threaded fasteners:

- Retainer rings
- Pins
- Keys
- Rivets

1.5.1 Retainer Rings and Lock Rings

Retainer rings and lock rings (*Figure 37*) are used for both internal and external fastening to a shaft or pin. Some retainer rings are seated in grooves. Others are self-locking and do not require a groove. Special pliers are used to remove internal and external rings. Self-locking rings snap in place.

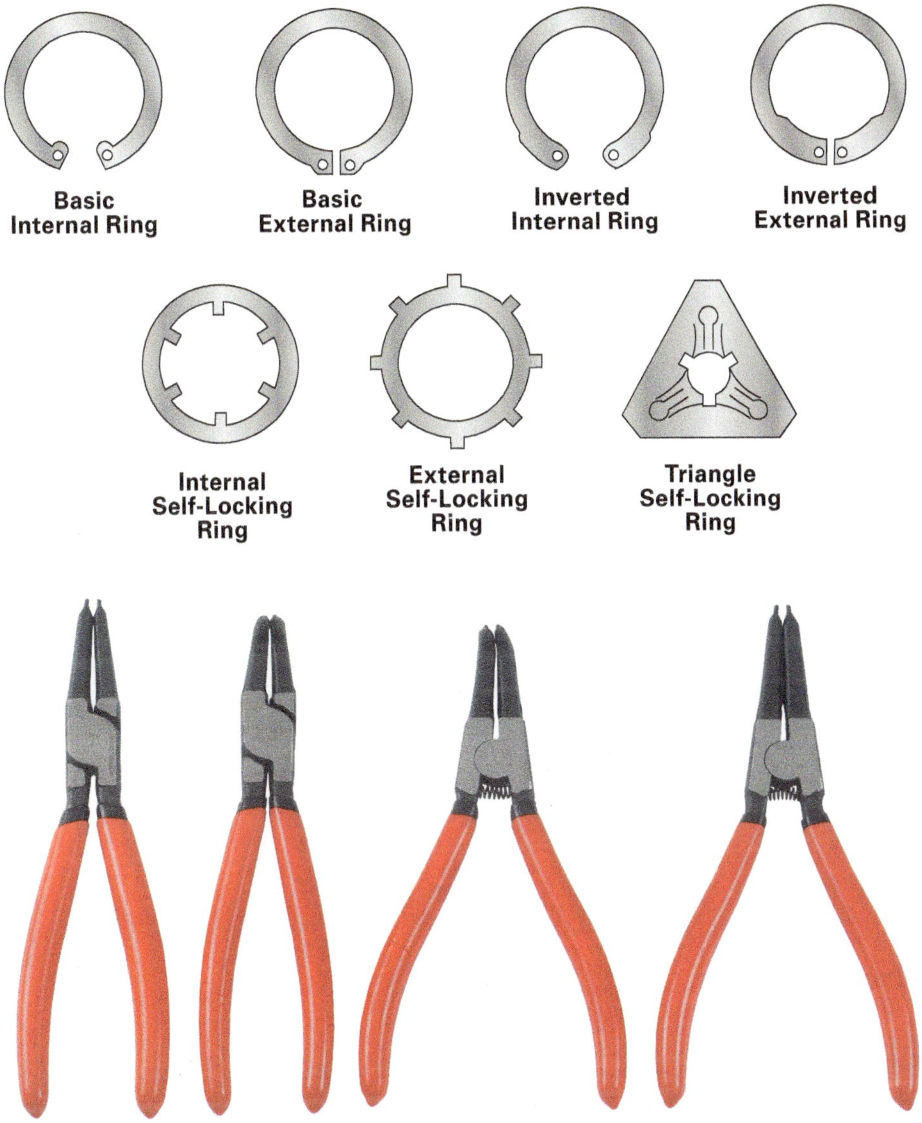

Figure 37 Retainer rings, lock rings, and lock ring pliers.
Source: Courtesy of Stanley Black & Decker, Inc.

1.5.2 Pins

Pin fasteners (*Figure 38*) are used to align mating parts, hold gears and pulleys on shafts, and secure slotted nuts.

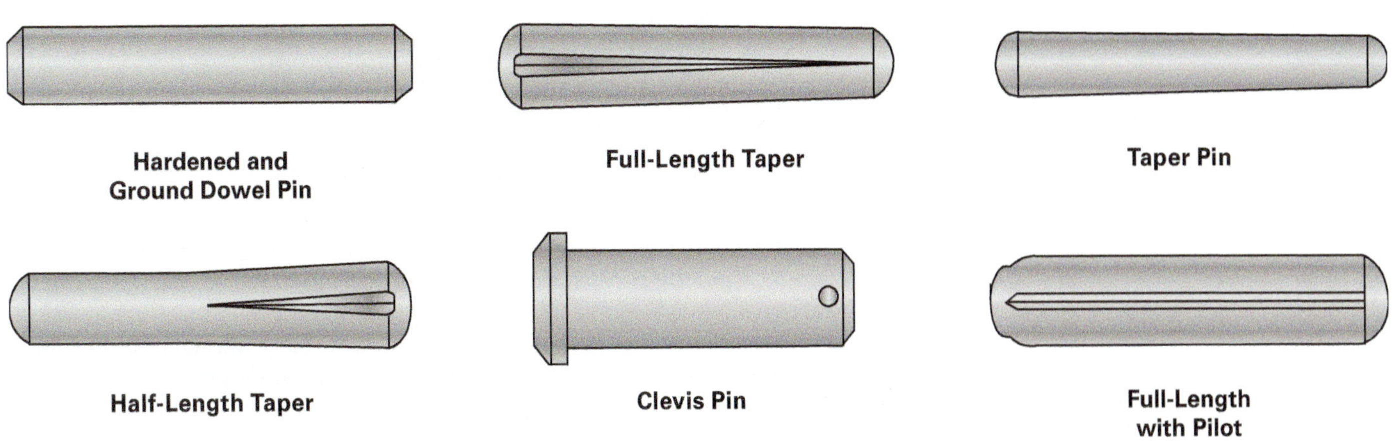

Figure 38 Pin fasteners.

Some common pins and their uses are as follows:

- *Dowel pins* — These are fitted into reamed holes to position mating parts. They also support a portion of the load placed on the parts.
- *Taper pins* — These are used to fasten gears, pulleys, and collars to a shaft.
- *Clevis pins* — These are used to secure components between the head and end of the pin. They are used where perfect alignment is not necessary, as they fit loosely and are designed for easy installation and removal. For example, they are used to secure lifting hardware such as shackles and connect towbars to a truck or tractor chassis.
- *Cotter pins (not shown)* — These are fitted into a hole drilled through a shaft or bolt shank. Cotter pins are used to prevent parts from slipping off the shaft. They are also installed to secure slotted nuts from working loose and to secure clevis pins in an assembly.

1.5.3 Keys

Keys (*Figure 39*) are metal parts used to prevent a gear or pulley from rotating on a shaft. One half of the key fits into a seat or groove on a shaft. The other half fits into a keyway in the hub of a gear or sheave.

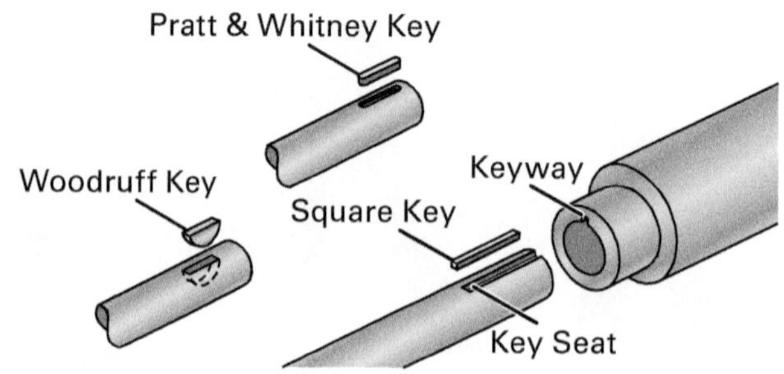

Figure 39 Keys.

1.5.4 Rivets

Rivets are used to semi-permanently join two pieces of material. One common type of rivet is the solid rivet (*Figure 40*).

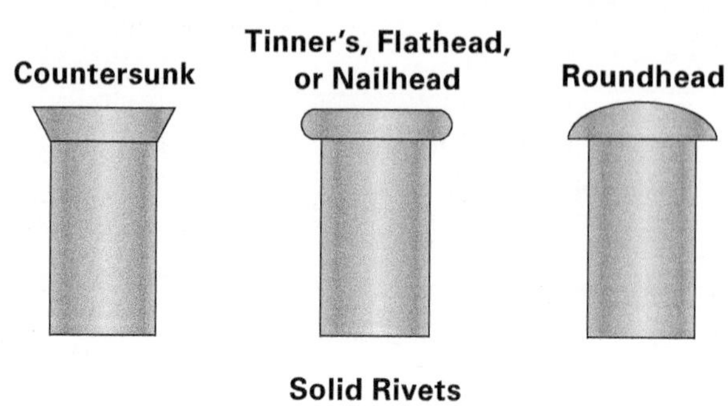

Figure 40 Solid rivets.

To install a solid rivet, the rivet is inserted into a drilled or punched hole slightly larger than the rivet shank. The small end of the rivet is then hammered into the form of a head.

Solid rivets are selected by material of construction, shank diameter, head type, and length. They are often sold in ounces or pounds per 1,000 rivets. For example, a 12-ounce rivet means that 1,000 rivets that size weigh 12 ounces. Aluminum rivets, of course, weigh far less than steel rivets.

Another popular rivet, often used in HVACR sheet metal work, is the pop rivet, also known as a *blind rivet*. A pop rivet tool (*Figure 41*) is used to install the

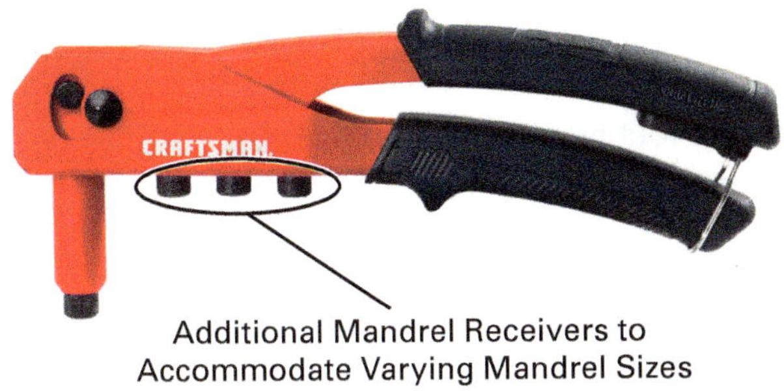

Figure 41 Pop rivet tool.
Source: Courtesy of Stanley Black & Decker, Inc.

rivets quickly; no hammering required. Pop rivets are used to fasten sheet metal, fiberglass, and plastics. They are available in various lengths and diameters.

Unlike solid rivets, pop rivets can be used when the joint can only be reached from one side. To install a pop rivet, insert the rivet mandrel (*Figure 42*) all the way into the tool's receiver. Then push the rivet head into the pre-drilled hole and pull the handles together. The opposite end will expand, and then the mandrel will break (pop) off when the stress overcomes it.

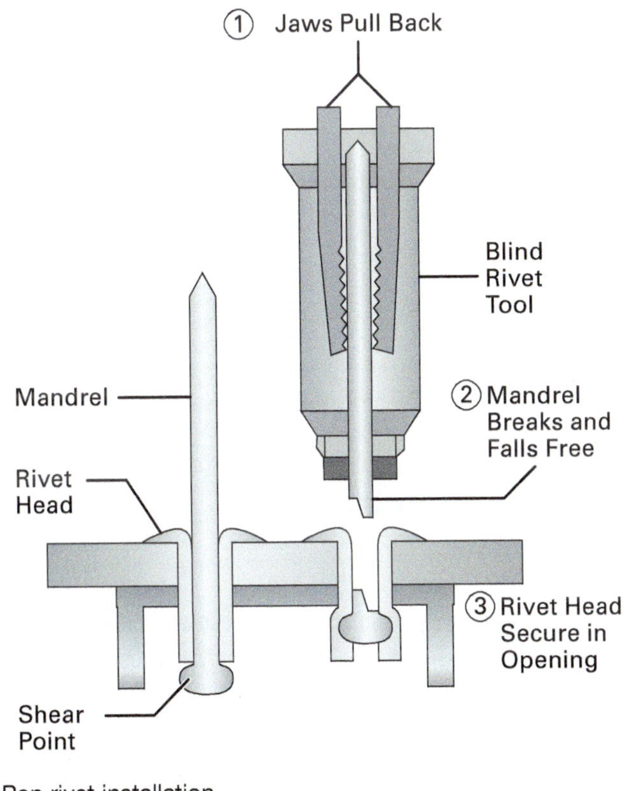

Figure 42 Pop rivet installation.

Pop rivets are made with several different mandrel sizes. To accommodate that, most pop rivet tools are provided with several interchangeable mandrel receivers, as shown in *Figure 41*.

Removing Rivets

Chisel or grind off the heads of solid rivets or steel blind rivets. Remove the shaft by striking through with a punch. Drill through pop or countersunk rivet heads using a high-speed electric drill, then punch out the shaft.

1.0.0 Section Review

1. Bolts used in all routine applications have ______.
   a. coarse threads
   b. fine threads
   c. extra fine threads
   d. metric fine threads

2. How many taps do you need to properly tap a blind hole?
   a. One
   b. Two
   c. Three
   d. Four

3. When using a properly selected torque wrench, the amount of torque required for your application will fall in a range of ______.
   a. 40% to 60% of full scale
   b. 75% to 100% of full scale
   c. 25% to 75% of full scale
   d. 50% to 80% of full scale

4. Non-expanding concrete anchors are held in place with ______.
   a. folding wings
   b. wedges
   c. retaining rings
   d. grout or epoxy

5. To keep it from loosening, a slotted nut is designed to accept a ______.
   a. snap ring
   b. cotter pin
   c. spring pin
   d. dowel pin

2.0.0 Vibration Isolators

Performance Tasks

There are no Performance Tasks in this section.

Objective

Identify and describe the installation of various types of vibration isolators.

a. Identify and describe vibration isolators used to support and suspend equipment and piping.
b. Identify and describe methods used to restrain equipment during seismic events.

HVACR equipment contains components such as fans, blowers, and compressors, all of which are driven by electric motors. The electric motors and their loads generate noises and vibrations that can be objectionable. If equipment that generates vibrations is in direct contact with other objects, those vibrations are transferred to them, and on to pipes and ducts connected to the equipment.

For example, an attic-mounted air handler that is installed directly on the attic floor can transmit vibrations to the structural members of the floor, allowing the vibrations to be heard throughout the structure. To reduce the transmission of unwanted noise and vibration, various kinds of **vibration isolators** can be used to lessen or eliminate the vibration transmission.

Vibration isolators: Devices that help prevent or reduce the transmission of vibrations from machinery to the surrounding structure.

2.1.0 Types of Vibration Isolators

Vibration isolators can be grouped based on the type of vibration they are designed to reduce or eliminate. They are broken into the following four distinct groups, based on the application:

- Vibration caused by compressors, fans, and blower motors
- Vibration transmission between objects
- Vibration transmission through ducts
- Vibration transmission through pipes

2.1.1 Compressor, Fan, and Blower Motor Vibration Isolators

Welded hermetic compressors installed in packaged units and split systems are mounted on grommet-like vibration isolators made of rubber (*Figure 43*). A bolt with an integral washer passes through the center of the grommet and holds the compressor to the unit's base pan. The rubber prevents metal-to-metal contact between the base pan and compressor, keeping any vibrations the compressor generates from being transferred to the rest of the unit.

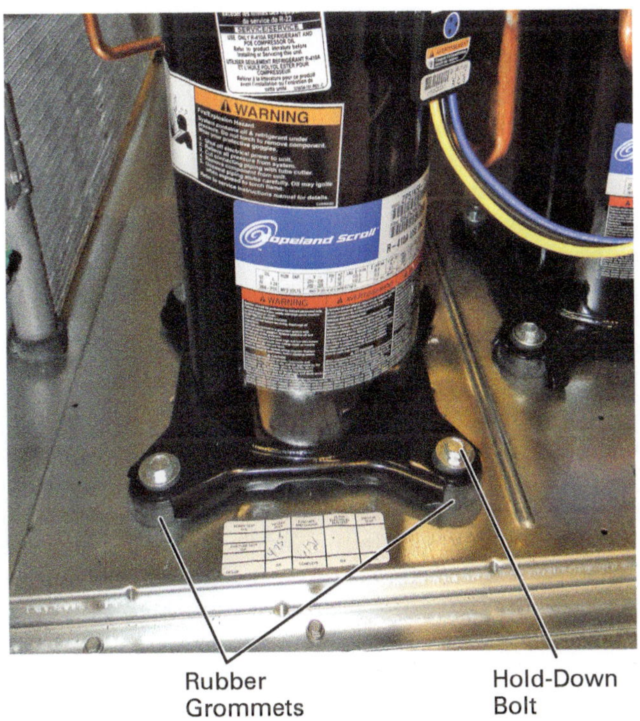

Figure 43 Hermetic compressor vibration isolators.

It is very important to note that isolators used in this way are of little or no value if the hold-down bolt is over-tightened. Compressor hold-down bolts are tightened at the factory for shipment of the unit only. However, during the initial startup, the bolts should be loosened so that they are not applying pressure to the feet. The bolts are there to simply keep the compressor from moving side to side. Compressing the rubber grommet by tightening the bolt eliminates the isolation characteristics.

A similar arrangement is used to isolate a direct-drive blower motor from the blower housing (*Figure 44*). Three rubber grommets are attached to the three motor mounts that are located 120 degrees apart around the motor. These bolts must be tight to hold the motor in place, but not tight enough to damage the isolator.

Heavier semi-hermetic compressors are mounted on spring isolators (*Figure 45*). These isolators must also be decompressed before start-up by backing the mount bolts out while leaving them in place for future use. Even though the compressor

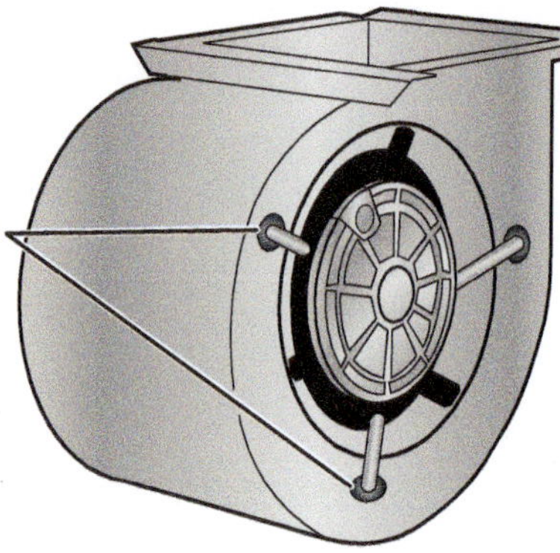

Figure 44 Direct-drive blower vibration isolators.

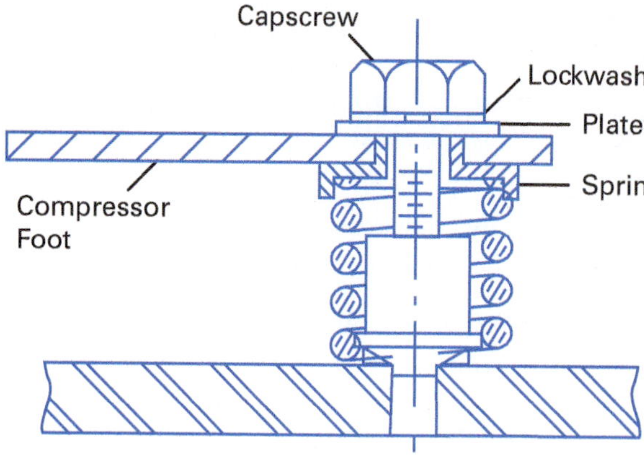

Figure 45 Semi-hermetic compressor vibration isolator.

rides on spring isolators, equipment manufacturers sometimes recommend that the entire unit be mounted on isolators to further reduce the transmission of vibrations to the structure. Rooftop-mounted chillers are a good example.

2.1.2 Vibration Isolators that Reduce Vibration Between Objects

Figure 46 Vibration-isolating pad.

Equipment vibration can be transmitted between the equipment and the object to which it is mounted. For example, a roof-mounted packaged unit can transmit vibrations to the structural members of the roof. Relatively lightweight residential air handlers installed directly on ceiling joists in an attic can be isolated by placing simple vibration-isolating pads (*Figure 46*) under key locations of the unit. This product is constructed of two layers of grooved rubber sheet, separated by a cork center. The grooves on each side travel in opposing directions.

When working with all vibration isolators, it is essential to ensure that the isolator is applied within its designed load range. The pad shown in *Figure 46*, for example, is rated at 50 lb per square inch (in^2). That means that only 4 in^2 of pad area is needed to properly support and isolate a 200-lb load.

If too much weight is applied to it, the pad will crush the cork center and reduce its effectiveness. If it is significantly underloaded, it will simply fail to provide isolation. This concept is essential to the proper application of all vibration isolation products. To perform the desired function, the isolators must be properly loaded.

Residential air handlers installed in attics can also be suspended from rafters using multiple threaded rods that are coupled through vibration-isolating hangers (*Figure 47*). Each threaded rod contains its own vibration isolator (*Figure 48*). This prevents vibration from being transmitted through the threaded rods to the roof structure.

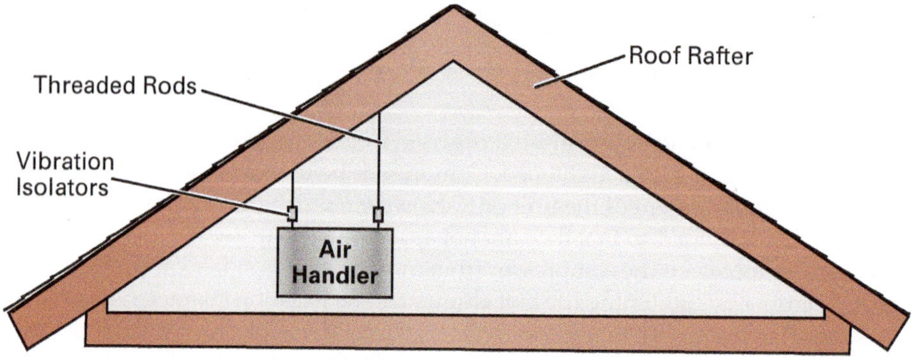

Figure 47 Air handler suspended with vibration isolators.

In *Figure 48*, note the spring deflection scale. The scale provides a quick visual means of determining how much the spring is deflected under the load. Loading a spring isolator to somewhere near mid-range is ideal. For example, if a 50-lb load on each corner of a suspended load is expected, vibration isolators with a maximum loading of roughly 100 lb would likely be chosen.

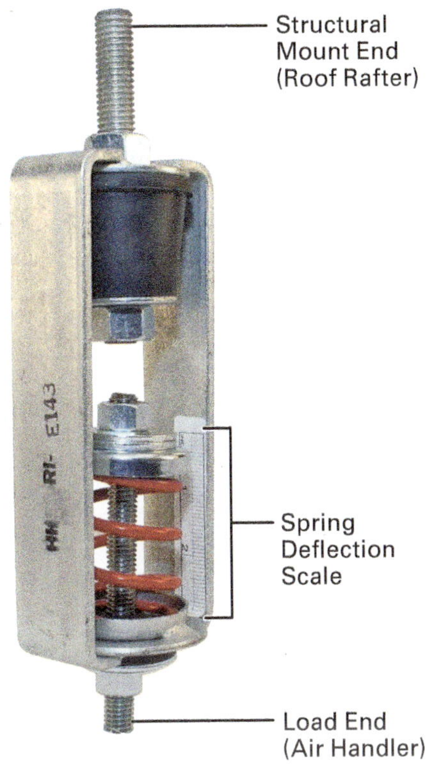

Figure 48 Vibration-isolating hanger.
Source: California Dynamics Corporation

Heavy HVACR equipment, like chillers, are often mounted on heavy-duty spring-mounted vibration isolators (*Figure 49*). Although it may appear in the photo that the upper pad of the isolator is touching the top of the base, it is riding on the spring. Because the corner and side-to-side loads of a large piece of equipment are different, it is not unusual to see different ratings on the vibration isolators supporting a unit. They are matched to the load at their location, based on load data provided in the installation instructions.

Figure 49 Heavy-duty vibration isolator.

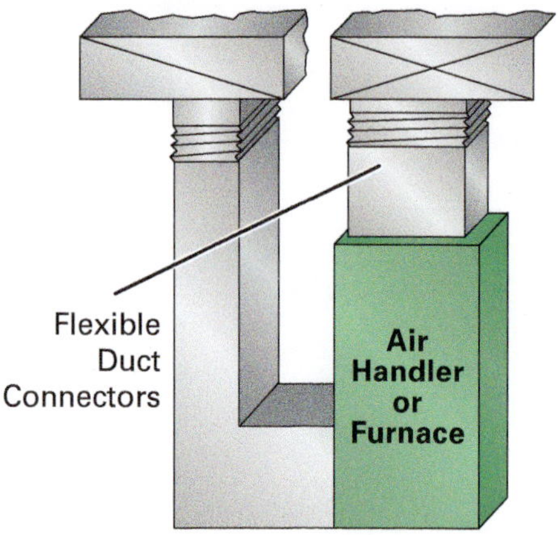

(A) Vibration/Noise Control at Air Handler

(B) Flexible Duct Connector

Figure 50 Ductwork vibration control.

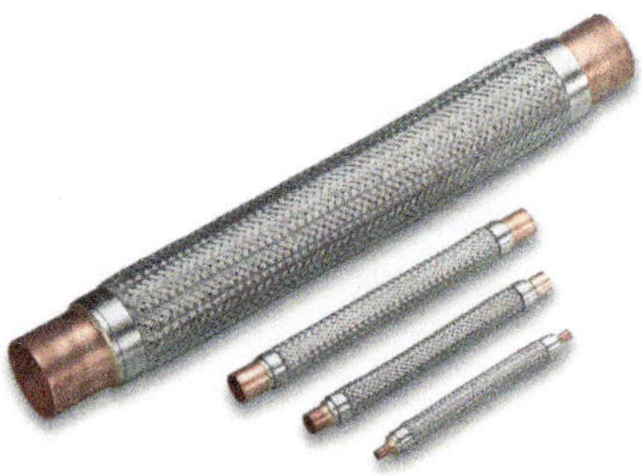

Figure 51 Flexible piping vibration eliminators.

Source: Courtesy of Sporlan Division—Parker Hannifin Corporation

2.1.3 Isolators for Ductwork

In metal duct systems, any noise or vibration generated by the equipment can be transmitted throughout the structure. Flexible duct and duct made with rigid fiberglass board do not easily transmit vibrations and do not require isolation.

The simplest way to prevent vibrations from the air handler traveling into a metal duct system is to install flexible (nonmetallic) connectors in the supply and return air ducts as they enter/leave the air handler (*Figure 50*). Flexible connectors can be installed in vertical and horizontal sections of either duct. In commercial applications, they are often found adjacent to air handling equipment of all sizes. Lining sheet metal ducts with fiberglass duct liner can also reduce the transmission of noise and vibrations.

2.1.4 Isolators for Piping

Pipes, like metal duct systems, can transmit any noise or vibration generated by the equipment throughout the structure. Vibrations can be reduced at the source (a pump or compressor) and in the piping system itself.

Compressors, especially larger semi-hermetic compressors, can generate noise and vibration at start-up and during operation. The starting torque of the motor can cause the compressor to move as it starts. If rigid pipes are connected to the compressor, they can be stressed by the starting torque and transmit noise and vibrations through the connected piping. To reduce this problem, a flexible metal vibration eliminator (*Figure 51*) can be installed in the refrigerant suction and/or discharge lines of the compressor.

This type of vibration eliminator is made of corrugated stainless steel with a braided stainless-steel jacket and copper ends for brazing. Bronze versions are also available. The corrugations in the body of the device make it flexible and able to absorb equipment vibrations. Note that the isolators in *Figure 51* are designed specifically for refrigerant piping and are selected to match the line size.

Piping suspended by threaded rods (*Figure 52*) can also include vibration isolators. Again, sizing each vibration isolator for the expected load is essential to get the desired performance.

Figure 52 Steam condensate piping suspended with vibration isolators.

2.1.5 Selecting Vibration Isolators and Materials

The selection of the correct vibration isolator or **seismic restraint** depends on the type of equipment, its weight, and how and where it is to be installed. Local codes often dictate what devices or materials are required for a particular application. Job specifications also dictate the type of vibration isolation that an application requires.

Seismic restraint: A device or cable arrangement used to control the movement of equipment when the release of geologic stress causes seismic events.

Manufactured vibration isolators and seismic restraints that are readily available for purchase include the following:

- Suspension-type vibration isolators
- Spring-type vibration isolators
- Grommet-type vibration isolators
- Vibration-isolating pads
- Flexible, vibration-isolating refrigerant piping connectors
- Factory-built seismic roof curbs

Manufacturers of these products and devices provide performance data so that the correct device can be selected for any application. In addition to common information such as dimensions, they provide weight or range of movement limitations.

Table 7 shows a standard vibration isolator with its performance and selection data. Note that many different springs can be selected for the same isolator. The spring design and construction characteristics determine its weight-bearing capacity. The springs are color-coded, making it easy to visually confirm which spring has been provided in an installation. Note that the isolators at the lower end of the table, with the highest weight rating, are equipped with two springs. In this case, a core spring is provided inside the main, outer spring to accommodate heavier weights.

Vibration isolators and seismic restraints may need to be fabricated by the installer using common hardware items such as the following:

- Angle iron
- Metal straps
- Steel cable
- Threaded rod
- Nuts, bolts, washers, etc.

The size and quality of the materials used are usually spelled out in local codes or job specifications.

Vibration isolators and seismic restraints use similar devices and methods to achieve similar results. An excellent resource for learning what to use for a particular application can be found in VISCMA 412, *Installing Seismic Restraints for Mechanical Equipment*, published by the Vibration Isolation and Seismic Control Manufacturing Association. Its tables offer suggestions on how to properly install various pieces of equipment.

For example, assume an air handler is to be mounted on a concrete floor. For simple vibration isolation, open- or housed-spring devices can be used. If the installation must be resistant to **seismic activity**, housed-spring isolators cannot be used because they cannot resist uplift. If open-spring devices are used and the installation must be seismic-compliant, snubbers or bumpers must be used to restrict back-and-forth and side-to-side motion. Equipment manufacturers may provide specific instructions and/or kits for installing their equipment to be compliant with seismic safety regulations.

Seismic activity: Movement of the earth's crust caused by the release of volcanic or other geologic stress.

TABLE 7 Vibrator Isolator Performance Data: Type 1" Deflection C, CI, CIP, & CIW Ratings

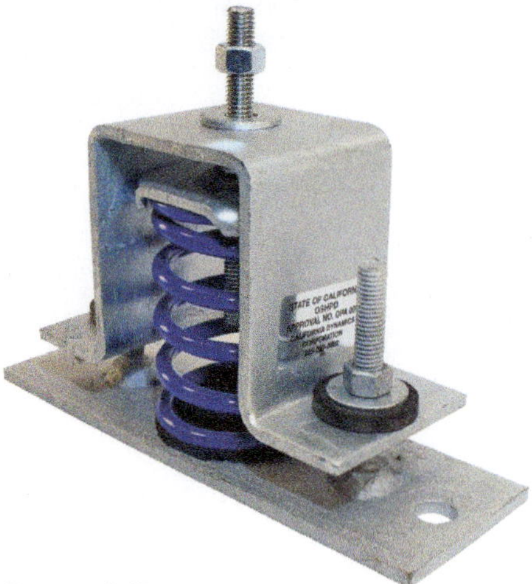

Source: CalDyn

Size	Rated Capacity (lb)	Rated Deflection (in)	Mount Constant (in/lb)	Spring Color/Stripe
A-45	45	1.60	28	Blue
A-75	75	1.50	50	Orange
A-125	125	1.33	94	Brown
A-200	200	1.15	174	Black
A-310	310	1.00	310	Yellow
A-400	400	1.00	400	Green
A-510	510	1.00	510	Red
A-625	625	1.00	625	White
B-65	65	2.10	31	Brown
B-85	85	2.10	40	White/Black
B-115	115	2.00	57	Silver
B-150	150	2.00	75	Orange
B-280	280	1.60	174	Green
B-450	450	1.31	344	Red
B-750	750	1.12	670	White
B-1000	1000	1.00	1000	Blue
B-1250	1250	1.00	1250	Gray
B-1650	1650	1.00	1650	Black
C-1000	1000	1.00	1000	Black
C-1350	1350	1.00	1350	Yellow
C-1750	1750	1.00	1750	Black
C-2100	2100	1.00	2100	Yellow*
C-2385	2385	1.00	2385	Yellow**
C-2650	2650	1.00	2650	Red*
C-2935	2935	1.00	2935	Red**
D-3500	3500	1.00	3500	Black*
D-4200	4200	1.00	4200	Yellow*
D-4770	4770	1.00	4770	Yellow**
D-5300	5300	1.00	5300	Red*
D-5870	5870	1.00	5870	Red**
E-7000	7000	1.00	7000	Black*
E-8400	8400	1.00	8400	Yellow*
E-9540	9540	1.00	9540	Yellow**
E-10600	10600	1.00	10600	Red*
E-11740	11740	1.00	11740	Red**
F-12600	12600	1.00	12600	Yellow*
F-14310	14310	1.00	14310	Yellow**
F-15900	15900	1.00	15900	Red*
F-17610	17610	1.00	17610	Red**
G-18900	18900	1.00	18900	Yellow*
G-21465	21465	1.00	21465	Yellow**
G-23850	23850	1.00	23850	Red*
G-26415	26415	1.00	26415	Red**

*with RED core spring
**with GREEN core spring

2.2.0 Seismic Restraints

In areas with high seismic activity, local codes often require that HVACR equipment be fitted with some form of seismic restraint to prevent significant movement if an earthquake occurs. The reason for this is quite simple. If gas-fired equipment moves too much, it can break gas and electrical connections, resulting in a fire. If heavier equipment breaks loose, it can cause additional structural damage and possibly injure bystanders, in addition to the fire hazard. Equipment can be restrained using angle iron, straps, bumpers, vibration-isolating devices, cables, and other methods.

The variety of seismic restraint requirements for all the different types of HVACR equipment is beyond the scope of this module. For more detailed information on this subject, refer to VISCMA 412, *Installing Seismic Restraints for Mechanical Equipment*. This guide is easy to access via the internet and more user-friendly than one might expect, using simple language and diagrams to provide valuable information.

Did You Know?

The Great San Francisco Earthquake

In 1906, a devastating earthquake struck the city of San Francisco. Many of the city's water mains were broken, resulting in a loss of water pressure. With little or no water pressure available to fight the many fires that started after the earthquake, fires raged out of control, destroying the city.

2.2.1 Seismic Restraints for Residential Equipment

There are several important requirements for equipment installed in a residential environment where seismic disturbances are a consistent concern. For example, residential gas furnaces and water heaters with input ratings less than 100,000 Btuh must use flexible connectors between any rigid gas piping and the appliance. Furnaces and water heaters can be secured with metal straps bolted to the structure. Residential split-system condensing units should be bolted to a concrete pad using concrete anchors.

WARNING!

If you work in an area subject to seismic disturbances and governed by seismic codes, you must become familiar with the requirements to ensure your work is done in accordance with the applicable guidelines. Failure to take the required steps to secure and isolate equipment can lead to serious personal injury and major property damage.

Bumpers can be used to prevent lateral (horizontal) movement on equipment that is not top-heavy and prone to tipping over. For equipment that has the potential to tip over, both bumpers and other seismic restraints may be required. The bumpers shown in *Figure 53* are bolted to concrete.

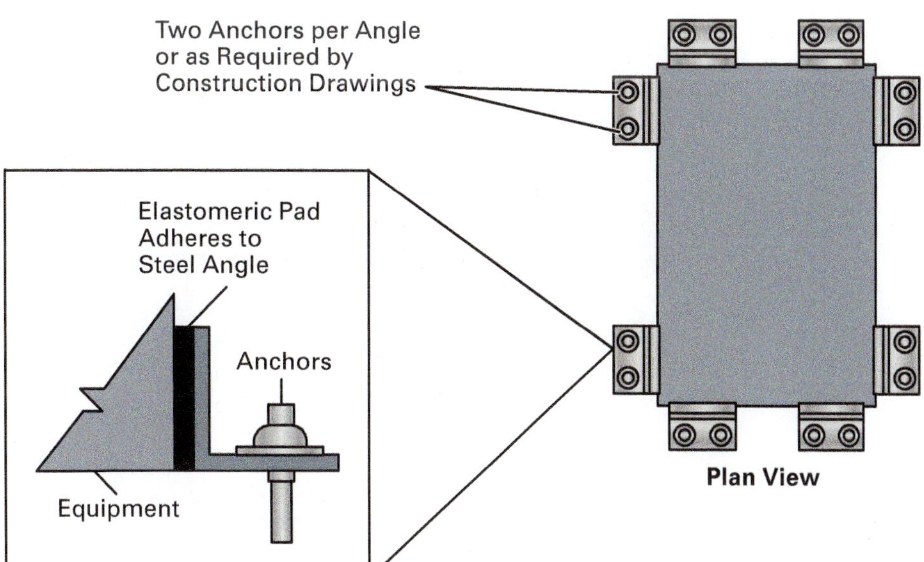

Figure 53 Bumpers.

Seismic Gas Shut-Off Valve

The danger of fire when gas lines rupture during an earthquake is obvious. A new gas shut-off valve is now available to help reduce this danger. Installed in the main gas line after the gas meter, this valve shuts off gas flow to the entire structure if an earthquake of magnitude 5.4 or above is detected. If shutoff occurs, the valve must be manually reset.

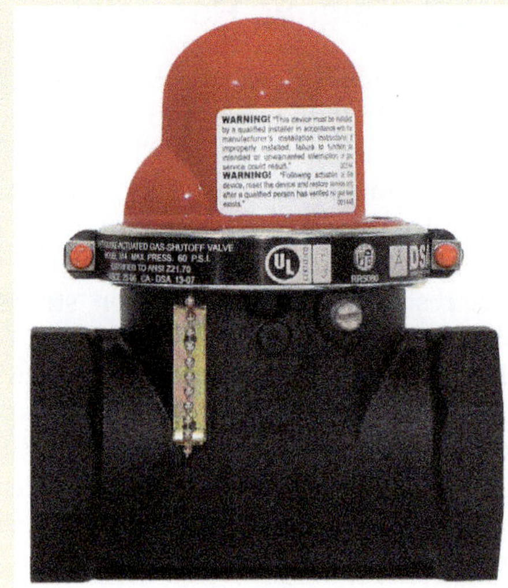

Source: Pacific Seismic Products

2.2.2 Seismic Restraints for Light Commercial Equipment

Packaged rooftop units that are curb-mounted can be converted in the field to be resistant to seismic events, or the unit can be installed on a factory-built seismic curb like the one shown in *Figure 54*. Equipment such as ceiling-mounted air handlers can be suspended and constrained with a combination of threaded-rod hangers with vibration isolators and steel cables (*Figure 55*).

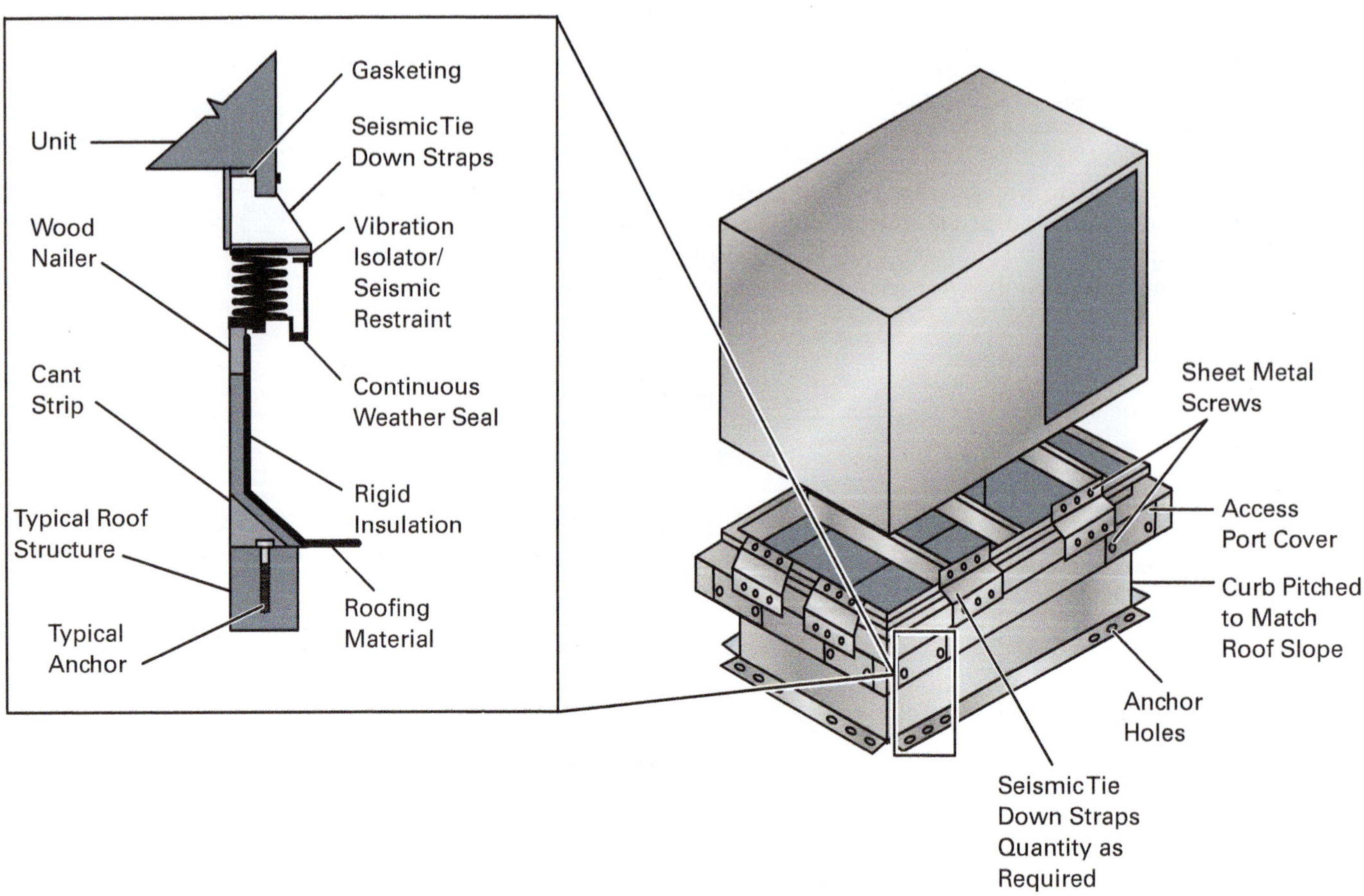

Figure 54 Factory-built seismic curb.

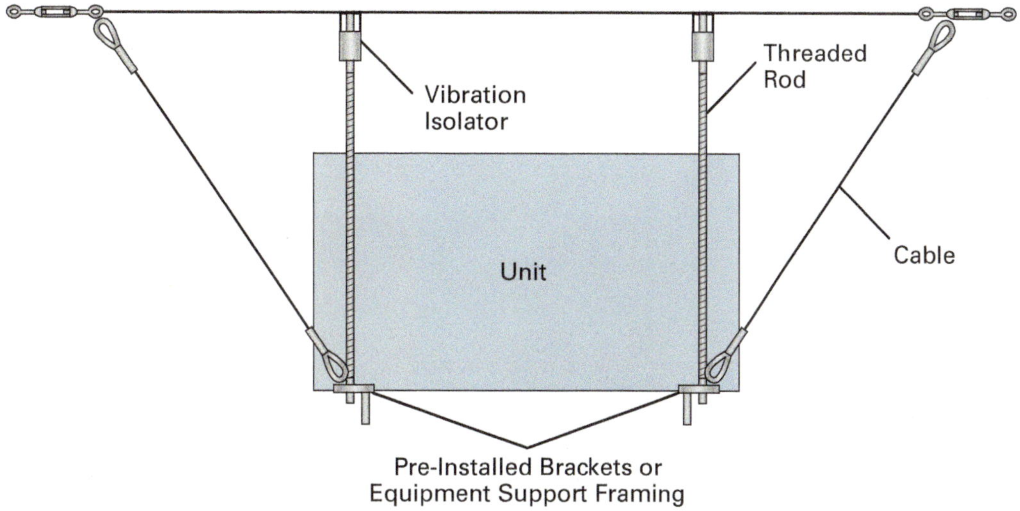

Figure 55 Suspended-equipment seismic restraints.

2.2.3 Seismic Restraints for Other Equipment

Pumps used in hydronic systems are restrained using a variety of devices. In *Figure 56*, vibration isolators and snubbers keep the pump baseplate on its

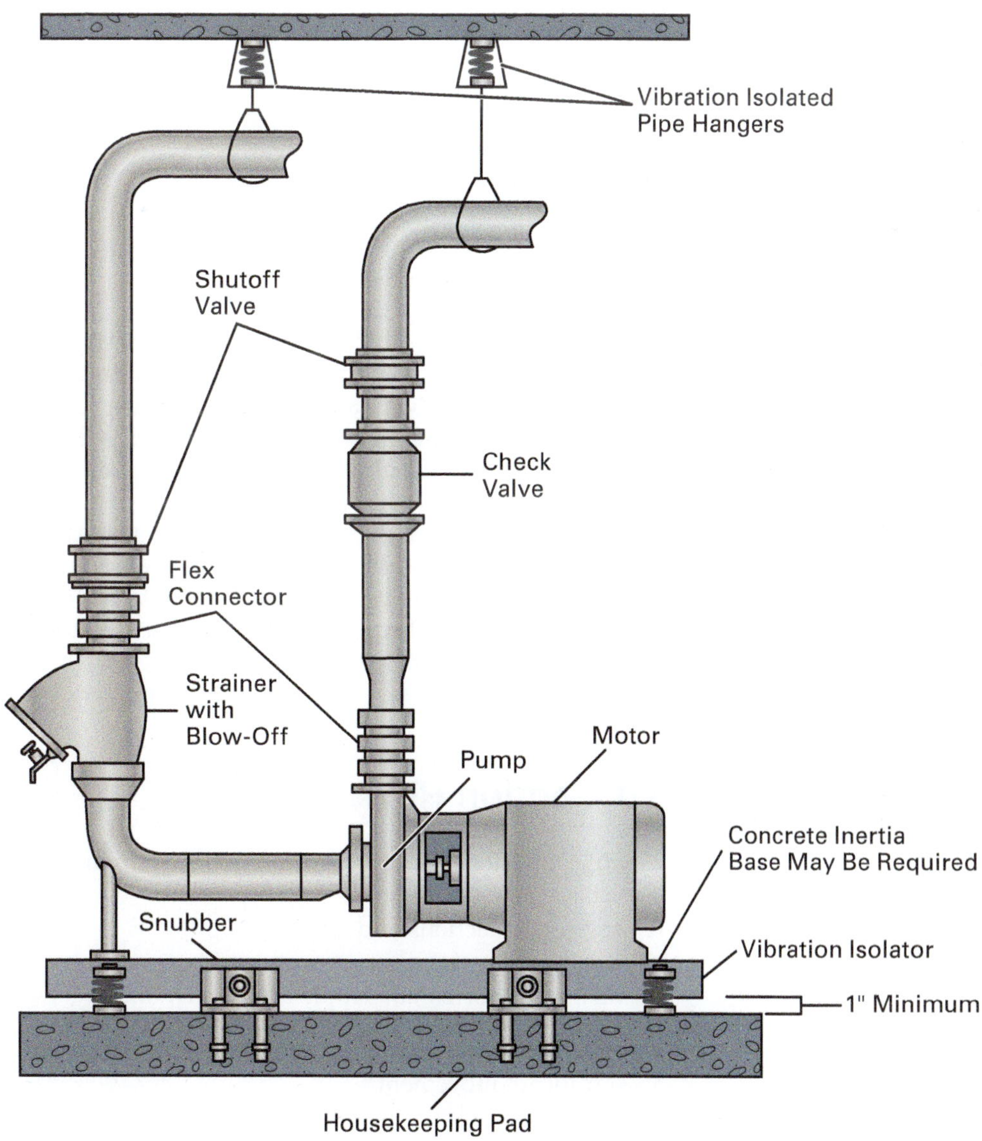

Figure 56 Typical pump seismic restraints.

Did You Know?

Earthquake-Resistant Buildings

Many new skyscrapers built in earthquake zones around the world are being built on giant suspension springs. This allows the building to sway instead of collapse when an earthquake occurs.

foundation. Flexible connectors in the pump's supply and return water lines, along with vibration-isolated pipe hangers, help prevent damage in case of seismic movement.

Figure 57 shows a combination coil-spring isolator and seismic restraint that can be used for various kinds of heavy, floor-mounted mechanical equipment.

Figure 57 Combination vibration isolator and seismic restraint.
Source: Courtesy of Kinetics Noise Control

There are many types of isolators and seismic restraints, and many more are constructed specifically for unique applications. In new construction, follow the plans and specifications and ensure that each isolator is installed in the right place. When replacing a worn or damaged isolator or seismic restraint, always use one that has the same basic characteristics and ratings as the original.

Remember that a properly piped hydronic system does not place weight on the connected equipment. Excess weight on the equipment from poorly supported piping overloads isolators, causing them to bottom out or fail altogether. When inspecting isolated equipment, ensure the isolators are actively supporting the equipment and are free to move.

2.0.0 Section Review

1. Flexible vibration isolators installed in ducts are mounted in ______.
   a. vertical sections of duct
   b. horizontal sections of duct
   c. supply air ducts only
   d. vertical or horizontal supply and return ducts

2. Seismic restraints are placed on HVACR equipment to prevent ______.
   a. significant movement
   b. vibrations
   c. expansion
   d. noise transmission

Module 03313 Review Questions

1. In a situation where the stress on a fastener is extremely high, which grade of bolt is *most likely* to be used?
   a. Grade 1
   b. Grade 2
   c. Grade 5
   d. Grade 8

2. A fastener with a more attractive head that is used where a finished appearance is important is called a ______.
   a. machine bolt
   b. lag screw
   c. cap screw
   d. stud bolt

3. A jam nut is used ______.
   a. when frequent removal is required
   b. to prevent another nut from loosening
   c. when appearance is important
   d. to secure a split washer

4. When threading a blind hole, the final tap to use is the ______.
   a. pipe tap
   b. plug tap
   c. bottoming tap
   d. taper tap

5. True or False: All lubricants perform equally well for thread tapping.
   a. True
   b. False

6. Sometimes during tightening, a fastener seems to seize as it approaches the recommended torque, often emitting a popping sound. This indicates a condition referred to as ______.
   a. stretching
   b. bonding
   c. rundown resistance
   d. set

7. The force required to loosen a fastener is called the ______.
   a. run-down torque
   b. service torque
   c. break-away torque
   d. loosen-up torque

8. How are concrete inserts installed?
   a. They are secured with epoxy.
   b. They are secured with grout.
   c. They are screwed directly into concrete.
   d. They are embedded in wet concrete.

9. True or False: Tapcon® screws require a pilot hole to be drilled before they are installed.
   a. True
   b. False

10. True or False: Pop rivets can be used when a joint can only be accessed from one side.
    a. True
    b. False

11. During the installation process, the hold-down bolts for a compressor in a new unit should be ______.
    a. checked to ensure they are properly torqued
    b. loosened but left in place
    c. removed completely
    d. removed, lubricated, then reinstalled and torqued

12. Vibration-isolating pads are rated in ______.
    a. pounds
    b. inch-pounds
    c. pounds per square inch
    d. angle of deflection

13. Flexible duct connections for vibration isolation can be installed in ______.
    a. return ducts only
    b. vertical ducts only
    c. horizontal ducts only
    d. both horizontal and vertical ducts

14. Residential gas furnaces and water heaters must have flexible connections on the gas lines if the input rating is ______.
    a. 50,000 Btuh or less
    b. 75,000 Btuh or less
    c. less than 100,000 Btuh
    d. less than 150,000 Btuh

15. Bumpers are used as seismic restraints on equipment to prevent ______.
    a. tip over
    b. up and down movement
    c. lateral movement
    d. swaying

Answers to Odd-Numbered Module Review Questions are found in *Appendix A*.

Answers to Section Review Questions

Answer	Section	Objective
Section One		
1. a	1.1.1	1a
2. c	1.2.2	1b
3. c	1.3.3	1c
4. d	1.4.2	1d
5. b	1.5.2	1e
Section Two		
1. d	2.1.3	2a
2. a	2.2.0	2b

MODULE 03206

Alternating Current

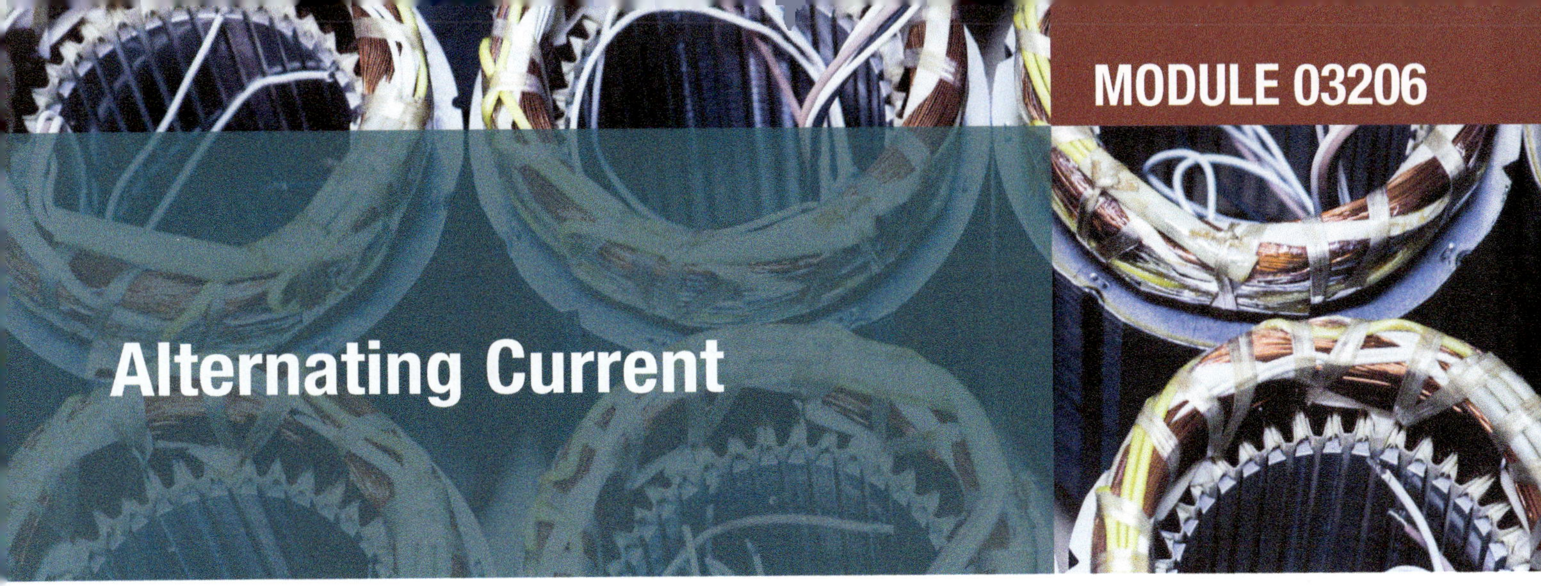

Source: Smoczyslaw/Getty Images

Objectives

Successful completion of this module prepares you to do the following:

1. Explain how alternating current is generated and how it is applied.
   a. State the basic concepts of power generation and the characteristics of alternating current.
   b. Explain how single- and three-phase AC power is provided for practical use.
   c. Identify resistive and inductive circuits.
2. Explain how transformers operate and describe various transformer types.
   a. Explain how transformers operate.
   b. Describe various types of transformers and their applications.
3. Describe various types of induction motors and explain how they operate.
   a. Describe how induction motors operate.
   b. Describe single-phase motor starting circuits.
4. Explain how to test various AC-powered components and terminate wires.
   a. Identify basic electrical safety guidelines.
   b. Explain how to use meters to test various electrical devices and motors.
   c. Identify common electrical termination hardware and explain how to properly terminate wiring.

Performance Tasks

Under supervision, you should be able to do the following:

1. Wire a multitap control transformer for the available primary voltage, energize the transformer, and measure the primary and secondary voltage.
2. Identify the start and run windings of an AC motor with an ohmmeter.
3. Beginning with raw wire, fabricate crimped connections and terminate both low- and line-voltage wiring on a compressor contactor.

Overview

The electrical power used to operate HVACR equipment is alternating current (AC). Compressor motors, fan motors, and transformers, along with many of the low-voltage control circuits used in managing HVACR systems, depend on AC to function. As an HVACR service technician, you'll work with and troubleshoot AC circuits and devices daily. For that reason, it is important to understand the fundamental characteristics of AC and begin learning how to test motors and similar components.

Digital Resource for HVACR

Scan this code using the camera on your phone or mobile device to view the digital resources related to this craft.

SCAN ME

NCCER Industry-Recognized Credentials

If you are training through an NCCER-accredited sponsor, you may be eligible for credentials from NCCER. The ID number for this module is 03206. Note that this module may have been used in other NCCER curricula and may apply to other level completions. Contact NCCER at 1.888.622.3720 or go to **www.nccer.org** for more information.

You can also show off your industry-recognized credentials online with NCCER's digital credentials. Transform your knowledge, skills, and achievements into credentials that you can share across social media platforms, send to your network, and add to your resume. For more information, visit **www.nccer.org**.

1.0.0 AC Power Generation and Uses

Performance Tasks

There are no Performance Tasks in this section.

Objective

Explain how alternating current is generated and how it is applied.

a. State the basic concepts of power generation and the characteristics of alternating current.
b. Explain how single- and three-phase AC power is provided for practical use.
c. Identify resistive and inductive circuits.

Most of the electricity used to power homes and businesses is produced by utility-owned power generating stations like the one shown in *Figure 1*. Renewable resources, such as the sun, wind, and hydroelectric power, are consistently being added to the electrical grid as well. All these sources distribute alternating current (AC) to users. Direct current (DC), when needed, is obtained by converting AC power to DC at the point of consumption. Batteries are also a common, but limited, source of DC voltage.

Figure 1 Nuclear power plant.
Source: Ververidis Vasilis/Shutterstock

1.1.0 Power Generation

Most electricity is produced by turbine-driven generators like the one shown in *Figure 2*. The turbine itself may be driven by water, steam, or its own internal combustion process. It is connected to the generator by a drive shaft. Generators can be designed to produce DC or AC power.

Figure 2 Utility-scale steam turbine and generator.

Electrical energy is generated by passing coils of wire through a magnetic field. The operating principles of a DC generator are illustrated in *Figure 3*. A simple horseshoe magnet forms a magnetic field as magnetic lines of force travel from one pole to the other. A loop of wire supported on a shaft is rotated within this field. As the loop turns, an electric current is generated and flows to terminals (segments) on the **commutator**. Two stationary contacts, called *brushes*, touch these commutator segments as the loop of wire rotates. Current generated in the loop is carried by these brushes to an external circuit, where it can be consumed.

Commutator: The movable contact surface on an electric generator or motor.

The commutator consists of two halves. When the loop of wire rotates one-half turn, each brush contacts the other half of the commutator. This is necessary to keep the current flowing in the same direction in the circuit.

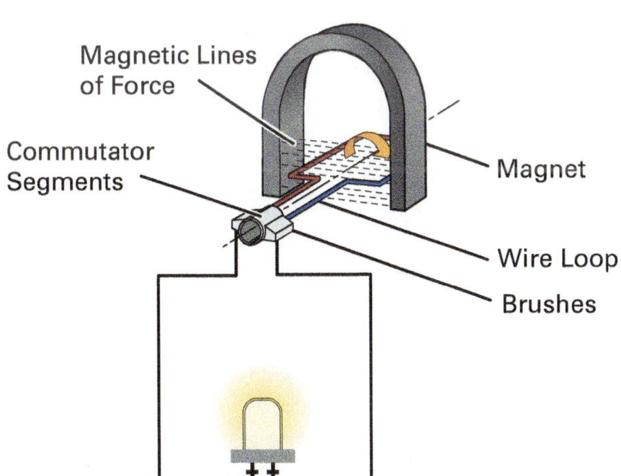

Figure 3 DC power generation.

Frequency: The number of complete cycles of an alternating current, sound wave, or vibrating object that occur over a predetermined time.

Hertz (Hz): The unit of measure for the frequency of alternating current. One Hertz equals one cycle per second.

Sine wave: A waveform that displays periodic oscillations in the generation of AC power.

Oscillation: The repetitive and/or periodic variation of some measure in relation to a central point.

Alternator: A device that generates alternating current through conductors rotating in a magnetic field.

1.1.1 Alternating Current

Alternating current is current that flows first in one direction, then in the opposite direction. The direction of current flow reverses at precisely established intervals. The number of complete cycles that occur each second is known as the **frequency**. The standard frequency for AC systems in the United States is 60 **Hertz (Hz)**, or 60 cycles per second. A single cycle of AC voltage looks like the **sine wave** shown in *Figure 4*. This periodic reversal of current at a precise interval is referred to as **oscillation**.

An alternating-current generator, frequently called an **alternator** (*Figure 5*), operates on much the same principle as the DC generator. The rotation of the conductor through magnetic lines of force generates current in the conductor.

However, instead of being connected to segments on a commutator, the conductor ends are each connected to *slip rings*. The slip rings are fastened to the shaft and rotate with it. Stationary brushes are in contact with the slip rings. The generated current is carried through the brushes to the load.

During one revolution of the conductor, each side passes through the magnetic lines of force, first in one direction and then in the other. Therefore, current flows first in one direction and then the other, thus producing alternating current.

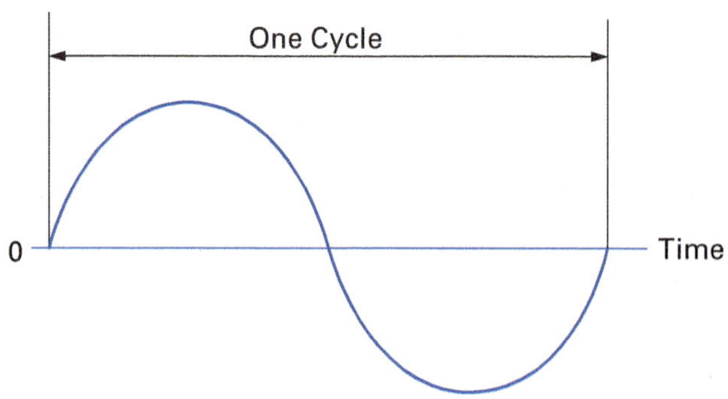

Figure 4 Alternating-current sine wave.

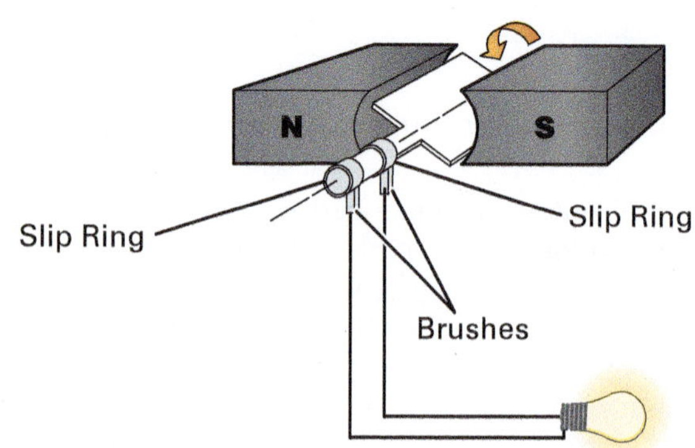

Figure 5 AC power generation.

Armature: The rotating component of a generator.

Induction: The generation of a current in a conductor by placing it in a moving magnetic field.

1.1.2 AC Sine Wave Generation

Figure 6 shows how a sine wave is generated. A voltage is *induced* as the conductor, called the **armature**, is moving through the magnetic field inside the generator. The amount of voltage induced depends on the strength of the magnetic field as well as the angle and speed at which the conductor cuts the magnetic lines of force. **Induction** is an electrical phenomenon you will hear about often, as it is also related to motor theory and operation.

In *Figure 6*, the armature has been divided into a dark half and a light half to make our discussion easier. No electrical load is shown in this example. Throughout this discussion, remember that the armature is not stopping at any of the positions shown. It is rotating at a consistent speed.

In the A position, the armature loop is positioned such that the magnetic lines of force are unbroken. The armature is not cutting through any lines of force, so no voltage is induced at that moment.

As the armature rotates toward the B position, it cuts more and more lines of force per second, and an increasingly strong voltage is induced. By the time it reaches the B position, the maximum voltage is attained because the armature has cut through the maximum number of lines of force per second.

As rotation continues toward the C position, the number of lines of force cut per second is again maximized. By the time it reaches the C position, though, the induced voltage decreases to zero, as it was when the armature was at the A position.

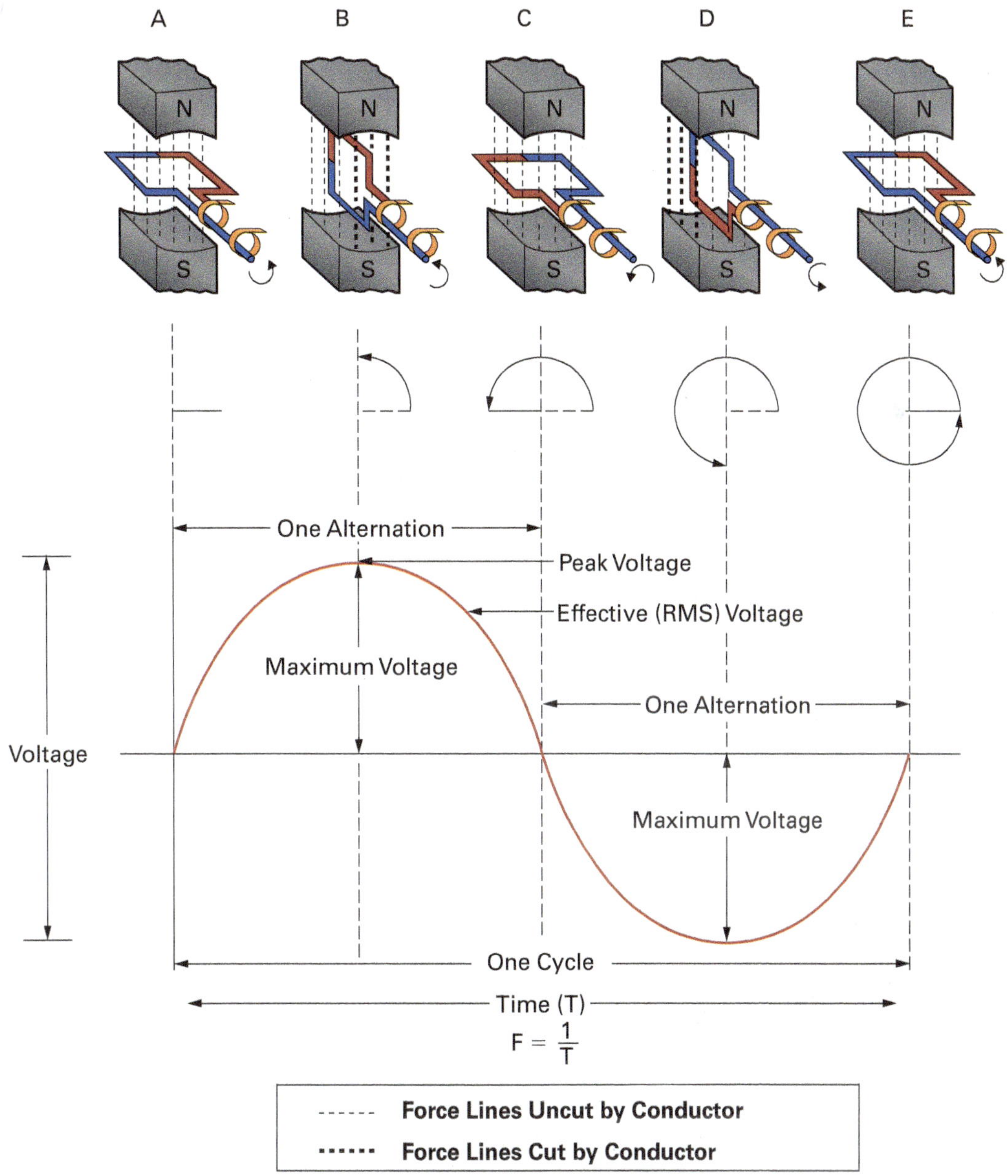

Figure 6 Sine wave generation.

When the armature reaches the C position, it has completed one-half of a full rotation. The first half of the sine wave in *Figure 6* shows the varying strength of the voltage during this half turn. The A position represents the zero point (0 degrees) of a 360-degree rotation. Once the armature reaches the C position, it has rotated 180 degrees, or half of a full circle. The voltage generated during that period is one alternation of the sine wave. Half of one cycle has been completed.

The armature continues rotating, through another 180 degrees and back to its starting point. A second alternation of voltage is generated. In this basic generator scheme, with one north and one south magnetic pole affecting the armature, one complete 360-degree rotation will always be required to generate a pair of alternations. Two alternations equal 360 degrees of rotation, or one complete rotation. Since we operate at a frequency of 60 Hz, the armature makes 60 complete rotations per second.

The direction of the voltage induced in this generator, known as its *polarity*, reverses every alternation, or 180 degrees of rotation. This also means electrical current flows through a load first in one direction, then in the other direction. The top half of the sine wave curve represents flow in one direction. The bottom half of the curve, lying under the horizontal axis, represents the flow of voltage in the opposite direction.

Amplitude: The maximum height and depth reached by the sine wave in a cycle, and the point on the wave at which the highest voltage is being generated.

NOTE

If you know the effective voltage by measuring it with a voltmeter and want to know the peak voltage, multiply the effective voltage by 1.414, which is also the square root of 2.

Effective voltage: The value of AC voltage that will produce as much power when connected across a load as an equivalent amount of DC voltage. Also known as root-mean-square (RMS) voltage.

Because the **amplitude** of the voltage waveform varies between zero and its peak, the amount of voltage available to do work—the **effective voltage**—is less than the peak voltage. The effective voltage, also known as *root-mean-square (RMS) voltage*, is 0.707 times the peak voltage. Therefore, a peak amplitude of 169.71 volts produces a working voltage of 120 volts.

$$\text{Peak voltage} \times 0.707 = \text{RMS (effective) voltage}$$
$$169.71 \text{ volts} \times 0.707 = 120 \text{ volts}$$

In HVACR service work, digital multimeters (DMM) like the one shown in *Figure 7* are widely used to measure the effective voltage. That's the voltage we interface with daily. Unless stated otherwise, the voltages stamped on the nameplates of HVACR equipment and other devices refer to effective voltage. A *true RMS* voltmeter is a more sophisticated instrument that can read the actual RMS voltage. Most voltmeters only estimate effective voltage because the actual RMS is difficult to measure if the AC current sine wave is imperfect in some way. A true RMS voltmeter can measure the voltage of imperfect sine waves. However, HVACR techs rarely need to work at that level of accuracy in AC voltage measurement.

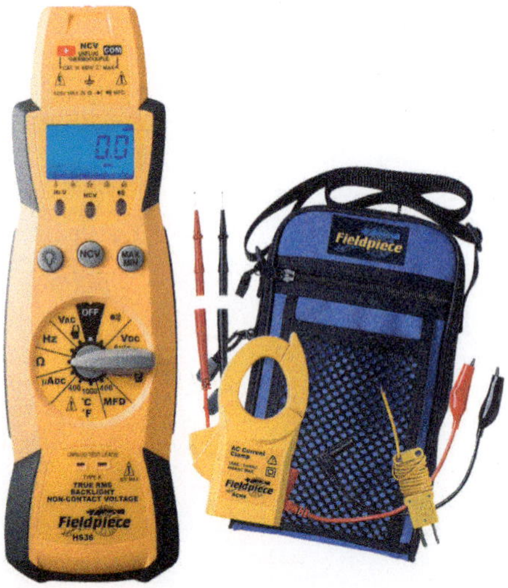

Figure 7 Standard DMM for HVACR service work.
Source: Courtesy of Fieldpiece Instruments

Did You Know?

Current Wars

After inventing the electric light bulb in 1879, Thomas Edison began work on a system for delivering electricity to homes and businesses. His system relied on DC—electric current that always flows in one direction. However, DC transmission over long distances proved to be impractical. Transmitting DCV at the low voltages useful for lighting or motor operation required the use of thick, expensive copper wire. In fact, the service areas of Edison's DC generating stations were limited to about a square mile and mainly served the downtown areas of large cities.

While Edison was pioneering DC distribution, electrical engineers in Europe were experimenting with alternating current (AC), which reverses direction at regular intervals. American businessman George Westinghouse saw the value in AC too. High-voltage AC power could be distributed over longer distances using thinner, less expensive copper wires. Experts theorized that special devices (transformers) could step the voltage up or down as needed. Increasing the voltage level would allow it to be transmitted over longer distances, while reducing the voltage near the point of use would enable the current to be used safely in homes and shops.

Westinghouse hired a young engineer named William Stanley, Jr., who developed the first effective transformer and demonstrated the first AC lighting system. Around the same time, inventor Nikola Tesla filed patents for other devices run by AC. After Westinghouse bought these patents, a full-scale industrial war known as the *Current Wars* erupted. At stake was whether Edison's DC or Westinghouse's AC would electrify America.

Edison claimed that alternating current was extremely dangerous and called for outlawing the high-voltage AC transmission lines. He wasn't wrong about that, as AC is dangerous. Westinghouse countered that transformers safely reduced AC voltages before they entered buildings. Within ten years, the value of the AC system had been convincingly demonstrated. In the years since, we have learned to safely harness AC power using purpose-built safety components and by developing safe work practices.

Eventually, Edison was forced to admit that AC power had won the battle, and General Electric, the company he founded, began building and installing high-voltage AC transmission systems.

1.1.3 Frequency

The speed at which an armature rotates affects the frequency of the voltage. If the rotational speed of the armature increases, so does the frequency. If a single loop rotates in a two-pole field, as shown in *Figure 6*, the current flows once in each direction to complete one pair of alternations, or one cycle, as the armature turns through 360 degrees (*Figure 8*).

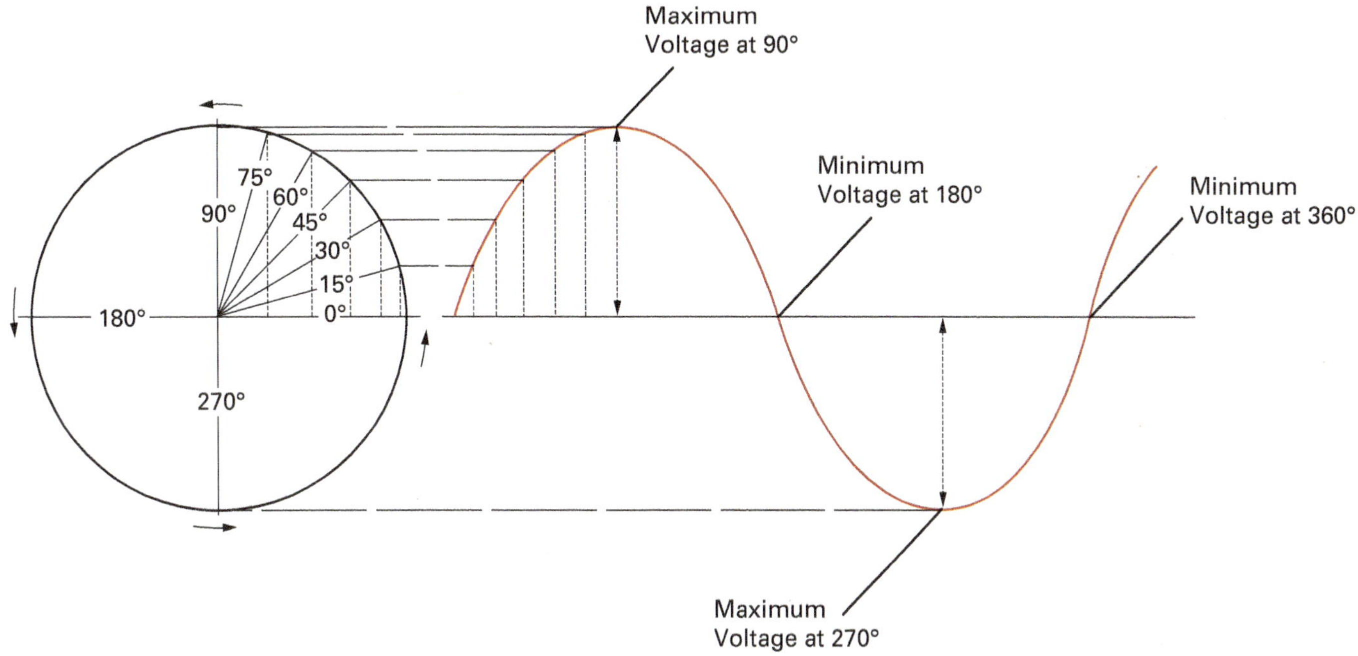

Figure 8 Sine wave plot.

If there is one complete rotation per second, the frequency is 1Hz. If the armature makes two complete rotations in one second, the frequency is 2Hz (*Figure 9*). For three rotations per second, the frequency will be 3Hz, etc. As stated earlier, in the United States, commercial power is supplied at a frequency of 60Hz.

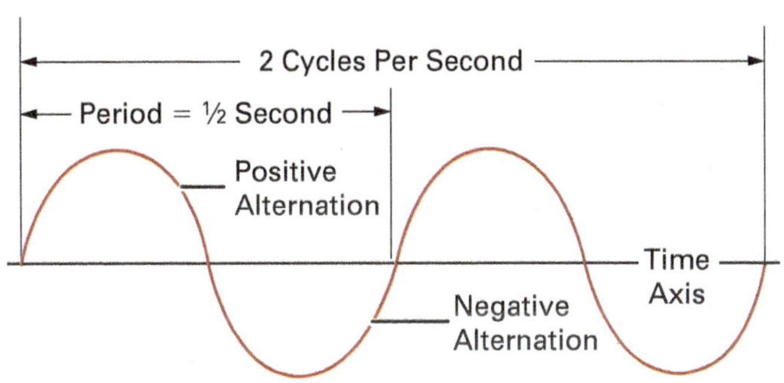

Figure 9 2Hz sine wave.

The more pairs of poles added to a generator, the higher the frequency will become, given the same armature speed. Generators that operate at a low speed need more pairs of poles than high-speed generators to provide voltage at the same frequency.

> **Around the World**
>
> **Electrical Frequency**
>
> While North America uses a 60Hz AC frequency as a standard, many other countries in Europe, Asia, and other regions provide power at 50Hz. Most of those, if not all, also do not use 120VAC power. Instead, the voltage provided at common receptacles in a home is about 230VAC @ 50Hz. Japan, however, uses both 50Hz and 60Hz power frequencies, depending upon the region.

1.2.0 Single-Phase and Three-Phase Power

Power produced by the utility is transmitted at a very high AC voltage, then stepped down to usable levels using large transformers (*Figure 10*). *Figure 11* shows the generation and distribution path of single-phase power. The single-phase voltage received at the service entrance panel of a residence is usually about 240VAC. However, the voltage provided to homes and businesses does vary somewhat from one area to another. The generated, transmitted, and distributed voltage upstream of the user also varies depending on the utility and the source of the generated power.

Figure 10 Substation transformer.

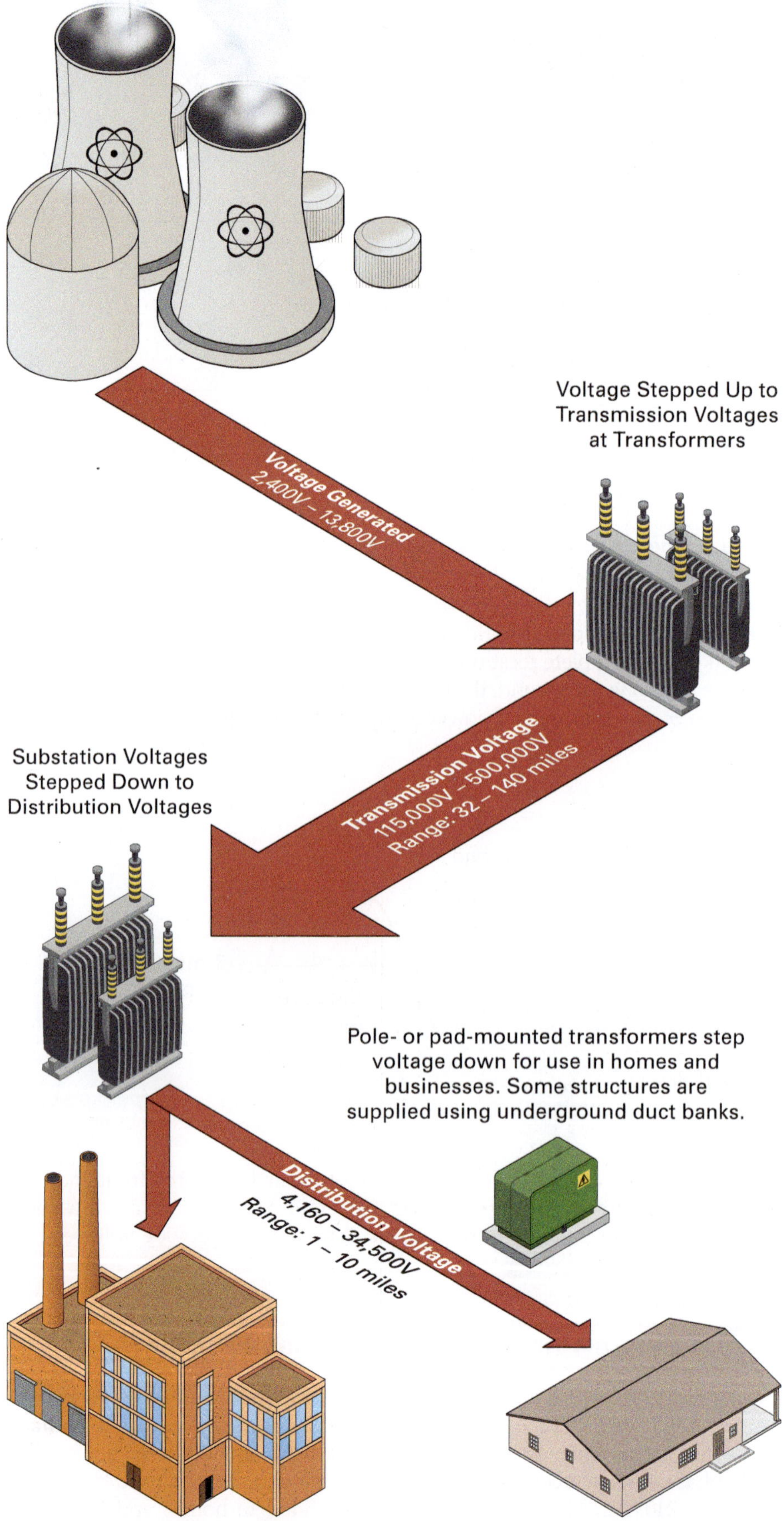

Figure 11 AC power generation and distribution path.

1.2.1 Single-Phase Power

An *Edison hookup* (*Figure 12*) is a common wiring arrangement for a power transformer that delivers 120/240VAC power to a residence. These transformers are located outside of the residence, usually on a pole, above ground in an enclosure, or below ground. Each of the two power wires delivering the 240V can alternatively be paired with a neutral line to create a 120V circuit. The neutral line is connected to the building ground, which may be an underground copper pipe or a copper rod that has been driven into the ground.

NOTE

Unless otherwise noted, all uses of the abbreviation V through the remainder of the module, representing voltage, refer to voltage alternating current (VAC).

Figure 12 Edison hookup.

The two hot legs (L) are commonly referred to as L1 and L2. The secondary winding of the transformer is center tapped. This leg is called the neutral (N) leg. If the voltage is measured between L1 and L2, it would be 240V, whereas the voltage between either L1 or L2 and N would be 120V. The neutral is electrically grounded at the power transformer. Other grounding points may exist at the electric meter and the main circuit panel base, depending on local codes.

After being stepped down, the power enters the structure through the service entrance. Wiring from the service entrance passes through a meter that records the power consumed and then continues to the circuit breaker panel (*Figure 13*). The circuit breaker panel, if it is the first enclosure the power source encounters after leaving the meter, may also be referred to as the *service entrance panel*. Power is distributed to a collection of branch circuits within the panel. The 240V branch circuits (*Figure 14*) serve major appliances, including most HVACR cooling equipment. Fossil fuel furnaces typically require 120V to operate.

Both legs of power in the 240V branch circuits are protected with circuit breakers rated for the current capacity of the wiring and the connected appliance. The wire connecting these circuits to the appliance must be sized to carry the load, as determined by the *National Electric Code*® (*NEC*®).

The 120V branch circuits (*Figure 15*) are also protected by circuit breakers. Each 120V branch is protected on the hot leg only. The 120V branches are taken

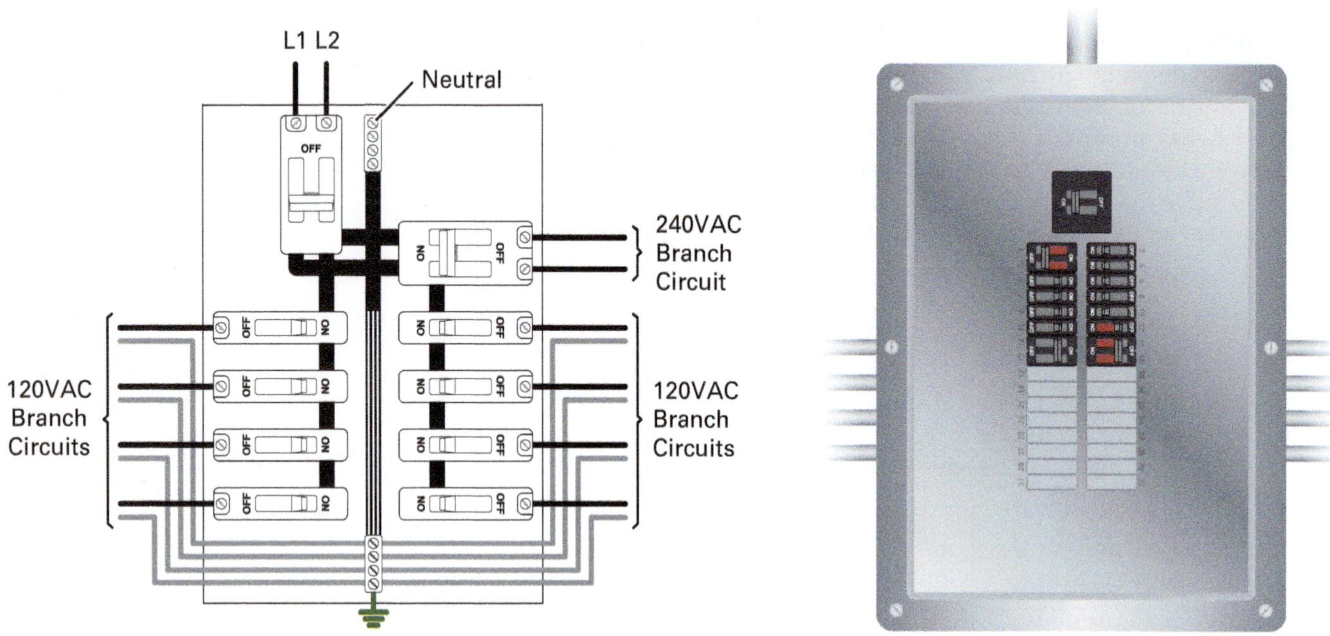

Figure 13 Circuit breaker panel.

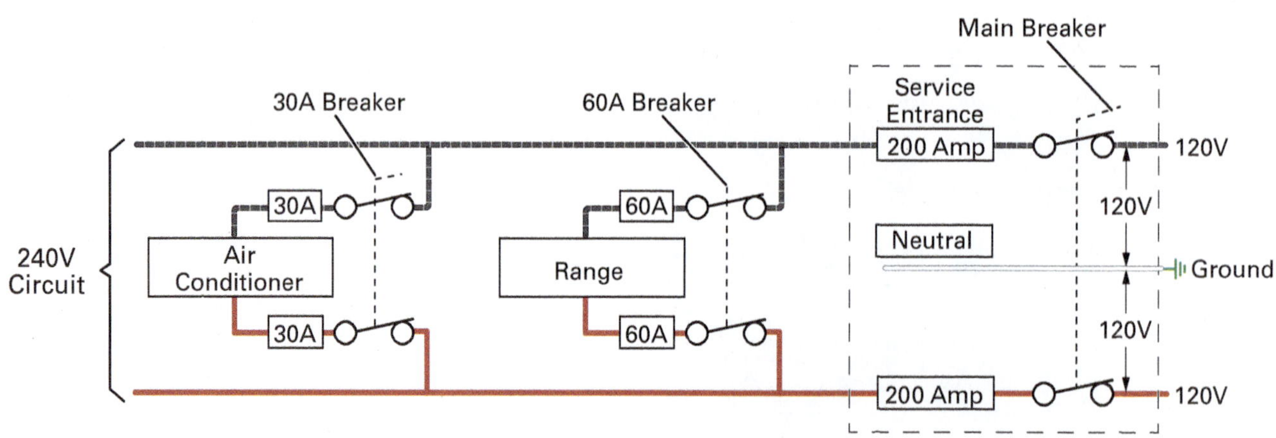

Figure 14 240V branch circuits.

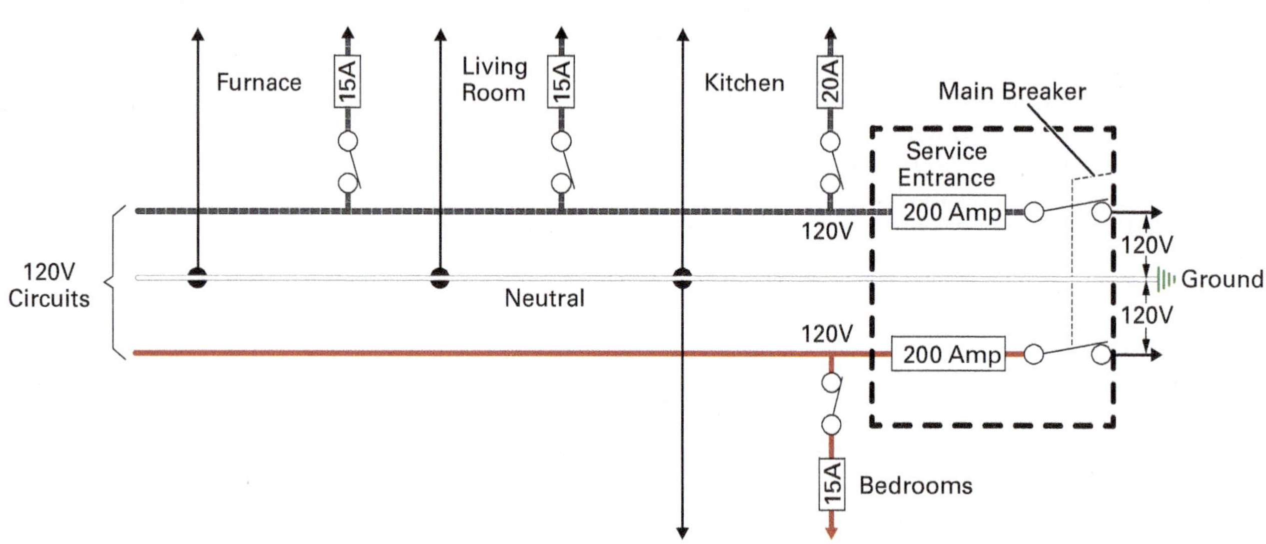

Figure 15 120V branch circuits.

from each of the 240V legs to balance the loads as closely as possible. The second wire of a 120V branch circuit is connected to the neutral terminal.

Figure 16 shows how the voltage from a 240V branch circuit might be distributed to an air conditioning compressor and its control circuit. Note that the compressor (and other line-voltage loads not shown) receives the full 240V, but the control circuit requires 24V. As shown in *Figure 16*, a transformer reduces the line voltage to 24V.

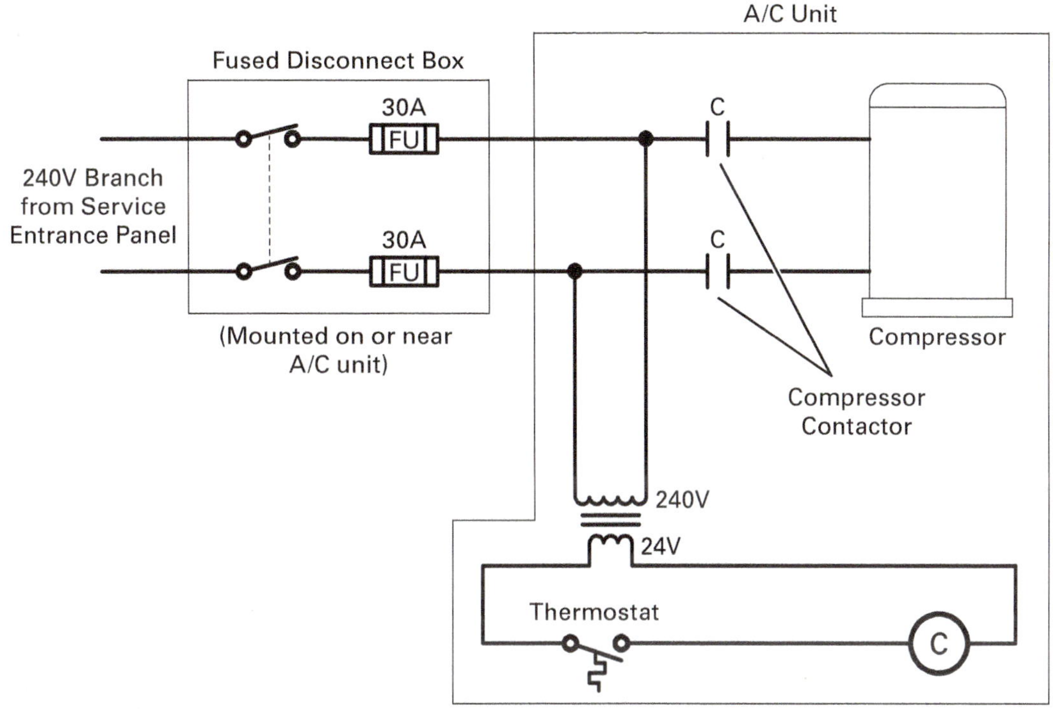

Figure 16 Example of an air conditioner branch circuit.

The *NEC*® requires that a power disconnect be installed on or within sight of the unit and be readily accessible (*Figure 17*). The disconnect may have an external handle that can be used to de-energize the circuit. Some may have a circuit breaker inside that can serve that purpose or a removable plug that disconnects power when it is pulled out. Some disconnects have fuses inside, while others do not.

Figure 17 Power disconnect mounted near a condensing unit.

Electrical Upgrades

The electrical panel, also known as a *breaker box* or *distribution board*, divides the incoming power from the utility into individual protected circuits that feed power to the building's electrical loads, such as lighting, outlets, and the HVAC system.

Before replacing fossil-fuel heating equipment with a heat pump, a thorough assessment of the electrical service and panel must be completed. The assessment should include an evaluation of the existing panel's capacity as well as its condition. The structure's existing electrical loads must be calculated to determine the impact of the additional load a heat pump may impose.

If a structure already has air conditioning, the electrical load of the heat pump may not change the situation substantially. On the other hand, the addition of supplemental electric heat can increase the electrical load considerably. An electrical assessment can determine if any panel or service upgrades are needed, including whether a subpanel may be a practical approach.

Electrical evaluations must be conducted by licensed and qualified electricians due to the liabilities involved.

1.2.2 Three-Phase Power

Single-phase power is adequate to power the common loads found in residences and small commercial businesses. The power demand is much greater in commercial and industrial applications, especially where large electric motors are used. Three-phase power is generally required for HVACR units that exceed 60,000 Btuh (5 tons) of capacity.

Thus far, the discussion of power generation has focused on a single conductor rotating in a magnetic field. If three rotating conductors are placed 120 degrees apart from each other, three equal voltages are generated at the same time. As shown in *Figure 18*, they occur 120 degrees apart in time; in other words, they are 120 degrees out-of-phase with one another. The three sine waves generated can be overlaid on each other, as shown.

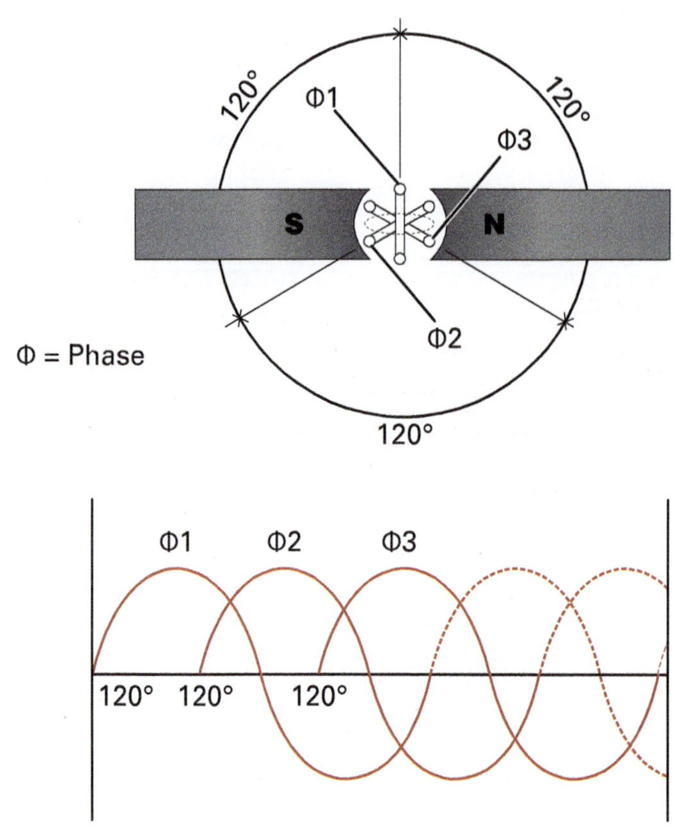

Figure 18 Three-phase voltage power generation.

There are several ways in which three-phase power sources can be connected, depending on the voltage(s) and the amount of current required. For example, a *four-wire closed delta configuration* (*Figure 19*) uses three transformers to provide 120V single-phase, 208V single-phase, and 240V three-phase power.

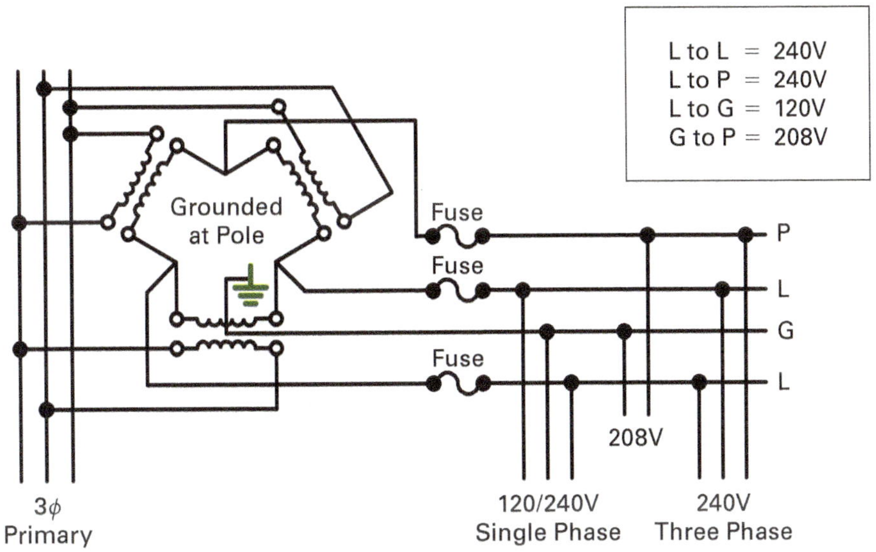

Figure 19 Four-wire closed delta configuration.

The four-wire open delta configuration (*Figure 20*) uses two single-phase transformers to produce 240V three-phase power, as well as 120V, 240V, and 208V single-phase power. This connection method is not as stable as the closed delta; thus, it is more difficult to keep the current in the three legs balanced. However, an industrial electrician might temporarily reconfigure a closed delta configuration to an open delta arrangement to accommodate a failed transformer. Only two transformers are required to make this configuration work.

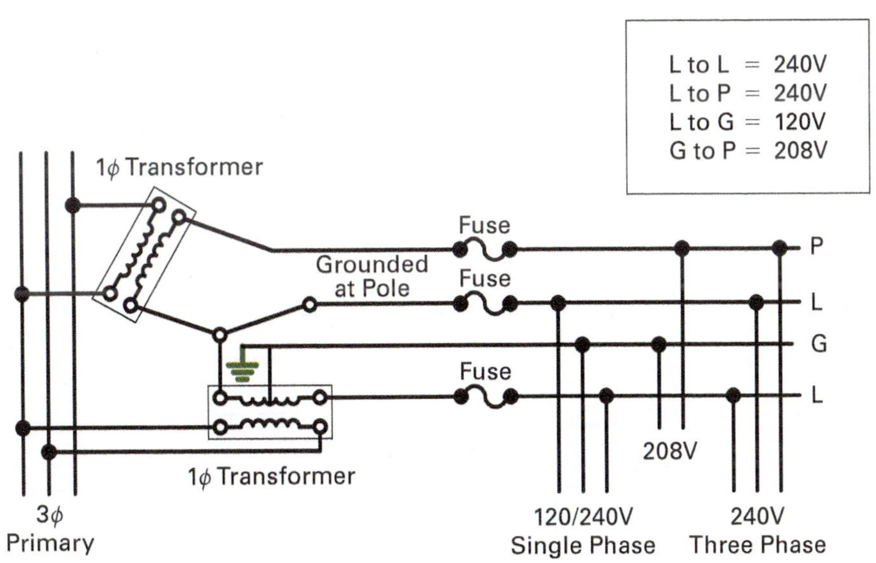

Figure 20 Four-wire open delta configuration.

A four-wire wye system (*Figure 21*) may be used to supply power for industrial use where high voltage is required to run large machines. This arrangement produces 480V three-phase power as well as the 277V single-phase power required for expansive fluorescent lighting systems.

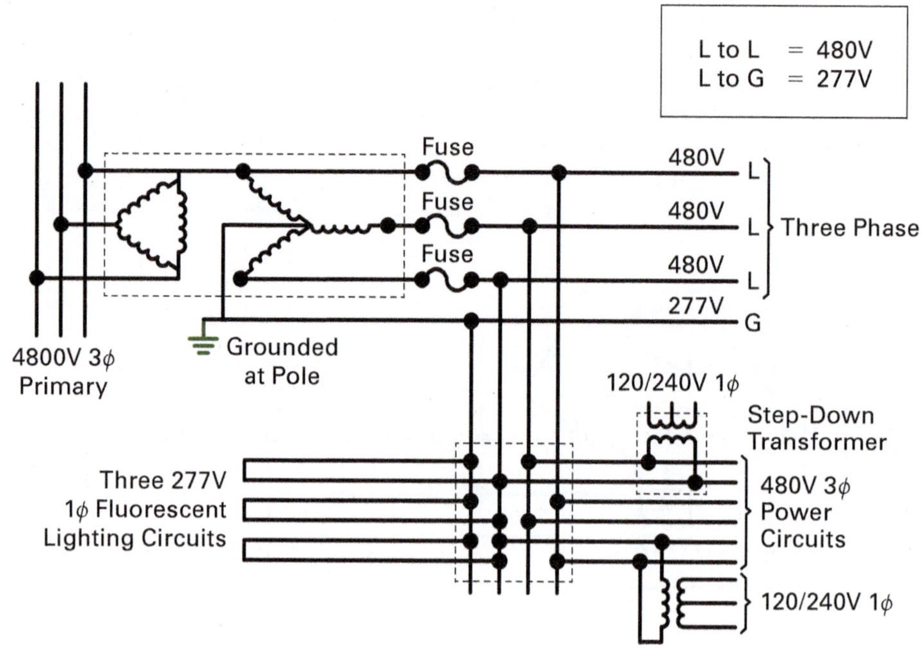

Figure 21 Four-wire wye configuration.

Figure 22 shows a wye-connected power source supplying power to a three-phase compressor circuit. Three-phase compressor motors are used for systems with a capacity greater than 5 tons, although three-phase compressors of lesser capacity are also produced in small numbers for unique situations.

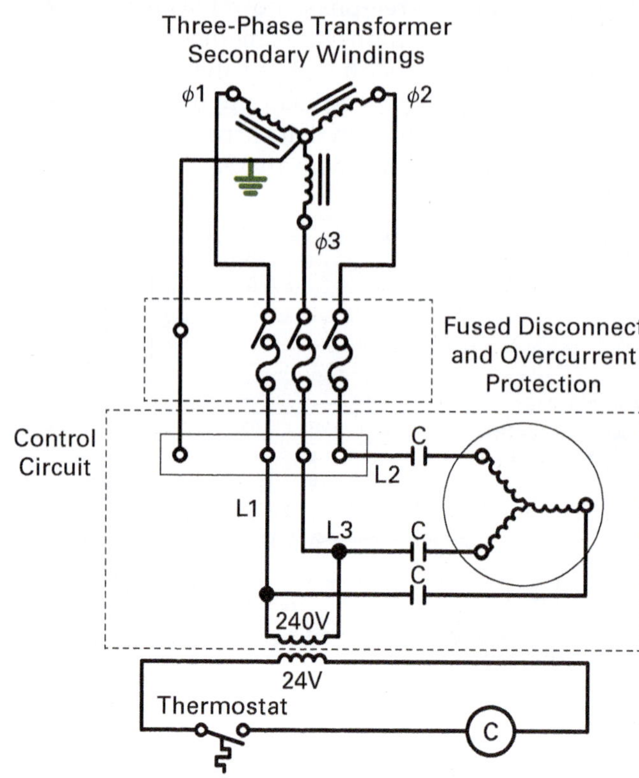

Figure 22 Example of a three-phase compressor circuit.

Voltage imbalance: For a three-phase power supply, the maximum deviation of the voltage of any phase from the average voltage of the three phases, divided by the average voltage. It is generally expressed as a percentage.

Current imbalance: For a three-phase power supply, the maximum deviation of the current in any phase from the average current in the three phases, divided by the average current. It is generally expressed as a percentage.

Voltage and Current Imbalance

Voltage imbalance and **current imbalance** are very important factors to consider when working with three-phase systems. The voltage and current on each phase serving a load need to be reasonably balanced.

A small imbalance in phase-to-phase voltage can result in a serious current imbalance. A current imbalance increases the heat generated in the motor windings. Both current and heat can cause nuisance overload trips and lead to motor failure. For this reason, the voltage imbalance between any two legs of the supply voltage applied to a three-phase system or motor should not exceed 2%. If a voltage imbalance of more than 2% exists at the input to the HVACR equipment, the equipment should be shut down and the problem corrected. Note that this is likely the responsibility of the local utility.

The voltage imbalance is determined in a three-phase system using the following formula:

$$\text{Voltage imbalance \%} = \frac{\text{maximum deviation from average}}{\text{average voltage}} \times 100$$

Refer to *Figure 23* for an example. Follow these steps to calculate voltage imbalance:

Step 1 Measure the voltage on each phase with the equipment in operation.

Step 2 Add the three voltage readings together and divide by three to determine the average voltage.

NOTE

Note that various standards place the allowable voltage imbalance as low as 1% and as high as 3%.

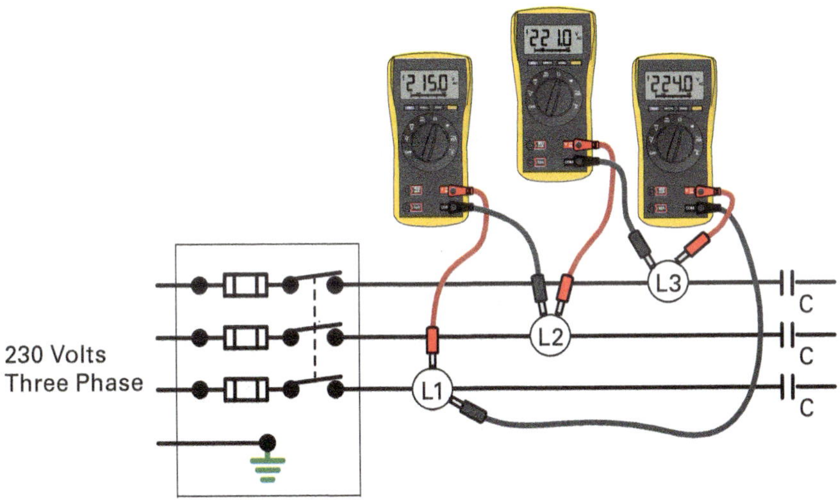

1) **Phase** **Measured Reading**

Phase	Measured Reading
L1	215V
L2	221V
L3	224V

2) $\text{Average Voltage} = \frac{215 + 221 + 224}{3} = 220\text{V}$

3) Individual Phase Imbalance from Average

L1 to L2 = 220 − 215 = 5V
L2 to L3 = 221 − 220 = 1V
L3 to L1 = 224 − 220 = 4V

4) 5V = Maximum Imbalance

5) $\text{\% Imbalance} = \frac{\text{Maximum Imbalance}}{\text{Average Voltage}} \times 100$

$\text{\% Imbalance} = \frac{5\text{V}}{220\text{V}} \times 100 = 2.27\%$ **(Out of Balance)**

Maximum voltage imbalance between any two legs must not exceed 2%.

Figure 23 Three-phase voltage imbalance calculations.

Step 3 Determine how far each of the three voltage readings deviates from the average.

Step 4 Identify which of the three readings shows the largest deviation.

Step 5 Use the formula to determine the percentage of imbalance.

In this example, the voltage imbalance is 2.27%. An imbalance of this magnitude should be investigated further and resolved.

The current imbalance in any phase of a three-phase system should not exceed 10%. Note that a 1% voltage imbalance alone can cause a current imbalance ranging from 6% to 10%. A current imbalance can develop without a voltage imbalance in the power source, although a voltage imbalance is a common cause. But a current imbalance can also result from loose or burned electrical contacts and terminations, which increase the resistance of the circuit. Since current follows the path of least resistance, the current in the other two legs will increase, causing more heat to be generated in those motor windings.

Refer to *Figure 24* for an example. The current imbalance in a three-phase system is determined using the same formula, but voltage measurements are replaced by current measurements:

$$\textit{Current imbalance \%} = \frac{\textit{maximum deviation from average}}{\textit{average current}} \times 100$$

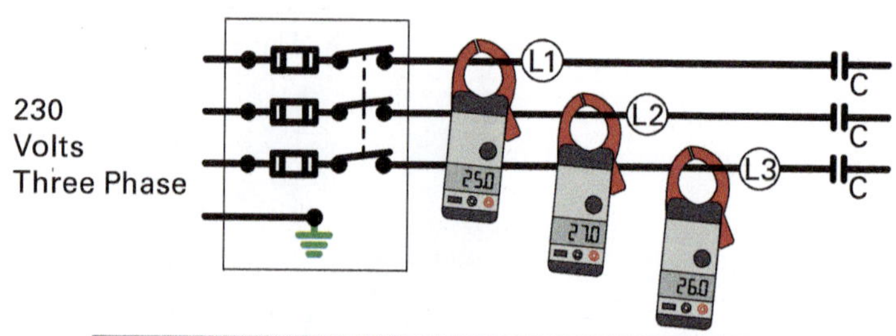

1)

Phase	**Measured Reading**
L1	25A
L2	27A
L3	26A

2) $\text{Average Current} = \frac{25 + 27 + 26}{3} = 26\text{A}$

3) Individual Phase Imbalance from Average:
- L1 to L2 = 25 − 27 = 2A
- L2 to L3 = 27 − 26 = 1A
- L3 to L1 = 26 − 25 = 1A

4) 2A = Maximum Imbalance

5) $\% \text{ Imbalance} = \frac{\text{Maximum Imbalance}}{\text{Average Current}} \times 100$

$\% \text{ Imbalance} = \frac{2\text{A}}{26\text{A}} \times 100 = 7.7\%$ **(In Balance)**

Maximum current imbalance between any two legs must not exceed 10%.

Figure 24 Three-phase current imbalance calculations.

Step 1 Measure the current on each phase serving an operating load.

Step 2 Add the three current readings together and divide the total by three to determine the average current.

Step 3 Determine how far each of the three current readings deviates from the average.

Step 4 Identify which of the three readings shows the largest deviation.

Step 5 Use the formula to determine the percentage of imbalance.

In the example shown in *Figure 24*, the 7.7% current imbalance is not enough to cause significant concern.

1.3.0 Resistive and Inductive Circuits

One important characteristic that distinguishes AC from DC is that AC acts differently in different kinds of circuits. HVACR components such as motors take advantage of this characteristic. This section will present two different types of load circuits—*resistive circuits* and *inductive circuits.*

The Power Formulas

As you work through this module, you may want to refer to the power formulas that were introduced in NCCER Module 03106, *Basic Electricity.* Remember, though, that these formulas are only valid for resistive loads, such as electric heat strips and crankcase heaters.

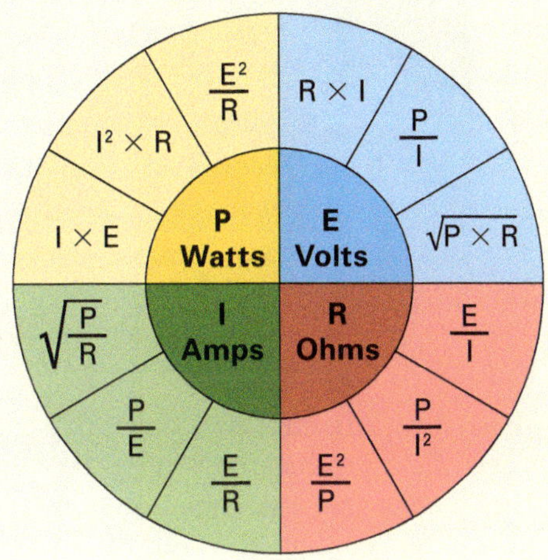

1.3.1 Resistive Circuits

An electric heating element is an excellent example of a resistive load. In a resistive circuit, the *current waveform,* created as power is consumed, occurs in phase with the voltage sine wave (*Figure 25*). The amount of power consumed by the load is determined by the following formula:

$$\text{P (power)} = \text{E (voltage)} \times \text{I (current)}$$

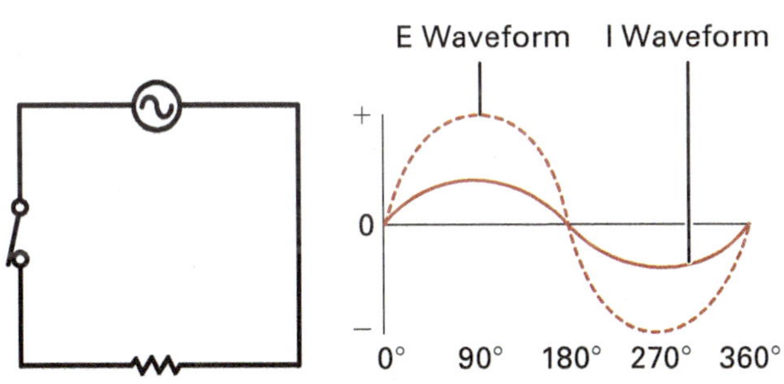

Figure 25 Resistive-circuit current and voltage sine waves.

All electrical loads generate some heat. Electric heaters, such as those used with heat pumps, along with stove burners and toasters, take advantage of this fact. One unit of measure for heat is the British thermal unit (Btu). One watt (W) can produce 3.414 Btus per hour (Btuh). Thus, a heater rated at 1,000W produces 3,414 Btuh. It is important to understand this principle because the heating load of a building is usually stated in Btuh, but the capacity of an electric heater is generally stated in watts. Larger heating elements, such as those used with a heat pump for supplemental heat, are usually rated in kilowatts (kW).

NOTE

For quick field calculations, it is generally safe to use 3.4 Btuh per watt. The resulting error is not significant.

Resistive Electric Heating Elements

Resistive electric heating elements like the ones shown here are typical of those used in HVAC equipment.

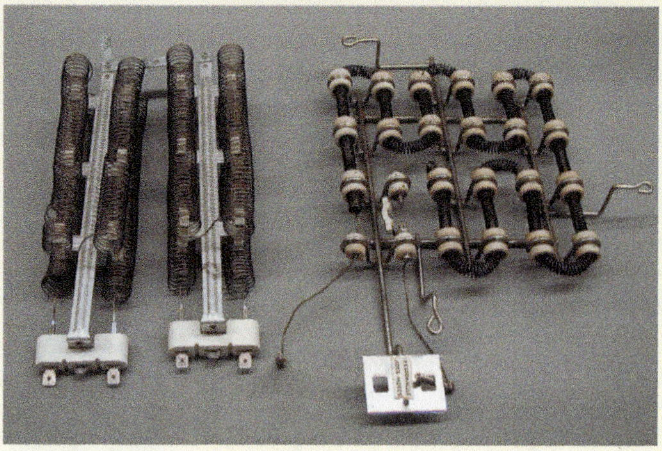

1.3.2 Inductive Circuits

When alternating current flows through a coil of wire, such as a motor winding, the magnetic field produced by one turn of the wire induces a voltage in adjacent turns. This means that a voltage is generated in an adjacent coil of wire without there being direct contact between the two coils.

Since the voltage is induced rather than physically conducted, it is known as an *inductive circuit*. The voltage induced in this manner is opposite in polarity to the applied voltage and therefore opposes current flow. For that reason, the current through a coil lags the voltage. In a purely inductive circuit, the current waveform will lag the voltage waveform by 90 degrees, as shown in *Figure 26*.

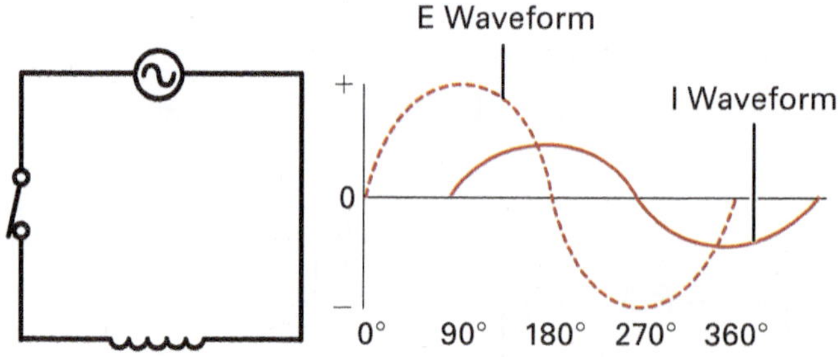

Figure 26 Inductive-circuit current and voltage sine waves.

1.0.0 Section Review

1. Compared to high-speed generators, what do low-speed generators require to provide power at an equal frequency?
   a. More transformers
   b. More pairs of poles
   c. More pairs of armatures
   d. More commutators

2. True or False: The transformer that steps the voltage down for residential use is located inside each home's service panel.
   a. True
   b. False

3. In a purely inductive circuit, the current waveform will lag the voltage waveform by ______.
   a. 45 degrees
   b. 90 degrees
   c. 180 degrees
   d. 360 degrees

2.0.0 Transformers

Objective

Explain how transformers operate and describe various transformer types.

a. Explain how transformers operate.
b. Describe various types of transformers and their applications.

Performance Task

1. Wire a multitap control transformer for the available primary voltage, energize the transformer, and measure the primary and secondary voltage.

A simple transformer consists of two or more coils of wire wound around a laminated iron core. An unbroken coil of wire is called a *winding*. One winding is connected to the power source and is called the *primary winding*. Separate windings can be placed around the opposite side of the same iron core. They are called *secondary windings* (*Figure 27*).

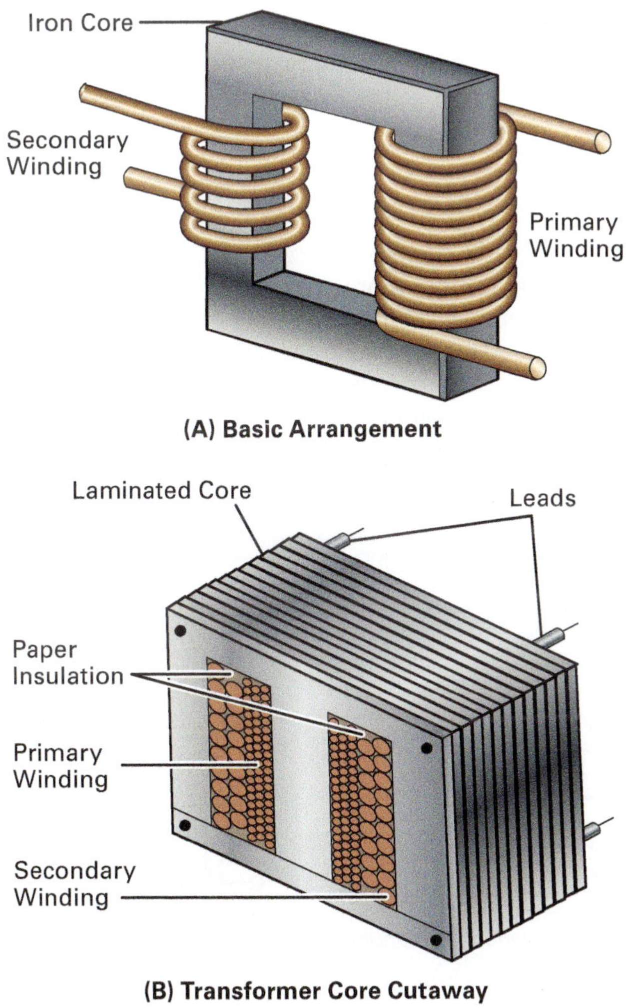

Figure 27 Basic components of a transformer.

There is no physical electrical connection between the primary and secondary windings. Electrical energy is transferred from the primary winding to the secondary winding by induction. Transformers related to HVACR work have no moving parts or fluids and therefore require no specific maintenance.

2.1.0 Transformer Operation

In operation, the primary winding consumes power from a source to induce electrical energy into the secondary winding. The secondary winding delivers the induced power to connected loads, usually at a different voltage. The power transferred from the primary winding to the secondary winding depends on the current flowing in the primary circuit.

The AC voltage induced in the secondary winding is directly related to the number of turns of wire in the secondary winding as compared to the primary winding. The output voltage from the secondary winding of a transformer is calculated by comparing the number of turns of wire in the primary winding to the number of turns in the secondary winding.

For example, if the number of turns in the secondary winding is twice that of the primary winding, its output voltage is twice that of the primary. 120V can be transformed to 240V this way. This identifies it as a *step-up transformer.* If the secondary winding, though, has half the number of turns of the primary winding, its output voltage is half that of the primary voltage. This identifies it as a *step-down transformer.* This relationship is stated by the formula:

$$\frac{E_P}{E_S} = \frac{N_P}{N_S}$$

Where:

E_P = voltage applied to the primary
E_S = voltage induced in the secondary
N_P = number of turns in the primary
N_S = number of turns in the secondary

Turns ratio: The ratio between the number of turns in a transformer's primary winding and a secondary winding.

This relationship between the primary and secondary windings is called the **turns ratio**. It is used to determine the output of both step-up and step-down transformers.

Some transformers contain more than one secondary winding, as shown in *Figure 28*. They are designed so that more than one voltage can be obtained from the secondary. The transformer is said to have multiple taps and is referred to as a *multitap transformer.* This is because most have only one winding that has been "tapped into" at different points between the beginning and end. This adds to its flexibility for use in different applications.

Transformers might also have multitap primary windings, allowing them to be used with different input voltages (*Figure 29*).

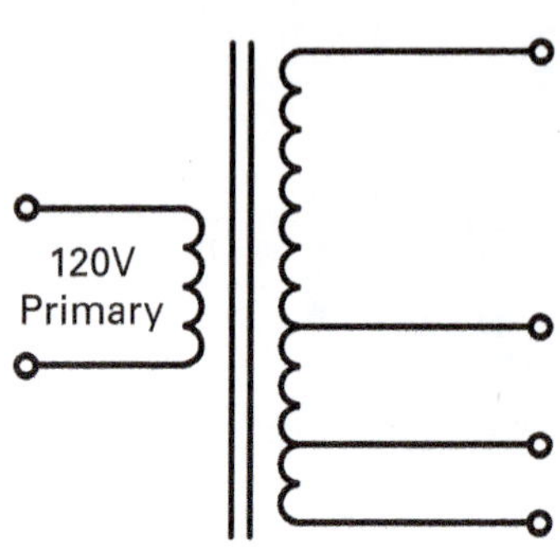

Figure 28 Multitap secondary winding.

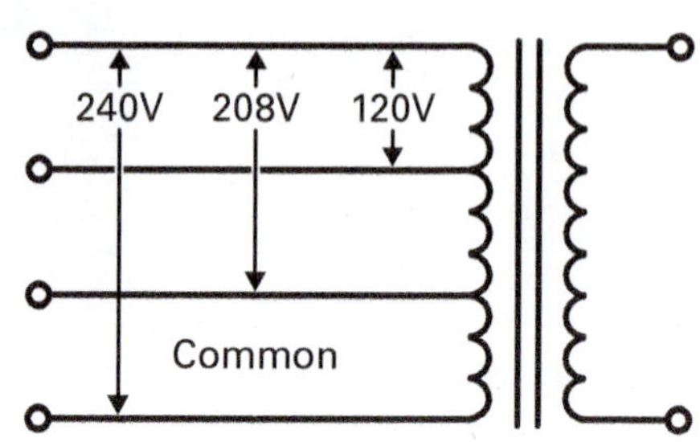

Figure 29 Multitap primary winding.

2.1.1 Transformer Capacity Rating

The current capacity of a transformer's secondary winding changes inversely with the voltage. For example, if the voltage across a step-up transformer is doubled from 120V to 240V, the current it can support is halved. Understanding a transformer's current capacity is essential to making a proper selection.

Transformers are rated or sized according to the amount of power the secondary winding can support. This power capability is expressed in volt-amps, or VA. This means that the VA rating is simply the voltage of the secondary winding multiplied by the amps of current it can support.

For example, if the secondary voltage of a transformer is 24V, and it can safely support 2 amps (A) of current, the VA rating of that transformer is 48VA (24V × 2A = 48VA).

Small control transformers rated at 40VA are very common in residential and light-commercial HVAC equipment. Assuming the rated output voltage is 24V, you can easily determine that it can handle 1.67A by dividing the VA rating by the voltage:

$$40\text{VA} \div 24\text{V} = 1.67\text{A}$$

You can determine how much current is flowing through the secondary winding with a clamp-on ammeter. You'll need to know this if you plan to add another load device to the control circuit, such as the coil of a relay. Then you can determine if there is enough capacity in the transformer to handle the additional load. Adding switches into the control circuit will make no difference, since they do not represent an electrical load.

NOTE

Never replace a failed control transformer with one of a lower VA rating unless it is known to be significantly larger than necessary. Using a transformer with insufficient capacity will result in a blown control fuse or transformer failure under load. A control transformer can, however, be replaced by one with a higher VA rating.

Some control transformers have an external fuse or manual-reset circuit breaker to protect the secondary winding from excessive current demand (*Figure 30*). Many others, especially small transformers, have a glass fuse buried beneath the casing and/or insulation soldered into the winding. They are not considered field serviceable. Larger power transformers have fuse blocks where larger, line-voltage fuse classes can be installed for protection.

Figure 30 Control transformer with an integral circuit breaker.
Source: Courtesy of Veris Industries

NOTE

Technically, any voltage above 50VAC is accepted as "line voltage" and hazardous. Voltages below 50VAC are considered low voltages. However, it is rare to encounter an AC voltage in HVACR work that falls between a nominal 24VAC and 120VAC by design.

2.2.0 Types of Transformers

When you need to select a transformer, the factors must be considered carefully. Errors usually result in transformer failure.

Inductive devices, such as motors, **contactor** coils, relay coils, and solenoid coils require more power to get started than they do during steady operation. Thus, for a short period after being energized, they draw extra current. This is referred to as **inrush current**, and you'll be exposed to the term again in a deeper discussion of motors. A transformer is often powering inductive loads, so the inrush current, although short-lived, must be considered in transformer selection.

Contactor: A control device consisting of a magnetic coil and one or more sets of contacts used as switching devices in high-voltage, high-current circuits, such as those serving compressors.

Inrush current: The maximal, short-lived current drawn by an inductive electrical device when first energized.

2.2.1 Single-Phase Transformers

Most of the transformers you'll be exposed to in HVACR work are single-phase transformers. Most residential and light-commercial equipment require a 24V power source to operate the controls. That source is a step-down transformer, appropriately sized for the loads. The transformer is usually powered by the same power source as the equipment it serves.

The simple cooling-system control circuit shown in *Figure 31* provides a good example of this. L1 and L2 represent the line-voltage power supply. It branches off to provide power to a compressor and the two other motors. Farther down the ladder, you'll see the transformer's primary windings powered by the same source. Beneath the transformer are the control devices in the circuit. The transformer's loads include the compressor contactor coil (C) and the indoor fan relay (IFR) coil. The other devices below the transformer are switches for various purposes, but switches do not represent loads.

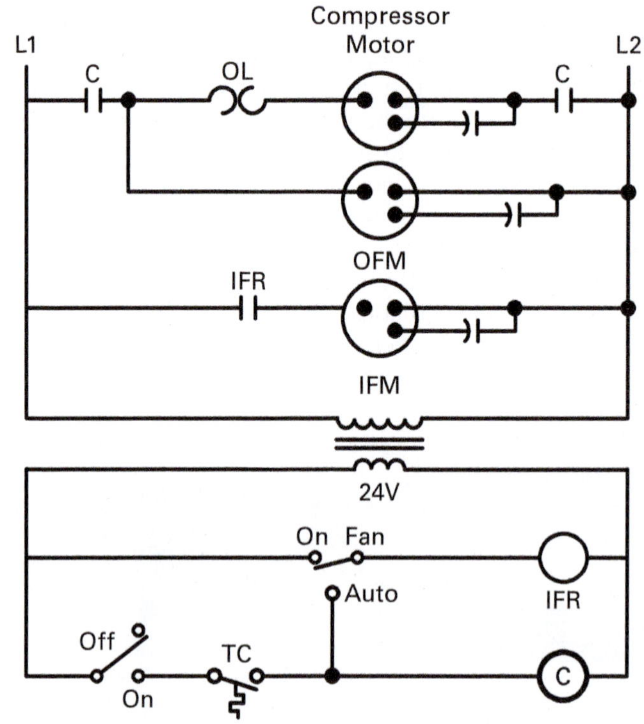

Figure 31 Example of a cooling-system electrical circuit.

NOTE

Although 24V is the standard for most residential and light commercial HVACR systems, some HVACR control transformers are rated to output as much as 27V. The higher voltage generally has no negative effect on the devices in the circuit, but the higher voltage is not generally needed. In some unique situations, though, they can better support a control circuit that has more **voltage drop** than desired.

Voltage drop: A decrease of electrical potential along the path of a current flowing in an electrical circuit, often caused by insufficient wire sizing for circuit length and poor wiring connections.

Transformers are available as step-up and step-down transformers. As noted earlier, a step-up transformer increases the voltage applied to the primary winding. Very large step-up transformers are used at power stations to increase the generated voltage before it is transmitted over long distances.

In HVACR work, step-up transformers are rarely encountered, although they are used in gas- and oil-furnace ignition systems. Step-down transformers are far more common. Most often, step-down transformers reduce an HVACR unit's line voltage down to 24V for the controls.

The transformer shown in *Figure 30* is a typical control transformer, although the winding insulation on many transformers is exposed rather than encased. This model has a multitap primary winding, and can be powered by a 120V, 240V, 277V, or 480V power supply. It produces only one secondary voltage, 24V. Rated at 75VA, it would most likely be found in a commercial HVAC system. With a rating of 75VA, we can determine that it supports 3.125A of current:

$$75\text{VA} \div 24\text{V} = 3.125\text{A}$$

The integral circuit breaker shown in *Figure 30* is set for 3.2 amps to protect the transformer against overloads.

Why Do We Care about Voltage Drop?

Many craftworkers understand what voltage drop is and how to look for it. Think of it as a plumbing issue. You may need to deliver 100 gallons per minute (gpm) of fluid to a process, but if the piping from the pump is too small, it can't be done unless you increase the size of the pump and create a larger pressure difference. A similar thing happens with voltage and wiring.

When a voltage is applied to conductors, it is there and available for electrical loads. If there is no demand for power, the voltage at the end of a long, undersized wire will consistently measure the same as it does at the source. It is only when there is a demand for power and current begins to flow that voltage drop manifests itself. Thus, voltage drop can only be detected on a circuit under load.

But why do we care? In a resistive circuit, such as one serving an electric heating element, a reduction in the voltage causes a similar reduction in current flow, and the element produces less heat. Not a good result, but perhaps not a serious problem.

However, in an inductive circuit, such as one serving a motor, voltage drop results in an *increase* in current flow. If the voltage drop is significant, the motor will exceed its amperage rating and eventually fail. If the voltage drop in the circuit is too severe, there may not be enough power available to even make the motor rotate. At that moment, the current flow is many times higher than normal, and the possible results include a tripped circuit breaker, blown fuses, and a smoking, destroyed motor.

If you aren't thinking about voltage drop as a factor, the problem can be difficult to identify. Again, the voltage will seem just fine when there is no load. You might find yourself saying, "The voltage to the motor is exactly what it should be," and move on to other possibilities. You must monitor the voltage as the load is applied to catch it in the act!

Transformers with multitap primaries are valued by technicians due to their installation flexibility. By providing additional taps along the primary winding, a single transformer may be used with a variety of common voltages. This allows technicians to carry fewer replacement parts. Multiple wires are provided to connect the primary winding to the power supply. One wire is identified as Common and is always used as one power lead. The other leads are labeled or color-coded to indicate which voltage they are tapped to accept. One of them must be used along with Common to energize the transformer. All the unused wires should be rendered safe with wire nuts or **crimped** butt connectors.

Crimped: Formed through compression, such as a connector on the end of an electrical conductor.

If the incorrect primary wire is chosen and connected, the secondary voltage will be incorrect. For example, if the power supplied to the transformer is 208V, but the wire identified for use with 240V is used instead, the result will be a secondary voltage less than the needed 24V. When a new transformer is installed, always measure both the primary and secondary voltage with a meter before considering the installation complete.

Large commercial and industrial equipment may have multiple transformers. For example, the voltage supplied to the unit may be 480V. The unit may have a few loads that require a 120V input. One transformer would then step down the voltage to 120V for those loads and perhaps provide a 120V input to a 24V control transformer. Alternatively, the control transformer can be powered by the main 480V power supply.

You may occasionally encounter a special transformer known as an **isolation transformer**. Isolation transformers have a 1:1 turn ratio, which means they deliver an output voltage equal to the input voltage. This special type of transformer is intended to either provide electrical isolation for certain components or systems, or to minimize electrical interference (noise) in electronic equipment. However, their use is very limited in the HVACR field.

Isolation transformer: A transformer with a one-to-one turn ratio, used to isolate electrical circuits or prevent electrical interference.

2.2.2 Three-Phase Transformers

For larger electrical loads, such as those used in commercial buildings and factories, three-phase power is required. Three-phase transformers have three primary windings and three secondary windings. However, three-phase transformers are almost always used in power distribution circuits, and not within HVACR equipment. Qualified electricians are typically responsible for their service and testing.

There are several ways in which three-phase transformers can be connected, depending on the voltage and current capacity required. The actual voltages produced depend on the way the transformers are connected. The four-wire wye-wye arrangement (*Figure 32*) is by far the most common in utility-class power distribution systems. It typically produces secondary voltages of 120V/208V or 277V/480V. A 277V power supply is used for fluorescent lighting in large commercial and industrial buildings. The closed delta-delta arrangement shown in *Figure 33* can provide 120V/240V or 240V/480V. A delta-wye configuration (*Figure 34*) can produce 120V/208V or 277V/480V.

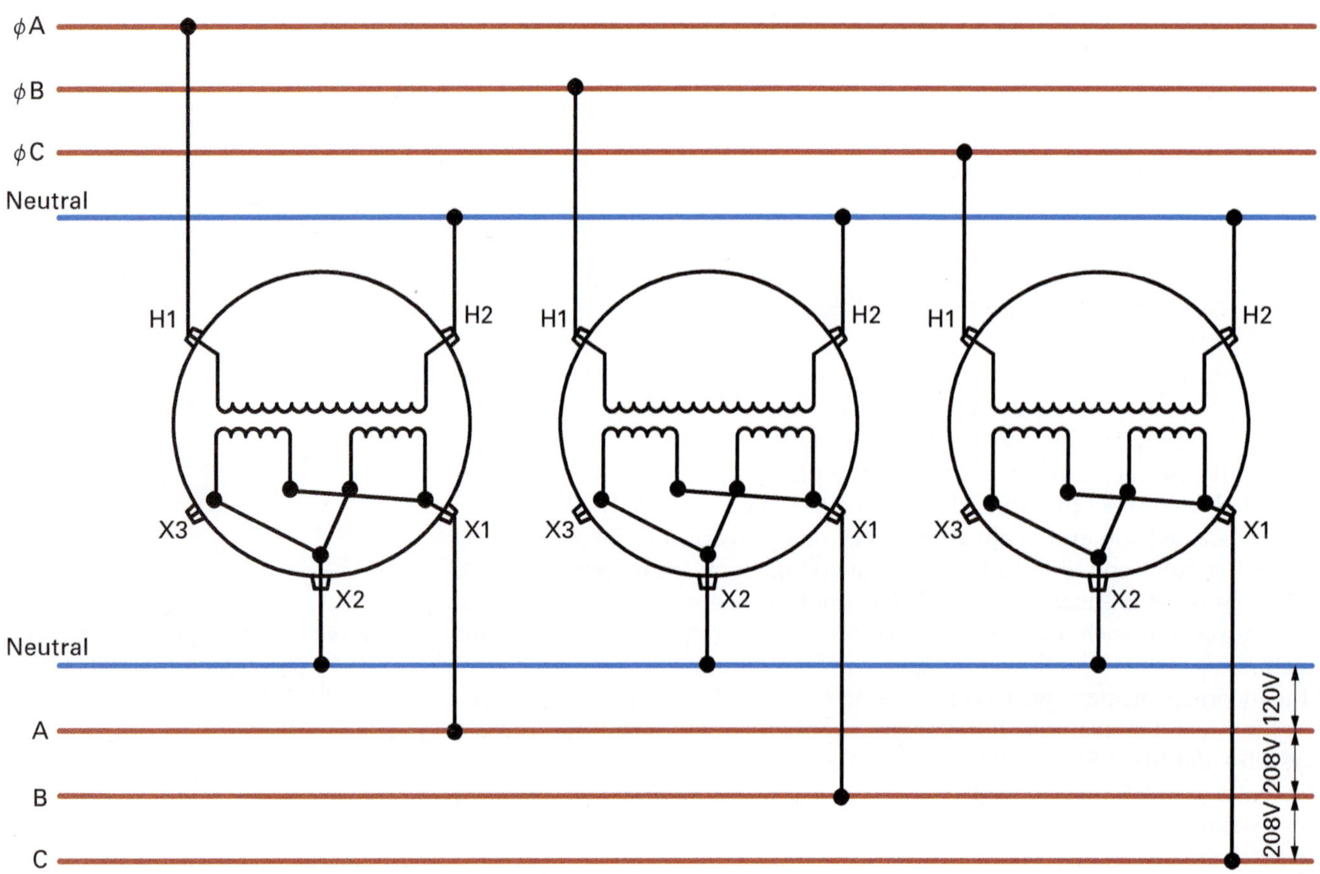

Figure 32 Four-wire, wye-wye transformer hookup.

WARNING!

The three-phase transformer hookups shown here are for example only. Three-phase power transformer hookups, as well as their service and maintenance, must be done by properly trained, qualified electricians.

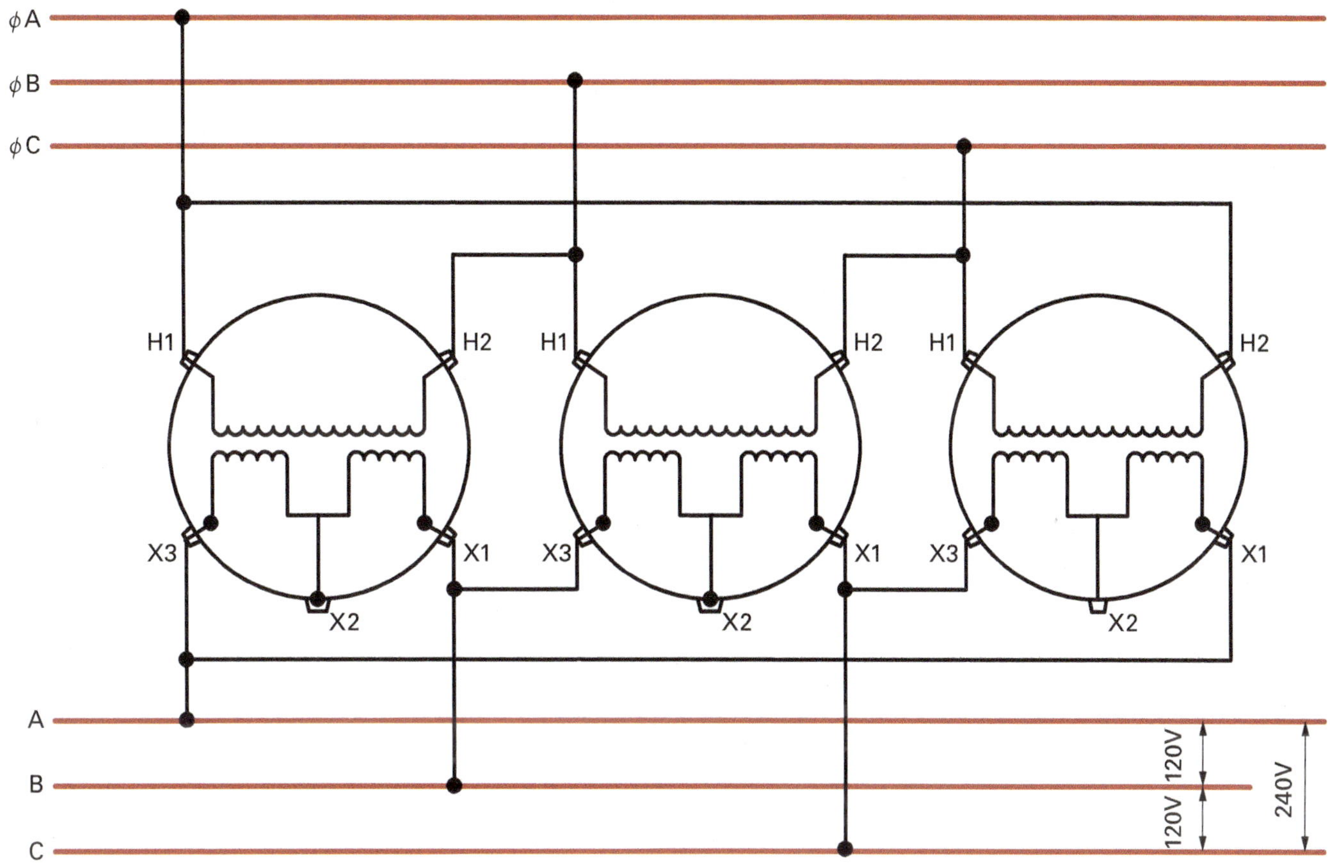

Figure 33 Closed delta-delta transformer hookup.

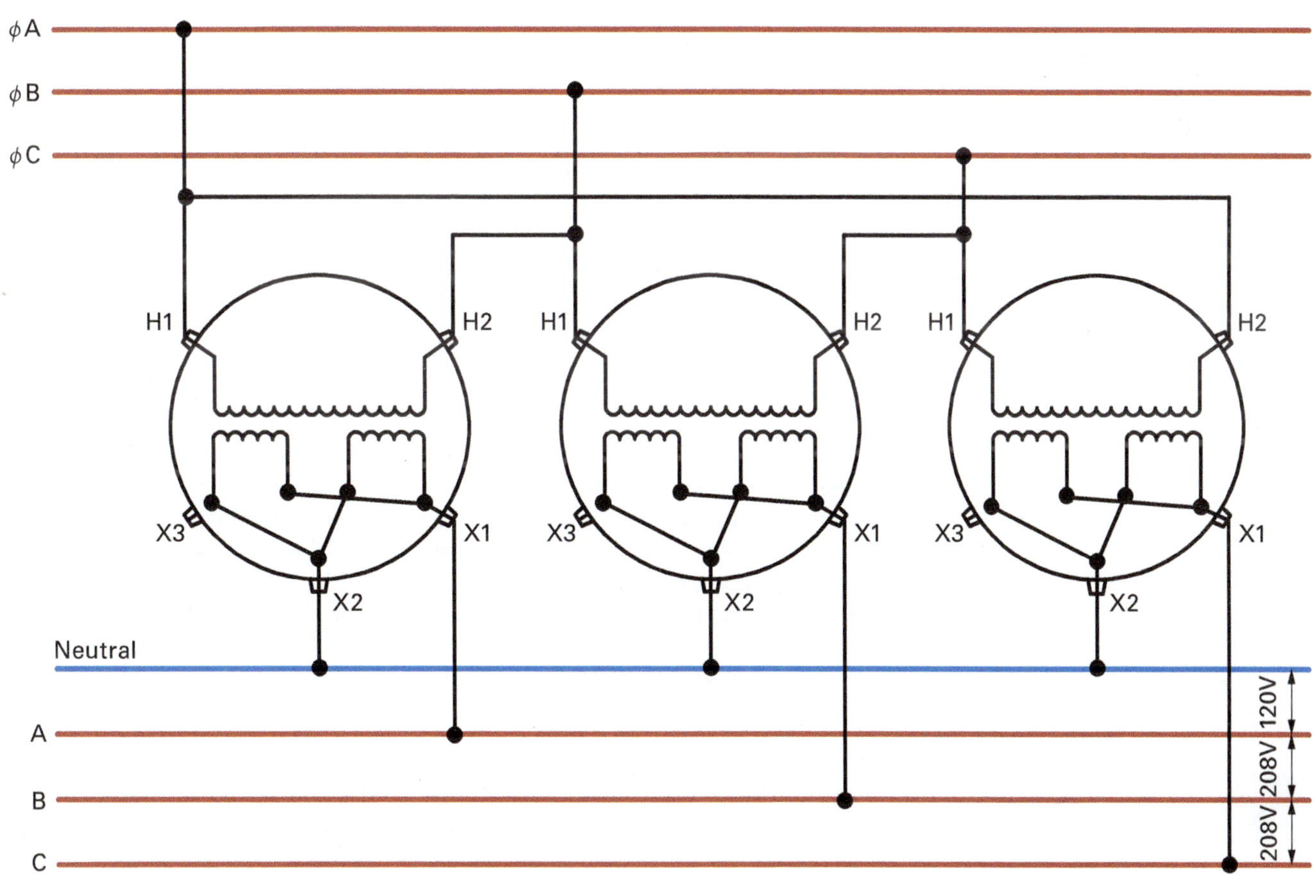

Figure 34 Delta-wye transformer hookup.

Step-Up Transformers

Transformers come in all sizes. The giant transformer shown here is used to step up the voltage produced by a power station to a very high voltage that will be carried by transmission lines. Voltages up to 765,000V, or 765 kilovolts (kV), are routinely carried by transmission lines. Substations along the way reduce the voltage to distribution-level voltages (such as 4kV), which is then sent along distribution lines to smaller transformers that reduce the voltage to a usable level for the consumer.

2.2.3 Control Transformer Replacement

When replacing a failed control transformer that is known to be original equipment, try to replace it with one of the same form and ratings. In this case, *form* refers to the way it mounts.

If it cannot be confirmed that it is original equipment, or at least has the same ratings as the original, make sure that the replacement will safely and efficiently handle the primary and secondary voltage and current demands of the load. If additional loads have been added to the circuit that were not part of the original equipment design, they must be considered in the selection process.

If someone has previously replaced a transformer with one of greater capacity than the original, there may be a good reason for that. Other loads may have been added, such as additional relays or solenoid valve coils. On the other hand, it is also possible that it was replaced with a larger size because it was what another technician had on the truck in a time of need.

If the new transformer is not equipped with a serviceable fuse or integral circuit breaker, it is never wrong to install an external fuse holder to protect the secondary (*Figure 35*). Consider that transformers are very reliable if the load is not excessive, but it failed for some reason. A shorted contactor coil, for example, could cause that to happen. Inserting a fuse sized to handle the current rating of the transformer is excellent insurance. If there is a problem in the control circuit, the fuse will protect the new transformer. Fuse holders come in various forms. The one shown in *Figure 35* is designed to simply be spliced into the secondary wiring as it leaves the transformer.

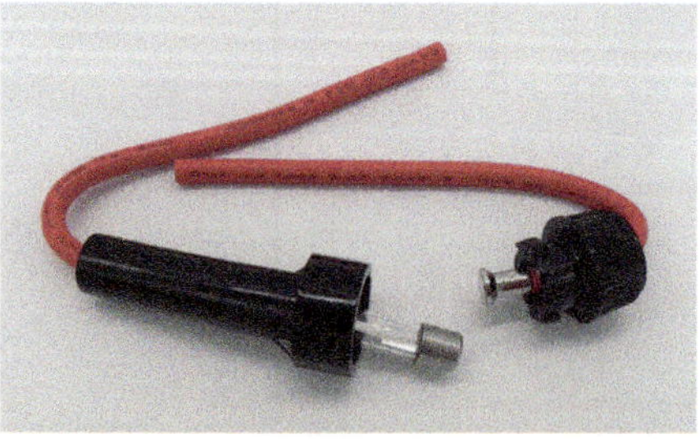

Figure 35 Inline fuse holder.

For control transformers, the following are the most important characteristics to consider:

- *Primary voltage* — You must match the listed input voltage with reasonable precision. Providing 208V to a transformer primary designed for 240V, for example, will significantly reduce the voltage provided by the secondary winding. Two or three volts of difference between the two won't likely cause a problem. However, a wider difference will reduce or increase the secondary voltage.

- *Secondary voltage* — Most control transformers produced for HVACR work are designed to provide 24V. A few are rated to provide as much as 27V, which can be useful if you are dealing with a circuit that has a significant amount of voltage drop. A long run of undersized thermostat wire is a common cause of voltage drop in control systems.
- *VA rating* — In most cases, matching the VA rating of the failed transformer will work. If you plan to add loads, you'll need to increase the VA rating to accommodate the added demand.
- *Mounting* — Transformers take many forms and can be secured in different ways. Some, especially smaller transformers, come with accessories that provide several different mounting schemes.

Before replacing the transformer, examine and record the existing wiring connections. Be sure that you know what each wire is for and ensure that the new transformer is wired accordingly. There are at least four connections to be made—two wires for the primary winding (input) and two wires for the secondary wiring (output).

It is very important to measure both the primary and secondary voltage and current once the transformer is energized but before activating anything in the control circuit. Then measure the voltage and current again while the transformer is under load. You should also measure the voltage at several 24V control devices while they are energized to ensure there is no significant voltage drop. Ensure that all the measurements fall within expectations. Otherwise, you will likely find it necessary to replace the transformer again.

2.0.0 Section Review

1. A transformer has 200 turns in its primary winding and receives a 120V input. The secondary winding has 400 turns. The resulting voltage from the secondary winding is _____.
   a. 60V
   b. 120V
   c. 240V
   d. 400V

2. True or False: In HVACR work, step-up transformers are most often found in fossil-fuel ignition systems.
   a. True
   b. False

3.0.0 Induction Motors

Performance Task	Objective
2. Identify the start and run windings of an AC motor with an ohmmeter.	Describe various types of induction motors and explain how they operate. a. Describe how induction motors operate. b. Describe single-phase motor starting circuits.

Induction motors: Electric motors that operate through induction.

Horsepower: A unit of measure for power equal to 550 foot-pounds per second, or 746 watts.

Torque: A measure of force that causes an object, such as a motor shaft, to rotate.

The primary load devices in HVACR equipment are AC **induction motors**. They range in size from a small fraction of one **horsepower** to hundreds of horsepower. They drive compressors, pumps, blowers, and many types of fans, among other things. Electromagnetism and induction are the keys to the operation of an AC induction motor.

There are several types of single-phase AC motors used in HVACR equipment. They are selected for their starting **torque** and running characteristics. Three-phase motors are used where more torque and/or horsepower is required.

3.1.0 Induction Motors

Both single-phase and three-phase motors are used in HVACR applications. Since three-phase power is not available at most residences, three-phase motors are found only in larger systems serving the commercial and industrial markets. Single-phase motors vary in design and are more complex than three-phase motors. For that reason, most of this section focuses on the various single-phase motors you will encounter.

Rotor: The rotating component of an induction motor.

Stator: The stationary portion of a motor, primarily consisting of the windings.

The two primary components of a single-phase motor are the **rotor** and the **stator** (*Figure 36*). The stator is fixed in place, while the rotor turns. The motor accomplishes work by converting electrical energy into mechanical energy. Mechanical energy is delivered by attaching a device such as a fan blade or pump shaft to the rotor shaft of the motor.

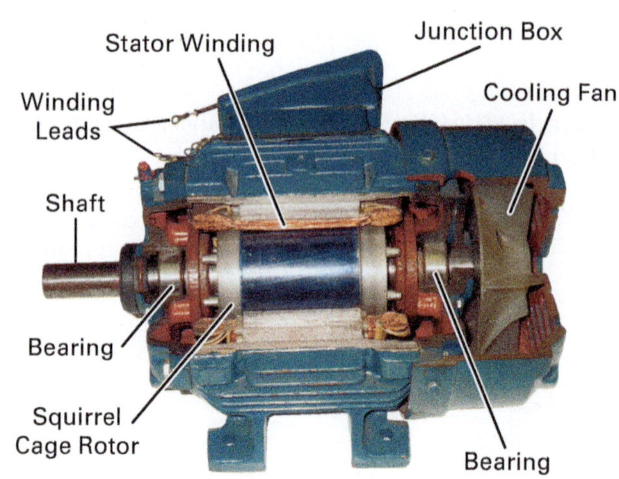

Figure 36 Parts of a motor.

To turn the rotor, voltage is applied to the stator winding (*Figure 37*), which contains many turns of wire. The magnetic field created by the current through the stator winding attracts or repels the rotor, causing it to turn. Because the polarity of the AC voltage applied to the stator is constantly changing, the relationship of the north and south magnetic poles of the stator and rotor constantly changes. This creates a rotating magnetic field that continuously pulls and pushes the rotor and keeps it turning.

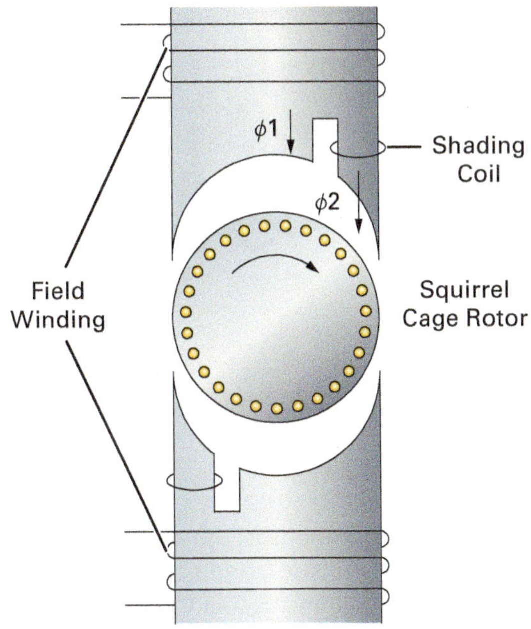

Figure 37 Basic AC motor.

NOTE

Motors, motor starting circuits, and motor troubleshooting are presented in greater detail in NCCER Module 03317, *Control Circuit and Motor Troubleshooting*.

The rotational speed of a motor, measured in revolutions per minute (rpm), is determined by the number of stator windings and the frequency of the applied voltage. The speed at which the magnetic field rotates is referred to as its **synchronous speed**. Synchronous motors, those that rotate at precisely the same speed as the magnetic field, are relatively rare in HVACR equipment. They are most often used in applications where precise rotational speed is required, such as a timer or tape recorder.

Synchronous speed: The speed at which the magnetic field in a motor rotates, determined by the number of stator windings and the frequency of the applied voltage.

Standard induction motors experience some *slippage*, or inefficiency, in the operation of the motor. Therefore, the actual speed of standard induction motors is about 95% to 97% of their synchronous speed.

3.1.1 Single-Phase Multispeed Motors

The speed of some motors can be changed using taps on the stator winding (*Figure 38*). They are referred to as *multispeed motors*. Note that their speed is not considered *variable*, as their speed can't be modulated using winding taps. It is important to use the terms *multispeed* and *variable speed* correctly when referring to a motor. Multispeed motors offer a selection of fixed speed choices.

Instead of using the entire stator coil, only a portion of it is energized—the larger the portion, the lower the speed. This creates more **impedance**, resulting in lower current, which reduces the power available. This method allows a single motor to be used to do varying levels of work. In many furnaces, for example, the controls automatically select a lower fan speed for heating and a higher fan speed for cooling.

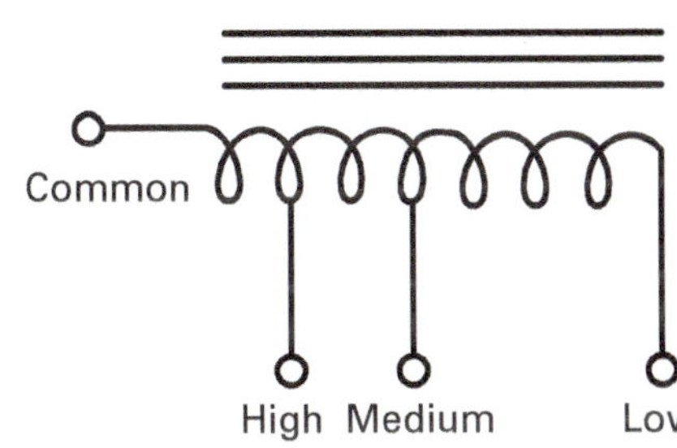

Figure 38 Multispeed motor speed taps.

Impedance: The effective resistance of a circuit or component to the flow of current.

3.1.2 Electronically Commutated Motors

Electronically commutated motors (ECMs) are now widely used in residential and commercial HVACR equipment. ECMs are three-phase, DC-powered motors that are *brushless*. They are introduced here because they are powered by the same AC power source serving the equipment in which they are installed. The AC input is converted to DC inside the motor assembly. ECMs are rapidly replacing fractional- and small-horsepower induction motors in a majority of HVACR applications.

Most ECM motors consist of two sections: a motor section and a control section (*Figure 39*). The control section has two functions. It first takes the single-phase AC input and converts it to DC. This is called *rectification*. The control section then converts the DC power to three-phase power before applying the voltage to the motor.

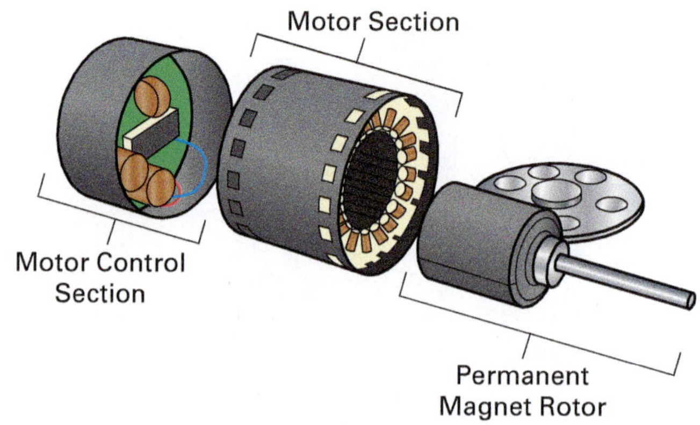

Figure 39 Major components of an ECM.

The motor control section also houses the electronics that control motor speed. ECMs are considered *variable speed motors*, since the speed can vary across a wide range. The permanent magnet rotor is turned by sequentially energizing the three sets of windings in the stator to move the rotor in the desired direction.

ECMs offer several benefits to the end user, including the following:

- *Increased energy savings* — ECMs consume a fraction of the power of conventional motors.
- *Improved comfort* — When coupled with modulating or multistage heating and cooling products, they can modulate the fan speed to match the heating or cooling capacity. Fan speeds can also be slowed down to allow for better dehumidification without overcooling the conditioned space.
- *Flexibility* — The operating characteristics of an ECM can be tuned to better meet the needs of each unique installation.
- *Quieter operation* — ECMs slowly ramp up to the required fan speed from a stop. When the thermostat is satisfied, they slowly ramp down their speed before shutting off. Continuous fan operation can modulate down to a very low speed, allowing for good air circulation without excess air movement and noise.

Rock and Roll!

When an ECM motor is first energized, it may rock back and forth a few times, appearing confused about what to do. This is normal operation, as the motor works to determine which direction is which and then chooses a direction in which to rotate.

3.1.3 Three-Phase Motors

Three-phase motors are used primarily in large commercial systems. They require three-phase AC power, which is not available in most residential areas. Although it may seem that three-phase motors would be more complex than single-phase motors, in fact they are much simpler to test and troubleshoot. Three-phase motors offer several other important advantages over single-phase motors:

- They offer excellent starting torque and require no special starting equipment, such as capacitors and start relays. Because the three stator windings are 120 degrees out of phase, at least one of the windings is always applying torque to the rotor. This feature also means a three-phase motor generates less vibration and pulsations than many single-phase motors.
- The direction of rotation can easily be reversed by reversing any two of the three supply voltage connections to the motor.
- They run very smoothly.

Internally, three-phase motor stators are connected in either a delta or wye configuration, depending on the application (*Figure 40*). A delta-connected motor provides more current, and therefore more starting torque. However, it also consumes more power than a wye-connected motor.

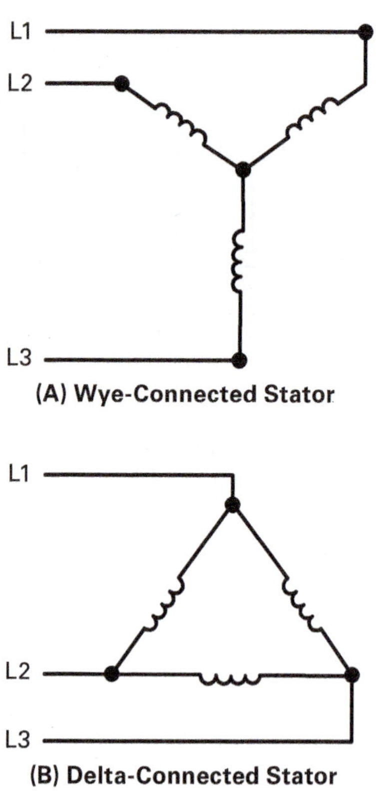

Figure 40 Three-phase motor stator windings.

3.2.0 Single-Phase Motor Starting Circuits

The problem with single-phase motors is getting them started. First, it takes extra energy (torque) to overcome **inertia** and force a rotor at rest to turn. An inductive load like a motor draws a great deal more current as it starts than it does when it is running at full speed. This is true regardless of the load that may be placed on the motor.

Inertia: The tendency of a body in motion to remain in motion and a body at rest to remain at rest.

Second, if the rotor comes to a stop at a position where the stator field is exerting equal pushing and pulling forces, the motor will not restart when power is reapplied. Various ways of dealing with these problems are incorporated into the motor design.

3.2.1 Split-Phase Motors

In the split-phase motor (*Figure 41*), an additional winding known as the **start winding** is added to the stator. The start winding contains many turns of very fine wire. Therefore, the current increase is slower than that of the **run winding**, which has fewer turns of heavier wire. The phase difference between the two windings creates torque that starts the rotor turning. A centrifugal switch opens to remove the start winding from the circuit once the motor has started and reached a speed that ensures the run winding will keep it turning without further assistance.

Start winding: The stator winding of a motor that is used to provide starting torque.

Run winding: The stator winding of a motor that draws current during the entire running cycle of the motor.

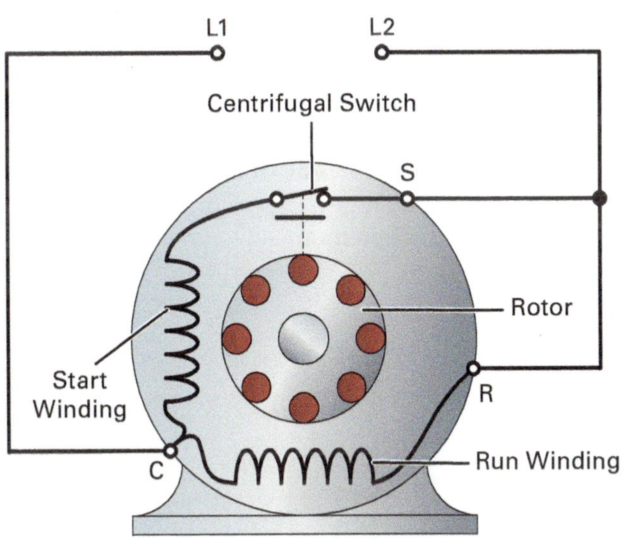

Figure 41 Split-phase motor schematic.

The rotor of a split-phase motor consists of conductive bars set in an iron core and connected at the ends by a copper or aluminum ring. This type of rotor is called a *squirrel cage rotor* because the arrangement of the copper bars resembles a cage. Split-phase motors can be used to power pumps, oil burners, and similar applications requiring $\frac{1}{3}$ horsepower (hp) or less.

Figure 42A shows a centrifugally actuated switch that is closed when the motor is at rest. This switch is wired in series with the start winding. The switch is opened and closed by the movement of an actuator disk against a contact lever. When the motor is at rest, springs retract the weight and cause the actuator disk to move toward the bearing at the end of the shaft. This pushes on the contact lever, closing the switch contacts.

After the motor starts and reaches about 75% of its rated speed, the weights swing out, overcoming the spring tension, and retract the contact disk from the lever. This opens the switch, removing the start winding from the circuit.

If the switch fails to open for some reason, the motor will operate at a reduced speed until it overheats the start winding, unless a *thermal overload* in the motor senses the condition and opens the circuit. *Figure 42B* shows the motor start and run windings.

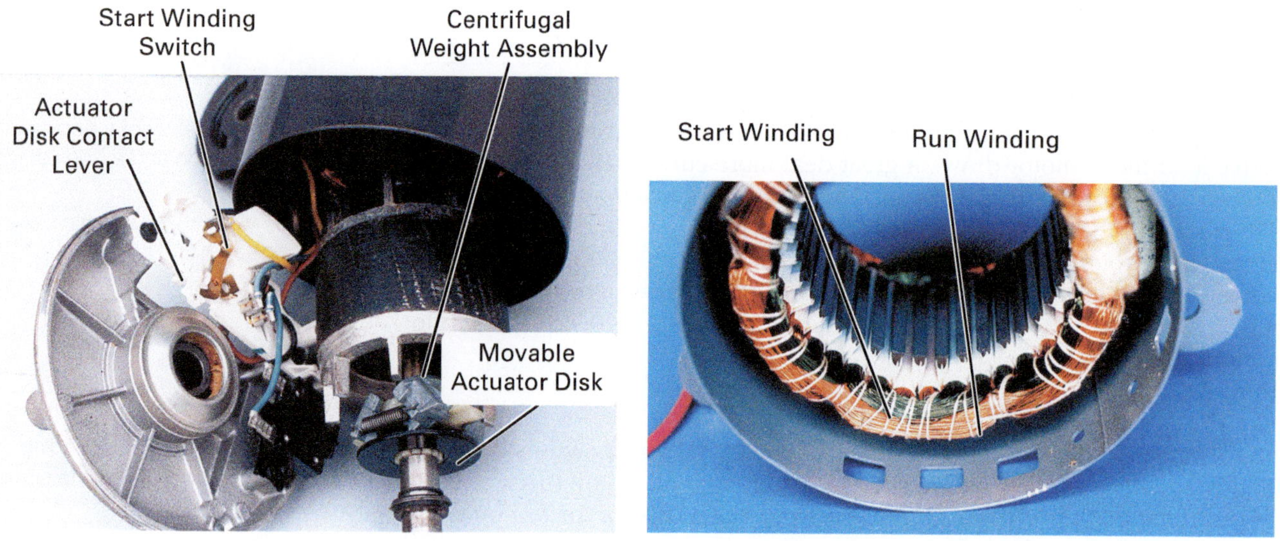

(A) Components (B) Windings

Figure 42 Split-phase motor components.

3.2.2 Capacitors

Capacitor: A passive electrical energy storage device containing two metal plates separated by an insulating (dielectric) material.

Before discussing other types of single-phase motors, it is essential to understand the function of a **capacitor** in single-phase motor operation. Split-phase motors do not incorporate a capacitor, but most of the motors you work with do require one.

A capacitor (*Figure 43*) is an electrical storage device that charges and discharges as the applied voltage changes. Capacitors play an important role in the operation of single-phase induction motors.

(A) Start Capacitor (B) Run Capacitor (C) Dual Run Capacitor

Figure 43 Examples of start and run capacitors.
Source: (A) Elcap via Wikimedia; (B) Winai_Tepsuttinun/Getty Images; (C) Ohms1999/Shutterstock

Dielectric: A material that strongly resists the passage of current.

Microfarad (mfd): A unit of measure for capacitance, equaling one-millionth of a farad.

A capacitor consists of two metal plates separated by an insulating material known as a **dielectric** (*Figure 44*). Its capacity, which is measured in **microfarads (mfd)**, is determined by the size of the plates, the distance between the plates, and the type of dielectric material used. A *farad* is a unit of measure for capacitance, named for English physicist Michael Faraday. One microfarad represents one-millionth of a farad.

Current will not flow through dielectric materials. When an AC voltage is applied across the plates of the capacitor, electrons will flow from one plate, through the load, and collect on the other plate, creating a charge with the same polarity as that of the input waveform. When the AC waveform changes direction (alternates), the capacitor discharges; that is, the charge built up on one plate flows rapidly through the load to the other plate.

Because the voltage across the capacitor is created by current flowing from one plate to the other, the voltage across the capacitor lags the current. In a

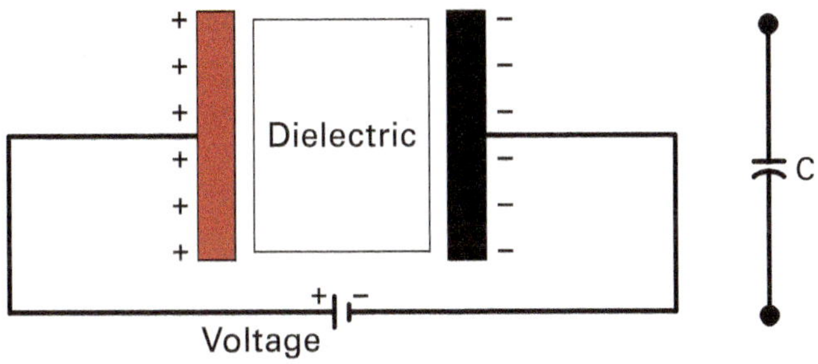

Figure 44 Basic capacitor construction and wiring diagram symbol.

purely capacitive circuit, voltage lags current by 90 degrees. This change in sine waves is known as a *phase shift*. The phase shift here is the opposite of that in an inductor, where voltage leads the current. *Figure 45* shows the phase relationships in resistive, inductive, and capacitive circuits. The phase shift is how a capacitor helps get a motor started.

In HVACR applications, there are two common types of capacitors. In an *oil-filled capacitor,* paper soaked with a special fluid serves as the dielectric. This capacitor has large plates and a large amount of dielectric oil to dissipate heat. This allows it to remain in the circuit continuously. Motor run capacitors are constructed this way. When a run capacitor is used in a motor circuit, it must be able to remain energized while the motor runs.

Run capacitors are available in single and dual configurations, as shown in *Figure 43.* A dual run capacitor has a Common terminal and two others. The capacitor is designed to support both a condenser fan motor and a compressor in residential and light-commercial equipment. To accommodate the different motors, the capacitor provides two microfarad values. For example, a dual run capacitor may be rated for 25+5 mfd. The 25-mfd side is wired to the larger motor (compressor), and the 5-mfd side is wired to the fan motor.

An *electrolytic capacitor* is often smaller and contains less dielectric material. It will overheat and be damaged if it is left in the circuit for very long. This approach is used for *start capacitors.* Start capacitors must be switched out of the circuit soon after being energized.

Capacitors and the phase shift they provide play an important role in helping to start single-phase induction motors and make them operate more efficiently.

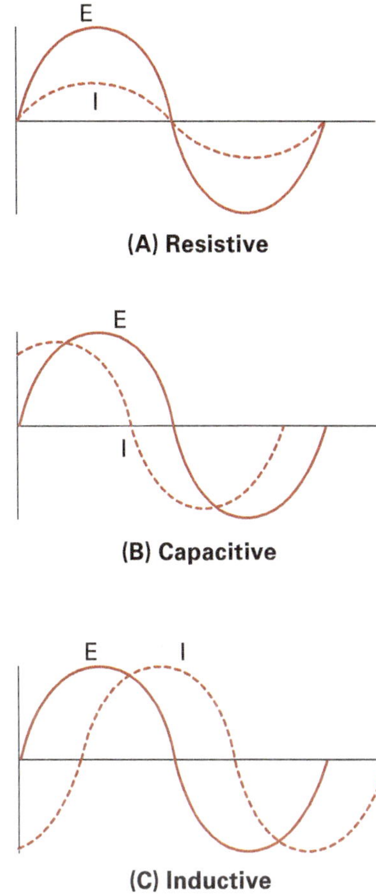

Figure 45 Phase relationships.

WARNING!

Some older capacitors may contain polychlorinated biphenyl (PCB), which was at one time widely used as a dielectric in capacitors and power distribution transformers. PCB is known to cause cancer in humans. If you have a defective oil-filled capacitor or transformer, do not attempt to pry it open or puncture it. If it is leaking, don't touch the oil or breathe the fumes. Dispose of it in accordance with applicable local or national codes.

Note that electrolytic start capacitors, typically constructed with a phenolic case instead of metal, do not contain oil.

3.2.3 Permanent Split Capacitor Motors

Permanent split capacitor (PSC) motors are easily the most common type used in HVACR equipment over the years, although they are steadily being replaced by ECM motors. They are used to drive blowers, fans, and compressors in residential and light commercial applications. They are also commonly used for condenser fans on much larger equipment.

CAUTION

It is very important that capacitors be sized correctly. If you must install or replace them, use the capacitance value specified by the motor manufacturer. However, a capacitor with a higher voltage rating than the original can be used if necessary.

In a PSC motor (*Figure 46*), a run capacitor provides a phase shift between the run and start windings that helps start the motor, then it remains in the circuit to improve running efficiency. A start capacitor is not generally required.

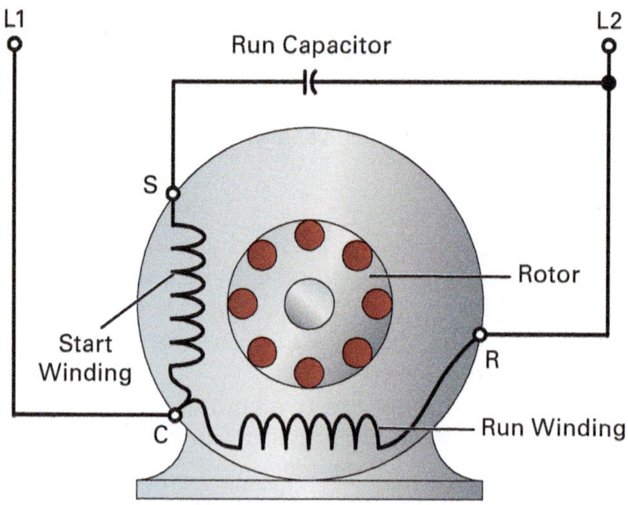

Figure 46 PSC motor schematic.

3.2.4 Capacitor-Start Motors

Capacitor-start motors range in capacity from fractional horsepower to as high as 10 hp. The high starting torque they provide makes them suitable for powering fans and even some refrigeration compressors.

The start capacitor, wired in series with the start winding, provides plenty of starting torque with a relatively low starting current. It is an electrolytic capacitor and must, therefore, be switched out of the circuit as soon as the motor reaches around 75% of its rated speed. The start winding itself must also be de-energized. Two methods are commonly used to do this. In *Figure 47*, a centrifugal switch is used, as described earlier for split-phase motors.

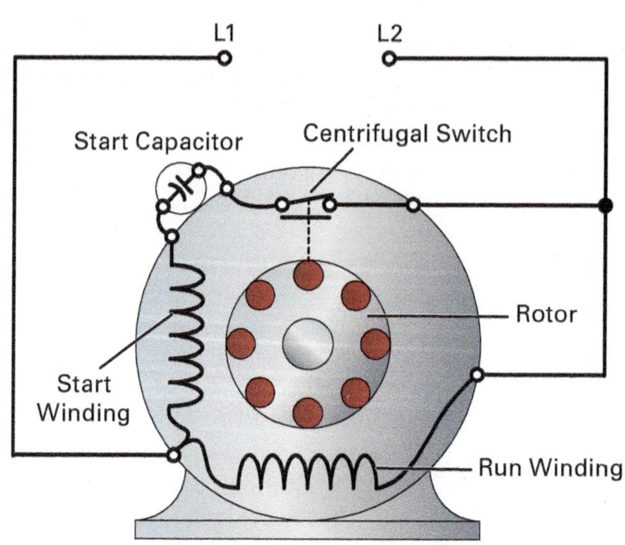

Figure 47 Schematic of a capacitor-start motor with a centrifugal switch.

A more common approach is shown in *Figure 48*. A start relay is used to enable and disable the start capacitor. In this case, the relay shown is called a **potential relay**. The name refers to the fact that the coil is energized by a voltage with more "potential" than the line voltage.

Potential relay: A relay energized by back EMF voltage generated during startup by the start winding in a capacitor-start motor.

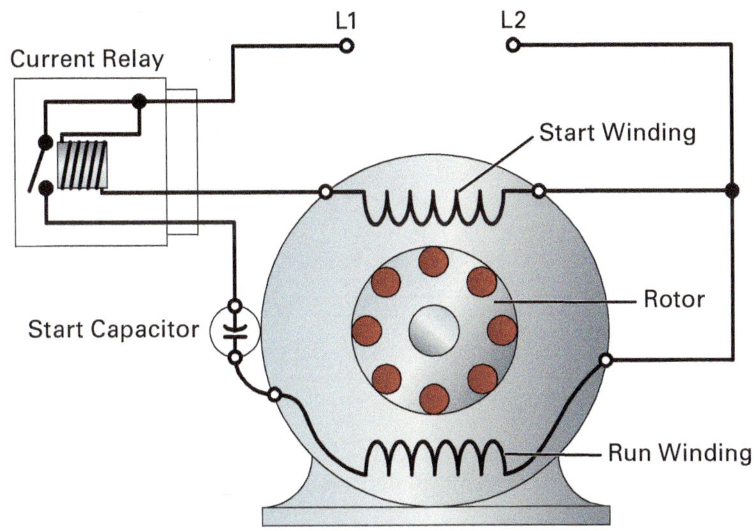

Figure 48 Schematic of a capacitor-start motor with a start relay.

While in operation, all electric motors have some capacity to generate a voltage. This results from induced voltages being generated in the motor stator windings by the motion of the rotating rotor. The voltage generated is known as *back electromotive force*, or back EMF.

It is important to understand that the back EMF generated in a motor normally has a much higher potential than the line voltage applied to the motor. For example, a motor being driven by a 230V line source can generate a back EMF voltage exceeding 400 volts.

In a capacitor-start motor using a potential start relay, the back EMF voltage, not the line voltage, energizes and actuates the start relay. By design, the start relay will energize at a specific back EMF voltage that is higher than the line voltage. The contacts of a potential relay are normally closed.

When the motor is first energized, the start relay remains de-energized. The start capacitor and start winding are connected in the circuit through the normally closed contacts of the start relay. As the motor picks up speed, the back EMF in the start winding increases. Eventually, it rises high enough to energize the start relay coil. This opens the relay contacts and disconnects both the capacitor and start winding from the circuit. As long as the motor continues to run at or above the current speed, the back EMF generated in the start winding keeps the start relay contacts open.

CAUTION

It is very important to use the correct start relay. If you must install or replace a relay, use the relay specified by the motor manufacturer.

3.2.5 Capacitor-Start, Capacitor-Run Motor

The capacitor-start, capacitor-run (CSCR) motor is often used to drive refrigeration compressors (*Figure 49*). It combines high starting torque with smooth, quiet, and efficient operation. However, CSCR motors are not quite as efficient as PSC motors. The substantial torque developed, though, provides a better chance for a successful start when the refrigerant circuit pressures are not fully equalized. Like the capacitor-start motor, the start capacitor is switched out of the circuit when the motor comes up to speed. The motor then operates like a PSC motor.

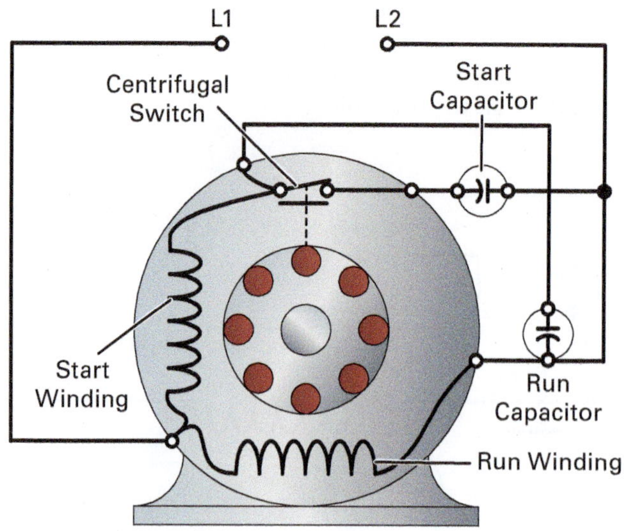

Figure 49 Capacitor-start, capacitor-run motor schematic.

NOTE

Motors, motor starting circuits, and motor troubleshooting are presented in greater detail in NCCER Module 03317, *Motor Testing and Troubleshooting*.

3.2.6 Shaded-Pole Motor

In a shaded-pole motor (*Figure 50*), the rotor turns within two pairs of stator windings. A groove in the stator pole separates a small portion of the stator from the rest. A metal band, called a *shading coil*, is placed around the smaller portion. This band causes a slight phase shift, which is enough to provide the torque to start the motor. However, it does not develop enough torque to start the motor under a significant load.

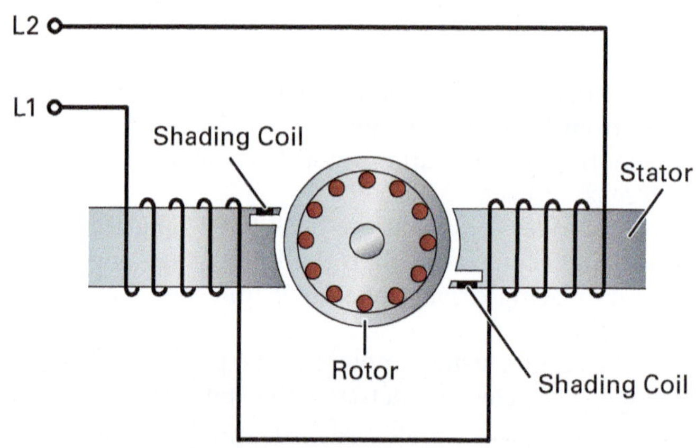

Figure 50 Shaded-pole motor schematic.

Shaded-pole motors are only used to drive the smallest fans and pumps, since they have very little starting torque. However, you will encounter them from time to time.

3.0.0 Section Review

1. The speed of a single-phase motor is determined by the number of stator windings and the _____.
   a. diameter of the armature
   b. number of rotor windings
   c. resistance of the run winding
   d. frequency of the applied voltage

2. A start relay may be used to control power to the start winding in a _____.
   a. three-phase motor
   b. shaded-pole motor
   c. capacitor-start motor
   d. PSC motor

4.0.0 Basic Component Testing and Wiring Termination

Objective

Explain how to test various AC-powered components and terminate wires.

a. Identify basic electrical safety guidelines.
b. Explain how to use meters to test various electrical devices and motors.
c. Identify common electrical termination hardware and explain how to properly terminate wiring.

Performance Tasks

Wire a multitap control transformer for the available primary voltage, energize the transformer, and measure the primary and secondary voltage.

3. Identify the start and run windings of an AC motor with an ohmmeter.
4. Beginning with raw wire, fabricate crimped connections and terminate both low- and line-voltage wiring on a compressor contactor.

Most components of AC circuits can be tested with the standard electrical test meters—the voltmeter, ammeter, and ohmmeter. A multimeter, or VOM, will handle most voltage and resistance readings and a clamp-on ammeter will suffice for current readings. In addition to these instruments, a capacitor tester and a **megohmmeter (megger)**, will sometimes be needed. A capacitor tester may also be incorporated into better multimeters.

Megohmmeter (megger): A test instrument used to test high-resistance circuits using a high-voltage test signal.

This module will present some basic electrical troubleshooting procedures. Note that more advanced troubleshooting techniques and opportunities to practice are presented in NCCER Module 03314, *Control Circuit and Motor Troubleshooting*. In electrical troubleshooting, practice and safe experimentation cannot be overdone.

4.1.0 Electrical Safety

Alternating current can be a deadly force if carelessly handled. One tenth of an ampere of alternating current flowing through a vital organ can prove fatal. Therefore, safety precautions must be observed whenever working with electrical circuits and devices.

The amount and duration of the current flow, the parts of the body involved, and the frequency of the current determine the potential for injury. The most serious injuries occur when the current flow is through or near nerve centers and vital organs.

4.1.1 Electrical Safety Hazards

People differ in their resistance to electric shock. A current level that may cause only a painful shock to one person might be fatal to another.

Table 1 presents the effect of 60Hz AC flowing through the body from hand-to-hand or hand-to-foot. The table shows that at approximately 1 milliampere (mA), or 0.001A, an electrical current can be felt. A 10mA shock is typically enough to prevent voluntary control of the muscles. This often leads to other physical injuries, especially if working on a ladder.

As the current increases beyond 20mA, muscle contractions can prevent the victim from pulling away from the current source. At 50mA and above, the shock can be fatal if it continues for more than a second. Respiratory paralysis may result in suffocation.

TABLE 1 Effects of Current on the Human Body

Current Value	Typical Effects
1mA	Perception level. Slight tingling sensation.
5mA	Slight shock. Involuntary reactions can result in other injuries.
6mA to 30mA	Painful shock. Loss of muscular control.
50mA to 150mA	Extreme pain, respiratory arrest, severe muscular contractions. Death possible.
1,000mA to 4,300mA	Ventricular fibrillation, severe muscular contractions, nerve damage. Typically results in death.

Did You Know?

Electrical Mishaps

Approximately 30,000 nonfatal and 1,000 fatal electrical shock accidents occur each year. However, most hospital admissions related to electrical accidents are from arc flash burns, not from shocks. An arc flash releases thermal energy at temperatures up to 35,000°F (19,427°C).

Current levels above 150mA may cause the heart to beat in an abnormal rhythm. This condition is fatal unless the heart rhythm is corrected using a medical device called a *defibrillator* (*Figure 51*). Defibrillators are possibly better known by their familiar acronym, AED, which stands for *automated external defibrillator.* Current levels of 4A or more may stop the heart altogether, resulting in death unless immediate care is provided. Remember, voltage doesn't kill you; it is the current that kills.

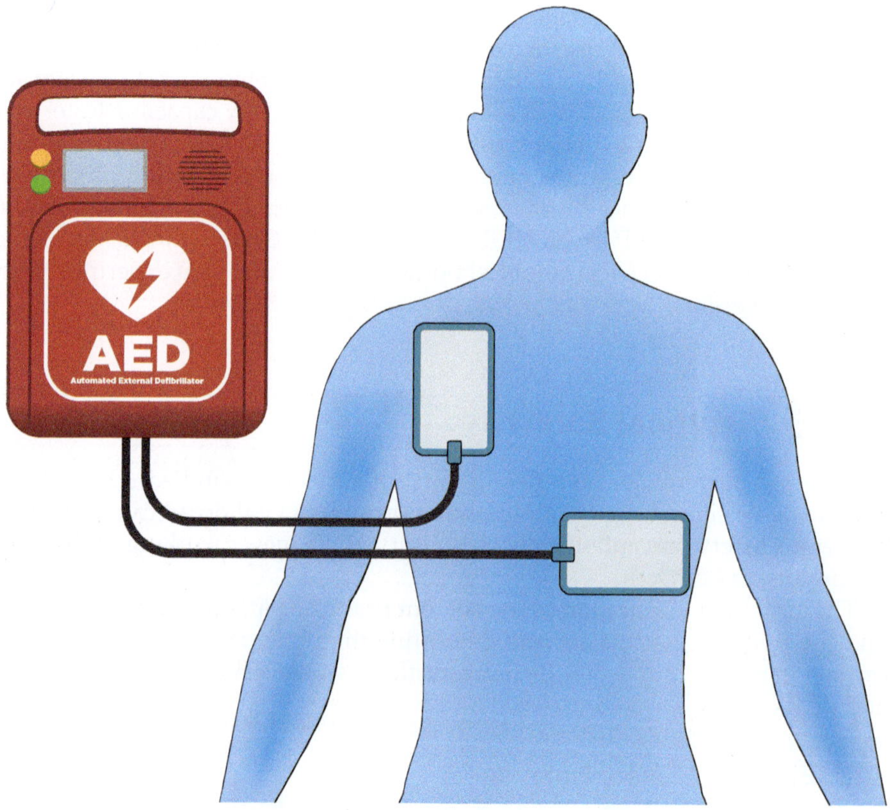

Figure 51 Automated external defibrillator (AED).
Source: chombosan/Getty Images

Other possible results of electrical shock include entry and exit wounds from high-voltage contact, and thermal burns from a current flow of a few amps and up. Thermal burns are often not immediately apparent. However, the tissue in the current path may be destroyed, causing the skin to die from the inside over time.

Current will follow any available path back to its source, but it prefers the path of least resistance. Because of this, you can prevent yourself from being shocked by being electrically isolated from ground. This means that no part of your body touches a grounded point. Another way to prevent electrical shock is to place one end of a conductor on the part of the circuit you are working on and the other end onto a grounded point. A conductor has almost no resistance, so current would likely flow through the conductor to ground instead of your body.

Two conditions must be present for an electric current to flow through the body and cause electric shock. First, the body, or some part of the body, must be part of a closed circuit. Second, there must be a voltage somewhere in the closed circuit. Thus, one way to prevent electric shock is to avoid making your body part of a closed circuit. Keeping your body well-insulated from the ground is an excellent preventive measure.

Ground Fault Circuit Interrupters

Ground fault circuit interrupters (GFCIs) are no joke. They are not just a gimmick—a GFCI-equipped extension cord can save your life. Always minimize your risk of shock by using extension cords equipped with a GFCI. If there is any current leakage to ground, the GFCI will trip long before a conventional circuit breaker would and before any potentially harmful levels of current are reached.

4.1.2 Avoiding Electrical Hazards

Most electric shocks result from human error, rather than equipment failure. Nearly all deaths due to electrical shock result from a worker's failure to observe safety precautions, failure to repair electrical defects, or failure to remedy all defects that are discovered.

It is part of your job to know and follow your employer's safety rules, policies, and procedures. It will take time for you to become familiar with all safety guidelines, but until then you can use these 10 general guidelines to develop good safety habits:

1. Always wear your hard hat and other safety equipment where required.
2. Use your tools, equipment, and PPE the way they were designed to be used.
3. Inspect your tools, PPE, and equipment daily before use, and never use damaged or unsafe equipment.
4. Perform only those tasks for which you have been trained.
5. Take personal responsibility for your safety as well as act in the best interests of your co-workers.
6. Work toward understanding and embracing your company's safety rules and policies.
7. Follow your employer's policy for tagging out, repairing, or discarding unserviceable tools and equipment.
8. Immediately correct or report all unsafe conditions to your supervisor.
9. Get involved with your company safety program.
10. When in doubt, stop and ask!

Electrical safety is largely about avoiding actions that allow you to make contact with energized circuits. Follow these basic safety precautions on the job to keep yourself and others safe from electrical shock:

- Never cut off the ground prong of a power plug or use a power cord that has had the prong removed.
- Never touch any electrical wire without ensuring that it is de-energized.
- Never switch an appliance on or off while standing in water or touching a wet surface.
- Always turn off power at the main disconnect before working on an electrical circuit or device. Once the equipment is de-energized, lockout/tagout devices must be attached to the source to prevent someone from accidentally energizing the circuit.
- Replace all worn or damaged power cords.
- Unplug cords by pulling on the plug head; do not pull the cord.
- If it becomes necessary to work on energized circuits, try to work with only one hand inside the panel at a time. If you are shocked while using only one hand, current will probably flow through the hand and arm, then through the feet toward the ground. If a shock is conducted through both hands and arms, the electrical path would be through the heart and lungs.
- Use protective equipment, such as rated electrical rubber gloves with leather protectors and insulated boots.
- Use well-insulated hand tools, and make sure the insulation isn't split or otherwise damaged.
- Remove metal jewelry, such as rings and watches, when working on or near energized circuits. Tie back long hair.

4.1.3 Arc Flash Hazards

Any time the electrical contacts of a component make or break, such as a switch or a contactor, an arc can occur. That is, current arcs between the open contacts as they approach or separate from each other. An arc can also occur between a live circuit and a nearby conductor, even a finger. An arc will certainly occur if a fuse is pulled from an energized and electrically loaded circuit.

The magnitude of the arc and the distance it travels are functions of the voltage in the circuit. If the voltage is high enough and the arc is sustained, an *arc flash* can occur, resulting in severe burns and damage to equipment. When very high voltages are involved, a tremendous amount of damaging energy is released in the form of an *arc blast*.

OSHA issues regulations addressing safety issues in the workplace, including electrical safety issues. While OSHA is a federal agency and its regulations are law, OSHA also relies on national consensus standards to establish certain requirements. For electrical safety, OSHA recognizes several standards from the National Fire Protection Association (NFPA). *NFPA 70®, National Electrical Code® (NEC®)*, provides requirements for electrical installations. NFPA 70E®, *Standard for Electrical Safety in the Workplace*, provides practical safe working requirements related to hazards arising from the use of electricity.

Qualified person: One who has the required skills and knowledge related to the construction and operation of electrical equipment and installations and has received relevant safety training to recognize and avoid the hazards involved.

Special safety procedures and PPE are required when working on or near circuits with voltage levels of more than 50V line-to-line. Remember that distance is the best protection against electrical hazards. Only a **qualified person** may work in areas with unguarded, uninsulated energized lines or equipment operating at 50V or more. An unqualified worker must remain well out of the danger zone at specified distances as follows:

- There is no distance specified for voltages less than 50V.
- 10' (3.05 m) for voltages from 50V to 750V.
- 20' (6.1 m) for voltages greater than 750V.

Industry standards have established specific limits of approach to exposed energized parts. These limits are for personal protection and are called *approach boundaries*. These boundaries include the following:

- An *arc flash boundary*
- *Shock protection boundaries*, including a *limited approach boundary* and a *restricted approach boundary*
- A *minimum approach distance*, or MAD

Only qualified persons are allowed within flash and shock protection boundaries and MADs. Unqualified personnel must be trained to stay away from areas designated as such. Depending on the level of voltage involved, flame-resistant (FR) PPE and other safety gear may be required.

When working with electrical equipment, always assume that the equipment is energized until you have personally verified otherwise. You should assume that all electrical equipment presents potential shock, flash, and blast hazards regardless of the voltage until you have verified otherwise.

Ideally, work on or near electrical equipment would always be performed with the circuit de-energized. This represents what is known as an *electrically safe work condition*. But that is not always possible. Therefore, specially trained workers analyze equipment installations to identify the hazards and the various boundaries.

Every possible electrical hazard within a work area must be analyzed and documented. This is called a *flash hazard analysis*. These areas must be clearly marked with appropriate signs (*Figure 52*) indicating the hazard and the boundaries. Signs must also provide information about the PPE that must be worn in the area. Specific PPE for a given situation is based on the information gathered from the hazard analysis. Personnel, both qualified and unqualified, working in the area must be trained to recognize and avoid the identified hazards. Only qualified persons using all required PPE are allowed to enter and work inside the hazard boundaries.

Figure 52 Electrical hazard warning sign.

4.2.0 Testing Electrical Devices and Motors

The most significant inductive load you'll work with regularly is the electric motor, in one form or another. Stator windings typically have a very low resistance; only a few ohms (Ω) of resistance are common. The resistance of the windings, though, will vary between motors. The larger the motor, the smaller the resistance of the winding is likely to be.

There are times when you need to differentiate between the start winding and the run winding. Remember this: The start winding of a single-phase motor will have a higher resistance than the run winding. The resistance of the start winding may be as much as three to four times higher.

The electrical resistance of motor windings is not something you'll see listed on the motor nameplate. Knowing the precise resistance of a winding, in ohms, is rarely important. What is important, though, is to know how to tell the start winding from the run winding. The actual numbers are not significant. But you may often need to sort the windings out, and the start winding will always have the greater resistance.

Measuring the resistance of either winding means placing one ohmmeter probe on the Common (C) terminal or wire and the other probe on either the Start (S) or Run (R) terminal (*Figure 53*).

When the terminals are unmarked, consider them as Terminal 1, 2, and 3. Take resistance measurements from 1 to 2, 1 to 3, and 2 to 3. The two terminals with the highest resistance are the Start and Run terminals. Now you know the remaining terminal is Common, and you can identify the other two from there.

An *open winding* will be readily apparent, since your meter will continue to read OL (open loop) on the display. When a winding is shorted, you won't be reading the full resistance of the winding, and the resistance will likely be lower than you expect (*Figure 54*).

However, when windings are shorted, they are usually shorted to ground. An important test to complete any time you are testing motor windings is to determine if they are grounded.

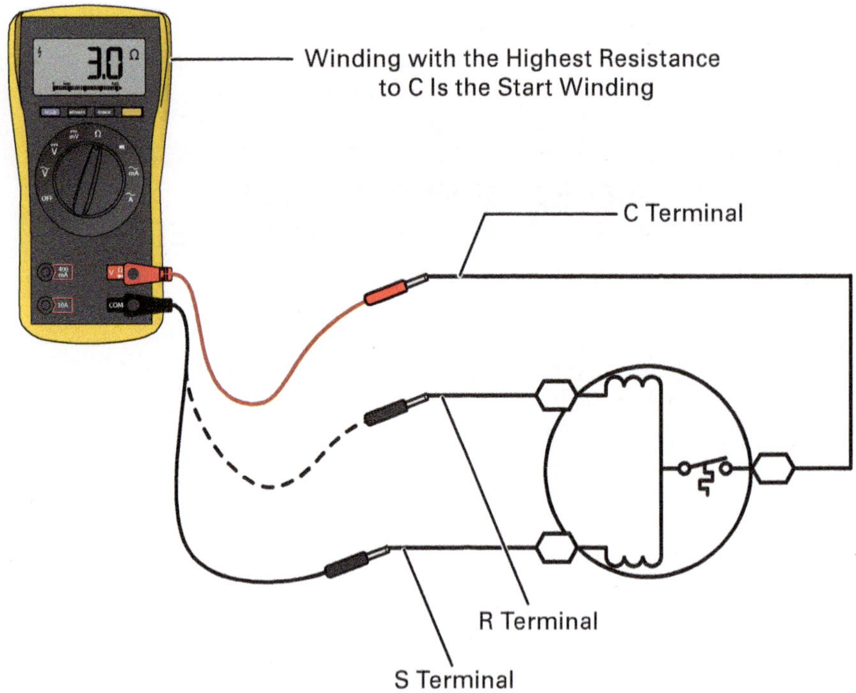

Figure 53 Identifying single-phase motor windings.

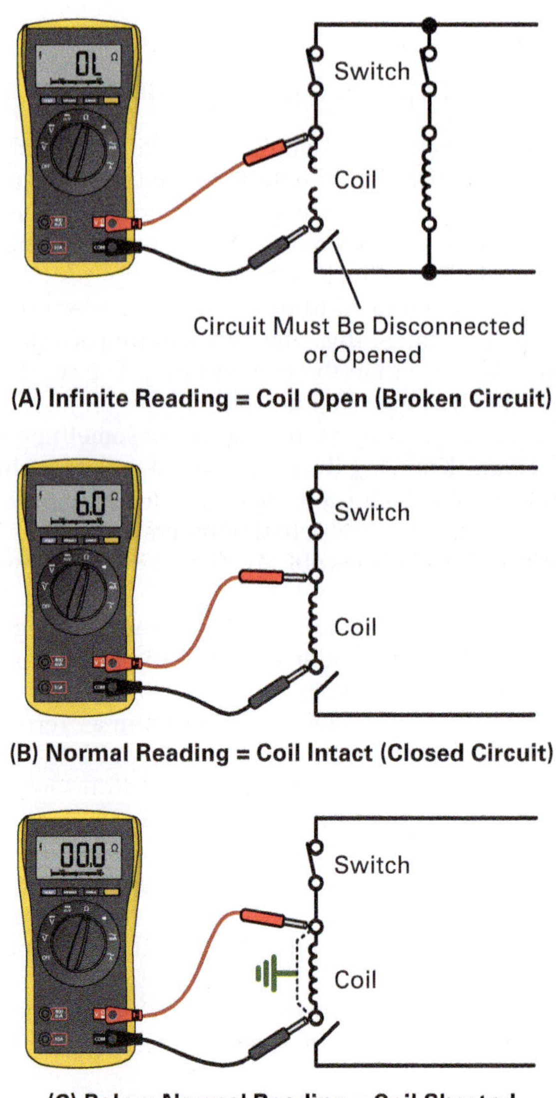

Figure 54 Testing for an open winding or relay coil.

4.2.1 Checking a Motor for Grounded Windings

If a circuit breaker serving HVACR equipment keeps tripping, you might suspect a grounded motor, or even a control device, as the source of the problem. Motors, however, due to the work they perform, are far more likely to experience a short to ground.

The following is a quick way to check an electric motor for a grounded winding using a multimeter:

WARNING!

Ensure that all sources of electrical power have been disabled and locked out. Test the circuit with a voltmeter before touching any components and testing with an ohmmeter. Test the voltmeter on a known voltage source first, test the de-energized circuit, and then retest the meter on the original known source to ensure it is functioning properly. Wear the appropriate PPE when working with line voltage circuits.

Step 1 Shut off power to the equipment and discharge any capacitors in the motor circuit. Verify with a voltmeter that the equipment is de-energized.

Step 2 Disconnect the motor leads to isolate the motor from the rest of the circuit.

Step 3 Set the ohmmeter range at its highest setting or enable the automatic ranging function.

Step 4 Place one ohmmeter probe on a ground source (usually the motor case) and the other probe on one of the motor winding leads. Be sure to avoid painted or coated surfaces when connecting the probe to ground. Firm metal to metal contact is required.

Step 5 Check the resistance of each of the remaining windings to ground. When using a common ohmmeter, any reading of continuity to ground indicates a failure.

Remember, when resistance and continuity checks are made, one end of the target component or series of components must be disconnected from the circuit. Otherwise, the ohmmeter might read the resistance of a parallel circuit. In the upper section of *Figure 54*, for example, note the open switch in the circuit. If it were closed, the meter would read the resistance of the parallel path, because it offers much less resistance than the open coil. You would not see that the winding is open.

Using a Megohmmeter (Megger)

Normally, an ohmmeter or multimeter is not used to measure extremely high resistances, such as those involving conductor insulation, insulation between motor or transformer windings, and similar components. In the exercise above, it is quite possible that you were unable to read the actual resistance between the winding and ground. More than likely, unless the winding was grounded, you would simply detect an open circuit between each winding and ground.

To adequately test motor winding resistance to a ground source, it is necessary to use a much higher electrical potential than is furnished by the small battery of an ohmmeter. Much higher test voltages, typically up to 5,000VDC, can be supplied by a megger. Note that the test voltage is DC, not AC.

Meggers measure resistance in *megohms*. One megohm (MΩ) equals 1,000,000Ω. There are small, handheld models that can provide around 1kVDC for testing (*Figure 55*). They are sufficient for most HVACR work. Larger battery-operated meggers and those that require an AC power source can develop much higher test voltages. Hand-cranked models that allow the user to develop the test voltage by cranking a tiny generator also remain available, but they are rapidly being replaced by advanced battery-operated models.

Figure 55 Examples of meggers. The model on the right can develop a 5kVDC test voltage.
Source: Ridge Tool Company

WARNING!

When using a megger, you can be injured or damage the equipment due to the extreme voltage. Follow these safety precautions at a minimum:

- Use meggers for high-resistance measurements only (such as insulation measurements or to check two separate conductors in a cable). Do not use a megger for general troubleshooting.
- Never touch the test leads.
- De-energize the circuit to be tested and verify that the power is off before connecting the meter.
- Fully disconnect the tested component from other electric circuits before using the meter.
- Never test a motor, such as one contained in a compressor, while the component is in a vacuum. This can cause internal high-voltage arcs that seriously damage the motor windings.

Meggers can perform motor insulation tests that can reveal deterioration of the motor winding insulation. The probes are used just as you would when using a standard ohmmeter (*Figure 56*). When a motor winding is grounded enough to cause a circuit breaker or fuse to open, any ohmmeter can likely detect the grounded condition, as the resistance will be relatively low. Acid in a refrigerant circuit, though, can slowly damage the insulation coating on the windings, reducing its strength. A megger can detect the reduction in insulation quality, while a common ohmmeter cannot. After a few periodic tests with a megger, you can see if the insulation is steadily deteriorating.

Traditionally, megohmmeter testing was reserved for large, expensive compressors and motors, and those that perform critical functions. These motors are tested periodically. The readings are documented so that trends can be observed over time. However, today's handheld meggers allow for insulation testing on all motors at a reasonable cost, although the test voltage may only be around 1kVDC.

Megger readings of an important motor's insulation resistance should be taken when the motor is first installed, and at least semiannually thereafter. These tests should also be applied to the resistance between the individual windings as well as between each winding and ground.

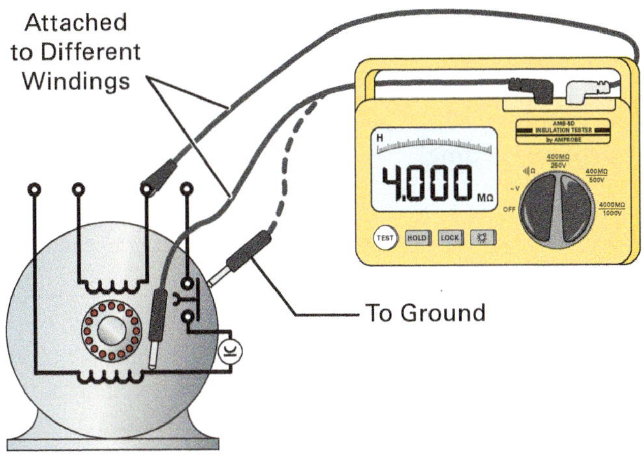

Figure 56 Using a megger.

A motor generally needs to be replaced or repaired if the megger reading drops below the recommended minimum resistance for the motor. Undamaged windings generally have a resistance to ground of at least 1,000Ω per volt of operating voltage. For example, the windings of a 230V motor should have a minimum resistance to ground of 230,000Ω (230V × 1,000Ω = 230,000Ω). A higher resistance reading indicates a sound motor. A resistance below that threshold is cause for concern. Note that a motor with sound winding insulation can have a resistance reading many times its minimum acceptable resistance.

Multimeters for Today's Technician

Your multimeter is one of the most important pieces of gear you'll ever own. A meter that is designed for the abuse that comes with field service, as well as offering features that make your job easier, will be a joy to use.

Multimeters like the Fieldpiece model shown here provide all the functions an HVACR technician will typically need, including capacitor testing and three-phase motor phasing for proper rotation. However, in most cases, you must observe a meter directly to see the display.

Bluetooth wireless technology and innovative software development have changed that. The meter shown here can connect wirelessly to a mobile device and display measured values within an unobstructed 35' radius. The meter, as well as others in the model family, is compatible with both Fieldpiece's Job Link® System App and the measureQuick® HVACR wireless tool platform. Now you can walk away and do other things while consistently monitoring power consumption, voltage, and current. It also allows loads to be energized and monitored from a safe distance.

Source: Courtesy of Fieldpiece Instruments

4.2.2 Checking Fuses

The safest way to test a fuse is by testing for continuity with an ohmmeter. The most important step is to ensure that the circuit is de-energized before a fuse is removed for a continuity test. Always use a voltmeter to verify the circuit is de-energized before proceeding!

WARNING!

Many fuses you encounter will be housed inside a disconnect switch with an external On-Off handle. There have been cases when the handle is pulled down to de-energize the circuit, but one of the blades inside the disconnect fail to separate from the contacts due to problems with the mechanical linkage. Whenever opening a disconnect switch of this type, visually verify that all the blades (*Figure 57)* have separated from the contacts.

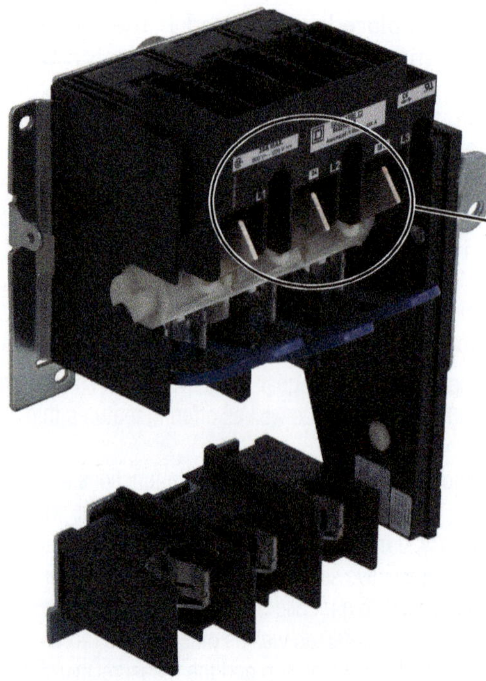

Figure 57 Disconnect switch blades.
Source: Schneider Electric

To safely check fuses, de-energize the circuit, open the unit disconnect switch or control panel, and remove the fuses using an insulated fuse puller (*Figure 58*). Test the fuses for continuity using an ohmmeter or multimeter (*Figure 59*). The fuses can also be checked with the ohmmeter while they remain in the disconnect.

If continuity is indicated across the fuse with 0Ω displayed, it means the fuse is good (intact). If an open circuit exists and OL is displayed on the meter, the fuse is open (blown). A blown fuse is usually caused by some abnormal overload condition, such as a short or grounded circuit within the equipment.

Figure 58 Fuse puller.
Source: Dave Willman/Dreamstime

WARNING!

Replacing a blown fuse without locating and correcting the cause can result in injury to personnel and damage to the equipment.

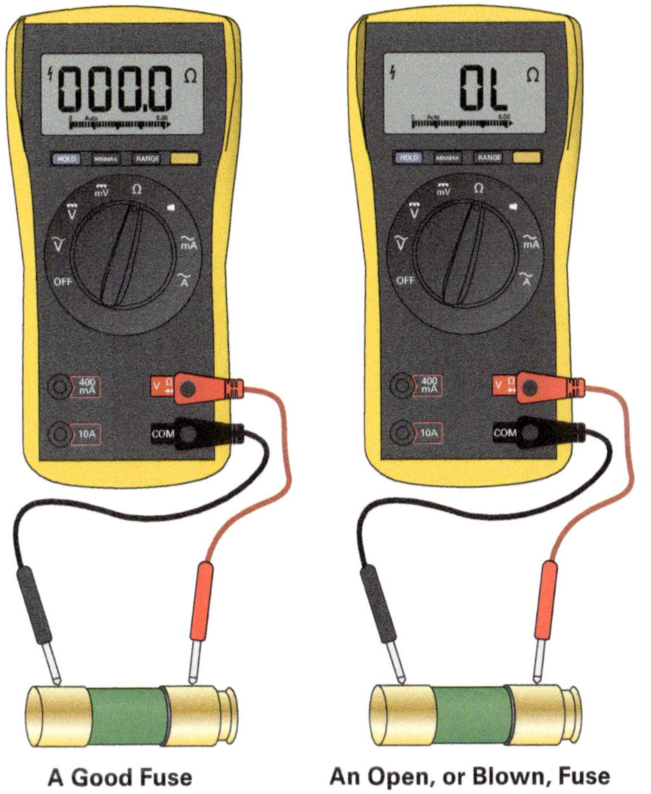

Figure 59 Testing fuses.

Fuses can also be tested with the circuit energized. Working with energized systems is sometimes a necessity, but it is best avoided whenever possible. Use the appropriate electrical PPE, including rated gloves, for the voltage present. Follow these basic steps to test fuses in an energized circuit:

Step 1 Set the voltmeter to measure AC voltage on the highest range or enable the auto-ranging function.

Step 2 Use the voltmeter first to confirm the presence of power across the top of both fuses (assuming there are two). The line voltage input in a typical disconnect switch, for example, should always be at the top of the fuses. The bottom is where the load connections are made.

Step 3 Refer to *Figure 60*. Place one of the meter probes on the line side of one fuse. Touch the other probe to the load side of another fuse. If voltage is measured between these two points, the fuse is intact. If the line voltage is not displayed on the meter, the fuse is open. Note that the fuse being tested is the one with the meter lead at the bottom. This method tests one fuse at a time.

Step 4 Repeat this procedure so that all fuses are measured with one test lead on the load side and the other test lead on the line side of a different fuse.

4.2.3 Capacitance Testing

A capacitor tester is a special-purpose meter used to test capacitors. It can test capacitors for current leakage, insulation breakdown, shorts, and opens. In addition, it can report the value of the capacitor, in microfarads. However, there are only a few capacitor testers on the market today.

Many of today's digital multimeters are equipped for capacitor testing. This is a valuable feature to look for when purchasing a multimeter for field use.

A capacitor tester is needed to effectively test a capacitor. If used correctly, a standard ohmmeter can be used to determine if the capacitor is open or shorted, but it can be tricky, and it cannot measure the actual capacitance. A capacitor tester (or equal function in a multimeter) is far more reliable and produces better information.

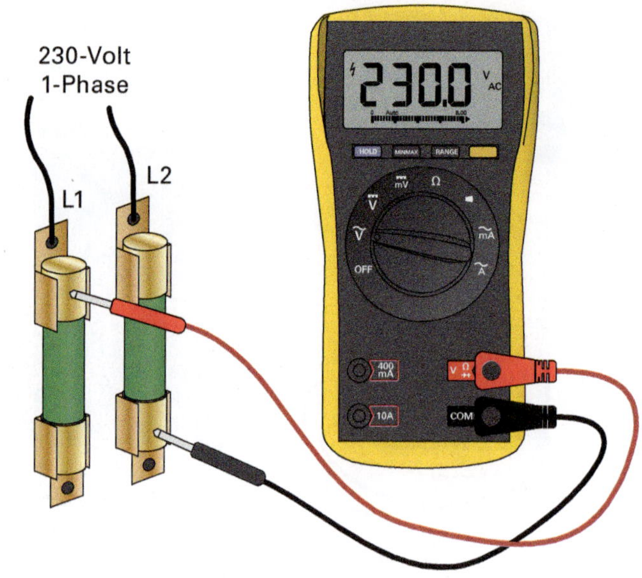

**Meter reads line voltage if the fuse on the right is intact.
If the fuse on the right is open, it will read 0V.**

Figure 60 Testing fuses on an energized circuit.

Before connecting a capacitor tester, de-energize the equipment and bleed off (discharge) any remaining capacitor charge.

Did You Know?

Hand Tool Electrical Insulation

When you see some rubbery or plastic material on the handles of a pair of pliers or similar hand tool, it doesn't mean that the tool is electrically safe. But there are several ways that you can determine if a tool is rated for protection against incidental contact with energized circuits up to 1,000V.

The first thing to look for is the symbol used to identify a tool that has been properly tested to a relevant standard. The symbol, shown here, should be found on the insulation of the tool, usually on the inside where it is somewhat protected. You'll also see references to the standard(s) used for testing, such as International Electrotechnical Commission (IEC) 60900.

The handles of electrically rated hand tools are double layered, and the two layers are usually different colors. This makes it easier to tell when the insulation has been damaged enough to render the tool unsafe. If you can see the color of the lower layer through a cut or scar, it has been compromised. The handles should also be chunky and be equipped with guards that prevent your hand from slipping down toward the metal portion.

Bleeding the Charge from a Capacitor

Capacitors can store a significant amount of energy—enough to injure you if the stored charge isn't released. Unlike a battery, though, the charge stored in a motor capacitor can be discharged in just a moment.

WARNING!

Never place your fingers across the capacitor terminals. The residual electrical charge on the capacitor can deliver a substantial shock.

Before handling a start or run capacitor, bleed off any remaining charge as follows. You'll need some needlenose pliers with electrically insulated handles and a resistor with a value between 10kΩ and 20kΩ. Refer to *Figure 61*:

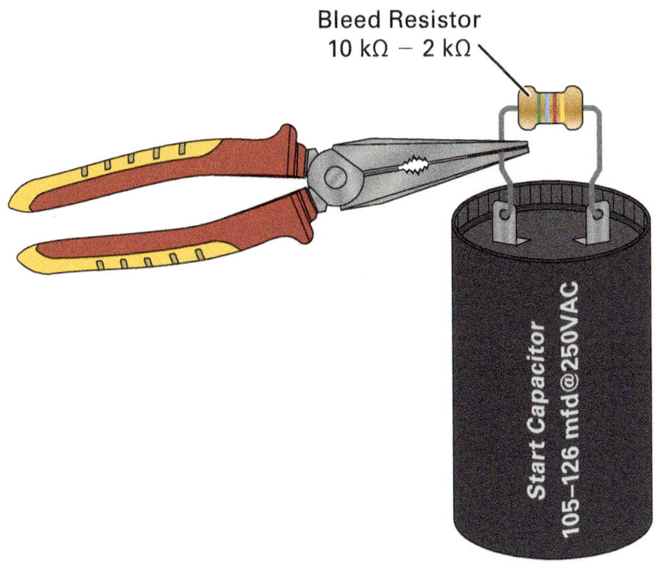

Figure 61 Discharging a capacitor.

Step 1 Use a voltmeter to ensure the circuit is de-energized.

Step 2 Hold the resistor with the pliers, keeping your hands away from the metal portion of the handles.

Step 3 Touch the ends of the resistor firmly against the terminals. If there are more than two terminals, make sure all are touched from one to another. The capacitor should now be safely discharged.

NOTE

Several tool manufacturers now offer convenient capacitor discharge tools.

Some start capacitors are equipped with bleed resistors. Once de-energized, they immediately discharge any remaining charge through the resistor. However, it is always a good idea to wait several minutes after the capacitor is de-energized before handling it.

Testing a Capacitor

Testing a capacitor is easy. First, ensure the capacitor has been safely discharged. Carefully disconnect the wires and pay close attention to them. An error in the wiring connections during reassembly can result in an electrical short, arcing, and significant personal injury.

When testing a start capacitor with a permanent bleed resistor installed, the resistor must be disconnected before testing. Otherwise, the small electrical charge developed by a capacitor tester will simply flow through the resistor rather than through the capacitor.

After connecting the two leads to the terminals, press the button to complete a test. A stand-alone capacitor tester will typically respond as follows:

- If the capacitor is open, it will display OPEN.
- If the capacitor is shorted, SHRT will be displayed.
- If the capacitor is intact, the microfarad value will be displayed.

When you are using a multimeter with a capacitor testing function, follow the guidance provided by the manufacturer. Many will provide similar responses.

The microfarad value of the capacitor should be within the tolerance allowed. This information is often shown on the capacitor label. An example would be "+10% to –5%." If there is no tolerance shown, use the following criteria:

- Run capacitors should test within ±10% of their labeled rating.
- Start capacitors should test within their labeled range.

Replace any capacitor that does not test within the specified range.

Versatile Capacitors

Capacitors have many possible values. As a technician, it is impossible to carry every possible size on the truck. Carrying a multirange, or universal, capacitor like the one shown here can help get equipment running faster. The desired microfarad value is determined by installing jumpers in a specific configuration. In many cases, it is best to use this type of capacitor temporarily until you can return with a capacitor of the proper value, although it can remain in place for long-term service. Both start and run multirange capacitors are available.

Source: Motors & Armatures, Inc. (MARS)

4.3.0 Terminating Wires

There are many different types of wiring terminations. The selection of the proper method for an application often depends on how and where the termination is used. Terminations are commonly made with a solder-less pressure connector.

The basic requirements for a good electrical termination include the following. These characteristics should last throughout the life of the connection:

- It must be both mechanically and electrically secure.
- It must be insulated as well as or better than the existing insulation on the conductors.

After finishing a wiring termination, evaluate it with these two important considerations in mind.

Did You Know?

Conductor Terminations and Splices

Poor electrical connections are responsible for a large percentage of equipment damage and fires. Many of these failures are the direct result of improper terminations, poor workmanship, and using the wrong hardware.

4.3.1 Wiring Termination Hardware

A variety of wire terminals for stranded wire are shown in *Figure 62*. These connectors are available in various sizes to accommodate wire sizes No. 22 American Wire Gauge (AWG) and larger. They can be installed in the field with a manual crimping tool. Thus, they are often referred to as *crimp connectors.* The optimal range of wire sizes is normally stamped on the tongue of each terminal.

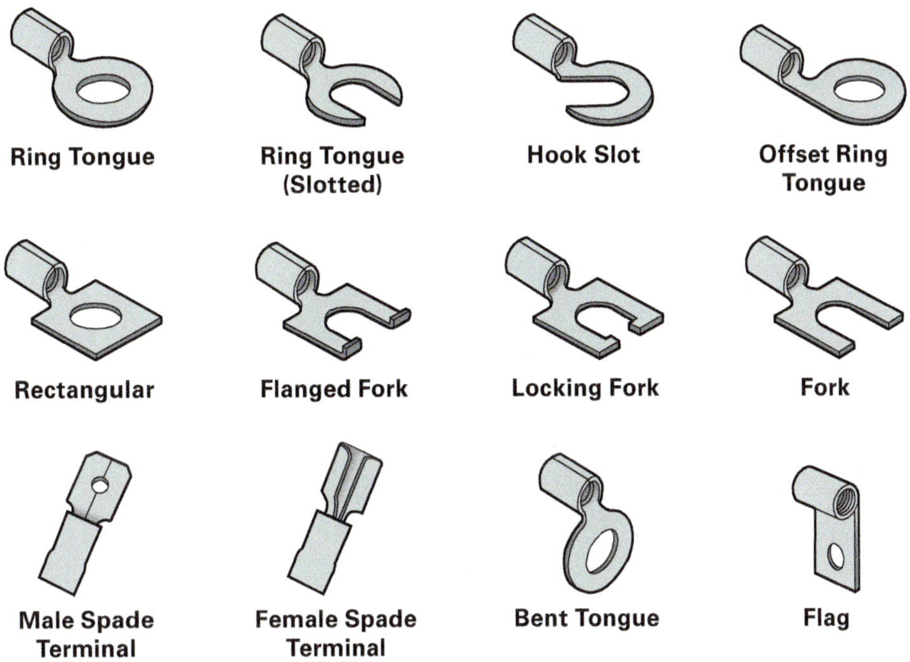

Figure 62 Crimp-on terminals for stranded wire.

Crimp connectors for wires smaller than No. 8 AWG are normally made to accept at least two different wire sizes. Those that are insulated are often color-coded. For example, one manufacturer's color code is as follows:

- Red for No. 18 and No. 20 AWG wire
- Blue for No. 16 and No. 14 AWG wire
- Yellow for No. 12 and No. 10 AWG wire

Termination hardware for wire sizes No. 8 AWG and larger, commonly called **lugs**, are made to accept one specific conductor size or multiple smaller conductors. In HVACR work, lugs are commonly found on components such as larger compressor contactors and fuse blocks, which conduct the highest currents in a system (*Figure 63*). A termination is made by inserting the wire beneath the screw and tightening. More than one wire may also be slipped inside a crimp connector, but this is rarely necessary.

Figure 64 shows the basic structure of a crimp connector. The crimp barrel receives the wire and is crimped to secure it in place. Wrinkles or dimples inside the barrel improve wire-to-terminal contact and increase the tensile strength of the connector. Most crimp connectors have nylon or vinyl insulation covering the barrel to reduce the possibility of shorting to adjacent terminals. An inspection hole is provided at the end of the barrel to allow visual inspection of the wire position inside.

CAUTION

Crimped terminals such as those shown in *Figure 60* should never be used with solid wire.

Lugs: Wiring terminals used on wire sizes No. 8 AWG and larger; also, a term used to describe terminals that apply pressure to a stripped conductor with a screw.

4.3.2 Fabricating Terminations

The correct connector for the task must first be selected. Match the wire size to the connector's range. Ring terminals require that the screw used to secure it be completely removed, then reinserted through the hole. Forked and slotted connectors can be slipped beneath a loosened screw. Also ensure you select a connector rated for the voltage it will carry. While many are rated for use up to 600V, some have a lower rating.

The conductor must then be stripped to the proper length to suit the connector, without exposing more wire than necessary.

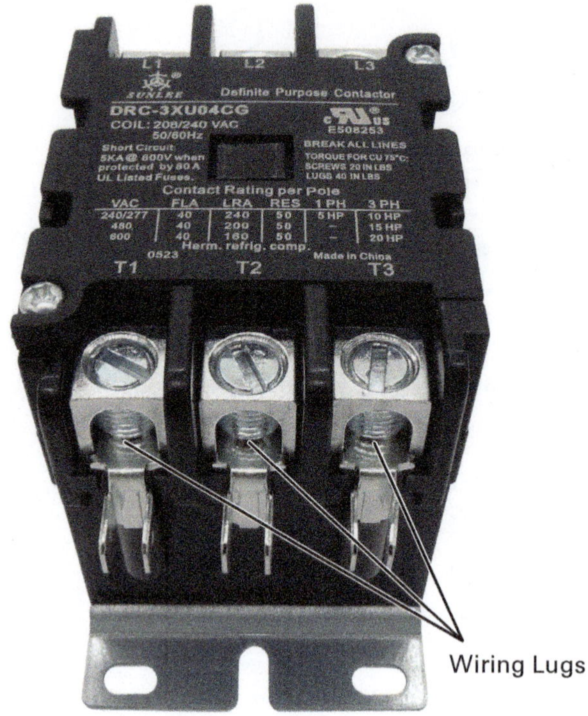

Figure 63 Three-phase compressor contactor.

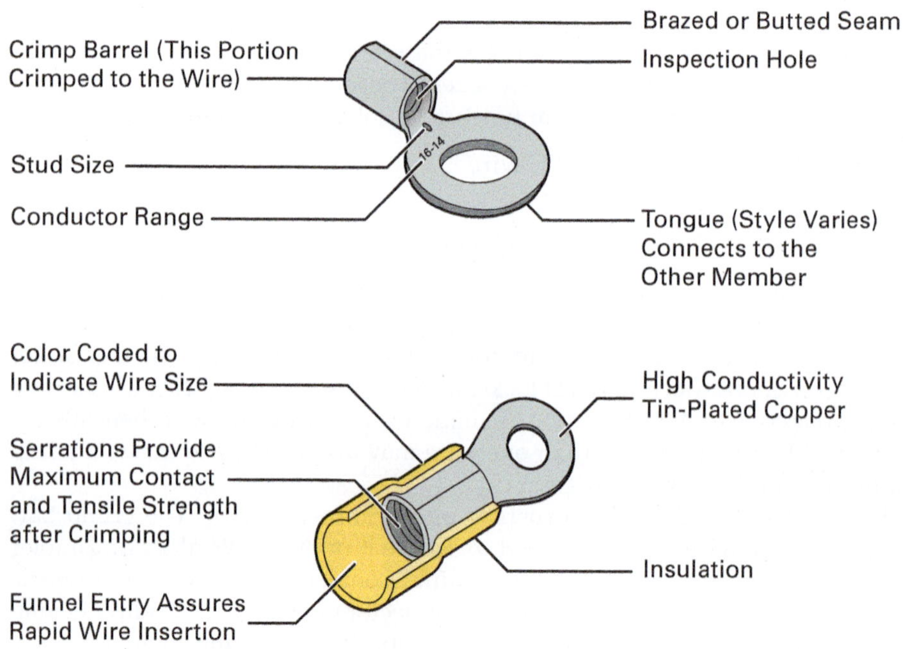

Figure 64 Basic crimp connector structure.

Preparing Conductors for Termination

Before any termination can be assembled, the end of the conductor must be properly cleaned and stripped. *Stripping* refers to the removal of the insulation from the end of the conductor. Conductors should only be stripped using an appropriate stripping tool. Never strip wire using a pocketknife or similar tool. Using a knife often results in significant cuts, both in the conductor and the user.

WARNING!

Never use a pocketknife or similar tool to strip conductors. Serious personal injury can result, and the knife blade can easily nick or damage the conductor.

Poorly stripped wires, where the conductor is nicked or gouged, result in weak, deficient connections. Heat, rapid temperature changes, vibration, and motion aggravate the damage. Lost strands—those accidentally severed during stripping—reduce the effective size of the conductor. Leaving any strands exposed creates a safety hazard.

Wires should only be stripped using an appropriate tool. *Figure 65* shows two common tools used for smaller conductors. The combination tool cuts, strips, and crimps. You must use the correct opening for stripping to ensure the conductor isn't damaged. Various crimping shapes are also offered.

(A) Combination Stripper/ Crimper/Cutter

(B) Self-Adjusting Stripper

Figure 65 Wire stripping and crimping tools.
Source: Courtesy of Stanley Black & Decker, Inc.

The other tool is only capable of stripping wire, but it automatically adjusts itself to fit the wire size. A separate tool for crimping is required, like the one shown in *Figure 66*.

Figure 66 Wire crimper.
Source: Southwire

Figure 67 shows how a wire looks in the connector barrel when stripped to the proper length. The conductor insulation must be inside the belled mouth of the terminal. This relieves mechanical stress on the wire and increases the strength of the connection. If the conductor protrudes through the connector barrel more than $\frac{1}{32}$", it may interfere with the terminal screw. It's best to allow it to protrude only enough to be seen. When stripping the wire for a lug, the insulation should touch or be very close to the lug when the wire is secured, and the wire should just slightly protrude through the opposite side.

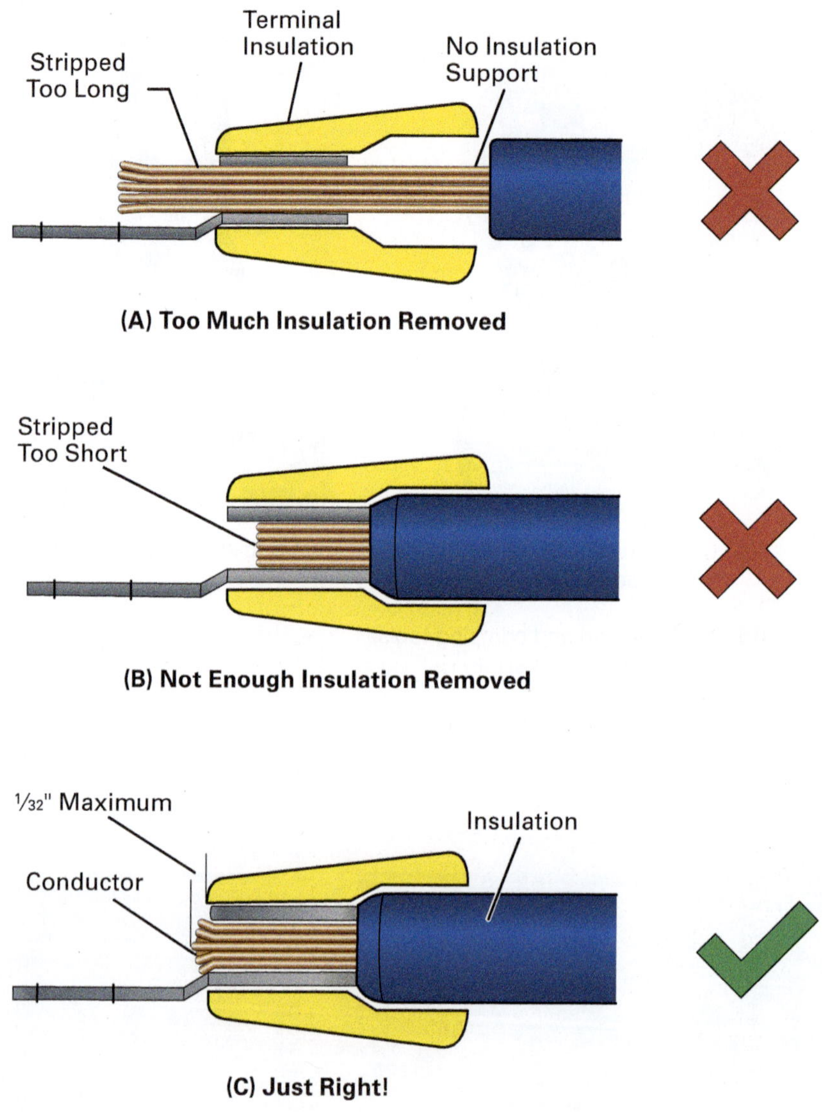

Figure 67 Proper stripping length.

To strip a wire using the combination tool shown in *Figure 65*, insert the conductor into the proper size stripping groove, squeeze the tool handles together, and pull the end of the wire through. The length of the strip is regulated by the amount of wire extending beyond the tool when inserted in the stripping groove. The tool cuts through the insulation only, allowing the conductor to be pulled out while the waste insulation falls away (*Figure 68*).

If you choose a stripping groove that is smaller than it should be, the conductors beneath will be damaged. If the conductor has fine strands, several of the strands will likely be cut along with the insulation.

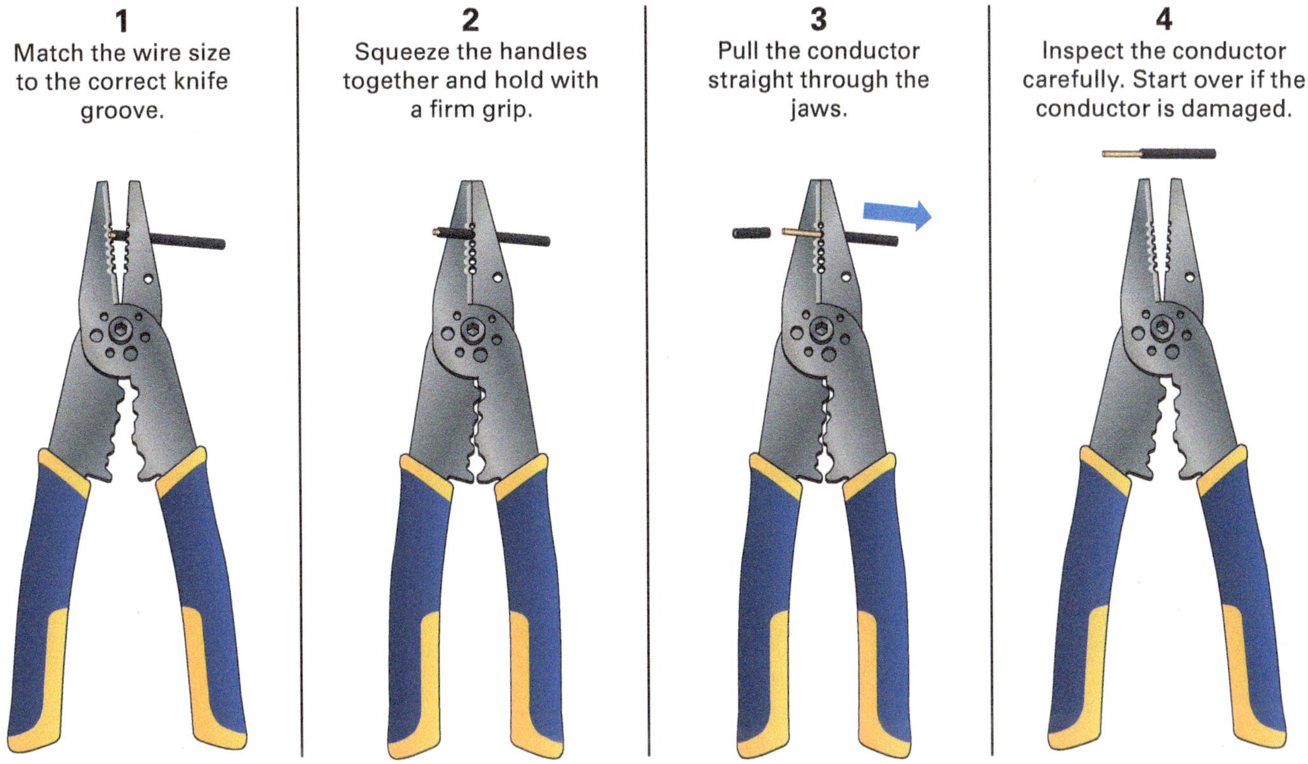

Figure 68 Stripping a wire.

Installing a Crimp Connector

To install a crimp connector, first insert the stripped end of the wire completely into the connector until the insulation is seated inside against the barrel. Position the crimping tool in the center of the connector barrel, then squeeze the tool firmly to crimp the connector. Make sure that the crimping tool jaws come together when squeezing.

CAUTION

Crimp connectors are not designed for use with solid wire. Solid wiring is most often and best terminated at equipment with a lug or coiled around a screw. Runs may be joined by a wire nut.

4.0.0 Section Review

1. The minimum approach distance for an unqualified worker where voltages between 50V and 750V is present is _____.
   a. 5'
   b. 10'
   c. 25'
   d. 100'

2. When a fuse is checked using an ohmmeter, the display should read _____.
   a. the line voltage
   b. infinity (OL)
   c. at least 10Ω
   d. 0Ω

3. True or False: Common crimped terminals can be used on all types and sizes of wire, including both solid and stranded conductors.
   a. True
   b. False

Module 03206 Review Questions

1. The number of times that a complete AC sine wave occurs each second is referred to as its ______.
   a. period
   b. frequency
   c. cycle
   d. armature

2. The effective (RMS) voltage of an AC sine wave with a peak voltage of 339V is ______. (Round your answer to the nearest whole number.)
   a. 111V
   b. 120V
   c. 192V
   d. 240V

3. To operate, fossil-fuel furnaces typically require a voltage of ______.
   a. 24V
   b. 120V
   c. 240V
   d. 460V

4. The cause of current imbalance in a three-phase system must be corrected if the current is out of balance by more than ______.
   a. 10%
   b. 15%
   c. 20%
   d. 25%

5. The input to output ratio of an isolation transformer is ______.
   a. 3:1
   b. 2:1
   c. 1:2
   d. 1:1

6. True or False: Multitap transformers are designed so that, if one primary winding tap fails, you can use another tap to get the same voltage without replacing the transformer.
   a. True
   b. False

7. How does the actual rotational speed of an induction motor relate to its synchronous speed?
   a. An induction motor always rotates at its published synchronous speed.
   b. The actual rotational speed is about 95% to 97% of its synchronous speed.
   c. An induction motor's synchronous speed is precisely 50% of its actual rotational speed.
   d. The synchronous speed is calculated by multiplying the rotational speed by 0.707.

8. Which type of motor will reverse its direction of rotation by simply swapping any two power connections?
   a. Split-phase motor
   b. Capacitor-start, capacitor-run motor
   c. Permanent split capacitor motor
   d. Three-phase motor

9. The start winding of a split-phase motor ______.
   a. has fewer turns of wire than the run winding
   b. remains in the circuit as long as it is running
   c. has more turns of wire than the run winding
   d. is 120 degrees out of phase with the run winding

10. A permanent split capacitor motor is equipped with ______.
    a. a run capacitor only
    b. a start capacitor only
    c. a synchronous capacitor
    d. both run and start capacitors

11. A shaded-pole motor is aided in starting by a ______.
    a. run capacitor
    b. shading coil
    c. start capacitor
    d. centrifugal switch

12. When testing the windings of a single-phase induction motor, the resistance of the start winding will be ______.
    a. about the same as that of the run winding
    b. about half that of the run winding
    c. several times greater than that of the run winding
    d. too low to read on a standard ohmmeter

13. A high-voltage motor winding resistance test is performed using a(n) ______.
    a. ohmmeter
    b. multimeter
    c. wattmeter
    d. megger

14. The proper and safe method of discharging a capacitor is to ______.
    a. span the terminals with a screwdriver
    b. connect a jumper wire across the terminals
    c. span the terminals with a resistor
    d. attach a capacitance tester

15. Crimp connectors for wires smaller than No. 8 AWG are typically made to accept ______.
    a. only one wire size
    b. at least two different wire sizes
    c. at least four different wire sizes
    d. any wire size; one connector fits all

Answers to Odd-Numbered Module Review Questions are found in *Appendix A*.

Answers to Section Review Questions

Answer	Section	Objective
Section One		
1. b	1.1.3	1a
2. b	1.2.1	1b
3. b	1.3.2	1c
Section Two		
1. c	2.1.0	2a
2. a	2.2.1	2b
Section Three		
1. d	3.1.0	3a
2. c	3.2.4	3b
Section Four		
1. b	4.1.3	4a
2. d	4.2.2	4b
3. b	4.3.1	4c

MODULE 03302

Compressors

Source: kadmy/Getty Images

Objectives

Successful completion of this module prepares you to do the following:

1. Identify and describe the operating characteristics of various compressors.
   a. Identify and describe hermetic, semi-hermetic, and open-drive compressors.
   b. Identify and describe reciprocating compressors.
   c. Identify and describe rotary compressors.
   d. Identify and describe scroll compressors.
   e. Identify and describe rotary screw compressors.
   f. Identify and describe centrifugal compressors\
2. Identify and describe various approaches to compressor capacity control.
   a. Identify and describe capacity control methods for reciprocating and scroll compressors.
   b. Identify and describe capacity control methods for rotary screw and centrifugal compressors.
3. Describe the common causes of compressor failures.
   a. Describe compressor failures related to the refrigerant circuit.
   b. Describe compressor failures related to electrical issues.
4. Identify and describe the operation of various compressor protection devices.
   a. Identify and describe the operation of various motor overload protection devices.
   b. Identify and describe the operation of other compressor protection devices.
5. Explain how to analyze the operation of a hermetic compressor.
   a. Explain how to evaluate the mechanical operation of a hermetic compressor.
   b. Explain how to evaluate the electrical operation of a hermetic compressor.

Performance Tasks

Under supervision, you should be able to do the following:

1. Use an acid/moisture test kit to test a refrigerant circuit.
2. Measure and record the electrical and other operating parameters of a running compressor.

Overview

The compressor is the heart of an HVACR system, providing the pressure differential that moves refrigerant through the system and taking advantage of the refrigerant's thermodynamic properties. It is essential to understand how compressors work, allowing technicians to service and troubleshoot as necessary. Although today's compressors are durable, external factors such as the state of the

Digital Resources for HVACR

NCCER

SCAN ME

Scan this code using the camera on your phone or mobile device to view the digital resources related to this craft.

refrigerant circuit and the quality of the power supply typically contribute to their failure. This module will help you recognize whether a compressor is the source of a problem or merely a symptom of a problem located elsewhere in the system.

> **NCCER Industry-Recognized Credentials**
>
> If you are training through an NCCER-accredited sponsor, you may be eligible for credentials from NCCER. The ID number for this module is 03302. Note that this module may have been used in other NCCER curricula and may apply to other level completions. Contact NCCER at 1.888.622.3720 or go to **www.nccer.org** for more information.
>
> You can also show off your industry-recognized credentials online with NCCER's digital credentials. Transform your knowledge, skills, and achievements into credentials that you can share across social media platforms, send to your network, and add to your resume. For more information, visit **www.nccer.org**.

1.0.0 Types of Compressors

Performance Tasks

There are no Performance Tasks in this section.

Objective

Identify and describe the operating characteristics of various compressors.

a. Identify and describe hermetic, semi-hermetic, and open-drive compressors.
b. Identify and describe reciprocating compressors.
c. Identify and describe rotary compressors.
d. Identify and describe scroll compressors.
e. Identify and describe rotary screw compressors.
f. Identify and describe centrifugal compressors.

The compressor is the heart of an HVACR system, providing the pressure differential that moves refrigerant through the system and takes advantage of the refrigerant's thermodynamic properties. As shown in *Figure 1*, the refrigerant flows through the components of the system in the direction indicated by the arrows. The basic refrigeration cycle is the foundation of heat transfer in a refrigerant circuit.

The evaporator receives low-temperature, low-pressure liquid refrigerant from the expansion device. The evaporator coil exposes cool liquid refrigerant to the warm air passing over the coil and tubing. Heat from the warm air is transferred to the cooler refrigerant, causing it to boil and vaporize. Superheated, low-temperature, low-pressure refrigerant vapor flows through the suction line from the evaporator coil to the inlet of the compressor.

Compression: The reduction in volume of a vapor by mechanical means.

Compression ratio: For refrigerant compressors, the ratio of discharge pressure to suction pressure. Note that *compression ratio* has a different meaning when applied to internal combustion engines.

Through the process of **compression**, the compressor converts the low-temperature, low-pressure vapor into a high-temperature, high-pressure vapor that flows to the condenser via the hot gas line. The **compression ratio** refers to the ratio of discharge pressure to suction pressure. For example, a compressor operating with 180 psig of discharge pressure and 60 psig of suction pressure has a compression ratio of 3:1.

Like the evaporator, the condenser coil contains tubing through which the refrigerant flows. As cooler outside air moves across the condenser coil, the hot refrigerant vapor gives up accumulated superheat and cools. As the refrigerant vapor continues to give up heat to the air, the temperature drops to the saturation point and the refrigerant begins to condense into a liquid.

Once it has completed the change of state to a liquid, additional cooling occurs and subcooling begins. This means that the liquid refrigerant is cooling below its saturation temperature. The resulting medium-temperature, high-pressure liquid then flows through the liquid line to the input of the expansion device.

The expansion device regulates the flow of refrigerant to the evaporator. The expansion device converts the medium-temperature, high-pressure refrigerant

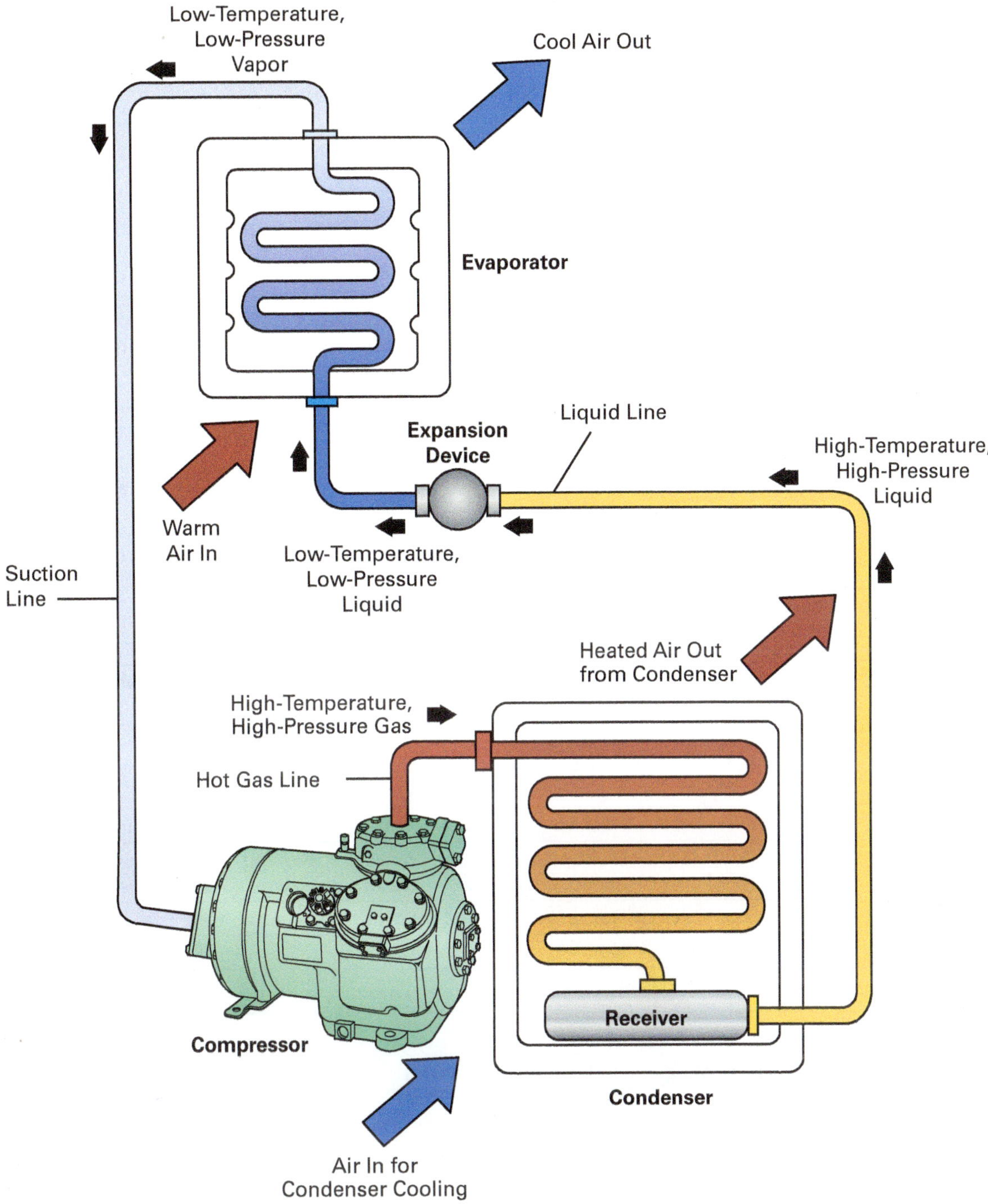

Figure 1 Typical refrigerant circuit.

from the condenser into the low-temperature, low-pressure refrigerant liquid/vapor mixture needed to absorb heat in the evaporator.

The compressor draws the refrigerant vapor from the evaporator coil, compresses it, and then sends the vapor downstream to the condenser. During this process, the following important conditions are created:

- The pressure and temperature of the refrigerant in the evaporator are lowered, allowing the refrigerant to boil and absorb heat from its surroundings.
- The pressure and temperature of the refrigerant in the condenser are raised, allowing the refrigerant to give up heat at existing temperatures to whatever medium (air or water) is used to absorb the heat.
- A pressure difference is maintained between the high- and low-pressure sides of the system. This pressure difference causes the refrigerant to flow through the system.

NOTE

Expansion devices are presented in detail in NCCER Module 03303, *Metering Devices*.

The variety of refrigerants available, as well as the application, location, and size of systems are just some of the factors that create the need for different compressor designs.

The following compressor types are common in HVACR systems:

- Reciprocating
- Rotary
- Scroll
- Rotary screw
- Centrifugal

Table 1 summarizes the uses and characteristics of these compressor types. Each type will be explored in the sections that follow. *Table 2* shows the applications for which each type is best suited.

TABLE 1 Compressor Comparison Chart

	Reciprocating	Rotary	Scroll	Rotary Screw	Centrifugal
Usage	Refrigeration and air conditioning, chillers, modular chillers, heat pumps, and transportation	Refrigerators, room air conditioners	Central systems for refrigeration, air conditioning, modular chillers, and heat pumps	Refrigeration, air conditioning, chillers, and heat pumps	Chillers
Capacity	Fractional tonnage through 150 tons	5 tons or less	1.5 to 70 tons	50 to 750 tons	100 to over 10,000 tons
Sub-Types	Open, semi-hermetic, and hermetic	Sliding vane and rolling piston type, hermetic only	Compliant, noncompliant, and digital hermetic	Rolling rotor, open, and hermetic	Single and multi-stage open and hermetic
Displacement	Positive	Positive	Positive	Positive	Non-positive
Typical Capacity Control	Multispeed and cylinder unloaders		Multispeed, variable speed, unloaders	Variable speed and intake slide valve	Variable speed and inlet guide vanes
Suction Valves	Yes	No	No	No	No
Discharge Valves	Yes	Yes	No	No	No

TABLE 2 Compressor Applications

	Reciprocating	Rotary	Scroll	Rotary Screw	Centrifugal
Domestic Refrigeration	✓	✓			
Room Air Conditioners	✓	✓			
Comfort Cooling (Direct)	✓		✓		
Comfort and Process Cooling (Chilled Water)	✓		✓	✓	✓
Heat Pumps	✓	✓	✓	✓	
Commercial Refrigeration	✓		✓	✓	
Industrial Refrigeration	✓			✓	

1.1.0 Hermetic, Semi-Hermetic, and Open-Drive Compressors

An electric motor typically drives a compressor. Very large compressors, especially those operating in remote locations, may be driven by internal combustion engines or steam turbines. Compressors can be divided into three categories based on the way they are joined to their drive component. The categories are: *hermetic, semi-hermetic,* or *open drive.*

1.1.1 Hermetic Compressors

Hermetic compressor housings are assembled and welded closed at the factory and cannot be serviced internally in the field. The compressor and motor have a common drive shaft. Most hermetic reciprocating, rotary, and scroll compressors (*Figure 2*) have the motor and drive shaft mounted vertically in the shared housing.

Figure 2 A hermetic scroll compressor.

Hermetic compressors (sometimes referred to as *tin cans*) are more compact, quieter, and require less maintenance than other types because they have no belts or drive couplings to wear out and very little potential for leakage. However, because they are factory sealed, the entire unit must be replaced when mechanical or electrical failures occur.

1.1.2 Semi-Hermetic Compressors

The semi-hermetic, or *serviceable hermetic*, compressor and motor also share the same housing and a common drive shaft (*Figure 3*). The drive shaft is usually horizontal. However, access to the compressor valves and a few other components for field repair is possible by removing the heads and various access plates.

Figure 3 A semi-hermetic compressor cutaway.

Going Green

Semi-Hermetic Compressors

Semi-hermetic compressor housings are very durable and can be reused numerous times. They are either *rebuilt* or *remanufactured.* All the moving parts in a remanufactured compressor are typically replaced. A rebuilt compressor often retains some of its moving parts if the parts meet specifications. Many semi-hermetic compressors replaced each year are rebuilt or remanufactured units that are less expensive than fully new models.

When a semi-hermetic compressor is purchased, the failed compressor should be traded in against the replacement. If not, a *core charge* is applied to the invoice, charging the buyer for the value of the core, or housing. This charge is waived when a trade-in is offered.

Semi-hermetic compressors are very sturdy and much heavier than hermetic models because the housing is usually made from cast iron. These compressors can be fully disassembled and rebuilt to their original specifications by the manufacturer or an independent rebuilder.

1.1.3 Open-Drive Compressors

The compressor and motor do not share an enclosure in an open-drive compressor. One end of the compressor's horizontal crankshaft extends outside the housing. A mechanical seal on the rotating shaft prevents refrigerant leakage. A motor drives the compressor through a flexible coupling (*Figure 4*) or a belt.

(A) Open-Drive Refrigeration Compressors

(B) Drive Couplings and Inserts

Figure 4 Open, direct-drive compressors with example shaft coupling and coupling inserts.
Source: (A) kadmy/Getty Images; (B) Photo courtesy TB Woods, http://www.tbwoods.com.

The compressor runs at the same speed as the drive motor when connected to it directly. Modulating the compressor speed is possible only if the motor is designed for multispeed or variable speed operation. The couplings shown in *Figure 4* are examples of flexible couplings, but some compressors require heavier, all-metal couplings. Belt-driven arrangements allow the compressor to run at a different rotational speed than the motor. The proper combination of pulleys produces the desired compressor speed.

Hermetic compressors are welded and can potentially leak only at the electrical terminals or line connections. Leaks are very rare as a result. Semi-hermetic compressors have additional ports and many gasketed joints that increase the likelihood of leaks, but they still hold up well. Leaks remain rare and are usually addressed easily enough with new gaskets and properly fitted plugs and caps.

Open-drive compressors, on the other hand, must have a leakproof seal where the rotating crankshaft passes through the housing. While sealing stationary components is relatively simple, maintaining a reliable seal around a rotating shaft is much more challenging.

Shaft seals (*Figure 5*) prevent refrigerant and oil from leaking out around the crankshaft. However, shaft seals must do so without applying significant friction to the shaft. Issues related to reliable shaft seals and refrigerant leakage were primary motivators for the development of hermetic and semi-hermetic compressors many years ago.

Some low-temperature systems, depending on the refrigerant, operate in a vacuum (below 0 psig) on the low-pressure side of the refrigerant circuit. In those cases, a leaking seal allows air and water vapor to enter the refrigerant circuit. This causes serious performance problems and eliminating the air and moisture requires all the refrigerant to be recovered and a system evacuation with a vacuum pump. The work is both expensive and time-consuming.

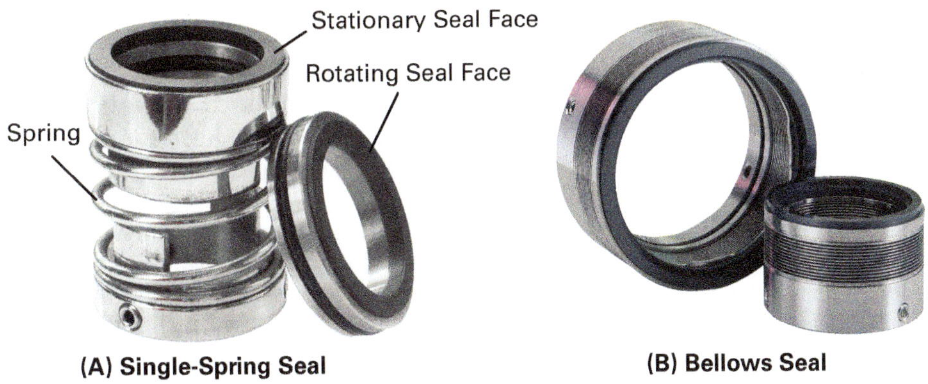

Figure 5 Common shaft seals for open-drive compressors.
Source: (A) nayladen/Shutterstock; (B) Flex-A-Seal Inc.

Single-spring seals and bellows seals are commonly used in compressors. Single-spring seals use two precisely mated friction surfaces. One surface turns with the crankshaft and is sealed to the shaft with an O-ring or synthetic material. The other surface is stationary and is mounted on the housing with leakproof gaskets. A bellows seal consists of a bellows assembly, carbon ring, and cover plate, installed in that order on the compressor crankshaft. The sealing surface is between the nose of the bellows and a shoulder on the crankshaft.

When installed correctly, shaft seals can last for years. For long seal life in a direct-drive unit, the compressor crankshaft and motor drive shaft must be perfectly aligned. Belt-driven units do not require the same level of precision in pulley alignment, but the belts must be adjusted for the proper tension. Excessive belt tension can damage the seal as well as the bearings.

Compressors on the Move

Open-drive compressors are common in industrial stationary air conditioning and refrigeration systems. However, a variation is also used in mobile applications, such as automobile air conditioners and refrigerated trailers. In these applications, a belt-driven, open-drive compressor is coupled to the vehicle's engine through a magnetic clutch mechanism in the compressor drive pulley hub. The magnetic clutch, controlled by the vehicle's thermostat or a refrigerant pressure switch, engages or disengages the compressor drive.

Source: snapper8S8/Shutterstock

1.2.0 Reciprocating Compressors

Reciprocating compressors are available in any of the three configurations mentioned previously hermetic, semi-hermetic, or open drive. Reciprocating compressors can have from one to as many as 12 pistons moving back and forth within their individual cylinders.

The primary mechanical parts include a cylinder, piston, connecting rod, crankshaft, cylinder head, and suction and discharge valves (*Figure 6*). The reciprocating motion of each piston within its cylinder is created by the eccentric lobes of the rotating crankshaft.

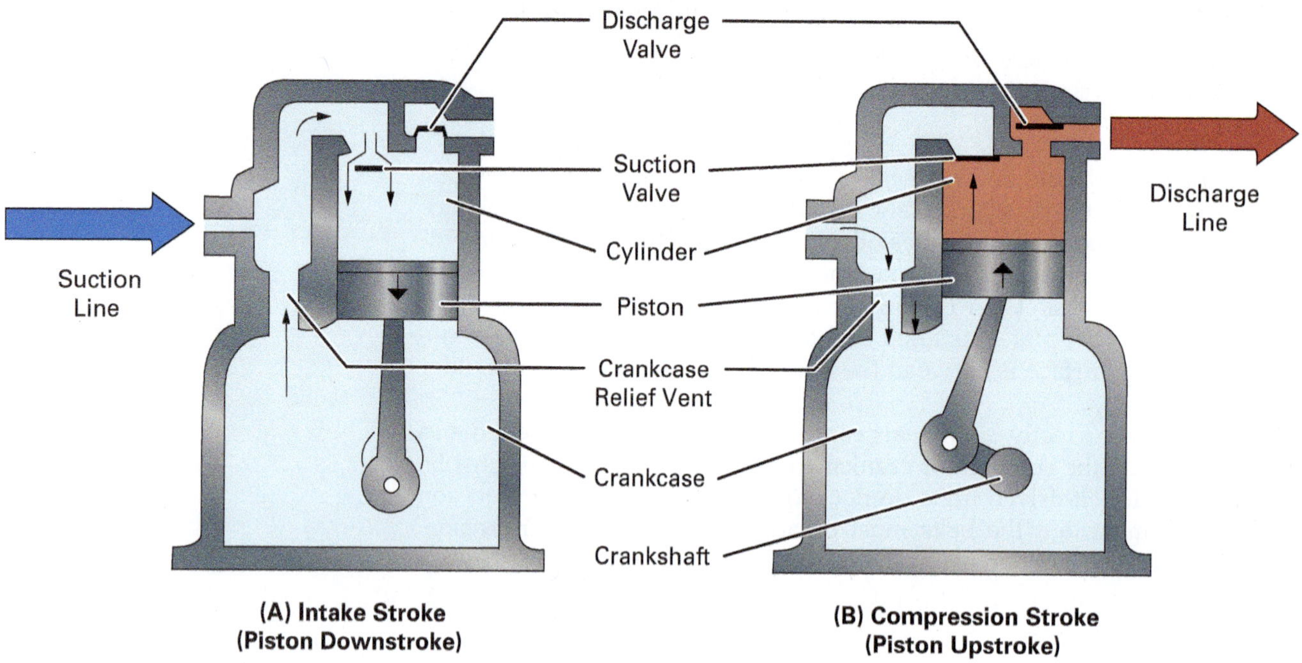

Figure 6 Reciprocating compressor operation.

The suction and discharge valves control refrigerant movement in and out of the cylinder. The crankcase contains the crankshaft and stores oil that is used to lubricate the moving parts. *Figure 6* shows the intake (suction) and compression (discharge) strokes for one cylinder of a reciprocating compressor. The following events take place during the intake stroke:

1. At the start of the intake stroke, both the discharge and suction valves are fully closed. As the piston starts moving down, a low-pressure area is formed under the suction valve. When the pressure falls below the suction line pressure, the suction valve opens, and the cylinder begins to fill with vapor from the suction line.
2. As the piston continues toward the bottom of the intake stroke, the cylinder continues to fill. At the bottom of the stroke, there is a very slight time lag as the crankshaft rotates the connecting rod to complete the stroke. During this time, the cylinder continues to fill with suction vapor until the pressure is equal to that in the suction line.
3. When the piston reaches its lowest point of travel, called *bottom dead-center*, the equal pressures in the cylinder and suction line allow the suction valve(s) to close.

During the compression stroke, the following sequence of events takes place:

1. At the start of the compression stroke, both the suction and discharge valves are closed. As the piston begins its upward stroke, compression of the vapor begins, and the pressure rises. Compressing vapors also produces heat. In hermetic and semi-hermetic compressors, the motor heat is transferred to the vapor as well. The suction valve remains closed by the pressure in the cylinder while the discharge valve remains closed due to the pressure in the discharge line.

2. As the piston continues its upward travel, the discharge valve remains closed until the piston approaches the top of the cylinder. At some point, the pressure in the cylinder exceeds the pressure in the discharge line, and the discharge valve opens. This allows the now high-temperature, high-pressure vapor to exit the compressor through the discharge line.
3. At the top of its travel, called *top dead-center*, the piston stops moving and leaves a small amount of vapor in a space at the top of the cylinder. This space is called the **clearance volume** (*Figure 7*). Although the space is very small, it is necessary to ensure that the piston does not strike the upper portion of the cylinder. This small amount of vapor, which is now at the discharge pressure, will re-expand when the piston starts back down at the start of the next intake stroke. It contributes to some inefficiency in compressor operation, because a small amount of vapor must be expanded and compressed again during every cycle.

Clearance volume: The amount of clearance between the cylinder head and the piston when it is at the top dead-center position of its travel.

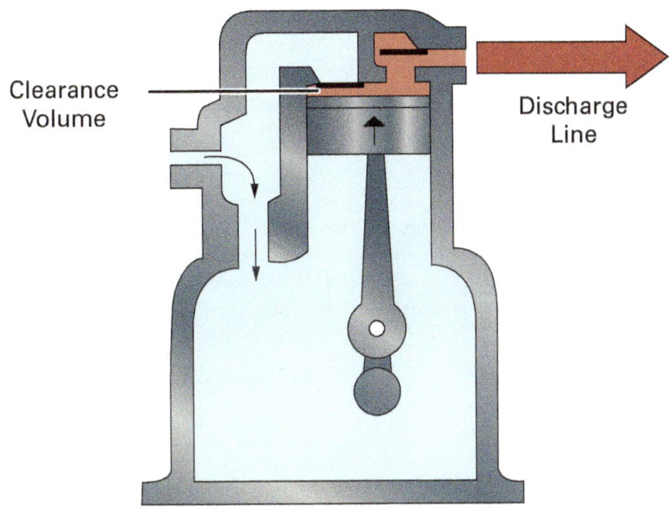

Figure 7 Clearance volume.

Hermetic Compressors

In the past, the limited capacity of hermetic compressors also limited their application. In older applications, hermetic compressors were typically found in systems with a capacity of less than 10 tons. Systems over that capacity tended to use semi-hermetic compressors.

Today, hermetic compressors with greater capacity are possible. More capacity is also possible through a modular arrangement, with more than one compressor installed on the same refrigerant circuit. This requires careful oil management to ensure each compressor receives its share of the lubricant in circulation. Even rotary screw compressors with significant capacity in a hermetic design are available today. Rotary screw compressors are typically semi-hermetic.

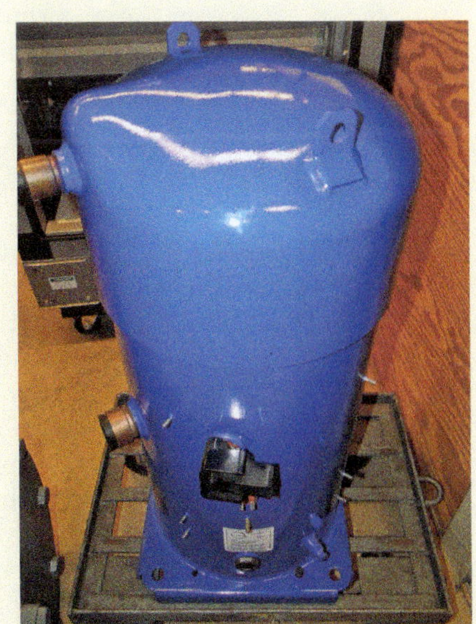

1.2.1 Pistons and Piston Rings

Pistons compress the refrigerant vapor using the mechanical force applied by the crankshaft. During the compression stroke, high-pressure vapor develops above the piston with low-pressure vapor below it. To prevent the high-pressure vapor from leaking past the piston, rings are inserted into grooves provided in the piston body.

Two types of rings are usually mounted on a compressor piston. The upper rings are called *compression rings*. The compression rings prevent the high-pressure discharge vapor from leaking past the piston into the low-pressure area of the crankcase.

The rings mounted on the piston below the compression rings are called *oil rings*. The oil rings function to prevent oil from flowing around and past the piston. Those familiar with automotive engines will recognize the similarity.

Note that some small reciprocating compressors have no rings. In these compressors, a film of refrigerant oil provides the only seal.

1.2.2 Compressor Cylinder Valves

Two common types of reciprocating compressor valves are: *reed*, or *flapper*, and *ring* (*Figure 8*). Both types can be applied to either suction or discharge ports.

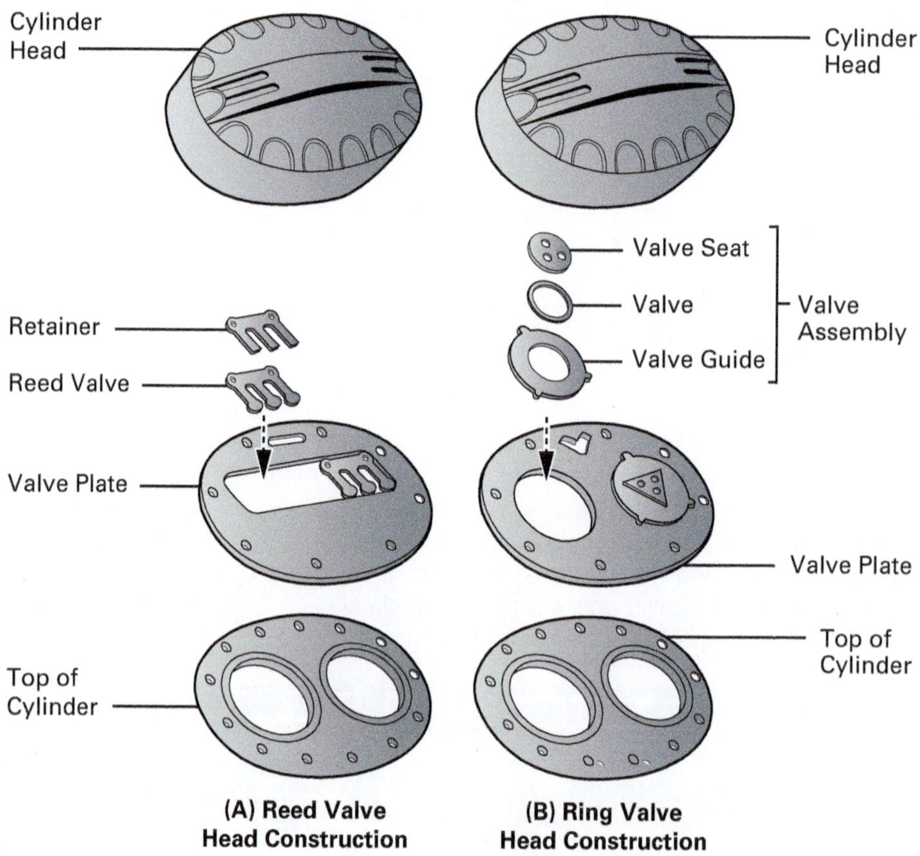

Figure 8 Common compressor valve plate assemblies.

The reed valve is a thin, flexible piece of spring steel mounted on a valve plate. The valve plate mounts between the head of the compressor and the top of the cylinder wall. A portion of the reed covers the suction or discharge port.

The natural spring tension of the reed keeps the valve closed. It is forced open by pressures that overcome the tension. Reed valves have a variety of shapes, with each compressor manufacturer having their own version.

Note the retainer on top of the reed valve in *Figure 8A*. The retainer is a much thicker piece of metal of roughly the same shape as the reed, bent up slightly to provide clearance for the valve to open. The retainer prevents the reed from opening too wide. The valve can only open as far as the retainer allows. Otherwise, the thin reed valve would break quickly under stress.

Ring valves are also mounted on valve plates. The assembly consists of valve seats, circular ring valves, and valve guides, as shown in *Figure 8B*. The valve plate is mounted at the top of the cylinder, with the suction valve below the plate and the discharge valve above the plate. The ring valves are held in the closed position by a group of small springs installed in equally spaced holes around the guide. When the pressure difference across the discharge or suction valve is greater than the tension of the valve springs, the valve moves from its seat, allowing vapor to flow.

The operation and condition of the valves are important. Broken or bent valves and leaking valve plate gaskets affect the seal between the high-pressure and low-pressure sides of the compressor. High-pressure, high-temperature vapor can leak into the suction side of the compressor through worn valves. The inability of a compressor to produce its normal pressure differential is a common symptom of damaged or worn valves. It can also result in reduced cooling performance and an overheated compressor. Significant valve damage as well as broken piston rods and other components can occur due to liquid refrigerant entering the compressor. However, excessive oil returning from circulation can also damage the valves. Liquids in any form cannot be compressed.

1.2.3 Lubrication Systems

Two lubrication methods are used with reciprocating compressors to lubricate the moving parts. One is the **splash lubrication system**, shown in *Figure 9A*. In this arrangement, crankcase oil is splashed onto the cylinder walls and bearing surfaces of the compressor during each revolution of the crankshaft while the compressor is running.

Splash lubrication system: A compressor lubrication system that relies on oil splashed up from the crankcase during each revolution of the crankshaft to provide lubrication to the internal parts.

Splash lubrication is used in hermetic compressors. Most have *oil scoops*, which are like small spoons attached to the lower ends of the connecting rods. The oil scoops scoop up the oil and splash it on the other parts.

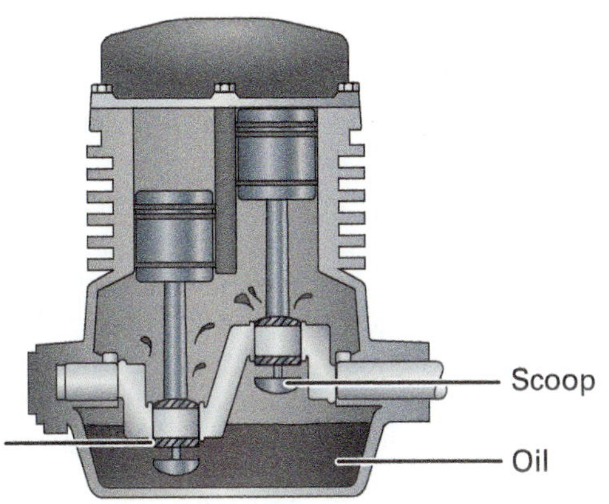

(A) Splash Lubrication

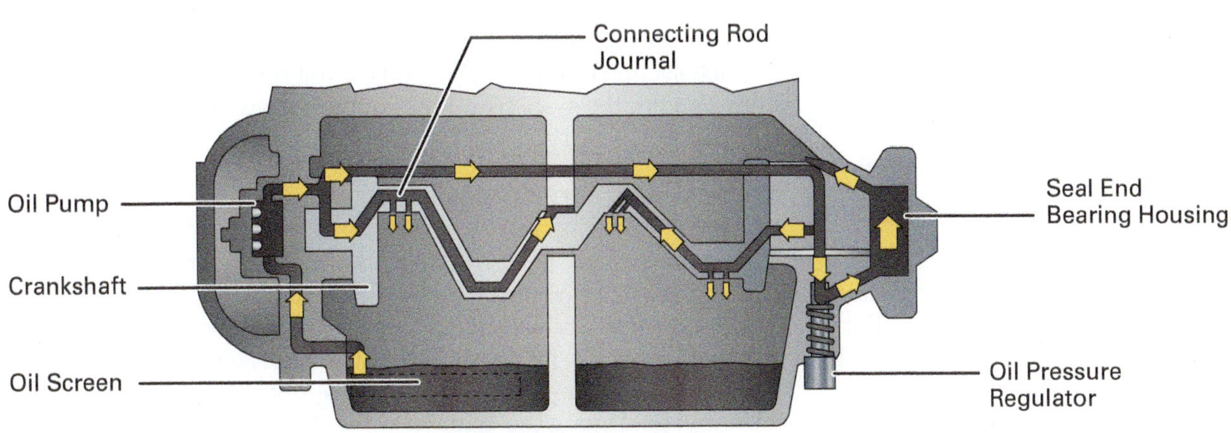

(B) Pressure Lubrication

Figure 9 Reciprocating compressor lubrication systems.

Going Green

Oil-Free Semi-Hermetic Compressor

Improved compressor efficiency translates into reduced energy consumption. In the HVACR compressor world, the introduction of the oil-free semi-hermetic compressor represented a significant performance improvement over a standard semi-hermetic compressor at full and part load. Magnetic bearings virtually eliminate friction, saving energy. Impellers, much like those found in a centrifugal compressor or pump, provide compression. A variable-speed drive is also incorporated into the motor to match the capacity to the load and provide maximum application flexibility.

Source: Danfoss Turbocor

Pressure (force-fed) lubrication system: A compressor lubrication system that uses a crankshaft-driven oil pump to draw oil from the crankcase and move it to the internal parts through the lubrication circuit.

The second and most popular method of lubrication for semi-hermetic reciprocating compressors is the **pressure (force-fed) lubrication system**, shown in *Figure 9B*. It relies on an internal oil pump mounted on the end of the crankshaft. The pump forces oil through the crankshaft to the bearing surfaces. In some cases, the connecting rods are drilled so the oil can access the pins that attach the pistons to the piston rods. An oil screen in the bottom of the crankcase strains the oil before it enters the pump.

An oil pressure regulator controls the oil pressure in some models. Excessive oil pressure leads to increased energy consumption and possible compressor damage. The arrows in *Figure 9B* show the direction of flow through the compressor.

Remember that some volume of compressor lubricating oil is normally entrained in the refrigerant being circulated through a system. The uniform movement of this oil through the entire refrigerant circuit and back to the compressor depends on proper piping practices, including attention to the refrigerant velocity in each section of piping. If the oil cannot return to the compressor, it will remain trapped in the system and the compressor will quickly run dry.

1.3.0 Rotary Compressors

Smaller rotary compressors are hermetic. The motor and drive shaft are mounted vertically in the compressor housing with the motor above the compressor.

Rotary compressors are usually one of two types: *stationary-vane* or *rotary-vane*. The rotary-vane type is far more common in refrigeration applications. Its vanes, assisted by spring tension, slide in and out of their retaining slots to provide a continuous seal along the cylinder walls (*Figure 10*).

Rotary-vane compressors have a rotor centered on the drive shaft. However, the drive shaft itself is positioned off-center in the cylinder. Two or more vanes

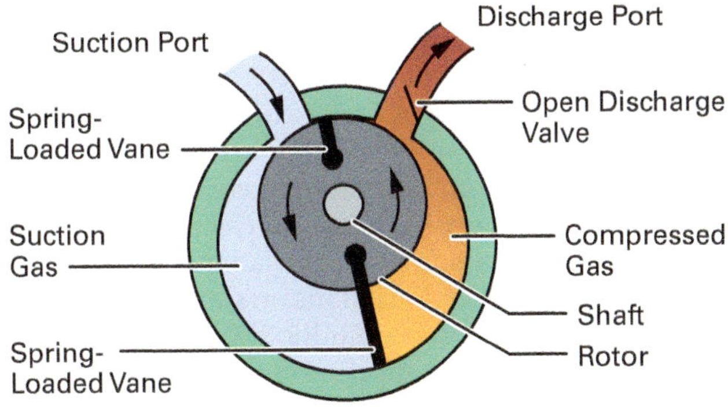

Figure 10 Rotary-vane compressor operation.

that slide in and out to follow the shape of the cylinder are mounted on the rotor. As the rotor turns, low-pressure vapor from the suction line is drawn into the cylinder behind the vanes. The vapor trapped ahead of the vane is consistently compressed against the cylinder wall until it is forced out of the discharge opening. The vanes provide separation between the high- and low-pressure areas.

Rotary compressors do not have suction valves, but they do have a discharge valve to prevent the backflow of refrigerant into the compressor when it is not running. A check valve is typically found in the intake to prevent compressor oil from being forced back into the suction line and into the evaporator.

Lubrication in a rotary compressor can be provided by either a splash system or a force-fed system. Regardless of the method, compressor performance and reliability depend on a continuous film of oil being maintained on the cylinder walls and vane surfaces.

On Site

Rotary-Vane Compressors

Rotary-vane compressors must be installed in clean systems. Years ago, rotary compressors were used in some split air conditioning and heat pump systems. However, dirt and contamination introduced during installation resulted in a high failure rate.

If small particles of debris are present in the refrigerant circuit, this debris can damage the compressor. For this reason, rotary-vane compressors are mainly used in window units and packaged terminal air conditioners (PTACs) where the cleanliness of a factory-sealed system is an advantage. Rotary-vane compressors typically start and run quietly because there are no pistons or any reciprocating motion. This makes rotary-vane compressors a good choice for systems that must operate near room occupants.

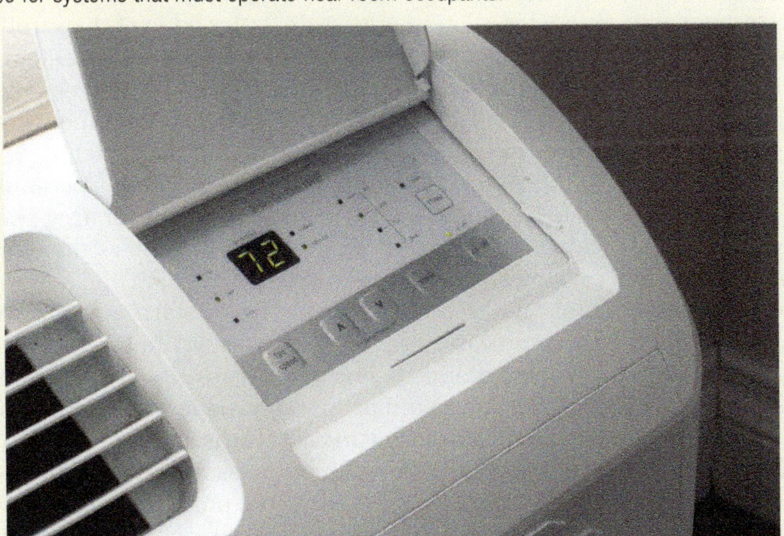

Source: Tada Images/Shutterstock

1.4.0 Scroll Compressors

Scroll compressors are almost exclusively hermetic, but a few semi-hermetic models have been introduced. Of all the compressor types, scroll compressors have the fewest moving parts. Scroll compressors operate efficiently even in applications that require a higher compression ratio, such as commercial refrigeration.

Scroll compressors and their numerous variations currently dominate the residential and commercial market. Variations on standard fixed-capacity models include variable speed operation and multiple stages of compression.

The scroll compressor develops compression using two spiral-shaped parts, called *scrolls* (*Figure 11*). The upper scroll is stationary, while the lower scroll is driven by the motor and moves inside the stationary scroll with an orbiting motion. The motor is usually mounted beneath the compression section. Slight contact exists between the two scrolls. Scroll compressors do not require suction or discharge valves because the design of the scroll minimizes refrigerant flow when they are not running.

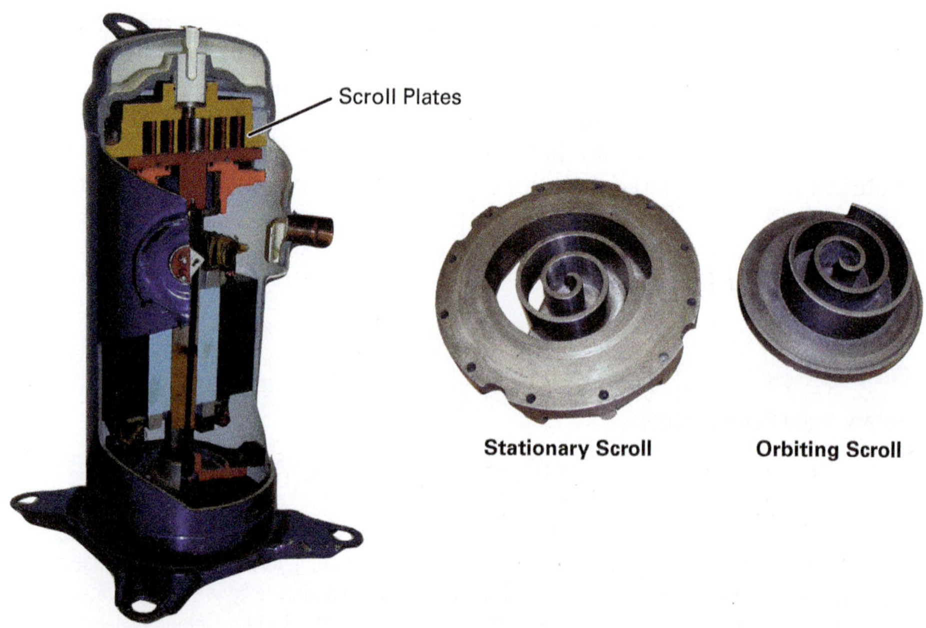

Figure 11 Scroll compressor.
Source: Emerson Climate Technologies

Refrigerant vapor enters the suction port at the outer edge of the scroll and, after compression, is forced out through the discharge port at the center of the stationary scroll. The orbiting action draws vapor into pockets between the two spirals. As the scroll rotates, the vapor inlet is sealed off and the vapor is forced into progressively smaller pockets as it progresses toward the center.

Compliant Scroll™ compressor: A version of the scroll compressor designed to allow the orbiting scroll to temporarily shift from its normal operating position, allowing liquid refrigerant that may enter to pass through without causing significant damage.

A version of the scroll compressor, called a **Compliant Scroll™ compressor**, allows the orbiting scroll to temporarily shift from its normal operating position if liquid refrigerant accidently enters the compressor. This allows the liquid to pass through without causing significant damage.

Scroll compressors often make unique sounds when starting up, running, and shutting down. For example, scroll compressors may run backwards for a few moments when the drive motor is de-energized, causing an unusual noise. The metallic sound is obvious, and a similar sound is made when a three-phase model is improperly phased, causing it to run backwards.

Unlike reciprocating compressors, scroll compressors only work in one direction of rotation. This is only a concern for three-phase models. However, brief periods of incorrect rotation rarely cause any damage. When starting a three-phase scroll compressor for the first time, the sound will usually indicate whether it is running in the proper direction.

Reversing any two of the three power wires changes the direction of rotation. If not familiar with these sounds, a technician would think something is wrong internally. Various compressor manufacturers and equipment manufacturers have training programs and materials available that help familiarize technicians with the normal and abnormal sounds made by scroll compressors.

Scroll compressors typically have protection against excessive discharge temperatures that can damage the compressor and/or burn the oil. A discharge temperature safety switch for this purpose is commonly wired into the compressor control circuit.

However, a discharge temperature safety valve may also be added. When the discharge temperature is excessive, the device opens an orifice that redirects hot discharge vapor to the motor's internal overload protection. Exposing the valve directly to hot vapor causes it to react and open quickly. This protects the compressor, but it can be misleading during troubleshooting if the technician is not familiar with the device.

1.5.0 Rotary Screw Compressors

Rotary screw compressors are used in commercial and industrial applications requiring 50 to 750 tons of capacity. Most models used for air conditioning and chilled water applications are semi-hermetic, while those used in industrial refrigeration often have an open drive (*Figure 12*). Hermetic versions are also available in the smaller sizes.

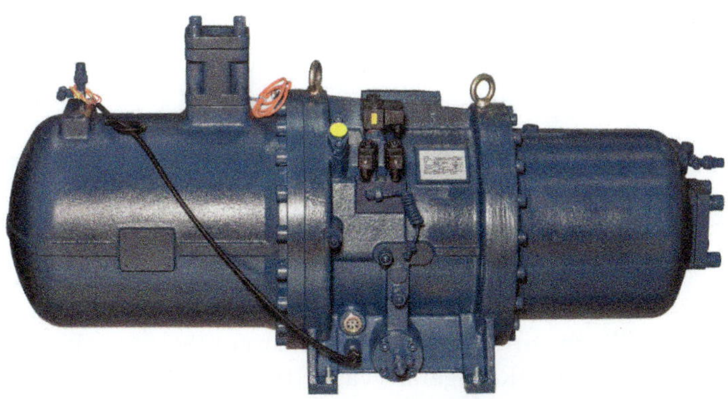

(A) Semi-Hermetic Screw Compressor

(B) Open-Drive Screw Compressor

Figure 12 Semi-hermetic and open-drive rotary screw compressors.
Source: (A) Sergii_Petruk/Shutterstock; (B) Milan Sommer/Shutterstock

Rotary screw compressors use a matched set of screw-shaped rotors enclosed within a cylinder (*Figure 13*). One rotor is designated male or referred to as the *drive rotor*. The other is designated female or referred to as the *driven rotor*. The motor turns the male (drive) rotor that in turn drives the second rotor. The two rotors turn in opposing directions. The drive rotor typically turns faster than the driven rotor because it has fewer *lobes*. Think of the lobes as threads on a screw. Typically, the drive rotor has fewer lobes than the driven rotor. As the rotors turn, the vapor is trapped between the lobes and compressed as it is pushed into a progressively smaller space.

Figure 13 Rotary screw compressor rotors.

A thin oil film fills the tiny void between the rotor lobes to minimize leakage and enhance compression. Refrigerant vapor that is virtually free of oil is sent to the condenser, thanks to a high-efficiency oil separator. Oil separated from the discharge vapor stream collects in an oil sump that is maintained at the condensing pressure while the compressor runs. Thus, oil consistently moves out of the sump and into areas where the pressure is lower; no pump is required. Oil coolers are sometimes needed, especially when the compressor must operate at a high compression ratio that increases heat production.

1.6.0 Centrifugal Compressors

Centrifugal compressors are also made in both open and semi-hermetic designs. Typically, centrifugal compressors are used on commercial/industrial cooling systems requiring compressors larger than 100 tons. Standard sizes are available up to roughly 10,000 tons of capacity. Custom-built models can exceed 20,000 tons.

Centrifugal compressors (*Figure 14*) use a high-speed impeller with many blades that rotate in a spiral-shaped housing. The impeller is driven at speeds around 10,000 rpm inside the compressor housing.

Refrigerant vapor enters the compression area through the center of the impeller. The impeller throws this incoming vapor in a circular path outward from between the blades and into the compressor housing (*Figure 15*). This action, an example of *centrifugal force*, pressurizes the high-velocity vapor and forces it out of the discharge port.

Figure 14 Centrifugal compressor.
Source: alifsufri/Shutterstock

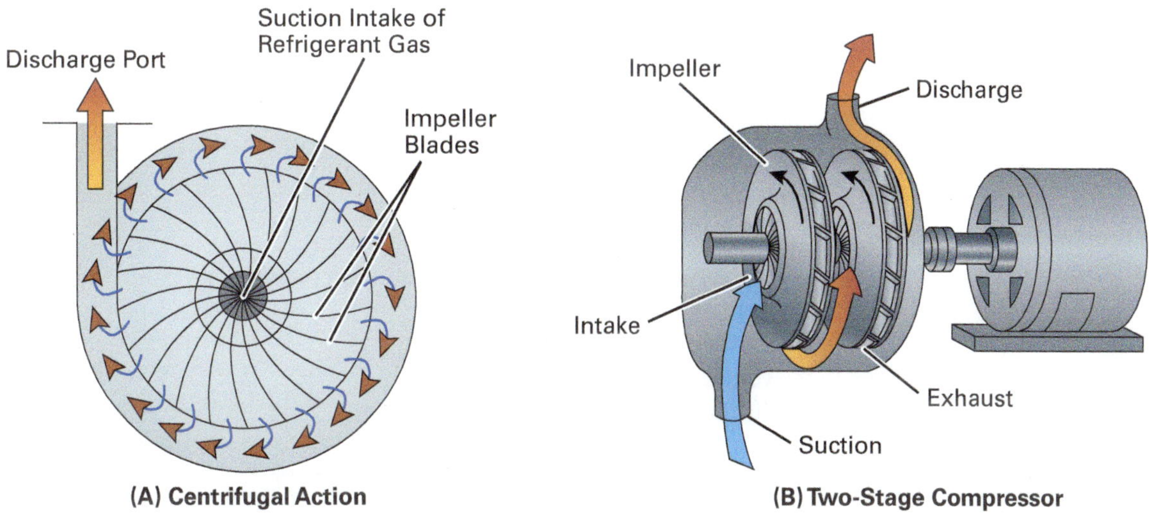

Figure 15 Centrifugal compressor operation.

Several impellers can be placed in series to create a greater pressure difference and/or to move a greater volume of vapor. A compressor with one impeller is a *single-stage compressor*; two or more impellers create a *multi-stage compressor*. When more than one stage exists, the discharge from one stage is fed into the inlet of the next stage, where it is further compressed.

1.0.0 Section Review

1. An advantage of a semi-hermetic compressor over a hermetic compressor is that it is ______.
   a. lighter in weight
   b. more energy efficient
   c. repairable
   d. permanently sealed

2. In a reciprocating compressor, the up and down movement of the piston is created by the ______.
   a. cylinder
   b. suction and discharge valves
   c. crankshaft lobes
   d. cylinder head

3. Rotary compressors do *not* ______.
   a. have a discharge valve
   b. have a suction valve
   c. compress vapor
   d. need a drive motor

4. Which of the following types of scroll compressors allow the rotating scroll to shift from its position to accommodate liquid entry?
   a. Semi-hermetic
   b. Modulating
   c. Liquid-safe
   d. Compliant

5. A rotary screw compressor would most likely be found in a(n) ______.
   a. industrial application
   b. window air conditioner
   c. packaged heat pump
   d. residential split system

6. The capacity of a centrifugal compressor is typically ______.
   a. larger than 100 tons
   b. less than 50 tons
   c. never larger than 500 tons
   d. unlimited

2.0.0 Compressor Capacity Control

Performance Tasks

There are no Performance Tasks in this section.

Objective

Identify and describe various approaches to compressor capacity control.

a. Identify and describe capacity control methods for reciprocating and scroll compressors.
b. Identify and describe capacity control methods for rotary screw and centrifugal compressors.

Capacity control: The manipulation of compressor operating characteristics to reduce the volume of vapor compressed to better match the load conditions. Other methods of capacity control can also be applied to a system without affecting compressor operation or the volume of vapor moving through it.

When a cooling or refrigeration system is operating properly, the load imposed on the compressor is often less than the capacity of the compressor. The load imposed on cooling and refrigeration systems can vary significantly.

Some systems make use of compressor and/or system **capacity control** to match the capacity with greater precision load. Capacity control can help maintain more stable temperatures, as well as conserve energy.

Various methods of capacity control can be applied to all systems. Capacity control has been common in commercial and industrial applications for many years. Today, many residential systems also take advantage of improved technology to control capacity and increase efficiency.

2.1.0 Capacity Control for Reciprocating and Scroll Compressors

A variety of methods are used to control the capacity of reciprocating and scroll compressors or the systems they serve. The methods discussed in this section include the following:

- On/off cycling
- Multiple compressors
- Cylinder unloading
- Compressor speed control
- Hot-gas bypass

2.1.1 On/Off Cycling

On/off control can be considered a simple form of capacity control. In a common comfort cooling system, a room thermostat closes when cooling is required. This energizes the compressor, allowing it to run until the temperature in the room reaches the thermostat's setpoint. The compressor then cycles off until the thermostat closes again. On/off cycling is commonly used in many residential and commercial systems.

On/off cycling works well in situations where the load remains consistent, and where materials or processes are not sensitive to minor temperature changes. However, if the compressor capacity is significantly greater than necessary for the load, the compressor may have to cycle on and off too often.

Short cycling occurs when a compressor runs for a short time, shuts down, and then restarts after a brief rest. Short cycling can result from poor thermostat control, rapid changes made by the user, or from the operation of some safety controls. Compressors and their drive motors experience more stress during startup than at any other part of their operating cycle.

Short cycling: A condition in which a compressor is restarted very soon after it has shut down, possibly accompanied by very short run times as well.

2.1.2 Multiple Compressors

The use of multiple compressors is another effective approach to capacity control. Multiple compressors can be applied to a single piece of equipment in two common methods.

The first method is to install two compressors that operate on a common refrigeration circuit. The two compressors are often referred to as *tandem compressors.* The compressors share a common evaporator, condenser, and refrigerant piping. The compressors are essentially connected in parallel. When the demand is low, only a single compressor is required to handle the load. As the load increases, the second compressor is started. However, remember that the refrigerant piping must be sized to handle the total refrigerant flow for all the connected compressors.

As a result, when only one compressor is running, the flow rate and velocity of the refrigerant is greatly reduced. The reduction can result in oil being trapped within the system's coils and piping sections. Oil that is unable to return to the compressors reduces the oil level in one or more compressors. The oil will begin to return when the system next operates at maximum capacity, but problems can and do occur if full-capacity operation does not happen often enough. Trapped oil may return at a faster rate than desired when the compressor begins operation at its full capacity, resulting in **oil slugging**.

For the oil that returns as expected, some method of balancing the oil level between the compressors is needed. An *oil equalizer line* is typically installed between the compressors and located at the normal oil level. When the compressors are off and the pressures equalize, the oil levels will also equalize with the help of gravity.

Another design feature is used to balance the oil levels while the compressors operate. When two compressors are connected in parallel and the upstream (relative to the suction line) compressor runs, most of the oil will return to that compressor first. The use of a restrictor in the downstream suction line provides a small pressure drop when the downstream compressor is running (*Figure 16*).

On Site

Tandem Compressors

Manufacturers of tandem compressors often require the replacement of both compressors when one fails.

Oil slugging: The entrance of oil in circulation into a compressor cylinder during operation in an unmanageable volume, leading to the potential for serious mechanical damage if it enters the compressor cylinder.

This drop causes the pressure in the crankcase of the downstream compressor to be slightly lower than the pressure in the upstream crankcase. This small pressure differential allows excess oil in the upstream compressor to move through the equalizer line to the downstream compressor during operation.

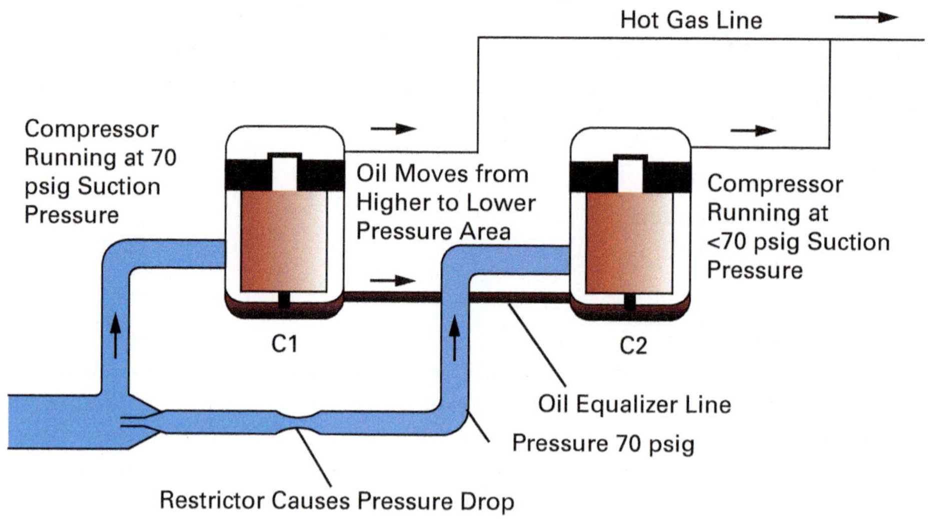

Figure 16 Oil management through pressure differential.

On Site

Active Oil Management

In commercial refrigeration and industrial environments, systems may have many compressors tied to a single refrigerant circuit. Passive oil management doesn't work well in these applications.

Active oil management involves the separation of oil from the vapor stream, collecting it in a sump, and then redistributing the oil to any number of compressors. As the example shows, each compressor is fitted with an independent oil level control that admits oil into the crankcase as needed.

While conventional oil separators that rely solely on centrifugal force to separate liquid from vapor can capture roughly 85% of the oil from the vapor stream, *coalescing oil separators* can capture as much as 99% of the oil. Less oil moving through the circuit also improves heat transfer at the evaporators and condensers.

Low-Pressure Oil System with TraxOil Controls

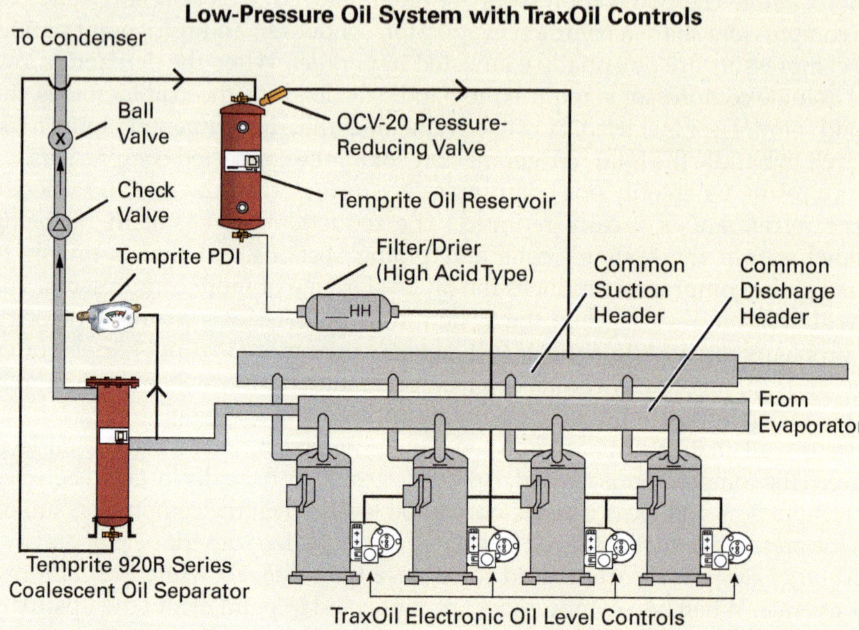

Source: Temprite

Multiple compressors can also be applied by providing an independent refrigerant circuit for each one. In this case, each compressor has its own condenser, metering device, evaporator, and piping. Compressors are cycled on and off as necessary to manage the cooling load.

One advantage to this approach is that it eliminates oil management and distribution issues. In addition, any refrigerant circuit problems that arise only affect a single compressor. However, for the system to operate effectively, the supply airflow (or water flow in a chilled water application) must be carefully balanced across the multiple evaporators.

2.1.3 Cylinder Unloading

Cylinder unloading is a common method of capacity control used with reciprocating compressors that have multiple cylinders. Reciprocating compressors with multiple cylinders are typically semi-hermetic or open drive.

A control device inside the compressor blocks or reroutes the flow of vapor to one or more cylinders of a multi-cylinder compressor. The pistons continue to move up and down, but no compression takes place in the unloaded cylinders. Auto engines have also taken advantage of a similar approach, referred to as *cylinder deactivation*, to improve fuel efficiency.

The primary types of cylinder unloading methods and other mechanical approaches to capacity control used today include the following:

- Suction-bypass unloading
- Suction-cutoff unloading
- Hydraulic unloading
- Scroll compressor mechanical unloading

Suction-Bypass Unloading

The suction-bypass unloading method uses a special cylinder head and a pressure- or electrically actuated unloader control valve. When the cooling load drops, the pressure and temperature at the compressor suction inlet also drop. The unloader control device senses the decrease in pressure (pressure-actuated valve) or temperature (electrically actuated valve) and actuates the unloader. A piston in the unloader retracts to a preset point and unloads the cylinder, causing the suction vapor to bypass the unloaded cylinder. *Figure 17* shows a pressure-actuated bypass unloader operating in the unloaded position.

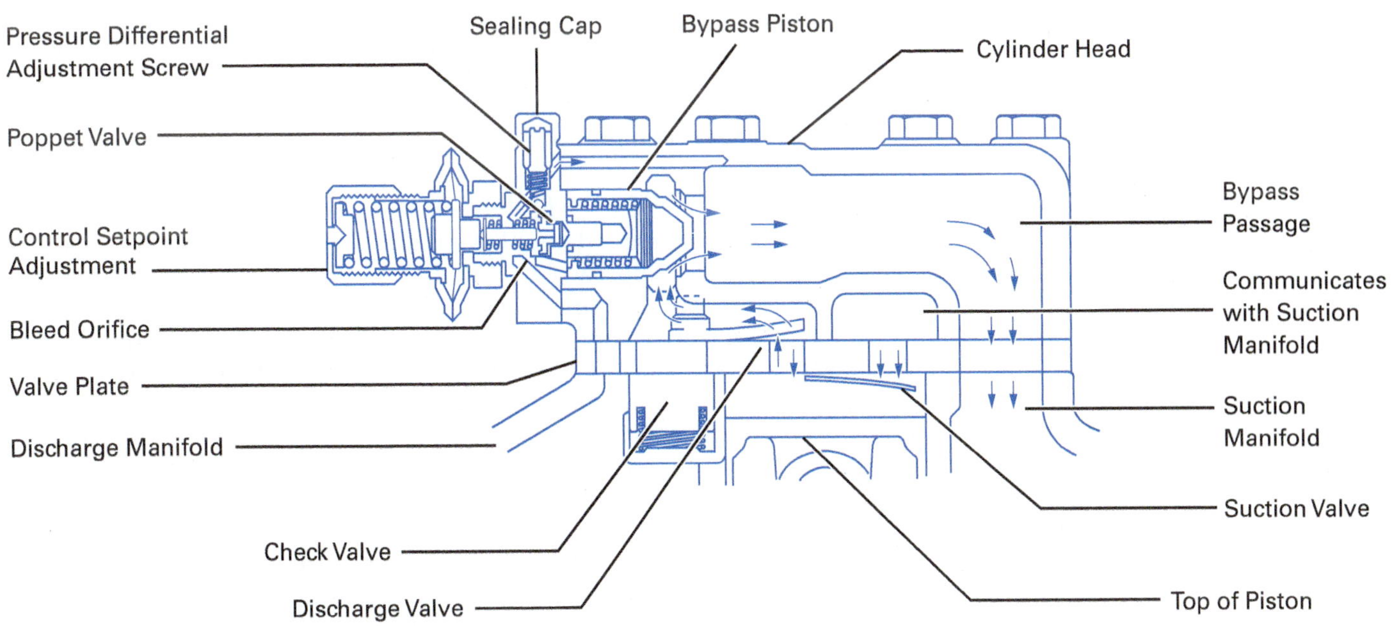

Figure 17 Suction-bypass unloader.

Suction-Cutoff Unloading

The suction-cutoff unloading method (*Figure 18*) is actuated in a similar manner. With a suction-cutoff unloader, the suction vapor is prevented from entering the unloaded cylinder at all, rather than being recirculated through it.

Both the suction-bypass and suction-cutoff unloaders have adjustable setpoints. The *control setpoint* determines the pressure setting for the cylinder to be loaded. The *differential setpoint* determines how far from the control setpoint the pressure must fall before the cylinder unloads. However, some unloaders may have fixed setpoints for specific applications.

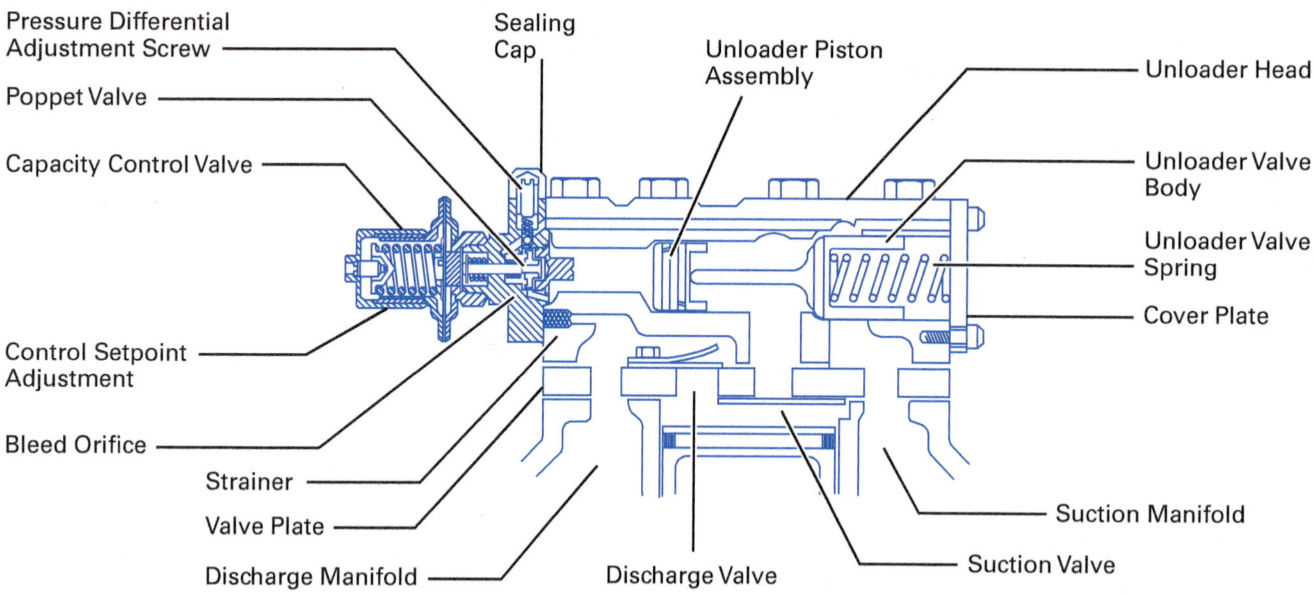

Figure 18 Suction-cutoff unloader.

Hydraulic Unloading

The hydraulic unloader unloads a compressor by holding the suction valve open during operation, thus preventing compression of the vapor. Compressor oil pressure powers the unloader valve mechanism. When capacity reduction is needed, oil pressure is relieved from a piston located in the capacity-reduction unit. The piston forces a mechanical linkage to lift the suction valve from its seat. The compressor piston is then no longer able to compress refrigerant and simply travels up and down within the cylinder. Like the other unloader types, the hydraulic unloader has adjustments to set the operating point at which the cylinders are loaded and unloaded.

Scroll Compressor Mechanical Unloading

Because scroll compressors do not have traditional cylinders like reciprocating compressors, another approach is needed for mechanical capacity control. One approach used for scroll compressors is referred to as *vapor bypass*.

An internal solenoid controls vapor bypass ports. At low loads, the solenoid is de-energized and the bypass ports remain open. As the load increases, the solenoid is energized, and the bypass ports are closed. The opening and closing of the bypass ports effectively changes the internal volume of the area where the refrigerant is compressed. This form of control typically results in two steps of capacity: 100% capacity with the bypass ports closed, and 67% capacity when the ports are opened.

One manufacturer of scroll compressors has introduced another means of mechanically modulating compressor capacity. Recall that a scroll compressor uses a fixed upper plate and an orbiting lower plate. Scrolls considered *compliant* can lift the upper plate when placed under stress by liquid refrigerant entering the scrolls and thereby prevent mechanical damage to the compressor.

Capacity control in a scroll compressor can be achieved by taking this feature one step further. In timed cycles, the two plates can be separated to cease compression (unloaded) and then be brought together for normal compression (loaded) (*Figure 19*). The unloaded cycle time is managed to precisely track the load. With this strategy, scroll compressor capacity can be modulated between 10% and 100%, resulting in reduced operating costs, consistent load control, and fewer compressor starts. This type of compressor is referred to by the manufacturer as a *digital scroll compressor.*

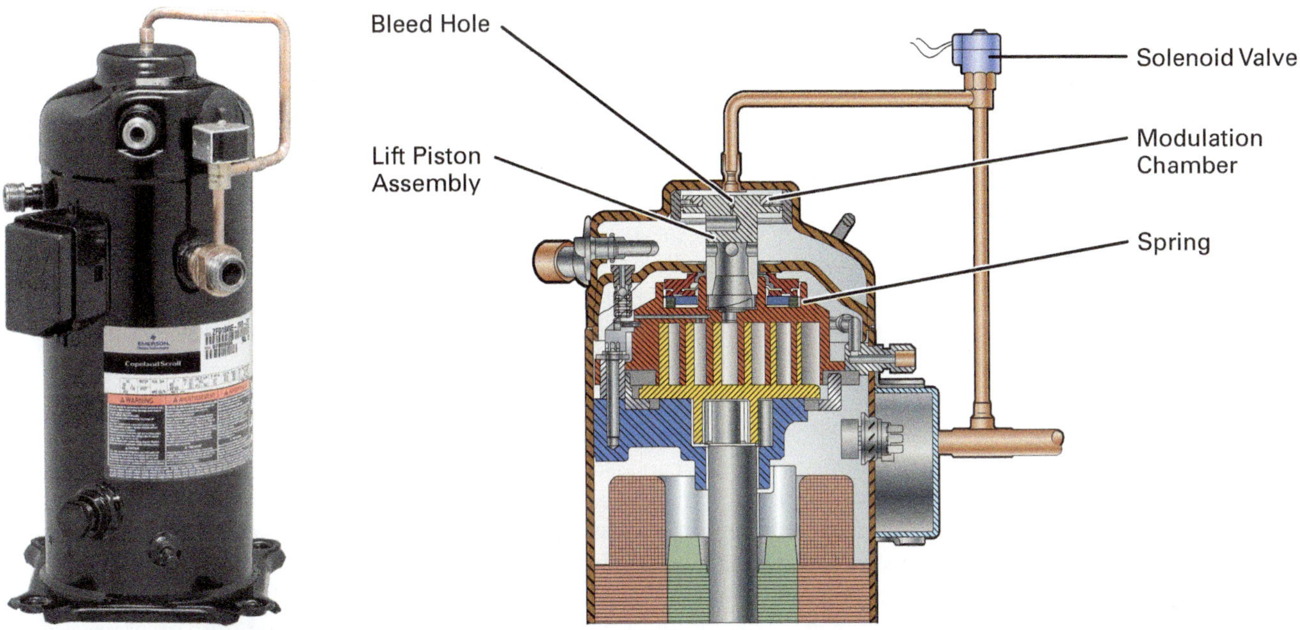

Figure 19 Digital scroll compressor capacity control.
Source: Emerson

2.1.4 Compressor Speed Control

All compressors, except the centrifugal compressor, are **positive displacement compressors**. A positive displacement compressor is one in which the pumping action is created by moving pistons or moving chambers.

Positive displacement compressors: Compressors that capture vapors in a chamber, then the size of the chamber is mechanically reduced to increase the pressure.

When they are operated at a constant speed, positive displacement compressors move a consistent volume of vapor. Compressors with greater displacement have higher capacities. The term *displacement* refers to the volume of vapor moved over time. For a reciprocating compressor, the total displacement is calculated as follows:

$$\text{Compressor displacement (cubic inches, or in}^3) = (\pi D^2 L n \div 4) \times \text{rpm}$$

Where:

$\pi = 3.1416$
D = diameter of the cylinder bore, in inches
L = length of piston stroke, in inches
n = number of cylinders
rpm = compressor revolutions per minute

From this equation, it should be noted that the displacement of a compressor, and therefore its capacity, is proportional to its speed. Thus, speed is a viable approach to capacity control.

Multispeed Capacity Control

Multispeed motors can be used to drive compressors and achieve fixed levels of capacity control. Electronic controls that automatically select the proper motor speed for the load conditions are more effective.

However, the power wiring and contactor arrangement for a multispeed motors can be complex. Multispeed motors contain dedicated sets of motor windings used to develop specific motor speeds. Care must be taken in the control circuits to ensure that windings for different speeds are not energized simultaneously, which could destroy the motor. Separate contactors are used to energize the different windings. The contactors are mechanically and/or electrically interlocked to prevent simultaneous operation.

Multispeed operation generally applies only to two speeds, referred to as *low speed* and *high speed*. Also note that the compressor must shut down before a speed change can be made.

> **On Site**
>
> **Centrifugal Compressors**
>
> Centrifugal compressors are classified as *dynamic displacement compressors*. Dynamic displacement compressors significantly increase the velocity of the vapor and then restrict its discharge, causing the pressure to increase. However, the generated pressure differential is significantly less than that developed by positive displacement compressors.

Variable Speed Capacity Control

Variable frequency drive (VFD): An approach to motor speed control that manipulates the voltage and frequency of power delivered to an AC motor to fully modulate its speed.

Compressor motors can also be driven at variable speeds, meaning that the speed is not restricted to fixed values. Today's variable speed technology can support motor operation at virtually any speed. The use of **variable frequency drive (VFD)** technology, which has long been used with commercial/industrial fans and pumps, has become very popular for HVACR compressors. By changing the frequency of the power to the motor, compressor speed can be controlled with great precision.

VFDs, which may also be referred to as *variable speed drives (VSDs)*, first change the incoming AC power from the utility to DC. This process is known as *rectification*. The power sent to the motor is simulated AC with the frequency of the signal now controllable. Returning the power to an AC signal the motor can use is called *inversion*. For this reason, VFDs are sometimes referred to as *inverter drives*. Note that the motor itself is still AC powered. Electronically commutated motors (ECMs), which are commonly used today as blower and condenser fan motors, are DC motors.

In *Figure 20*, the gray dashed line represents the temperature swings that occur when a standard compressor is cycled on and off to maintain the room temperature at a given setpoint. The blue line represents a compressor operating with an inverter drive. This demonstrates the consistent level of comfort that can be achieved.

Although it may seem like the intermittent off cycles would generate greater energy savings, many tests and real-world applications have proven that keeping the compressor running and modulating its speed consumes less energy. Modulating the compressor speed also significantly reduces the number of starts. Motors generally suffer more wear and tear during startup than when running normally, so reducing the number of starts is a significant advantage.

The technology is available to drive compressor motors at very low speeds, even down to 1% of their maximum speed. However, the basic refrigeration cycle fails to produce any reasonable cooling effect when the speed is too low. For that reason, most variable speed compressors have a minimum speed that ranges from 20% to 40% of their maximum speed.

Compressors using VFD technology are brought up to speed gradually. Soft starting reduces the starting current substantially, reducing the stress applied to the contactor and other line-voltage control devices.

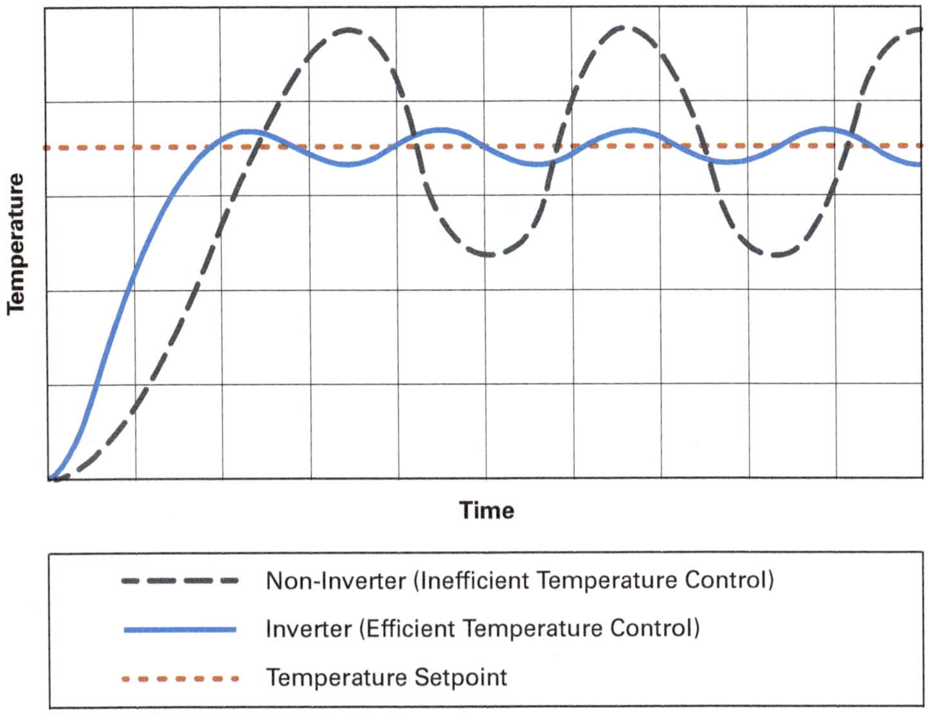

Figure 20 Inverter versus non-inverter temperature control.

2.1.5 Hot-Gas Bypass

The hot-gas bypass method of capacity control is not directly related to compressor operation, but is worth noting. The hot-gas bypass method can be used to maintain extreme consistency in temperature for applications that require precise control.

System capacity is reduced by mixing some of the compressor discharge vapor with liquid/vapor mixture leaving the metering device. The hot vapor is typically, and ideally, introduced between the metering device and the evaporator inlet, as shown in *Figure 21*. This forces the boiling refrigerant in the evaporator to also de-superheat the introduced vapor, reducing its capacity to

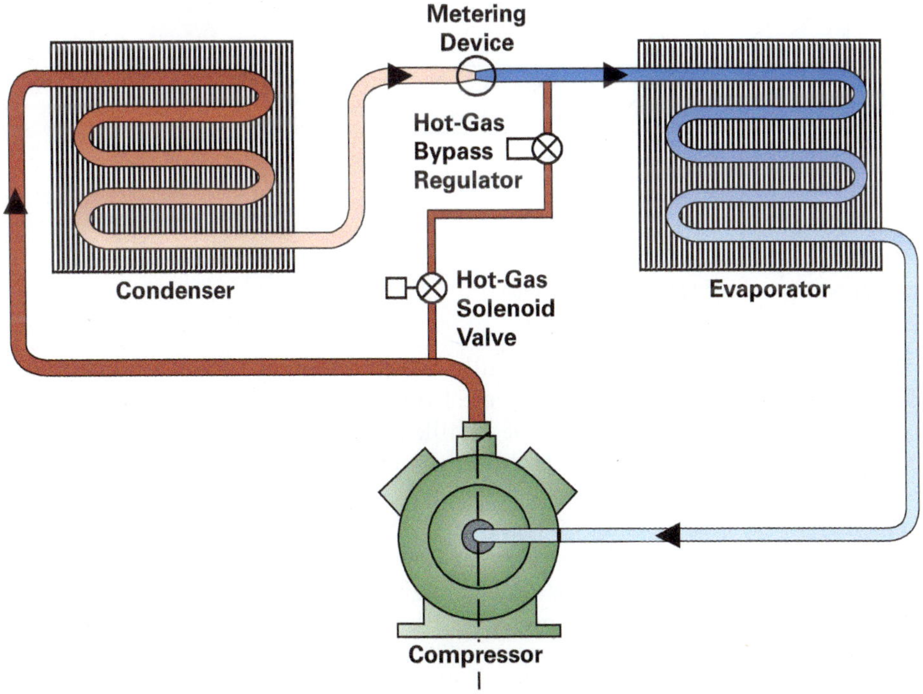

Figure 21 Basic hot-gas bypass schematic.

cool the air flowing across the evaporator. A solenoid valve provides on-off control of the hot gas, while the hot-gas bypass regulator reduces the pressure to a value closer to that normally found in the evaporator.

Throughout this scenario, the compressor continues to operate normally and at full speed. Although very precise temperature control can be achieved this way, it is not an energy efficient approach. When practical, other strategies for controlling system capacity are preferred today.

2.2.0 Capacity Control for Rotary Screw and Centrifugal Compressors

Due to their designs, screw and centrifugal compressors require different approaches to capacity control.

2.2.1 Rotary Screw Compressor Capacity Control

To mechanically control the capacity of a rotary screw compressor, controlling the amount of compression allowed is the current approach for most of those used in HVACR applications. A slide valve is used to block or open a path of escape for vapor entering the compression area between the lobes (*Figure 22*).

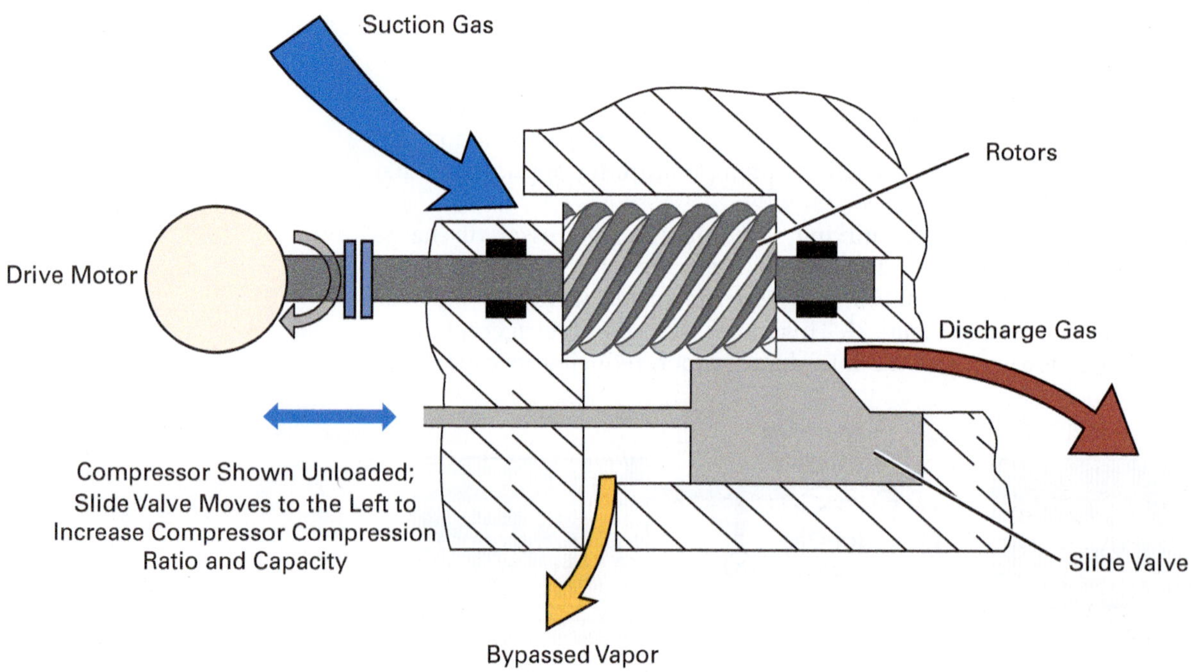

Figure 22 Rotary screw compressor slide valve operation — compressor unloaded.

The slide valve position between fully loaded and fully unloaded is infinite. The needed capacity can be tracked electronically and the valve positioned to closely track the load. In the case of a chiller, for example, the slide valve may be positioned based on the temperature difference between the returning water temperature and the leaving water setpoint. A secondary factor may be the rate of change in the return water temperature.

Some rotary screw compressors in service may use slightly different approaches to capacity control, such as using a slide valve to vary the size of the suction inlet to control vapor flow. Regardless of the compressor type or age, always review the manufacturer's documentation on the specific model being serviced before beginning any service or maintenance activities.

2.2.2 Centrifugal Compressor Capacity Control

Centrifugal compressor capacity control is generally achieved by one of three methods:

- Motor speed control
- Variable-pitch inlet guide vanes
- Suction or discharge throttling valves

As with other compressors, changing the speed of a centrifugal compressor drive motor affects its capacity. Machines equipped with a variable speed drive respond to load changes by ramping the motor speed up or down. Microprocessor-based controls provide load signals to the VFD that reacts by changing speed.

Variable-pitch inlet guide vanes (*Figure 23*) are also widely used in centrifugal compressors to control capacity. The guide vanes are mounted in front of the impeller vapor inlet. Controlling the volume of vapor entering the impeller controls capacity while allowing the impeller to continue rotating at its optimum speed.

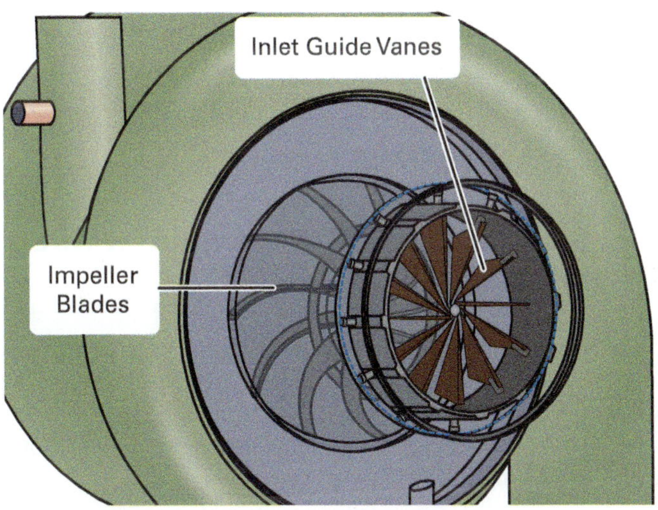

Figure 23 Inlet guide vanes on a centrifugal compressor.

2.0.0 Section Review

1. The most basic form of compressor capacity control is _____.
   a. hot-gas bypass
   b. cylinder unloading
   c. on/off compressor cycling
   d. compressor speed control

2. Which of the following is used for capacity control of a rotary screw compressor?
   a. Inlet guide vanes
   b. Cylinder unloading
   c. Hydraulic unloading
   d. Slide valve

3.0.0 Common Causes of Compressor Failures

Performance Tasks

1. Use an acid/moisture test kit to test a refrigerant circuit.

Objective

Describe the common causes of compressor failures.

a. Describe compressor failures related to the refrigerant circuit.
b. Describe compressor failures related to electrical issues.

Slugging: The entrance of excess liquid refrigerant and/or oil into a compressor cylinder during operation.

Floodback: The return of liquid refrigerant or a liquid refrigerant/oil mixture to the crankcase of an operating compressor.

Flooded starts: Compressor startups during which slugging, foaming, and inadequate lubrication occur due to oil in the compressor having absorbed condensed refrigerant during the off cycle, or due to the presence of liquid refrigerant at the compressor suction inlet. When the compressor crankcase pressure drops quickly at startup, any liquid refrigerant present boils rapidly, causing foaming and potentially washing the oil out of the crankcase.

Problems in the refrigerant and electrical circuits often result in compressor damage or failure. If you assume that the cause of a compressor failure is a deficiency within the compressor, you will likely make the first of many unnecessary compressor replacements. If the compressor fails and the cause has not been identified and resolved, the replacement compressor is also subject to failure. While compressors do occasionally fail due to an internal fault or normal wear, it is often an external factor at work, and sometimes more than one.

Common causes of compressor failure include the following:

- **Slugging** of liquid refrigerant and/or oil
- **Floodback** of liquid refrigerant into the compressor crankcase
- **Flooded starts** when liquid refrigerant has migrated back to the compressor crankcase or near the suction inlet during the off cycle
- Contamination of the refrigerant circuit with air, moisture, or dirt
- Overheating
- Electrical irregularities
- Incorrect installation of system components, piping, or accessories

3.1.0 Mechanical Causes of Refrigerant Circuit Compressor Failures

Part of the difficulty in identifying the true source of a compressor failure is that the system is inoperable. Without a functional compressor, refrigeration circuit operation and performance cannot be analyzed. For this reason, it is essential that system operation be carefully analyzed when a new or replacement compressor is started for the first time.

3.1.1 Slugging

Slugging occurs when a compressor tries to compress liquid refrigerant, oil, or both. Liquids are not compressible, and serious mechanical damage is often the result when a significant amount of liquid enters a cylinder or other compression area.

Slugging can occur at startup or as the result of rapid changes in system operating conditions while the compressor is running. Slugging is often indicated by a loud knocking sound from the compressor.

When a compressor tries to compress a liquid refrigerant or oil, pressures exceeding 1,000 psi (6,900 kPa) can be reached in the cylinder. The extreme pressure can result in blown cylinder heads and valve plate gaskets, damage to the compressor pistons and valves, and broken crankshafts or piston rods.

Common causes of slugging include the following:

- A significant overcharge of refrigerant.
- An overcharge of oil.
- An oversized or failed metering device that allows too much refrigerant to pass through.
- Refrigerant condensing in the evaporator or any portion of the refrigerant circuit between the evaporator and the compressor during the off cycle. Buried refrigerant lines or lines that pass through cold spots in refrigeration applications are common locations for vapor to condense when the system is idle. In winter conditions especially, refrigerant can condense in the suction line

and even inside the compressor itself. While crankcase heaters help prevent refrigerant from condensing in the compressor, heaters can and do fail.

- Slugs of oil trapped in the suction line or other components due to low refrigerant velocity. In most cases, the oil and refrigerant mix and travel reasonably well together. The oil often moves through the low-pressure side of the system as small droplets being pushed along by the velocity of the refrigerant vapor. If oil gets trapped in the system piping (such as when the compressor is operating in an unloaded condition) and then returns all at once, oil slugging occurs.

3.1.2 Floodback

Floodback refers to the return of liquid refrigerant to the compressor crankcase while it is operating. Floodback usually dilutes the oil, resulting in foaming and loss of lubrication. Oil foam is easily swept out of the crankcase and into the system piping. When floodback is severe and sustained, it can cause damage to the pistons, rings, and valves.

The following conditions can result in floodback:

- Oversized metering device admitting too much refrigerant into the evaporator.
- A broken or poorly mounted thermostatic expansion valve sensing element.
- Low superheat setting on an expansion valve.
- A lightly loaded evaporator due to restricted airflow, failure of the indoor fan or blower, or operation at indoor temperatures that are lower than recommended. Reduced airflow can cause frost to develop on the coil quickly, adding to the problem.
- An overcharge of refrigerant, especially in systems that use a fixed-orifice metering device.

On Site

Refrigerant Piping and Slugging

Refrigerant piping installations where the evaporator is placed lower than the condensing unit require a properly sized oil trap at the evaporator coil suction line connection. This trap should be shallow to minimize the volume of oil that can collect. The purpose of the trap is to create turbulence in the vapor passing through, increasing the likelihood of oil being caught in the vapor stream and carried up the riser.

In many cases, the suction leaving the evaporator and the riser must be reduced in size to ensure the velocity is high enough to carry the oil uphill under all operating conditions. Once the line changes to horizontal travel, the line size can be increased to reduce the pressure drop.

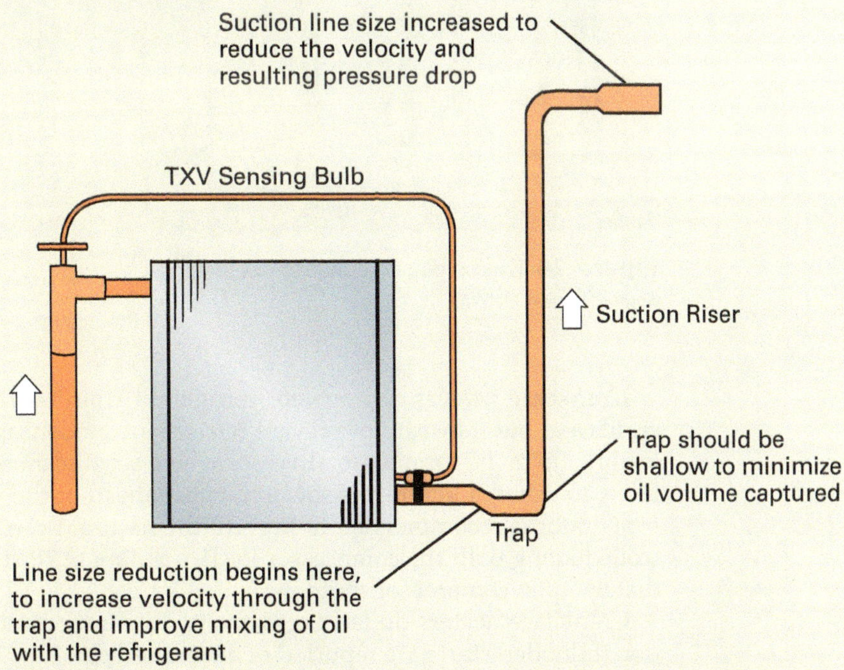

3.1.3 Flooded Starts

Flooded starts often result from refrigerant migrating to, and condensing in, the compressor crankcase and suction line. When the compressor is not running, refrigerant vapor tends to migrate to the coldest part of the system, where the vapor condenses to a liquid. For example, if it is cold outdoors and warm indoors, refrigerant vapor will migrate outdoors to the condensing unit. The crankcase can then become a convenient liquid-refrigerant storage vessel. Now add the sudden need for cooling, despite the season due to a house full of guests, and the compressor starts with liquid refrigerant in the crankcase.

At startup, the crankcase pressure drops sharply and the refrigerant in the oil will boil violently. Some of the liquid refrigerant will flash from a liquid to a vapor, causing the oil to foam. The oil-rich foam moves through the compressor, providing marginal or inadequate lubrication and carrying an excessive volume of oil into the system.

Crankcase heaters are used to keep the crankcase warm during the off cycle, preventing refrigerant from migrating to the compressor crankcase. Crankcase heaters are available in a variety of shapes and voltages (*Figure 24*). Some are mounted to the outside of the crankcase, while others are inserted into a well in the crankcase. Most hermetic compressors equipped with a heater use a flexible type that is wrapped around the base of the compressor shell, as shown in *Figure 25*. The heaters are referred to as *belly-band heaters*. Some have a self-regulating feature that modulates the heat produced as needed. Crankcase heaters can also be thermostatically controlled.

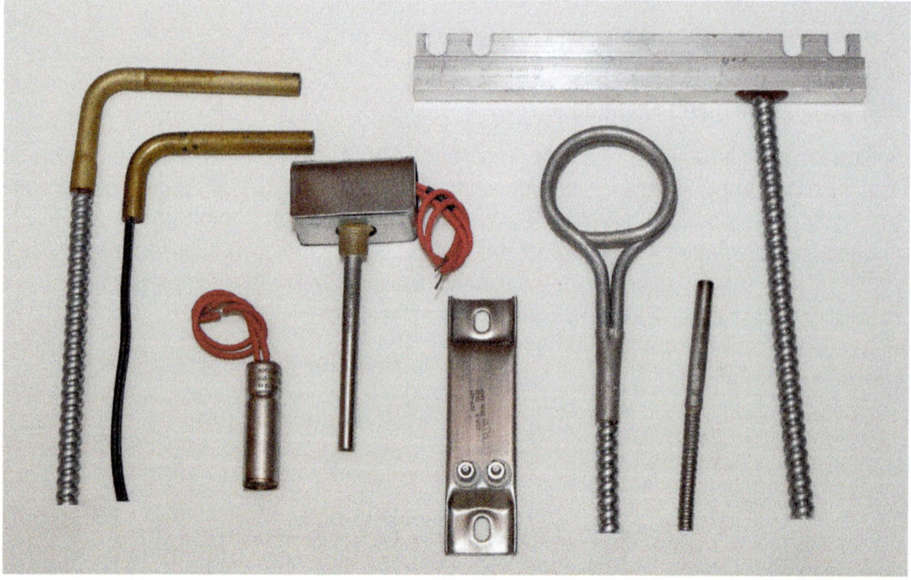

Figure 24 Examples of crankcase heaters.
Source: Backer HotWatt

Crankcase heaters can prevent refrigerant from condensing directly in the crankcase, but they cannot prevent refrigerant condensation in the suction line leading up to the compressor. In cases where significant refrigerant migration is likely to occur, a liquid-line solenoid valve ahead of the metering device at the evaporator is recommended to prevent the natural flow. The valve can be controlled along with the compressor itself, opening at startup and closing immediately upon compressor shutdown.

Crankcase heaters do fail occasionally. It is important to check their operation if flooded starts are reported or encountered.

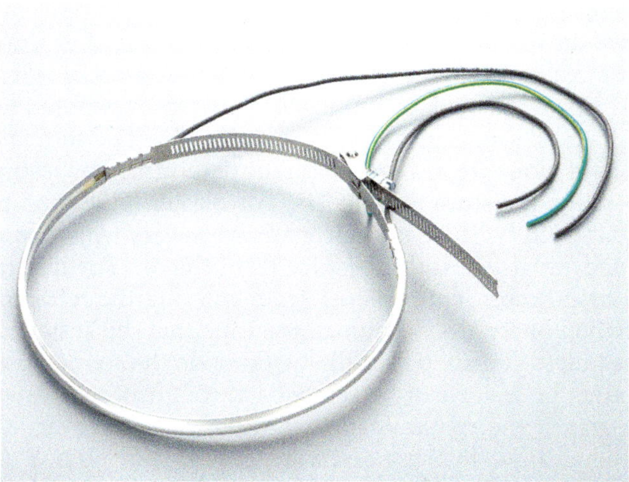

Figure 25 Belly-band crankcase heater.

3.1.4 Contamination

Refrigerant circuits ideally contain only refrigerant and oil. The circuit must be free of air and water. Anything other than refrigerant and oil in the circuit is considered a contaminant. Common contaminants include air, water, acids, dirt, and copper oxides.

Air, moisture, dirt, and copper oxides enter the system during installation or service procedures. Acid, soot, varnish, and hard carbon are typically created inside the system either by chemical reactions with contaminants or from a compressor motor **burnout**.

Burnout: A condition in which the failure or degradation of the motor winding insulation causes the windings to electrically short to ground.

Air

Air usually contains some moisture, and the air takes up space in the system because it does not condense at the conditions present in a refrigerant circuit. Air generally accumulates in the condenser reducing the condenser volume available for refrigerant condensing and increasing the condensing temperature.

To ensure that air is removed from a system, evacuate the refrigerant circuit using the proper procedures after installation and any time the refrigerant circuit is opened for service.

On Site

Installing Split Systems

Split system refrigerant circuits are easily contaminated during the installation process because the two sections of the system must be connected by field-installed refrigerant piping. Here are some tips to help minimize the possibility of system contamination:

- Use only ACR or refrigeration grade tubing. Tubing kits for residential systems are constructed from the proper tubing and protected with caps or plugs on the ends.
- Keep the ends of the tubing tightly capped or plugged until you are ready to make the final connection to the equipment. If you are using hard-drawn copper tubing and fitting a new refrigerant circuit, always maintain protection on the open ends and use a nitrogen purge when brazing.
- Ensure that the equipment is furnished with a liquid-line filter drier in the outdoor unit. If not, install a properly sized filter-drier in the liquid line.

Moisture

Moisture is a common contaminant. Under the heat of compression, it may react with the refrigerant to form acids. These acids cause corrosion and break down the insulating coating that protects the motor windings. The compressor motor will eventually be shorted or grounded.

Moisture in the refrigerant can also cause *oil sludge*, which reduces the lubricating properties of the oil and plugs oil passages and screens. Moisture can freeze at the metering device, especially in low-temperature refrigeration applications and in heat pumps where the outdoor coil metering device is exposed to low winter temperatures. Ice blocks the orifice and creates intermittent performance problems.

The presence of moisture in a system can be determined by testing the refrigerant and/or with a moisture test kit. The polyolester (POE) and polyalkylene glycol (PAG) oils used with today's refrigerants are very **hygroscopic**, meaning they readily and rapidly absorb moisture from the air. For this reason, great care must be taken to prevent moisture from entering the refrigeration circuit during installation or repairs. Minimize the time that the system is open, and cap open lines unless you are physically working on the connection.

Hygroscopic: A tendency to absorb water and water vapor readily and consistently, e.g., desiccants.

NOTE

Oils and their properties are presented in NCCER Module 03301, *Refrigerants and Oils*.

Small amounts of water can be removed through proper evacuation procedures. However, if the contamination is significant, the evacuation process requires additional time. In these cases, the system may be evacuated several times. Between evacuation cycles, the vacuum in the system can be broken with dry nitrogen that will absorb some of the water. The nitrogen is then released, carrying the moisture with it, and the evacuation process begins again.

The installation of a filter-drier is also important. When moisture is known to be a problem, you may need to use a larger size to increase the volume of the desiccant.

Acid

Acid is formed inside an operating system by the reaction of air or moisture with the refrigerant and/or oil. Acid is produced in great quantities by a compressor when it experiences a severe burnout, due to the electrical surge and the breakdown of the winding insulation. If acid remains in the system after the failed compressor is replaced, it will break down the winding insulation on the new compressor, causing another motor burnout.

The presence of acid in a system can be determined by testing the refrigerant and/or oil with an acid test kit. Simple kits are readily available. The tester shown in *Figure 26* is designed to take a tiny sample of refrigerant vapor from

Before

After (Positive Result)

Figure 26 Refrigerant acid test kit for vapor in use.
Source: QuikProducts, Mainstream Engineering Corporation

the unit service valve by depressing the Schrader core. If acid is present, the strip inside the tube will turn red. Other acid test kits are made specifically for testing refrigerant oil for the presence of acid (*Figure 27*).

Figure 27 Acid test kit for refrigerant oil.
Source: Nu-Calgon

On Site

Compressor Oil Analysis

It is a common practice to send oil samples from centrifugal and other large compressors to an independent laboratory for detailed analysis. An analysis is usually done on a regular basis, such as annually or after a certain number of operating hours. An analysis can help identify problems in the system before they cause permanent damage or an unexpected failure and is referred to as *predictive maintenance*.

Foreign Materials

Foreign material such as copper filings, copper oxides, and dirt get into the system during installation and service tasks. The particles can clog small oil passages in the compressor as well as the metering device.

Copper oxide is a black, flaky contaminant that forms rapidly when brazing joints. These same oxides form whenever copper is exposed to oxygen and are responsible for the darkening of copper surfaces. However, the pace of natural formation is rather slow. Due to the heat applied when brazing a copper joint, copper oxides form very quickly, both inside and outside the tubing and fittings, and in high volume.

Loose copper oxide particles are easily washed from the inner tubing surfaces when the system operates, allowing it to circulate with the refrigerant and oil. For that reason, it is critical to prevent the formation of copper oxides within the tubing while brazing, and to remove oxides that do form. Proper brazing techniques include purging the refrigerant lines with nitrogen while brazing to displace the oxygen.

NOTE

For detailed coverage of brazing copper and establishing a nitrogen purge, refer to NCCER Module 03104, *Soldering and Brazing*.

3.1.5 Compressor Overheating

Compressors are expected to generate heat and are designed to handle it. One cause of compressor overheating is excessive compression ratio. The compression ratio is usually highest in low-temperature applications, where a greater difference between the suction and discharge pressures exists. In this case, a high compression ratio is normal and expected. For some low-temperature compressors, an additional form of cooling, such as a cooling fan mounted on the compressor head, is needed.

As discussed previously, excessive superheat, lubrication problems, condenser fouling, and electrical problems can all cause overheating. Temperatures between 275°F and 300°F (135–150°C) in the discharge line cause oil and refrigerant to break down. Lubricating oils at these temperatures begin to burn, fouling the rest of the oil. Laboratory tests have shown that for each 18°F (10°C) rise above normal in the discharge temperature, the rate of chemical reactions in the system doubles.

WARNING!

Certain parts of operating compressors can be very hot under normal operating conditions. Wear gloves and other personal protective equipment to avoid being burned when working with and around operating compressors. Do not touch an operating compressor without first measuring the temperature.

Remember that the maximum discharge temperature for most HVACR compressors is 225°F (107°C). A discharge temperature of 180°F (82°C) is a reasonable target.

3.2.0 Compressor Electrical Failures

Electric motors driving compressors operate on either single-phase or three-phase AC power. Power wiring to HVACR equipment must be done in accordance with the latest edition of the *National Electrical Code®* (*NEC®*) as well as local codes. The wire size and overcurrent protection device are based on the equipment nameplate full-load amperage (FLA) or rated-load amperage (RLA), any listed *NEC®* derating factors, and any increase in wire size needed to prevent excessive voltage drop on long runs. These calculations are typically the responsibility of the electrical craft. However, it is important for HVACR technicians to ensure the power source is compatible with the electrical characteristics found on the equipment nameplate.

Know the Code

NEC Article 430, *Motors, Motor Circuits, and Controllers* applies to HVACR equipment. ***NEC Article 440***, *Air Conditioning and Refrigerating Equipment*, provides additional details and some exceptions to the rules shown in ***NEC Article 430***. However, open drive compressors, with motors that are external, are not affected by ***NEC Article 440***.

NEC Article 440 contains the following parts:

- Part I — *General*
- Part II — *Disconnecting Means*
- Part III — *Branch-Circuit Short-Circuit and Ground-Fault Protection*
- Part IV — *Branch-Circuit Conductors*
- Part V — *Controllers for Motor-Compressors*
- Part VI — *Motor-Compressor and Branch-Circuit Overload Protection*
- Part VII — *Provisions for Room Air Conditioners*

Open-drive compressors and their motors must be in a well-ventilated area to aid motor cooling. Unlike hermetic and semi-hermetic compressors, motors for open-drive compressors are cooled only by the surrounding air. Some motors have ventilation openings on their shells to allow air to pass through for cooling. This type of motor, referred to as an *air-over (AO) motor*, must be used in a relatively clean environment to prevent fouling.

Other motors used with open-drive compressors are totally enclosed to prevent air from freely entering the motor housing. This motor depends on heat radiating from the motor's shell to cool the windings and is known as a *totally enclosed, non-ventilated (TENV) motor*. A motor may also have a small fan mounted to the motor's shaft to aid in cooling where the environment allows it. This type of motor is referred to as a *totally enclosed, fan-cooled (TEFC) motor*.

The motor for a hermetic or semi-hermetic compressor is enclosed in the same housing. Thus, cool refrigerant vapor from the evaporator passes over and cools the motor windings before entering the compressor. If the refrigerant charge is low or other issues cause abnormally warm vapor to return to the compressor, hermetic and semi-hermetic compressors can overheat.

On Site

Compressor Performance Charts

The nameplate on a unit usually contains very little information about the compressor beyond its RLA and LRA. For detailed information regarding compressor performance, record the compressor model number and visit the manufacturer's website. Manufacturers provide extremely detailed performance information across a range of conditions.

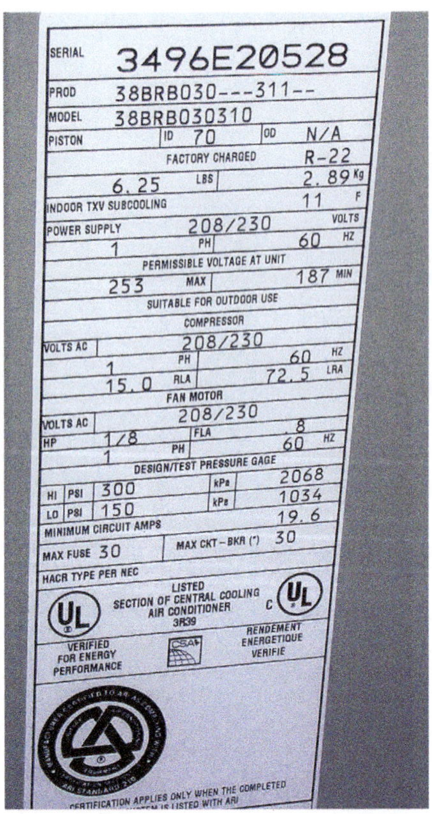

Figure 28 Typical unit nameplate.

3.2.1 Nameplate Information

Valuable information about a unit's operating voltage and current can be obtained from the unit information plate, referred to as a *nameplate* or *dataplate* (*Figure 28*). The nameplate is attached to the unit in a clearly visible location.

The nameplate contains a lot of helpful information. Note, however, that all manufacturers have their own unique patterns and codes for model and serial numbers:

- *Unit model number* — The unit model number typically identifies the family or series of equipment, the unit nominal capacity, and the voltage. For example, in *Figure 28*, the numbers 030 indicate that the unit has a nominal capacity of 2½ tons. The number 3 following the capacity indicates the unit requires a 208/230VAC, single-phase power supply.

- *Unit serial number* — The serial number is generally coded to show when a unit was manufactured, and it often shows where it was made. In *Figure 28*, the first four digits, 3496, indicate the unit was manufactured in the 34th week of 1996. The letter E refers to a specific manufacturing plant. The rest of the numbers are assigned sequentially to provide a unique number for each piece of equipment.
- *Refrigerant* — The type of refrigerant the unit requires is very important. Many refrigerants are currently in use, and more possibilities are in development. Always check this information before connecting your gauges to a system and be certain about which refrigerant you are working with.
- *Power supply and minimum/maximum voltage* — Most units today have a dual voltage rating of 208/230VAC. This provides flexibility, given that different voltages are used in different areas. The minimum and maximum voltage represent a range of ±10%. On a dual voltage unit, the 10% is added to the higher voltage and subtracted from the lower voltage.
- *Compressor electrical data* — The rated load amperage (RLA) or full load amperage (FLA) is used to identify the current consumption of an electric motor when fully loaded. Nameplates for hermetic compressors built after 1972 are marked with the RLA. However, you might see the FLA used for other motors. The locked rotor amperage (LRA) represents the current when the compressor is energized but the rotor fails to move.

The manufacturer determines the RLA for a hermetic compressor by testing the compressor in actual operating conditions. The manufacturer first determines the compressor's maximum continuous current (MCC). The MCC is the maximum current value that the compressor's motor protection device will endure without opening, which de-energizes the compressor. Any additional current consumption above the MCC will cause the protection device to open. Sometimes the MCC value is also marked on the unit nameplate. The RLA is important because it is used to size conductors and contactors.

NOTE

The compressor RLA is not used to size the unit power supply. The power supply must be based on the electrical characteristics of the entire unit. Many units, including the one shown in *Figure 28*, provide the minimum circuit ampacity (MCA) for the unit. This value is used to size the conductors. The maximum fuse and/or circuit breaker rating is also provided.

On Site

Crankcase Pressure Regulators

Compressors in low-temperature applications may be equipped with a crankcase pressure regulator (CPR) to prevent the pressure in the compressor crankcase from rising to a level that will overload the compressor. This can happen following a defrost cycle, for example, when the evaporator coil is well above freezing when the system restarts. For a short time, the load may be outside of the compressor's normal operating range.

The RLA value is used when adjusting the CPR setting. The valve setting must be adjusted while the compressor is under an excessive load following a defrost cycle, and while watching an ammeter. As the valve is adjusted towards the closed position, the current falls as the load on the compressor is reduced. The proper setpoint is just below the compressor's RLA. The valve allows the compressor to regain control of the load without overloading.

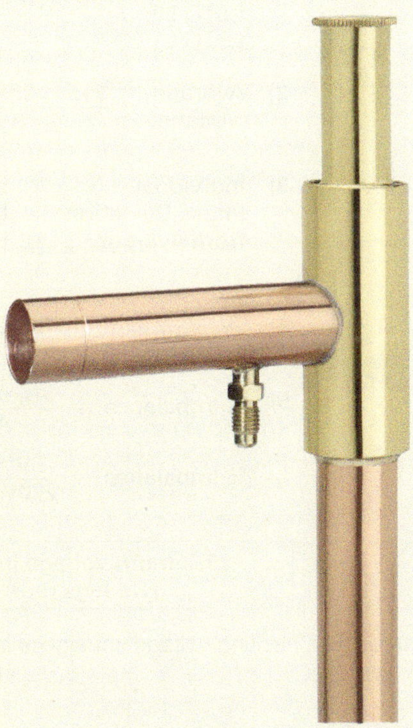

Source: Courtesy of Sporlan Division-Parker Hannifin Corporation

3.2.2 Motor Operating Voltage Ranges

A high or low operating voltage can cause a compressor motor to overheat. Operating voltages must be maintained within the limits posted on the motor's nameplate. If the voltage falls outside the limits, the problem should be corrected before starting the system. The problem may be with the building distribution system or with the power supply to the building. The AC voltage tolerances typically applied to motors are as follows:

- *Motors rated at a single voltage* — The input supply voltage should be within ±10% of the nameplate voltage. For example, a motor with a nameplate voltage rating of 230V should have an input voltage that ranges between 207V (–10% of 230V) and 253V (+10% of 230V).
- *Dual-voltage rated motors* — A motor with a nameplate dual-voltage rating of 208/230V should have an input voltage that ranges between 187V (–10% of 208V) and 253V (+10% of 230V).

3.2.3 Voltage Imbalance on Three-Phase Motors

The voltage imbalance between any two legs of the supply voltage applied to a three-phase motor should not exceed 2%. A small imbalance in the input voltage results in a considerable amount of heat being generated by the motor windings. A 2% imbalance typically causes a temperature rise in the motor windings 8% above normal. At a 5% imbalance, the winding temperature can increase to 50% above normal.

Figure 29 shows an example of how to calculate the voltage imbalance in a three-phase system. The process begins by taking a reading of the voltage between L1 and L2, L2 and L3, and L3 and L1. Note the three readings shown in the example are all well within the operating range individually. However, sufficient deviation exists in the voltage between phases to create an unacceptable voltage imbalance.

1)

Phase	Measured Reading
L1 TO L2	215V
L2 TO L3	221V
L3 TO L1	224V

2) $$\text{Average Voltage} = \frac{215 + 221 + 224}{3} = 220\text{V}$$

3) Individual Phase Deviation from Average:
L1 TO L2 = 220 – 215 = 5V
L2 TO L3 = 221 – 220 = 1V
L3 TO L1 = 224 – 220 = 4V

4) 5V = Maximum Imbalance

5) $$\%\ \text{Imbalance} = \frac{\text{Maximum Imbalance}}{\text{Average Voltage}} \times 100$$

$$\%\ \text{Imbalance} = \frac{5\text{V}}{220\text{V}} \times 100 = 2.27\%$$ (Out of Balance)

Maximum voltage imbalance between any two legs must not exceed 2%.

Figure 29 Checking voltage imbalance in a three-phase system.

3.2.4 Current Imbalance on Three-Phase Motors

Current imbalance between any two legs of a three-phase motor should not exceed 10%. While a voltage imbalance will always produce some current imbalance, a current imbalance can easily occur without a voltage imbalance present. This can happen when an electrical terminal or contact set becomes loose or corroded, causing a higher resistance in one leg of power. Because current follows the path of least resistance, the current in the other two legs will increase. Calculate the current imbalance in a motor using the following steps:

Step 1 Measure the current to the motor on all three phases.

Step 2 Add the three values together, then divide by three to determine an average current. For example, if the three current readings are 54A, 48A, and 50A, the average would be 50.7A, rounded to one decimal place.

Step 3 Select the one current value of the three phases that deviates the most from the average. The 54A value is the farthest of the three from the average, and therefore is selected for the rest of the calculation.

Step 4 Determine the difference between the selected value and the average. In this example, the difference is 3.3A (54A – 50.7A = 3.3A). Then divide this value by the average current:

$$33\text{A} \div 50.7\text{A average} = 0.065\text{A}$$

Step 5 Multiply the result by 100 to determine the percentage of imbalance. This result is within the 10% tolerance for current imbalance:

$$0.065\text{A} \times 100 = 6.5\%\ \text{current imbalance}$$

Single-phasing of a three-phase motor is an example of a severe case of current imbalance. Single-phasing occurs when voltage from one of the three phases is missing. If the compressor is operating when one phase is lost, the motor may continue to run. The other two phases will attempt to carry the load, resulting in a current increase to about 1½ times the normal running current.

Single-phasing: A condition in which a three-phase load receives power from only two of the three phases (phase loss).

Single-phasing will probably cause an overload device to trip if it is equipped with a current overload protection device. Otherwise, the motor will continue running until it overheats. Eventually, temperature limits or other safety controls step in to shut the compressor down, or worse, the compressor fails.

Once the motor stops, it cannot be restarted again until all three phases of power have returned. Without all three phases, there is not enough torque to initiate rotation. If repeated attempts to start are not prevented, the motor will likely sustain permanent damage.

3.0.0 Section Review

1. Which of the following conditions is *most likely* to cause slugging?
   a. Low refrigerant charge
   b. Buried refrigerant lines
   c. Compressor running loaded
   d. High refrigerant velocity

2. The motor in a hermetic compressor is primarily cooled by _____.
   a. liquid coolant
   b. air
   c. a cooling fan
   d. returning refrigerant vapor

4.0.0 Compressor Protection Devices

Performance Tasks

There are no Performance Tasks in this section.

Objective

Identify and describe the operation of various compressor protection devices.

a. Identify and describe the operation of various motor overload protection devices.

b. Identify and describe the operation of other compressor protection devices.

A number of devices are used to protect compressors. The size and type of compressor often determines which safety devices are needed and installed.

4.1.0 Motor Protection Devices

The most common causes of motor failure are current overloads and overheating. A current overload condition occurs when the current exceeds the motor's RLA. An overload condition can also result in external problems, such as melted conductor insulation.

Overheating can occur regardless of the operating current. One problem unique to three-phase motors is single-phasing. The resulting current imbalance from single-phasing rapidly increases the temperature in the remaining two windings, causing the motor to overheat. Devices designed to prevent single-phasing of a three-phase motor consist of an electronic module with current sensors that monitor current in each phase. If current flow ceases in one of the three windings, the module opens the control circuit to the compressor contactor coil.

It is important to protect all motors from both current overloads and overheating. To accomplish this with compressors, current- and/or heat-sensing devices are used to open the circuit before damage occurs. A compressor protection device can be classified as either a **line-duty device** or a **pilot-duty device** (*Figure 30*).

Line-duty device: An electrical component designed to handle the voltage and current of load devices such as motors.

Pilot-duty device: An electrical component designed only to handle small current values such as those found in a control circuit.

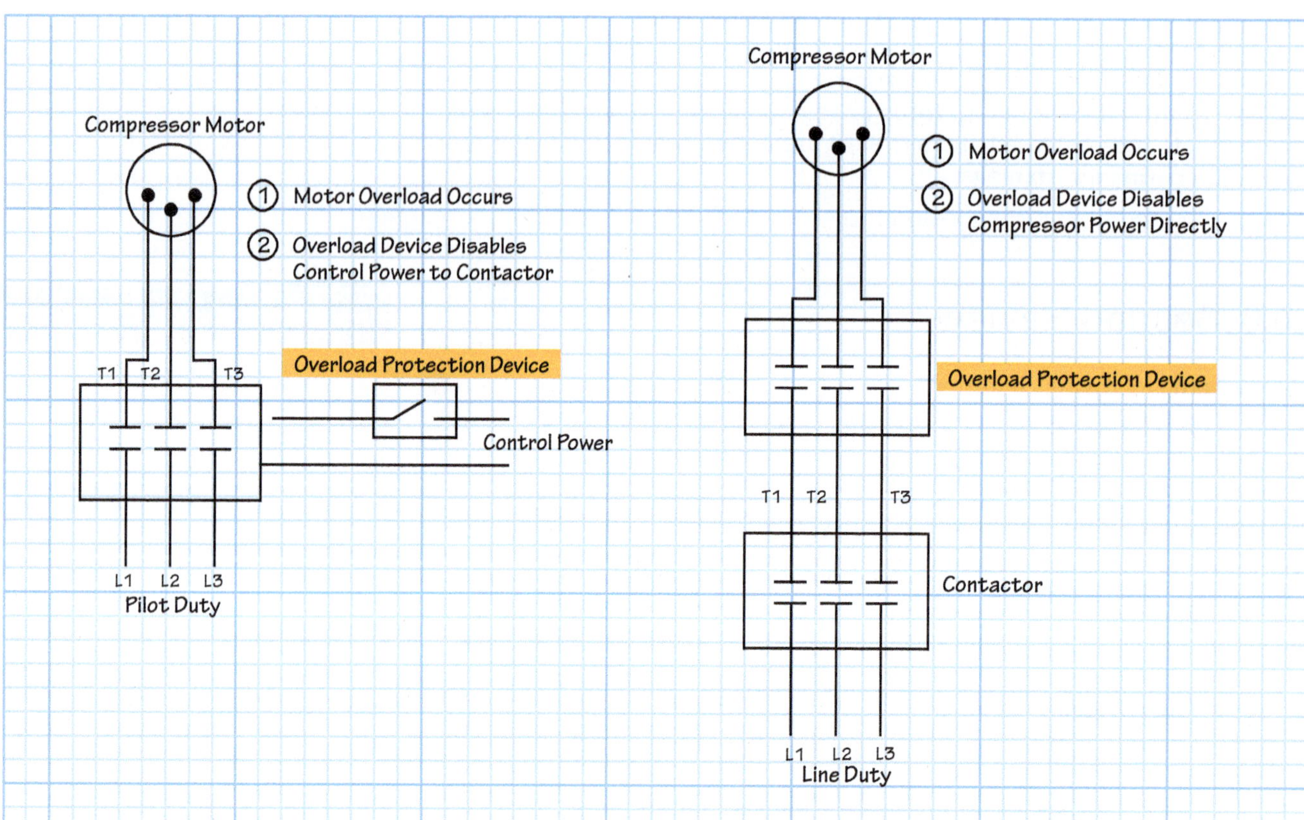

Figure 30 Pilot-duty and line-duty devices.

Pilot-duty devices sense a current overload or high temperature and open the contactor or motor-starter control circuit to shut down a compressor. When line-duty devices sense a current overload or high temperature, they either open the motor winding connections internally or open the line-voltage power between the contactor and the compressor.

On Site

Pilot-Duty and Line-Duty Devices

Pilot-duty overload devices are rarely needed for small hermetic compressors. Instead, they typically use an internal line-duty device for motor protection. Pilot-duty devices are more often used with semi-hermetic compressors and the motors of open-drive units.

Once an overload device has opened a circuit, it must be reset before the motor can operate again. A manual-reset overload device must be physically reset once it has tripped. An automatic-reset overload device will automatically close after the overcurrent or over-temperature condition has passed. The reset time of automatic devices is based on recovery from the condition, and can vary from seconds to minutes. This is especially true of devices that protect against overheating, as they must cool down before they can be reset.

Many types of motor protection devices are available, including the following:

- External line-break overloads
- Internal line-break overloads
- Motor thermostat overloads
- Electronic overloads
- Three-phase overloads
- Current-monitoring devices
- Voltage-monitoring devices

4.1.1 External Line-Break Overloads

The external line-break overload, often called a Klixon®, is a single-phase motor overload protector. Some types provide current protection, others are sensitive to temperature, and still others can monitor both values.

These devices depend on a bimetal warped-disc switch. Generally, the external line-break device is inside a compressor terminal box and connected for line duty in series with the motor's Common terminal.

Figure 31 shows an example of an automatic external line-break overload. External line-break overloads are commonly found on compressors in room air conditioners and small refrigeration units.

Some versions are used in pilot-duty applications. In this case, the contacts are connected in series with the contactor control circuit. Both devices reset automatically after recovery from the overload condition.

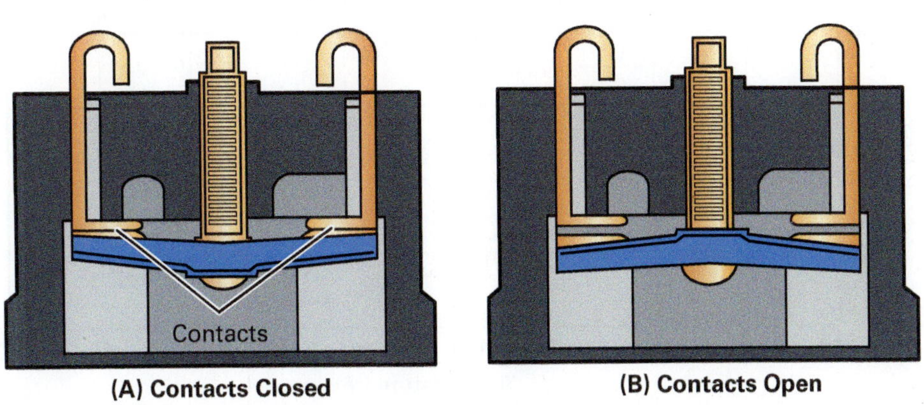

Figure 31 Example of external line-break overload.

4.1.2 Internal Line-Break Overloads

The internal line-break overload is often used in hermetic compressor motors. This protector is also a line-duty device, but it is located inside the motor (*Figure 32*). The protector is placed in series with the Common winding and opens when high temperature, excessive current, or a combination of both is sensed. It resets automatically after recovery from the overload condition and a cooling-off period.

This type of protector is located inside the compressor motor, so it is impossible to remove and inspect. If the overload itself fails to close after a fault condition, the compressor must be replaced. However, this is not a common occurrence. When the compressor motor windings are tested with a meter, a failed internal overload shows no continuity through the Common winding.

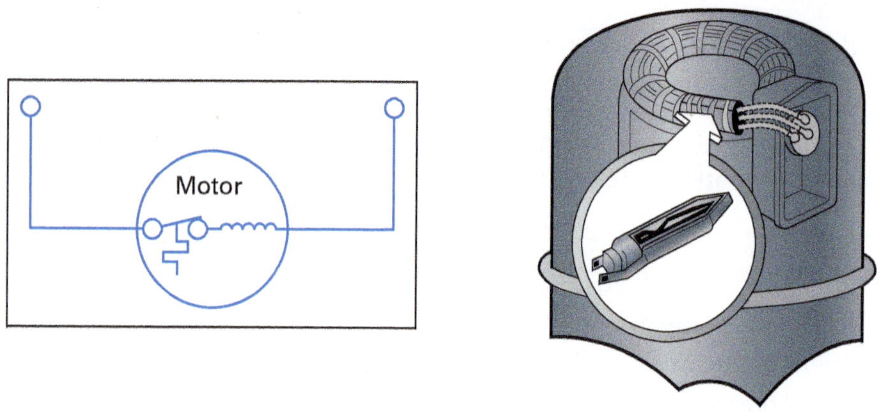

Figure 32 Example of an internal line-break overload.

4.1.3 Electronic Protection Modules

Thermistors are resistors that vary their resistance in response to temperature changes. Thermistors can decrease resistance as temperature increases (negative temperature coefficient thermistor, or NTC), or increase resistance as temperature increases (positive temperature coefficient thermistor, or PTC). Thermistors can be embedded in a motor winding to sense temperature.

An electronic control connected to internal thermistors can react to excessive temperatures by opening the control circuit. Manufacturers typically supply resistance versus temperature charts (*Figure 33*) that are useful in determining if the thermistors are accurate. Note that the thermistors are not field accessible or replaceable.

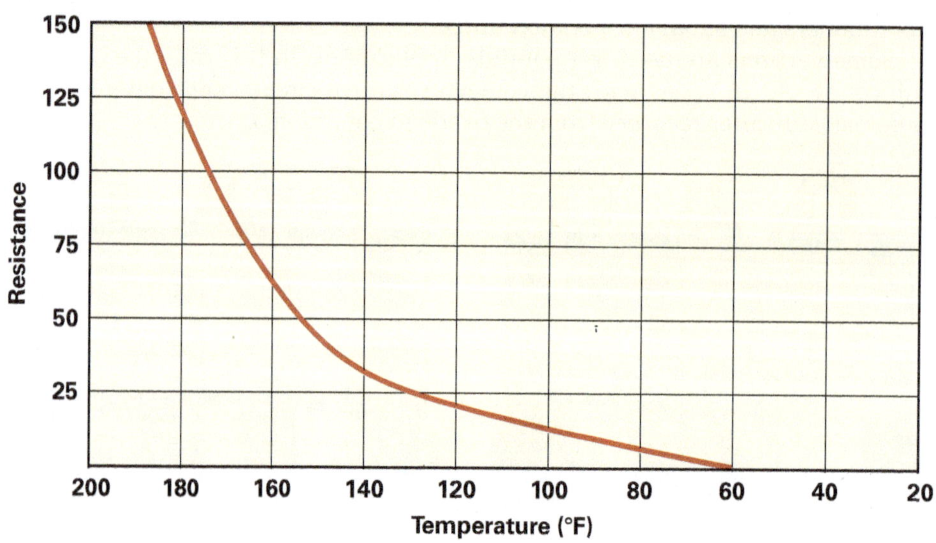

Figure 33 Example thermistor resistance versus temperature chart.

One common protection device for large compressors is a single-module, three-sensor system. The thermistors are inserted in the hermetic motor winding to provide maximum protection (*Figure 34*).

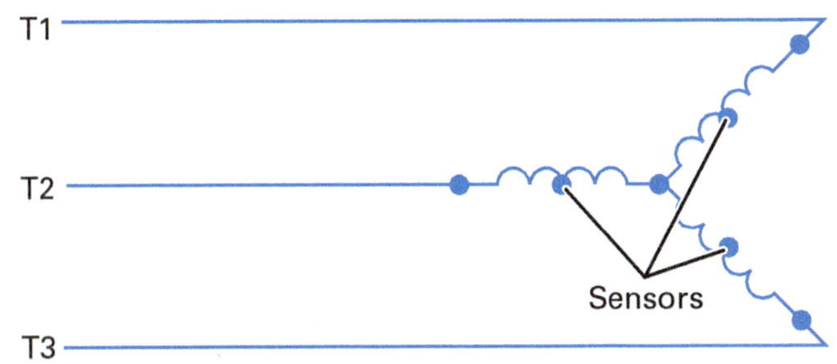

Figure 34 Sensor locations in a three-phase motor.

The module requires a power supply, sensor connections, a control circuit interlock, and a manual reset mechanism and/or terminals to connect a remote reset device. *Figure 35* shows an example of the connections required.

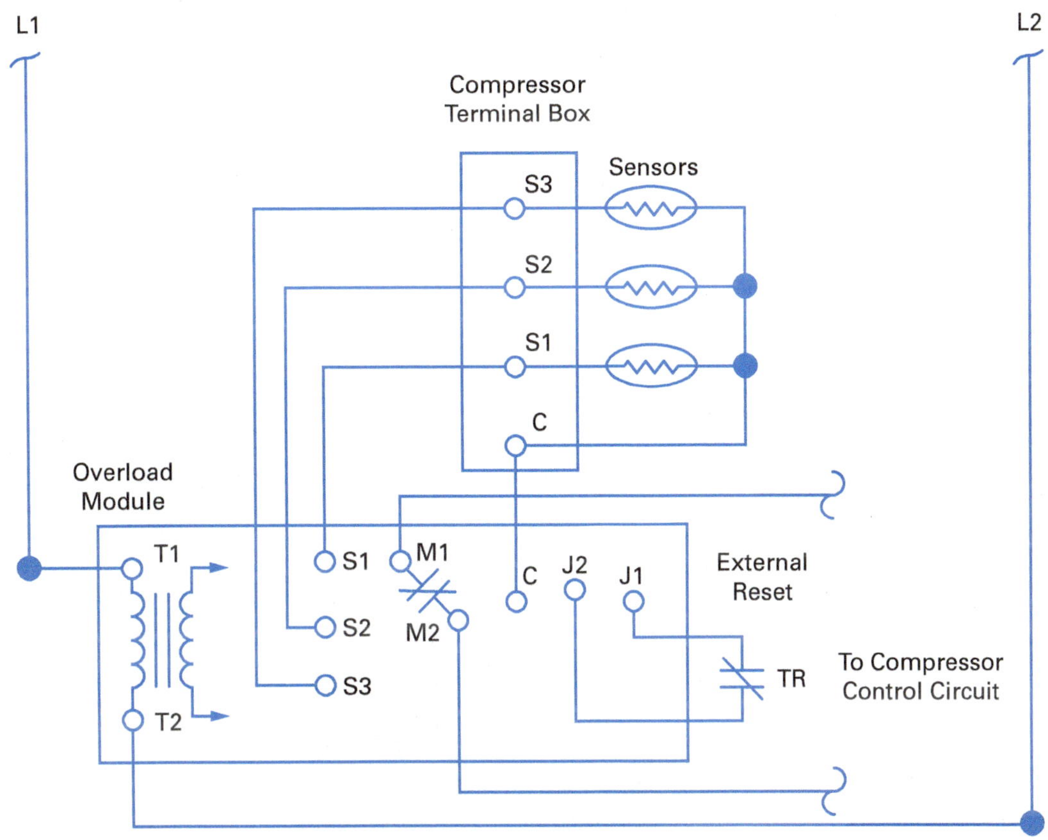

Figure 35 Example of wiring connections for an overload module.

4.1.4 Current Monitoring Devices

Current transformers (CTs) are often used to accurately monitor current in compressor power leads (*Figure 36*). This type of CT is basically a current-sensing wire loop with a predictable response to current flow. CTs may also be referred to as *current traps*.

CTs operate in the same way as a clamp-on ammeter to sense current. If the current setpoint of the relay is exceeded, the control opens the control circuit.

One or more compressor power leads are routed through CTs. Only one lead can be routed through an individual CT, so a separate CT must be provided for each lead to be monitored. Controls for single-phase compressors monitor current only on the Common lead while three-phase controls monitor current on all three legs.

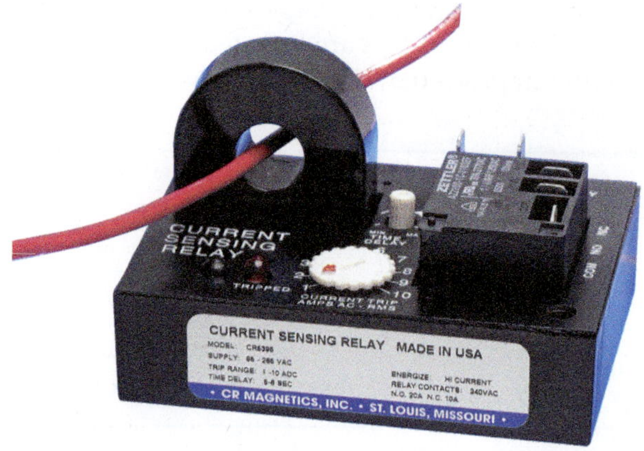

Figure 36 Current transformer (CT) relay.
Source: CR Magnetics

CTs can also be used to ensure a compressor starts as expected. If the control circuit for a compressor is energized, but the CT relay does not sense any current in the compressor lead(s), the relay opens and typically locks out the compressor control circuit until action is taken and the control power is cycled. Its purpose is based on the premise that, if the compressor is not running as expected, there is a problem somewhere. Note that the CT is not monitoring the precise current in this case; it is only monitoring for the presence of current. These devices are a bit simpler than CT relays that monitor with greater precision.

4.1.5 Voltage Monitoring Devices

Voltage monitoring devices help protect against flaws in the power source. In addition, many also protect against power-related flaws within the unit itself.

For example, the compressor contactor may experience damage to one or more of its contacts. The damage can cause a significant voltage drop on that leg of power, or even the complete loss of the affected phase. Because the incoming power to the contactor is unaffected, this condition can only be sensed by also monitoring the voltage on the load side of the contactor. Many voltage monitors, such as the one shown in *Figure 37*, monitor the voltage on both the line side and load side of the contactor.

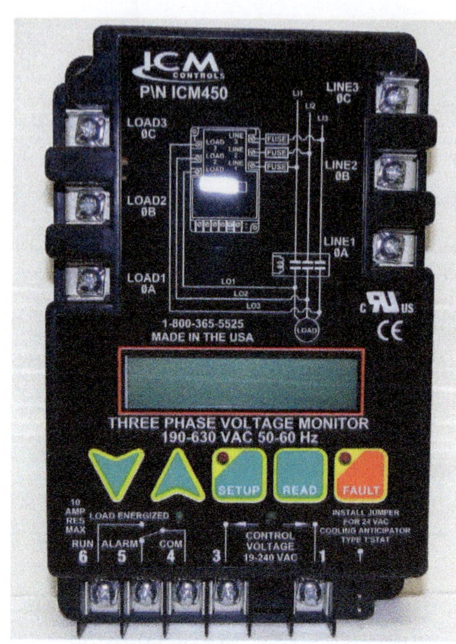

Figure 37 Three-phase voltage monitor.

This type of protection is generally not provided with a new unit. It is occasionally offered by an equipment manufacturer as an option. Most voltage monitors are aftermarket additions to protect critical systems. Voltage monitors are far more valuable in the protection of three-phase systems, but single-phase systems that regularly suffer from brownout conditions can also be protected.

Voltage monitors can react to a problem very quickly. The reaction time can be less than 0.1 second. Many are also capable of storing faults so that technicians can see what happened after they arrive on the scene.

The monitor shown in *Figure 37* monitors for under- or over-voltage conditions on both the line side and load side of the compressor contactor. It can be programmed to monitor any voltage from 80V to 300V and operate on any input voltage from 18V or 240V. It can also monitor and react to voltage imbalance, phase loss, and phase reversal. The monitor also acts as an anti-short-cycle protector and stores the most recent 25 faults in memory.

In addition to the normally closed (NC) contacts that break the circuit when a problem is identified, a second set of contacts can be used to set off a remote visual or audible alarm.

On Site

The Reality of Compressor Protection

Many of the compressor protection devices discussed herein are not found on common equipment. Residential comfort-cooling equipment, for example, usually has few protective devices. A high-pressure switch is common, and a discharge-temperature safety switch is common for scroll compressors. Beyond that, compressor protection consists of internal thermal protection, a pressure relief device, and the equipment fuses and/or circuit breaker. Most heat pumps are also equipped with a loss-of-charge switch that opens when a substantial amount of refrigerant has been lost.

More protective devices are added as compressors get larger and more expensive. Critical applications also get added protection, often through the installation of aftermarket devices. For example, a cooling unit that protects life-saving blood or antivenom supplies may be small in capacity, but maximum protection, as well as a means for instantaneous notification of a problem, is needed.

4.2.0 Other Protection Devices

Compressors also require protection from system related issues, such as excessive pressures and rapid cycling. These additional protective devices include the following:

- High- and low-pressure switches
- Oil pressure safety switch
- Short-cycling time-delay relay

Some safety controls, once activated, must be reset manually before the compressor attempts to start again. Some high-pressure safety switches, for example, have a reset button that must be pressed to close the open contacts. Other safety controls may reclose automatically once the condition has passed. Some systems may be equipped with a lockout circuit that opens when any safety device is activated.

Compressor lockout protection is used to prevent the compressor from repeatedly cycling on and off due to the opening and closing of an automatic-reset safety control. In a low-pressure condition, for example, the compressor control circuit is de-energized when a low-pressure safety switch opens. Moments later, after the compressor shuts down, the pressure increases, the switch automatically recloses, and the compressor is again commanded to start. This sequence of events can repeat rapidly, over and over, in the absence of a lockout circuit.

Figure 38 shows the wiring of a lockout relay. It may be referred to on diagrams as a compressor lockout (CLO). Note that the relay coil has a uniquely high resistance and is referred to as an *impedance relay* for this reason. Due to its

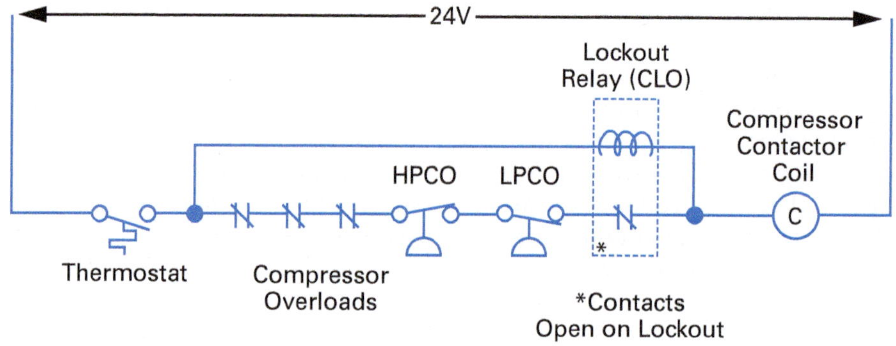

Figure 38 Compressor lockout relay wiring.

high resistance, it is not energized during normal operation; the easier electrical path is through the pressure switches and overloads. However, when any one of the safety controls opens the circuit to the contactor coil, current takes the available remaining path through the lockout relay coil. The CLO is then energized, and the contacts open.

The relay remains energized, keeping the contactor control circuit open until control power is interrupted. Cycling the line-voltage power to the equipment containing the control transformer, or simply turning the thermostat to Off and back to On, resets the circuit.

Proper operation of this circuit depends on the resistance of the lockout relay coil being much greater than the resistance of the compressor contactor coil. If the lockout relay must be replaced, replace it with an exact duplicate to ensure the proper balance in circuit resistance remains.

4.2.1 Pressure Protection

Compressors require protection from unusual and potentially damaging pressure conditions. Some require protection from a loss of oil pressure. Several different approaches to pressure protection are presented here. In commercial and industrial systems, all approaches may be applied to a single system.

Internal Pressure-Relief Valves

Many compressors are equipped with internal pressure-relief valves. Internal relief valves are designed to be the last means of defense against excessive pressure. The setting for internal relief valves is higher than the appropriate setting for a high-pressure safety switch. When an internal relief valve opens, it bypasses discharge vapor directly back to the suction side of the compressor. As a result, the refrigerant pressures suddenly come much closer together—the suction pressure rises while the discharge pressure falls. The compressor does not shut down unless another safety control, such as an internal thermal-protection device, is activated. That activation is likely, as the compressor will overheat quickly when it is bypassing vapor internally.

The relief valve will close after the pressure falls substantially. However, if the source of the problem remains, the valve will soon reopen. Eventually, the compressor will overheat, and the thermal safety devices should open the motor windings.

Most manufacturers state that the internal relief valve is not designed for repetitive use, and that it may eventually fail to reclose. If the valve does not close, the compressor becomes useless because the valve is inaccessible. Note that the compressor makes a lot of noise when the relief valve opens.

High-Pressure and Low-Pressure Safety Switches

High-pressure and low-pressure safety switches are a very common form of protection. While some pressure switches are used to control other components and actions, such as the operation of condenser fans, safety switches are in the circuit strictly for protection.

A high discharge pressure is a particular concern, as it increases the compression ratio and the compressor's load. Of course, extremely high pressure can also cause tubing or refrigerant circuit components to rupture unexpectedly. Although a low-pressure condition is less stressful on a compressor in some ways, the compression ratio is still negatively affected.

As shown in *Figure 39*, the high- and low-pressure switches are wired in series with the compressor contactor coil and other protective devices. High- and low-pressure switches for safety purposes are usually small, compact models with fixed settings. Note that there is little difference between the two switches visually. You must pay strict attention to the numbers and ranges provided on the switch label. In some cases, high-pressure safety switches have red leads, and also a red button to provide a manual reset function.

Figure 39 Pressure safety switches.

This type of safety switch mounts directly to the refrigerant line on a ¼" male flare fitting. Adjustable pressure switches are usually surface mounted and are connected to the line by a copper capillary line or a refrigerant hose.

The NC contacts inside are kept closed by spring pressure. If the pressure in the system rises or falls beyond the range of the spring, the contacts open and break the control circuit. The high-pressure switch opens on a rise in pressure, while the low-pressure switch opens on a reduction in pressure. In *Figure 39*, the appearance of the pressure switch symbols indicates the action of each switch.

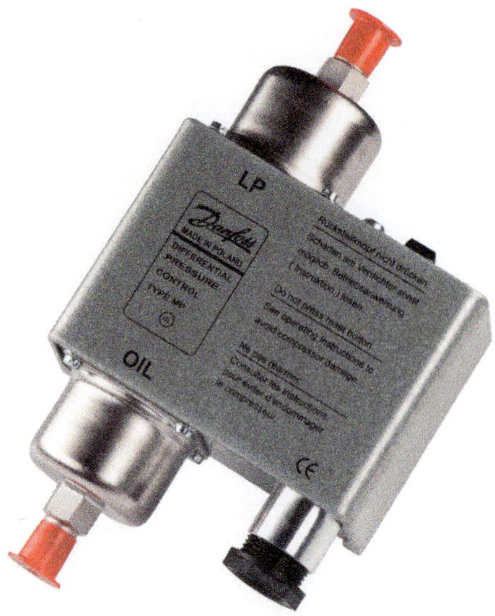

Figure 40 Oil-pressure safety switch.
Source: Danfoss Turbocor

Oil-Pressure Safety Switches

Semi-hermetic and open-drive compressors with oil pumps often incorporate oil pressure protection. Oil-pressure safety switches, like the one shown in *Figure 40*, are *differential-pressure switches*. This means that the switch accepts two pressure inputs and monitors the difference.

One capillary tube or hose is connected to a point on the compressor at suction pressure, which is the same pressure available at the inlet of the oil pump. The other is connected to the discharge side of the oil pump. Actual, or *net*, oil pressure is determined by subtracting the suction pressure from the discharge pressure.

On Site

Low-Pressure Switches

The loss-of-charge (LOC) switch used in a heat pump has a much lower setpoint than a normal low-pressure safety switch. The reason: The outdoor coil on a heat pump serves as the evaporator in the heating mode, and it does so at very low outdoor temperatures. This results in low suction pressures that are normal for the heating mode, but not for the cooling mode. The opening pressure of the switch must be low enough to allow normal heating operation at low outdoor temperatures. This is why they are referred to as loss-of-charge switches rather than low-pressure switches.

The following information is common to many oil-pressure safety switches, although the terminal markings may differ. Refer to *Figure 41*, which represents the wiring for one common safety control. An oil-pressure safety switch incorporates a time delay to allow the compressor time to develop oil pressure at startup:

- The control uses a small heater, labeled TD HTR, to warm a bimetallic switching element. Once the heater warms enough to change the element's shape, it snaps to a new position and cycles the device contacts. By design, the timing is predictable. Common controls allow a compressor 90 seconds to develop enough oil pressure.
- Resistor R, shown in the lower right corner of *Figure 41*, is not a heater, but a *dropping resistor* that allows the control to be powered by 120VAC or 240VAC.
- The control receives operating power on terminal 2 from L1 of the compressor power circuit and on the 240VAC terminal from L2. Energizing the control initiates a timing cycle and the heater begins to warm up. Note that these two terminals are powered from the load side of the contactor and are not energized unless the compressor is energized.
- The control circuit of the compressor contactor flows through the L and M terminals. They are connected by the time-delay relay (TD) through the closed TD1 contacts. This enables the compressor contactor to remain energized during the timing cycle.
- The TD2 contacts are normally open (NO) and enable an alarm output.
- The double-pole relay PA is pressure actuated—a pressure switch. An increase in pressure above the setpoint cycles the relay, and the contacts change position. On startup, the PA1 contacts remain closed, energizing the time-delay heater.

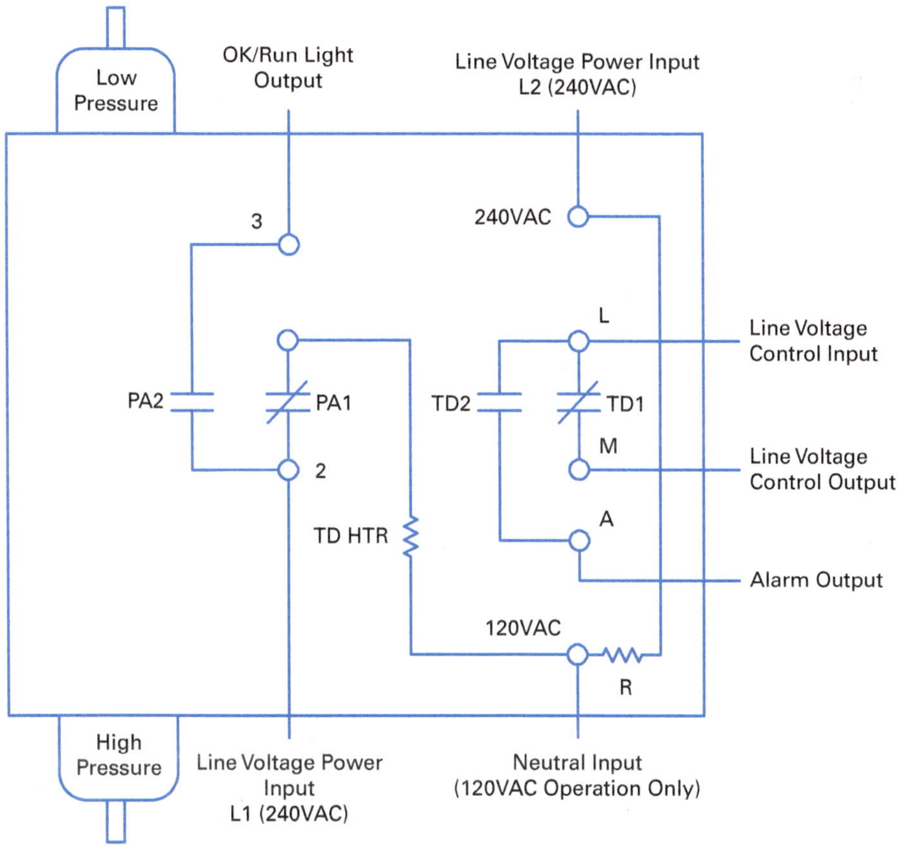

Figure 41 Oil-pressure safety switch wiring example.

- If sufficient oil pressure develops in the allotted time, relay PA changes position, the heater is de-energized, and the timing cycle stops. Power is also enabled through closed contacts PA2 to power an optional indicating light on a panel, confirming that the oil pressure is within range.
- If sufficient oil pressure does not develop, the heater eventually causes relay TD to cycle. Contacts TD1 open, breaking the compressor control circuit. Contacts TD2 close, enabling power to an external alarm circuit, such as a light or horn.
- Most oil pressure safety controls require a manual reset using a button on the enclosure.

Oil-pressure safety switches are commonly available with fixed differentials to satisfy the design requirements of specific compressor manufacturers. The design oil pressure varies by compressor model and manufacturer, but there are similarities among families of compressors. Some switches have an adjustable pressure setting and can be applied to a variety of compressors. When using the adjustable models, the correct oil pressure for the compressor being serviced must be known to properly set the control.

Making the pressure connections using flare tees that provide an adjacent service port is best. This allows a gauge manifold to be connected at the same point for troubleshooting or measuring the oil pressure during inspections and troubleshooting.

For compressors fitted with pressure transducers, electronic oil-pressure safety switches can be used. The copper capillary tubes or hoses are replaced by sensor leads, eliminating potential leaks.

4.2.2 Short-Cycling Protection

Before restarting a reciprocating compressor after an operating cycle, allow the suction and discharge pressures to equalize. Short cycling is a condition in

> **NOTE**
>
> Scroll compressors, due to their design and lack of suction and discharge valves, are not affected by unequal pressures when started.

which the compressor attempts to restart quickly after an operating cycle, and sometimes in rapid succession. Working against a pressure differential places far more strain on the motor. The strain can also cause the circuit breaker or overload protection to open if the starting current is sufficient and sustained long enough.

Short cycling can happen due to a momentary interruption of power to the compressor or in its control circuit. This might occur when the power is out for a few seconds during a storm, or when the thermostat setting is rapidly changed up and down. With older mercury-bulb thermostats, erratic movement of the control lever would make the mercury bounce back and forth, causing the circuit to open and close a few times. A battle over setpoint at the thermostat in a home or business might result in several rapid setpoint changes.

Short-cycling protection can be provided by placing a solid-state electronic timer in the contactor control circuit. The device shown in *Figure 42* is NC. When power is applied, the contacts remain closed. When the circuit is later de-energized, the contacts open and remain open for the set delay. Once they close again, the compressor can be restarted.

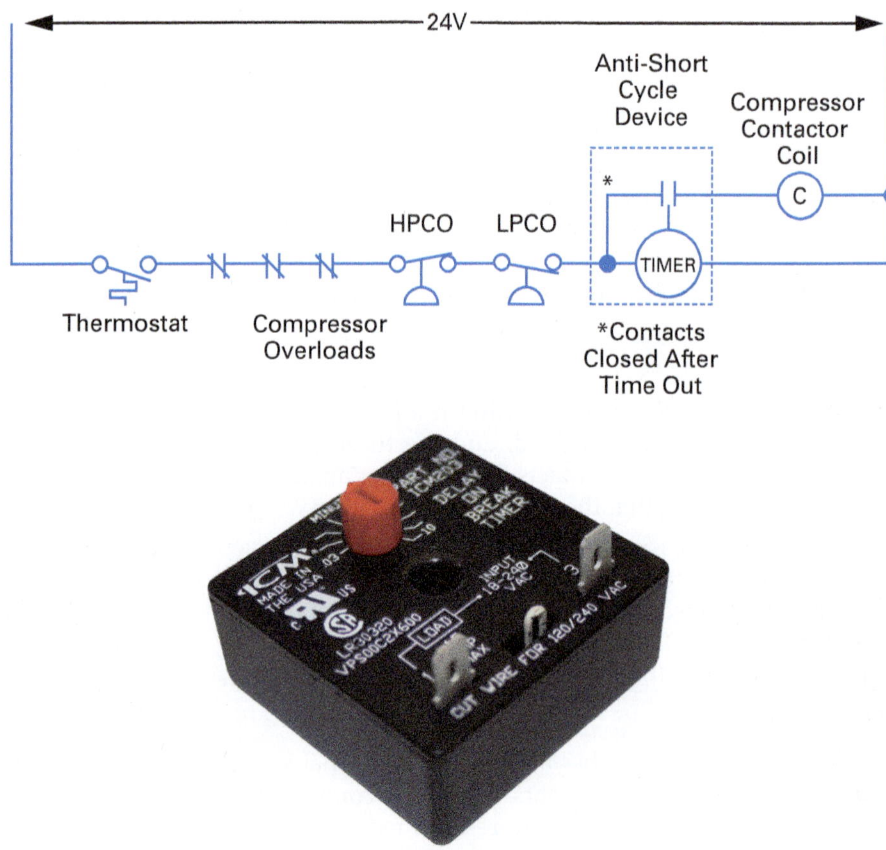

Figure 42 Compressor short-cycle timer and simplified circuit.
Source: ICM Controls

This type of relay is called a *delay-on-break* relay. The relay may have an adjustable range of about one second up to five minutes. A setting of three to five minutes is common for HVACR compressor applications.

While a delay-on-make relay can be used, it is NO and begins the timing period when power is applied. Thus, the compressor will not start until the set delay has elapsed, even if it has been idle for hours, needlessly delaying the restart.

Today, most electronic room thermostats, even inexpensive models, contain a built-in compressor delay timer to prevent compressor short cycling. Once the thermostat has satisfied and opens the circuit to the compressor, its internal time delay begins. The timer must complete its cycle before the thermostat energizes the control circuit for the next cycle.

When applying compressor short-cycle protection, be aware of how the operation of other system controls may affect the operation. Avoid using multiple devices for short-cycle protection, as this may cause some discomfort due to extended delays.

Installer programming is usually required to enable, disable, or change the time value of the thermostat's short-cycle protection. *Figure 43* shows an example of an installer's menu where the feature can be adjusted or turned off completely. The number on the right shows the length of the time delay. Checking the thermostat programming is a very important step during both start-up and service visits.

Figure 43 Example of installer's menu for short-cycle protection programming.

Before leaving this topic, it is important to note that short run times can also be a problem for compressors. A compressor that repeatedly starts and then stops after only two or three minutes of operation will eventually lose lubrication. Without some extended run times, oil in circulation may fail to return. Because there is no safety control on many systems to address this problem, it is up to HVACR technicians to understand the importance of extended runs to enhance oil return, and to inform users when they encounter a system that is operating this way.

4.0.0 Section Review

1. Which listed device works most like a CT?
   a. Clamp-on ammeter
   b. Rectifier
   c. Time-delay relay
   d. Ohmmeter

2. Once a reciprocating compressor shuts down, what needs to happen before a restart is attempted?
   a. The incoming voltage must stabilize.
   b. The motor must cool off significantly.
   c. The suction and discharge pressures need to equalize.
   d. Any refrigerant remaining in the liquid line must vaporize.

5.0.0 Hermetic Compressor Analysis

Performance Task

2. Measure and record the electrical and other operating parameters of a running compressor.

Objective

Explain how to analyze the operation of a hermetic compressor.

a. Explain how to evaluate the mechanical operation of a hermetic compressor.
b. Explain how to evaluate the electrical operation of a hermetic compressor.

Many functional compressors are replaced because a technician was unable to accurately analyze a system problem. Before condemning the compressor, certain procedures should be performed to identify the source of the problem. Even if the compressor has failed, these checks must be made to find out why it failed after a new compressor is installed. Remember that, if a compressor fails and the cause has not been determined and corrected, the replacement compressor is also likely to fail.

5.1.0 Hermetic Compressor Mechanical Analysis

A basic mechanical review of a system is primarily visual, although a few things may require basic tools and instruments.

WARNING!

Ensure all electrical power to the equipment is turned off whenever it is not required for troubleshooting and testing. Open, lockout, and tag disconnect switches or circuit breakers. Be aware of pressurized and hot components. Follow all safety instructions labeled on the equipment and in the manufacturer's service manual for the equipment.

5.1.1 Preliminary Analysis

Before operating the compressor, a preliminary inspection is performed as follows:

Step 1 Turn off the equipment. Lockout and tag the power source.

Step 2 Look for an evaporator or condenser mounted above the compressor that might affect how the system should be piped to avoid liquid floodback, slugging, and similar problems.

Step 3 Look for the following piping problems:

- Improperly sized lines that increase pressure drop (too small) or decrease velocity below the minimum needed for reliable oil return (too large)
- Long or uninsulated suction line that might affect superheat
- Liquid line running through an unconditioned space (hot or cold) that might affect subcooling
- Buried lines that might allow refrigerant to condense in the suction line
- Very long refrigerant lines that hold an excessive amount of refrigerant
- Suction line risers without traps at the base, or traps that are too deep

Step 4 At the indoor and outdoor coils, check for the following:

- Coil fin collars corroded
- Coil fins dirty or badly damaged
- Air filters dirty or missing
- Supply air plenum dirty
- Blowers and fans dirty
- Drive belt loose

Step 5 Check the following at the compressor (if applicable):

- Ensure both service valves are fully open.
- Ensure the compressor mount bolts are loosened or tightened per the manufacturer's instructions. Note that the hold-down bolts on most compressors must be loosened when the unit is installed. This allows freedom

NOTE

A damaged or dirty coil reduces heat transfer. In the cooling mode, a fouled outdoor coil results in higher head pressure. This increases the compressor operating temperature. A fouled indoor coil can result in liquid floodback. Both conditions degrade performance and efficiency. Dirty or plugged filters, dirty indoor blower wheels, and loose drive belts all impede airflow across the indoor coil and have the same negative effect.

of movement for the vibration isolators. The bolts are tightened by the manufacturer to secure the compressor for shipping.
- Ensure the crankcase heater is functional. Although the power may be off, the area around the heater may still be warm enough to verify it is working. If in doubt, check the continuity of the heater with an ohmmeter to ensure it is electrically intact.
- Inspect the compressor cylinder heads to see if they are scorched or blistered from excessive heat.
- Check the oil level in the sight glass (if applicable).
- Ensure the compressor electrical terminals are in good condition and there is no evidence of refrigerant or oil leakage around them.

WARNING!

Crankcase heaters can be extremely hot. Do not casually touch the heater itself even when the power is off. Check to see if it is producing heat by touching nearby areas of the compressor housing first.

The listed piping issues can cause several problems. Refrigerant lines that are too small increase pressure drop and reduce efficiency. Excess superheat can cause the compressor to overheat. Refrigerant condensing in buried suction lines can result in slugging at startup. A technician must be aware of the characteristics of the piping system and understand how they might affect compressor and system operation.

For example, if a **pump-down** is attempted on a system with a very long line set, the condenser and/or receiver may not be adequate to hold the entire refrigerant charge. This may cause the system to shut down due to a high-pressure condition before the pump-down cycle is complete. This does not necessarily represent a hazard to the compressor, but it can affect the subsequent service procedures.

Pump-down: Operating a compressor to clear most or all refrigerant from the low-pressure side of the system or a portion thereof. A pump-down cycle may be applied at the end of an operating cycle, or as a means of transferring refrigerant into the condenser for service work.

The preceding checks are simple to perform but are often overlooked, especially if the compressor appears to be operating normally. For example, signs of overheating or condensation in the wrong places suggest a problem that may eventually lead to compressor failure. Finding the source of the problem and correcting it quickly will extend the life of the compressor. The best way to accomplish this is to be highly observant, both before and after a compressor is replaced.

CAUTION

Scroll compressors should not be pumped down below a suction pressure of 7 psig (48 kPa). Allowing it to operate below this pressure for more than a few seconds may cause overheating and damage to the scroll plates. Testing to see if a scroll compressor can reduce the suction pressure to or below 0 psig serves no purpose and can damage the compressor. This test was often done with reciprocating compressors in the past to check the integrity of the internal valves, but it should not be applied to scroll compressors.

5.1.2 Operating Analysis

If the compressor is operable, important operating parameters should be measured and documented. The current operating conditions can then be compared against previous records. *Figure 44* shows a typical comfort cooling refrigerant circuit with temperatures and pressures that are considered normal under common conditions. *Table 3* summarizes temperature and pressure readings from *Figure 44*.

The importance of properly recording the operating conditions cannot be overstated. The record serves as a historical document that can be used for future comparison. The recorded data can also serve as a system commissioning record to verify compliance with contract specifications. Keep in mind that your readings will vary from those shown in *Table 3* for a variety of reasons. Also do not assume that the figures provided represent perfect operation; specific operating conditions and equipment characteristics are infinitely variable. With experience, you will develop a feel for the proper temperatures and pressures appropriate for various systems and climates.

To create your own document similar to *Table 3* for an operating system, use the following steps:

WARNING!

Avoid rotating, pressurized, or hot components. Follow all labeled safety instructions as well as those in the manufacturer's equipment service manual. Failure to do so could result in personal injury.

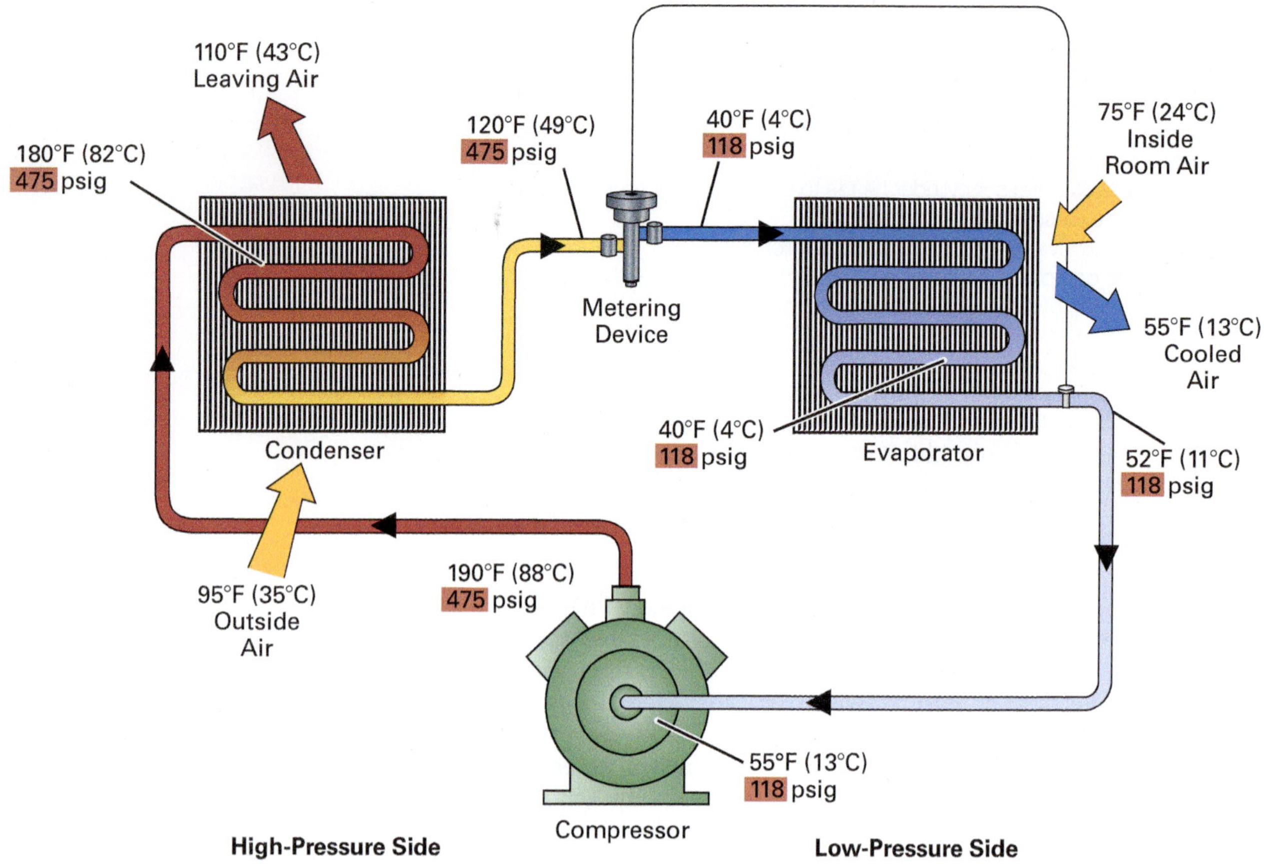

Temperature °F	Pressure HFC-410A
27	91.0
28	92.9
29	94.9
30	96.8
31	98.8
32	100.9
33	102.9
34	105.0
35	107.1
36	109.2
37	111.4
38	113.6
39	115.8
40	118.1
41	120.3
42	122.7
43	125.0
44	127.4
45	129.8
46	132.2
47	134.7
48	137.2
49	139.7

Temperature °F	Pressure HFC-410A
50	142.2
55	155.5
60	169.6
65	184.5
70	200.4
75	217.1
80	234.9
85	253.7
90	273.5
95	294.4
100	316.4
105	339.6
110	364.1
115	389.9
120	416.9
125	445.4
130	475.4
135	506.9
140	540.1
145	575.1
150	611.9

Temperature/Pressure Chart

Figure 44 Typical operating temperatures and pressures for HFC-410A refrigerant in a comfort cooling application.

TABLE 3 Temperature-Pressure Data Summary

System Parameter	Comfort Cooling System Operating with HFC-410A
Suction Pressure	118 psig (814 kPa)
Suction Line Temperature at Evaporator Outlet	52°F (11.1°C)
Saturated Suction Temperature	40°F (4.4°C)
Superheat – TXV	12°F (6.7°C)
Discharge Pressure (psig)	475 psig (3,275 kPa)
Discharge Line Temperature (°F)	190°F (87.8°C)
Saturated Discharge Temperature (°F)	130°F (54.4°C)
Liquid Temperature Leaving Condenser (°F)	120°F (48.9°C)
Liquid Line Temperature Entering Metering Device (°F)	120°F (48.9°C)
Subcooling (°F)	10°F (5.6°C)
Air Temperature Entering Evaporator (DB°F)	75°F (23.9°C)
Air Temperature Leaving Evaporator (DB°F)	55°F (12.8°C)
Air Temperature Drop Across Evaporator (DB°F)	20°F (11.1°C)
Outdoor Temperature Rise Across Condenser (DB°F)	15°F (8.3°C)
Compressor Compression Ratio (Absolute Discharge Pressure ÷ Absolute Suction Pressure)	3.72 to 1

Step 1 Connect a gauge manifold to the system gauge ports. Connect thermometers or temperature sensors on the refrigerant lines in the system. Use *Figure 44* as a reference when placing temperature sensors. Note that placing sensors in the middle of the coil to measure refrigerant temperature isn't practical; the sensors are shown on the diagram for reference only. However, you may wish to measure the suction line temperature at both the evaporator outlet and the compressor inlet to see if it changes point-to-point. Checking the change in temperature between the compressor discharge and the condenser inlet is also a good exercise. Of course, if you are working with a classroom system that is close coupled, no change is likely.

Step 2 Place a clamp-on ammeter around the Common lead to the compressor, on the load side of the contactor. Set the ammeter to measure and remember the peak current at startup if the instrument has that function. The electrical analysis of a compressor is covered in sections that follow.

Step 3 Restore power to the system and adjust the thermostat to call for cooling. Start the system and monitor its operating characteristics:

- Note the inrush current as the compressor starts and watch the current as it drops and begins to stabilize.
- Listen and look for excessive vibration of the compressor, piping, motors, and fans.
- Listen for knocks or rattling noises from the compressor that may indicate liquid is being drawn into the cylinders or scrolls. If this condition continues for more than a few seconds, shut down the system and look for the cause of slugging. If the sounds don't begin until minutes after startup, floodback is more likely than slugging. Although the two are similar in effect, they have different sources.

Step 4 Check the compressor oil sight glass (if applicable). Any significant foaming should diminish within 10 minutes. If it does not, an excessive amount of refrigerant may have condensed in the oil. Some foaming during startup, however, is not unusual.

WARNING!

If the compressor terminals are damaged and the system is under pressure, the terminals can be violently blown from the compressor. Disturbing those terminals to take electrical measurements is unnecessary. When using a meter on a pressurized system, take measurements at or near the contactor, terminal boards, and other test points away from the compressor. If applying a meter directly to the compressor terminals is necessary, do not stand in front of or face the terminals directly. Safety glasses should be worn whenever you are working with refrigerants under pressure.

Step 5 Ensure all related fans, blowers, and pumps are operating and providing air or water flow.

Step 6 Measure or calculate and record the following information related to the compressor's mechanical performance:

- Suction pressure
- Corresponding saturated suction temperature
- Suction line temperature
- Calculate superheat
- Discharge pressure
- Corresponding saturated discharge temperature
- Discharge line temperature
- Calculate subcooling
- Liquid-line temperature entering the metering device
- Compression ratio

Once the information is documented, compare it against previous results to determine if there is a problem.

As noted earlier, readings vary for many reasons. Always refer to the manufacturer's service documentation to evaluate a system's performance under the current operating conditions.

Based on the analysis, identify and investigate any problems. When there is doubt about the mechanical condition of the compressor, contact the manufacturer if necessary to discuss the recorded figures. There are two potential problems that are directly related to the compressor:

- *Low compression ratio* — A low ratio between the suction and discharge pressures, coupled with low current, often reflect compressor valve or scroll damage. With reciprocating compressors, suction valves tend to be damaged more often than discharge valves. When this occurs, the suction pressure will remain abnormally high while the discharge pressures may appear near normal or somewhat low.
- *No pressure changes through the compressor; current very low, and little or no sound from the compressor* — This usually indicates a broken crankshaft and/or piston rod. It may surprise you to record a tiny operating current, indicating the compressor motor is running, but with little or no sound or pressure changes present. Mechanical noise may accompany this problem, but the compressor may also be whisper quiet.

5.2.0 Hermetic Compressor Electrical Analysis

The electrical system should be evaluated to ensure reliable compressor operation. Analyzing an operating compressor involves checking voltage and current to ensure both are within expected ranges. The entire electrical system of the unit in which the compressor is installed should also be checked to ensure other electrical deficiencies are not a factor in system performance.

5.2.1 Preliminary Electrical Inspection

Like mechanical inspections, an electrical inspection of the compressor begins with the senses of sight, sound, touch, and smell to identify problems. Note that the electrical and mechanical integrity of the system can be evaluated at the same time.

WARNING!

Be sure the equipment is de-energized when power is not required for troubleshooting and testing. Open, lock, and tag disconnects. Be aware of pressurized and hot components. Follow all safety instructions labeled on the equipment and in the manufacturer's service manual for the equipment.

The preliminary electrical inspection is performed as follows:

Step 1 De-energize, lockout, and tag the equipment.

Step 2 Check the following:

- Electrical connections at the unit, contactor, and compressor. Look for terminals that are discolored, indicating the terminals have overheated.
- Ground connections are present and intact.

5.2.2 Energized Testing

Electrical testing of a running compressor is performed as follows:

Step 1 Remove the electrical lockouts and restore power to the equipment.

Step 2 Measure and record the input voltage at the line side of the compressor contactor.

- The measured voltage should be within ±10% of the compressor and/or unit nameplate value. Remember that the compressor and motors may be dual voltage, expanding the range of acceptable voltage.
- On a three-phase system, calculate the voltage imbalance. Remember that the voltage imbalance between phases cannot exceed 2%.

Step 3 Check the unit nameplate to find and record the RLA and LRA values for the compressor.

Step 4 Place a clamp-on ammeter on the Common lead of the compressor (single-phase compressors) as it leaves the contactor (*Figure 45*). For a three-phase compressor, choose any one of the three leads. If a recording or Maximum Hold function is available, set the ammeter accordingly.

Figure 45 Checking compressor current.
Source: David Spates/Shutterstock

Step 5 Set the thermostat to call for cooling when ready. Measure and record the compressor peak current at startup and its running current after allowing time for the system to stabilize.

- Compare the running current to the compressor RLA. The actual value on a normally functioning compressor will likely be something less than its RLA, depending on the operating conditions. On a three-phase compressor, measure the current on each phase. Calculate the current imbalance to ensure it is within 10%.

Step 6 Check the voltage on the line side of the contactor with the compressor running.

- Compare the voltage to the reading taken before the compressor was started. There should be no more than a 3% voltage drop. If the drop exceeds 3%, de-energize and secure the system, then inspect all the wiring connections between the unit and the circuit breaker panel. Also confirm that the size of the conductors is adequate for the load.
- Calculate the voltage imbalance and ensure that it is within the 2% maximum.

Step 7 Carefully take a voltage reading across the contacts of the contactor. This measures the voltage drop across the contacts alone, and the reading will likely be in millivolts. Ideally, no voltage drop occurs, but it is not unusual to see around 100 to 200 millivolts (mV). A significant voltage drop across the contactor produces a lot of heat. A voltage drop of 500 mV or more typically indicates it is time to clean the contacts or replace the contactor.

Step 8 Before completing the work, check the control voltage at the contactor and ensure it is within the required range.

5.0.0 Section Review

1. Foam remaining in a compressor's sight glass after 10 minutes of operation often indicates ______.
   a. excessive refrigerant dissolved in the oil
   b. oil contamination from water
   c. a low refrigerant charge
   d. a defective hot-gas bypass valve

2. The voltage drop in the circuit powering a running compressor should not exceed ______.
   a. 1%
   b. 3%
   c. 5%
   d. 10%

Module 03302 Review Questions

1. Which compressor type is limited to 5 tons of cooling capacity or less?
   a. Rotary
   b. Screw
   c. Centrifugal
   d. Scroll

2. Most hermetic reciprocating, rotary, and scroll compressors have the motor and drive shaft mounted ______.
   a. horizontally
   b. vertically
   c. at a 45-degree angle
   d. outside the compressor housing

3. Just before the intake stroke of a reciprocating compressor begins, the suction valve(s) are ______.
   a. fully open
   b. partially open
   c. fully closed
   d. nearly closed

4. Which type of compressor has the *fewest* number of moving parts?
   a. Rotary screw
   b. Scroll
   c. Centrifugal
   d. Reciprocating

5. Which of the following statements about centrifugal compressors is *correct*?
   a. Centrifugal compressors are limited to roughly 200 tons of capacity.
   b. Centrifugal compressor impellers spin much more slowly than a common water pump.
   c. Refrigerant vapor enters the compression area through the center of the impeller.
   d. Centrifugal compressors only work well with refrigerants that operate at very high pressures.

6. What issue that is unique to the arrangement must be considered when tandem compressors share a common refrigeration circuit?
   a. Balancing the voltage
   b. Balancing the oil level
   c. Excess sound and vibration
   d. Metering device design

7. When a hot-gas bypass strategy is used to reduce system capacity, where is the hot gas best introduced into the refrigerant circuit?
   a. Between the metering device and the evaporator inlet.
   b. Directly into the suction inlet of the compressor.
   c. Between the condenser outlet and the expansion device.
   d. Near the center of the condenser.

8. Which of the following is *not* a means of controlling the capacity of a centrifugal compressor?
   a. Speed control
   b. Variable-pitch inlet guide vanes
   c. Internal vapor bypass
   d. Suction or discharge throttling valves

9. How can slugging in a hermetic compressor be detected?
   a. A knocking sound
   b. High superheat
   c. High subcooling
   d. Liquid-line sight glass bubbles

10. If an oil product is hygroscopic, ______.
    a. it contains hydrogen
    b. molecules of the oil are very small
    c. it is water-soluble
    d. it readily absorbs moisture

11. When installing a dual-voltage 208/230V compressor, the supply voltage cannot be any *lower* than ______.
    a. 187V
    b. 208V
    c. 219V
    d. 230V

12. A 2% voltage imbalance in a three-phase motor can result in a temperature rise in the motor windings of ______.
    a. 3% above normal
    b. 5% above normal
    c. 8% above normal
    d. 12% above normal

13. What does the term *single-phasing* mean?
    a. Using a single-phase motor in place of a three-phase motor.
    b. A three-phase motor is trying to run on one leg of power.
    c. A three-phase motor is trying to run on two of three phases.
    d. A single-phase motor is being operated on three-phase power.

14. A technician can tell when the internal line-break overload of a single-phase compressor is remaining open when ______.
    a. there is no continuity through the Common winding
    b. the windings show continuity to ground
    c. the compressor is noisy but isn't pumping
    d. the compressor is silent but draws a small amount of current

15. A compressor internal pressure-relief valve is usually set ______.
    a. by adjusting it while watching an ammeter
    b. through the installer's menu at the thermostat
    c. to open at a higher pressure than the high-pressure safety switch
    d. to close when the refrigerant circuit goes into a vacuum

16. A time-delay relay for compressor short-cycle protection is installed in ______.
    a. series with the compressor Common lead
    b. parallel with the compressor contactor
    c. the contactor control circuit
    d. parallel with the room thermostat

17. Refrigerant lines that are buried underground can cause ______.
    a. the copper tubing to collapse in just a few years
    b. the pressure to drop significantly
    c. refrigerant to condense in the suction line
    d. vapor to be trapped in the evaporator coil

18. If a system's refrigerant lines are long, ______.
    a. the compressor may not be able to complete a pump-down cycle
    b. the compressor cannot overcome the pressure drop
    c. more time should be allowed between compressor starts
    d. the system pressures will never equalize

19. Scroll compressors should *not* be pumped down below a suction pressure of ______.
    a. 35 psig (241 kPa)
    b. 20 psig (138 kPa)
    c. 7 psig (48 kPa)
    d. 0 psig (0 kPa)

20. On which lead of a single-phase compressor should the compressor current be measured?
    a. Neutral
    b. Start
    c. Run
    d. Common

Answers to Odd-Numbered Module Review Questions are found in *Appendix A*.

Answers to Section Review Questions

Answer	Section	Objectives
Section One		
1. c	1.1.0	1a
2. c	1.2.0	1b
3. b	1.3.0	1c
4. d	1.4.0	1d
5. a	1.5.0	1e
6. a	1.6.0	1f
Section Two		
1. c	2.1.1	2a
2. d	2.2.1	2b
Section Three		
1. b	3.1.1	3a
2. d	3.2.0	3b
Section Four		
1. a	4.1.4	4a
2. c	4.2.2	4b
Section Five		
1. a	5.1.2	5a
2. b	5.2.2	5b

MODULE 03303

Metering Devices

Source: Suranto W/Shutterstock

Objectives

Successful completion of this module prepares you to do the following:

1. State the function of refrigerant metering devices and describe their effect on refrigerants.
   a. State the function of metering devices.
   b. Describe how refrigerants react as they pass through a metering device.
   c. Describe refrigerant distributors and explain how they affect metering device and coil performance.
2. Identify and describe the operation of fixed metering devices.
   a. Identify and describe the operation of fixed-orifice metering devices.
   b. Identify and describe the operation of capillary tubes.
   c. Describe common problems associated with fixed metering devices.
3. Identify and describe the operation of expansion valves.
   a. Identify and describe the operation of manual expansion valves.
   b. Identify and describe the operation of automatic expansion valves.
   c. Identify and describe the operation of thermal expansion valves.
   d. Identify and describe the operation of electric and electronic expansion valves.
   e. Describe common problems associated with expansion valves.
4. Explain how thermal expansion valves are selected and installed.
   a. Explain how thermal expansion valves are selected for a given application.
   b. Describe installation practices related to thermal expansion valves.

Performance Tasks

Under supervision, you should be able to do the following:

1. Remove and replace a fixed-orifice piston.
2. Install an externally equalized expansion valve, correctly placing the sensing bulb and equalizer tube.
3. Measure superheat in an operating system with an expansion valve and adjust it to change the superheat.

Overview

The metering device controls refrigerant flow and provides the needed reduction in pressure to cause a change of state as the refrigerant enters the evaporator coil. Different approaches exist to create the pressure drop, but all metering devices provide a similar function. HVACR technicians

Digital Resources for HVACR

Scan this code using the camera on your phone or mobile device to view the digital resources related to this craft.

SCAN ME

must be able to identify various metering devices and understand how they affect the refrigeration cycle. Different metering devices require different service procedures and significantly affect how refrigerant is added.

> **NCCER Industry-Recognized Credentials**
>
> If you are training through an NCCER-accredited sponsor, you may be eligible for credentials from NCCER. The ID number for this module is 03303. Note that this module may have been used in other NCCER curricula and may apply to other level completions. Contact NCCER at 1.888.622.3720 or go to **www.nccer.org** for more information.
>
> You can also show off your industry-recognized credentials online with NCCER's digital credentials. Transform your knowledge, skills, and achievements into credentials that you can share across social media platforms, send to your network, and add to your resume. For more information, visit **www.nccer.org**.

1.0.0 Introduction to Metering Devices

Performance Tasks

There are no Performance Tasks in this section.

Objective

State the function of refrigerant metering devices and describe their effect on refrigerants.

a. State the function of metering devices.
b. Describe how refrigerants react as they pass through a metering device.
c. Describe refrigerant distributors and explain how they affect metering device and coil performance.

Metering devices are essential to the operation of HVACR systems. Metering devices are considered one of the four primary components in a working refrigerant circuit. The type of metering device used also has a major impact on how a system performs, cycle after cycle.

1.1.0 The Function of Metering Devices

The metering device in a refrigerant circuit performs two important functions. First, the metering device attempts to match the rate of refrigerant flow entering the evaporator with the rate at which liquid is boiling and changing to its vapor state. Second, the metering device provides a significant pressure drop that separates the high-pressure side of the system from the low-pressure side. The low-pressure side of the system is where heat is transferred from the load to the refrigerant.

A variety of metering devices are used in modern systems to control refrigerant flow. These devices include capillary tubes, fixed orifices, pressure-operated valves, thermal expansion valves (TEVs or TXVs), float valves, and electronic expansion valves (EEVs). Other components, such as *distributors* and *equalizer lines*, are also used to support and enhance their performance. Metering devices are also referred to as *expansion devices* because they are part of a **direct-expansion (DX) cooling system**.

Direct-expansion (DX) cooling system: A system dependent on a refrigerant circuit in which the cooling effect is achieved by the expansion of a refrigerant as it changes from a liquid to a vapor state.

To transfer heat effectively, a DX refrigerant circuit must operate at differing pressures and temperatures. The refrigerant temperature entering an air-cooled condenser, for example, must be higher than the outdoor temperature for heat to be rejected to the air. If not, then heat would instead be added to the refrigerant. The increase in temperature is accomplished by compressing refrigerant vapor to increase the pressure and temperature. By making the refrigerant warmer than the outdoor air, we can take advantage of the air as a heat sink.

The refrigerant entering the evaporator, on the other hand, must be at a lower temperature than the indoor air or other load so that heat can be transferred to the refrigerant. To reach a lower temperature, the refrigerant pressure must be significantly reduced. While the system compressor is responsible for increasing the pressure, the metering device is responsible for decreasing the pressure.

The metering device is installed between the condenser and the evaporator, usually at the evaporator inlet (*Figure 1*). The metering device controls the volume of refrigerant that enters the evaporator. Ideally, only as much refrigerant needed for effective system operation at a given set of conditions is allowed to pass. It must pass enough refrigerant to support the cooling demand on the evaporator. At the exit of the evaporator coil, only refrigerant vapor should be present, with some degree of superheat present to confirm that no liquid remains.

NOTE

Many of today's ductless split systems place the metering device in the outdoor unit, rather than at the evaporator coil inlet. This means that both the suction line and the liquid-vapor line leaving the metering device must be insulated between the indoor and outdoor unit.

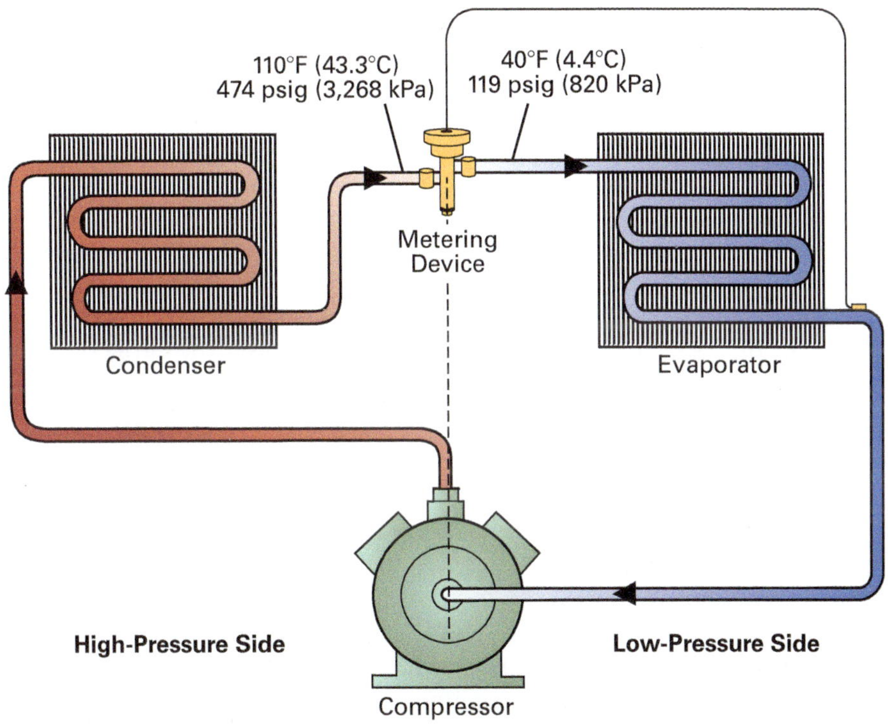

Figure 1 Metering device position in the refrigerant circuit (HFC-410A pressures and temperatures shown).

When the amount of refrigerant entering the evaporator is not sufficient, the liquid boils away too quickly and much of the coil surface becomes ineffective. To maximize heat transfer in the coil, the internal surfaces must be bathed in boiling refrigerant liquid. Once the liquid has completed the change of state to a vapor, it is virtually useless as a heat transfer medium.

Conversely, if too much liquid enters the evaporator, some of it will still be in the liquid state when it reaches the end of the evaporator. The liquid will then enter the suction line and return to the compressor as *floodback*. Floodback can wash away and/or dilute the lubricating oil in the compressor, and liquid refrigerant can destroy a compressor if it enters the cylinder or other compression area. Because liquids cannot be compressed, floodback often results in severe mechanical damage in the form of broken valves, pistons, connecting rods, scroll plates, and crankshafts. Thus, the metering device must perform these two vital functions:

- Restrict and control the refrigerant flow to produce the low-pressure, reduced-temperature refrigerant required for cooling.
- Admit only the volume of refrigerant needed in the evaporator to satisfy the cooling load, without allowing a surplus of refrigerant to pass.

Ideally, a metering device can effectively achieve these goals under every set of operating conditions. As you will see, some metering devices are far more capable than others.

1.2.0 Metering Devices and the Refrigerating Effect

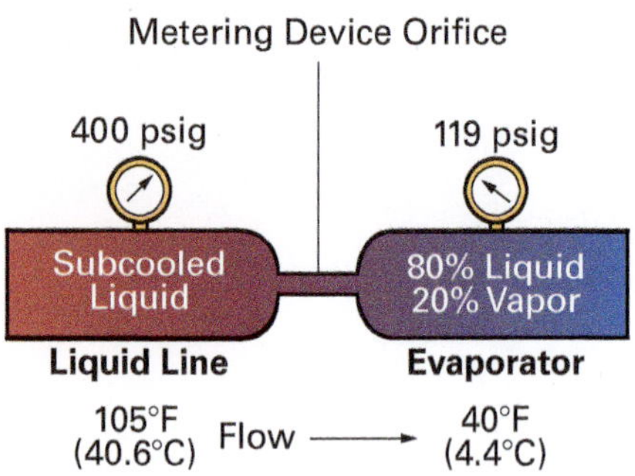

Figure 2 Example HFC-410A refrigerant pressure and temperature drop after passing through a metering device.

Saturation: The condition of a refrigerant when it exists in both the liquid and vapor states. Saturation is the point at which a refrigerant can be in transition from one state to the other or at a stable ratio at a fixed temperature and pressure. The ratio of liquid to vapor is not a consideration; as long as both liquid and vapor are present in some amount, the refrigerant is considered saturated.

Flash gas: Refrigerant vapor that forms spontaneously due to a rapid change in pressure that forces the refrigerant to boil.

NOTE

Terms used here such as *high pressure* and *high temperature* are relative to the system characteristics. Most refrigerant circuits do not operate at truly high pressures or high temperatures. Much higher pressures and temperatures are encountered in hydraulic and steam systems, for example. Here, these terms simply help differentiate between the two "sides" of a refrigerant circuit.

The following discussion is an example of how refrigerant reacts as it passes through a metering device in a comfort cooling application. Remember, though, that a similar reaction occurs in all DX systems and with all conventional metering devices.

The metering device receives moderate-temperature, high-pressure, subcooled liquid refrigerant from the condenser. The metering device opening, or *orifice*, is tiny when compared to the refrigerant lines. As a result, a large pressure drop occurs as the liquid passes through the orifice. Essentially, the metering device converts high-pressure, high-temperature liquid refrigerant to a low-temperature, low-pressure mixture of liquid and vapor.

For example, HFC-410A refrigerant may enter a metering device at 400 psig (2,758 kPa) and a temperature of 105°F (40.6°C). Refer to *Figure 2*. On the opposite side of the metering device, the pressure has been reduced to 119 psig (820 kPa). According to the pressure-temperature (P-T) chart for HFC-410A, refrigerant in **saturation** at this pressure should be at a temperature of 40°F (4.4°C).

To satisfy this new pressure condition while in saturation, the refrigerant temperature must drop 65°F (36°C) instantly. But where can the heat go so quickly?

Recall that a change of state in a substance consumes large amounts of latent heat. Therefore, for the liquid refrigerant to rid itself of the heat quickly, a latent process must take place. A portion of the refrigerant leaving the metering device flashes instantly from a liquid to a vapor. The latent process uses heat from the liquid to flash a portion of it to a vapor, until the remaining refrigerant reaches 40°F. The vapor now mixed with the liquid refrigerant is known as **flash gas**.

The refrigerant leaves the orifice of the metering device as a low-pressure, low-temperature mixture of liquid and vapor. The refrigerant mixture now entering the evaporator is roughly 80% liquid and 20% vapor, as shown in *Figure 2*. This is a common ratio for comfort cooling systems. However, the ratio varies widely depending on the operating conditions. One major factor affects it. Before presenting that factor, it is crucial to understand the importance of this ratio.

As the refrigerant enters the evaporator and is exposed to the heat load, the remaining liquid refrigerant continues to boil. Once again, it is the latent heat exchange that consumes most of the heat. The liquid refrigerant in the evaporator will remain at 40°F (4.4°C) until all the liquid has boiled to its vapor state, but a great deal of heat is being transferred by the latent process.

Once the refrigerant has completely changed to the vapor state, the only additional heat exchange that occurs is a sensible rise in the vapor temperature. The amount of heat required to raise the temperature of vapor is insignificant. Therefore, vapor inside the evaporator is useless for effective heat transfer. Only boiling liquid, while a change of state is taking place, has any real cooling value in the refrigerant circuit.

Therefore, any vapor mixed with the liquid refrigerant as it enters the evaporator is of no real value. If 20% of the liquid has changed to a vapor before ever entering the evaporator, then 20% of the refrigerant is not enhancing the cooling process.

Consider a low-temperature refrigeration application that requires the refrigerant entering the evaporator to be as low as –25°F (–31.7°C). In this case, the liquid refrigerant temperature must be reduced by an astounding 130°F (72.1°C). Obviously, a great deal of liquid refrigerant must flash to vapor to create this cooling effect.

In most DX systems, it is a necessary and unavoidable loss. However, minimizing this loss can significantly increase the efficiency of the system. Remember this important fact—exposing the maximum amount of internal evaporator surface area to boiling liquid refrigerant maximizes the cooling effect.

The major factor mentioned earlier that affects the liquid/vapor ratio is refrigerant *subcooling*. Further subcooling the liquid refrigerant before it enters the metering device reduces the amount of flash gas. Imagine subcooling the liquid all the way down to 40°F before it reaches the metering device (*Figure 3*). Theoretically, the refrigerant entering the evaporator would still be 100% liquid and all of the refrigerant would be contributing to the cooling effect.

Although subcooling to that level is impractical for many applications, it demonstrates the value and importance of subcooling to system performance. The type of metering device used is not a factor in this topic; subcooling provides a significant advantage at the evaporator, regardless of the type of metering device used.

Consider the opposite situation, which happens quite often. Instead of having subcooled liquid entering at 105°F, the refrigerant from the condenser has no subcooling at all. In other words, it has not fully condensed to a liquid. Some liquid likely exists, but a great deal of vapor also remains. Perhaps the condenser is clogged with dust and debris.

If any refrigerant condensed as it flowed through the condenser, it is saturated because both liquid and vapor are present. This means that the refrigerant temperature will precisely match a specific pressure on the PT chart. As noted in *Figure 4*, there is no way to know the liquid/vapor ratio entering the evaporator. When a refrigerant is in its saturated state, it can be 98% liquid and 2% vapor, or 2% liquid and 98% vapor. In this scenario, the ratio is probably closer to the worst case, and heat transfer is seriously diminished.

To summarize, when the refrigerant entering the metering device is not subcooled to some degree, the cooling capacity of the system is significantly compromised. Again, this happens often in the field. It can be the result of problems such as a refrigerant undercharge; one or more failed condenser fan motors; or a condenser clogged with dust and debris as mentioned earlier. These problems and deficiencies often confront HVACR service technicians, and it is our job to identify and resolve them.

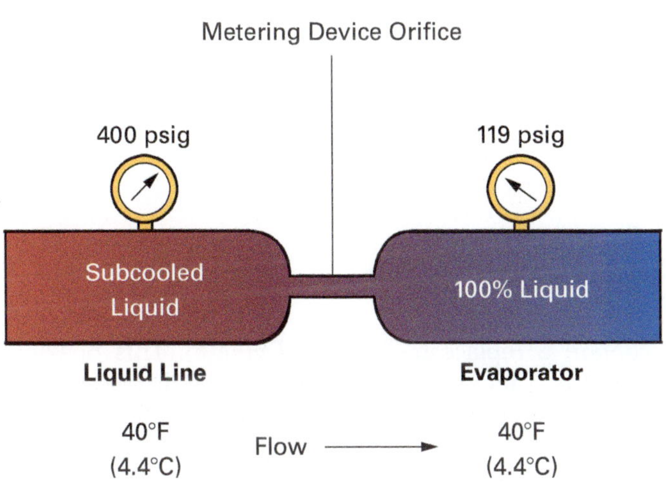

Figure 3 The positive effect of deeply subcooled liquid.

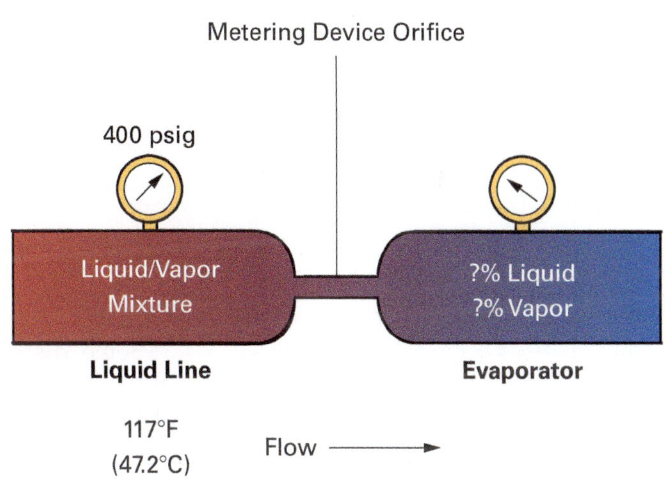

Figure 4 Refrigerant with no subcooling passing through the metering device.

1.3.0 Distributors

Before discussing metering devices in more detail, the effect of another device associated with metering devices should be understood.

For best system performance, the liquid/vapor mixture exiting the metering device must be distributed evenly throughout the evaporator coil. Thus, the refrigerant is rarely fed into a coil at only one point, except in small systems.

To simplify many illustrations here and in other publications related to metering devices, the details of refrigerant distribution at the evaporator are often omitted. However, to ensure that the entire coil is actively transferring the maximum amount of heat, refrigerant is typically introduced to the coil at multiple points. As the name implies, a *distributor* is used to disperse the refrigerant through multiple coil circuits (*Figure 5*).

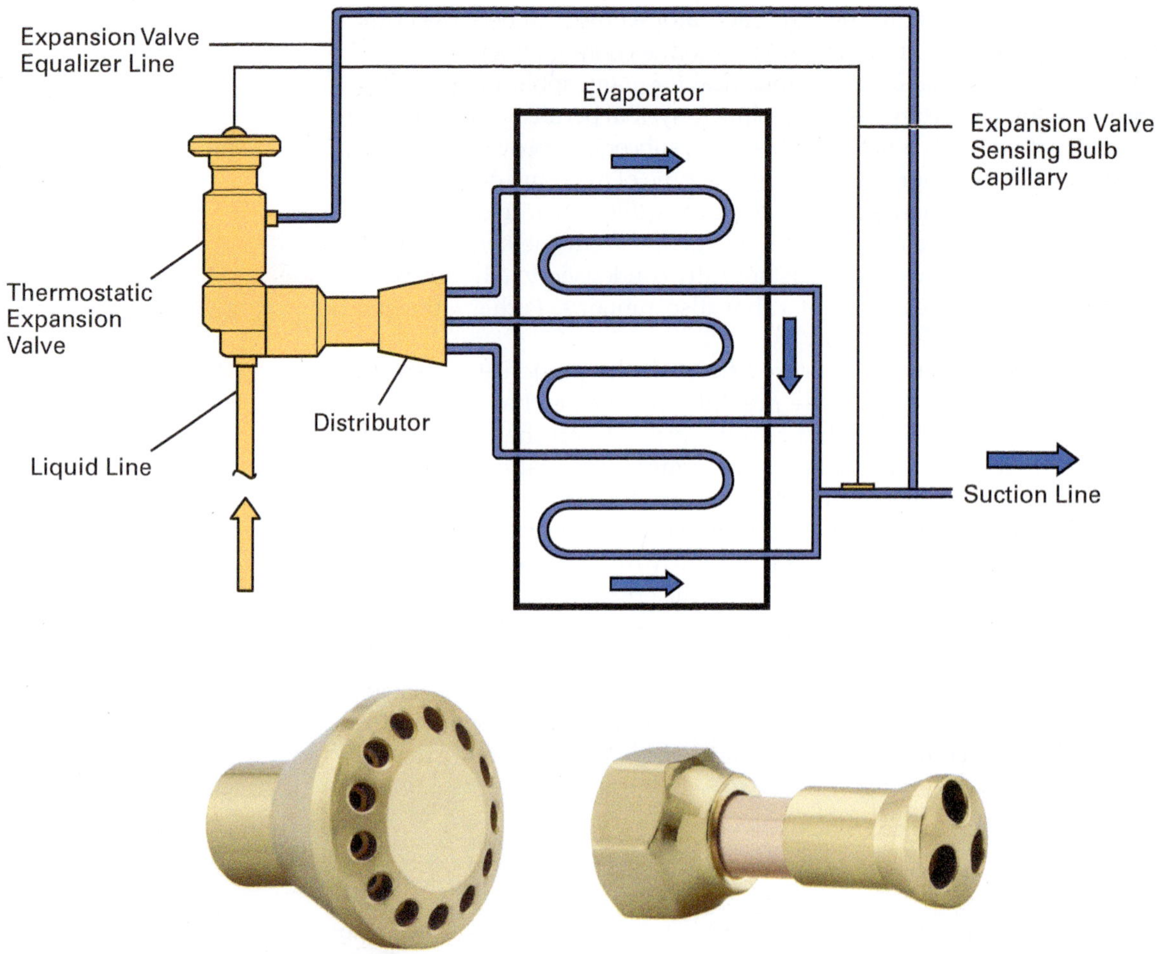

Figure 5 Distributors and coil piping schematic.
Source: Courtesy of Sporlan Division—Parker Hannifin Corporation

Figure 6 Replaceable distributor nozzle.
Source: Courtesy of Sporlan Division—Parker Hannifin Corporation

Some distributors contain a replaceable nozzle (*Figure 6*). This nozzle is quite large when compared to a metering device orifice. The distributor nozzle helps the vapor and liquid to mix thoroughly. This ensures that each coil circuit receives liquid and vapor in equal proportions. The nozzle size can be changed to allow for the use of the same distributor body with systems of differing capacities. Smaller distributors have a one-piece design with no replaceable nozzle.

Distributors vary by the number of outlets and by refrigerant line size (both inlet size and the size of the multiple outlets). Installing a large distributor with many outlets can be very challenging under field conditions. Brazing the individual tubes that are very close together while avoiding leaks requires skill and patience. As a result, manufacturers often make complex distributors available with factory-installed outlet tubes. Each tube is then field cut to length as needed.

Because the pressure drop through a distributor can be significant, it must be considered in the metering device selection process. It is critical that the pressure drop through the complete evaporator circuit be held within predictable and reasonable limits. When distributors are encountered, they are usually fed by expansion valves that accommodate the total pressure drop through the distributor and coil using an *equalizer line*. However, occasionally distributors are used with fixed metering devices as well.

Systems exceeding 10 tons of capacity may have several expansion valves, with each expansion valve equipped with a distributor. However, multiple expansion valves are not usually applied to a single evaporator coil. These systems may have two or more evaporator coils instead. An example would be a 20-ton packaged comfort cooling unit, with two independent 10-ton refrigerant circuits. Each circuit would likely have its own compressor(s), expansion valve, and distributor.

Note that distributors are almost never the source of a problem in the refrigerant circuit. Distributors have no moving parts, and the internal openings are much larger than the metering device orifice. As a result, they are not likely to become blocked or fouled. When the metering device is chosen by a manufacturer, the resultant pressure drop must be considered.

NOTE

Equalizer lines are introduced in Section 3.3.3, *TXV Valve Equalization*.

1.0.0 Section Review

1. True or False: Liquid refrigerant that exits the evaporator and returns to the compressor is referred to as *floodback*.
   a. True
   b. False

2. A common ratio of liquid-to-vapor leaving the metering device in a comfort cooling application would be _____.
   a. 80% liquid, 20% vapor
   b. 60% liquid, 40% vapor
   c. 40% liquid, 60% vapor
   d. 20% liquid, 80% vapor

3. True or False: The pressure drop through a distributor is insignificant and need not be considered in metering device selection.
   a. True
   b. False

2.0.0 Fixed Metering Devices

Objective

Identify and describe the operation of fixed metering devices.

a. Identify and describe the operation of fixed-orifice metering devices.
b. Identify and describe the operation of capillary tubes.
c. Describe common problems associated with fixed metering devices.

Performance Tasks

1. Remove and replace a fixed-orifice piston.

Fixed metering devices include *fixed-orifice* and *capillary tube* metering devices. Both types have a fixed opening through which the refrigerant from the condenser passes, but the two devices create the pressure drop in slightly different ways.

2.1.0 Fixed-Orifice Metering Devices

One of the most popular metering devices used in recent years is the fixed-orifice metering device. Its popularity stems from the fact that it is very simple, inexpensive, and easily serviced or replaced. It is also less likely to clog or foul than a capillary tube. Further, its use eliminates a great deal of additional piping and components from a heat pump system. It consists of a removable **piston** with a precisely sized orifice (*Figure 7*).

Note the small fins positioned around the piston and the tapered shape on one end. The piston can slide back and forth within its housing. Movement in one direction creates a seal around the perimeter of the device as two machined surfaces meet, forcing all refrigerant to pass through the center orifice. In this position, the piston acts as a metering device.

Piston: In this context, the component of a fixed-orifice metering device that contains the orifice. The piston is free to move back and forth inside the housing. At one end of its travel, the piston meters refrigerant flow through the orifice; at the other end, it allows refrigerant to flow both through the orifice and around its outside perimeter.

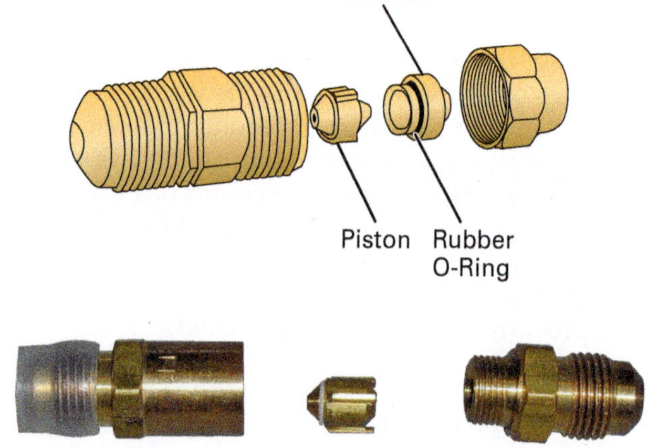

Figure 7 Fixed-orifice metering device assembly.

When the flow is reversed, the piston slides to the opposite end of the housing. Here the orifice is not mated to a matching seat and refrigerant can flow both through, and around, the piston. The turbulence does create a small amount of pressure drop, but not enough to significantly affect the refrigerant.

The easy reversal of refrigerant flow supports heat pump operation perfectly. Reversing the direction of refrigerant flow and the role of the two coils is unique to a heat pump. With a fixed-orifice device located at the condenser and evaporator coils, the two coils can easily swap roles. For this reason, fixed-orifice metering devices are widely used in heat pump systems and made heat pump systems simpler and less expensive to manufacture. However, these devices are also used in cooling-only systems because today's residential evaporator coils are designed for use in both heat pump and cooling-only systems. The manufacturer chooses which orifice size to provide with a coil based on the capacity of the matching outdoor unit.

On Site

Servicing and Cleaning Fixed-Orifice Metering Devices

When servicing fixed-orifice metering devices, be aware that both right-hand and left-hand threads may be encountered when disassembling the housing.

Once removed, fixed-orifice metering devices are easy to clean, but use caution. The piston and housing are usually made from brass. If the orifice is enlarged or scratched, system operation will be affected. After disassembling the housing, use a small bent wire to gently pull the piston out if it does not slide out freely.

Deposits resulting from compressor failures or other contaminants may cause the piston to seize in place. If a piston does stick occasionally or consistently, it will cause immediate and serious performance problems.

The piston can be removed easily from the housing and replaced with a different size. In a matter of minutes, a system can be fitted with the correct orifice for systems ranging from one to five tons in capacity. The installation labor is significantly reduced as a result; no brazing is required. A single housing can accommodate a compatible piston, regardless of the orifice size.

Because of their low cost, fixed-orifice metering devices are also used on many light commercial systems. Rather than use an expansion valve with a distributor, a number of manufacturers have used multiple fixed-orifice devices at the evaporator coil(s) of $7\frac{1}{2}$- to 20-ton packaged units (*Figure 8*). In these applications, the orifice is not free to move back and forth. Thus, this arrangement is not used on heat pumps. This resulted in a significant cost reduction for manufacturers.

As shown in *Figure 8*, the pistons are brazed into a fixed position inside the coil header. Because there are no moving parts, the approach is generally reliable. However, because the orifice size is smaller than the average single orifice, the

Figure 8 Multiple fixed-orifice metering devices in an evaporator coil header.

orifices tend to foul more easily. Further, these orifices are difficult to access for cleaning or inspection, so cutting and brazing is required. To inspect or clean the orifices following a compressor failure, the entire coil header must be removed. This often leaves the copper tubing weak and brittle from excess heat. Removing and replacing the entire coil header as an assembly is often the most practical approach.

Although the simplicity of the device is attractive, the disadvantage associated with all fixed metering devices is that they cannot adjust to accommodate changes in the load. The flow rate through them is completely dependent on the pressure differential across them. For example, as the high-side pressure increases relative to the low-side pressure, the flow rate increases. Fixed metering devices are best suited for smaller systems that operate at a reasonably predictable load.

On Site

Sight Glasses

Sight glasses are often placed in the liquid line allowing technicians to see the condition of the refrigerant as it flows through. Flashing and bubbling can indicate an undercharge condition on some systems. However, that is not always the case; other operating conditions can create the same effect. Sight glasses are of limited value in systems with fixed metering devices because these systems are charged to a superheat value rather than to a subcooling value.

Sight glasses are often installed on split systems at the outdoor condensing unit. Although this may be convenient and not necessarily incorrect, viewing the condition of the refrigerant as it enters the metering device is far more informative to a technician. For that reason, sight glasses are best installed just ahead of the metering device.

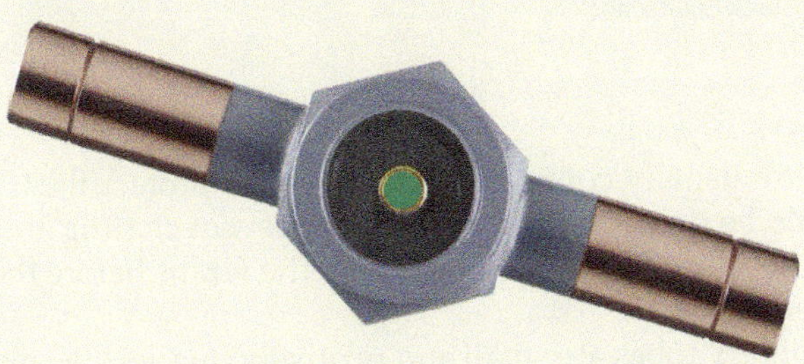

Source: Courtesy of Sporlan Division—Parker Hannifin Corporation

2.1.1 Fixed-Orifice Metering Device Selection

In a split system, the proper fixed orifice size is determined by the capacity of the system and the characteristics of the coil. The condensing unit capacity is typically the primary factor. Condensing units are typically matched with evaporator coils that mirror their capacity precisely. However, a larger evaporator coil is sometimes selected, further increasing system efficiency. A small increase in total capacity often results as well.

It is essential to follow the manufacturer's recommendations for any equipment combination. Do not assume that every combination of 3-ton equipment, for example, uses the same orifice size. If a condensing unit and coil combination has not been tested by the manufacturer and a piston size identified, it is probably not an approved combination.

It is common for the manufacturer to ship a piston with the condensing unit. The evaporator coil is also shipped with a piston installed. In many cases, the piston shipped with the evaporator should be replaced with the piston supplied with the condensing unit. Again, refer to the manufacturer's installation instructions. Ensuring the piston size is correct from the start helps avoid costly rework and refrigerant recovery to access the piston.

2.2.0 Capillary Tubes

A capillary tube (*Figure 9*), often referred to as a *cap tube,* is a slender copper tube with a small, precise inside diameter. They typically range in length from 20" to 140". The combination of diameter and length are selected to provide the needed pressure drop for the application at design conditions.

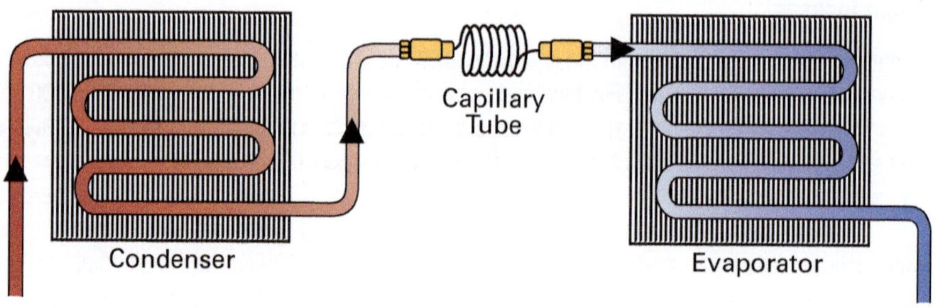

Figure 9 Capillary tubing.
Source: (Top) pisanstock/Shutterstock

Cap tubes are usually coiled to save space and protect them from damage. The coils should be tightly bound together to prevent chafing and leaks. A bead of silicone caulk is sometimes applied across the wraps before the coil is bound to help dampen vibration.

Because capillary tubes are small and relatively fragile, they must be handled with care. Bending the tube at a sharp angle will cause a permanent kink that renders it useless. Dents in the side can also ruin a cap tube, since the inside diameter is very small.

Because of the risk of restricting the ends and the difficulty of effectively reaming the tiny opening, cap tubes are not cut with regular tubing cutters or saws. One approach is to deeply score the tube all around the outside with a tri-corner file. Flexing the tube a few times at the weakened spot will result in it separating without damaging the end. A cutting tool specifically designed for cutting capillary tubes without damage is also available if you work with the material regularly (*Figure 10*). Always examine the tube ends for burrs or debris before installing a cap tube.

Figure 10 Capillary tube cutter.
Source: Ritchie Engineering Co., Inc./YELLOW JACKET

The cap tube is usually selected and installed at the factory because its length and diameter are critical. If the tube is too long or the diameter is too small, the evaporator will be starved for refrigerant. The result will be high superheat, high subcooling, and inadequate cooling. The same result can be caused by a restricted capillary tube. Because the tube is easily fouled, a fine-mesh liquid strainer or filter-drier is often provided at the cap tube inlet (*Figure 11*). A filter-drier can be used to protect any metering device, but strainers are often used in small systems. The style shown is called a *spun-copper strainer.*

Figure 11 Spun-copper refrigerant strainer.
Source: Courtesy of Sporlan Division—Parker Hannifin Corporation

If the capillary tube is too short or the inside diameter is too large, excess liquid will be fed to the evaporator. This can cause refrigerant floodback.

When replacing a capillary tube, it is crucial that the new tube be the same precise length and diameter as the original tube. Even the smallest deviation will result in different system operating characteristics. Capillary tubes for a specific piece of equipment are best ordered from the manufacturer.

2.3.0 Common Fixed Metering Device Problems

Fixed metering devices have been used extensively in the HVACR industry for many years. Numerous advantages are clear, however, these devices do have one very significant disadvantage. Because the orifice size is fixed, they cannot respond in any way to accommodate changes in the load and regulate refrigerant flow.

The flow rate of any fixed metering device depends solely on the pressure difference across it. When the system is not operating, the pressure is the same on each side of the orifice. The pressures on the high and low side of the system are balanced. Once the compressor starts, the pressure begins to fall on the low side of the system and rise on the high side. Refrigerant flow through the metering device begins as soon as there is a difference in pressure. The flow rate through it continues to increase as the pressure difference increases. Eventually, the system pressures and the flow rate reach equilibrium.

Assuming that the refrigerant charge is correct, and the heat load is within the design limitations of the system, some level of superheat should be present. For the purposes of this discussion, assume that the superheat on a capillary-tube equipped system is 20°F (8.9°C). This approximates the superheat for a residential comfort-cooling system using a fixed metering device when the return air temperature is around 78°F (26°C).

However, the superheat produced by a fixed metering device consistently changes as the load and pressure difference changes. As the home cools and the air temperature falls, the heat load is reduced. As a result, the superheat will begin to gradually drop as well. With a reduced heat load, the boiling refrigerant travels farther through the coil before becoming fully vaporized. By the time the temperature in the home has dropped to around 73°F (23°C), the superheat may have fallen to 14°F (7.8°C). This is quite normal for a system using a fixed metering device. The concern here is that, if the heat load becomes very low, as would be the case if the thermostat is set to a very low temperature, eventually there is no superheat and floodback occurs.

On Site

Refrigerant Charging

Another issue related to the use of fixed metering devices is the refrigerant charge. The volume of refrigerant charged into these systems is far more important than in systems using expansion valves. In fact, the systems in which they are installed are sometimes referred to as *critically charged systems.*

To ensure proper operation, the volume of refrigerant in the system needs to be just right. There is significantly more room for error without causing problems when expansion valves are installed. Although this concept is discussed in depth in other modules, always remember that systems with fixed metering devices are charged using superheat values. Systems with expansion valves are primarily charged to subcooling values.

Conversely, if the heat load begins to increase while the system is running, the superheat will begin to rise. The metering device will continue at roughly the same flow rate, and the refrigerant will complete the change of state in the coil at an earlier point. As the heat load increases, the suction pressure may rise. This reduces the pressure difference between the high and low side of the system, and the flow rate falls a little more.

Consider this scenario—one you undoubtedly encounter. A home system was left off on a very hot day, or perhaps there was an extended power outage. The temperature in the home may have risen to 95°F (35°C) or more. When the system is started, an enormous heat load is present. Once the system reaches equilibrium, the superheat may be as high as 40°F (22.2°C) or even higher. The fixed metering device has no way to increase the flow of refrigerant to the coil to accommodate the load and reduce the superheat. A great deal of the evaporator coil is filled with vapor only and is inactive. Thus, the cooling process is impeded and the home cools very slowly. At a time when maximum system capacity is needed, the capacity is minimized.

This demonstrates the important lesson that fixed metering devices do not provide optimum cooling. Although most systems operate without serious problems, this characteristic can result in poor performance under certain conditions and decrease system efficiency. You will notice that systems with the highest efficiency ratings do not rely on fixed metering devices.

As government mandates continue to increase the required equipment efficiency ratings, fixed metering devices become less popular. Generally, only new systems rated at lower, yet legally acceptable, efficiency levels are equipped with fixed metering devices today. More efficient systems, which are also more expensive, are equipped with an expansion valve. Of course, many fixed restrictor systems made in prior years remain in service today.

Causes of poor system performance that are related to fixed metering devices include the following:

- Moisture in the system freezing at the metering device.
- Dirt or other contaminants obstructing the flow of refrigerant through the metering device. Note that a strainer or filter-drier can also become clogged if it traps an unusual amount of debris.
- Wax from lubricating oils collecting and solidifying at the metering device outlet (rare, and usually only occurs in low-temperature applications such as freezers).
- Undersizing or oversizing the metering device.

Problems related to freezing moisture and wax can often be hard to diagnose. These conditions often develop gradually through an operating cycle. Before the metering device can be removed for inspection, the problem is gone. The gradual onset of the problem during a cooling cycle is often the best clue. The superheat would slowly but gradually rise while the suction pressure gradually drops.

Of course, this problem is far more likely to occur on systems designed to refrigerate below freezing. Comfort-cooling systems rarely suffer from moisture freezing at the metering device. Wax is also not a typical problem for comfort-cooling applications. In low-temperature systems however, wax from lubricants may solidify and collect at the metering device.

Remember that many problems can be eliminated if the system is kept clean, dry, and in a proper state of charge, regardless of what type of metering device is used.

2.0.0 Section Review

1. The reversal of refrigerant flow through a refrigerant circuit is a feature unique to _____.
   a. heat pumps
   b. refrigeration systems
   c. water-cooled systems
   d. direct expansion (DX) systems

2. Capillary tubes can be cut without damage using a _____.
   a. round file
   b. tri-corner file
   c. hack saw
   d. four-wheeled tubing cutter

3. True or False: When using a fixed metering device, if the heat load increases during an operating cycle, the superheat increases.
   a. True
   b. False

3.0.0 Expansion Valves

Objective

Identify and describe the operation of expansion valves.

a. Identify and describe the operation of manual expansion valves.
b. Identify and describe the operation of automatic expansion valves.
c. Identify and describe the operation of thermal expansion valves.
d. Identify and describe the operation of electric and electronic expansion valves.
e. Describe common problems associated with expansion valves.

Performance Tasks

2. Install an externally equalized expansion valve, correctly placing the sensing bulb and equalizer tube.
3. Measure superheat in an operating system with an expansion valve and adjust it to change the superheat.

The heat load on an evaporator varies according to changes in the temperature or volume of the cooled medium (usually air or water). Unlike fixed metering devices, most expansion valves have some ability to respond to these changes. An ideal metering device reacts quickly and regulates the flow of refrigerant into the evaporator to precisely match the conditions.

Several types of expansion valves exist. The types to be explored here include the following:

- Manual expansion valves
- Automatic expansion valves
- Thermal expansion valves
- Electronic expansion valves

3.1.0 Manual Expansion Valves

A manual expansion valve (*Figure 12*) is a hand-operated valve with a tapered or pointed valve stem. The valve tip fits the seat precisely. Like a fixed metering device, these very simple valves have no ability to adjust themselves. However, they can be manually adjusted, and, therefore, are not considered fixed metering devices. From the outside, the valves do not appear to be significantly different from other common hand valves and are usually distinguished by a bar-shaped and/or yellow handle.

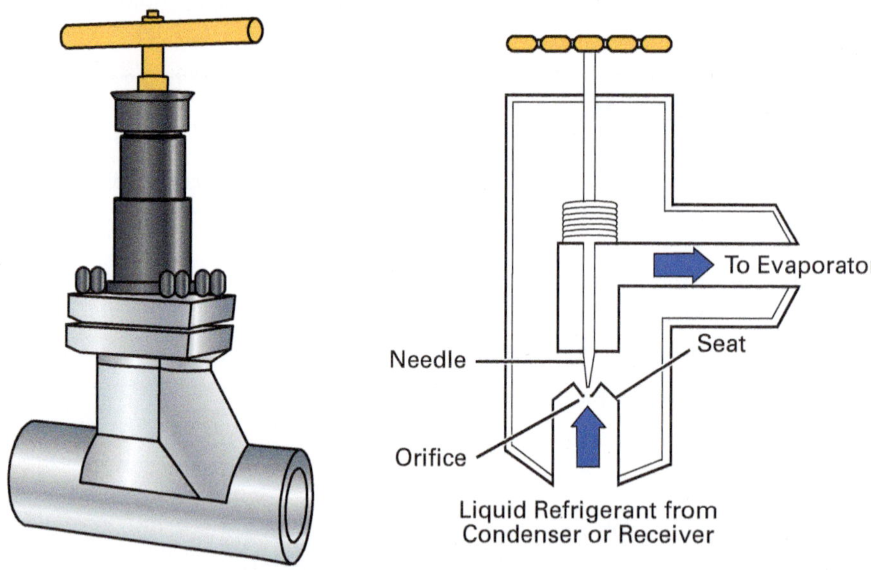

Figure 12 Manual expansion valve.

Manual expansion valves are adjusted by hand to match the present load and are only suitable in systems with a reasonably constant load. Even then, the system requires constant monitoring to adjust for changing conditions, to ensure the appropriate amount of refrigerant is flowing to the evaporator.

This type of expansion valve is not used for comfort cooling or even in common refrigeration applications today. However, manual expansion valves can still be found on large industrial ammonia-based refrigeration systems or subsystems.

3.2.0 Automatic Expansion Valves

An automatic expansion valve (AEV), also known as a *constant pressure valve*, is designed to maintain a constant evaporator pressure (*Figure 13*). It cannot react to temperature directly. However, by maintaining a consistent evaporator pressure with the refrigerant in a saturated condition, it can maintain a reasonably stable evaporator temperature. An AEV relies on a spring-loaded diaphragm to position the valve, thus regulating the refrigerant flow.

The spring pushing down on the diaphragm acts as an opening force. In other words, the spring works to move the valve off the seat and allow refrigerant to flow. Pressure at the outlet of the valve is the same as the pressure of the evaporator coil. This pressure, working with a spring at the bottom of the valve, represents a closing force.

The pressure exerted by the upper spring can be adjusted. For this discussion, assume that the pressure setting is 60 psig (414 kPa).

In operation, the opening and closing forces applied by the spring and the evaporator pressure become balanced. As pressure is increased on the spring, the

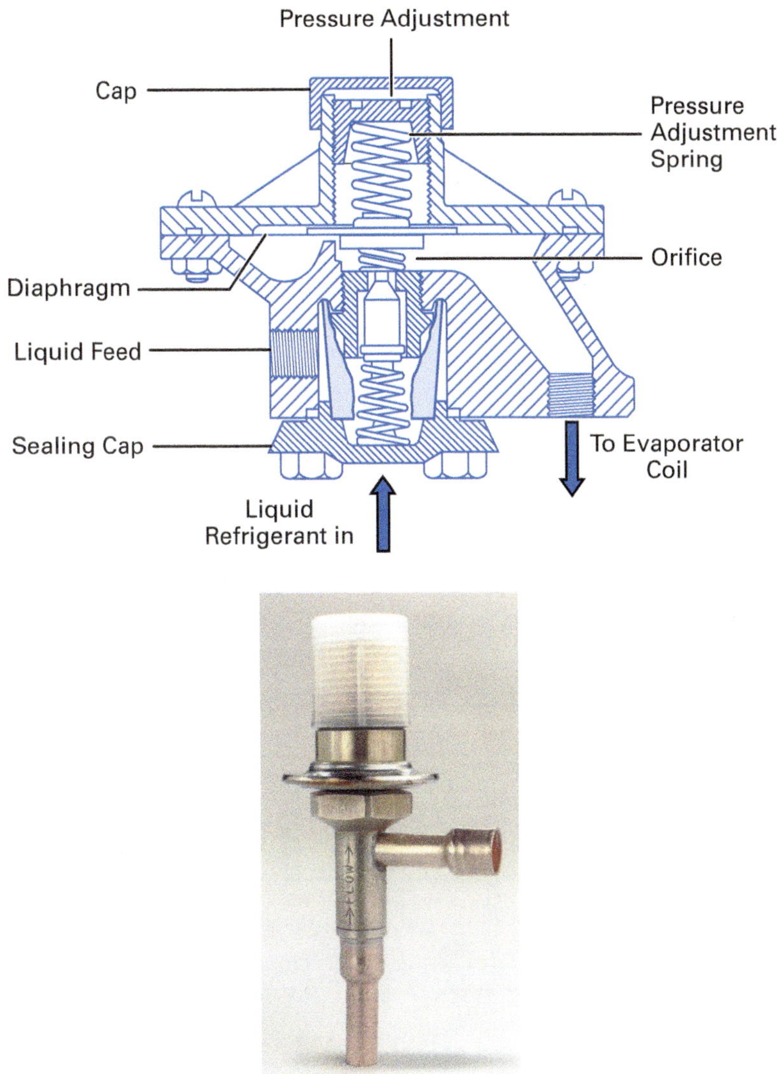

Figure 13 Automatic, or constant pressure, expansion valve.
Source: Courtesy of Sporlan Division—Parker Hannifin Corporation

valve opens slightly until the pressure in the evaporator rises an equal amount. The two forces are then in equilibrium. A reduction in spring pressure causes the valve to close slightly until the evaporator pressure falls proportionately.

Once the valve is set, the valve opens and closes in response to pressure changes in the evaporator. As the evaporator pressure changes in response to a changing load, the diaphragm works to open and close the valve accordingly.

The automatic expansion valve, however, responds to load changes in opposition to what logic might dictate. When there is an increase in the heat load on the evaporator coil, the suction pressure rises. This causes the automatic expansion valve to reduce, or throttle, the amount of refrigerant being fed to the evaporator. A larger increase in the heat load will cause an even larger reduction in refrigerant flow.

Similarly, when there is a decrease in the heat load being handled by the evaporator, the suction pressure goes down. This causes the valve to open and feed more refrigerant to the coil. If the heat load continues to drop, the coil may become flooded with liquid refrigerant resulting in floodback to the compressor. The valve has no way of sensing or preventing inadequate or excessive refrigerant flow, since it responds only to pressure changes.

Despite their unique operation, automatic expansion valves are well-suited for small capacity equipment with a consistent load, such as refrigerators, freezers, and some window air conditioners. AEVs are simple to install because the only connections to make are the two refrigerant lines. They can be installed in any position because their operation is not affected by gravity.

On Site

High- and Low-Side Float Valves

High- and low-side float valves are also metering devices. They are used in large systems, such as centrifugal chilled water and industrial refrigeration applications.

The high-side float valve reacts to the level of liquid from the condenser, opening and closing the orifice in response to changes in the liquid level. The amount of liquid leaving the condenser is driven by the amount of vapor produced in the evaporator. This, in turn, is a function of the cooling load on the evaporator. As the cooling load increases, more liquid leaves the condenser. This extra liquid raises the float, lifting the orifice to send more refrigerant to the evaporator.

The low-side float valve is the best control device for use with flooded evaporators. It may be installed directly in the evaporator or in a separate float chamber outside the evaporator. Note that it is installed on the low side of the metering orifice, while the high-side float valve is installed on the high side. As the load on the evaporator increases, more liquid boils off, reducing the amount of liquid in the float chamber. The reduced liquid level lowers the float, opens the orifice, and supplies more refrigerant to the evaporator.

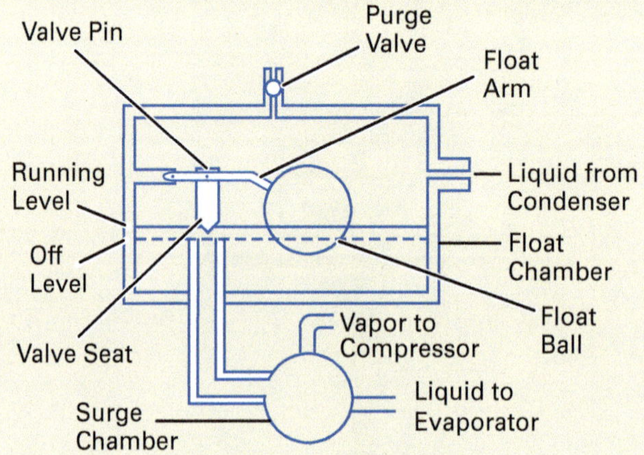

High-Side Float Valve

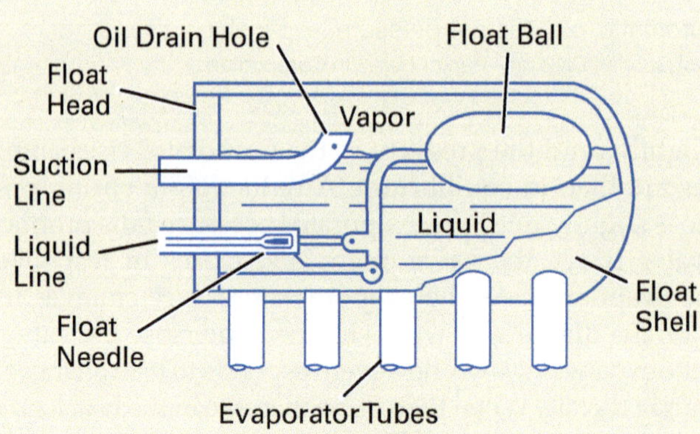

Low-Side Float Valve

3.3.0 Thermal Expansion Valves

Thermostatic element: An assembly filled with refrigerant in saturation with an attached sensing bulb that responds to the temperature of the suction line. The pressure from the element acts as an opening force in a thermostatic expansion valve. It is also referred to as a *power element*.

The thermal expansion valve (TXV or TEV) is a metering device with an external sensing bulb mounted to the suction line at the evaporator outlet. The bulb is part of the **thermostatic element** and is connected to it by a slender copper tube (*Figure 14*). The thermostatic element provides a pressure input that varies with the temperature of the suction line. The thermostatic element is also referred to as a *power element* or *power head*.

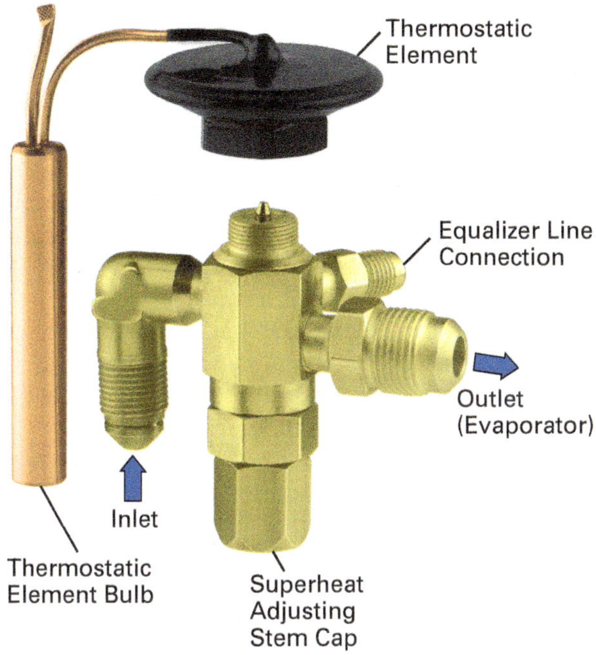

Figure 14 Thermal expansion valve.
Source: Courtesy of Sporlan Division—Parker Hannifin Corporation

A TXV works to maintain a consistent superheat value at the evaporator outlet. The TXV should not be confused with the thermal-electric expansion valve (TEEV/THEV) or the electronic expansion valve (EEV), both of which use one or more electrical inputs to operate. TXVs are not powered.

3.3.1 Operating Principles

TXV control is primarily based on the amount of superheat in the refrigerant as it leaves the evaporator. Changes in the temperature of the refrigerant leaving the evaporator cause the valve orifice to open or close as needed.

Refer to *Figure 15*. The refrigerant liquid/vapor mixture leaves the valve at Point A and continues to boil as it absorbs heat in the coil. By the time it reaches Point B, all the liquid has vaporized, and the change of state is complete. Now that only vapor is present, the temperature begins to increase. In other words, superheating of the vapor has begun.

It must be noted that Point B in *Figure 15* is simply a graphic representation for the sake of discussion. It is understood that the change of state needs to occur before the refrigerant reaches the end of the coil. The point at which the

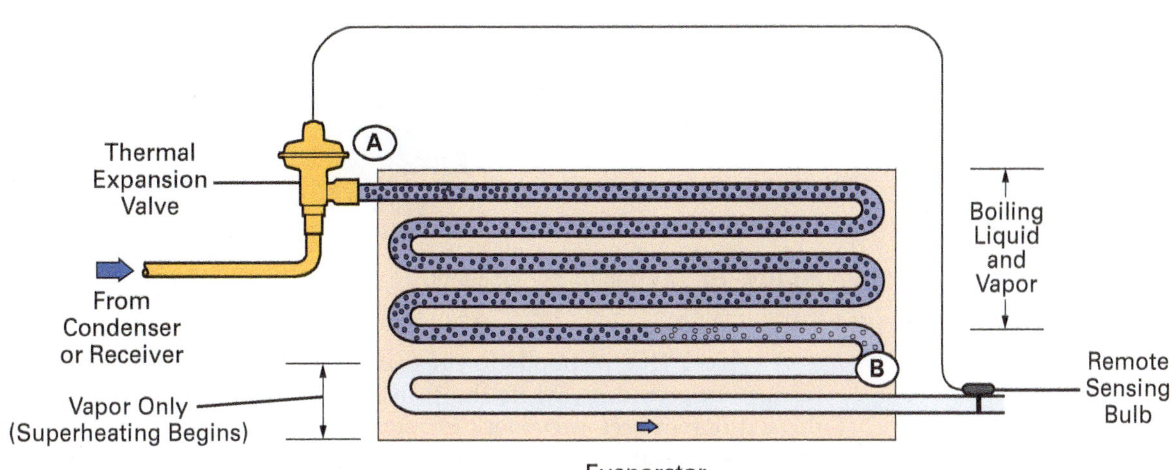

Figure 15 Refrigerant flow in the evaporator.

change of state is complete will move back and forth in the evaporator as the heat load changes and the valve modulates the flow of refrigerant. In common comfort-cooling systems, Point B ideally occurs somewhere in the last 20% of the evaporator coil. A low superheat setting moves the point closer to the end of the coil, while a high superheat setting moves the point toward the coil inlet.

The expansion valve modulates the refrigerant flow as the heat load changes. The goal is to keep the superheat value consistent. The specific superheat target differs from system to system and depends significantly on the metering device used. More reliable and precise refrigerant control allows lower superheat settings, more heat transfer, and improved efficiency.

The superheat setting needs to be relatively low to maximize heat absorption in the evaporator. At the same time, the setting must be high enough to ensure liquid refrigerant does not reach the compressor as the heat load changes. Small variations in the superheat value are expected when using a TXV, so it is not wise to push the issue in either direction. However, TXVs are very reliable and capable of maintaining the superheat within 2°F degrees of their setpoint under most conditions.

The parts involved in the movement of the TXV are shown in *Figure 16*. The diaphragm on the top of the valve is part of the thermostatic element. As the diaphragm moves downward in response to a pressure increase from the sensing bulb, it pushes against the operating pins. The operating pins in turn push down on the pin carrier assembly. The tapered valve tip is mounted to the pin carrier, and points upward towards its seat in the orifice. When the diaphragm moves down, it pushes the valve tip away from the seat, opening the valve. Thus, an increase in pressure at the sensing bulb, indicating an increase in superheat, results in more refrigerant entering the coil.

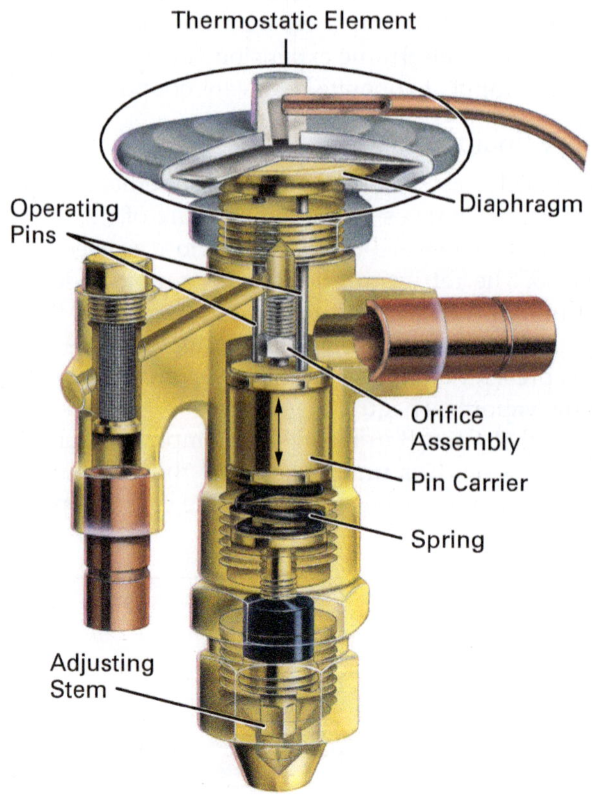

Figure 16 TXV control components.
Source: Courtesy of Sporlan Division—Parker Hannifin Corporation

A typical TXV responds to pressure from three different sources. Two are refrigerant pressures, while the third is a spring. *Figure 17* shows how these three different pressures work to control refrigerant flow and maintain a consistent superheat.

The first pressure, shown at Point 1 of *Figure 17*, is the pressure from the thermostatic element sensing bulb. The bulb is tightly clamped to the outside of the suction line. Inside the bulb, there is a refrigerant in saturation—both liquid and vapor are present. In some cases, a mixture of refrigerants is used to charge the bulb. The charge of saturated refrigerant in the bulb has a strict pressure-temperature relationship, as all compound refrigerants do. As the temperature at the bulb rises, the pressure inside the bulb and thermostatic element rises in response. The increased pressure pushes down against the diaphragm, providing an opening force.

It is important to note that the refrigerant in the bulb and thermostatic element is segregated from the refrigerant in the system itself. No direct connection exists between the two, and it is unlikely that the two refrigerants are even the same.

The second pressure, shown at Point 2 in *Figure 17*, represents the pressure of the evaporator. This pressure comes from the outlet of the valve (the inlet to the evaporator). This pressure is pushing up on the diaphragm, opposing the force from the thermostatic element and providing a closing force.

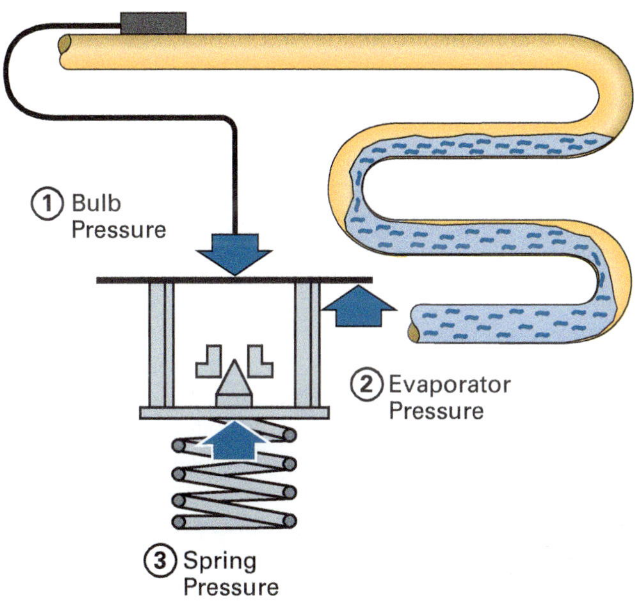

Figure 17 Pressures that influence TXV position.
Source: Courtesy of Sporlan Division—Parker Hannifin Corporation

The third pressure, shown at Point 3, is the mechanical pressure applied by the spring. It is also a closing force, working along with the evaporator pressure to oppose the force from the thermostatic element. The spring and evaporator pressure work against the single opening force of the thermostatic element.

If the heat load increases, the refrigerant in the evaporator will boil faster. As noted in *Figure 15*, Point B will begin to move toward the inlet of the coil. The suction line temperature and superheat value increase, warming the thermostatic element bulb and increasing the pressure inside. The additional pressure pushes the diaphragm down and the valve opens in response. Moments later, all three forces are again in balance. Point B moves back to where it was, or very nearly so, unless the change in heat load was dramatic.

If the heat load decreases, which happens as a room becomes cooler, the opposite sequence takes place. The suction line becomes cooler, the superheat value drops and the pressure inside the thermostatic element bulb drops in response. The spring and evaporator pressure applied to the bottom of the diaphragm are then able to move the valve slightly toward the closed position against the reduced pressure in the thermostatic element.

3.3.2 TXV Superheat Settings

Many of the expansion valves used in residential or light commercial HVACR equipment are not adjustable. The manufacturer of the equipment selects a TXV with the correct characteristics and installs it at the factory or provides it with the equipment order for field installation. The manufacturer may also offer a TXV as a desirable alternative to a fixed metering device, at an additional cost. In either case, the TXV manufacturer presets the desired superheat value for the application. The setting is usually 10°F to 12°F (5.5°C to 6.7°C) of superheat.

Of course, many other TXVs are adjustable. The adjustment is made using the adjustment stem shown in *Figure 16*. As a rule, turning the stem clockwise, or in towards the valve, increases the spring pressure and results in a higher superheat (increases the closing force). Turning it counterclockwise decreases spring pressure, resulting in a lower superheat (decreasing the closing force). Always consult the manufacturer's instructions for a particular product to ensure you are clear on its operation and adjustments.

Table 1 provides some suggested superheat settings for common applications. Note that the settings become lower as the design evaporator temperature drops. In refrigeration applications, the temperature difference (ΔT or TD) between the coil and the cooled medium is generally lower.

TABLE 1 Suggested Superheat Settings for Different Applications

Application [Saturated Suction Temperature]	Comfort Cooling [40°F to 50°F (4.5°C to 10°C)]	Commercial Refrigeration [0°F to 40°F (–18°C to 4.5°C)]	Low Temperature Refrigeration [–40°F to 0°F (–40°C to –18°C)]
Typical Superheat Values	10°F to 14°F (4.4°C to 6.6°C)	6°F to 10°F (3.3°C to 4.4°C)	4°F to 8°F (2.2°C to 3.3°C)

For example, in a comfort cooling system, a 40°F (4°C) evaporator temperature is a common target. The air temperature in a home is often around 75°F (24°C), which creates a 35°F (19°C) temperature difference between the coil and the air by design. However, the coil to air ΔT in a low-temperature application may be only 10°F (5.5°C). This slows the heat transfer process.

To compensate, the design superheat value is lowered in refrigeration applications to maximize the evaporator capacity. However, these systems often use accumulators and other accessories to provide physical protection from floodback.

Note that the values shown in *Table 1* are typical for many applications. However, there are exceptions. Also note that these values apply to TXVs. When more sophisticated metering devices are used that can more reliably maintain a low superheat setting, manufacturers often design the equipment to operate at a lower superheat. Today's chillers equipped with electronic expansion valves, for example, often operate at a superheat as low as 4°F (2.2°C).

3.3.3 TXV Valve Equalization

Along with spring pressure, the evaporator inlet pressure acts on the bottom of the diaphragm as a closing force. The sensing bulb pressure acts on the top of the diaphragm. These three forces interact to position the TXV.

As is true for any section of piping, some pressure drop occurs as refrigerant flows through a distributor and/or evaporator. This fact is often discounted in many discussions. However, if there is a significant refrigerant pressure drop through the evaporator, the pressure on the bottom of the valve diaphragm (a closing force) will be higher than the pressure exerted on the top (an opening force). This can result in the valve maintaining a higher superheat than desired. In some cases, it is better to acknowledge and compensate for the pressure drop through the evaporator.

TXVs that use a pressure input from the valve outlet (the evaporator inlet) are referred to as **internally equalized** valves. See *Figure 18A*. Internally equalized valves are generally used in small systems where the pressure drop through the evaporator coil is not excessive, such as a kitchen reach-in refrigeration unit. The allowable pressure drop through the coil will vary with the refrigerant and the evaporating temperature, ranging from 0.75 psig to 3 psig (5 kPa to 27 kPa). Remember that, on an internally equalized TXV, evaporator inlet pressure is fed to the bottom of the diaphragm through an internal passage.

Internally equalized: Describes an expansion valve that uses the pressure value at its own outlet to serve as a closing force. This effectively disregards the pressure drop through the evaporator.

If the pressure drop through an evaporator is greater than 3 psig, the TXV should be **externally equalized** (*Figure 18B*). Most evaporators of substantial size are externally equalized. Evaporators with a refrigerant distributor should always be fitted with an externally equalized valve due to the extra pressure drop they create.

Externally equalized: Describes an expansion valve that has a pressure connection to the suction line downstream of the evaporator coil to serve as a closing force. External equalizer lines are used when there is a significant pressure drop through the evaporator and/or distributor.

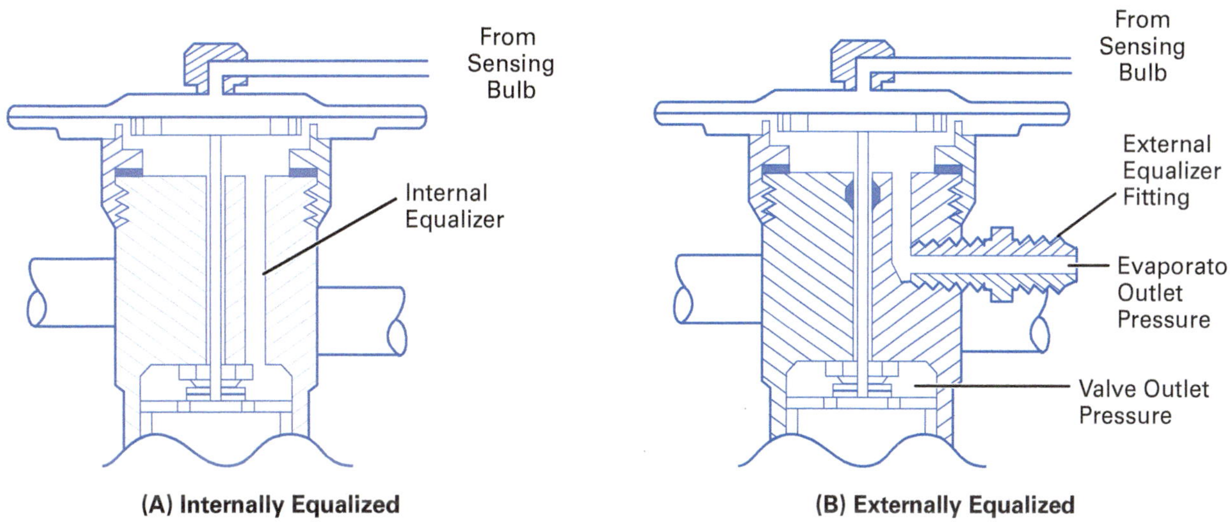

Figure 18 Internally and externally equalized TXVs.

In an externally equalized valve, the TXV diaphragm is isolated from the evaporator inlet pressure. The evaporator outlet pressure is routed to the valve from a location near the outlet of the coil. The pressure is then fed back through an equalizer line to the bottom of the diaphragm. This allows the TXV to better compensate for the pressure drop and operate more accurately.

Figure 19 shows the routing of an external equalizer line. The line is constructed from ¼" OD copper tubing or smaller. The line can be small since there is no refrigerant flow through it; it merely provides a pressure input to the valve. It is connected to the valve and/or the suction line by a flare fitting, solder joint, or brazed joint. The line is generally connected just past the sensing bulb mounting location. The equalizer line should never be capped, plugged, or eliminated, as the valve cannot operate accurately without it.

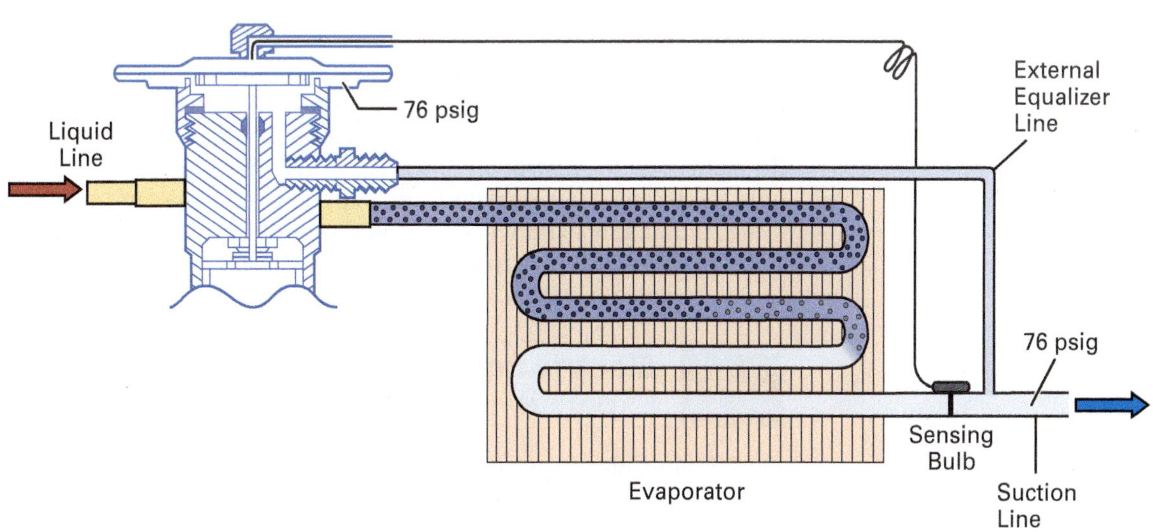

Figure 19 Externally equalized TXV installation.

3.3.4 Special TXV Features

The following sections introduce a few special features associated with TXVs that are important to understand.

Bi-Flow TXVs

The first heat pumps introduced used TXVs at the indoor and outdoor coils. To allow for flow reversal, an external check valve was placed in parallel with each valve. When the direction of refrigerant flow changed, refrigerant could move freely around one TXV while being metered through the other TXV.

Bi-flow TXVs (*Figure 20*) have been developed that incorporate a check valve into the TXV valve body. This feature eliminates the external check valve and the associated piping. Visually, bi-flow valves resemble a conventional TXV.

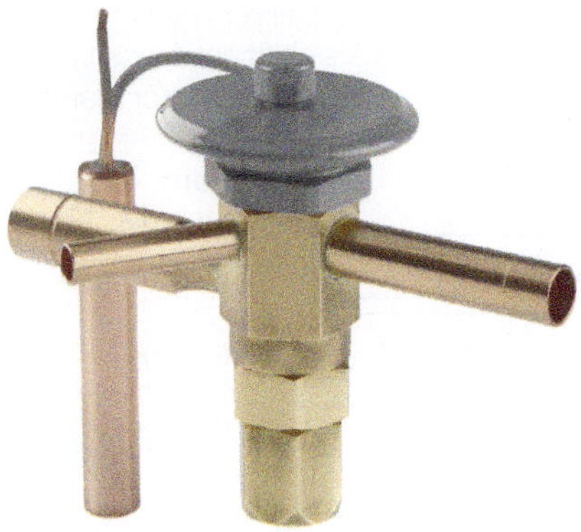

Figure 20 Bi-flow TXV.
Source: Courtesy of Sporlan Division—Parker Hannifin Corporation

Rapid Pressure Balancing

Most TXVs have a small bleed port inside that allows the low- and high-side pressures to equalize when the system shuts down. If the pressures do not equalize, the compressor may attempt to start against a large pressure differential. This can place a tremendous amount of stress on the compressor motor, potentially shortening its life. Reciprocating compressors with low starting-torque motors may be unable to start. The normal bleed port in TXVs may not equalize pressures quickly enough for these compressors.

To resolve this problem, TXVs are available with a feature that allows the system pressures to equalize quickly. Note that this feature is not recommended for most systems. This type of valve should not be used randomly. Always check with the manufacturer before considering a TXV with this feature.

MOP Valves

Maximum operating pressure (MOP) expansion valves operate in much the same way as a conventional TXV, with one added function. The MOP feature allows the valve to limit the evaporator pressure downstream. Once the pressure limit is reached, the valve begins to throttle the flow of refrigerant, preventing the pressure from rising any higher. The primary reason for this is to limit the suction pressure returning to the compressor and prevent it from overloading. This feature is used on refrigeration systems almost exclusively—mainly those that have a defrost cycle.

Following a defrost cycle, the evaporator and the refrigerated area are much warmer than usual. Since the compressor was selected for low-temperature use, the operating current can be excessive immediately following a defrost cycle. The MOP feature helps control the flow of refrigerant back to the compressor until the heat load has returned to normal. The MOP feature is also valuable when a refrigerated enclosure is initially started up, because the temperature inside is abnormally high.

Balanced Port Valves

TXVs are generally constructed such that the direction of refrigerant flow through the orifice is toward the movable part of the valve. In other words, the refrigerant flow is in a direction that tends to act as an opening force. This has little effect on the operation of a system that operates under stable and predictable conditions. However, the added opening force may influence TXVs used on systems that may experience wide variations in loads and pressures. As a result, the superheat may vary from the desired setting. The balanced port design helps the valve avoid being influenced by the liquid flow as an opening force.

3.4.0 Electric and Electronically Controlled Expansion Valves

Electric and electronically controlled expansion valves (EEVs) have become increasingly popular. Although TXVs have been used successfully for many years with excellent performance and reliability, EEVs offer unparalleled flexibility and accuracy. Several different types of electric and electronic valves will be introduced here.

3.4.1 Analog and Pulsing Valves

Some of the early electric metering devices were based on the same design as a common refrigerant solenoid valve. In a solenoid valve, an energized electric coil creates a magnetic field to attract and lift the stem of the valve plunger. When the coil is de-energized, the valve plunger returns to its previous position using spring pressure. Solenoid valves are very common accessories used to start and stop the flow of refrigerant in refrigeration systems, but they are rarely used in residential and commercial comfort cooling systems.

This basic concept can also be used to modulate the flow of refrigerant. By opening and closing the valve rapidly at a controlled frequency, refrigerant flow control can be achieved. The valve is still either open or closed, with no position in between, much like a solenoid valve. But by changing the frequency at which the valve is cycled, the flow rate can be modulated. These valves are still in use, although typically with a very narrow range of control.

The valves are controlled by an external electronic device through a process called *pulse-width modulation* (PWM). A signal is repeatedly sent to the valve (pulsation), with each signal instructing it to open. When the signal stops, the valve closes. The controller, based on temperature and/or pressure inputs from sensors on the suction line, determines the timing of the pulses. *Figure 21* shows a solenoid valve controlled by PWM.

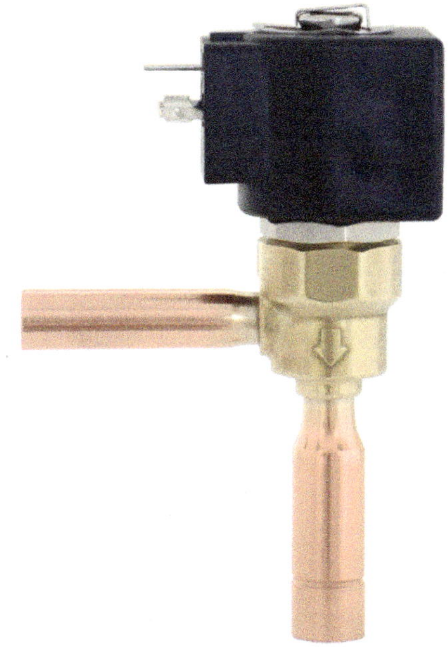

Figure 21 PWM-controlled solenoid valve.

Source: Courtesy of Sporlan Division—Parker Hannifin Corporation

Analog valves are also based on the construction of solenoid valves. Instead of rapidly cycling from open to closed, analog valves use a variable-strength magnetic field to lift and hold the valve plunger, modulating the size of the opening. The valve plunger moves up and down in response to the intensity of the magnetic field. Unfortunately, the valve plunger does not repeat well. This means that, when provided with an identical signal repeatedly, the valve plunger does not achieve the same position each time.

3.4.2 Thermal-Electric Expansion Valves

Thermal-electric expansion valves (TEEVs), represent an entirely different approach to expansion valve control (*Figure 22*). They are also referred to as *heat motor valves*, although no motor is involved.

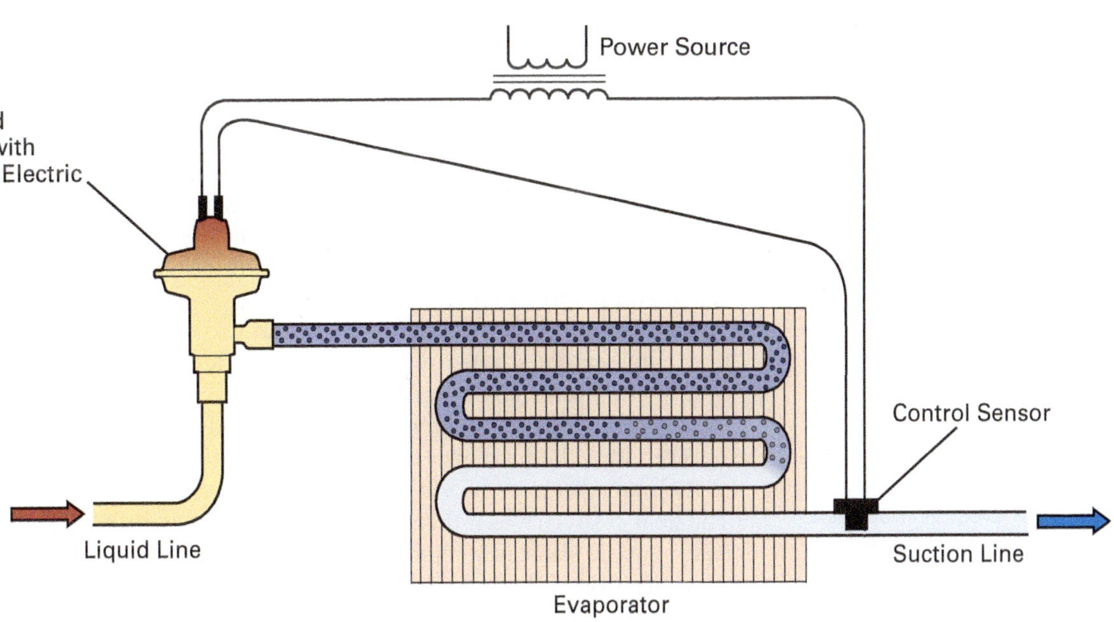

Figure 22 Thermal-electric, or heat motor, expansion valve.

The operator of the valve is filled with a fluid. A tiny electric heater is immersed in the fluid. The amount of power sent to the electric heater is modulated based on the temperature at the suction line sensor location. As the fluid expands and contracts in response to temperature change from the heater, it changes the position of the valve plunger via a diaphragm.

3.4.3 Electronically Controlled Expansion Valves

Stepper motor: A type of motor used to open and close an electronically controlled expansion valve. The motor rotates in small increments called *steps*, based on signals received from an external controller.

An EEV (*Figure 23*) is a microprocessor-controlled device that is driven by a miniature DC **stepper motor**. The microprocessor control is not part of the valve itself. The stepper motor does not spin continuously, as do most familiar motors. Instead, the motor turns only a tiny fraction of a revolution when it receives a single signal from the controller. Each of these small movements is referred to as a *step*.

Figure 23 Electronically controlled expansion valves (EEVs).
Source: Courtesy of Sporlan Division—Parker Hannifin Corporation

Each time the stepper motor turns, the valve plunger moves in or out of the orifice to control refrigerant flow. While earlier versions of the stepper motor had less than one hundred individual steps, valves are now available with thousands of steps. The distance the valve plunger travels per step can be as little as 0.00008" (0.002 mm). The controller can typically issue hundreds of step instructions per second to the motor. Thus, EEVs offer the advantage of incredible precision and response time.

The EEV plunger must move back and forth to open or close the refrigerant port. Since motor shafts rotate on an axis, a device is needed to change the rotation of the motor shaft into linear, or push/pull motion. In addition, the tiny motor is not designed to generate a great deal of force.

Digital linear actuator (DLA): A device that changes the rotational force of an EEV stepper motor to a push/pull (linear) motion needed to open and close the valve. The geared portion of the device also serves to increase the torque of the tiny motor.

A device called a **digital linear actuator (DLA)** is used to change the direction of the motor's force, and to increase that force through a gear train (*Figure 24*). The gear train drives a threaded shaft. The valve plunger is threaded onto this shaft. As the shaft is rotated by the DLA, the plunger is in turn moved up and down as it rides the threaded shaft.

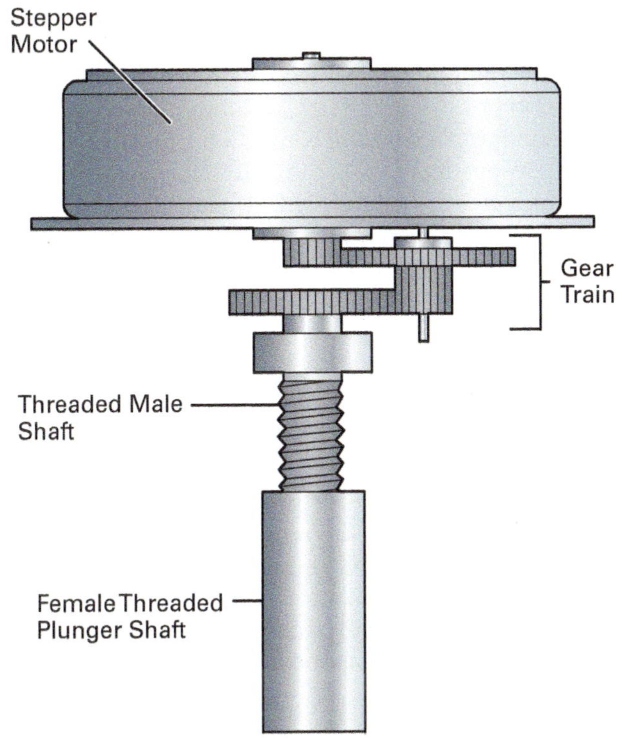

Figure 24 EEV digital linear actuator (DLA) assembly.

On Site

Low-Ambient Head Pressure Control

When refrigeration units must operate at low outdoor air temperatures, the pressure differential across the TXV is reduced. In cold weather, the system head pressure can be very low. The lack of pressure can result in insufficient refrigerant flow through the metering device.

The head pressure can be maintained by cycling and controlling the condenser fan motor(s), but only to a point. In extreme cold, even a condenser with no fans running at all may be too effective. Cycling fans on and off also results in a constantly changing head pressure.

A more elaborate but better method of head pressure control is to use the refrigerant circuit itself. The device used to accomplish this is called a *head pressure control valve*. When the high-side pressure falls to the valve's factory setting, the valve goes to work. The valve restricts the liquid leaving the condenser, mixing hot gas with the liquid on its way to the receiver. This causes liquid to back up into the condenser, reducing its capacity by reducing its effective surface area. This in turn increases pressure in the high side of the system.

For the system to work properly, the refrigerant charge in the system must be sufficient to partially fill the condenser and still leave some in the receiver.

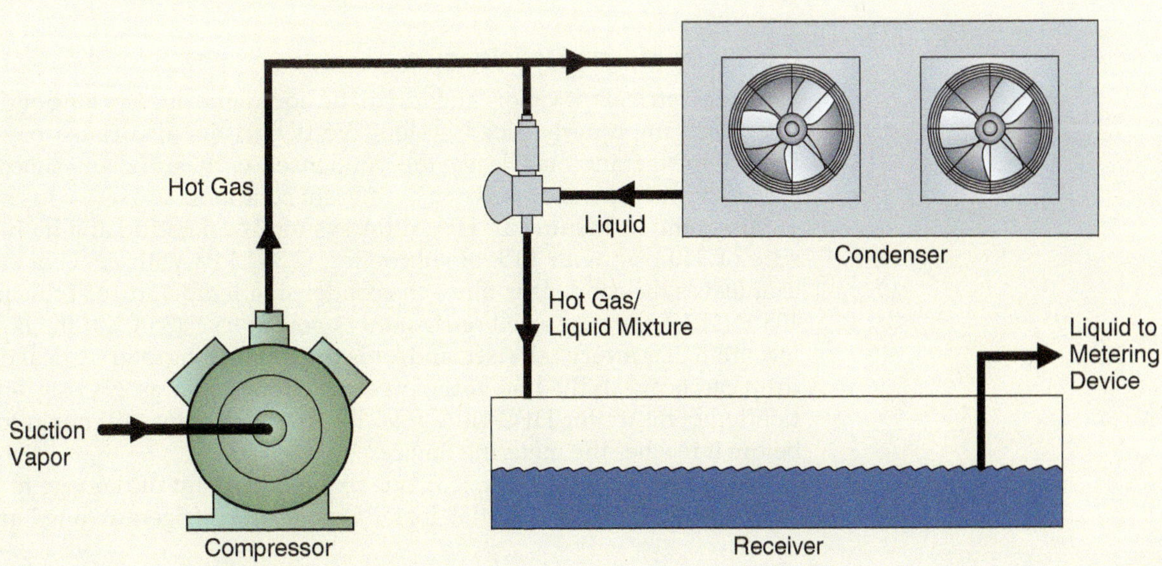

The DLA maintains its position when the motor is idle. When a system shuts down, the stepper motor is usually powered to the closed position by its controller. This is determined by the controller programming which varies from one application to another.

When power is first applied to the controller, the valve is often sent several closing commands to ensure it is at the fully closed position. Once the controller is certain the valve is fully closed, it uses it as a reference point from which to move. Although the stepper motor receives power only when it is commanded to move, power to the valve controller is constant, even during the off cycle.

3.5.0 Common Expansion Valve Problems

Metering devices are often blamed for problems that can have many sources. Many functional expansion valves are replaced simply because technicians do not fully understand them. Problems associated primarily with TXVs and EEVs are the primary focus of this section.

Some problems associated with expansion valves are the same as those that affect fixed metering devices, such as the following:

- Dirt or other contaminants obstructing the flow of refrigerant through the valve port
- Improper sizing
- Wax from oil that builds up on the needle valve and seat (low temperature systems with mineral oil)
- Moisture frozen in the valve port (low temperature systems)

3.5.1 Refrigerant Charge Issues

The system refrigerant charge must be within a reasonable range for an expansion valve to operate properly, but it can compensate for an overcharge. If the refrigerant charge is too low, proper condensing will not occur and the valve will not receive subcooled liquid refrigerant. This causes a high superheat, even though the valve tries to compensate by opening the orifice wider.

If the refrigerant charge is a little higher than necessary though, subcooling is not adversely affected and the valve can operate properly. With fixed metering devices, liquid floodback is far more likely. An expansion valve, on the other hand, can control the refrigerant flow to maintain the superheat as set.

3.5.2 Liquid Line Deficiencies

Note that refrigerant vapor can form in the liquid line due to a long or undersized line or an improperly piped system. Vertical risers also cause pressure loss. Liquid refrigerant may leave the condenser with sufficient subcooling but arrives at the expansion valve in a different condition.

For example, assume an HFC-410A system is operating at a high-side pressure of 340 psig with 10°F of subcooling (2,344 kPa with 5.5°C of subcooling) as it leaves the condenser. Since the condensing temperature at this pressure is 105°F (41°C), the subcooled refrigerant is now at 95°F (35°C). If the liquid line is too small, has a very tall riser, and/or is unusually long, a great deal of pressure drop can occur in the line. If the pressure drops as far as 294 psig—the saturation temperature for HFC-410A at 95°F—the refrigerant will begin to vaporize before it reaches the metering device.

This same problem can occur due to a single point of damage to the liquid line. Especially in residential installations, the refrigerant lines are usually exposed and at ground level. The lines can easily be damaged by a lawnmower wheel, a baseball, or anything else that delivers a significant blow to the soft copper tubing. A significant dent or crimp in the liquid line can act as a metering device by causing a significant pressure drop.

If the damage is serious enough, there will be a temperature drop immediately downstream of the problem. The liquid line in this area may even develop condensation on the outside if it falls below the dew point. A badly restricted filter-drier, clogged with collected debris, also exhibits a measurable temperature drop and possible condensation.

Note that performance problems under these circumstances are not the fault of the metering device. However, if technicians fail to consider these possibilities and do not fully troubleshoot the problem, the expansion valve will likely be suspected.

3.5.3 Thermostatic Element and Equalizer Line Deficiencies

A TXVs failure is quite often related to a loss of charge from the thermostatic element. The tiny tube that connects the thermostatic head and diaphragm to the sensing bulb is relatively fragile. If the sensing bulb loses its refrigerant charge due to a broken or damaged sensing tube, this removes the primary opening force. As a result, the valve will often remain closed. The symptoms are an extremely low suction pressure and a very high superheat. Very little, if any, cooling will be provided, and the compressor will likely overheat if it runs for very long.

Many TXVs have a replaceable thermostatic element. Adjustable TXVs are more likely to have a replaceable element. Smaller valves and those used by equipment manufacturers may not have replaceable elements. In those cases, the entire valve must be replaced if the thermostatic element is damaged or has lost its charge.

TXVs are also affected by problems related to the equalizer line, such as the following:

- Misapplication of an internally equalized valve in place of an externally equalized valve, or poor placement of the equalizer line connection.
- Restricted, plugged, or capped external equalizer line.

3.5.4 Hunting

Hunting is a term commonly used to describe excessive and repetitive changes in refrigerant flow as an expansion valve tries to find a stable position. A TXV may seem unable to balance the refrigerant flow with the load. The symptom is a refrigerant superheat value that consistently rises and falls. Hunting can also be observed by watching the system suction pressure rise and fall. Even the sound of the compressor may change as the pressures rise and fall when hunting occurs.

Hunting: Describes the operation of an expansion valve opening and closing excessively and repeatedly, unable to stabilize to a consistent flow rate.

Oversized expansion valves are a common cause of hunting. Low load conditions significantly affect an oversized TXV. With a reduced load on the evaporator, the TXV must operate with the valve plunger very close to the seat. Further, with the refrigerant flow rate reduced, the distributor may not be able to provide a consistent mixture of liquid and vapor to all parts of the coil. The liquid may go to the lower circuits of the evaporator while the vapor goes to the upper circuits. One way to diagnose this problem is to check for high temperatures at the top of the evaporator coil and low temperatures at the bottom.

The sensing bulb location, position, and quality of contact with the suction line can also cause hunting. The bulb must be secure, properly positioned on the line, and insulated. The thermostatic element must also have the proper

NOTE

Refer to Section 4.2.2, *Sensing Bulb Location*, for more information about bulb installation.

charge for the application. Because thermostatic elements are available with different charges to match the application, it is very important to select the proper element.

Hunting is not always caused by a TXV deficiency. In some cases, hunting results from problems with the evaporator. For example, if the evaporator is clogged with debris due to a lack of proper filtration, heat transfer will be very poor, and the valve will function as if it is oversized for the application. Duct insulation or other foreign material may loosen and block a portion of the coil with the same result. In chilled water systems, an evaporator tube bundle can become fouled due to poor water quality or sediment. All these conditions can result in hunting.

Hunting can be diagnosed by making several suction line temperature measurements over time. A repetitive pattern such as that shown in *Figure 25* indicates hunting.

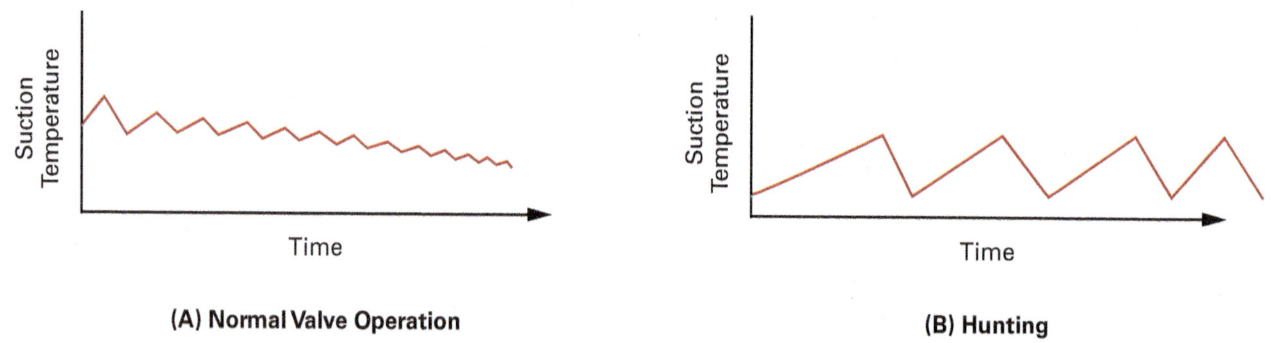

Figure 25 Suction temperature variations indicate hunting.

Before troubleshooting a TXV, establish a baseline superheat reading. The operation of a TXV can then be checked by removing the sensing bulb and holding it tightly in your hand, or by holding the bulb next to a warm heat source. The system should respond to the increase in bulb temperature by feeding more refrigerant, lowering the superheat. Next, place the sensing bulb in a glass of ice water. The system superheat should increase, as the drop in bulb temperature should force the TXV to throttle the refrigerant flow. If the TXV does not respond to these changing conditions, it is likely damaged or defective and must be serviced or replaced.

3.5.5 Corrosion

Corrosion is rarely a problem for TXVs. However, significant corrosion on or around the sensing bulb can interfere with heat transfer between the bulb and the suction line.

However, EEVs do occasionally suffer from corrosion between the operating coils and the plunger that controls the flow rate. Expansion valves of all types generate condensation on the body due to the cold refrigerant at the outlet and the surrounding humidity. As a result, they should be properly insulated to prevent water from dripping where it doesn't belong.

In the case of EEVs, the condensation can cause significant corrosion to occur in and around the moving parts. The primary symptom is a failure to move and reposition itself consistently, despite the proper electrical inputs being provided. In the worst case, the plunger seizes in position and refuses to move. The user may not be aware that a problem is developing, as the system may continue to provide reasonable comfort for some time. All EEVs should be tested using the techniques provided by the manufacturer.

3.0.0 Section Review

1. True or False: Manual expansion valves are used often in comfort cooling applications.
   a. True
   b. False

2. In an automatic expansion valve, the spring pressure pushing down on the diaphragm ______.
   a. cannot be adjusted
   b. is a closing force
   c. is an opening force
   d. offsets the pressure at the valve inlet

3. A typical TXV responds to ______.
   a. two different influences
   b. three different influences
   c. four different influences
   d. five different influences

4. The type of expansion valve that uses a tiny electric heater in the valve operator is a(n) ______.
   a. thermal-electric expansion valve
   b. thermostatic expansion valve
   c. electronically controlled expansion valve
   d. automatic expansion valve

5. With a TXV, if the refrigerant charge is slightly higher than necessary, _____.
   a. liquid floodback is likely to occur
   b. a high superheat value will result
   c. the valve is still able to maintain the proper superheat
   d. the refrigerant will not be properly subcooled

4.0.0 Thermal Expansion Valve Selection and Installation

Objective

Explain how thermal expansion valves are selected and installed.

a. Explain how thermal expansion valves are selected for a given application.
b. Describe installation practices related to thermal expansion valves.

Performance Tasks

2. Install an externally equalized expansion valve, correctly placing the sensing bulb and equalizer line.
3. Measure superheat in an operating system with an expansion valve and adjust it to change the superheat.

This section discusses TXV selection and common installation practices. Proper installation of a TXV is critical for it to operate correctly.

4.1.0 TXV Selection

It is important to select a TXV with the correct characteristics for peak system performance. Manufacturers provide detailed selection tables to assist. They also provide a wealth of additional information related to TXV operation to ensure that users fully understand their operation. Study the information manufacturers provide at length. This is one of the best ways to advance your skills and enhance your understanding of the refrigerant cycle.

TXV selection tables contain data relating to pressure drop across the valve, liquid refrigerant temperature entering the valve, evaporator temperature, and the valve capacity (in tons) at these conditions. Other characteristics that must be chosen include the thermostatic element charge, type of equalizer, valve body style, and adjustability.

As you will see, looking only at the label of a TXV to determine its capacity will not result in a good selection. The label information only reflects the nominal tonnage of a valve at a standard set of conditions.

4.1.1 Pressure Drop

The amount of pressure drop across the TXV is a significant factor in the selection process. As the refrigerant flow rate through a valve increases, the amount of pressure drop increases as well. Therefore, the pressure drop provides a means of comparing the refrigerating capacity of valves on an equal basis.

On a TXV selection chart, the pressure drop must be the same for every valve shown to allow a meaningful comparison. But note that the actual pressure drop across the metering device in a system needing a valve is often different from that shown on the chart. To determine the actual capacity of the valve for a specific use, a mathematical adjustment must be made. These adjustments are referred to as *correction factors*.

In *Figure 26*, note the correction factors shown in the bottom right-hand corner. For a 40°F (4.4°C) evaporator temperature, which is common in air conditioning applications, the correction factor is 1.00 at a 160-psi pressure drop. That means that the capacities of the valves shown on the upper left-hand chart are based on a 160-psi pressure drop. If the pressure drop for the application is higher, then the valve's capacity will be higher as well.

Air Conditioning and Heat Pump Applications

Valve Types	Nominal Capacity	Refrigerant 410A Recommended Thermostatic Charge ZCP180, ZGA ① Evaporator Temperature °F		
		40°	20°	0°
ER, RC	1	1.2	1.3	1.2
ER, RC	1-1/2	2.1	2.3	2.0
ER, RC	2	2.7	3.0	2.7
ER, RC	3	3.8	4.1	3.7
ER, RC	4	5.0	5.4	4.9
ER, RC	5	5.9	6.5	5.8
ER, RC	6	7.1	7.8	6.3
ER	8	9.5	10.4	9.3
ER	12-1/2	12.3	13.5	12.1
ER	15	14.3	15.7	14.0

Valve Types		Nominal Capacity	Refrigerant 410A Recommended Thermostatic Charge ZCP180, ZGA Evaporator Temperature °F		
			40°	20°	0°
OZE-20		20	19.7	21.6	20.4
OZE-25		25	23.7	25.9	24.5
OZE-35		35	32.6	35.6	33.6
OZE-50		50	49.3	54.0	51.0
OZE-60		60	59.2	64.8	61.2
Valve Types	Cartridge Type				
BQ, EBQ, SBQ	AAA	1/3	0.4	0.5	0.4
BQ, EBQ, SBQ	AA	3/4	0.9	1.0	0.9
BQ, EBQ, SBQ	A	1-3/4	1.9	2.1	1.8
BQ, EBQ, SBQ	B	3-1/2	3.4	3.7	3.2
BQ, EBQ, SBQ	C	6	6.2	6.8	5.9

Refrigerant	Liquid Temperature Entering TEV °F										
	40	50	60	70	80	90	100	110	120	130	140
	Correction Factor, CF Liquid Temperature										
410A	1.39	1.31		1.17	1.12	1.06	1.00	0.94	0.88	0.82	0.76

Evaporator Temperature °F	Pressure Drop Across TEV (psi)							
	80	120	160	200	240	280	320	360
	Correction Factor, CF Pressure Drop							
40°	0.71	0.87	1.00	1.12	1.22	1.32	1.41	1.50
20° & 0°	0.63	0.77	0.89	1.00	1.10	1.18	1.26	1.34

① The RC valve for R-410A applications is only available with a ZGA element.

Figure 26 Typical TXV selection chart for HFC-410A.
Source: Courtesy of Sporlan Division—Parker Hannifin Corporation

For example, if the pressure drop expected is 200 psi, then the capacity shown in the table for each valve would be multiplied by 1.12. Therefore, an ER- or RC-type valve with a nominal capacity of 4 tons would have an actual capacity of 5.6 tons at a 200-psi pressure drop. This was determined by multiplying the 5.0-ton capacity shown by 1.12.

This example demonstrates that the nominal capacity of the valve on a selection chart should not be used blindly as a final selection factor. The nominal capacity of a TXV is provided as a general reference; a place to start. In this example, the 4-ton ER or RC-type valve could properly serve a system that produces 5 tons of cooling.

4.1.2 Liquid Temperature Entering the TXV

The liquid refrigerant temperature entering the TXV affects its capacity. Cooler refrigerant is better and always provides more capacity. Again referring to *Figure 26*, note in the bottom row on the left side that the correction factor is 1.00 at a 100°F entering temperature. Therefore, the valve capacities shown in the table at the upper left are based on a liquid refrigerant temperature of 100°F. (37°C).

If the liquid temperature is expected to be higher or lower, the appropriate correction factor must be applied to the capacity. A liquid refrigerant temperature of 100°F is a common design factor. However, technicians in the coolest and hottest areas of the country may need to apply these correction factors.

4.1.3 Evaporator Temperature

When using a TXV selection chart, it is crucial that the column showing the capacity at the proper evaporator temperature be used. Looking at the wrong column will provide incorrect values. In *Figure 26*, note that the capacity for the ER and RC valve types with a nominal rating of 4 tons has more capacity when used at 20°F (–6.7°C) instead of 40°F (4.4°C). This is because a 20°F evaporator will operate at a lower pressure. As a result, the pressure drop through the valve at the same discharge pressure will be higher. In addition, a correction factor will need to be applied to compensate for the higher pressure drop. This will further increase the capacity rating of the valve.

Figure 26 provides the valve capacity for three different HFC-410A evaporator temperatures.

- The 40°F evaporator temperature is common for comfort cooling applications.
- The 20°F temperature is common for most commercial refrigeration applications where the product temperature remains above freezing.
- The 0°F temperature is common for commercial freezers that maintain product temperatures at or near 10°F (–12°C).

4.1.4 Thermostatic Element Charge

Manufacturers also provide charts to assist in selecting the proper thermostatic charge for the application. Using the correct thermostatic element is essential. When the chosen element is incorrect, the results are unpredictable. *Figure 27* shows one such chart. Whenever there is doubt about the thermostatic element that should be used for a given application, consult the manufacturer. Most manufacturers are very helpful and are truly interested in their products being applied correctly.

Also remember that thermostatic elements can be ordered with longer sensing bulb capillaries if necessary.

4.1.5 Equalizer Type

It is also important to apply the correct equalization method (internal or external) for the application. Remember that internally equalized valves should only be used on single-circuit evaporators that have a small pressure drop. This primarily limits their use to small refrigeration systems. When selecting a valve for replacement, select the same equalizing method used by the equipment manufacturer or consult the manufacturer for guidance. Most comfort cooling applications require a TXV that is externally equalized.

Sporlan Selective Charges Engineered for Peak Performance for Each Specific Application

Application	Refrigerant: 12, 409A	22, 422D, 407A	410A	134a	401A	402A	404A, 408A	407C	502	507	Actual Thermostatic Charges
Air Conditioning	FCP60	–	–	JCP60	XCP60	–	–	–	–	–	JCP60
	–	VCP100	–	–	–	–	–	NCP100	–	–	VCP100
	–	–	ZCP180	–	–	–	–	–	–	–	ZCP180
	–	VGA	–	–	–	–	–	NGA	–	–	VGA
	–	–	–	–	–	–	SCP115	–	RCP115	–	SCP115
	–	–	ZGA	–	–	–	–	–	–	–	ZGA
Commercial Refrigeration 50°F to −10°F	FC	–	–	JC	XC	–	–	–	–	–	JC
	–	VC	–	–	–	–	–	NC	–	–	VC
	–	–	–	–	–	–	SC	–	RC	–	SC
	–	–	–	–	–	LC	–	–	–	PC	PC
Low Temperature Refrigeration 0°F to −40°F	FZ	–	–	–	–	–	–	–	–	–	JZ
	FZP	–	–	–	–	–	–	–	–	–	JZP
	–	VZ	–	–	–	–	–	–	–	–	VZ
	–	VZP40	–	–	–	–	–	–	–	–	VZP40
	–	–	–	–	–	LZ	SZ	–	RZ	PZ	SZ
	–	–	–	–	–	LZP	SZP	–	RZP	PZP	SZP
Extreme Low Temperature Refrigeration −40°F to −100°F	–	VX	–	–	–	–	–	–	–	–	VX
	–	–	–	–	–	LX	SX	–	RX	PX	SX

***Applicaton Factors:**

1. The Type ZP charges have essentially the same characteristics as the Type Z charge with one exception: they produce a pressure limit Maximum Operating Pressure (MOP). ZP charges are not intended as replacements for Z charges. Each should be selected for its own unique purpose.
2. All air conditioning and heat pump charges are intended for use with externally equalized valves.
3. Type L liquid charges are also available for most commonly used refrigerants in most element sizes.
4. **The Type X charges are not to be used with "EBS" and "0" valves.**
5. If in doubt as to which charge to use, review the section on thermostatic charges in Bulletin 10-9 or contact Sporlan Division of Parker, Washington, Missouri with complete system data.
6. **For dual temperature applications, use the "C" charge.**
7. **The "C" charge may be used on applications down to -30°F on R-22, R-404A and R-507.**

Figure 27 Thermostatic charge selection chart.

Source: Courtesy of Sporlan Division—Parker Hannifin Corporation

4.1.6 Body Style

TXVs are typically available in straight-through or angled body styles. Those considered to be straight-through usually do have some offset in the line connections, as shown in *Figure 28.*

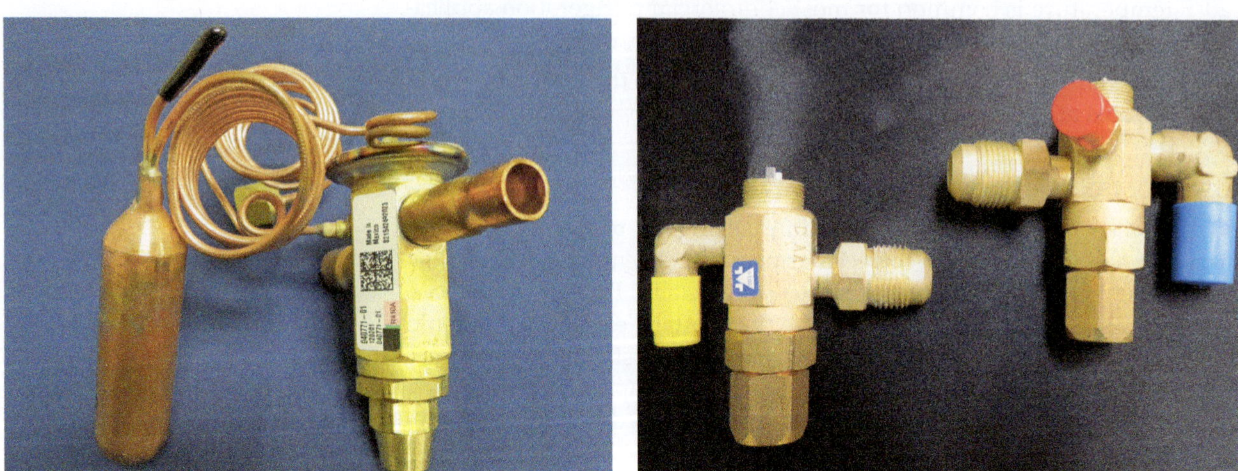

Figure 28 Straight-through and angled valve bodies.

Most TXVs are available in a limited selection of line sizes. The outlet is often larger than the inlet. Note that a TXV should never be selected by its line size. The size of the connections on a valve is not a reflection of its capacity. A proper selection process must be done first, and then the line size available that best suits the installation is chosen.

Some TXVs offer modular construction for versatility (*Figure 29*). The valve is composed of several user-chosen parts. The upper module of the valve can be fitted with the desired thermostatic element and orifice assembly required. The orifice assembly may be referred to as a *cage assembly*. The lower module of the valve provides multiple options in body style and line size.

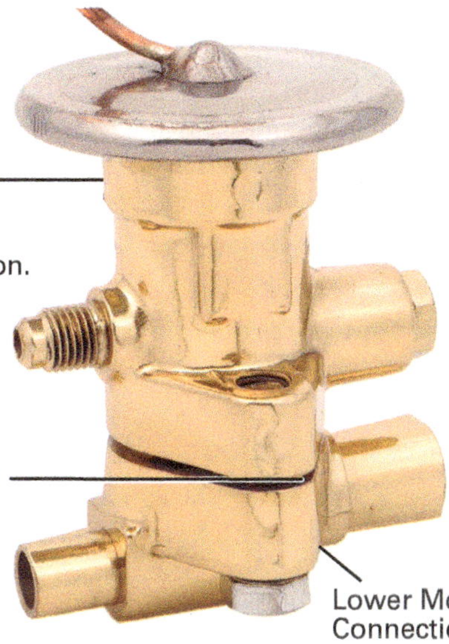

Figure 29 Modular TXV.
Source: Courtesy of Sporlan Division—Parker Hannifin Corporation

Valves are also available with different line connection styles. Both flared and soldered/brazed connection styles are commonly used, but rarely in the same model of valve. Some are also available in a relatively new style of fitting called a *Chatleff connector.* This type of connection is becoming more common, and kits are available to convert from older styles to the Chatleff style. The Chatleff connector uses an O-ring as a seal.

4.1.7 Adjustability

TXVs may be non-adjustable or adjustable. Some HVAC equipment manufacturers may special order a large quantity of valves that are non-adjustable for their equipment. Using the results of intense product testing, manufacturers can determine the exact characteristics needed for a system. By eliminating adjustability, manufacturers save money. When replacing a valve, as long as all the other operating requirements are met, an adjustable valve can replace a non-adjustable valve. Most technicians prefer adjustable valves for field replacement, but it is not an essential feature in every case.

4.2.0 TXV Installation Considerations

A TXV's type of line connection determines the steps required to physically connect the valve. The standard practices for those methods, such as soldering a joint or forming a flare, will apply. When soldering or brazing is to be conducted, the valve can be disassembled to avoid damaging internal components. However, disassembling the valve is not always necessary for soldering.

When soldering or brazing, the flame should always be directed away from the valve body. Using heat sinks and wet rags to protect the valve body is recommended. Some valves have synthetic materials inside that can be damaged.

Using a nitrogen purge is always recommended while soldering or brazing to minimize copper oxide formation. Even the light flow of nitrogen through the valve will also help carry some of the heat away.

4.2.1 Location

TXVs typically function properly when mounted in any position. However, the following are factors that affect system performance and are related to the valve location and position:

- The distributor functions best if it is installed directly to the expansion valve outlet. Gravity still affects a closed system; the liquid in the mixture will tend to fall to the bottom of the tube if the refrigerant leaving the valve travels a significant distance horizontally. If some horizontal travel is required before entering the distributor, maintaining good velocity helps keep the liquid and vapor well-mixed. Elbows in the TXV outlet piping should be avoided.
- The best performance is obtained when the valve feeds vertically up or down into the distributor. Doing so helps eliminate any liquid/vapor separation caused by gravity.
- Larger valves should be properly supported to prevent stress on the refrigerant lines.

4.2.2 Sensing Bulb Location

The location and installation of the sensing bulb is a critical part of the process. Even when installed properly, oxide deposits can form between the bulb and suction line over time, impeding heat transfer. This is a condition to look for if a valve is not properly controlling the superheat.

In general, the sensing bulb needs to be near any liquid refrigerant that might be in the suction line, so that it can sense it and respond. However, oil in circulation often carries a small amount of liquid refrigerant along with it. This is not considered floodback and it is not desirable for it to influence TXV operation. Therefore, the sensing bulb is not placed at, or even immediately adjacent to, the bottom of a suction line.

The following are common guidelines for installing the TXV sensing bulb:

- The bulb should be mounted on a horizontal section of the suction line near the evaporator outlet and always upstream from any traps in the suction line (*Figure 30*). If this is not possible, the bulb can be mounted on a vertical section of the line that is descending; do not mount it on an ascending suction line.
- On suction lines that are $^7/_8$" OD and larger, the bulb should be mounted at the 4 o'clock or 8 o'clock position (*Figure 31*). This avoids any refrigerant-carrying oil traveling at the bottom of the line, but keeps the bulb near the bottom in case too much refrigerant is being fed. For suction lines smaller than $^7/_8$" OD, any position at or above the 4 o'clock or 8 o'clock positions is acceptable.

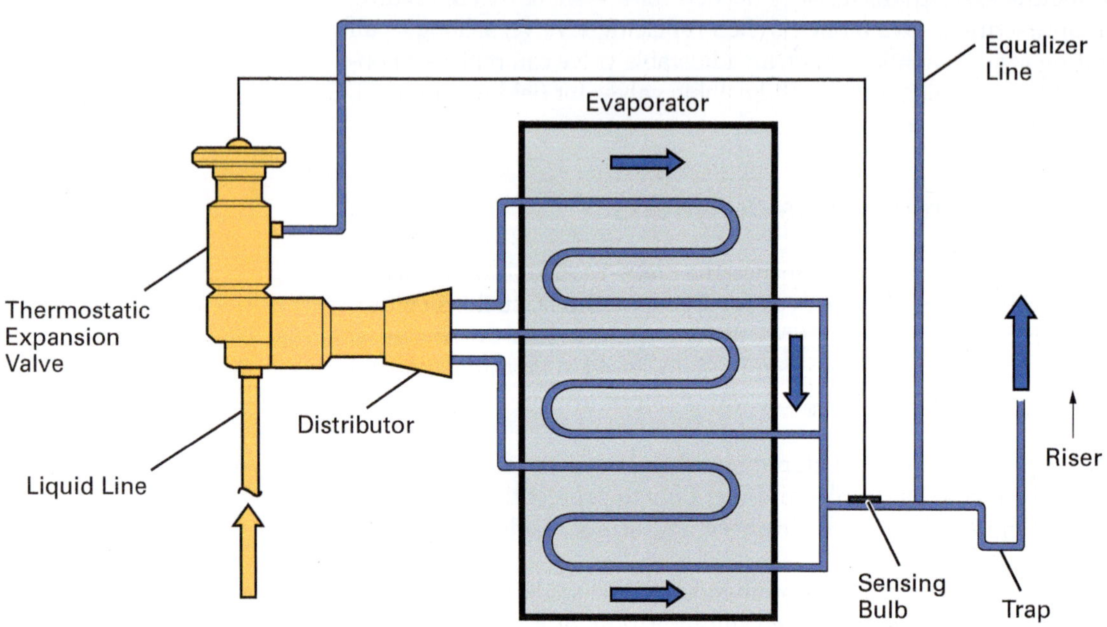

Figure 30 Sensing bulb location.

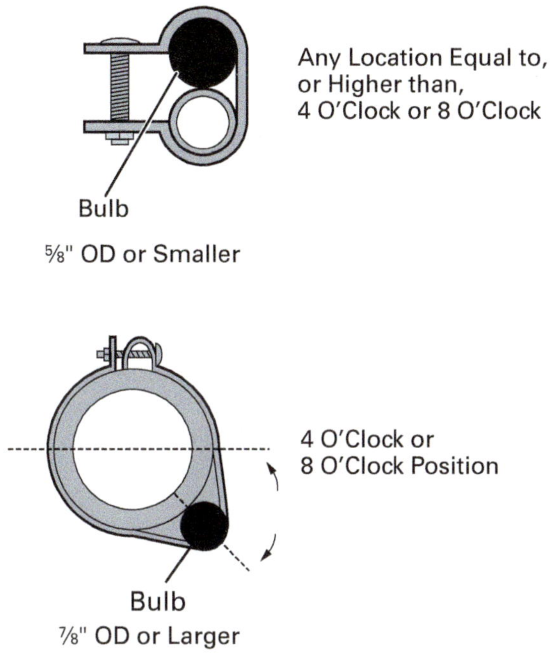

Figure 31 Sensing bulb positioning.

- TXVs are usually provided with two clamps to mount the bulb to the suction line. Both clamps should be used to ensure good contact along the full length of the bulb. Poor thermal contact will cause the valve to respond slowly and/or feed too much refrigerant.
- Ideally, the bulb should be located where it is exposed to the same temperature as the evaporator during the Off cycle. If this is not possible, the bulb and the surrounding area should be well insulated. Make the insulation water-tight and water-repellent when working with low temperature systems. Otherwise, moisture that enters or condenses beneath the insulation may freeze there and affect valve performance.
- If the sensing bulb is located in a significantly warmer location than the head of the thermostatic element, refrigerant in the bulb may migrate to the diaphragm area during the Off cycle and condense there. This could happen where the expansion valve is in the same area as a freezer coil, but the sensing bulb is outside the coil enclosure. To minimize migration in these situations, it is recommended that the thermostatic element be insulated.

On Site

Suction Line Pitch

As the suction line leaves the evaporator, it should be pitched slightly downward to allow for proper drainage of oil and any liquid refrigerant to move away from the coil and TXV sensing bulb.

4.2.3 External Equalizer Line

The external equalizer line should be connected to the suction line near the sensing bulb, where it can sense the line pressure at the sensing bulb location. The connection is usually made just downstream of the TXV sensing bulb, but this specific position is not essential. In unusual situations, the equipment manufacturer may connect it just upstream of the sensing bulb, for example. The equalizer line should never be capped or plugged, as the valve will not operate properly without it.

A refrigerant pressure regulator or similar flow control may be found on some evaporators, such as those installed on the many evaporators of large refrigeration systems serving food processing plants. When present, connect the equalizer line between the evaporator and the regulator.

4.2.4 TXV Adjustment

Adjustable TXVs allow the superheat setting to be changed as needed for proper performance. All TXVs are tested and set at the factory prior to shipment. The factory superheat setting will generally be suitable for the valve application; no further adjustment is required in most cases.

However, do not assume the setting is correct. Always check the superheat after a TXV is installed. The application may also call for a different superheat setting. Refrigeration systems, for example, often run at a lower superheat than comfort cooling systems.

To begin, a superheat reading must be taken. The superheat can be measured using the following procedure:

Step 1 Start the system and allow it to fully stabilize before taking any readings.

Step 2 Accurately record the temperature of the suction line at the position where the sensing bulb is attached. Note that infrared thermometers may not be accurate enough for this purpose. A direct measurement using a contact thermometer or temperature sensor firmly attached to the line is always best.

Step 3 Obtain the suction pressure in the suction line at the bulb location. Connect a gauge in the external equalizer line (if the unit is equipped with a fitting for this purpose) or record the gauge pressure at the first available point downstream. If the pressure is obtained from a downstream location, an estimated suction line pressure drop between the sensing bulb location and the reading location must be added to the gauge reading. The sum of the gauge reading and estimated pressure drop should be equal to the suction line pressure at the sensing bulb.

Step 4 Convert the recorded pressure to a saturated refrigerant temperature using a P-T chart. Most refrigeration gauges have the corresponding temperatures for this conversion listed on the gauge face, for your convenience. Digital gauge sets also provide this information.

Step 5 Subtract the saturated suction temperature from the line temperature at the bulb location. The difference is the superheat value.

To find the system superheat rather than the evaporator superheat, measure the suction line temperature near the compressor instead and follow *Step 2* through *Step 4*. Depending on the length of, and the environment around, the suction line, the superheat reading may be higher or lower. In most cases, the superheat will be higher. The system superheat combines the superheat added in the evaporator with the superheat added as the vapor travels through the suction line to the compressor.

On Site

Suction Pressure and Pressure Drop

The vast majority of systems in service do not have a low-side gauge port in the equalizer line or near the evaporator. In many cases, the compressor or condensing unit may be the only location with a low-side gauge port. When replacing a TXV with an external equalizer, a good practice is to install a gauge port in the equalizer line during the process.

Proper piping practices demand that the pressure drop in the suction line be kept very low. Most properly installed systems, unless other refrigerant control devices are in the suction line, have no more than 3 psi (21 kPa) of pressure drop between the evaporator and the compressor.

If the superheat setting needs to be changed, it must be done while the system is operating. The following discussion of changing the superheat setting of a TXV is general in nature. For proper details related to a specific valve model, consult the manufacturer's literature. Remember that your adjustment changes the spring pressure that is working against the diaphragm.

Step 1 Remove the adjusting stem cap.

Step 2 If the adjusting stem has a packing nut at the base, it may have to be loosened slightly before the stem will turn easily. Be sure to retighten the packing nut after changes are made to avoid refrigerant leaks, but do not overtighten. Packing nuts like these do not require a lot of torque to effectively seal.

WARNING!

Do not remove the packing nut from the adjusting stem. The packing nut should only be loosened slightly. Removing the packing nut while the valve is under pressure could result in a sudden refrigerant release. Always retighten the packing nut after any adjustments are complete.

Step 3 To reduce the superheat setting, turn the stem counterclockwise (out). The valve will feed more refrigerant at the new setting. To increase the setting, turn the valve clockwise (in). This will cause the valve to feed less refrigerant. The best way to remember this is to note that it is the same as any common faucet or water valve; they are opened by turning the handle counterclockwise.

Step 4 Adjust the valve no more than one complete turn at a time. Keep track of the number of turns made, and in which direction, during the entire process. This makes it easier to return to the previous setting.

Step 5 Once the adjustment is made, the superheat may change dramatically at first. However, the valve will soon begin to find its new balance point and the superheat will stabilize. The valve may hunt temporarily, but it should stabilize in just a few minutes.

Step 6 Continue adjusting as necessary to achieve the desired superheat value and operating stability.

4.0.0 Section Review

1. Refer to Review Question *Figure SR01*. At a pressure drop of 200 psi, an entering liquid temperature of 100°F, and an evaporator temperature of 20°F, the actual capacity of an ER or RC-2 expansion valve is ______.
   a. 2.70 tons
   b. 3.00 tons
   c. 3.02 tons
   d. 3.36 tons

Air Conditioning and Heat Pump Applications

Valve Types	Nominal Capacity	Refrigerant 410A Recommended Thermostatic Charge ZCP180, ZGA ① Evaporator Temperature °F		
		40°	20°	0°
ER, RC	1	1.2	1.3	1.2
ER, RC	1-1/2	2.1	2.3	2.0
ER, RC	2	2.7	3.0	2.7
ER, RC	3	3.8	4.1	3.7
ER, RC	4	5.0	5.4	4.9
ER, RC	5	5.9	6.5	5.8
ER, RC	6	7.1	7.8	6.3
ER	8	9.5	10.4	9.3
ER	12-1/2	12.3	13.5	12.1
ER	15	14.3	15.7	14.0

Valve Types		Nominal Capacity	Refrigerant 410A Recommended Thermostatic Charge ZCP180, ZGA Evaporator Temperature °F		
			40°	20°	0°
OZE-20		20	19.7	21.6	20.4
OZE-25		25	23.7	25.9	24.5
OZE-35		35	32.6	35.6	33.6
OZE-50		50	49.3	54.0	51.0
OZE-60		60	59.2	64.8	61.2
Valve Types	Cartridge Type				
BQ, EBQ, SBQ	AAA	1/3	0.4	0.5	0.4
BQ, EBQ, SBQ	AA	3/4	0.9	1.0	0.9
BQ, EBQ, SBQ	A	1-3/4	1.9	2.1	1.8
BQ, EBQ, SBQ	B	3-1/2	3.4	3.7	3.2
BQ, EBQ, SBQ	C	6	6.2	6.8	5.9

Refrigerant	Liquid Temperature Entering TEV °F										
	40	50	60	70	80	90	100	110	120	130	140
	Correction Factor, CF Liquid Temperature										
410A	1.39	1.31	1.23	1.17	1.12	1.06	1.00	0.94	0.88	0.82	0.76

Evaporator Temperature °F	Pressure Drop Across TEV (psi)							
	80	120	160	200	240	280	320	360
	Correction Factor, CF Pressure Drop							
40°	0.71	0.87	1.00	1.12	1.22	1.32	1.41	1.50
20° & 0°	0.63	0.77	0.89	1.00	1.10	1.18	1.26	1.34

① The RC valve for R-410A applications is only available with a ZGA element.

Figure SR01
Source: Courtesy of Sporlan Division—Parker Hannifin Corporation

2. The best location for the distributor is ______.
   a. mounted directly to the TXV outlet
   b. mounted directly to the TXV inlet
   c. at least 12" away from the TXV inlet
   d. at least 12" away from the TXV outlet

Module 03303 Review Questions

1. A metering device is a refrigerant circuit device that _____.
   a. converts high-pressure, high-temperature liquid refrigerant to a low-temperature, low-pressure vapor
   b. converts high-pressure, high-temperature liquid refrigerant to a low-temperature, low-pressure mixture of liquid and vapor
   c. is located at or near the evaporator outlet
   d. always contains a feature to automatically adjust the refrigerant flow

2. Which of the following statements about metering devices is *correct*?
   a. Roughly 80% of the refrigerant entering the evaporator is vapor.
   b. More subcooling improves the efficiency and capacity of the system.
   c. In a low-temperature evaporator, there is probably less flash gas present.
   d. Low subcooling can be caused by a failed evaporator fan motor.

3. When a refrigerant is saturated, indicated by a pressure and temperature match on a PT chart, _____.
   a. at least 25% of it is in the liquid state
   b. at least 50% of it is in the liquid state
   c. at least 75% of it is in the liquid state
   d. the percentage of liquid to vapor cannot be determined

4. The primary purpose of a distributor in an air conditioning system is to _____.
   a. feed equal amounts of refrigerant to the evaporator circuits
   b. make sure the evaporator receives only liquid refrigerant
   c. make sure the evaporator receives enough refrigerant
   d. distribute refrigerant from the condenser to the various metering devices

5. The primary factor in determining the size of a fixed-orifice metering device is typically the _____.
   a. suction line size
   b. liquid line size
   c. condenser coil capacity
   d. condensing unit capacity

6. Which of the following statements about capillary tubes is *correct*?
   a. Capillary tubes are all the same length but have different diameters.
   b. Capillary tubes are usually made from plastic tubing.
   c. Capillary tube length and diameter are both critical characteristics.
   d. Capillary tubes are never used on direct expansion systems.

7. The condition *most likely* to cause floodback when a fixed metering device is in use is _____.
   a. starting a system when the heat load is extremely high
   b. using an orifice that is too small
   c. setting the room thermostat abnormally low
   d. operating the system when it is cold outdoors

8. A clue to the problem of moisture freezing at a metering device is a _____.
   a. gradual increase in suction pressure and a decrease in superheat
   b. gradual decrease in suction pressure and an increase in superheat
   c. sudden increase in suction pressure and a decrease in superheat
   d. sudden decrease in suction pressure and an increase in superheat

9. In which of these applications would a manual expansion valve be a good choice?
   a. A large ammonia system with a fairly constant load
   b. A system that experiences frequent load changes
   c. A lobby in a busy hotel
   d. An automotive service shop

10. Another name for an automatic expansion valve is ______.
   a. thermal expansion valve
   b. constant pressure valve
   c. constant temperature valve
   d. heat motor expansion valve

11. True or False: A thermostatic expansion valve responds to two opening forces and a single closing force.
   a. True
   b. False

12. One distinguishing feature of the TXV is that it ______.
   a. has a sensing bulb filled with refrigerant or other thermostatic fluid
   b. needs to be manually adjusted as the cooling load changes
   c. uses a thermistor as a sensing device
   d. uses a float to maintain a constant liquid level in the evaporator

13. Superheat settings for low-temperature systems are typically lower to compensate for the ______.
   a. chances of liquid floodback are reduced
   b. refrigerant entering the expansion valve is deeply subcooled
   c. temperature difference between the coil and cooled medium being higher
   d. temperature difference between the coil and cooled medium being lower

14. Which of the following is *best suited* for a system with a large pressure drop across the evaporator?
   a. An internally equalized TXV
   b. An externally equalized TXV
   c. An automatic expansion valve
   d. A manual expansion valve

15. Maximum operating pressure (MOP) expansion valves are primarily used on ______.
   a. refrigeration systems with a defrost cycle
   b. refrigeration systems without a defrost cycle
   c. residential comfort cooling systems
   d. large commercial comfort cooling systems

16. Pulsing and analog EEVs are special designs based on common ______.
   a. thermostatic expansion valves
   b. thermostatic distributors
   c. solenoid valves
   d. automatic expansion valves

17. The device that converts an EEV's motor rotation to a back-and-forth motion is known as a(n) ______.
   a. digital linear actuator
   b. stepper actuator
   c. heat motor actuator
   d. electronic controller

18. If a system with a TXV constantly exhibits a very low suction pressure and a very high superheat, the *most likely* metering device-related cause is a(n) _____.
    a. clogged distributor
    b. oversized valve
    c. missing strainer
    d. loss of charge in the thermostatic element

19. True or False: The thermostatic element for TXVs is the same for all applications—only the means of attachment changes.
    a. True
    b. False

20. The correct location for a TXV sensing bulb is _____.
    a. at the condenser outlet
    b. on a suction riser
    c. at the TXV inlet
    d. at the evaporator outlet

Answers to Odd-Numbered Module Review Questions are found in *Appendix A.*

Answers to Section Review Questions

Answer	Section	Objective
Section One		
1. a	1.1.0	1a
2. a	1.2.0	1b
3. b	1.3.0	1c
Section Two		
1. a	2.1.0	2a
2. b	2.2.0	2b
3. a	2.3.0	2c
Section Three		
1. b	3.1.0	3a
2. c	3.2.0	3b
3. b	3.3.1	3c
4. a	3.4.2	3d
5. c	3.5.1	3e
Section Four		
1. b	4.1.1	4a
2. a	4.2.1	4b

MODULE 03301

Refrigerants and Oils

Source: chalermphon_tiam/Shuttestock

Objectives

Successful completion of this module prepares you to do the following:

1. Identify and classify refrigerants and match them to specific applications.
   a. Identify the desirable characteristics of refrigerants.
   b. Explain how refrigerants are chemically classified.
   c. Describe the various ways that refrigerants are identified and categorized.
   d. Match common refrigerants with cooling and refrigeration applications.
   e. Describe the characteristics of various refrigerant cylinders.
   f. State environmental concerns related to refrigerants.
2. Describe refrigerant blends and explain how blend chemistry affects pressure-temperature relationships.
   a. Identify and describe blended azeotropic, zeotropic, and near-azeotropic refrigerants.
   b. Explain how to use PT charts for compound, azeotropic and near-azeotropic refrigerants.
   c. Explain how to use PT charts for zeotropic refrigerants.
3. Describe various lubricating oils used in refrigerant circuits and identify considerations related to refrigerant and oil conversions.
   a. List and describe important refrigeration oil properties.
   b. Compare mineral-based and synthetic oils.
   c. Describe oil contaminants and testing methods.
   d. State common practices associated with handling, charging, and removing oils.
   e. Identify considerations and describe common practices related to refrigerant and oil conversions.

Performance Task

Under supervision, you should be able to do the following:

1. Identify contained refrigerants by temperature and pressure using a refrigerant gauge manifold and a PT chart.

Overview

HVACR technicians must be thoroughly familiar with the refrigerants and oils used in today's systems. Laws originally adopted years ago mandated the elimination of many popular refrigerants and replacement with products that were considered more environmentally friendly. Today, new refrigerants continue to enter the market and new mandates regarding use are the norm rather than the exception.

Digital Resources for HVACR

NCCER

SCAN ME

Scan this code using the camera on your phone or mobile device to view the digital resources related to this craft.

Many new refrigerants are not compatible with lubricants comprised of mineral oil, which were used almost exclusively with refrigerants that have been phased out. This resulted in the development of new synthetic lubricants that perform well with today's refrigerants. This module introduces refrigerants and familiarizes you with the characteristics of oils in common use today.

NCCER Industry-Recognized Credentials

If you are training through an NCCER-accredited sponsor, you may be eligible for credentials from NCCER. The ID number for this module is 03301. Note that this module may have been used in other NCCER curricula and may apply to other level completions. Contact NCCER at 1.888.622.3720 or go to **www.nccer.org** for more information.

You can also show off your industry-recognized credentials online with NCCER's digital credentials. Transform your knowledge, skills, and achievements into credentials that you can share across social media platforms, send to your network, and add to your resume. For more information, visit **www.nccer.org**.

1.0.0 Refrigerant Characteristics and Applications

Performance Tasks

There are no Performance Tasks in this section.

Objective

Identify and classify refrigerants and match them to specific applications.

a. Identify the desirable characteristics of refrigerants.
b. Explain how refrigerants are chemically classified.
c. Describe the various ways that refrigerants are identified and categorized.
d. Match common refrigerants with cooling and refrigeration applications.
e. Describe the characteristics of various refrigerant cylinders.
f. State environmental concerns related to refrigerants.

A *refrigerant* is a substance in a refrigerant circuit that is used to transfer heat. A refrigerant often changes state (liquid or vapor) when heat is added or taken away. When the pressure remains constant while the temperature changes, a change of state is likely to occur. If, however, the pressure changes as well as the temperature, a change in state does not necessarily occur. As you have learned from previous modules, a much larger heat exchange is required to change the state of a substance than to simply change its temperature.

In the condenser, a refrigerant vapor releases heat, allowing it to condense to a liquid state. In the evaporator, liquid refrigerant boils (vaporizes) as it absorbs heat. This rejection and absorption of heat by the refrigerant as it changes state is the underlying principle that allows mechanical refrigeration systems to function. Heat is absorbed by the refrigerant from an area where it is not wanted and transferred to an area where the heat is not objectionable.

Oils are used in refrigerant circuits to lubricate the moving parts of the compressor. The oil must be compatible with the refrigerant used in the system, since the two coexist in the refrigerant circuit. The oil must also be compatible with all the other materials used in refrigerant circuits.

On Site

The Early Days

The principles of mechanical refrigeration were put to practical use more than 170 years ago when American doctor John Gorrie patented a mechanical refrigeration system that compressed air, then released it into a chamber where it rapidly expanded. With water circulating around the chamber, heat was transferred from the water to the cooler chamber.

Essentially, he had created a chiller. He circulated the chilled water to cool sick patients. He was eventually able to create enough cooling effect to chill the water below freezing and make ice. Several years later, mechanical refrigeration systems using ammonia as a refrigerant were developed to produce far more ice.

In the early twentieth century, relatively hazardous refrigerants such as sulfur dioxide and methyl chloride were in widespread use. It was not until the 1920s that modern, less hazardous refrigerants were developed around **halogens**, such as fluorine and chlorine. These refrigerants enabled the development of safer domestic refrigerators and freezers, and the further development and widespread use of comfort cooling systems.

By the 1980s, scientists realized that the chlorine used in refrigerants was a major contributor to atmospheric **ozone** depletion. International treaties were enacted to restrict and eventually phase out the offending refrigerants. Refrigerants that did not interact with and destroy atmospheric ozone were developed to replace those products.

However, the growing issue of **global warming** further complicated the environmental initiative. Although the replacement refrigerants do not damage the ozone layer, the impact on global warming was even greater. This discovery prompted more research, leading to more new refrigerants and a global effort to eliminate refrigerants that have any significant environmental impact.

Halogens: The elements chlorine, fluorine, bromine, iodine, and astatine. Halogens are reactive nonmetallic elements that form strongly acidic compounds and form simple salts when combined with metals such as calcium.

Ozone: A highly reactive colorless or pale blue toxic gas with a pungent odor and strong oxidizing properties. Ozone is composed of three oxygen atoms (O_3) and is slightly soluble in water.

Global warming: A measurable increase in the temperature of earth's atmosphere, primarily attributed to the greenhouse effect from carbon dioxide (CO_2) and other materials released into the atmosphere that impede the movement of heat away from the planet.

1.1.0 Desirable Refrigerant Characteristics

Many substances can be used to transfer heat. However, many of them possess certain characteristics, such as toxicity or extreme flammability, that render them undesirable or unsafe. So, what does an ideal, or at least acceptable refrigerant look like? What characteristics does it need to have?

First, a substance must support good heat transfer in practical use and perform satisfactorily under widely varying operating conditions. In addition, a refrigerant needs the following characteristics:

- Stable
- Nontoxic
- Nonflammable/nonexplosive
- Noncorrosive
- Easily detectable leaks

1.1.1 Stability

Refrigerants must perform under a wide variety of temperature and pressure conditions while being in contact with various metals and nonmetallic substances commonly found in the refrigerant circuit. Under all expected conditions, the refrigerant ideally remains stable and does not break down into its component elements or combine with other substances to form harmful byproducts.

1.1.2 Toxicity

Modern refrigerants, for the most part, are nontoxic. Of course, any vapor or gas that substantially displaces breathable air from a confined space must be considered a hazard. However, this characteristic alone does not define a toxic substance.

Ammonia, which is used in large industrial refrigeration systems, is more toxic than other refrigerants and irritates the lungs at high concentrations. However, ammonia systems must be designed with special safety and alarm systems that alert personnel if a leak develops. With the proper precautions in place, ammonia is not considered to be a significant hazard.

On the other hand, sulfur dioxide, used in home refrigerators and similar applications many years ago, is quite toxic. Sulfur dioxide as a refrigerant was cast aside many years ago due to its toxicity. A refrigerant leak, especially indoors, could cause serious harm.

WARNING!

A confined space, by definition, has limited access, often has poor ventilation, and is not designed for continued human occupancy. Confined spaces may require a permit and/or steps to provide proper ventilation before entry is authorized. Never enter a confined space without taking the proper precautions and without proper authorization.

1.1.3 Flammability

Ideally, a refrigerant is not flammable or explosive. The normal operation of HVACR electrical devices often produces electrical arcs. In addition, service and repair procedures that require a torch could ignite leaking refrigerants that are flammable or explosive.

It is important to note that several flammable products are currently in use as refrigerants and refrigerant blend components. The interest in using flammable products stems from the fact that these refrigerants are more environmentally friendly than most traditional refrigerants and perform well in the primary role of transferring heat. The use of refrigerants with some level of flammability is rapidly expanding on a global scale, and will continue to grow for those reasons. Thus, special service techniques and procedures must be applied to minimize the risks when working with flammable products, such as HC-290 (refrigerant-grade propane) currently used in many small refrigeration systems.

1.1.4 Corrosive Properties

Refrigerants must not cause corrosion or the degradation of refrigerant circuit materials. Some refrigerants, notably ammonia, are corrosive to copper and bronze in the presence of moisture. For that reason, ammonia refrigerant piping is made of iron or steel. Aluminum is also an acceptable metal for use with ammonia, although its use is typically limited to the construction of coils because it is far less durable than iron and steel. However, most of today's refrigerants are compatible with copper and similar nonferrous metals.

1.1.5 Leak Detection

Despite your best efforts as an HVACR technician, refrigerant leaks will occur. Small or large, no leak can be repaired until it is located. Modern electronic leak detectors draw in an air sample and check for the presence of specific elements or **compounds** (*Figure 1*). When using an electronic leak detector, ensure that it can detect the type of refrigerant in use. However, several ways exist to locate leaks that do not rely on chemical sensitivity.

Compounds: Elemental combinations in a fixed ratio that exhibit their own unique characteristics. Refrigerants are typically compounds or blends of compounds.

We will likely be in pursuit of the perfect refrigerant for many years to come. No ideal refrigerant can possibly be identified today. In refrigerant selection, negative and positive characteristics must be traded off. As a result, the environmental benefits and performance offered by some refrigerants must be traded for minor safety concerns. Knowledge and experience often lead to a change in priorities.

NOTE

Refrigerant leak detection is presented in detail in NCCER Module 03205, *Leak Detection, Evacuation, Recovery, and Charging*.

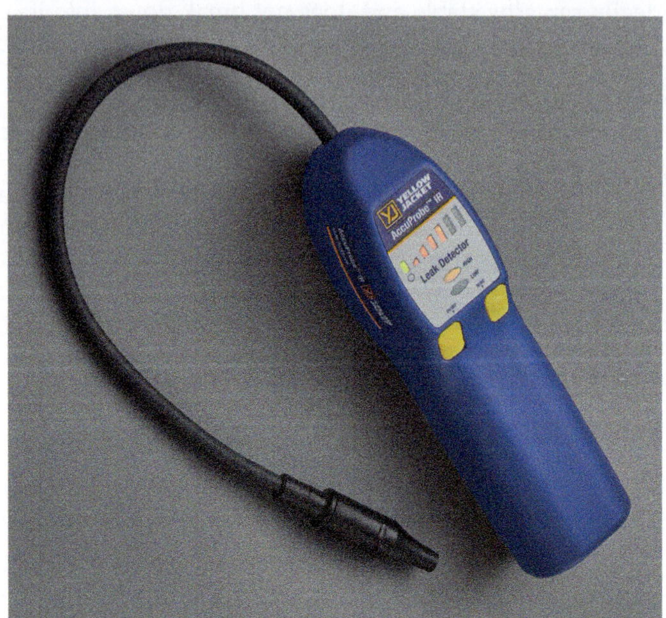

Figure 1 Electronic infrared leak detector.
Source: Ritchie Engineering Co., Inc./YELLOW JACKET

1.2.0 Refrigerant Chemistry

Refrigerants may consist of or contain **hydrocarbons (HCs)**, which are compounds composed only of carbon and hydrogen. When some or all the hydrogen atoms in a hydrocarbon are combined with a halogen, such as chlorine, fluorine, or bromine, the result is a compound known as a *halocarbon*. Halogens have replaced all the hydrogen in a *fully halogenated hydrocarbon*. At least one hydrogen atom remains in a *partially halogenated hydrocarbon* (*Figure* 2).

Both refrigerants shown in *Figure 2* are no longer produced in the United States, largely because of the volume and type of halogens they contain. Some halogens create problems in the atmosphere.

Hydrocarbons (HCs): Compounds comprised of only carbon and hydrogen atoms. When some or all the hydrogen atoms in a hydrocarbon are combined with a halogen, the result is a compound known as a *halocarbon*. Hydrocarbon refrigerants are typically flammable.

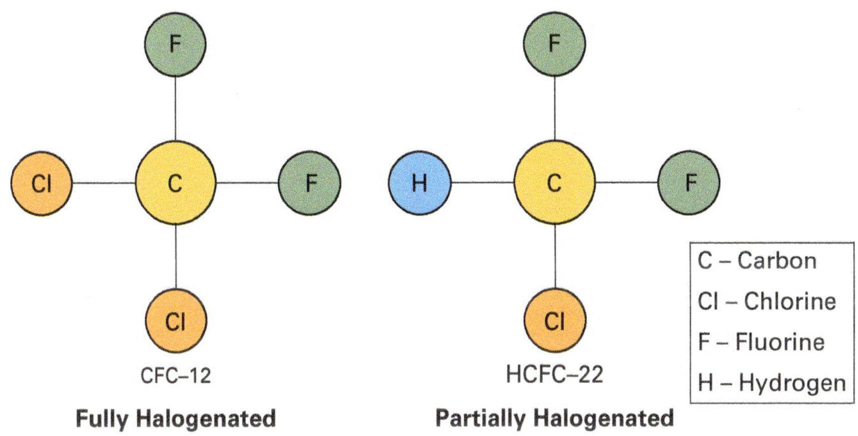

Figure 2 Halogenated hydrocarbons.

The basic combination of hydrogen and carbon supports the composition of many refrigerants. Today's refrigerants fall into the following major chemical classifications:

- *Chlorofluorocarbons (CFCs)* — These are comprised of chlorine, fluorine, and carbon. Hydrogen is not a component.
- *Hydrogenated chlorofluorocarbons (HCFCs)* — HCFCs are constructed of hydrogen, chlorine, fluorine, and carbon.
- *Hydrogenated fluorocarbons (HFCs)* — HFCs do not contain chlorine atoms; only hydrogen, fluorine, and carbon atoms.
- *Hydrofluoroolefins (HFOs)* — Much like HFCs, HFOs contain hydrogen, fluorine, and carbon. One or more of the carbon atoms are *double-bonded*, rather than *single-bonded* like those of HFCs. HFOs are relatively stable compounds but are more reactive than their HFC cousins due to the double (carbon–carbon) bond.
- *Hydrocarbons (HCs)* — Basic HCs are used as refrigerants. As mentioned above, they contain only hydrogen and carbon and are often referred to as *natural refrigerants*. Most are quite flammable, but blends containing very small amounts of HCs are not typically flammable.

NOTE

The environmental concerns related to refrigerants are presented in this module's Section 1.6.0, *Refrigerants and the Environment*.

Refrigerants can be a pure compound, or they can be blended to obtain the desired characteristics. For example, HFC-134a is a compound. HFC-410A, on the other hand, is a 50-50 blend by weight of HFC-32 and HFC-125—two different compounds that do not create a new compound when blended. A variety of compound and blended refrigerants will be discussed in this module. As you will see, some blends exhibit unique behavior in the refrigerant circuit that technicians must understand and consider when working with them.

1.2.1 CFC Refrigerants

Chlorofluorocarbon (CFC) refrigerants were the most damaging to the ozone layer because they contain a significant amount of chlorine. The production of CFCs in the United States ended in 1995, and global production in developing countries ended in 2010. One example of a CFC is CFC-12, which was used for many years in both refrigeration equipment and automotive air conditioning systems.

Although a few systems using CFCs may still be in service today, they are primarily mentioned here to provide some history on the topic. While CFCs performed admirably as refrigerants, the negative environmental impact led to their demise.

1.2.2 HCFC Refrigerants

Hydrogenated chlorofluorocarbon (HCFC) refrigerants also contain chlorine, but CFCs generally had more. HCFCs are kinder to the ozone layer than CFCs as a result, which allowed the phase out to progress over a longer period. HCFC refrigerants are no longer produced in most developed countries, but some are still produced by developing countries that are unable to afford the price of conversion as easily as developed countries.

The most common HCFC refrigerant was HCFC-22, used in most residential and commercial comfort-cooling systems worldwide until the 1990s.

1.2.3 HFC Refrigerants

Hydrogenated fluorocarbon (HFC) refrigerants do not contain chlorine. That makes this group of refrigerants more ozone-friendly than the CFCs and HCFCs (chlorinated halocarbons). However, HFCs contribute to global warming in a very significant way. HFC-410A is the best example, as it is already familiar to all HVACR technicians working in the craft today.

HFCs were long ago selected to replace some of the most popular CFCs and HCFCs. Thus, their presence in the market and across the world increased in sudden and dramatic fashion. Many of today's comfort cooling and refrigeration systems are supported with HFCs. However, the topic of global warming has advanced in an equally dramatic fashion since their introduction. Although you will encounter many systems using HFCs as a service technician, these refrigerants are being steadily phased out.

On Site

HFC Phaseout

The United States formally made HFCs a target for phaseout in 2020 with the passage of the American Innovation and Manufacturing (AIM) Act. The AIM Act reduced HFC production by 10% in 2022.

At this writing, the production of HFCs in the US will be reduced by an additional 30% in January 2024. This increases the importance of managing the current stock of HFCs and recycling or reclaiming these products at every opportunity. Some in the industry expect a shortage of HFC refrigerants to develop as a result.

Beginning in January 2025, a ban on the use of HFCs in new equipment takes effect. Equipment manufacturers must have made decisions on what refrigerant will be used in new equipment long before that date arrives.

1.2.4 HFO Refrigerants

HFOs contain the same primary elements as HFCs—hydrogen, fluorine, and carbon. However, they are derived from *olefins* rather than *alkanes* (saturated hydrocarbons that include methane, ethane, and propane). One or more of the carbon atoms are double-bonded rather than single-bonded like those of HFCs.

Current HFOs are mildly flammable and require different service procedures to ensure technician safety as a result. HFO-454B is an example of an HFO, and one which is being used by some manufacturers to replace HFC-410A in new lines of equipment. The European market has moved towards these substances at a more rapid pace than the United States, as they are considered more environmentally friendly than some older refrigerants. Systems using HFO-based refrigerants already number in the millions globally.

1.2.5 HC Refrigerants

Hydrocarbon, or HC, refrigerants are becoming increasingly popular as replacements for both HCFC and HFC refrigerants. Although HFC refrigerants are not known to harm the ozone layer, most do contribute to global warming at some level. HC refrigerants do not contribute to either of these environmental problems. The most common HC refrigerant currently is refrigerant-grade propane, designated as HC-290. Propane is also used as a fuel to heat homes and cook.

Common HC refrigerant blend components are propane and butane. You will recognize both as flammable fuels, with propane being a common heating fuel for furnaces and boilers in rural areas. Indeed, all present HC refrigerants in use are flammable. While non-toxic, non-corrosive, and considered environmentally safe, the flammability of HC refrigerants cannot be ignored.

CAUTION

HC-290 is a grade of propane that is refined to a higher degree of purity than propane sold as a fuel. Do not use fuel-grade propane in the refrigerant circuit.

As a result, working with these refrigerants represents a significant new safety hazard for technicians.

HC refrigerants are not currently approved for use in comfort-cooling equipment in the United States. They are approved for use in some industrial applications, and more recently for use in domestic appliances and some packaged commercial refrigeration systems. However, the size of the refrigerant charge in those applications is significantly limited.

HC refrigerants are reasonably compatible with most synthetic oil types. In some cases, the refrigerant is too soluble in the oil. This results in oil thinning and a lower effective **viscosity**. Thus, the use of a higher viscosity oil may be directed for HC-based equipment.

It is important to note that some refrigerant blends have a small amount of HC refrigerant added to enhance oil return in a system. However, in these cases, the HC is limited to a maximum of 3% of the blend components. The resulting blend is not considered significantly flammable.

CAUTION

The propane used as a refrigerant has been purified to a higher level than common heating fuel. In addition, the odorant added to propane fuels is not present in refrigerant-quality propane. As a result, you cannot rely on the familiar odor to detect a leak.

Viscosity: The resistance of a fluid to flow, often related to its thickness. The viscosity of a fluid typically varies, depending on its temperature. For refrigeration oils, the chosen viscosity must allow the oil to flow and provide proper lubrication over the system's range of operating temperatures.

CAUTION

The use of HC refrigerants in existing equipment not specifically designed for it is prohibited. Any new equipment based on an approved HC refrigerant in the United States must be designed and specifically tested for that purpose, must operate in limited conditions, and have a limited refrigerant charge. Note that no refrigerant should be used to replace another in existing equipment without the approval of the manufacturer.

1.3.0 Refrigerant Identification and Categorization

The American Society of Heating, Refrigeration, and Air Conditioning Engineers (ASHRAE) has developed a standardized method of identifying refrigerants. This method is largely based on the number of atoms of the various elements contained in the refrigerant.

The basic concept of the numbering system is outlined below, although there is no need to commit this information to memory. Note that the numbering system begins on the right rather than the left. This is contrary to our normal way of reading information:

- The first number on the right indicates the number of fluorine atoms.
- The second number from the right indicates the number of hydrogen atoms, plus one.
- The third number from the right indicates the number of carbon atoms, minus one. It is not used when equal to zero.
- The fourth number from the right indicates the number of double or triple carbon bonds. It is not used when equal to zero.
- An alpha suffix is found on many refrigerant identifiers. The suffix usually consists of uppercase or lowercase letters. A number may also follow a letter. A suffix consisting of a capital B and a number, for example, indicates the number of bromine atoms, when used. Although it is helpful for technicians to understand the unique properties that suffixes identify, it has little value in field service, other than to distinguish one refrigerant from another.

A lower-case alpha suffix indicates the refrigerant is an *isomer*. Isomers are substances that share the same atomic components and molecular weights, but have different molecular structures and measurable properties. The lower-case letter identifies which isomer, or molecular assembly of the atomic components, is present. HFC-134a is an example of a refrigerant with a lower-case suffix.

An uppercase alpha suffix alone indicates that there are different percentages of refrigerants comprising the blend. For example, HCFC-402A and HCFC-402B are blends comprised of the same three refrigerants, but the two products have different percentages of each refrigerant.

Each refrigerant number is more casually preceded by the letter R, such R-410A and R-290, in most publications. While this is very common and acceptable, in this book you will see them identified using their appropriate ASHRAE prefixes to help you commit the different types to memory.

1.3.1 ASHRAE Refrigerant Classification

ASHRAE Standard 34, *Designation and Safety Classification of Refrigerants* outlines the safety classification system for refrigerants. Refer to *Figure 3*.

Refrigerants fall into one of eight classifications, based on their level of toxicity (A or B), and their level of flammability (1, 2, 2L, or 3). As shown in *Figure 3*, a refrigerant classified as A1 is not flammable or toxic. A refrigerant classified as B3 would be quite flammable and significantly toxic.

	Lower Toxicity	Higher Toxicity
Higher Flammability	A3	B3
Lower Flammability	A2	B2
Mildly Flammable	A2L	B2L
No Flame Propagation	A1	B1
	No known toxicity ≤400 ppm	*Toxicity noted at <400 ppm*

Figure 3 ASHRAE refrigerant safety classifications.

As flammable refrigerants were being developed and tested, it was apparent the refrigerants exhibited varying levels of flammability. Flammability, in this case, reflects the speed at which the flame travels. The three designations for flammability did not accurately address these differences. Thus, ASHRAE developed the L suffix to further differentiate between levels of flammability. A2L refrigerants are less flammable than A2 refrigerants, as they burn more slowly.

Table 1 shows the safety classification of some common refrigerants as well as some newer refrigerants entering the market.

TABLE 1 Refrigerant Safety Classification Examples

Refrigerant	ASHRAE Safety Classification
HFC-134a	A1
HFC-404A	A1
HFC-407C	A1
HFC-410A	A1
HFO-1233zd(E)	A1
HFC-32	A2L
HFO-454B	A2L
HFO-1234yf	A2L
HFO-1234ze	A2L
HC-290 (propane)	A3
HCFC-123	B1
R-717 (ammonia)	B2L

Note that most gases can reach what might be perceived as a toxic level. Any space that is filled with a gas that displaces breathable oxygen becomes a space that cannot support life. However, this hazard is related to the absence of breathable air and is not related to the toxicity of the substance displacing the air. For example, nitrogen gas is not toxic, but if it is allowed to displace all the breathable air in a room, suffocation results.

1.3.2 Refrigerant Labeling Requirements

The OSHA Hazard Communication Standard (HCS) includes portions of the Globally Harmonized System of Classification and Labeling of Chemicals (GHS). The GHS attempts to reduce errors in readily identifying chemical hazards.

Figure 4 shows a sample label that incorporates the required elements. OSHA now requires six specific elements on refrigerant packaging, as follows:

- Product identifier
- Signal word
- Hazard statement(s)
- Pictogram(s)
- Precautionary statement(s)
- Name, address, and telephone number of the manufacturer, importer, or other responsible party

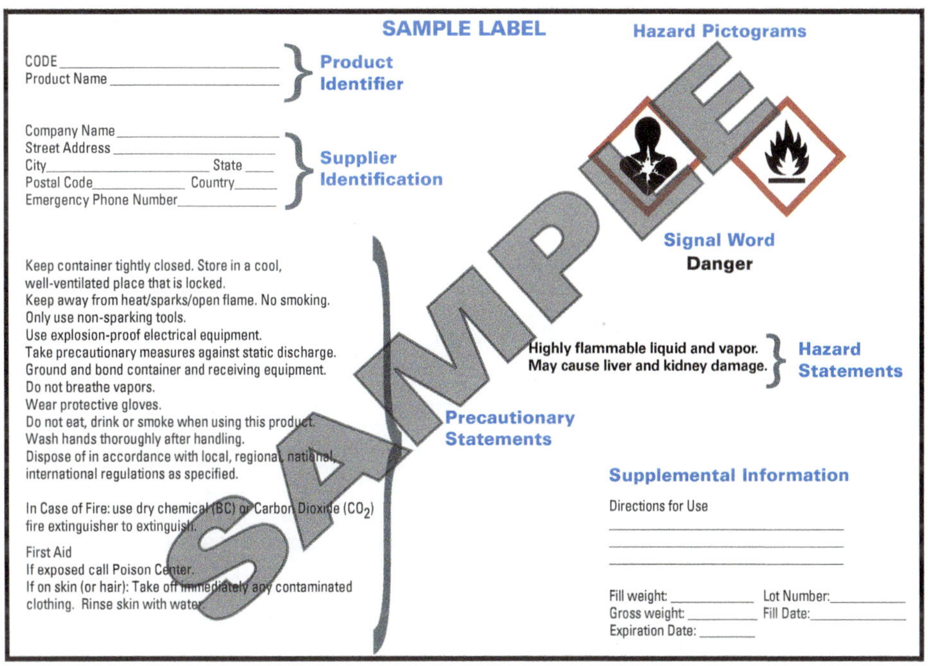

Figure 4 OSHA chemical label sample.

Source: Occupational Safety and Health Administration, U.S. Department of Labor

1.4.0 Common Refrigerants and Applications

There are many HVACR applications, and the characteristics of each refrigerant suit only a few of those applications. A refrigerant is selected based on the conditions under which the system must operate; the physical design of the system, including the materials used in the piping, coils, and accessories; and the cost and availability of the refrigerant.

The temperature at which the evaporator operates is always a significant factor. For example, the evaporator temperature for a typical air conditioner is around 40°F (4.4°C). HFC-410A has the physical properties that make it a good choice for this application.

HFC-407C is a refrigerant with similar temperature and pressure characteristics as HCFC-22 and can be used in its place with minor equipment modifications.

CAUTION

Refrigerants should never be mixed. Mixed refrigerants are not easily separated and must be incinerated when they are received and detected by an authorized reclaimer. The cost of incineration is the responsibility of the owner or organization submitting the refrigerant.

HFC-407C has been referred to as a *drop-in replacement* for HCFC-22. This term has long been applied to refrigerants that require minimum changes to the system for it to be used. A drop-in refrigerant does not require the refrigerant circuit to be flushed to remove the current oil. As you will see in the section related to refrigerant oils, many refrigerants are not compatible with traditional mineral oils used in HVACR systems. Thus, few drop-in replacements exist.

Over the years, the performance of many systems using a drop-in replacement has been less than ideal. Some manufacturers today prefer not to use the term to avoid confusion about the requirements, and to ensure technicians do not assume that old and retrofit refrigerants can be mixed. Refrigerants should never be mixed for any reason.

HFC-134a has properties that make it a better choice for high- and medium-temperature refrigeration applications where evaporator temperatures are lower. It is also used in many automobiles and other mobile applications. However, at temperatures below –10°F (–23°C), HFC-134a loses capacity and efficiency, making it a poor choice for very low-temperature applications.

In another example, the low boiling point of ammonia (–28°F (–33°C)) makes it ideal for use in low-temperature applications. At low temperatures, the evaporator pressure is still positive. Many refrigerants would need to operate in a vacuum to perform at the same temperatures. Ammonia is a very efficient refrigerant (in terms of heat transfer per pound of refrigerant flow), relatively inexpensive, and environmentally friendly. Many large food processing and cold storage facilities continue to rely on ammonia for their refrigeration needs.

Some common refrigerants and their applications are shown in *Table 2*. The refrigerant that a system uses is determined by the equipment manufacturer. Manufacturers seek out and test refrigerants with the needed pressure and temperature characteristics. However, manufacturers must also navigate the constantly changing mandates and guidelines from the EPA, ASHRAE, and other organizations.

TABLE 2 Common Refrigerants and Applications

Refrigerant	ASHRAE Safety Classification	Use	Comments
HFC-32	A2L	HFC-410A replacement	Represents 50% of the HFC-410A blend; more efficient than HFC-410A but mildly flammable.
HCFC-123	B1	CFC–11 replacement	Use restricted after 2020; currently being phased out.
HFC-134a	A1	CFC–12 replacement	Used in domestic refrigeration and automotive air conditioning; currently being phased out.
HFC-404A	A1	CFC–502 and HCFC–22 replacement	Equipment modification required; currently being phased out.
HFC-407C	A1	HCFC–22 replacement	Equipment modification required; currently being phased out.
HFC-410A	A1	HCFC–22 replacement	Higher pressures than HCFC-22; currently being phased out.
HFO-454B	A2L	HFC-410A replacement	Along with HFC-32, appears to be a popular choice for new comfort-cooling systems.
HFO-1233zd(E)	A1	HCFC-123 replacement	Now approved for use in the U.S. in new equipment requiring a low-pressure refrigerant.
HFO-1234yf	A2L	HFC-134a replacement	Mildly flammable; used in automotive and refrigeration applications.
HFO-1234ze	A2L	HFC-410A replacement	Mildly flammable; air conditioning, chillers, and similar applications.
HC-290 (propane)	A3	Commercial and domestic refrigeration	Flammable. Performance is comparable to several very popular refrigerants, but with a much lower flow rate required.
R-717 (ammonia)	B2L	Industrial refrigeration	Biodegradable; excellent heat transfer properties and environmentally friendly.

Note that research and testing of refrigerants continues daily. New refrigerants are constantly being introduced, and new regulations or standards for their use soon follow. Reviewing all current and potential refrigerants is beyond the scope of this module. All technicians have a responsibility to research and understand any refrigerant they handle to ensure their personal safety and the safety of their clients, as well as the proper operation of the systems they service.

The world of refrigerants is more complicated than it ever has been. While the world transitions away from what are now traditional refrigerants, a great deal of research and development must still be done before there are any final answers. Thus, while we are currently transitioning to new refrigerants, many are thought of as *transitional refrigerants*. Transitional refrigerants will become the standard for a period of years, until better options have been identified and approved for use. It is important to recognize how these changes and those to come affect the cost of HVACR equipment. Research and development costs are typically paid by the end user in the cost of the equipment.

The following sections introduce some of the refrigerants shown in *Table 2*.

1.4.1 HFC-32

As an HFC, HFC-32 has been scheduled for phaseout by the EPA. However, it has become an acceptable option for many manufacturers in the shorter term. HFC-32 combines good efficiency with a lower environmental impact.

Due to its flammability, this refrigerant was previously discounted as an option. Blending it with HFC-125 to create HFC-410A reduced its flammability, but at the cost of efficiency and environmental impact. Today, it has been embraced by several manufacturers for comfort cooling applications, at least until better options are available. HFC-32 operates at slightly higher pressures than HFC-410A.

Daikin® refers to HFC-32 as "the most balanced refrigerant for stationary air conditioners and heat pumps." As of December 2022, it is estimated that there are already more than 200 million HFC-32 units on the market globally. Units designed around HFC-32 were first introduced in Japan in 2012.

1.4.2 HFC-134a

HFC-134a is widely used in domestic and commercial refrigeration units as well as in automotive air conditioning systems. However, HFC-134a will be phased out. Future alternatives include HFO-1234yf and HC-290.

The use of HFC-134a in new light-duty vehicles was prohibited beginning in model year 2021. Many of today's new vehicles are charged with HFO-1234yf.

1.4.3 HFC-410A

HFC-410A, a blend of HFC-32 and HFC-125 in equal proportions, has been the primary refrigerant used in new air conditioning systems in the United States for roughly 20 years. However, HFC-410A cannot be used as a drop-in replacement for HCFC-22 due to its elevated operating pressures and mineral oil incompatibility.

Some equipment manufacturers are clearly committed to using HFO-454B in new cooling equipment, with new equipment already entering the market. HFC-32 will likely be used by others.

1.4.4 HFO-454B

HFO-454B is a blend of HFC-32 and HFO-1234yf. It can be used in many HVAC applications, including split and packaged cooling systems, heat pumps, some chillers, and window units. Several major manufacturers have committed to its use and are currently building equipment using HFO-454B.

1.4.5 HFO-1233zd(E)

This refrigerant has been available in European centrifugal chillers since 2014 and is now approved for use in the US for installations such as ice-skating rinks, cold storage warehouses, and industrial process cooling. However, HFO-1233zd(E) is not considered an alternative to replace HCFC-123 in existing chillers.

1.4.6 HFO-1234yf

HFO-1234yf was primarily developed to answer the need for a refrigerant to replace HFC-134a in the European market. HFO-1234yf is primarily used for automotive air conditioning systems at this writing, including vehicles sold in the United States. Its use in stationary refrigeration applications is likely to increase.

1.4.7 HFO-1234ze

This HFO refrigerant was also developed for the European market to replace HFC-410A. It is well suited for medium- and high-temperature refrigeration. Also approved for use in the United States, HFO-1234ze offers good energy efficiency, and is cost effective. Like HFO-1234yf, it is mildly flammable.

Note that you will see some derivatives, such as HFO-1234ze(E), on the market; always ensure you have the correct refrigerant, as the two refrigerants are not precisely the same. HFO-1234ze and HFO-1234ze(E) will also be used as propellants in inhalers and in similar applications.

1.4.8 HC-290

HC-290 is a purer form of propane than that used as a fuel gas. It is quite flammable as a result. At present, its use is restricted to small domestic and light commercial refrigeration systems in the United States, and there are limitations on the size of the refrigerant charge. Equipment using this refrigerant must be clearly marked with warning placards. Although the EPA does not require specialized training to service such systems, proper training is highly recommended.

WARNING!

HC-290 refrigerant is a purified form of propane and is highly flammable. Extreme caution must be used when handling this refrigerant. Note that the familiar odorant added to propane fuel gases is absent. Open flames and smoking in the service area cannot be allowed. In addition, many electrical devices, such as relays, motors, and temperature controls, must be designed for use in equipment bearing this refrigerant. Do not replace these and similar components with generic equivalents. Any electrical components replaced must be provided, or approved, by the equipment manufacturer to ensure safe operation.

Going Green

Carbon Dioxide as a Refrigerant

Early in the twentieth century, carbon dioxide (CO_2) was widely used as a refrigerant. The introduction of CFC and HCFC refrigerants quickly displaced CO_2 in the market.

Today, engineers are again embracing CO_2. It is non-flammable, non-toxic, inexpensive, and does not harm the ozone layer. Although CO_2 in large quantities has a proven greenhouse effect, the amount used in refrigeration applications is miniscule in comparison to what is released to the atmosphere from other processes. Actually, the CO_2 needed can be taken from the atmosphere.

The major disadvantage is its very high operating pressures for common applications. CO_2 has, however, been used as the low-temperature refrigerant in cascade systems with success. With a saturation pressure of 815 psig (5,619 kPa) at 68°F (20°C), a lot will need to change about system design if it is to be used for common applications in the future.

1.5.0 Refrigerant Cylinders

Refrigerant cylinders have historically been color-coded to make them easy to visually identify. Cylinder color codes are established by Air Conditioning, Heating, and Refrigeration Institute (AHRI) Guideline N, *Guideline for Assignment of Refrigerant Color Codes.*

Table 3 shows the color codes traditionally used on disposable cylinders for popular refrigerants. For example, disposable cylinders of the refrigerant HFC-410A are a rose color, while HFC-134a cylinders are light blue.

TABLE 3 Common Refrigerant Cylinder Color Codes

Refrigerant	Cylinder Color
HCFC-22	Light green
HFC-32	Light blue-green
HCFC-123	Blue-gray
HFC-134a	Light blue (Sky)
HFC-404A	Orange
HFC-407C	Medium brown
HFC-410A	Rose
HFO-454B	Light color with red shoulder band
HFO-513A	Sky blue
HFO-1234yf	Light color with red shoulder band
Reusable/recovery cylinder	Gray with yellow top

Cylinders containing flammable refrigerants have a red band around the upper portion of the cylinder to represent the flammability hazard (*Figure 5*). Alternatively, the entire upper portion of the cylinder may be red. Note that cylinders containing refrigerants with some level of flammability, such as HFO-454B, require an adapter to connect standard refrigerant hoses, as the

Figure 5 Disposable cylinder containing a flammable refrigerant.
Source: FluoroFusion

cylinder fitting has left-hand threads. This helps ensure that you are aware of the nature of its contents.

Going Green

Non-refillable refrigerant cylinders are also referred to as *disposable cylinders*. However, don't take that term to mean you can't recycle them. They are much better recycled than slowly decomposing in a landfill.

Technicians have been able to readily identify new refrigerant by cylinder color for many years. However, the introduction of so many new refrigerants largely exhausted the supply of colors that were easily distinguishable. In 2017, AHRI made changes to Guideline N that changed the default cylinder color for all refrigerants to light gray-green (*Figure 6*). The Occupational Safety and Health Administration (OSHA), as well as the Department of Transportation (DOT), directs that refrigerants be clearly identified by the labeling, regardless of color. In other words, only the labeling should be used to positively identify a refrigerant today. The color of the cylinder may be helpful, but it is not a reliable indicator. Specific colors most likely will not be assigned to refrigerants in the future.

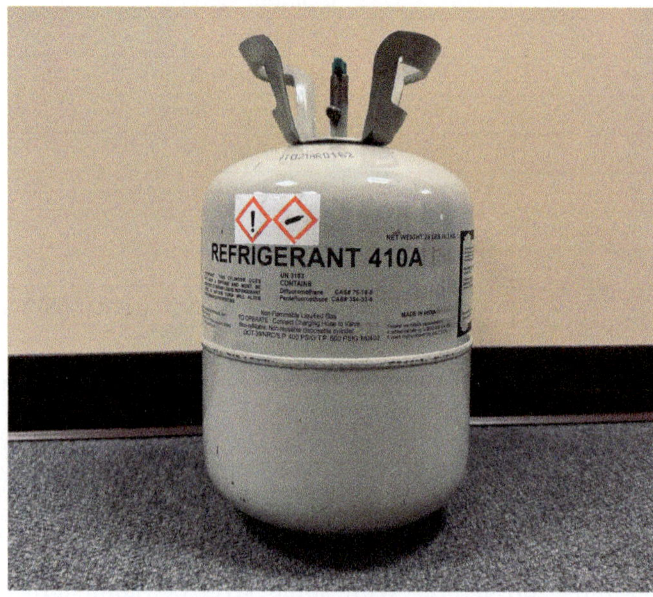

Figure 6 Disposable refrigerant cylinders meeting the new AHRI guidelines.
Source: Ritchie Engineering Co., Inc./YELLOW JACKET

These changes were not mandated. Manufacturers were not required to repaint, or even to fully adopt the new color guidelines. In *Figure 6*, note that the cylinders are white rather than light gray-green. As time passes, you will see more cylinders with little or no distinguishing color.

Given the number of refrigerants in use today, the significance of the damage that can be done by charging systems with the wrong refrigerant, and the introduction of flammable products, it is more important than ever for technicians to positively verify what they purchase and what they reach for to charge a system. Make a habit of reading the label before you even grasp the handles of a cylinder.

1.5.1 Refrigerant Cylinder Use and Handling

You will encounter two primary types of refrigerant cylinders: *disposable* (non-refillable and non-reusable) and *returnable/reusable*. Each type has its own unique handling requirements to ensure your safety.

Recovery: The act of transferring refrigerant from a system or other vessel into a reusable recovery cylinder for future use, recycling, or reclamation.

Disposable cylinders are single-use cylinders. It is against the law to reuse them in any way, including as a **recovery** cylinder. Break off the valve on the cylinder or puncture the *rupture disk* before recycling or disposing of the cylinder (*Figure 7*).

WARNING!

Before removing the valve or puncturing a cylinder, be sure that it is at atmospheric pressure. Any liquid or vapor remaining should be recovered before removing the cylinder from service.

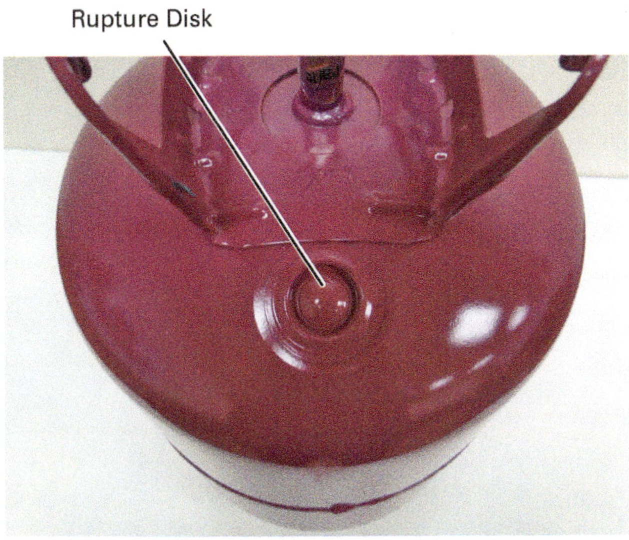

Figure 7 Disposable refrigerant cylinder rupture disk.

The rupture disk in a disposable cylinder is designed to fracture if the cylinder pressure exceeds the setpoint. Note that this can result in an unexpected release of refrigerant, although this is preferred to an explosive release of any kind. However, once a rupture disk has fractured, it does not reclose. All pressure in the cylinder will be released until it reaches atmospheric pressure.

Cylinders containing flammable refrigerants are equipped with a resettable relief valve. After the valve opens to relieve a high-pressure condition, it will reclose after the pressure has been sufficiently relieved to a safe pressure.

To prevent damage to the cylinder during transport, keep it in its cardboard shipping container when practical. This cushions the cylinder and keeps it from rolling around. If the cardboard container is destroyed, restrain the cylinder in the upright position with light chains or bungee straps.

Returnable/reusable cylinders can be returned to the distributor and/or refrigerant manufacturer to be refilled (*Figure 8*). Returnable/reusable cylinders are also used in the field for refrigerant recovery. The 30# and 50# models are commonly used for refrigerant recovery efforts.

These cylinders are colored gray with a yellow top. Another common characteristic of reusable cylinders is that they typically have two separate valves (*Figure 9*).

Figure 8 Returnable/reusable cylinders.
Source: Ritchie Engineering Co., Inc./YELLOW JACKET

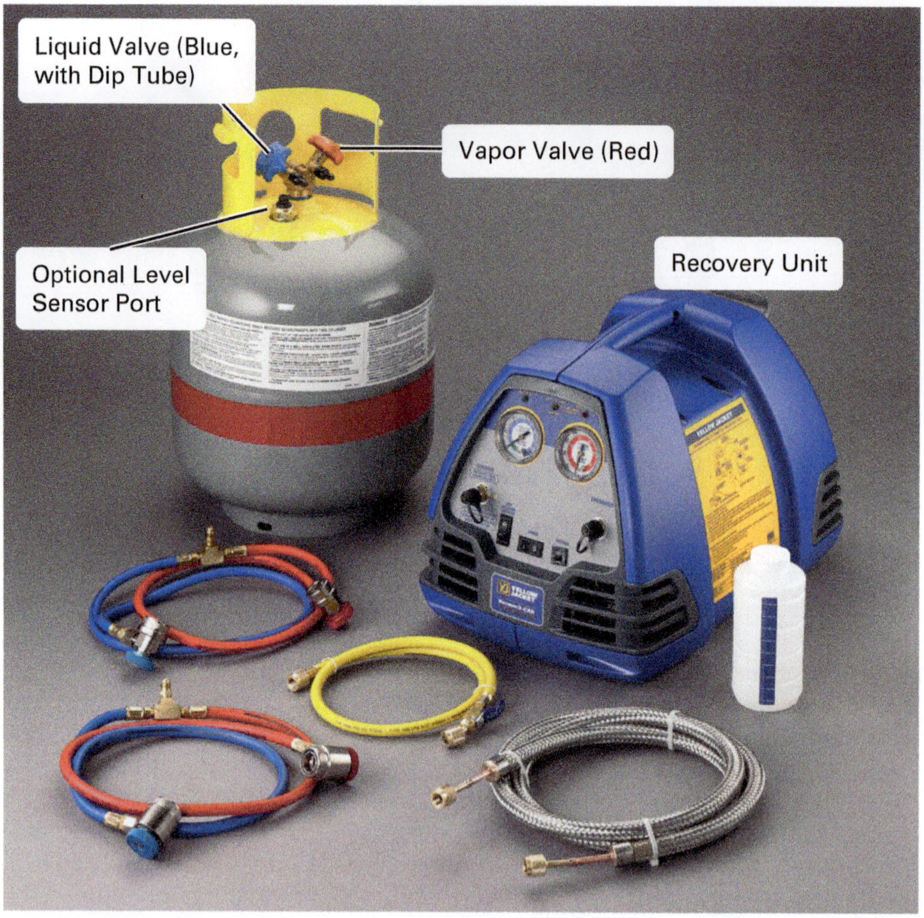

Figure 9 Recovery cylinder and common equipment.
Source: Ritchie Engineering Co., Inc./YELLOW JACKET

In two-valve models, one of the two valves is equipped with a dip tube to allow liquid to be drawn from the bottom. The other valve has no dip tube, and therefore provides access to vapor at the top of the cylinder. Although it is not mandated, most manufacturers color code the valves—blue for liquid (dip tube attached), and red for vapor.

It is very important to note that non-refillable cylinders often have a dip tube as well, connected to the single valve provided. This is the case for **zeotropes** and **near-azeotropes** that must be charged into a system in the liquid state. For many years, technicians turned HCFC cylinders upside down when dispensing liquid was desired. Doing that with some cylinders today will result in vapor being dispensed, not liquid! Pay close attention to the information and diagrams shown on the cylinders.

Zeotropes: Blended refrigerants that are not compounds. As a result, one or more components vaporize and condense over a range of temperatures instead of at a single temperature/pressure combination.

Near-azeotropes: Zeotropes with azeotropic properties. Refrigerant HFC-410A is an example of a near-azeotrope. Like zeotropes, near-azeotropes do exhibit glide characteristics, but they are slight and not a factor technicians need to consider during handling and charging.

Never overfill a recovery cylinder. The maximum capacity of a recovery cylinder is 80% of the cylinder's gross weight (the total weight of cylinder and refrigerant). Some recovery cylinders are equipped with a float switch or level sensor that shuts off the recovery machine when the maximum safe level in the cylinder is reached, as shown in *Figure 9*. If the cylinder is not equipped with a switch that interlocks with the recovery equipment, the weight must be carefully monitored using a scale (*Figure 10*). Some manufacturers are now making cylinders without this fitting since the requirement for it has now been waived.

WARNING!

Overfilling a refrigerant cylinder can result in an unexpected release of refrigerant or an explosion as the liquid warms and expands. Refrigerant volume changes with changes in temperature. By law, cylinders must never be filled to more than 80% of their capacity, and the cylinder temperature must never be allowed to exceed 122°F (50°C). Overfilling a cylinder can result in catastrophic failure, injury, or death. Note that, even though the refrigerant itself may not be toxic, the released contents of a refrigerant cylinder can quickly displace the oxygen in an enclosed area, such as a service van.

NOTE

Zeotropic and near-azeotropic refrigerants are introduced and discussed in detail in this module's Section 2, *Refrigerant Blends*.

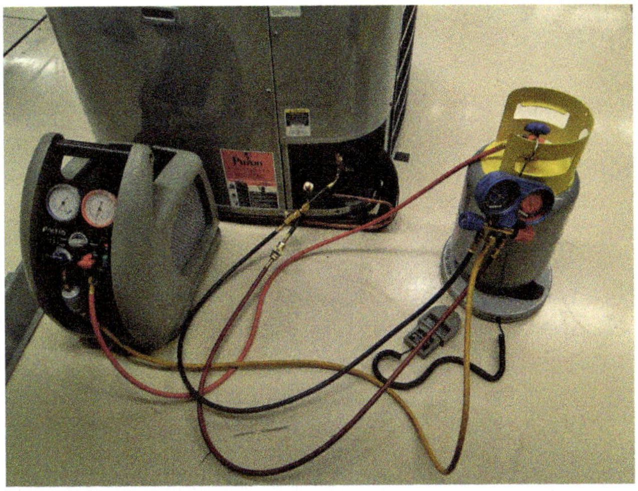

Figure 10 Connected condensing unit, recovery unit, and recovery cylinder on a scale.

NOTE

Refrigerant recovery procedures are presented in NCCER Module 03205, *Leak Detection, Evacuation, Recovery, and Charging*.

Did You Know?

Calculating the Maximum Safe Weight of a Refillable Refrigerant Cylinder

You can determine the maximum safe weight of a refrigerant cylinder using the following equation:

$$\text{Maximum Cylinder Weight} = (V \times D \times 0.8) + TW$$

Where:

V = Cylinder volume in cubic feet (ft^3)

D = Liquid density of the refrigerant at 130°F

TW = Tare (empty) weight of the cylinder

Begin by determining the cylinder volume. Examine the cylinder markings and find the water capacity (WC). Assume the water capacity is 47.7 pounds (lb). Since water has a density of 62.5 pounds per cubic foot (lb/ft^3), divide the cylinder's water capacity by the density of water to determine the cylinder volume as follows:

$$V = 47.7 \text{ lb} \div 62.5 \text{ lb/ft}^3$$
$$V = 0.763 \text{ ft}^3$$

Next, determine the liquid density of the refrigerant being recovered. For example, assume you are recovering HFC-134a, which has a liquid density of 67.46 lb at a saturation temperature of 130°F. Note that it has a higher density than water.

You will also find the tare weight of the cylinder stamped on it. The tare weight represents its empty weight. Assume it is a nominal 50-lb cylinder with a tare weight of 25.8 lb.

Now use the equation to determine the maximum safe weight as follows:

$$\text{Maximum Cylinder Weight} = (V \times D \times 0.8) + TW$$
$$\text{Maximum Cylinder Weight} = (0.763 \text{ ft}^3 \times 67.46 \text{ lb/ft}^3 \times 0.8) + 25.8 \text{ lb}$$
$$\text{Maximum Cylinder Weight} = 41.18 \text{ lb} + 25.8 \text{ lb}$$
$$\text{Maximum Cylinder Weight} = 66.98 \text{ lb}$$

Do not be tempted to exceed the safe weight of a cylinder. The risk is extreme, as an exploding cylinder of refrigerant can easily kill or cause devastating injuries.

Larger returnable/reusable cylinders are too heavy for most mobile service work. These cylinders are more often used in fixed locations where the cylinder can be moved short distances on a cylinder cart. When not in use, these cylinders must be secured in an upright position using light chains. The protective cap must be screwed on top of the cylinder to protect the valves from damage unless the cylinder is in use.

1.6.0 Refrigerants and the Environment

Ozone in our atmosphere prevents excessive ultraviolet (UV) radiation from the sun from reaching Earth where it has a negative effect on plants, animals, and humans. In other words, although ozone is an air contaminant near Earth's surface, its presence in the upper atmosphere protects us.

Refrigerants containing chlorine are chemically stable, remaining in the atmosphere for many years. These refrigerants rise into the upper levels of the atmosphere where UV radiation causes the refrigerant chemistry to break down, releasing the chlorine atoms. The chlorine atoms then combine with one of the three oxygen atoms that make up an ozone molecule. This results in the destruction of the ozone molecule. This initiates a series of chemical reactions that further break down ozone molecules.

For refrigerants, CFCs were the biggest offender in ozone depletion because they contain more chlorine in their makeup. For that reason, CFCs were the first refrigerants to have their use restricted and eventually eliminated in the United States. It should be noted that CFCs were also widely used in industrial processes and manufacturing, such as in the creation of foam insulation products.

Ozone depletion potential (ODP): A measure of a substance's ability to deplete atmospheric ozone, expressed as a number. The baseline ODP of 1 is assigned to refrigerant CFC-11 for practical comparison.

HCFC refrigerants have a lower **ozone depletion potential (ODP)** than CFCs. Thus, HCFC refrigerants were allowed for a longer period but are also being phased out to help protect the ozone layer. HFCs do not contain chlorine and are considered ozone friendly.

Refer to *Table 4*. CFC-11 and CFC-12 had ODPs of 1.0. HCFC-22 has an ODP of 0.05, while HCFC-123 has an ODP of 0.02. The lower the number, the lower the ODP. These values indicate that the effects on the ozone layer were minor when compared to the CFCs, but note that they were also being released in greater volume. All HFCs and HFOs have ODPs of zero because they do not contain chlorine.

On Site

SNAP

The Significant New Alternatives Policy (SNAP) Program is an EPA project. The program was established to evaluate and regulate new refrigerants that are potential replacements for refrigerants being phased out of use. The Clean Air Act authorizes the EPA to establish and publish lists of refrigerants deemed to be acceptable or unacceptable substitutes, and to establish guidelines for their use. Visit the SNAP website at **https://www.epa.gov/snap**.

TABLE 4 Ozone Depletion Potential of Refrigerants

Refrigerant	ODP
CFC-11	1.0
CFC-12	1.0
HCFC-22	0.05
HCFC-123	0.02
HFC-410A	0.0
HFC-134a	0.0

1.6.1 Global Warming

Greenhouse effect: The result of atmospheric gases such as CO_2 trapping heat in the atmosphere in the same way that a greenhouse captures and holds solar heat.

Global warming potential (GWP): A measure of a substance's contribution to global warming, expressed as a number. For practical comparison, carbon dioxide (CO_2), a known greenhouse gas, is assigned a GWP of 1. The higher the number, the greater the warming potential.

With the destruction of the ozone layer by refrigerants addressed, the world's attention soon turned to global warming. Global warming is related to the **greenhouse effect**, which is said to be a primary cause (*Figure 11*). Various pollutants in the atmosphere, including CO_2, trap the heat of the sun in the atmosphere similar to how glass in a greenhouse traps the sun's heat. Although refrigerants without chlorine do not damage the ozone layer, many have been found to have a high and unacceptable **global warming potential (GWP)**.

Table 5 shows the GWP for common refrigerants. The GWP shown is based on a span of 100 years, since the atmospheric life of each refrigerant also differs. Some linger for many years, while others are broken down and decompose in a relatively short period of time. Note that these figures change from time to time as new information is gathered. CO_2 was assigned the value of 1 to allow practical comparison. As the table shows, HFO refrigerants tend to have much lower GWP ratings than HFCs.

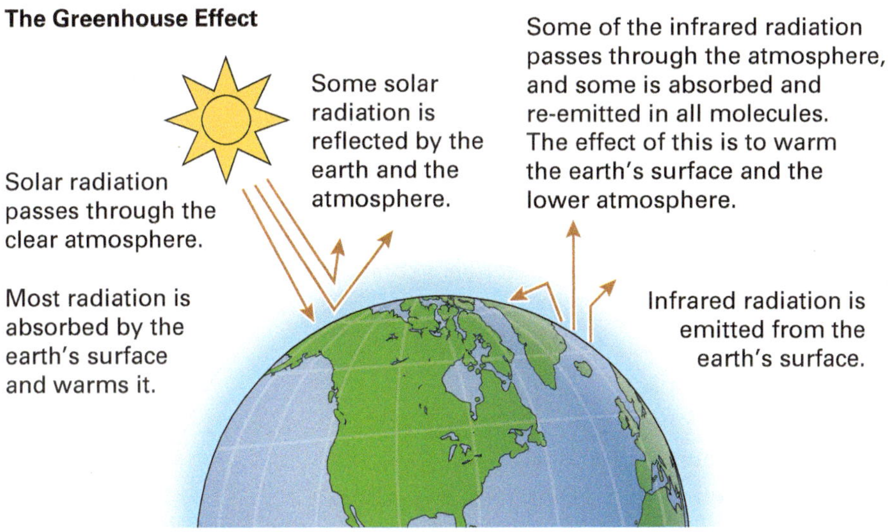

Figure 11 Greenhouse effect illustrated.
Source: Adapted from U.S. Environmental Protection Agency

TABLE 5 Global Warming Potential for Common Refrigerants

Refrigerant	GWP (100 Year)
CO_2 (baseline)	1
HFC-32	677
HCFC-123	79
HFC-134a	1,300
HFC-404A	3,922
HFC-407C	1,774
HFC-410A	2,088
HFO-454B	466
HFO-1233zd(E)	1
HFO-1234yf	4
HFO-1234ze	<1
HC-290 (propane)	<1
R-717 (ammonia)	0

1.6.2 Working with Refrigerants

As an HVACR technician working with refrigerants, it is your responsibility to prevent refrigerants from escaping into the atmosphere when servicing or repairing equipment. In the United States, all HVACR technicians must be certified by the EPA to handle refrigerants (*Figure 12*). The requirement is associated with Section 608 of the Clean Air Act. The certification card can take several forms, depending on the organization that proctors the test.

For clarity, the EPA defines a technician as an individual who performs any of the following activities:

- Attaching and detaching hoses and gauges to and from an appliance to measure pressure.
- Adding refrigerant to, or removing refrigerant from, an appliance.
- Any other activity that violates the integrity of a motor-vehicle air conditioner (MVAC), an MVAC-like appliance, or small appliance, other than for equipment disposal.

Note that registered apprentices are exempted from certification provided they are closely and continually supervised by a certified technician.

The four levels of certification are as follows:

- *Type I certification* allows technicians to work on small appliances containing less than 5 lb of refrigerant.

Figure 12 EPA certification card.

- *Type II certification* authorizes technicians to work on high-pressure and very-high pressure appliances, such as residential and commercial air conditioners, without a limitation on charge weight.
- *Type III certification* allows technicians to work on low-pressure chillers.
- *Universal certification* allows technicians to work on any systems covered by the Type I, II, or III certifications.

There are four sections to the test. All four sections must be passed to receive a Universal certification. The first section is called the Core. It covers topics such as but not limited to the following:

- The Clean Air Act and Montreal Protocol
- Environmental impacts
- Section 608 regulations
- Substitute refrigerants and oils
- Recovery, recycling, and reclamation procedures
- Safety

Safety is a potential topic in each section of the exam. Questions about the refrigerant circuit, pressure-temperature relationships, changes of state, and working with refrigerants in general may also be encountered in each section.

The Type I section of the exam covers more recovery requirements and techniques, and specifically those that are related to systems with less than 5 lb of refrigerant.

The Type II exam for high-pressure systems (no charge limit) tests this same information relevant to the larger systems. In addition, it covers leak detection and repair requirements.

NOTE

All EPA Section 608 exam candidates are encouraged to visit the EPA Section 608 website found at *https://www.epa.gov/section608* and take advantage of the resources provided to prepare.

The Type III section of the exam is related to low-pressure systems and refrigerants. The recovery and recycling procedures for these refrigerants differ significantly from the others, as do the laws governing leak repair. This section tests your knowledge of low-pressure refrigerant recovery, recycling, leak detection, and repair requirements, as well as your knowledge of the typical low-pressure refrigerant circuit.

Note that none of these certification levels is related to the use of refrigerants in mobile applications, such as automobiles. A separate certification is required for those who work with mobile systems, in accordance with Section 609 of the Clean Air Act.

You have already gained much of the knowledge needed to pass the EPA certification exam from modules that preceded this one. NCCER Module 03205, *Leak Detection, Evacuation, Recovery, and Charging* will provide information related to those four named topics that you'll need for the EPA exam.

1.0.0 Section Review

1. A combination of elements in a fixed ratio that exhibits its own unique characteristics describes a(n) _____.
   a. blend
   b. alloy
   c. compound
   d. mixture

2. Which refrigerant type contains chlorine?
   a. HCFC
   b. HFC
   c. HC
   d. HFO

3. ASHRAE Standard 34 is a resource for _____.
   a. all current EPA guidance
   b. the safety classification of refrigerants
   c. the color codes assigned to refrigerant cylinders
   d. refrigerant labeling requirements

4. Which of the following statements about HFO-454B is *correct*?
   a. It is currently prohibited for use in comfort cooling systems.
   b. It has an A1 safety classification.
   c. Several manufacturers have committed to its use in their products.
   d. It is mildly flammable.

5. Non-reusable refrigerant cylinders containing flammable refrigerants are equipped with _____.
   a. a hose fitting with left-hand threads
   b. multiple relief valves
   c. alarm circuitry
   d. a unique quick-disconnect fitting

6. HFCs are considered ozone friendly because they _____.
   a. don't cause global warming
   b. don't contain halogens
   c. don't contain chlorine
   d. aren't flammable

2.0.0 Refrigerant Blends

Objective

Describe refrigerant blends and explain how blend chemistry affects pressure-temperature relationships.

a. Identify and describe blended azeotropic, zeotropic, and near-azeotropic refrigerants.
b. Explain how to use PT charts for compound, azeotropic and near-azeotropic refrigerants.
c. Explain how to use PT charts for zeotropic refrigerants.

Performance Task

1. Identify contained refrigerants by temperature and pressure using a refrigerant gauge manifold and a PT chart.

Many of the refrigerants used today as well as those gaining popularity are blends rather than pure compounds. When a mixture of refrigerants does not result in a new compound, the components of the blend tend to maintain their own unique characteristics. As you will see, this has a significant effect on how blends behave in a refrigerant circuit and how to check and adjust the refrigerant charge.

2.1.0 Refrigerant Blends and Properties

Refrigerants are created in two ways. Chemists often develop a pure refrigerant that consists of a single compound. HCFC-22 was an example of a single compound refrigerant. In other cases, chemists will mix different compounds to obtain characteristics that differ from the characteristics of the individual refrigerants. However, in a blend, the components don't always act like their compound counterparts.

HFC-410A is a good example of a blended refrigerant as it is a 50-50 blend (by weight) of HFC-32 and HFC-125. Blends containing two refrigerants are called **binary blends**. A blend of three refrigerants is called a **ternary blend**. Blended refrigerants fall into three main categories: **azeotropes**, zeotropes, and near-azeotropes.

Binary blends: Blended refrigerants consisting of two components.

Ternary blend: A blended refrigerant consisting of three components.

Azeotropes: Blended refrigerants that are not true compounds but behave much like a compound. The refrigerant evaporates and condenses at specific pressure and temperature combinations. From the technician's perspective, working with an azeotropic blend is like working with a compound refrigerant.

2.1.1 Azeotropes

An azeotropic blend behaves like a compound, even though it is technically not one. All the refrigerant components vaporize and condense together at specific pressures and temperatures. From the technician's standpoint, working with an azeotrope is like working with a compound. Many of the older CFC refrigerants were azeotropes.

2.1.2 Zeotropes

Most zeotropes are ternary blends, making it less likely that all the components will behave the same. As a result, the components evaporate and condense over a range of temperatures, called the refrigerant's **glide**, at a constant pressure (*Figure 13*).

Glide: The temperature range across which a zeotropic refrigerant blend evaporates and condenses.

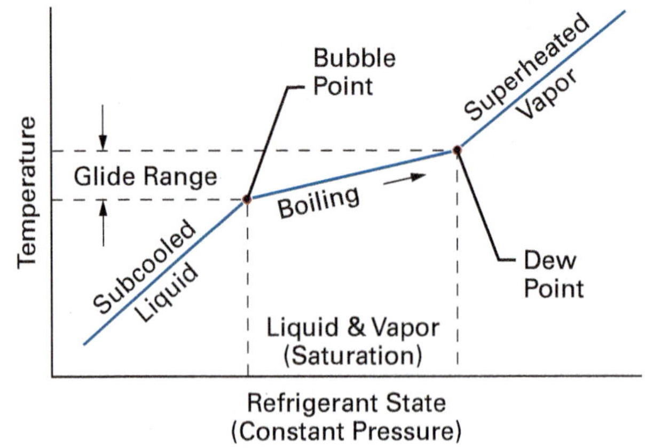

Figure 13 Temperature glide illustrated.

Refrigerant HFC-407C is an example of a zeotropic blend. When looking at a pressure-temperature (PT) chart for a zeotropic blend, two points are provided: the **bubble point** and the **dew point** (*Figure 14*). These are the points at which the blend begins vaporizing (bubble point) and finishes vaporizing (dew point) when heat is added. When heat is removed from the refrigerant, they represent the point where the blend begins condensing (dew point) and finishes condensing (bubble point).

Bubble point: The point at which a zeotropic refrigerant begins to vaporize as heat is added. It is also the point at which it completes the condensing process as heat is removed. See *dew point*.

Dew point: The point at which a zeotropic refrigerant stops vaporizing as heat is added. It is also the point that the refrigerant starts condensing when heat is removed. See *bubble point*.

The concept of refrigerant glide can be a little confusing, but it does not interfere significantly with refrigerant circuit operation. However, it is important to understand that some components of a zeotropic blend can be in the process of changing state while the others are not. Zeotropic blends vaporize and condense across a range of temperatures rather than at a single temperature/pressure combination.

NOTE

Some PT charts for zeotropic blends do not have dew point and bubble point column headings; the headings often differ. Columns labeled *Vapor* equate to the dew point, while columns labeled *Liquid* equate to the bubble point.

When heat is added to the refrigerant, the dew point is the point at which all the components of a blend have vaporized, and superheating begins. Any temperature rise measured above the dew point would indicate superheat. When heat is removed from the refrigerant, the dew point is seen as the point at which the vapor begins to condense to liquid.

Also, when heat is added to the refrigerant, the bubble point is the point at which the blend begins to vaporize. When heat is removed from the refrigerant, the bubble point is seen as the point where all the vapor has condensed to a liquid. Any temperature drop measured below the bubble point would indicate subcooling.

To summarize, when calculating superheat with a zeotropic blend, use the dew point column of a PT chart. When calculating subcooling, use the bubble point column.

Fractionation

Fractionation: A process in which one or more components of a blended refrigerant vaporize while its other components are not changing state.

Another problem with zeotropic blends is **fractionation**. Fractionation becomes a problem when a leak occurs, and it must be considered when charging a system. The different refrigerants in the blend boil off at different rates. Depending on the leak location, one or more components can leak out at a faster rate than others. This upsets the blend proportions, essentially creating a new blend.

For this reason, when you identify a leaking system using a zeotropic blend, you do not know the status of the remaining refrigerant. There is no

			MP-39 (R-401A)			MP-66 (R-401B)		R-409A		
°F	°C	R-134a	Bubble	Dew	500†	Bubble	Dew	Bubble	Dew	R-22
−40	−40.0	14.8	8.1	13.2	7.6	6.5	11.8	5.7	13.6	6
−35	−37.2	12.5	5.1	10.7	4.5	3.3	9.1	2.4	11.1	6
−30	−34.4	9.8	1.7	7.9	1.1	0.2	6.1	0.6	8.4	9
−25	−31.7	6.9	1.0	4.8	1.3	2.1	2.8	2.6	5.4	4
−20	−28.9	3.7	3.0	1.4	3.3	4.3	0.5	4.7	2.0	0.2
−15	−26.1	0.1	5.2	1.2	5.5	6.6	2.5	7.1	0.8	3.2
−10	−23.3	1.9	7.7	3.3	7.9	9.2	4.7	9.6	2.8	6.5
−5	−20.6	4.1	10.3	5.5	10.6	12.0	7.1	12.4	5.0	20.1
0	−17.8	6.5	13.2	8.0	13.4	15.1	9.7	15.5	7.5	24.0
5	−15.0	9.1	16.3	10.7	16.5	18.4	12.6	18.7	10.1	28.3
10	−12.2	11.9	19.7	13.7	19.9	22.0	15.8	22.3	13.0	32.8
15	−9.4	15.0	23.4	16.9	23.5	25.9	19.2	26.1	16.1	37.8
20	−6.7	18.4	27.4	20.4	27.4	30.1	23.0	30.3	19.5	43.1
25	−3.9	22.1	31.7	24.2	31.7	34.6	27.0	34.7	23.2	48.8
30	−1.1	26.1	36.4	28.3	36.2	39.5	31.4	39.5	27.1	55.0
35	1.7	30.4	41.3	32.8	41.0	44.8	36.1	44.6	31.4	61.5
40	4.4	35.0	46.6	37.6	46.2	50.4	41.1	50.1	36.1	68.6
45	7.2	40.1	52.4	42.7	51.8	56.4	46.6	56.0	41.0	76.1
50	10.0	45.4	58.5	48.2	57.7	62.8	52.4	62.2	46.3	84.1
55	12.8	51.2	65.0	54.1	64.0	69.6	58.7	68.9	52.0	92.6
60	15.6	57.4	71.9	60.4	70.7	76.9	65.4	75.9	58.1	101.6
65	18.3	64.0	79.3	67.2	77.8	84.7	72.5	83.4	64.7	111.2
70	21.1	71.1	87.1	74.4	85.4	92.9	80.1	91.4	71.6	121.4
75	23.9	78.7	95.4	82.1	93.4	101.6	88.2	99.8	79.0	132.2
80	26.7	86.7	104.2	90.2	101.9	110.9	96.8	108.7	86.9	143.6
85	29.4	95.2	113.6	98.9	110.8	120.7	106.0	118.1	95.2	155.7
90	32.2	104.3	123.4	108.1	120.3	131.0	115.6	128.0	104.1	168.4
95	35.0	113.9	133.8	117.9	130.3	141.9	125.9	138.4	113.4	181.8
100	37.8	124.2	144.8	128.2	140.8	153.4	136.8	149.4	123.4	195.9
105	40.6	135.0	156.4	139.1	151.9	165.5	148.2	161.0	133.8	210.8
100	43.3	146.4	168.5	150.6	163.5	178.3	160.4	173.1	144.9	226.4
115	46.1	158.4	181.3	162.8	175.7	191.6	173.1	185.9	156.6	242.8
120	48.9	171.2	194.7	175.6	188.6	205.7	186.6	199.2	168.9	260.0
125	51.7	184.6	208.8	189.0	202.1	220.4	200.7	213.2	181.8	278.0
130	54.4	198.7	223.6	203.2	216.2	235.9	215.6	227.8	195.5	296.9
135	57.2	213.6	239.1	218.1	231.0	252.0	231.2	243.1	209.8	316.7
140	60.0	229.2	255.3	233.8	246.5	268.9	247.6	259.1	224.8	337.4
145	62.8	245.7	272.2	250.2	262.7	286.6	264.8	275.7	240.5	359.0
150	65.6	262.9	289.8	267.4	279.7	305.0	282.8	293.1	257.0	381.7

Figure 14 PT chart for several refrigerants.

Source: Courtesy of Honeywell International Inc., Performance Materials and Technologies

way to tell if more of one blend component escaped than another. For some refrigerants, this means any remaining charge should be recovered and reclaimed, the leak repaired, and the unit evacuated and recharged with virgin refrigerant.

However, this is not always necessary. It depends on the leak location and the volume lost. A small leak on the liquid line of a unit that operates around the clock, for example, likely results in a loss of fully blended refrigerant in its liquid state; no harm is done to the remaining refrigerant. A leak in an evaporator coil, however, is a different matter. Where the refrigerant is normally in a saturated state around the leak, that is when the blend proportions are most likely to be disrupted. Unfortunately, when the system is not operating, it is difficult to predict what will leak out.

When charging a system, zeotropic blends must always be added in the liquid state. This ensures that the components are in proper proportions to each other. If the refrigerant is added to an operating system through the suction service port, the liquid may need to be passed through a metering orifice before it enters the unit, to prevent compressor damage. Always follow the equipment manufacturer's charging instructions. With care and experience, your refrigerant gauge manifold can serve as an expansion device during liquid charging.

2.1.3 Near-Azeotropes

Near-azeotropes can best be described as zeotropes with azeotropic properties. The refrigerant HFC-410A is an example of a near-azeotrope. Like zeotropes, near-azeotropes typically exhibit minimal glide characteristics, to the point where glide is not a factor for technicians to consider when charging a unit.

For example, the charging charts for HFC-410A have the same arrangement as a compound refrigerant, without bubble and dew point columns. *Figure 15A* and *Figure 15B* show a PT chart from a refrigerant manufacturer that compares HFC-410A and HCFC-22 in a single chart. Note that AZ-20, which appears on the chart, is the Honeywell® brand name for HFC-410A. There are no bubble points or dew points—only single pressures and temperatures that correspond to each other. However, like zeotropes, near-azeotrope refrigerants should also be charged into a system as a liquid. This assures the integrity of the blend.

	PSIG			PSIG			PSIG	
(°F)	AZ-20	R-22	(°F)	AZ-20	R-22	(°F)	AZ-20	R-22
-40	10.8	0.6	0	48.3	24.0	40	118.1	68.6
-39	11.5	1.0	1	49.6	24.9	41	120.3	70.0
-38	12.1	1.4	2	50.9	25.7	42	122.7	71.5
-37	12.8	1.8	3	52.2	26.5	43	125.0	73.0
-36	13.5	2.2	4	53.6	27.4	44	127.4	74.5
-35	14.2	2.6	5	55.0	28.3	45	129.8	76.1
-34	14.9	3.1	6	56.3	29.2	46	132.2	77.6
-33	15.6	3.5	7	57.8	30.1	47	134.7	79.2
-32	16.3	4.0	8	59.2	31.0	48	137.2	80.8
-31	17.1	4.5	9	60.7	31.9	49	139.7	82.4
-30	17.8	4.9	10	62.2	32.8	50	142.2	84.1
-29	18.6	5.4	11	63.7	33.8	51	144.8	85.7
-28	19.4	5.9	12	65.2	34.8	52	147.4	87.4
-27	20.2	6.4	13	66.8	35.8	53	150.1	89.1
-26	21.1	6.9	14	68.3	36.8	54	152.8	90.8
-25	21.9	7.4	15	69.9	37.8	55	155.5	92.6
-24	22.7	8.0	16	71.5	38.8	56	158.2	94.4
-23	23.6	8.5	17	73.2	39.9	57	161.0	96.1
-22	24.5	9.1	18	74.9	40.9	58	163.8	98.0
-21	25.4	9.6	19	76.6	42.0	59	166.7	99.8
-20	26.3	10.2	20	78.3	43.1	60	169.6	101.6
-19	27.2	10.8	21	80.0	44.2	61	172.5	103.5
-18	28.2	11.4	22	81.8	45.3	62	175.4	105.4
-17	29.2	12.0	23	83.6	46.5	63	178.4	107.3
-16	30.1	12.6	24	85.4	47.6	64	181.5	109.3
-15	31.1	13.2	25	87.2	48.8	65	184.5	111.2
-14	32.2	13.9	26	89.1	50.0	66	187.6	113.2
-13	33.2	14.5	27	91.0	51.2	67	190.7	115.3
-12	34.2	15.2	28	92.9	52.4	68	193.9	117.3
-11	35.3	15.9	29	94.9	53.7	69	197.1	119.4
-10	36.4	16.5	30	96.8	55.0	70	200.4	121.4
-9	37.5	17.2	31	98.8	56.2	71	203.6	123.5
-8	38.6	17.9	32	100.9	57.5	72	207.0	125.7
-7	39.8	18.7	33	102.9	58.8	73	210.3	127.8
-6	40.9	19.4	34	105.0	60.2	74	213.7	130.0
-5	42.1	20.1	35	107.1	61.5	75	217.1	132.2
-4	43.3	20.9	36	109.2	62.9	76	220.6	134.5
-3	44.5	21.7	37	111.4	64.3	77	224.1	136.7
-2	45.7	22.4	38	113.6	65.7	78	227.7	139.0
-1	47.0	23.2	39	115.8	67.1	79	231.3	141.3

Figure 15A Comparison of HFC-410A and HCFC-22 refrigerant pressures and temperatures (Part 1 of 2).
Source: Courtesy of Honeywell International Inc., Performance Materials and Technologies

	PSIG			PSIG			PSIG	
(°F)	AZ-20	R-22	(°F)	AZ-20	R-22	(°F)	AZ-20	R-22
80	234.9	143.6	104	334.9	207.7	128	463.2	289.2
81	238.6	146.0	105	339.6	210.8	129	469.3	293.0
82	242.3	148.4	106	344.4	213.8	130	475.4	296.9
83	246.0	150.8	107	349.3	216.9	131	481.6	300.8
84	249.8	153.2	108	354.2	220.0	132	487.8	304.7
85	253.7	155.7	109	359.1	223.2	133	494.1	308.7
86	257.5	158.2	110	364.1	226.4	134	500.5	312.6
87	261.4	160.7	111	369.1	229.6	135	506.9	316.7
88	265.4	163.2	112	374.2	232.8	136	513.4	320.7
89	269.4	165.8	113	379.4	236.1	137	520.0	324.8
90	273.5	168.4	114	384.6	239.4	138	526.6	329.0
91	277.6	171.0	115	389.9	242.8	139	533.3	333.2
92	281.7	173.7	116	395.2	246.1	140	540.1	337.4
93	285.9	176.4	117	400.5	249.5	141	547.0	341.6
94	290.1	179.1	118	405.9	253.0	142	553.9	345.9
95	294.4	181.8	119	411.4	256.5	143	560.9	350.3
96	298.7	184.6	120	416.9	260.0	144	567.9	354.6
97	303.0	187.4	121	422.5	263.5	145	575.1	359.0
98	307.5	190.2	122	428.2	267.1	146	582.3	363.5
99	311.9	193.0	123	433.9	270.7	147	589.6	368.0
100	316.4	195.9	124	439.6	274.3	148	596.9	372.5
101	321.0	198.8	125	445.4	278.0	149	604.4	377.1
102	325.6	201.8	126	451.3	281.7	150	611.9	381.7
103	330.2	204.7	127	457.3	285.4			

Figure 15B Comparison of HFC-410A and HCFC-22 refrigerant pressures and temperatures (Part 2 of 2).
Source: Courtesy of Honeywell International Inc., Performance Materials and Technologies

Going Green

Global Warming Potential

HFC refrigerants have been in use for some time now. Many were developed during the years when destruction of the ozone layer was the topic of worldwide conversation. Since those days, the hot topic has become global warming.

Unfortunately, many of the HFC refrigerants are major contributors to the greenhouse effect, thought to be the primary cause of global warming. Of course, there is no damage until they are released into the atmosphere. However, some can have a devastating effect. CO_2 has been assigned a global warming potential (GWP) of 1 as the baseline to measure other products. HFC-508B, an azeotropic blend developed for ultra-low temperature applications, has a GWP of 13,396. That means each pound of the refrigerant that is released to the atmosphere has the same greenhouse effect as 13,396 lb of CO_2. HFC-23, developed for the same applications, has an even higher GWP of 14,800. Refrigerants like these are on their way out as alternatives continue to be identified and developed.

2.2.0 Using PT Charts for Compound, Azeotropic and Near-Azeotropic Refrigerants

A PT chart shows the pressure and temperature relationships for a specific refrigerant in *saturation*. You'll recall that saturation is the point at which a refrigerant exists in both the liquid and vapor phase, but with no fixed proportion between the two phases.

For example, if an unidentified cylinder has some liquid refrigerant in it and the temperature and pressure is known, the refrigerant can generally be identified using a PT chart. What is not known is the ratio of liquid to vapor present. A cylinder that is 10% liquid and 90% vapor will have the same pressure and temperature characteristics when the ratio is reversed. If a temperature does not correspond to its listed pressure on a PT chart, the refrigerant is not in a saturated state. The refrigerant is exclusively subcooled liquid or superheated vapor.

Superheat is defined as heat added to a refrigerant after it is fully vaporized. Subcooling is defined as heat removed from a refrigerant after it has fully condensed to a liquid. A PT chart can be used to calculate both superheat and subcooling. At this point in your training, you have already practiced these calculations. However, the basic calculations are important and are reviewed here as a refresher before presenting superheat and subcooling calculations for zeotropic refrigerants.

Did You Know?

Refrigerant PT Apps

Some refrigerant and refrigeration product manufacturers now offer convenient apps that contain the PT relationships for all common refrigerants. The apps offer more information about each refrigerant than a pocket card can possibly offer. In addition, the user can select the desired pressure and temperature units and quickly convert from one to another.

Source: Courtesy of Danfoss

2.2.1 Calculating Superheat

The tools and instruments needed to calculate superheat include a gauge manifold, an accurate thermometer that can measure surface temperature on refrigerant lines, and a PT chart for the refrigerant in the system. Calculate superheat using the following steps:

Step 1 With power off, connect the gauge manifold to a low-pressure service port at the outlet of the evaporator. If there is no port at the evaporator coil to measure pressure, take the pressure measurement at the outdoor unit and adjust the reading for the pressure drop in the suction line. In a properly designed split system, the pressure drop should not exceed 3 psig. Attach the thermometer probe to the suction line where it exits the evaporator coil. Insulate the probe for an accurate reading.

Step 2 Turn the system on and let it run for 5 to 10 minutes to allow the temperatures and pressures to stabilize. Record the suction pressure and the line temperature at the outlet of the evaporator (indoor) coil. For example, assume an HFC-410A unit has a suction pressure of 123 psig (848 kPa) and the suction line temperature is 50°F (10°C).

Step 3 Consult the HFC-410A PT chart in *Figure 15A* and find the pressure closest to 123 psig. To the left of the pressure, read the corresponding saturation temperature. It is 42°F (5.5°C).

Step 4 Calculate the superheat by subtracting the saturation temperature from the measured temperature. The difference is the amount of superheat.

Superheat = Measured Temperature − Saturation Temperature

Superheat = 50°F − 42°F

Superheat = 8°F (4.4°C)

The superheat can also be calculated by taking both the pressure and temperature measurements at the condensing unit outdoors. However, you must remember that this will include the superheat gained by the vapor on its way back to the compressor.

Assume that this equipment has a fixed restrictor for a metering device and should operate at a superheat of 12°F (6.6°C) at the current conditions, according to the manufacturer. We calculated 8°F (4.4°C) of superheat. To obtain the required superheat of 12°F, refrigerant would need to be removed from the system until the superheat rises to 12°F. Of course, this assumes that all other factors, including the volume of airflow across the indoor coil, are as they should be.

NOTE

System charging is presented in detail in NCCER Module 03205, *Leak Detection, Evacuation, Recovery, and Charging*.

2.2.2 Calculating Subcooling

Calculate subcooling using the following steps:

Step 1 With power off, connect the gauge manifold to the high-pressure service port at the condensing unit. Attach the thermometer probe near the high-pressure service port. Insulate the probe for a more accurate reading.

Step 2 Turn the system on and let it run for 5 to 10 minutes to allow the temperatures and pressures to stabilize. Record the liquid line pressure and temperature. On our example HFC-410A unit, assume the high-pressure reading is 400 psig (2,758 kPa) and the liquid line temperature measures 107°F (42°C).

Step 3 Consult the HFC-410A PT chart in *Figure 15B* and find the pressure closest to 400 psig. In this case, it is 400.5 psig. To the left of the pressure reading, read the corresponding saturation temperature, which is 117°F (47°C).

Step 4 Calculate the subcooling by subtracting the measured temperature from the saturation temperature. The difference is the amount of subcooling.

Subcooling = Saturation Temperature − Measured Temperature

Subcooling = 117°F − 107°F

Subcooling = 10°F (5.6°C)

Subcooling can determine if some systems are charged properly. Subcooling values are primarily used for charging systems that use thermostatic expansion valves (TXVs) or similar modulating metering devices. Based on the measured subcooling, refrigerant may have to be added or removed from the system to obtain the correct charge. Generally, refrigerant is added to increase subcooling and removed to decrease subcooling.

On Site

Charging Analyzers

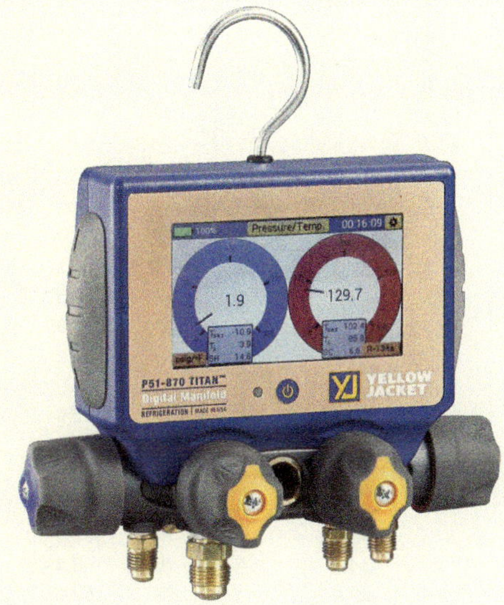

Source: Ritchie Engineering Co., Inc./ YELLOW JACKET

Technology has taken the industry forward through the development of gauge manifolds that eliminate the need for separate instruments and PT charts. The Yellow Jacket® Refrigeration System Analyzer shown here displays measurements on a backlit digital display and can be transmitted via Bluetooth® to a smartphone or tablet for further analysis. The local display of system pressures and temperatures, as well as superheat and subcooling values, provide instant analysis. When connected to a smartphone, target superheat and subcooling are easily attained, along with all the other capabilities of the ManTooth® app. The manifold features on-board data logging and standard vacuum sensor measurements (vacuum sensor included). At this writing, 126 different refrigerant profiles are stored in the instrument.

With resources like these, the most challenging part of charging a unit might be driving to the jobsite.

2.3.0 Using PT Charts for Zeotropic Refrigerants

The PT charts for zeotropic refrigerants are different from those used with other refrigerants. The charts must address glide, and therefore contain dew point and bubble point columns. On the PT chart for HFC-407C shown in *Figure 16*, dew point is shown in the Vapor Pressure column. The bubble point is shown in the Liquid Pressure column.

Calculating superheat and subcooling for a zeotropic refrigerant is similar to calculating those values for other refrigerants. The most important thing to learn when using these charts is which column to use for the task.

2.3.1 Calculating Superheat for Zeotropic Refrigerants

To calculate the superheat for zeotropic refrigerants, follow the same basic steps. It is always best to take the temperature and pressure readings at the same point. But when a refrigerant port is not available at the indoor coil, you can take the pressure reading at the outdoor unit and subtract the expected pressure drop from the reading to compensate.

After the system has been in operation for 5 to 10 minutes and has stabilized, record the suction pressure and the line temperature. For our example using HFC-407C, assume the suction pressure is 71 psig (490 kPa) and the suction line temperature is 50°F (10°C).

CAUTION

Use the Vapor Pressure or Dew Point column when calculating superheat.

Saturation Pressure-Temperature Data for R-407C (psig)*															
TEMP (°F)	PRESSURE		TEMP (°C)	TEMP (°F)	PRESSURE		TEMP (°C)	TEMP (°F)	PRESSURE		TEMP (°C)	TEMP (°F)	PRESSURE		TEMP (°C)
	LIQUID	VAPOR			LIQUID	VAPOR			LIQUID	VAPOR			LIQUID	VAPOR	
-49	*1.9*	*10.4*	-45.0	1	30.4	20.2	-17.2	51	99.8	80.5	10.6	101	228.8	199.3	38.3
-48	*1.2*	*9.8*	-44.4	2	31.4	21.0	-16.7	52	101.7	82.2	11.1	102	232.2	202.4	38.9
-47	*0.4*	*9.2*	-43.9	3	32.3	21.8	-16.1	53	103.7	84.0	11.7	103	235.6	205.6	39.4
-46	0.2	*8.6*	-43.3	4	33.3	22.6	-15.6	54	105.6	85.7	12.2	104	239.0	208.9	40.0
-45	0.6	*8.0*	-42.8	5	34.3	23.5	-15.0	55	107.6	87.5	12.8	105	242.4	212.1	40.6
-44	1.0	*7.3*	-42.2	6	35.3	24.3	-14.4	56	109.6	89.3	13.3	106	245.9	215.4	41.1
-43	1.4	*6.7*	-41.7	7	36.4	25.2	-13.9	57	111.7	91.2	13.9	107	249.5	218.8	41.7
-42	1.9	*6.0*	-41.1	8	37.4	26.1	-13.3	58	113.7	93.0	14.4	108	253.0	222.2	42.2
-41	2.3	*5.3*	-40.6	9	38.5	27.0	-12.8	59	115.8	94.9	15.0	109	256.6	225.6	42.8
-40	2.7	*4.6*	-40.0	10	39.5	27.9	-12.2	60	118.0	96.8	15.6	110	260.3	229.0	43.3
-39	3.2	*3.9*	-39.4	11	40.6	28.8	-11.7	61	120.1	98.7	16.1	111	263.9	232.5	43.9
-38	3.7	*3.2*	-38.9	12	41.7	29.8	-11.1	62	122.3	100.7	16.7	112	267.6	236.1	44.4
-37	4.1	*2.4*	-38.3	13	42.9	30.7	-10.6	63	124.4	102.7	17.2	113	271.4	239.7	45.0
-36	4.6	*1.6*	-37.8	14	44.0	31.7	-10.0	64	126.7	104.7	17.8	114	275.1	243.3	45.6
-35	5.1	*0.9*	-37.2	15	45.2	32.7	-9.4	65	128.9	106.7	18.3	115	279.0	246.9	46.1
-34	5.6	*0.1*	-36.7	16	46.3	33.7	-8.9	66	131.2	108.8	18.9	116	282.8	250.6	46.7
-33	6.1	0.4	-36.1	17	47.5	34.7	-8.3	67	133.5	110.9	19.4	117	286.7	254.3	47.2
-32	6.6	0.8	-35.6	18	48.7	35.7	-7.8	68	135.8	113.0	20.0	118	290.6	258.1	47.8
-31	7.2	1.2	-35.0	19	50.0	36.8	-7.2	69	138.2	115.1	20.6	119	294.6	261.9	48.3
-30	7.7	1.6	-34.4	20	51.2	37.9	-6.7	70	140.5	117.3	21.1	120	298.6	265.8	48.9
-29	8.3	2.1	-33.9	21	52.5	39.0	-6.1	71	142.9	119.5	21.7	121	302.6	269.7	49.4
-28	8.8	2.5	-33.3	22	53.8	40.1	-5.6	72	145.4	121.7	22.2	122	306.7	273.6	50.0
-27	9.4	3.0	-32.8	23	55.1	41.2	-5.0	73	147.8	124.0	22.8	123	310.8	277.6	50.6
-26	10.0	3.5	-32.2	24	56.4	42.3	-4.4	74	150.3	126.2	23.3	124	315.0	281.6	51.1
-25	10.6	3.9	-31.7	25	57.7	43.5	-3.9	75	152.8	128.6	23.9	125	319.2	285.7	51.7
-24	11.2	4.4	-31.1	26	59.1	44.7	-3.3	76	155.4	130.9	24.4	126	323.4	289.8	52.2
-23	11.8	4.9	-30.6	27	60.5	45.9	-2.8	77	158.0	133.3	25.0	127	327.7	293.9	52.8
-22	12.4	5.4	-30.0	28	61.9	47.1	-2.2	78	160.6	135.6	25.6	128	332.0	298.1	53.3
-21	13.1	5.9	-29.4	29	63.3	48.3	-1.7	79	163.2	138.1	26.1	129	336.4	302.4	53.9
-20	13.7	6.5	-28.9	30	64.7	49.6	-1.1	80	165.8	140.5	26.7	130	340.7	306.7	54.4
-19	14.4	7.0	-28.3	31	66.2	50.8	-0.6	81	168.5	143.0	27.2	131	345.2	311.0	55.0
-18	15.1	7.6	-27.8	32	67.7	52.1	0.0	82	171.3	145.5	27.8	132	349.7	315.4	55.6
-17	15.8	8.1	-27.2	33	69.2	53.4	0.6	83	174.0	148.1	28.3	133	354.2	319.8	56.1
-16	16.5	8.7	-26.7	34	70.7	54.8	1.1	84	176.8	150.6	28.9	134	358.7	324.2	56.7
-15	17.2	9.3	-26.1	35	72.2	56.1	1.7	85	179.6	153.2	29.4	135	363.3	328.8	57.2
-14	17.9	9.9	-25.6	36	73.8	57.5	2.2	86	182.4	155.9	30.0	136	368.0	333.3	57.8
-13	18.7	10.5	-25.0	37	75.4	58.9	2.8	87	185.3	158.5	30.6	137	372.7	337.9	58.3
-12	19.4	11.1	-24.4	38	77.0	60.3	3.3	88	188.2	161.2	31.1	138	377.4	342.6	58.9
-11	20.2	11.7	-23.9	39	78.6	61.7	3.9	89	191.1	163.9	31.7	139	382.1	347.3	59.4
-10	20.9	12.3	-23.3	40	80.2	63.2	4.4	90	194.1	166.7	32.2	140	387.0	352.1	60.0
-9	21.7	13.0	-22.8	41	81.9	64.6	5.0	91	197.1	169.5	32.8	141	391.8	356.9	60.6
-8	22.5	13.7	-22.2	42	83.6	66.1	5.6	92	200.1	172.3	33.3	142	396.7	361.7	61.1
-7	23.4	14.3	-21.7	43	85.3	67.6	6.1	93	203.2	175.2	33.9	143	401.6	366.6	61.7
-6	24.2	15.0	-21.1	44	87.0	69.2	6.7	94	206.3	178.1	34.4	144	406.6	371.6	62.2
-5	25.0	15.7	-20.6	45	88.8	70.7	7.2	95	209.4	181.0	35.0	145	411.7	376.6	62.8
-4	25.9	16.4	-20.0	46	90.6	72.3	7.8	96	212.5	184.0	35.6	146	416.7	381.1	63.3
-3	26.8	17.2	-19.4	47	92.4	73.9	8.3	97	215.7	186.9	36.1	147	421.9	386.8	63.9
-2	27.7	17.9	-18.9	48	94.2	75.5	8.9	98	219.0	190.0	36.7	148	427.0	392.0	64.4
-1	28.6	18.7	-18.3	49	96.0	77.2	9.4	99	222.2	193.0	37.2	149	432.2	397.2	65.0
0	29.5	19.4	-17.8	50	97.9	78.8	10.0	100	225.5	196.1	37.8	150	437.5	402.5	65.6

**Red Italics Indicate Inches of Mercury Below Atmospheric Pressure*

Figure 16 HFC-407C PT chart.
Source: Arkema

Consult the HFC-407C PT chart in *Figure 16* and find the pressure entry in the Vapor Pressure column closest to 71 psig. In this case, it is 70.7 psig. To the left of the pressure reading, read the corresponding saturation temperature from the Vapor Pressure column. It is 45°F (7.2°C).

Calculate the superheat by subtracting the saturation temperature from the measured temperature:

$$\text{Superheat} = \text{Measured Temperature} - \text{Saturation Temperature}$$
$$\text{Superheat} = 50°\text{F} - 45°\text{F}$$
$$\text{Superheat} = 5°\text{F}\ (2.8°\text{C})$$

2.3.2 Calculating Subcooling for Zeotropic Refrigerants

To calculate subcooling for zeotropic refrigerants, you must use the Liquid or Bubble Point column on the PT chart. For this example, assume a unit charged with HFC-407C has a high-side pressure of 250 psig (1,724 kPa) and the liquid line temperature is 97°F (36°C).

Consult the HFC-407C PT chart in *Figure 16* and find the liquid pressure closest to 250 psig. In this case, it is 249.5 psig. To the left of the pressure reading, read the corresponding saturation temperature from the Liquid Pressure column. It is 107°F (42°C).

Calculate the subcooling by subtracting the measured temperature from the saturation temperature:

Subcooling = Saturation Temperature − Measured Temperature

Subcooling = 107°F − 97°F

Subcooling = 10°F (5.6°C)

CAUTION

Use the Liquid Pressure or Bubble Point column when calculating subcooling.

2.0.0 Section Review

1. With what type of refrigerant blend is fractionation a significant issue?
   a. Zeotropic
   b. Ternary
   c. Binary
   d. Azeotropic

	PSIG			PSIG			PSIG	
(°F)	AZ-20	R-22	(°F)	AZ-20	R-22	(°F)	AZ-20	R-22
−40	10.8	0.6	0	48.3	24.0	40	118.1	68.6
−39	11.5	1.0	1	49.6	24.9	41	120.3	70.0
−38	12.1	1.4	2	50.9	25.7	42	122.7	71.5
−37	12.8	1.8	3	52.2	26.5	43	125.0	73.0
−36	13.5	2.2	4	53.6	27.4	44	127.4	74.5
−35	14.2	2.6	5	55.0	28.3	45	129.8	76.1
−34	14.9	3.1	6	56.3	29.2	46	132.2	77.6
−33	15.6	3.5	7	57.8	30.1	47	134.7	79.2
−32	16.3	4.0	8	59.2	31.0	48	137.2	80.8
−31	17.1	4.5	9	60.7	31.9	49	139.7	82.4
−30	17.8	4.9	10	62.2	32.8	50	142.2	84.1
−29	18.6	5.4	11	63.7	33.8	51	144.8	85.7
−28	19.4	5.9	12	65.2	34.8	52	147.4	87.4
−27	20.2	6.4	13	66.8	35.8	53	150.1	89.1
−26	21.1	6.9	14	68.3	36.8	54	152.8	90.8
−25	21.9	7.4	15	69.9	37.8	55	155.5	92.6
−24	22.7	8.0	16	71.5	38.8	56	158.2	94.4
−23	23.6	8.5	17	73.2	39.9	57	161.0	96.1
−22	24.5	9.1	18	74.9	40.9	58	163.8	98.0
−21	25.4	9.6	19	76.6	42.0	59	166.7	99.8
−20	26.3	10.2	20	78.3	43.1	60	169.6	101.6
−19	27.2	10.8	21	80.0	44.2	61	172.5	103.5
−18	28.2	11.4	22	81.8	45.3	62	175.4	105.4
−17	29.2	12.0	23	83.6	46.5	63	178.4	107.3
−16	30.1	12.6	24	85.4	47.6	64	181.5	109.3
−15	31.1	13.2	25	87.2	48.8	65	184.5	111.2
−14	32.2	13.9	26	89.1	50.0	66	187.6	113.2
−13	33.2	14.5	27	91.0	51.2	67	190.7	115.3
−12	34.2	15.2	28	92.9	52.4	68	193.9	117.3
−11	35.3	15.9	29	94.9	53.7	69	197.1	119.4
−10	36.4	16.5	30	96.8	55.0	70	200.4	121.4
−9	37.5	17.2	31	98.8	56.2	71	203.6	123.5
−8	38.6	17.9	32	100.9	57.5	72	207.0	125.7
−7	39.8	18.7	33	102.9	58.8	73	210.3	127.8
−6	40.9	19.4	34	105.0	60.2	74	213.7	130.0
−5	42.1	20.1	35	107.1	61.5	75	217.1	132.2
−4	43.3	20.9	36	109.2	62.9	76	220.6	134.5
−3	44.5	21.7	37	111.4	64.3	77	224.1	136.7
−2	45.7	22.4	38	113.6	65.7	78	227.7	139.0
−1	47.0	23.2	39	115.8	67.1	79	231.3	141.3

Figure SR01

2. Refer to *Figure SR01*. Assume an HFC-410A unit is operating at a suction pressure of 111 psig and the suction line temperature is 52°F. What is the superheat?
   a. 67°F
   b. 33°F
   c. 15°F
   d. 9°F

Saturation Pressure-Temperature Data for R-407C (psig)*

TEMP (°F)	PRESSURE LIQUID	PRESSURE VAPOR	TEMP (°C)	TEMP (°F)	PRESSURE LIQUID	PRESSURE VAPOR	TEMP (°C)	TEMP (°F)	PRESSURE LIQUID	PRESSURE VAPOR	TEMP (°C)	TEMP (°F)	PRESSURE LIQUID	PRESSURE VAPOR	TEMP (°C)
−49	*1.9*	*10.4*	−45.0	1	30.4	20.2	−17.2	51	99.8	80.5	10.6	101	228.8	199.3	38.3
−48	*1.2*	*9.8*	−44.4	2	31.4	21.0	−16.7	52	101.7	82.2	11.1	102	232.2	202.4	38.9
−47	*0.4*	*9.2*	−43.9	3	32.3	21.8	−16.1	53	103.7	84.0	11.7	103	235.6	205.6	39.4
−46	0.2	*8.6*	−43.3	4	33.3	22.6	−15.6	54	105.6	85.7	12.2	104	239.0	208.9	40.0
−45	0.6	*8.0*	−42.8	5	34.3	23.5	−15.0	55	107.6	87.5	12.8	105	242.4	212.1	40.6
−44	1.0	*7.3*	−42.2	6	35.3	24.3	−14.4	56	109.6	89.3	13.3	106	245.9	215.4	41.1
−43	1.4	*6.7*	−41.7	7	36.4	25.2	−13.9	57	111.7	91.2	13.9	107	249.5	218.8	41.7
−42	1.9	*6.0*	−41.1	8	37.4	26.1	−13.3	58	113.7	93.0	14.4	108	253.0	222.2	42.2
−41	2.3	*5.3*	−40.6	9	38.5	27.0	−12.8	59	115.8	94.9	15.0	109	256.6	225.6	42.8
−40	2.7	*4.6*	−40.0	10	39.5	27.9	−12.2	60	118.0	96.8	15.6	110	260.3	229.0	43.3
−39	3.2	*3.9*	−39.4	11	40.6	28.8	−11.7	61	120.1	98.7	16.1	111	263.9	232.5	43.9
−38	3.7	*3.2*	−38.9	12	41.7	29.8	−11.1	62	122.3	100.7	16.7	112	267.6	236.1	44.4
−37	4.1	*2.4*	−38.3	13	42.9	30.7	−10.6	63	124.4	102.7	17.2	113	271.4	239.7	45.0
−36	4.6	*1.6*	−37.8	14	44.0	31.7	−10.0	64	126.7	104.7	17.8	114	275.1	243.3	45.6
−35	5.1	*0.9*	−37.2	15	45.2	32.7	−9.4	65	128.9	106.7	18.3	115	279.0	246.9	46.1
−34	5.6	*0.1*	−36.7	16	46.3	33.7	−8.9	66	131.2	108.8	18.9	116	282.8	250.6	46.7
−33	6.1	0.4	−36.1	17	47.5	34.7	−8.3	67	133.5	110.9	19.4	117	286.7	254.3	47.2
−32	6.6	0.8	−35.6	18	48.7	35.7	−7.8	68	135.8	113.0	20.0	118	290.6	258.1	47.8
−31	7.2	1.2	−35.0	19	50.0	36.8	−7.2	69	138.2	115.1	20.6	119	294.6	261.9	48.3
−30	7.7	1.6	−34.4	20	51.2	37.9	−6.7	70	140.5	117.3	21.1	120	298.6	265.8	48.9
−29	8.3	2.1	−33.9	21	52.5	39.0	−6.1	71	142.9	119.5	21.7	121	302.6	269.7	49.4
−28	8.8	2.5	−33.3	22	53.8	40.1	−5.6	72	145.4	121.7	22.2	122	306.7	273.6	50.0
−27	9.4	3.0	−32.8	23	55.1	41.2	−5.0	73	147.8	124.0	22.8	123	310.8	277.6	50.6
−26	10.0	3.5	−32.2	24	56.4	42.3	−4.4	74	150.3	126.2	23.3	124	315.0	281.6	51.1
−25	10.6	3.9	−31.7	25	57.7	43.5	−3.9	75	152.8	128.6	23.9	125	319.2	285.7	51.7
−24	11.2	4.4	−31.1	26	59.1	44.7	−3.3	76	155.4	130.9	24.4	126	323.4	289.8	52.2
−23	11.8	4.9	−30.6	27	60.5	45.9	−2.8	77	158.0	133.3	25.0	127	327.7	293.9	52.8
−22	12.4	5.4	−30.0	28	61.9	47.1	−2.2	78	160.6	135.6	25.6	128	332.0	298.1	53.3
−21	13.1	5.9	−29.4	29	63.3	48.3	−1.7	79	163.2	138.1	26.1	129	336.4	302.4	53.9
−20	13.7	6.5	−28.9	30	64.7	49.6	−1.1	80	165.8	140.5	26.7	130	340.7	306.7	54.4
−19	14.4	7.0	−28.3	31	66.2	50.8	−0.6	81	168.5	143.0	27.2	131	345.2	311.0	55.0
−18	15.1	7.6	−27.8	32	67.7	52.1	0.0	82	171.3	145.5	27.8	132	349.7	315.4	55.6
−17	15.8	8.1	−27.2	33	69.2	53.4	0.6	83	174.0	148.1	28.3	133	354.2	319.8	56.1
−16	16.5	8.7	−26.7	34	70.7	54.8	1.1	84	176.8	150.6	28.9	134	358.7	324.2	56.7
−15	17.2	9.3	−26.1	35	72.2	56.1	1.7	85	179.6	153.2	29.4	135	363.3	328.8	57.2
−14	17.9	9.9	−25.6	36	73.8	57.5	2.2	86	182.4	155.9	30.0	136	368.0	333.3	57.8
−13	18.7	10.5	−25.0	37	75.4	58.9	2.8	87	185.3	158.5	30.6	137	372.7	337.9	58.3
−12	19.4	11.1	−24.4	38	77.0	60.3	3.3	88	188.2	161.2	31.1	138	377.4	342.6	58.9
−11	20.2	11.7	−23.9	39	78.6	61.7	3.9	89	191.1	163.9	31.7	139	382.1	347.3	59.4
−10	20.9	12.3	−23.3	40	80.2	63.2	4.4	90	194.1	166.7	32.2	140	387.0	352.1	60.0
−9	21.7	13.0	−22.8	41	81.9	64.6	5.0	91	197.1	169.5	32.8	141	391.8	356.9	60.6
−8	22.5	13.7	−22.2	42	83.6	66.1	5.6	92	200.1	172.3	33.3	142	396.7	361.7	61.1
−7	23.4	14.3	−21.7	43	85.3	67.6	6.1	93	203.2	175.2	33.9	143	401.6	366.6	61.7
−6	24.2	15.0	−21.1	44	87.0	69.2	6.7	94	206.3	178.1	34.4	144	406.6	371.6	62.2
−5	25.0	15.7	−20.6	45	88.8	70.7	7.2	95	209.4	181.0	35.0	145	411.7	376.6	62.8
−4	25.9	16.4	−20.0	46	90.6	72.3	7.8	96	212.5	184.0	35.6	146	416.7	381.1	63.3
−3	26.8	17.2	−19.4	47	92.4	73.9	8.3	97	215.7	186.9	36.1	147	421.9	386.8	63.9
−2	27.7	17.9	−18.9	48	94.2	75.5	8.9	98	219.0	190.0	36.7	148	427.0	392.0	64.4
−1	28.6	18.7	−18.3	49	96.0	77.2	9.4	99	222.2	193.0	37.2	149	432.2	397.2	65.0
0	29.5	19.4	−17.8	50	97.9	78.8	10.0	100	225.5	196.1	37.8	150	437.5	402.5	65.6

**Red Italics Indicate Inches of Mercury Below Atmospheric Pressure*

Figure SR02

3. Refer to *Figure SR02*. On an HFC-407C system, assume the high-side pressure is 271 psig and the measured liquid line temperature is 97°F. The amount of subcooling is ______.
   a. 0°F
   b. 5°F
   c. 16°F
   d. 24°F

3.0.0 Lubricating Oils and Refrigerant Conversions

Objective

Describe various lubricating oils used in refrigerant circuits and identify considerations related to refrigerant and oil conversions.

a. List and describe important refrigeration oil properties.
b. Compare mineral-based and synthetic oils.
c. Describe oil contaminants and testing methods.
d. State common practices associated with handling, charging, and removing oils.
e. Identify considerations and describe common practices related to refrigerant and oil conversions.

Performance Tasks

There are no Performance Tasks in this section.

Oil is required in a compressor to lubricate the moving parts. Oil also helps provide a seal around piston rings and other minute openings, and it helps dampen mechanical noise. Oil serves as a coolant and flushes tiny metal particles from moving parts and into the oil where they can collect and be drained away.

For many years, mineral oils were chosen for systems that used CFC and HCFC refrigerants, including the previously popular CFC-12 and HCFC-22. Additives used to enhance the lubricating qualities and the stability of mineral oils made them adequate for most applications.

The newer refrigerants that replaced CFC and HCFC refrigerants are not, however, compatible with mineral oils. As a result, various synthetic oils have been developed to be compatible with the new refrigerants.

All refrigerant oils must have good lubricating qualities and be **miscible** with the refrigerant with which it is being used. Miscibility allows the oil and refrigerant to dissolve within each other, allowing the mixture to move around the system together and ensuring the oil returns to the compressor. Additionally, oil must have a viscosity that allows it to flow under all the temperature conditions in which the system must operate. Like refrigerants, the oil must be free of water.

Miscible: Having the necessary properties to mix with or dissolve into another material. Many refrigerants, for example, are miscible in refrigeration oil.

3.1.0 Lubricating Oil Properties

All lubricating oils, whether mineral or synthetic, require properties that make them useful in a refrigerant circuit. The properties of greatest concern include the following:

- Stability
- Viscosity
- **Pour point**
- **Dielectric strength**
- **Floc point**
- **Flash point**
- Compatibility
- Tendency to foam

Pour point: An oil quality related to viscosity, defined as the temperature at which oil first starts to flow. See *viscosity*.

Dielectric strength: The ability of an electrical insulating material to resist breakdown in the presence of voltage.

Floc point: The temperature at which a 90%/10% mixture of oil and refrigerant forms a cloudy precipitate of wax in the mixture.

Flash point: The temperature at which heated oil vapors burst into flame. Flash point is a measure of a lubricant's flammability; higher flash points are best.

3.1.1 Stability

Refrigerant oil, like refrigerant, must be stable over the complete range of temperature and pressure conditions encountered in the circuit. It must not break down or combine with refrigerants or other substances in the system to form harmful byproducts.

3.1.2 Viscosity

Viscosity can be described as the resistance of a fluid to flow. Honey, for example, has a higher viscosity than water. As a result, honey has a harder time flowing from a container than water does. The viscosity of a fluid often varies with its temperature. Warm a jar of honey and it will flow from the jar much faster because it is less *viscous*.

The viscosity of a refrigeration oil must be low enough to enhance flow and provide proper lubrication over the complete range of temperatures encountered in refrigerant circuits. An oil that is too thick, or *viscous*, at cold temperatures can cause mechanical stress and impede lubrication.

3.1.3 Pour Point

The pour point is a quality related to viscosity. Pour point represents the temperature at which oil first begins to flow. In low-temperature refrigeration systems, it is important to use an oil with a low pour point to ensure that lubrication remains effective at low temperatures.

3.1.4 Dielectric Strength

Dielectric strength is the ability of oil or any material to resist breaking down in the presence of voltage. Refrigeration oil must have a dielectric strength high enough so that it does not break down or support an arc under normal operating voltage conditions. Water lowers the dielectric strength of the oil and causes it to break down, among other negative consequences.

3.1.5 Floc Point

The floc point of oil is the temperature at which a specific mixture of oil and refrigerant forms a cloudy, or *flocculent*, precipitate of wax in the mixture. Wax is generally undesirable in a refrigerant circuit, as it can plug orifices and prevent check valves from functioning properly if it collects and solidifies. Paraffin (wax) impurities in mineral oils that cannot be removed are the source. Synthetic oils do not contain paraffin, and therefore do not have a floc point.

The floc point of a mineral oil must be low enough to ensure wax does not form at the operating temperatures normally encountered in the system.

3.1.6 Flash Point

The flash point of an oil is the temperature at which heated vapors burst into flame. In other words, flash point is a measure of flammability. Ideally, compressor lubricant oils have a high flash point.

3.1.7 Material Compatibility

Compressor lubricating oils must be compatible with the various materials in the compressor and the refrigerant circuit, as well as with the refrigerant. Under normal operating conditions, lubricating oils do not initiate corrosion or otherwise react with materials, such as plastics and desiccants to break them down or form contaminants.

3.1.8 Foaming

Some foaming in the refrigeration oil is normal. Foaming is typically caused by mechanical movement of the compressor components and dissolved refrigerant boiling out of the oil. Excessive foaming is usually caused by vaporizing refrigerant.

Excessive foam reduces the lubricating qualities of the oil, leading to mechanical wear in the compressor. Foaming can be reduced by maintaining the oil at the correct level, and by using a crankcase heater to prevent refrigerant from condensing in the compressor crankcase during off cycles. Some refrigeration oils contain antifoaming additives.

3.2.0 Types of Refrigeration Oil

A variety of oils are used in HVACR systems today. The list expanded rapidly as new refrigerants were developed, many of which are incompatible with traditional mineral oils.

3.2.1 Mineral Oils

Mineral oils are used extensively in systems with CFC and HCFC refrigerants. Mineral oils have a paraffin base, meaning they are derived from the same crude oil from which gasoline and diesel fuel are refined.

Mineral oils can be classified as *naphthenic, paraffinic*, and *aromatic*. These categories refer to their chemistry. Generally, naphthenic mineral oils were chosen for use with the CFC and HCFC refrigerants.

Mineral oils are available in a wide range of viscosities and contain specific additives to make them compatible with the refrigerant and the application in which they are being used. However, mineral oils do not function well with HFCs and other new refrigerants, primarily due to low miscibility. Mineral oils and HFC refrigerants do not mix well, creating a problem with oil flow back to the compressor.

Since CFC and HCFC refrigerants are being phased out, fewer systems now contain mineral oil. Synthetic oils now dominate the market and are engineered to work better with new and unique refrigerants.

3.2.2 Synthetic Oils

The predominant synthetic oils in use are based on glycol, esters, ethers, or alkylbenzene. *Figure 17* shows a typical container of ester-based oil.

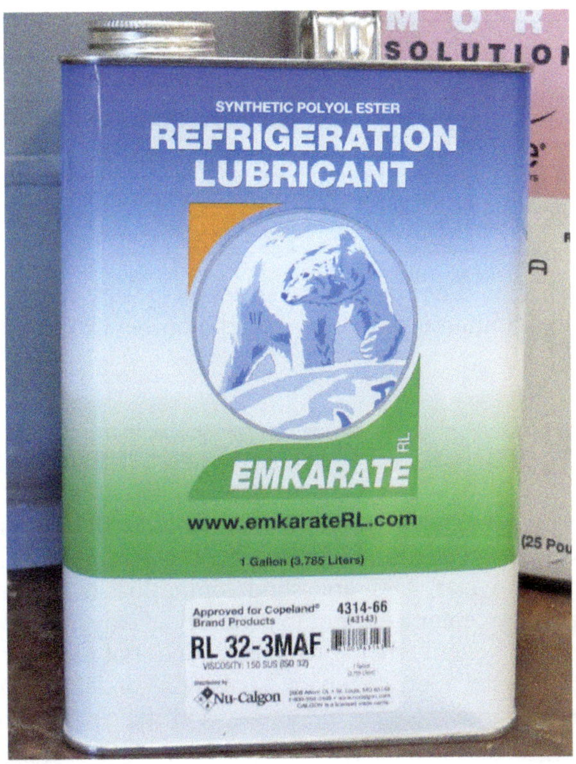

Figure 17 Synthetic ester-based refrigerant oil.

The following provides a brief introduction to the synthetic oils:

- *Alkylbenzene oil* is closely related to mineral oil and used in similar applications. It is compatible with mineral oil but has properties that improve its movement through the system. Alkylbenzene oil is widely used in low-temperature applications but it is not compatible with other synthetic oils. It works well with CFC and HCFC refrigerants.
- *Polyalkylene glycol (PAG) oil* is widely used in automotive air conditioning systems using HFC-134a refrigerant. As CFC-12 automotive air conditioners are converted to HFC-134a, the original oil must be drained and replaced with PAG oil. PAG oils are not compatible with mineral oil or alkylbenzene, and is the most **hygroscopic** type of oil of those listed here.
- *Polyolester (POE) oil* is compatible, and widely used, with most HFCs. Like PAG oils, POE is not compatible with mineral oil or alkylbenzene, and it is also hygroscopic.

Hygroscopic: Tending to absorb water and water vapor readily and aggressively, much like a desiccant.

- *Polyvinyl ether (PVE) oil* has properties similar to mineral oils, but can be used with HFC refrigerants. PVE oils are not as hygroscopic as POE and PAG oils.

Refrigerant compatibility with these oils varies. The equipment or refrigerant manufacturer must be consulted to determine the proper oil to use for an application. Common refrigerant and oil matches are shown in *Table 6*. For example, HCFC-123 is compatible with mineral oil, alkylbenzene oil, and synthetic oils such as POE and PVE. On the other hand, most HFCs such as HFC-410A and HFC-134a must be used only with a synthetic oil product.

TABLE 6 Matched Refrigerants and Oils

Refrigerant	Compatible Oils
HCFC-22	Mineral oil
HFC-32	Synthetic (POE, PVE)
HCFC-123	Mineral, alkylbenzene, and synthetic (POE, PVE)
HFC-134a	Synthetic (PAG, POE)
HC-290 (propane)	Synthetic (POE)
HFC-404A	Synthetic (POE, PVE)
HFC-407C	Synthetic (POE, PVE)
HFC-410A	Synthetic (POE, PVE)
HFO-454B	Synthetic (POE)
HFO-1234yf	Synthetic (PAG, POE)
HFO-1234ze	Synthetic (POE, PAG)

Note that *Table 6* should not be used to select oil for any specific application. Always follow the manufacturer's recommendation regarding the appropriate oil for all systems.

3.3.0 Refrigeration Oil Contamination and Testing

Ideally, the only two substances inside a refrigerant circuit are refrigerant and refrigerant oil. Anything else must be considered a contaminant. Typical contaminants include air, water, acid, and solid contaminants such as metal particles. Contaminants can enter a system in a variety of ways.

Air usually enters a system when a technician does not properly evacuate the system following an installation or repair. Air is non-condensable in the circuit, which causes the discharge pressure to increase and the circuit to operate erratically. Another problem with air is that it usually contains water vapor, a more harmful contaminant. The best way to avoid air in a system is to thoroughly evacuate a system following a leak test. A thorough evacuation removes both air and moisture.

The use of ACR-grade refrigerant tubing helps to ensure no moisture is in the piping materials when installed. Unused tubing should be tightly sealed immediately after being cut. Any time a refrigerant circuit is opened for service, the open circuit should be capped or plugged as soon as possible to keep moisture out. Multiple evacuations, with each vacuum broken with dry nitrogen, and repeatedly replacing the filter-drier can remove moisture from a contaminated system. This assumes that a small amount of moisture entered along with the air, perhaps on a very humid or rainy day. A system flooded with water requires a detailed analysis before acting. In many of these cases, the equipment must be replaced. Flooding does occur in chillers and water-cooled systems of all types.

Acid forms inside the system as the result of a compressor motor burnout or through substances combining with refrigerant, oil, and/or water at a high temperature. Once acid is in the system, it must be removed using filter-driers and/or by replacing the oil one or more times. Oil replacement is difficult with hermetic compressors. In addition to a liquid line filter-drier, a suction line filter-drier with charcoal to absorb acids in its core is often installed as well.

When the suction filter-drier must be replaced repeatedly to ensure a clean system, it is best to use a replaceable core filter-drier on larger systems (*Figure 18*).

Once the cleanup process is complete, the suction line filter-drier should be removed completely. Because suction line filter-driers are filtering vapor and oil, they must be larger to minimize the pressure drop through them. Once the cleanup is complete, it is best to eliminate the pressure drop they typically cause. If a replaceable core model is used, the shell can remain in the circuit; simply remove the core(s). A clean liquid line filter-drier should remain in the circuit.

Solids are usually metal particles resulting from normal wear, or leftovers from the manufacturing process. Other solid particles can come from improper repair procedures such as cutting tubing with a hacksaw (copper filings), or simply allowing dirt to get into a component while working on the ground. Solid particles can plug orifices or score piston cylinders.

Some compressors are equipped with magnetic drain plugs in the compressor sump that capture ferrous particles. Draining the compressor oil (when practical) helps remove solid contaminants from the sump. Filter-driers capture solid contaminants of all types as the system runs, but only if the contaminants remain in motion long enough to reach the filter.

(A) Shell (B) Charcoal-Infused Core

Figure 18 Replaceable core filter-drier shell and core.
Source: Courtesy of Sporlan Division—Parker Hannifin Corporation

On Site

Cheap Insurance

A liquid line filter-drier is a relatively inexpensive component. Make a habit of installing one in the liquid line before it enters the metering device. A filter-drier is cheap and appropriate insurance. It protects the system from contaminants and can make your job of cleaning up a system much easier if a compressor burnout occurs. Replace the filter-drier any time a refrigerant circuit is opened for any significant service.

3.3.1 Oil Testing

Oil and refrigerant can be tested for the presence of acid and moisture. An acid/moisture test kit (*Figure 19*) can be used to sample refrigerant vapor from a service port. Crystals in a sealed tube change color in the presence of acid and/or moisture.

CAUTION

Check the label of an oil test kit carefully and ensure that it is the correct type for the oil being tested. Using the wrong kit can cause misleading results.

Before After (Positive Result)

Figure 19 Acid/moisture test kit in use.
Source: QuikProducts, Mainstream Engineering Corporation

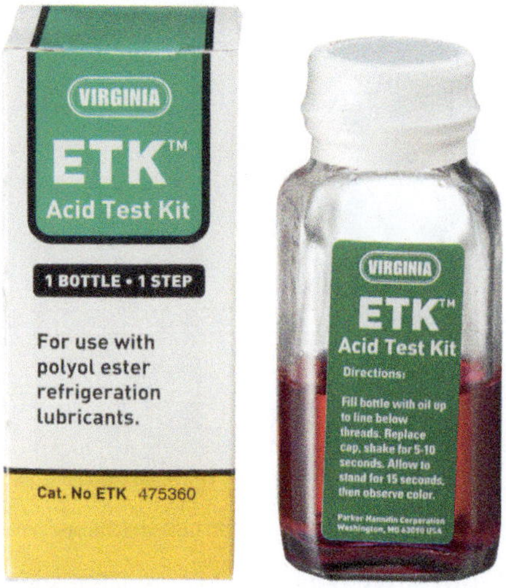

Figure 20 Acid test kit for POE refrigeration oil.

Source: Courtesy of Sporlan Division—Parker Hannifin Corporation

Vapor testing often reveals contamination, but most of the acid is generally entrained in the oil. Oil test kits (*Figure 20*) require an oil sample to be collected. When mixed with the chemicals in the kit, the oil will change color if contaminants are present. Note that there are different kits for different oils. Follow the kit directions carefully to ensure accurate results.

For large and/or critical compressors, oil samples may be taken periodically and sent to a laboratory for a thorough analysis. Wear metals are also measured during this type of testing, allowing the owner to monitor internal wear and establish trends.

On Site

Oil Samples

Obtaining an oil sample from a semi-hermetic compressor is relatively easy. Welded hermetic compressors present a bigger problem. The compressor must either be removed and the oil poured out (invasive and expensive), or a sample must be obtained through the compressor process tube (possible, but challenging). For that reason, an acid/moisture tester that checks the refrigerant is the easiest approach for residential and light commercial systems.

3.4.0 Oil Handling Guidelines

Refrigerant oils must be handled properly to avoid personal injury, damage to sensitive materials, and a negative environmental impact. Always consult the SDS for the oil to fully understand the product. The SDS lists all potential health hazards and provides the necessary precautions and handling requirements. *Appendix 03301A* shows an SDS for a typical POE refrigerant oil for your review.

WARNING!

When handling refrigerant oils, wear oil-resistant gloves and eye protection. Use caution when handling oil from a compressor that has experienced a burnout because the oil may contain harmful acids. Synthetic oils such as POE will irritate exposed skin. Use soap and water to wash the affected area if contact occurs. If a spill occurs, clean it up immediately to prevent slips and falls.

3.4.1 Working with Oils

Most refrigerant oils are hygroscopic to some degree. That is, the oils readily absorb moisture from the air. Because moisture is a major refrigeration system contaminant, refrigeration oil must be kept free of water.

That can be more challenging than it sounds. Synthetic oils such as POE and PAG are far more hygroscopic than mineral oils. In fact, many synthetic oils are so hygroscopic, these oils will not be found packaged in plastic containers; they are packaged in metal containers only (*Figure 21*). This is because synthetic oils can draw moisture from the air through the walls of a plastic container over time.

This concept is difficult for some to grasp. But it is an indication of the care that must be taken to prevent synthetic oils and moisture from being in the same place at the same time. Once moisture has contaminated refrigerant oil, the oil is virtually useless and must be disposed of properly.

It is also important to note that the bond that takes place between hygroscopic oils and water resists separation through evacuation. Filter-driers will capture the water over time, but less effectively as with mineral oils.

When working with synthetic oils, follow these guidelines:

- If a system must be opened for a repair, seal any openings immediately following separation.

Figure 21 Refrigeration oil containers.
Source: Courtesy of Sporlan Division—Parker Hannifin Corporation

- Avoid opening a system or an oil container on a rainy or foggy day whenever possible. Keep oil containers closed unless oil is being pumped. Ensure the neck of the container is sealed around the oil pump dip tube while in use.
- Use fresh oil from a new, unopened container when possible. Obtain a container of oil closest in size to the amount needed to minimize the storage of unused oil.
- Pump oil into the system; do not pour it in as this exposes the oil to air.
- When retrofitting a system that used mineral oil, purge all the oil from the system. Synthetic oils can usually tolerate up to 5% mineral oil in the system without problems. Oil test kits are available to measure the amount of mineral left in a system (*Figure 22*), but these kits may be ineffective until the system has operated for some time to mix any remaining oil with the new oil.

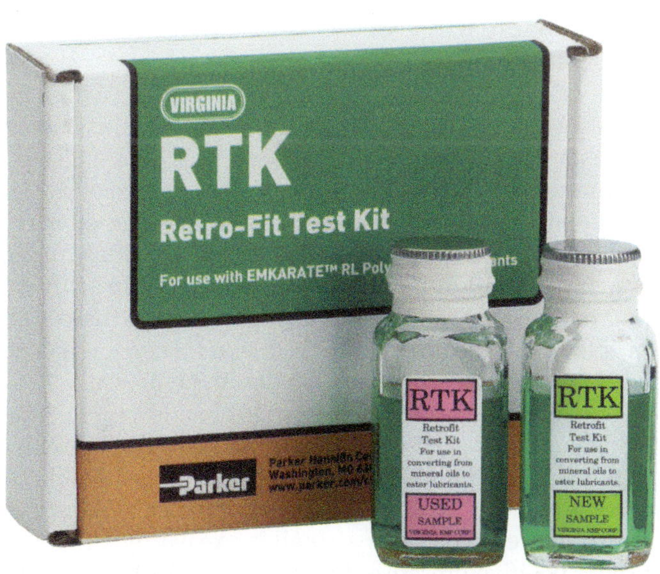

Figure 22 Residual mineral oil test kit.
Source: Courtesy of Sporlan Division—Parker Hannifin Corporation

CAUTION

Some roofing materials are sensitive to refrigeration oils and can be seriously damaged. When working on a rooftop, avoid allowing oil to drain or spill onto the roof. Thoroughly clean the roof following a spill of any kind.

Also remember that, like refrigerants, refrigerant oils should not be mixed. It is acceptable to use the same oil from different containers made by the same manufacturer.

- If a hygroscopic oil such as PAG or POE oil has absorbed some moisture, it cannot be removed effectively with a vacuum pump. The chemical bond is not that easily broken. An approved filter-drier offers the best performance. The desiccant of the drier does have a higher affinity for moisture than the oil does, but not generously so.
- Synthetic oil can act as a mild solvent. When the oil is changed to support a refrigerant retrofit, the oil can clean the inside of tubing and other components, possibly disturbing contaminants and placing them in circulation. For that reason, the installation of a fresh filter-drier is recommended when replacing the oil to collect any dislodged contaminants.
- Follow the equipment manufacturer's instructions to prevent damage to roofing materials. If protective measures are not taken, roof membranes can be damaged even when spilled oil is cleaned off the roof quickly.

3.4.2 Adding and Removing Oil from a System

Before adding or removing oil from a compressor, it is best to consider why the task is required. If the compressor is low on oil, it may indicate a piping design problem that is preventing oil from returning, or there may be a blockage in the system, such as an accumulator with a blocked oil-return orifice. If there is no evidence of leakage and the oil level was correct previously, then it is still somewhere in the system.

If the oil level is excessive, it may indicate that oil previously trapped in the piping has now returned to the compressor, and oil had been added when it wasn't necessary. To summarize, never casually add or remove oil from a system. Always make sure the reasoning is sound first.

Generally, oil is added or removed from semi-hermetic compressors that have an oil sight glass. Oil is not usually added to welded hermetic compressors because there is no reliable way of determining the oil level, short of removing the compressor and pouring the oil out—a costly and time-consuming process. Even then, there is no way to know how much oil is out in the system.

It is best to add oil to a compressor using a hand pump (*Figure 23*) and draw the oil directly from the original container. You can remove oil using a similar pump, a siphon, or by gravity when the circumstances allow it. The pump on the left in the figure can be used with a variety of containers, and to pump oil from a compressor crankcase. The pump on the right fits the neck of several standard containers.

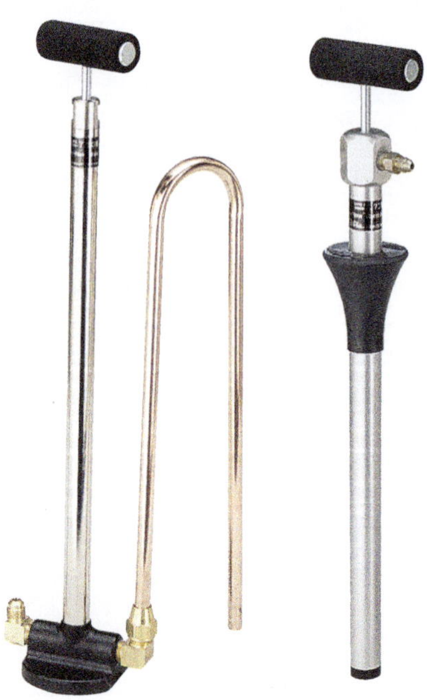

Figure 23 Refrigeration oil pumps.
Source: Ritchie Engineering Co., Inc./YELLOW JACKET

Before adding oil to a semi-hermetic compressor, reduce the crankcase pressure to zero or within 1 or 2 psig (7 to 14 kPa). To do this, close the suction service valve of an operating compressor. This will allow the compressor to pump down. Then close the discharge service valve to isolate the compressor as the suction pressure falls to zero.

Several cycles may be necessary to clear the crankcase as refrigerant vaporizes from the oil. Another way to reduce the pressure is to valve off the compressor and recover the refrigerant from it.

Oil is then pumped into the crankcase through a service port. *Figure 24* shows a cutaway of a semi-hermetic compressor crankcase. The oil service port and sight glass are both located on the far sidewall and are hidden by the crankshaft.

Oil pumps generally produce enough pressure to pump against a reasonable backpressure. Maintaining a slight positive pressure in the compressor helps keep air from entering. If the compressor is opened when it is at a negative pressure, air will be drawn in immediately to balance the pressure.

To remove oil from a semi-hermetic compressor, lower the crankcase pressure as previously described. If the compressor is equipped with an oil drain plug, remove the plug, and allow the oil to drain into a suitable container. For compressors that receive a high degree of maintenance as a critical system, the drain plug is usually replaced with a small back-seating valve to simplify the task. A very slight positive pressure (1 to 2 psig, or 7 to 14 kPa) in the crankcase will help push the oil out. The pressure can be provided by a nitrogen cylinder if needed.

Figure 24 Semi-hermetic compressor crankcase cutaway.

The semi-hermetic compressor shell usually has a removable plug or service valve above the oil level. Oil is usually added through this port, but it can also accommodate a siphon tube inserted into the oil sump to pump the oil out. This is a better option when the installation makes draining the oil from the plugged opening messy or impractical. Always follow the compressor manufacturer's instructions for adding or removing oil.

3.4.3 Waste Oil Disposal

Waste oil is not considered a hazardous waste unless it is mixed with other oils. Oil contaminated with acid, lead, or other heavy metals or contaminants that would make it flammable or corrosive is considered hazardous waste.

Take uncontaminated used oil to an oil recycler. Oils deemed hazardous must be taken to a hazardous waste disposal facility.

Federal and state regulations governing the handling of hazardous waste are constantly evolving. For that reason, technicians should check with local agencies and review the current regulations.

WARNING!

Wear rubber gloves and eye protection when handling waste oil of any kind.

3.5.0 Refrigerant and Oil Conversions

With CFC, HCFC, and HFC refrigerants being phased out, refrigerant manufacturers have developed more environmentally friendly refrigerants. Converting to a different refrigerant to replace a phased-out product can extend the life of equipment in good condition. However, the decision to do so cannot be made casually. Many issues need to be considered.

WARNING!

No current system using CFC, HCFC, or HFC refrigerants can be converted to a refrigerant that has a flammable ASHRAE rating, such as an A2L refrigerant.

NOTE

Consult the equipment manufacturer for guidance before attempting a refrigerant conversion. Manufacturers have knowledge that will help avoid problems and/or voiding of any remaining equipment warranty.

3.5.1 Refrigerant Conversion Considerations

Issues to be addressed during the conversion include the following:

- material and equipment compatibility
- refrigerant characteristics
- lubricant compatibility

Material and Equipment Compatibility

Materials used in a system designed for one refrigerant may not be compatible with another refrigerant or its required oil. In addition, equipment such as indoor and outdoor coils may not be compatible for other reasons, such as a pressure rating that is lower than that required for the new refrigerant.

Material compatibility is not a significant issue with metal parts. However, if the system contains natural or synthetic O-rings, gaskets, or similar parts, the system may not be compatible with the new refrigerant and/or oil. Leaks or component failure may result. When automotive air conditioners were converted from CFC-12 to HFC-134a, for example, O-rings in the compressor often had to be changed. Automotive supply houses often sell conversion kits designed for specific compressors and refrigerants.

The filter-drier must always be replaced when changing refrigerants and/or oils. The core material in a filter-drier is designated for use with certain refrigerants. Make sure the new filter-drier is compatible with the new refrigerant and oil. This information is usually available on the label, but it can always be acquired from the manufacturer's catalog data.

The new refrigerant often requires a different metering device. On systems using a fixed metering device, the size of the orifice will usually need to change. Thermostatic and electronic expansion valves may or may not be compatible with the new refrigerant. These devices need to be checked for compatibility.

The suitability of the equipment itself must also be considered. In the past, when a residential HCFC-22 system was being converted to HFC-410A, it was common to replace the outdoor condensing unit and retain the existing indoor unit and interconnecting tubing. Retaining the indoor coil, however, is not always an option. Two important issues must be considered:

- The existing indoor coil may not be rated for the pressures of the new refrigerant, posing a safety hazard. This was often the case when old systems were being converted from HCFC-22 to HFC-410A. The indoor coil must be checked to ensure it has the appropriate pressure ratings to handle the new refrigerant. The manufacturer can supply the needed information.
- The existing indoor coil may not allow the system to deliver the capacity, or the efficiency, expected from the new outdoor unit.

Refrigerant Characteristics

The new refrigerant will rarely have the same performance characteristics of the refrigerant it replaces. For example, HFC-407C is a popular replacement refrigerant for HCFC-22. Product data sheets state that HFC-407C is a reasonable performance match for HCFC-22. However, especially in low-temperature applications, decreased capacity is to be expected. The equipment manufacturer will know what to expect in terms of capacity for any approved refrigerant replacement.

Lubricant Compatibility

If the new refrigerant requires an oil that is not compatible with the current oil, every drop of old oil removed is important. Synthetic oils can usually tolerate no more than 5% contamination from mineral oil.

Procedures for removing oil from a semi-hermetic compressor were discussed earlier in the module. Hermetic compressors without access to the crankcase must be removed and the oil poured out. Even if every drop of the oil in the compressor is removed (unlikely), additional oil remains in the coils, piping, and accessories in the system. Components that normally trap and hold small amounts of oil, such as an accumulator or receiver, must also be removed to drain trapped oil.

To remove trapped oil in coils and piping, begin by purging the system with dry nitrogen (*Figure 25*).

CAUTION

Ensure that the nitrogen source is completely dry (contains no moisture). The definition of *dry nitrogen* can differ from vendor to vendor, so be certain that you are using nitrogen with the proper characteristics.

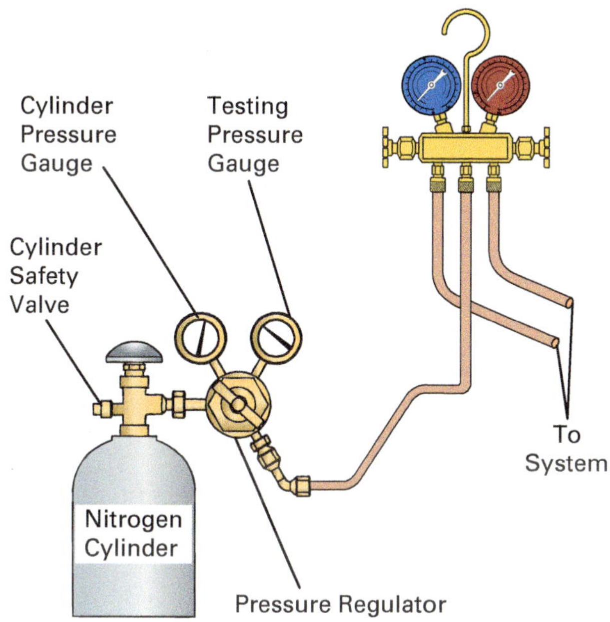

Figure 25 Connecting nitrogen to the system through a gauge manifold.

WARNING!

Never purge or pressurize a system with an unregulated nitrogen source. A pressure regulator is required. The cylinder pressure (typically over 2,000 psig) can rupture system components and cause personal injury. In addition, never pressurize or purge a system with oxygen or a flammable gas. Such a mixture can explode on contact with oil, causing personal injuries.

When a residential split-system is replaced with equipment that uses an incompatible oil, it is often unnecessary to replace the interconnecting lines. However, because the lines will contain some oil, they must be purged. Using nitrogen gas alone for this is not productive. Oil in trapped areas and clinging to the walls of the tubing is unlikely to move.

A liquid is usually circulated through the lines to pick up and carry the oil out. Special kits are available for this purpose. *Figure 26* shows one of the options, which relies on a solvent-based liquid that evaporates easily after use. A propellant or an external pressure source moves the product through the circuit.

Note that this approach can also be used to flush the indoor coil while it remains connected to the lines. However, remember that the metering device must be removed and bypassed. Otherwise, the velocity and turbulent flow through the system will be significantly reduced, defeating the purpose of the flush.

Follow these basic steps to use a flushing product. However, follow the manufacturer's guidance carefully for the product in use:

- The system refrigerant is recovered, and the old outdoor unit is removed. The metering device is removed, and a temporary bypass is installed in its place.
- An appropriate vessel is prepared at the end of one disconnected refrigerant line outdoors to capture the oil and materials as they exit the system. An empty one-gallon oil container will usually suffice.

Figure 26 Refrigerant line flushing kit.
Source: Rectorseal

- Nitrogen is used to provide the pressure needed to circulate the solvent if the product is not fully self-contained. The nitrogen source is connected to the product solvent tank.
- The piping and/or indoor coil are flushed for several minutes at 150 psig (1,034 kPa). The discharge rate from the solvent tank can often be throttled so the solvent stays in the system a little longer.
- The discharged product is visually checked to ensure it appears clear and free of oil and debris. The flushing continues if the discharge still appears to contain oil or contaminants.
- Once the flush is complete, the system is then evacuated. The solvent evaporates very quickly as the evacuation progresses. Monitor the vacuum pump for contamination from the solvent and change as necessary.

Several refrigerants are on the market that can be used to retrofit an existing system. Few, if any, should be considered drop-in replacements. In nearly every case, something, and often several things, need to change. As HCFC-22 was being phased out, there were a few refrigerants that could be used as interim replacements until an HFC-410A-based system was installed. Now that we are replacing HFCs, and many of the refrigerants considered as long-term replacements have some degree of flammability, fewer interim refrigerants will be available. It is also important to note that contractors assume a degree of risk when retrofitting a system with an interim refrigerant. The frequency at which you encounter retrofit projects will likely be affected by your employer's position on retrofitting existing systems and the liability for problems that they are willing to assume.

Retrofits and conversions require careful thought and research before moving forward. To summarize the issue of retrofits and conversions, there are two important points to remember that cannot be overemphasized:

- Do not plan a retrofit or conversion on your own. Contact the equipment manufacturer for guidance and support and follow their guidance carefully. This is essential if you expect to have their support if problems arise. Do not proceed with a retrofit or conversion without the support of the equipment manufacturer.
- Refrigerants with an ASHRAE flammability rating of 2, 2L, or higher, cannot be used in existing systems. These refrigerants must only be used in systems designed for them.

3.0.0 Section Review

1. Warming a lubricating oil generally reduces its ______.
   a. stability
   b. viscosity
   c. floc point
   d. dielectric qualities

2. Mineral oils are derived from ______.
   a. crude oil
   b. coal tar
   c. amino acids
   d. polypropylene

3. True or False: Once installed, a suction line filter-drier should always remain in the line.
   a. True
   b. False

4. Which gas can be used to safely push oil out of a semi-hermetic compressor crankcase?
   a. Acetylene
   b. Propane
   c. Oxygen
   d. Nitrogen

5. Which of these components must typically be removed to drain the oil out of it before a new oil is used in a system?
   a. Metering device
   b. Accumulator
   c. Indoor coil
   d. Refrigerant line set

Module 03301 Review Questions

1. What is likely to happen to a refrigerant when heat is added or taken away from it while it remains at a constant pressure?
   a. It expands.
   b. It condenses or vaporizes.
   c. It breaks down.
   d. It contracts.

2. A hydrocarbon that has some or all its hydrogen atoms replaced with chlorine atoms is called a(n) ______.
   a. azeotrope
   b. HFO
   c. halocarbon
   d. hydrofluorocarbon

3. The group of refrigerants that are *least* likely to damage the ozone layer are the ______.
   a. chlorinated halocarbons
   b. CFCs
   c. HCFCs
   d. HFCs

4. If you have a white refrigerant cylinder with a red band around it, which of the following refrigerants might it contain?
   a. HFC-410A
   b. HFO-1234yf
   c. HCFC-22
   d. HCFC-123

5. Which of the following refrigerants has the *highest* ozone depletion potential (ODP)?
   a. HFC-410A
   b. HCFC-22
   c. HCFC-123
   d. HFC-134a

6. What type of refrigerant do you have if its PT chart contains a bubble point and a dew point?
   a. HCFC
   b. Near-azeotrope
   c. Zeotrope
   d. Ternary

7. With what type of refrigerant must glide be considered when using a PT chart?
   a. Zeotropes
   b. Azeotropes
   c. Compounds
   d. Binary blends

8. Which of the following is *not* required to determine the subcooling for a system?
   a. Liquid line temperature
   b. Low-side pressure
   c. High-side pressure
   d. Refrigerant saturation temperature

9. The temperature at which an oil first begins to flow is known as the ______.
   a. floc point
   b. viscosity
   c. flash point
   d. pour point

10. A very low floc point for a mineral oil helps avoid the formation of ______.
    a. wax
    b. acid
    c. foam
    d. sludge

11. What type of compressor oil is likely to be found in an automotive air conditioner using HFC-134a?
    a. POE
    b. PVE
    c. PVC
    d. PAG

12. Acid can form inside a refrigerant circuit due to ______.
    a. low refrigeration loads
    b. a motor burnout
    c. a low oil floc point
    d. incorrect oil viscosity

13. A comprehensive review of health hazards for a given refrigeration oil can be found on ______.
    a. the unit dataplate
    b. the compressor data plate
    c. its SDS
    d. the related work order

14. Which of the following can interrupt the flow of oil through the refrigerant circuit as it travels back to the compressor?
    a. A plugged orifice in an accumulator
    b. A suction line pitched toward the compressor
    c. A high discharge pressure
    d. A high velocity in the suction line

15. True or False: Nitrogen alone is an ideal choice to flush refrigerant oil from existing refrigerant lines, coils, and other system components.
    a. True
    b. False

Answers to Odd-Numbered Module Review Questions are found in *Appendix A*.

Answers to Section Review Questions

Answer	Section	Objective
Section One		
1. c	1.1.5	1a
2. a	1.2.2	1b
3. b	1.3.1	1c
4. d	1.4.0; Table 1	1d
5. a	1.5.0	1e
6. c	1.6.0	1f
Section Two		
1. a	2.1.2	2a
2. c	2.2.1	2b

Superheat = Measured Temperature − Saturation Temperature
Superheat = 52°F − 37°F
Superheat = **15°F**

Answer	Section	Objective
3. c	2.3.2	2c

Subcooling = Saturation Temperature − Measured Temperature
Subcooling = 113°F − 97°F
Subcooling = **16°F**

Answer	Section	Objective
Section Three		
1. b	3.1.2	3a
2. a	3.2.1	3b
3. b	3.3.0	3c
4. d	3.4.2	3d
5. b	3.5.1	3e

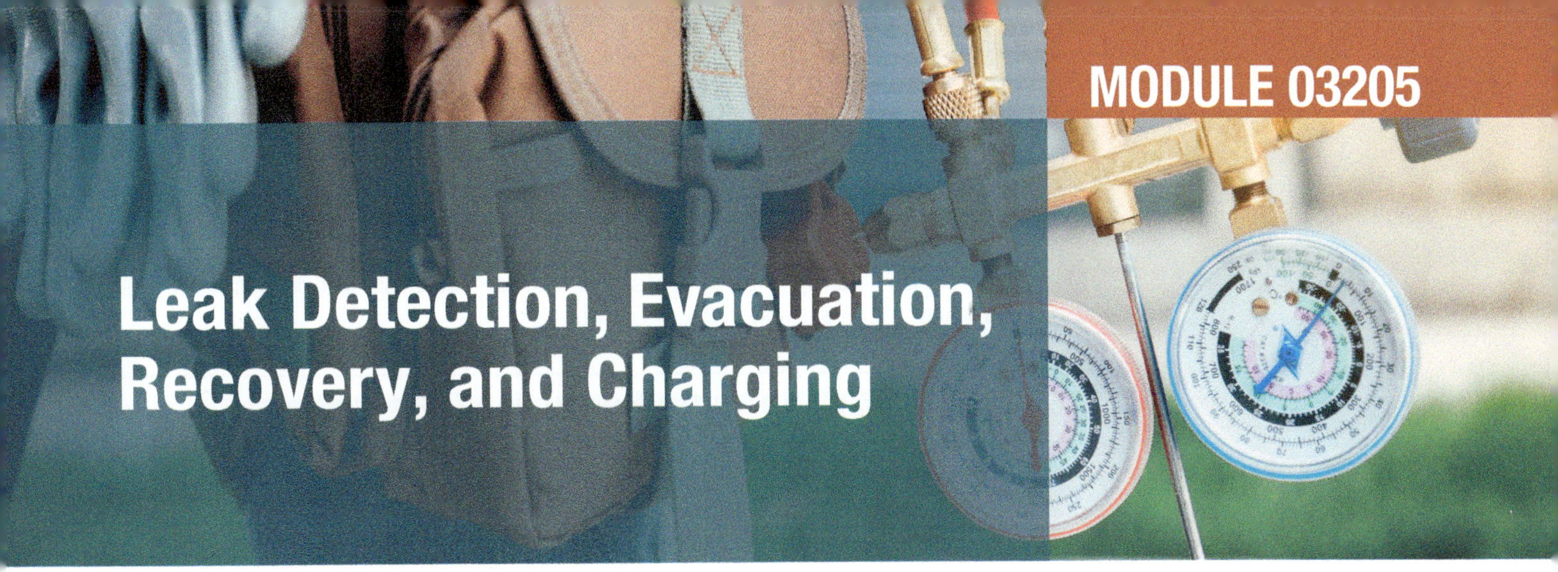

Source: I AM NIKOM/Shutterstock

MODULE 03205

Leak Detection, Evacuation, Recovery, and Charging

Objectives

Successful completion of this module prepares you to do the following:

1. Describe the equipment and approaches used to test refrigerant circuits for leaks.
   a. Describe various methods used to detect and locate refrigerant leaks.
   b. Describe how leak detection is approached based on a system's current refrigerant charge.
2. State refrigerant containment and recovery requirements and explain how to recover refrigerant.
   a. State the basic refrigerant containment requirements of Section 608 of the Clean Air Act.
   b. Describe refrigerant recovery and recycling equipment and explain how to recover refrigerant.
3. State the principles of evacuation and describe how to evacuate refrigerant circuits.
   a. State the principles of refrigerant circuit evacuation.
   b. Describe vacuum pumps and explain how to properly evacuate a system.
4. Explain how to properly charge refrigerant circuits using various techniques.
   a. State important concepts related to refrigerant charging and the related equipment.
   b. Explain how to charge a refrigerant circuit by weight.
   c. Explain how to charge a refrigerant circuit by superheat.
   d. Explain how to charge a refrigerant circuit by subcooling.

Performance Tasks

Under supervision, you should be able to do the following:

1. Demonstrate the ability to conduct a leak test on a pressurized refrigerant circuit using at least two of the following methods:
   - Electronic leak detection
   - Ultrasonic leak detection
   - Liquid bubble solution
   - Ultraviolet/fluorescent leak detection kits
2. Use a recovery unit to recover refrigerant from a system.
3. Evacuate a system using the triple-evacuation method and perform a vacuum leak test.
4. Demonstrate how to properly charge a refrigerant circuit using the following methods:
   - By weight
   - By superheat (fixed-orifice metering device)
   - By subcooling (expansion valve)

Digital Resources for HVACR

SCAN ME

Scan this code using the camera on your phone or mobile device to view the digital resources related to this craft.

Overview

The proper handling of refrigerants in accordance with EPA guidelines is one of the most important responsibilities of HVACR technicians. Technicians consistently perform the critical tasks of refrigerant circuit leak detection, evacuation, recovery, and charging. Work related to the refrigerant circuit is expensive and time-consuming, and must be done with attention to detail to ensure users have an efficient, reliable system at their disposal.

NCCER Industry-Recognized Credentials

If you are training through an NCCER-accredited sponsor, you may be eligible for credentials from NCCER. The ID number for this module is 03205. Note that this module may have been used in other NCCER curricula and may apply to other level completions. Contact NCCER at 1.888.622.3720 or go to **www.nccer.org** for more information.

You can also show off your industry-recognized credentials online with NCCER's digital credentials. Transform your knowledge, skills, and achievements into credentials that you can share across social media platforms, send to your network, and add to your resume. For more information, visit **www.nccer.org**.

1.0.0 Refrigerant Circuit Leak Testing

Performance Task

1. Demonstrate the ability to conduct a leak test on a pressurized refrigerant circuit using at least two of the following methods:
   - Electronic leak detection
   - Ultrasonic leak detection
   - Liquid bubble solution
   - Ultraviolet/fluorescent leak detection kits

Objective

Describe the equipment and approaches used to test refrigerant circuits for leaks.

a. Describe various methods used to detect and locate refrigerant leaks.

b. Describe how leak detection is approached based on a system's current refrigerant charge.

Service tasks related to the refrigerant circuit, such as leak detection, evacuation, recovery, and charging, are required when installing new systems and when servicing existing ones. Failure to perform any one of these procedures correctly results in poor or inefficient system operation and may even result in system failure. The work must also be performed in accordance with the requirements of the Environmental Protection Agency (EPA) as set forth in the Clean Air Act.

Refrigerants today must be contained and must not be released into the atmosphere, either voluntarily or involuntarily. Section 608 of the Clean Air Act requires the certification of all technicians who handle refrigerants and governs the use and handling of refrigerants relative to the environment. This section of the Clean Air Act was originally developed to support a global movement to protect the stratospheric ozone layer. Stratospheric ozone provides a level of protection against harmful ultraviolet (UV) rays from the sun.

In recent years, more attention has been placed on global warming. Many refrigerants in use today contribute to the *greenhouse effect* and global warming, although research and development to identify and implement environmentally safe products continues at a rapid pace. The EPA regulations include the following broad requirements, as well as many other detailed regulations:

- Refrigerant system service practices must be implemented that maximize the recovery, recycling, and reclamation of refrigerants. This includes all refrigerants except those that are specifically exempted. Notable exceptions include carbon dioxide (CO_2), ammonia (NH_3), propane (HC-290), isobutane (HC-600a), and water.

- Certification requirements exist for all persons working with refrigerants, recycling and recovery equipment, and HVACR systems that contain controlled refrigerants.
- Appliance disposal and recycling facilities must maintain documents that indicate refrigerant was recovered properly from refrigerant-bearing appliances prior to disposal. Records need to be maintained for three years. The documents include signed statements from those dropping the appliance indicating the refrigerant was recovered. The disposal facility is otherwise responsible for recovering the refrigerant to a specified level.
- Owners or operators must take corrective action when an appliance with a refrigerant charge of 50 pounds (lb) or more leaks an ozone-depleting refrigerant at an annual rate that exceeds the *trigger rate*. *Table 1* shows the trigger rate of leakage for various systems. Owners or operators of an appliance that is leaking an ozone-depleting refrigerant above the applicable trigger rate must do one of the following:
    - Repair leaks within 30 days from the date the leak was discovered, *or*
    - Develop a plan within 30 days to retrofit or retire the appliance and then complete the actions outlined by that plan within 1 year.
- Owners or operators of systems that exceed the trigger rate must also conduct initial and follow-up verification tests after leaks have been repaired.

NOTE

It is important to recognize that the detailed rules related to leak repair only apply to ozone-depleting refrigerants, as of April 2020. This, of course, does not mean you should ignore leaks in systems using HFCs and other refrigerants that have no *ozone depletion potential (ODP)*—refrigerants that do not contain chlorine. However, the detailed rules concerning leak repair and documentation do not currently apply to those systems. While this rule was far-reaching when CFC and HCFC refrigerants were prevalent and their phaseout began, at this writing few systems remain that contain ozone-depleting refrigerants.

TABLE 1 Trigger Rates for Ozone-Depleting Refrigerants

System Type	Annualized Leak Rate
Industrial process refrigeration	30%
Commercial refrigeration	20%
Comfort cooling	10%
All other systems containing 50 lb or more	10%

Going Green

Ammonia (NH_3)

Anhydrous ammonia (ammonia free of water) does not damage the ozone layer or contribute to global warming. This organic refrigerant, which remains popular in industrial refrigeration applications due to its high heat-transfer efficiency and low cost, does not require recovery under the Clean Air Act. However, this doesn't mean that ammonia can be randomly released in bulk, as it is considered a toxic gas by its ASHRAE designation.

Source: Milan Sommer/Shutterstock

1.1.0 Leak Detection Methods

Refrigerant leaks are a consistent problem in HVACR systems. Leaks must be detected and repaired to prevent refrigerant from being released into the atmosphere. Leaks also waste an expensive resource and can allow air and moisture to enter the refrigerant circuit. Refrigerant leaks are very likely the predominant cause of equipment failures that lead to discomfort and jeopardize refrigerated products.

Leak testing is normally performed under the following circumstances:

- When an operating system is found with a low refrigerant charge.
- Following assembly or installation, but before evacuating and charging a new system.
- After repairing a leak or replacing a component in a refrigerant circuit.
- When testing indicates air and/or moisture exists in a system using a low-pressure refrigerant that operates below atmospheric pressure (primarily centrifugal chillers).

It is not acceptable to consider the problem solved by simply topping off the refrigerant charge on a system if it is found to be low. The refrigerant circuit is a sealed system that should retain the charge unless there is a leak or a relief valve has opened. Remember that refrigerant never wears out or needs to be replaced due to its age. When you encounter a system with a low refrigerant charge, you must assume there is a leak unless you have reason to believe the system was not charged properly during previous service. A well-trained, competent technician knows how to avoid that type of error. Leaks need to be located and repaired.

In some cases, it is best to add the needed refrigerant to a system before the leak search begins. Not only does this increase the system pressures and likely make the leak easier to locate, it also allows the system to begin providing comfort or refrigeration while the search is conducted.

Several approaches to locating refrigerant system leaks exist. This section describes four common ways to search for leaks:

- Using electronic devices to detect the presence of a refrigerant at the point where it is escaping.
- Using sensitive sound-detection devices to hear escaping refrigerant.
- Using refrigerant circuit additives with a special light that detects the presence of the additives at the leak location.
- Using simple soap solutions to develop visible bubbles where refrigerant is escaping.

WARNING!

Leak detection devices that could provide a source of ignition must not be used near systems containing refrigerants such as, but not limited to, HC-290 (propane) and HC-600a (isobutane).

Your choice of a leak detection method depends on several factors, including the amount of refrigerant remaining in the system (if any); the surrounding environment, as it relates to sound and wind; and a technician's own preference based on experience.

1.1.1 Electronic Detectors

Electronic leak detectors are reasonably accurate and easy to use. Some reportedly detect leaks as small as 0.03 ounce (oz) per year. The following electronic detectors will be introduced in this section:

- Corona discharge detectors
- Heated-diode detectors
- Infrared detectors

- Ultrasonic detectors
- Combustible gas detectors
- Hydrogen detectors

Corona Discharge Leak Detectors

Corona discharge units (*Figure 1*) look for changes in the conductivity of the air that passes between two electrodes. A high-voltage DC arc (corona) that bridges the gap between the electrodes, without any vapors present, establishes the normal range. As the corona arc is sustained and encounters a different substance, changes in the arc's current indicate that an insulating gas has intervened. The change initiates a series of alarm lights that reflect the leak concentration, or an audible signal that indicates the concentration with an escalating tone.

Corona discharge units work best with refrigerants containing halogens such as chlorine and fluorine. This includes the CFCs, HCFCs, HFCs, and blends that contain halogens. However, refrigerants with a very small volume of halogens are harder to detect. These units are common but other designs are beginning to dominate the market.

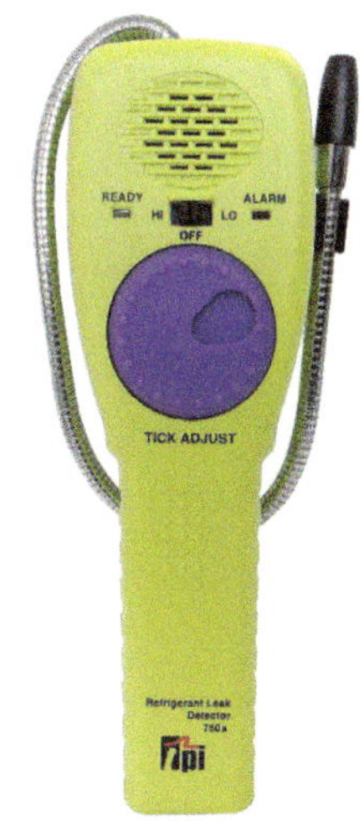

Figure 1 Corona discharge refrigerant leak detector.
Source: Test Products International

Heated-Diode Leak Detectors

A heated-diode leak detector (*Figure 2*) has a ceramic heating element that heats refrigerant molecules, breaking them apart. What remains are positively charged ions that are attracted to a collection point with a negative electrical charge. As the positively charged ions flow toward the collection point (usually a wire), a small current is generated. Once the current reaches a preset level, the instrument alerts the operator.

Heated-diode units tend to be reliable and accurate. These units are less likely to generate false alarm signals than corona discharge units. Heated-diode units can detect CFCs, HCFCs, HFCs, and HFOs, including those that are designated by ASHRAE as A2L (mildly flammable). However, like corona discharge units, detectable halogens must be present.

It is also important to note that heated-diode sensors as well as corona discharge sensors do not have unlimited life. While some may provide several hundred hours of service, others may need to be replaced after 50 hours of service. If operated in areas where the refrigerant is present in heavy concentrations, the service life is typically reduced.

Infrared Leak Detectors

Infrared (IR) leak detectors (*Figure 3*) have been increasing in popularity. One reason is that these detectors are not sensitive to specific types of refrigerants. Even though IR leak detectors are more expensive than most other types of detectors, users generally report good results.

Figure 2 Heated-diode refrigerant leak detector.
Source: Courtesy of Fieldpiece Instruments

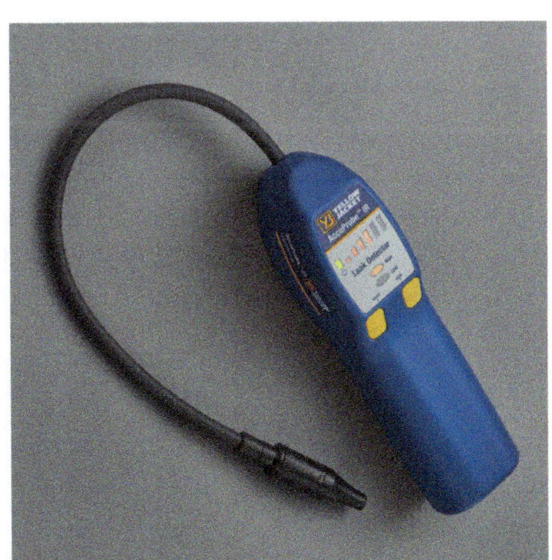

Figure 3 Infrared leak detector.
Source: Ritchie Engineering Co., Inc./YELLOW JACKET

An infrared leak detector draws air past an optical sensor. The sensor rapidly compares each sample with the next to determine how much IR radiation each sample has absorbed. The detector alarms when it finds anomalies in the analysis. What this means to the user is that it can't be steadily held in one place like a corona discharge or heated-diode unit. The probe must be consistently swept back and forth.

While this characteristic makes it more difficult to pinpoint the precise location of a leak, practice helps. However, these detectors offer excellent sensitivity and work with a wide variety of gases. The sensors also remain serviceable much longer, often providing around 1,000 hours of service.

Ultrasonic Leak Detectors

Ultrasonic leak detectors (*Figure 4*) are reasonably accurate and easy to use. Ultrasonic leak detectors rely exclusively on sound to detect a leak. The ultrasonic frequencies detected are beyond the range of human hearing. Because of the reliance on sound, their use is not dependent on any specific substances being present, which is an essential feature in some situations. These detectors are not affected by stray gases in the search area that can affect refrigerant-specific leak detectors.

Ultrasonic leak detectors detect the sound of a gas leaking from a pressurized system. As the gas passes through an opening, the flow becomes turbulent. This turbulent flow creates ultrasonic sound waves that are heard by the unit. These devices can also detect leaks in a system under vacuum because the device can hear the turbulence caused by air passing through the opening in either direction.

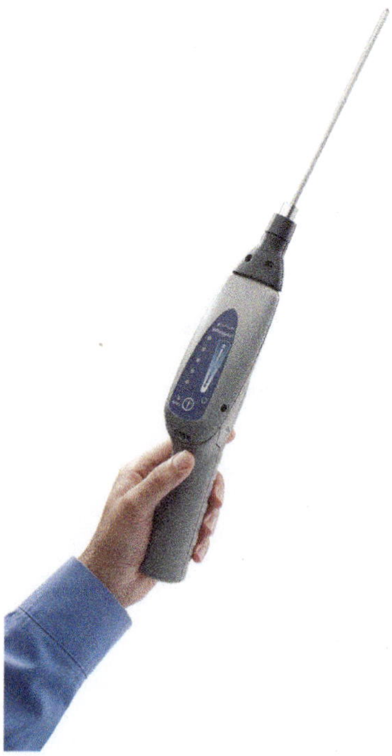

Figure 4 Ultrasonic leak detector.
Source: INFICON, Inc.

Because the ultrasonic detector senses sound waves rather than elements or compounds, it can be used to detect any type of escaping gas. This is a distinct advantage, due to the many types of refrigerant now in use. In addition, when you encounter a system that has lost all its refrigerant or you are testing a new installation, nitrogen alone can provide the needed pressure. It can also be used to detect electrical arcs, bearing wear and damage, and problems with steam traps.

One disadvantage of the ultrasonic detector is that sounds from other sources may affect its ability to detect a leak. Ultrasonic detectors do function in windy areas, but the wind can make sounds that might confuse the operator. The detectors do, however, work well in areas with stray gases present that might confuse other electronic detectors, depending on the composition of the stray gases. Technicians sometimes use cardboard or similar materials to shield the test area from wind.

Ultrasonic leak detectors include both a handheld detector and a headset for the operator. Typically, you don the headset, turn on the instrument, and begin listening for leaks as you slowly move the detector around the system. The ambient sound is processed by the instrument and becomes audible in the headset.

Combustible Gas Detectors

Combustible gas detectors (*Figure 5*) are used to detect propane, butane, methane, and other products that contain ethane. They are easy to use, with the audible alarm intensity changing to reflect the strength of the sensed concentration. This helps pinpoint the leak location.

Combustible gas sensors can typically provide up to 500 hours of service. Because other craftworkers often use these detectors for applications beyond HVACR work, there are more manufacturers in the market from which to choose.

Hydrogen Detectors

Corona discharge, heated-diode, and combustible gas detectors are sensitive only to specific substances. If that substance isn't present in the leaking contents, the detector will not detect a leak. In many cases, a technician must leak test a system without a refrigerant present.

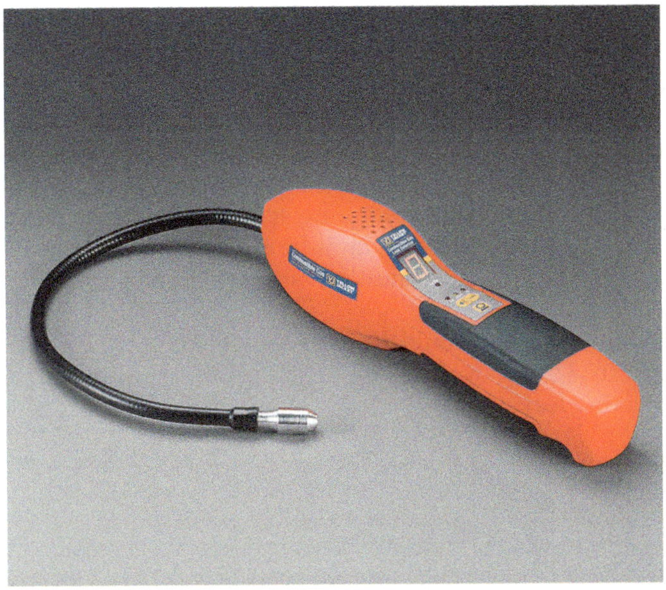

Figure 5 Combustible gas detector.
Source: Ritchie Engineering Co., Inc./YELLOW JACKET

Today, hydrogen is often used as *trace gas*. Hydrogen is the smallest atom of the elements and will escape through the smallest of openings. A mixture of 95% nitrogen and 5% hydrogen is commonly used to pressurize a system for leak testing. A hydrogen detector (*Figure 6*) is then used to search for leaks. After testing, the nitrogen/hydrogen mixture can be vented to the atmosphere. Commercially prepared mixtures are available from HVACR wholesalers and distributors.

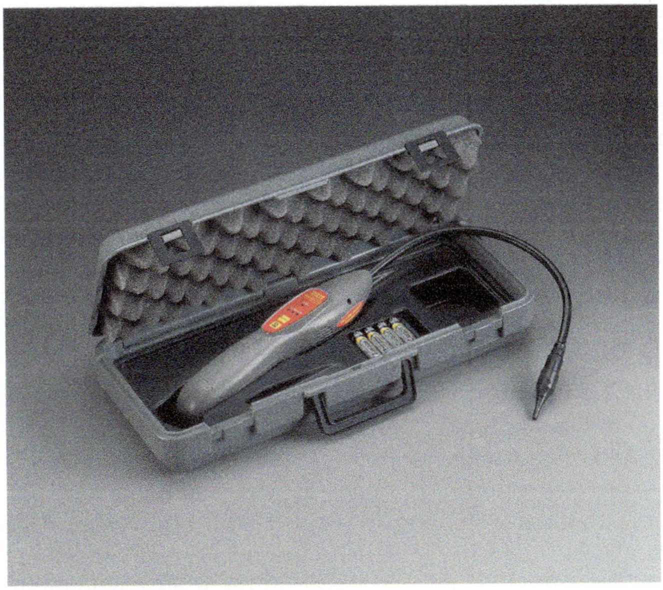

Figure 6 Hydrogen detector.
Source: Ritchie Engineering Co., Inc./YELLOW JACKET

Using and Maintaining Electronic Leak Detectors

Electronic leak detectors should be operated, maintained, and calibrated as directed by the manufacturer. Relatively delicate instruments, electronic leak detectors are unreliable if not properly maintained.

Since many of these detectors rely on a tiny pump to move air across a sensor, the air filter must be checked and replaced as needed. The tip is moved around and near the equipment and potential leak sites, so the tip is subject to fouling from oil and dirt. Keeping the unit clean is essential to protect the sensor and ensure reliable operation.

CAUTION

Never test a leak detector by exposing the sensor to a stream of refrigerant from a cylinder or system. This will typically saturate and degrade the sensor, shortening its service life and interfering with its ability to detect small leaks.

Many electronic detectors come with a substance, referred to as a *leak standard*, that can be used to quickly test the instrument to ensure it is working properly. It is usually contained in a tiny bottle.

The sensing tip is placed next to each joint and connection in the system, then moved slowly about while searching for the leak. Move at a rate of about 1" (2.5 cm) per second. Keep the tip close to potential leak sources. Pausing at a potential source is often necessary, except in the case of infrared devices. However, pausing in one spot with an infrared leak detector simply causes the unit to recalibrate itself for the conditions at that spot. Infrared detectors should remain in motion.

When possible, drafts should be reduced by shutting off fans and other devices that cause air movement. A breeze can quickly disperse refrigerant vapors from a tiny leak.

WARNING!

Do not use corona discharge, heated-diode, and similar leak-detecting devices in the presence of flammable refrigerants or explosive vapors unless the manufacturer clearly specifies the product is safe. Their sensors operate at high temperatures and can provide a source of ignition. Infrared detectors, ultrasonic detectors, and dedicated combustible gas detectors are safe for use with all refrigerants. Read and follow the manufacturer's instructions for each individual device, paying strict attention to the safety information.

Remember that many electronic leak detectors sense the presence of halogens. Halogens must be present for the electronic leak detectors to work. If a system has no refrigerant in it and is charged only with nitrogen for leak testing, an electronic refrigerant detector will not help. Ultrasonic and infrared leak detectors are not affected by the lack of halogens.

Section 608 of the Clean Air Act does allow one refrigerant to be used to help detect leaks. A small amount of HCFC-22 can be charged into a system, and then the pressure increased as necessary with nitrogen. The nitrogen provides the needed pressure, while the refrigerant provides halogens that can be detected with an appropriate device. It is also important to note that this HCFC-22 and nitrogen mixture is not expected to be recovered; it can be legitimately released into the atmosphere. However, HCFC-22 is very expensive and rarely found on today's service vehicles. Thus, other non-refrigerant gases such as hydrogen are often used today.

1.1.2 Ultraviolet/Fluorescent Dye Leak Detection

This method of leak detection requires a fluorescent dye to be injected into the refrigerant circuit of a functional system (*Figure 7*). The dye then exits the system through the leak along with the oil and refrigerant. The leak area then has a yellow-green glow when viewed under the beam of a high-intensity UV lamp. Note, however, that the system must be functional to allow the dye to circulate.

Kits generally include a high-intensity ultraviolet lamp, capsules filled with a premeasured amount of fluorescent dye, and the tools needed to inject the dye into the system. The amount of dye injected depends on the size of the system.

The dye can be injected into the system through any of the following methods:

- Use liquid refrigerant from a cylinder to move the dye into the low-pressure side of the system.
- Connect a compatible dye capsule between the high and low sides of the system and allow the high-side pressure to move the dye into the low-pressure side.
- Add the dye to refrigeration oil and then pump the oil into the system.
- Manually pump a premeasured solution into the circuit using a special injector, as shown in *Figure 7*. Note that these have limitations in system pressure against which they can work. Many cannot be used to inject dye into a charged HFC-410A system, for example.

CAUTION

Before injecting any dye or other fluid into a refrigerant circuit, ensure that the dye is compatible with the refrigerant and oil. Also check with the equipment manufacturer and make sure the substance is approved for use. Introducing a substance to the system that is not approved by the manufacturer may void the equipment warranty.

(A) Injecting Dye Into the Refrigerant Circuit

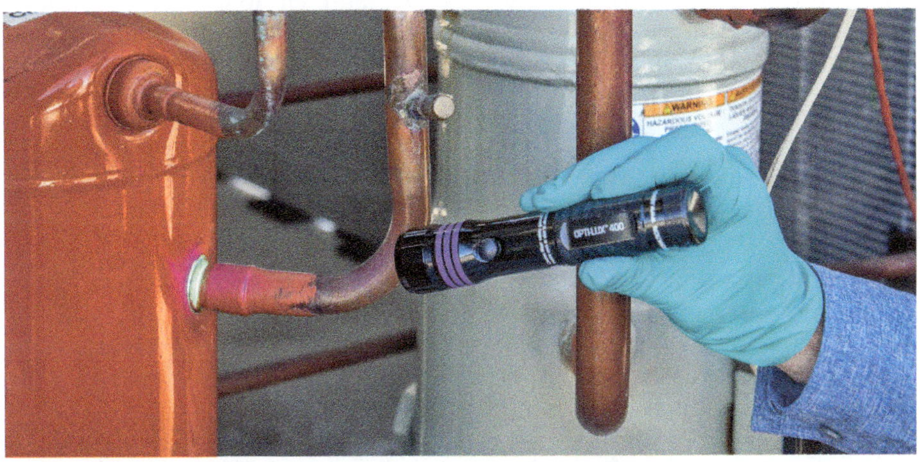

(B) Using the UV Light to Locate Leaking Dye

Figure 7 Using UV light and fluorescent dye for leak detection.
Source: Airgas, Inc.

After allowing time for the solution to circulate and mix thoroughly, the UV lamp is used to scan the system for leaks. At spots where refrigerant or oil is leaking, the fluorescent dye also leaks out. The bright yellow-green glow of the solution pinpoints the leak. Using a hood or other means of darkening the search area helps significantly.

After any leaks have been repaired, the visible dye can be removed from the leak site with a general-purpose cleaner suitable for removing oil residue. Use the light to ensure it is clean. The fluorescent solution remains in the system. This allows for future leak testing of the system. Even if the refrigerant charge is lost due to a leak, enough of the solution usually remains mixed with the compressor oil in the system to mark future leaks. However, a compressor change will significantly dilute its volume.

When leaks prove difficult to locate in a short time using other approaches, this method offers an alternate approach. For example, the solution may be charged into a system after a technician fails to locate a leak but must recharge and return the system to service quickly to preserve food or restore comfort. A leak search with a UV lamp can then be completed on a follow-up visit.

1.1.3 Liquid Bubble Solutions

Using a liquid bubble solution (*Figure 8*) is simple and non-invasive. A good solution can identify very small leaks in a refrigerant circuit. A bubble solution is often used to identify the exact point and size of a leak that was found using one of the other methods. These solutions are the most common way for technicians to test new joints quickly following a repair or system installation.

Figure 8 Liquid leak detector solution.
Source: Courtesy of Sporlan Division—Parker Hannifin Corporation

The soap-like solution is brushed onto the suspected area. Escaping gas at the leak causes bubbles to form, providing a visual indication. It may take a minute or so for foam to form on a very small leak, but bubbles often appear immediately. Tiny leaks slowly create a dense foam, while larger leaks tend to create larger bubbles. Always wipe or rinse off the residue when the search is over to keep the joints clean.

The main advantages of bubble solutions are low cost and ease of use. A disadvantage is that larger leaks may blow through the solution, causing no bubbles to appear. However, if the leak is large enough to do that, it will be noticed as soon as you touch the area with the applicator.

Commercial solutions are recommended because they are safer to use on all metals. These solutions also provide longer-lasting bubbles and the added colorants aid visibility. Leak testing using solutions made from common laundry or kitchen detergents should be avoided. Many of these detergents contain chlorides, which can corrode some metals. They also have a thinner consistency than commercial leak detection solutions, causing them to drip from surfaces too quickly. The product shown in *Figure 8* has a high viscosity and a fluorescent color that stands out.

On Site

Leak Repair

Leaks in the refrigerant circuit are repaired in a variety of ways. Some joints can be repaired by brazing and soldering, while others may require a new flare joint to be formed. But the location of some small leaks makes it challenging if not impossible to repair them through conventional approaches. Of course, you must also find the leak to repair it.

For small leaks that are difficult to access or locate, leak-sealing agents that can be injected into the refrigerant circuit are available. Similar products have been used in other industries for years. In the HVACR craft, the concern has long been that such products would clog small passages that are essential to system operation, and perhaps cause parts, such as the pushrods in an expansion valve, to bind.

Today's leak-sealing products remain inert inside the circuit until they reach a leak. Once the opening is reached, these leak-sealing products quickly react with air or water vapor to begin closing the leak. The process does take time, perhaps as much as a week, before a leak is completely closed. The product is often combined with UV dyes to advance the leak search if the sealant doesn't take care of the problem.

It is always best to consider conventional leak repair practices first, but when those practices aren't an option or the leak can't be located, leak-sealing agents offer another option.

Source: Nu-Calgon

1.2.0 Leak Testing and the Refrigerant Charge

Leak testing is performed on operational systems, on systems that have a partial refrigerant charge, and those with no refrigerant remaining. A leak test should always be performed before charging a new system or after making a refrigerant circuit repair on an existing system.

All leak testing is done with the refrigerant circuit under pressure. When pressurizing a system for testing, do not exceed the maximum pressure rating stamped on the equipment nameplate or listed in the manufacturer's service literature for the unit. The maximum pressure rating of equipment varies significantly. The established pressure is primarily based on the refrigerant for which it was designed.

Also note that the maximum pressure rating of indoor and outdoor components may differ. For that reason, always check the nameplate of each piece of equipment in the circuit to determine its maximum pressure rating. Do not assume that all the equipment in a system carries an equal pressure rating. Leaks are typically easier to find at higher pressures, but never exceed the maximum pressure rating of any component.

Leaks are often tiny and difficult to find, making leak detection one of the most challenging aspects of HVACR service. Some areas are often difficult to access. However, a competent technician must be able to conduct effective leak testing.

1.2.1 Operational Systems

If the system is operational, a visual check might reveal the source of a suspected leak. Because oil is mixed with refrigerant inside the system, the presence of oil around tubing joints, fittings, and on coil surfaces often indicates a leak. Vibration can loosen mechanical fittings over time so close examination is needed.

Use both your eyes and ears while looking for leaks. Larger leaks can often be heard. Leaks can also be found in unsuspected places, such as inside the bend of a 90-degree ell. Fittings can, and do, crack due to stress.

If the simple sight-and-sound method fails to identify the leak area, broaden the search, and use an appropriate tool for the task. When the system is running, an electronic leak detector or liquid bubble solution can be used. The noise level may be too high for an ultrasonic unit to be effective. Also remember that blowers and fans in operation can disturb the leak, allowing the escaping vapor to dissipate very quickly. It is often more productive, as well as safer, to search for leaks with the system off.

Remember that, when a system is in operation, the pressure on the low side of the system is at its lowest. If the leak is suspected to be on the low side, you can raise the test pressure by turning the system off. The increased pressure on the low side may help locate the leak. Conversely, searching for leaks on the high side of the system may be best done with the system running, since the high-side pressure is at its highest at that time.

1.2.2 Systems with a Partial Charge

When a system with a partial refrigerant charge has insufficient pressure to support a leak test, your options are limited. Nitrogen cannot be added to the system as is to increase the pressure. The resulting mixture cannot be released to the atmosphere by law (unless it happens to be an HCFC-22 system), nor can the mixture be recovered and reclaimed since the refrigerant would be contaminated with nitrogen.

The remaining charge must be recovered before nitrogen is added. The circuit can then be pressurized with nitrogen alone or a trace-gas mixture as previously described. If it is an HCFC-22 system, recover the refrigerant until the system has a slight positive pressure remaining. The system can then be pressurized with nitrogen while the remaining refrigerant serves as a trace gas. Otherwise, recover all the remaining refrigerant.

1.2.3 Systems Without a Refrigerant Charge

When a system has no refrigerant charge, pressure must be provided by another source. As mentioned previously, the EPA allows pressure testing with a mixture of nitrogen and HCFC-22. Only a small amount of HCFC-22 is added to act as a trace gas.

Other gases, primarily hydrogen, are most often used as a trace gas today. Mixtures of 95% nitrogen and 5% hydrogen are prepared and marketed for this purpose. A hydrogen detector works with these mixtures.

WARNING!

Hydrogen gas is highly combustible. Never use hydrogen gas alone as a leak-detection gas. Also never use oxygen, compressed air, or a flammable gas to pressurize a refrigerant circuit or conduct leak testing on any system or component.

To use an infrared or ultrasonic detector, nitrogen alone can be used to increase system pressure. Bubble solutions can also be used with dry nitrogen.

Figure 9 shows a typical setup for pressurizing a refrigerant circuit. The refrigerant gauge manifold is used to connect both the refrigerant and the nitrogen source or trace-gas mixture to the system.

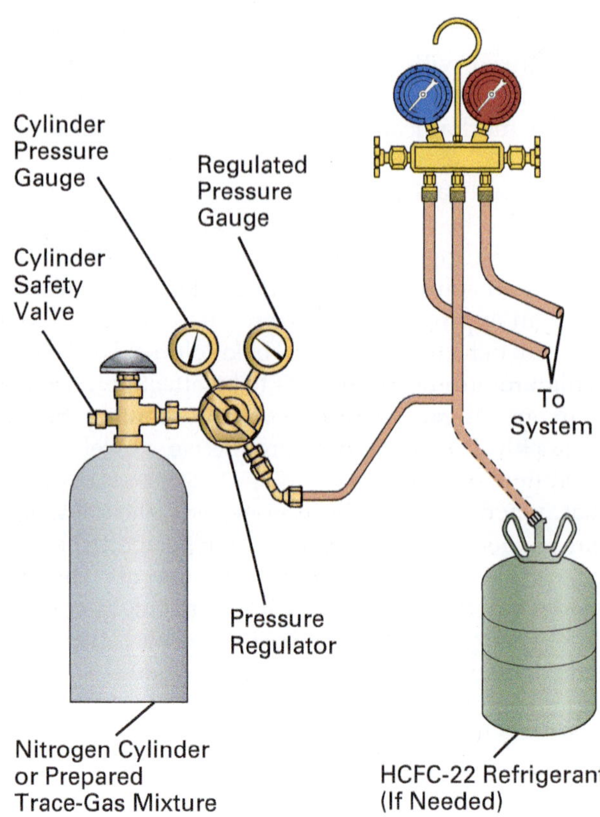

Figure 9 Equipment hookup for pressurizing a refrigerant circuit.

WARNING!

Nitrogen is stored in cylinders at a high pressure. At full pressure (typically 2,000 psig or more), system components can unexpectedly rupture. A pressure regulator must be used to limit and control the outlet pressure of the cylinder.

When using HCFC-22 as a trace gas, always add the refrigerant to the circuit first. Then turn the cylinder valve off and completely disconnect the cylinder from the gauge manifold before connecting the nitrogen cylinder. This ensures that an excessive pressure is not applied to the refrigerant cylinder, and that the two gases do not mix in the cylinder. Do not connect both pressure sources at the same time.

After locating and marking leaks in the system, the trace-gas mixture can be vented to the atmosphere. This remains true if HCFC-22 is properly used as a trace gas. The law currently allows for the mixture to be released, as the two gases cannot be separated if recovered.

CAUTION

Avoid connecting a refrigerant cylinder and any other source of pressure to a gauge manifold at the same time, to ensure the two gases cannot inadvertently mix.

1.2.4 Standing Pressure Tests

A standing pressure test using nitrogen alone is a common approach to testing a new system. This test involves pressurizing the refrigerant circuit with nitrogen to a specific pressure. The pressure is documented and then checked again after some time has passed. It is common to pressurize a system and leave it overnight to be checked the following day. If the pressure drops, then the system may have a leak somewhere.

The one problem with the test is that the nitrogen pressure changes along with the temperature. If the system is pressurized on a hot day, using nitrogen from a warm cylinder, the pressure will likely drop when left overnight. This may lead to the conclusion that the system is leaking when it may actually be free of leaks.

There is a way to determine how much pressure drop (or rise) is due to changes in the ambient temperature. The following equation is derived from what is referred to as the Ideal Gas Law and has been adjusted for working exclusively with nitrogen in a closed system that does not change in volume:

$$P_2 = T_2 \times (P_1 \div (T_1))$$

Where:

P = Absolute nitrogen pressure (psia)

T = Temperature in Rankin (°R)

P_1 represents the nitrogen pressure to which the system was charged. P_2 represents the new pressure, assuming there is no leakage, that results from a change in temperature. T_1 and T_2 represent the original ambient temperature and the new temperature, respectively.

For example, assume that you pressurize a system to 200 psig while the ambient temperature is 90°F. Overnight, the ambient temperature drops to 75°F. How far will the pressure drop due to the change in temperature?

In the first part of the equation, note that the pressure and temperature are being converted to the proper units. Complete the math as follows:

$$P_2 = T_2 \times (P_1 \div (T_1))$$
$$P_2 = (75° + 459.67°) \times ((200 \text{ psig} + 14.7) \div (90° + 459.67°))$$
$$P_2 = 534.67 \times (214.7 \div 549.67)$$
$$P_2 = 534.67 \times 0.391$$
$$P_2 = 209.06 \text{ psia}$$

Remember that the result shown represents absolute pressure. It must be converted back to gauge pressure by subtracting 14.7 psi:

$$P_2 = 209.06 \text{ psia} - 14.7 \text{ psi}$$
$$P_2 = 194.36 \text{ psig}$$

Thus, in this scenario, the pressure on the gauge should be 194.36 psig, with the loss of pressure from 200 psig attributed solely to the change in temperature. A lower pressure would indicate a leak.

Note that most refrigerant gauges only display the pressure in whole numbers; working to the same level of precision as the math is rarely possible. However, some digital manifolds today can make these nitrogen pressure calculations for you.

Although these calculations are necessary for an accurate pressure test using nitrogen, the challenge is found in accurately determining the ambient temperature to which the entire system is exposed. In a packaged unit, where the entire circuit is exposed to one stable temperature, the ambient temperature is easy to determine.

However, a large split system may be exposed to two or more different temperatures. The indoor unit may be in an attic or crawl space, for example, where the temperature is higher or lower than the outdoors. In this case, determining the temperature of the nitrogen in the refrigerant circuit becomes challenging, which affects the accuracy of the calculations.

During a standing pressure test, the gauge manifold usually remains connected to the system. It is important to remember that gauge manifolds and hoses also leak occasionally. Although it is best to leave the hoses and manifold connected, isolate them by front-seating the system's charging valves if possible.

1.0.0 Section Review

1. Owners and operators are required to take swift corrective action when their equipment leaks an ozone-depleting refrigerant if the equipment has a refrigerant charge of _____.
   a. 5 lb or more
   b. 20 lb or more
   c. 50 lb or more
   d. 100 lb or more

2. Nitrogen and hydrogen can be used to pressurize a system, providing a trace gas for electronic detection, in a nitrogen/hydrogen concentration of _____.
   a. 70%/30%
   b. 75%/25%
   c. 85%/15%
   d. 95%/5%

2.0.0 Refrigerant Recovery and Containment

Performance Task

2. Use a recovery unit to recover refrigerant from a system.

Objective

State refrigerant containment and recovery requirements and explain how to recover refrigerant.

a. State the basic refrigerant containment requirements of Section 608 of the Clean Air Act.
b. Describe refrigerant recovery and recycling equipment and explain how to recover refrigerant.

HVACR technicians must adhere to strict guidelines when handling refrigerants, in accordance with Section 608 of the Clean Air Act. The EPA requirements dictate how to conduct and document refrigerant recovery and recycling operations.

2.1.0 Section 608 Refrigerant Containment Requirements

All personnel who handle or purchase refrigerants must be certified through testing. Technicians must be certified in all equipment categories that they plan to service or install.

2.1.1 Certification Categories

The EPA refrigerant certification exam has 100 questions, randomly chosen from a bank of questions. Note that the first 25 questions of the exam are referred to as the Core. The three remaining sections of the exam each have 25 questions as well.

Technician certifications are broken into four separate categories. For any level of certification to be awarded, the candidate must first pass the Core. The types of certification are as follows:

- The *Type I Small Appliances* certification covers equipment that contains up to 5 lb of refrigerant. Examples include refrigerators, water coolers, and window air conditioning units.
- The *Type II High-Pressure Appliance* certification covers equipment charged with more than 5 lb of a high-pressure refrigerant. *High-pressure refrigerants* are defined as having a boiling point between −58°F (−50°C) and 50°F (10°C). This range covers most common refrigerants such as HCFC-22, HFC-410A, and many others.
- The *Type III Low-Pressure Appliance* certification covers systems using refrigerants that have a boiling point below −58°F (−50°C), regardless of the volume of refrigerant. Most low-pressure equipment, such as centrifugal chillers, require a substantial volume of refrigerant.
- The *Type IV (Universal)* certification is awarded to individuals that pass all four sections of the exam (including the Core). Universal certification, however, does not cover refrigerant handling in mobile applications, such as cars and trucks.

NOTE

For mobile and vehicular applications, a separate certification is required, as outlined in Section 609 of the Clean Air Act. Certification under Section 608 does not allow a technician to service the refrigerant circuit in mobile or vehicular applications.

Certification is achieved by passing the EPA's Section 608 Technician Certification Program Examination for each type, as proctored by an EPA-approved certifying agency and a proctor. Individual test proctors can also register with the EPA and proctor the exam. A complete list of EPA-approved testing agencies can be found on the EPA website at *www.epa.gov*.

2.1.2 A Technician Defined

The EPA offers a clear definition of a technician as it relates to the handling of refrigerants. A *technician* is defined as any person who conducts maintenance, repair, or service operations on HVACR equipment that could potentially result in the release of refrigerants to the atmosphere. Anyone who is engaged in the disposal of refrigerant-bearing equipment (except mobile systems) also falls under the description of a technician, if there is any chance of refrigerant being released.

Note that apprentices are exempt from the certification requirement, providing they are closely and continually monitored by an EPA-certified technician while they work. It is important to understand that the relevant federal regulation, found in 40 *CFR* Part 82, Subpart 152, defines an apprentice as "any person who is currently registered as an apprentice in maintenance, service, repair, or disposal of appliances with the U.S. Department of Labor's Office of Apprenticeship (or a State Apprenticeship Council recognized by the Office of Apprenticeship). A person may only be an apprentice for two years from the date of first registering with that office."

There are some activities that are excluded from the technician definition. Here are some examples of activities that do not require certification:

- Conducting work that is not reasonably expected to result in an accidental refrigerant release. Examples include painting, making wiring connections, and replacing air filters.
- Maintaining, servicing, or disposing of equipment when the refrigerant has already been recovered to the EPA-required levels.

Only an EPA-certified individual may purchase *bulk refrigerant*. Bulk refrigerant refers to containers holding 20 lb or more. However, companies can have drivers pick up and transport purchased refrigerants by prearranging it with a wholesaler. The wholesaler will require a copy of the EPA certification belonging to the person who is making the purchase. They will file this information to show the product was sold to a certified buyer. Fully assembled equipment, such as a window unit, refrigerator, or a water cooler are exempt and may be purchased by people who are not certified.

2.1.3 Containing Refrigerants During Service

Most refrigerants being handled during a service procedure must be recovered, recycled, reclaimed, or destroyed. The EPA has specific definitions for the terms *recovery*, *recycle*, and *reclamation*, as follows:

- *Recovery* — The removal and temporary storage of refrigerant in containers approved for that purpose. Recovery does not provide for any cleaning or filtration of the refrigerant.
- *Recycle* — The circulation of recovered refrigerant through filtering devices that remove moisture, acid, and solid contaminants. This does not mean that the recycled refrigerant meets the purity standards for new refrigerants. Recycling can be done in the field or in a shop environment.
- *Reclamation* — The remanufacturing of used refrigerant to return it to the standards of purity and cleanliness required of virgin refrigerant. These standards are outlined in AHRI Standard 700, *Specifications for Refrigerants*, which is incorporated into the law by reference. Reclamation is by no means a field-service procedure. It is a complex process conducted only by reprocessing and industrial facilities.

NOTE

Table 3 shows the required levels of recovery only for recovery equipment manufactured on or after November 15, 1993. On this date, the manufacturing standards for active recovery equipment were increased. The EPA allows a recovery percentage of 80% for all equipment, regardless of the compressor status, when using recovery equipment manufactured before this date. However, it is highly unlikely that any recovery equipment manufactured in 1993 or earlier and designed for field use remains in service. On the other hand, information relative to this date and recovery equipment performance remain a certification exam topic at this writing.

Refrigerant recovery is typically performed in the following situations:

- Before a refrigerant circuit is opened to make repairs
- Before pressurizing a refrigerant circuit for leak testing
- Before disposing of any HVACR system, appliance, or component
- To remove excess refrigerant from an overcharged system

Before opening a refrigerant circuit, the refrigerant must be recovered from the equipment to target pressures established by the EPA. *Table 2* shows the required evacuation levels for a variety of stationary equipment. The evacuation levels for small appliances containing 5 lb or less of refrigerant are shown in *Table 3*.

To use *Table 2*, you must also be able to distinguish between high-pressure, medium-pressure, and low-pressure appliances. The following information is found in 40 *CFR*, Part 82, Section 152 *Definitions*:

- A *high-pressure appliance* means an appliance that uses a refrigerant with a liquid phase saturation pressure between 170 psia and 355 psia at 104°F. Examples include but are not limited to appliances using R-22, R-407A, R-407C, R-410A, and R-502.
- A *medium-pressure appliance* means an appliance that uses a refrigerant with a liquid phase saturation pressure between 45 psia and 170 psia at 104°F. Examples include but are not limited to appliances using R-114, R-124, R-12, R-134a, and R-500.

TABLE 2 Required Levels of Evacuation for Stationary Appliances (Except for Small Appliances)

Type of Appliance	Inches of Hg Vacuum (Relative to Standard Atmospheric Pressure of 29.9" Hg)	
	Using pre-1993 recovery equipment	Using post-1993 recovery equipment
Very high-pressure appliance	0	0
High-pressure appliance, or isolated component of such appliance, with a full charge of less than 200 lb of refrigerant	0	0
High-pressure appliance, or isolated component of such appliance, with a full charge of 200 lb or more of refrigerant	4"	10"
Medium-pressure appliance, or isolated component of such appliance, with a full charge of less than 200 lb of refrigerant	4"	10"
Medium-pressure appliance, or isolated component of such appliance, with a full charge of 200 lb or more of refrigerant	4"	15"
Low-pressure appliance	25 mm Hg absolute	25 mm Hg absolute

TABLE 3 Small Appliance Recovery Requirements

Recovery Equipment in Use and Recovery Situation	Vacuum Level or Recovered Percentage
Active and passive recovery equipment made on or after November 15, 1993; serviced unit has an OPERABLE compressor	4" Hg or 90% of the charge
Active and passive recovery equipment made on or after November 15, 1993; serviced unit has an INOPERABLE compressor	4" Hg or 80% of the charge

- A *low-pressure appliance* means an appliance that uses a refrigerant with a liquid phase saturation pressure below 45 psia at 104°F. Examples include but are not limited to appliances using R-11, R-123, R-113, and R-245fa.

Note that the EPA consistently refers to *appliances*. Although many of us think of a domestic refrigerator or a dishwasher as an appliance, the EPA defines an appliance as "any device which contains and uses a Class I or Class II substance or substitute as a refrigerant and which is used for household or commercial purposes, including any air conditioner, motor vehicle air conditioner, refrigerator, chiller, or freezer."

NOTE

Class I and *Class II* substances refer to ozone-depleting substances, such as CFCs and HCFCs, listed in 40 *CFR*, Part 82, Subpart A, Appendix A and Appendix B.

When recovering refrigerant from equipment listed in *Table 2*, a recovery unit must be used that will reduce the system pressure to the level shown in the table. For equipment listed in *Table 3*, the refrigerant is considered completely recovered when 90% of the refrigerant by weight is removed if the appliance has a functional compressor, or 80% of the refrigerant by weight is removed if the unit has an inoperative compressor. For practical purposes, the refrigerant in small appliances can also be considered recovered when the pressure has been reduced to 4" Hg vacuum. The technician need only reach the correct weight or the required vacuum, but not both.

There are some exceptions to the EPA recovery levels. If evacuation to the specified levels is not achievable because of leaks in the system, you must perform the following steps:

- Isolate the leaking components from the rest of the system, if possible;
- Recover the refrigerant from non-leaking components to be opened or disposed of to the levels specified in *Table 2*;
- Recover the refrigerant from leaking components to the lowest level attainable without substantially contaminating the refrigerant but not below 0 psig, as this will allow air to enter and contaminate the refrigerant.

There are two types of refrigerant releases that are allowed under Section 608 regulations, since these releases are considered unavoidable:

- A release that is referred to as a **de minimis release**. This simply means that a very small volume of refrigerant released while connecting a refrigerant gauge manifold or purging hoses, for example, is acceptable.
- The release of a mixture comprised of nitrogen and HCFC-22 used exclusively for the purpose of leak detection.

De minimis release: *De minimis* is a Latin expression referring to small or insignificant things. In the HVACR craft, the term refers to a minute release of refrigerant that is unavoidable. The very brief, low-pressure purging of a refrigerant hose prior to connection is an example of a de minimis release.

2.1.4 Options for Recovered Refrigerant

Once the refrigerant has been recovered, technicians have several options. For example, assume a leak on a brazed fitting has been repaired. A leak test shows the repair was successful. Following an evacuation, the recovered refrigerant can be returned to the system as is.

If there is no indication that the recovered refrigerant is contaminated, it can be returned to the system directly. It is best to pass the recovered refrigerant through a filter-drier as it is returned to the circuit. It is also best to filter the refrigerant as it is recovered. Note that some recovery equipment manufacturers require a filter-drier to maintain the warranty.

NOTE

The evacuation process is presented in this module's Section 3.0.0, *Refrigerant Circuit Evacuation*.

Another option is to *reclaim* the refrigerant. Remember that reclamation means returning the refrigerant to the same standard of purity applied to virgin refrigerant. The recovered refrigerant is sent to an authorized facility for this purpose (*Figure 10*). Contractors may choose to do this for several reasons. One reason would be that the refrigerant is thought or known to be contaminated with acid. This could be the case after a compressor burnout, where acids have been released from the motor windings into the refrigerant and oil. Some contractors, however, prefer to reclaim any recovered refrigerants simply as a matter of policy. When the refrigerant is reclaimed, new refrigerant is used to charge the system.

Figure 10 A worker extracts refrigerant from cylinders at a modern reclamation facility.
Source: Airgas, Inc.

There is a very important regulation regarding the reuse of recovered refrigerant that technicians must be aware of. Refrigerant recovered from a system cannot be transferred to another owner. Regardless of the condition of the recovered refrigerant, it cannot be charged into equipment owned by others. It can, however, be used in another piece of equipment owned by the same person or entity, as-is or following recycling.

Technicians should always avoid mixing recovered refrigerants as well as mixing refrigerants with other gases. Mixed refrigerants and refrigerants contaminated with other gases cannot be reclaimed. When refrigerant arrives in a condition that prevents reclamation, the refrigerant must be incinerated by the reclaiming facility. This will usually result in the sender being charged to cover the cost of its incineration. Nitrogen and refrigerant mixtures for leak testing, done in accordance with the law, are released into the atmosphere.

2.1.5 Equipment Disposal

The refrigerant must be recovered from any equipment to be disposed of. The regulations require that most systems have a service *aperture*, or connection point, to allow the refrigerant to be removed. Small appliances, such as a refrigerator, need only have a *process tube* available, where a temporary valve can be installed for refrigerant removal. A process tube is simply a short section of copper tubing on a compressor or system tubing through which the manufacturer evacuated and charged the equipment (*Figure 11*).

Figure 11 Compressor process tube.
Source: Nor Gal/Shutterstock

The final individual in the disposal chain, such as a landfill operator, is required to ensure that the refrigerant was recovered and must maintain documentation to that effect. Refrigerant recovered from equipment taken out of service can only be used on equipment owned by the same entity, or it can be reclaimed.

Technicians recovering refrigerant from equipment for disposal purposes with a full charge of more than 5 lb but less than 50 lb of refrigerant must keep records for three years that document the following information:

- The company name, location of the equipment, date of recovery, and the type of refrigerant recovered;
- The total volume of refrigerant recovered from all discarded HVACR equipment in each calendar month; and
- The quantity of refrigerant, by type, sent for reclamation or destruction, the entity to whom it was transferred, and the date of transfer.

2.1.6 Refrigerant Recordkeeping

Companies and individuals must keep records documenting refrigerant use and management. The recordkeeping requirements include, but are not limited to, the following:

- When the equipment serviced contains 50 lb or more of refrigerant, technicians must provide documentation on the invoice showing how much refrigerant was added. However, most contractors place the information on every invoice for clarity.

- Reclaiming facilities must maintain a record of the refrigerant sent to them for reclamation. The information includes the name and address of the person or company that submitted it, as well as the amount submitted.
- Owners and operators of equipment containing refrigerant also have some responsibility. They must maintain records of service calls and the amount of refrigerant added, if any, if the equipment they possess contains 50 lb or more of refrigerant. Because residential systems typically contain less than 10 lb of refrigerant, homeowners are not affected by this requirement.
- Wholesalers selling refrigerant must maintain records of the refrigerant sold, the date, and the authorized purchaser. Remember that the purchaser must be EPA certified to handle refrigerants.
- Organizations operating certification programs must maintain a record of each person's name, address, and test score, as well as the date and location where the exam was proctored. They must also provide a semi-annual activity report to the EPA that shows the testing schedules and the pass-fail rates for the candidates.

It makes sense for contractors to be proactive in their documentation of work related to refrigerants. Doing so provides the needed information to defend against an accusation of a violation. The following recommendations apply to both contractors and technicians:

- Employers should have their technicians clearly document the status of all refrigerant that passes through their hands. It is either on the service truck, in storage, or its place of use has been documented.
- All refrigerant purchases should be documented. In addition, records should be kept of all refrigerant recovered, recycled, and reclaimed. It is important to document where recovered refrigerants went following the recovery.
- Any accidental venting of refrigerants should be reported to the EPA. This indicates a good faith effort on the part of the technician and/or the company. A full and accurate disclosure of the incident should be provided. Although the EPA is unlikely to tolerate careless disregard for the law, the EPA is also unlikely to prosecute anyone putting forth their best effort to comply with the laws and self-report an accidental release. If a change in work practices would prevent a similar accidental release in the future, it is best to document that in the report as well.

2.1.7 Enforcing the Law

Whenever there are laws, there are penalties for breaking laws. Both civil and criminal penalties can be incurred for failing to follow the regulations documented in Section 608 of the Clean Air Act. Typical violations include the failure to show ownership of an approved recovery unit while HVACR service work was ongoing; intentionally venting refrigerants to the atmosphere; and the failure to report an accidental refrigerant release.

Smuggling and illegal ventures are also occasionally discovered. A 2012 case was related to an individual selling HCFC-22 refrigerant in cylinders via a website and advertising that no certification was necessary to buy the product. Violations can and do occur at administrative, distribution, manufacturing, and user levels.

Penalties for failing to abide by the law include the following:

- Civil penalty fines up to $44,539 per day can be assessed per violation of the Clean Air Act. Note that the maximum civil penalty changes periodically.
- An individual's certification can be revoked for a variety of reasons. For example, if a technician shows that they are not familiar with the laws or are unable to demonstrate the skills to properly recover refrigerant, their certification would likely be revoked.

A reward of up to $10,000 can be given to individuals who report a violation or provide information leading to a conviction. The EPA can take several steps when a violation is suspected, ranging from a simple request for information to prosecution in federal court. These steps include the following:

- A letter referred to as a *Section 114 information collection request* can be sent to a suspected offender, outlining the possible violation. The letter requires a response and additional information on the matter from the recipient.
- A different document can be sent as notification that a known violation has occurred and the report is being filed for further review.
- An administrative order can be filed, which directs the violating party to comply with regulations or face additional punishment, such as the closing of the business operation.
- A field citation can be written, much like a speeding ticket, for minor violations.
- For more significant violations, an administrative-style legal hearing can be assembled.
- For the most serious violations, the case is prosecuted before a federal court.

2.2.0 Refrigerant Recovery Equipment

Refrigerant recovery usually requires a certified refrigerant recovery unit or combination recovery/recycling unit. Because the recovery process must be accomplished without releasing refrigerant into the atmosphere or contaminating the refrigerant, only certified technicians may perform this task.

2.2.1 Active and Passive Recovery

Active recovery can be defined as a recovery process using a self-contained refrigerant recovery unit. Most of today's recovery units rely on an oil-less pump, rather than a compressor that requires lubrication.

Active recovery: Recovery process that relies on a self-contained refrigerant recovery unit to draw or push the refrigerant out of a refrigerant circuit.

An active recovery does not require or rely on a functional compressor in the system. Most recovery units can recover both liquid and vapor. During a recovery, refrigerant is transferred into a reusable refrigerant cylinder, which should never be filled with liquid beyond 80% of its volume.

Passive recovery does not rely on a self-contained unit. The passive process uses the system's own pressure and/or the system compressor to push the refrigerant out. The refrigerant must be captured in a non-pressurized container. Although special bags designed for this purpose are available to contain a very small volume of refrigerant, very few are in use today. An evacuated reusable refrigerant cylinder is typically used. The vacuum in the cylinder helps draw refrigerant from the system. For very small equipment, like a domestic refrigerator, an evacuated cylinder alone is often sufficient to accomplish the task and draw the system pressure down to the EPA-required level.

Passive recovery: Refrigerant recovery process that makes use of system pressure and/or its compressor to push the refrigerant out; no dedicated recovery unit is used. The refrigerant is captured in a non-pressurized container, such as an evacuated refrigerant cylinder.

This module focuses on the active recovery process as it is the most common approach for HVACR technicians, but it is still important to be aware of the passive methods of recovery.

2.2.2 Self-Contained Recovery Units

Although large recovery units are available, portable units dominate the service industry. Large units are more likely to be found in the shop where they may also be used to effectively recycle refrigerant through heavy filtration and oil separation. Portable recovery units (*Figure 12*) are usually equipped with an oil-less pump to move the refrigerant.

Before using any recovery unit, make sure that it can safely recover the type of refrigerant present. This information will be found on its nameplate. Many models today can recover all common refrigerants, such as HFC-134 and HFC-410A, and are also rated to recover A2L refrigerants. Note that HC-290 (propane) is an A3 refrigerant and cannot be recovered using equipment rated for A2L refrigerants.

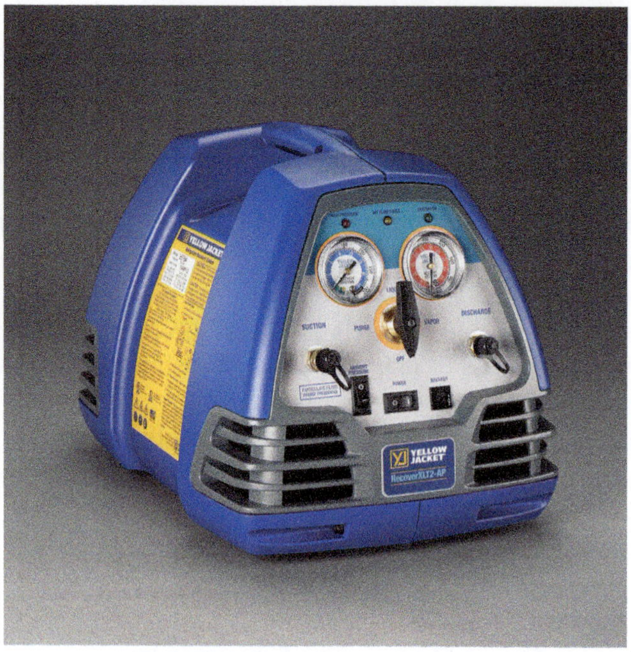

Figure 12 Oil-less portable recovery unit.
Source: Ritchie Engineering Co., Inc./YELLOW JACKET

WARNING!

Ensure that a recovery unit is designed for use with the refrigerant to be recovered. This is especially important when working with flammable refrigerants. Using equipment not designed for use with flammable products may result in fire or explosion.

Most units can recover refrigerant in either the vapor or liquid state. Pound for pound, transferring liquid is much faster than transferring vapor. For example, a unit may have a specification indicating that its liquid transfer capacity is 5 lb per minute, while the vapor capacity is 0.6 lb per minute. Therefore, it is best to evaluate the task and try to recover as much of the refrigerant as possible in its liquid form. However, the process normally must be completed by recovering the vapor that remains after the liquid has been recovered.

Recovery units may have a third capacity specification. It is known as the *push-pull* recovery rate. Normally, a recovery unit draws refrigerant from a system by creating a lower pressure with its pump, encouraging refrigerant to flow in and through. Many units though, can be configured to push the refrigerant out. The unit draws vapor refrigerant from the recovery cylinder, compresses it, and then pushes it into the system, forcing the liquid out.

Push-pull recovery rates vary significantly, but it is very efficient when it can be used. For example, it is not unusual for the push-pull recovery rate to be up to 10 times that of the specified liquid recovery rate. However, the push-pull method can only be used under the right conditions.

Each recovery unit model is somewhat unique, so it is important to follow the manufacturer's instructions carefully. The manufacturer also provides maintenance instructions. Operating the unit through a *purge cycle* is one common maintenance activity that is typically done following every recovery task. Recovery units should be purged of their contents to ensure that there is no possibility of contaminating the refrigerant of another system, and to avoid the possibility of mixing refrigerants.

Remember that certain vacuum levels must be achieved when recovering refrigerant. This is an indication that most of the refrigerant has been removed. Refer again to *Table 2* and *Table 3* for this information. However, on most occasions, the pressure immediately begins to rise in the system as soon as the recovery unit is turned off. This is due to the presence of small pockets of liquid refrigerant that remain and are boiling off, raising the pressure. Refrigerant is also dissolved in the oil and must vaporize before it can be recovered.

Thus, the recovery unit may need to be cycled several times for trapped refrigerant to finish vaporizing and exit the system. This issue affects operations far more in the winter than the summer because the refrigerant tends to vaporize slowly in a cold environment. As long as the pressure continues to rise when the recovery unit is turned off, refrigerant remains in the circuit. This assumes, of course, that the system is not leaking, allowing air to enter.

2.2.3 Recovery Safety

Before beginning a refrigerant recovery, note the following safety precautions:

- Wear safety goggles and appropriate gloves.
- Become thoroughly familiar with the unit operating manual. Ensure that the unit is appropriate for the refrigerant being recovered.
- Make sure the area is well ventilated. Refrigerants can displace breathable air quickly in the event of a hose rupture or other major leak.
- Inspect the power cord and ensure that the unit is connected to a power source protected by a ground fault circuit interrupter (GFCI). Ensure that any extension cords are properly sized for the load. Excessive voltage drop can damage the recovery unit.
- Do not operate the recovery unit near flammable liquids and vapors.
- Units charged with HC refrigerants, such as HC-290 (propane) and HC-600a (isobutane) must have red charging ports and/or process tubes. Be certain of the refrigerant in use before connecting a gauge manifold or any other equipment to a system.
- Inspect the refrigerant hoses to ensure they are in good condition. Most refrigerant manifolds are equipped with ¼" hoses; using ⅜" hoses will speed the process since ⅜" hoses offer reduced pressure drop. The larger hoses are also better for system evacuation.
- Ensure that the recovery cylinder has enough capacity to hold the expected refrigerant charge. Remember that cylinders should never be filled beyond 80% of the cylinder volume.
- Always remember that refrigerants in various states and at various pressures are present. Pay careful attention to ensure that the hoses are properly connected.
- Disconnect hoses carefully to prevent frostbite or other injury resulting from the sudden release of liquid refrigerant.

On Site

Power Cords

Recovery units, as well as vacuum pumps, typically operate on a 120VAC power supply. The motors consume a significant amount of power. An operating current of 9A–12A is common for this type of equipment.

Take care in the selection of power cords. If the cord length exceeds 5', power cords with No. 12 wire are generally required. If the wire size of the cord is too small for the load and cord length, the voltage can drop significantly under load. This results in an even higher operating current and leads to overheated power cords, tripped circuit breakers, and possible equipment damage. A No. 10 cord is needed for long runs to support significant current demands. Also, be sure to use a GFCI-protected circuit or cord on every occasion.

2.2.4 Liquid and Vapor Refrigerant Recovery

The following guidelines for recovering refrigerant are generic in nature. These guidelines will be helpful in most situations, but cannot be considered accurate for every situation and recovery unit. Always follow the manufacturer's guidance for the recovery unit in use.

As noted earlier, larger refrigerant hoses help accelerate the recovery. Most refrigerant hoses also have a valve core depressor built into the female connection. The depressor, as well as the valve core itself, creates a restriction to flow. It is best to use a tool that connects to the access fitting and can temporarily withdraw the valve core without refrigerant loss (*Figure 13*). The core is pulled back into a separate chamber of the tool, clearing the refrigerant flow path. The valve core depressor can then be removed from the end of the hose. Removing these restrictions accelerates the recovery process. The tool shown in *Figure 13* can be used for evacuation, charging, valve core removal, and recovery tasks.

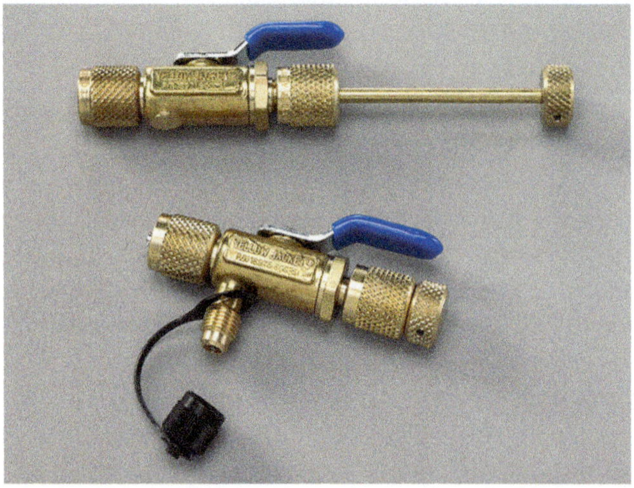

Figure 13 Four-in-one ball valve tool.
Source: Ritchie Engineering Co., Inc./YELLOW JACKET

Removing refrigerant from a system in its vapor state causes the pressure to drop rapidly. As a result, puddles of liquid refrigerant will begin to boil (vaporize). These areas can be identified by frost on the lines and cold, frosty components. Carefully warming these areas will help the refrigerant to vaporize more rapidly.

Figure 14 shows the appropriate hose connections for one model of recovery unit, but the connections are typical for most units. The discharge port of

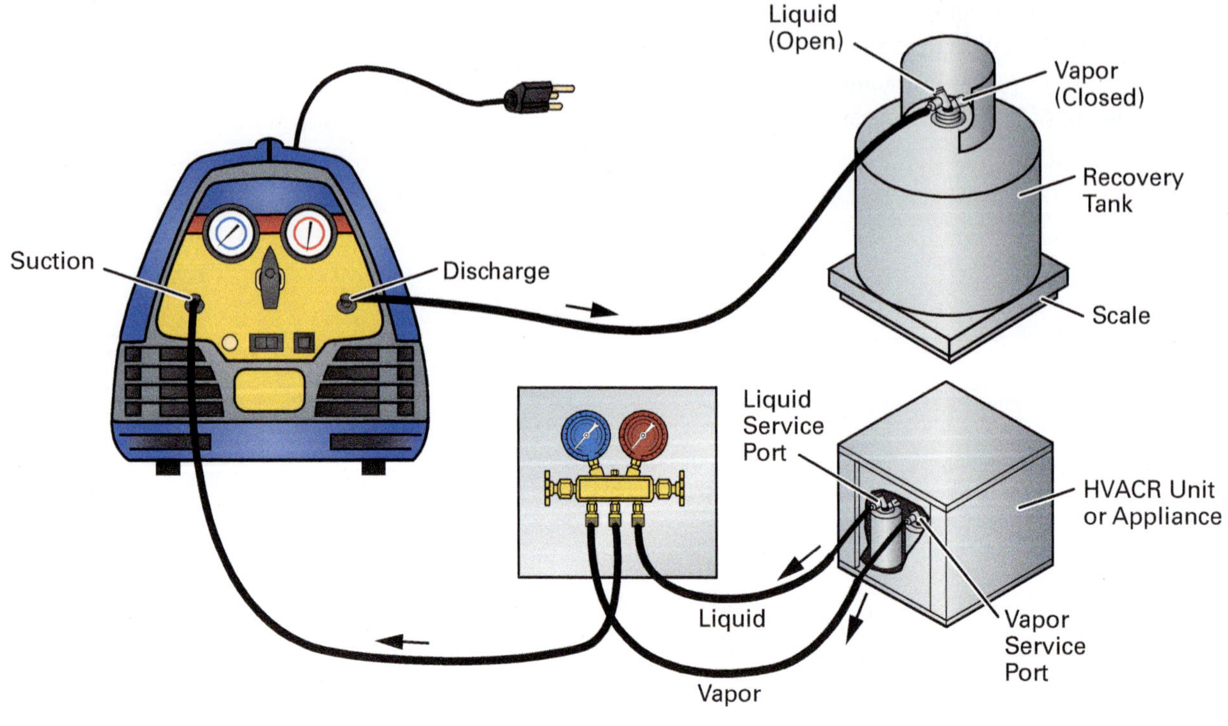

Figure 14 Typical recovery unit connections.
Source: Ritchie Engineering Co., Inc./YELLOW JACKET

the recovery unit is connected to the recovery cylinder liquid valve (the port equipped with a dip tube). The suction port of the unit is connected to the system through the center port of a gauge manifold, which provides additional control over the refrigerant flow.

The manifold is connected to both the low and high sides of the system. A scale is used to monitor the cylinder weight. If a recovery unit with a float switch is being used, the switch is connected directly to the recovery unit. It will automatically shut the recovery unit down when the cylinder has reached 80% capacity.

A properly connected recovery unit, cylinder, and scale are shown in *Figure 15*. Using a scale to monitor the weight of the cylinder is essential when a float switch is not used (*Figure 16*).

The following steps outline the process of recovering liquid refrigerant:

Step 1 Make the necessary hose connections per the manufacturer's operating instructions. Remember that $^3/_8$" hoses are recommended to reduce the pressure drop. Check the connections twice to ensure accuracy.

Step 2 Purge all the connected hoses with refrigerant before starting, to ensure air is not captured in the recovery cylinder along with the refrigerant. These are de minimis releases.

Step 3 Open the liquid valve on the recovery cylinder.

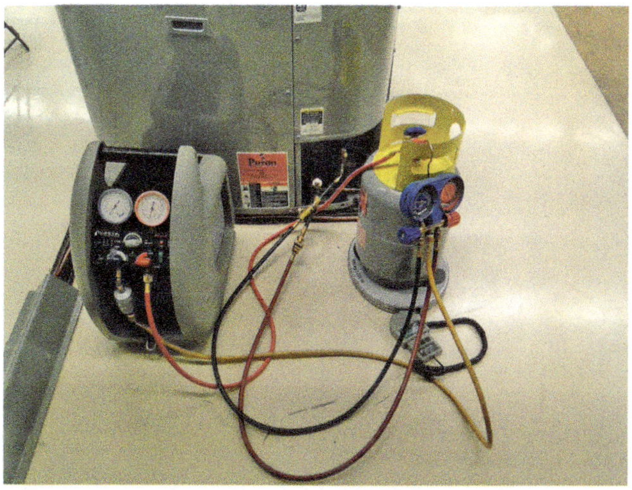

Figure 15 Recovery unit connected to a system.

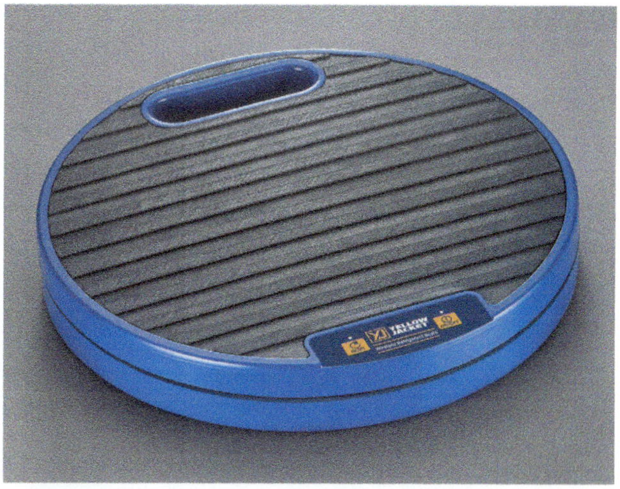

Figure 16 Wireless refrigerant cylinder scale.
Source: Ritchie Engineering Co., Inc./YELLOW JACKET

Step 4 Start the recovery unit and set the mode selector for liquid recovery as appropriate. Note that the mode selector is typically a manually operated valve with multiple positions.

Step 5 Open the high-side valve on the gauge manifold. This allows liquid from the high-pressure side of the system to enter the manifold. Also open the utility valve on the manifold, allowing refrigerant to flow to the suction port of the recovery unit. Monitor the weight shown on the scale to ensure the cylinder weight does not exceed 80% of its gross weight.

Step 6 Once the liquid has been removed, vapor recovery can begin. To switch to a vapor recovery mode, set the recovery unit mode selector to the appropriate position.

Step 7 Both the high- and low-side valves can be opened on the gauge manifold for vapor recovery. This allows vapor to enter the recovery unit from both sides of the system at the same time.

Step 8 When the required vacuum level has been achieved as indicated on the recovery unit suction gauge or the gauge manifold, the mode selector can be turned to the Off position.

Step 9 Monitor the pressure after the unit is switched off. Cycle the unit several more times as necessary until the pressure remains at the target pressure.

Step 10 The recovery unit can now be purged. Turn the mode selector to the Purge position. Monitor the suction gauge of the unit; when the desired vacuum level has been reached, the purge process is complete. The mode selector valve can be returned to the Off position.

Know the Code

Locking Valve Caps

The International Mechanical Code (IMC) now requires locking caps on refrigerant service valves that are located outdoors. This is to prevent unauthorized access into the system by making it difficult to remove the access port caps. Although the IMC has not been adopted in every area, using locking caps on refrigerant access ports is a good idea.

The caps protect homeowners and vandals alike from the risks associated with refrigerants, such as frostbite and inhalation hazards. Most locking caps require a special tool or key to remove them. Over time, you will likely collect several different types of cap keys.

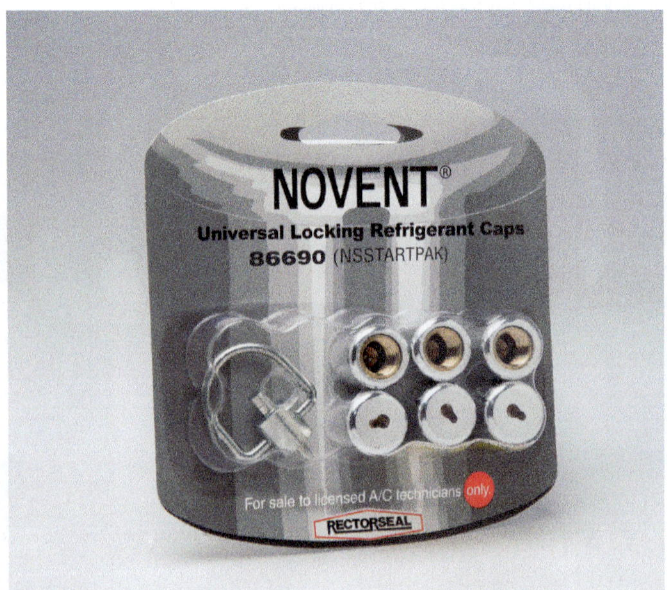

Source: RectorSeal

Note that some units do not need to be switched from a liquid recovery mode to a vapor recovery mode; some automatically change modes once the flow of liquid stops.

If a recovery unit is switched on while there is a significant pressure differential across the gauges, it may have trouble starting. Like a compressor in an HVACR unit, the motor may not have sufficient torque to start when it is working against significant pressure differences. Recovery units often require a simple procedure (open and closing valves) to equalize the pressures before the unit is restarted. It may be as simple as moving the mode selector to the Off position, allowing time for the pressures to balance, and then returning it to the proper position after the unit is started.

Always test the recovered refrigerant for the presence of acid and moisture before recharging it back into the system. When liquid refrigerant is being recovered, a filter-drier should be installed in the utility hose line of the gauge manifold connected to the recovery unit. Be sure to orient the filter-drier for the correct direction of flow. Small filter-driers with flare connections to fit the hoses are available for this purpose. It never hurts to ensure the refrigerant is as clean as possible, and the filter-drier also protects the recovery unit from debris. When the refrigerant is suspected to contain acid or moisture, it is best to recharge the system with virgin refrigerant and send the recovered refrigerant to be reclaimed.

Push-Pull Refrigerant Recovery

The push-pull recovery method is typically the fastest method. It is best suited for moving a large volume of refrigerant quickly. However, there are some conditions that prevent it from being an effective approach:

- Systems equipped with an accumulator or reversing valve (heat pumps) cannot be recovered through the push-pull method.
- Systems that contain less than 10 lb of refrigerant are not good candidates for push-pull recovery. The process works best with systems containing 20 lb or more, as a substantial volume of refrigerant is required to make the process work.

Like standard liquid recovery, some vapor will remain in the system after the push-pull process has removed as much liquid as possible. Using this method, you can monitor the flow of liquid through a sight glass as it exits the system. Once liquid refrigerant is no longer visible in the glass, the remaining refrigerant must be recovered as vapor.

To make the process work, a recovery cylinder containing about 10 lb of refrigerant is needed. A sight glass and an additional hose are also needed.

Figure 17 shows the typical connections for the push-pull method. Note that the discharge port of the recovery unit is now connected to the equipment suction line. The suction port of the recovery unit is connected to the vapor valve of the recovery cylinder.

CAUTION

The refrigerant needed to support the process in the recovery cylinder cannot come from a system owned by others unless it will all be reclaimed. It must either be virgin refrigerant or usable refrigerant that belongs to the same equipment owner. When virgin refrigerant is used, it is no longer considered virgin. This means that the entire volume of refrigerant can only be used on equipment owned by the same entity or be reclaimed. Also ensure that you do not mix contaminated refrigerant with clean refrigerant.

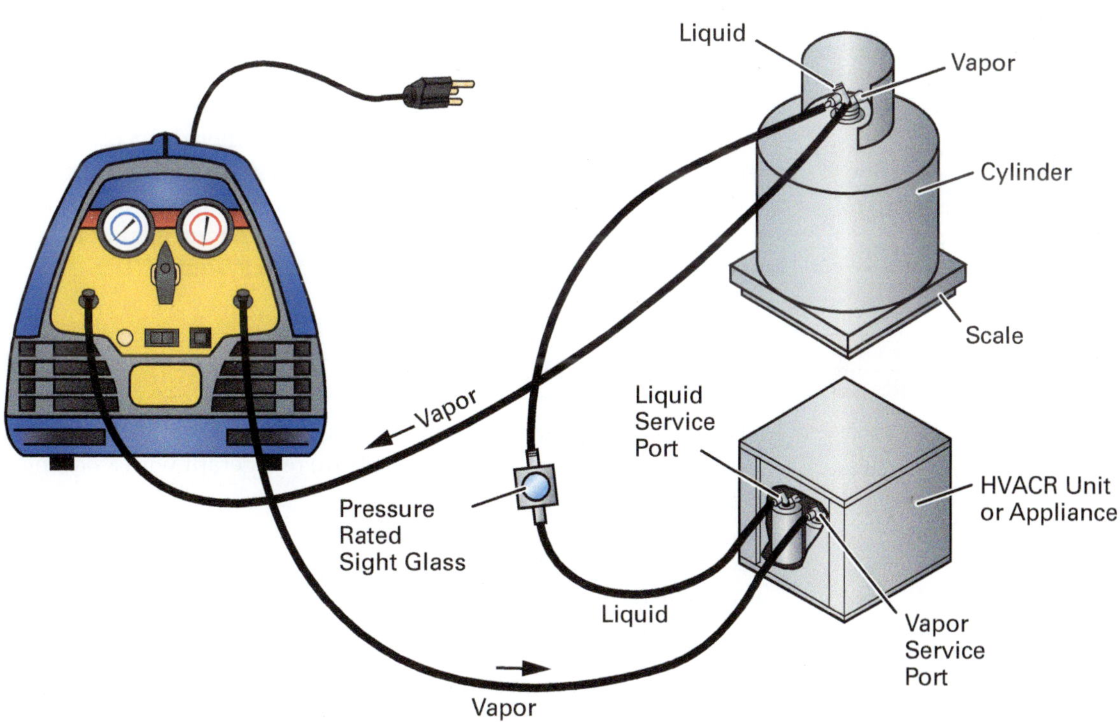

Figure 17 Push-pull recovery setup.
Source: Ritchie Engineering Co., Inc./YELLOW JACKET

The following steps outline the process of recovering refrigerant using the push-pull method. However, always follow the recovery unit manufacturer's guidance for all recovery operations:

Step 1 Make the necessary hose connections per the manufacturer's operating instructions. Remember that $\frac{3}{8}$" hoses are recommended to reduce the pressure drop. Check the connections twice to ensure accuracy.

Step 2 Purge all the connected hoses with refrigerant before starting, to ensure that air is not captured in the recovery cylinder along with the refrigerant.

Step 3 Open both valves on the recovery cylinder.

Step 4 Set the mode selector to the proper position as directed by the operator's manual and start the recovery unit.

Step 5 Monitor both the sight glass and the recovery cylinder weight. Refrigerant vapor is being drawn from the recovery cylinder, compressed to a higher pressure, and sent to the equipment. The high-pressure vapor pushes the liquid through the unit and out to the recovery cylinder, where the liquid falls to the bottom and remains.

Step 6 When no more liquid flow is observed through the sight glass, liquid recovery is complete. The hose connections are then changed as necessary to accommodate vapor recovery.

Figure 18 Shop refrigerant recycling unit with integral vacuum pump.
Source: CPS Products

2.2.5 Recycling Units

Recovered refrigerant can be recycled in the field or in the shop by a certified technician using a certified recycling unit (*Figure 18*). Make sure that the unit is designed for use with the refrigerant to be cleaned. Follow the recycling unit manufacturer's instructions carefully.

A recycling unit dehydrates and cleans recovered refrigerant. Filter-driers are primarily used for this process. The same types of filter-driers or filter-drier cores used in HVACR systems are used with recycling units. After recycling, the refrigerant is cleaner, but its purity is not the same as that of virgin or reclaimed refrigerant.

Acid and moisture testing of recycled refrigerant should be performed to verify the quality of the refrigerant before putting it back into a system. Again, recovered and/or recycled refrigerant can only be reused in the same system from which it was recovered, or another system owned by the same person or entity. It cannot be used in equipment owned by someone else. Recycling the refrigerant does not change the options for its reuse.

Many contractors have found that using a filter-drier on the recovery unit, rather than investing in equipment that is designed for more effective refrigerant cleansing, better suits their needs. If the refrigerant is contaminated, there is less risk and time involved in simply sending the refrigerant away for reclamation and using virgin refrigerant to recharge a system.

2.0.0 Section Review

1. Which of the following technician certifications is required to charge a refrigerant circuit containing a low-pressure refrigerant with a boiling point below −58°F (−50°C)?
   a. Type I
   b. Type II
   c. Type III
   d. Type IV

2. During a recovery in progress, spotting a patch of frost on the liquid line is an indication that ______.
   a. refrigerant vapor is condensing in that area
   b. liquid refrigerant is vaporizing in that area
   c. there is a refrigerant leak at or near that location
   d. the recovery unit is too small and is not recovering as it should

3.0.0 Refrigerant Circuit Evacuation

Objective

State the principles of evacuation and describe how to evacuate refrigerant circuits.

a. State the principles of refrigerant circuit evacuation.

b. Describe vacuum pumps and explain how to properly evacuate a system.

Performance Task

3. Evacuate a system using the triple-evacuation method and perform a vacuum leak test.

HVACR technicians must keep foreign substances out of refrigerant circuits. This includes air and moisture, both of which can be effectively removed through evacuation, except when very *hygroscopic* oils are used. Some synthetic oils such as polyalkylene glycol (PAG) form a chemical bond with water that is difficult to break through evacuation alone. This makes it even more important than ever to ensure moisture does not enter a refrigerant circuit.

Systems must be evacuated to draw out both air and moisture before refrigerant is added. However, it is important to point out that good service practices can prevent all but the smallest amounts of air and moisture to enter a system.

3.1.0 Evacuation Principles

Refrigerant circuits are intended to contain only refrigerant and oil. Anything else in a closed system is considered a contaminant. Contaminants such as air and moisture cause poor system performance and even compressor failures.

In conversation, air is often referred to as a *non-condensable* rather than a contaminant. Air generally migrates to and accumulates at or near the top of the condenser. This reduces the condenser internal volume available for condensing refrigerant. The result is an increase in the condensing pressure and temperature, making the system work harder. Think of it as having a smaller condenser coil than needed.

Moisture promotes the creation of acids through chemical reactions with the oil and refrigerant. These acids corrode metal and break down the insulation on the compressor's motor windings. Moisture in the refrigerant also causes sludge, which reduces the efficacy of the lubricant. There are many reports over the years of *green slime* being discovered in HVACR circuits. This unique sludge is created from moisture and oil, coupled with copper oxides (copper corrosion residue) from tubing. These problems are in addition to the potential for water to freeze at the outlet of the metering device, restricting refrigerant flow. It is much easier to keep water out of a refrigerant circuit than it is to remove it and its byproducts. A wise technician takes every precaution to prevent circuit contamination.

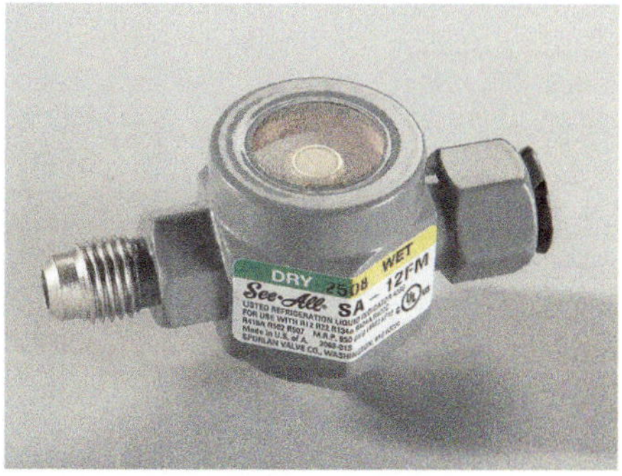

Figure 19 Refrigerant sight glass.
Source: Ritchie Engineering Co., Inc./ YELLOW JACKET

The presence of air can generally be determined from pressures, temperatures, and performance. The presence of moisture in a system can be confirmed using an acid/moisture test kit. Most sight glasses are also equipped with a material that changes color in the presence of moisture (*Figure 19*). However, very few residential systems are equipped with a sight glass. Test kits typically offer greater accuracy.

Generally, a system evacuation is required in the following scenarios:

- After a new system is assembled and pressure tested for leaks, but before refrigerant is added or released from the outdoor unit.
- After an existing system is opened to the atmosphere for service and subsequently reclosed.
- When moisture testing shows moisture is present. In this case, the refrigerant may need to be recovered and replaced with fresh refrigerant following a thorough evacuation. Thorough recycling of the refrigerant to clean it up before returning it to the system is also an option.

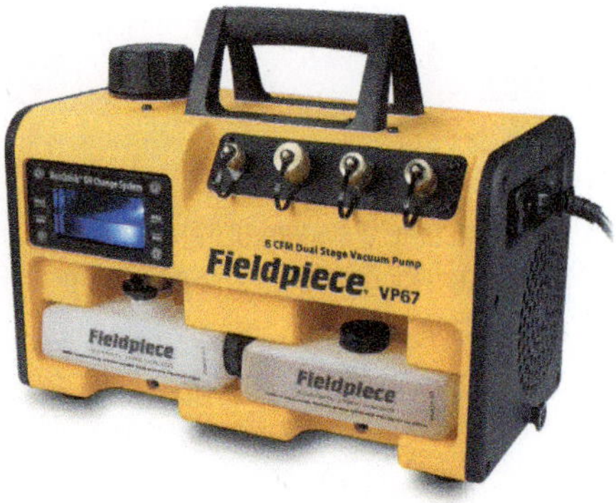

Figure 20 Vacuum pump.
Source: Courtesy of Fieldpiece Instruments

NOTE

A perfect vacuum is not technically possible. Even in the vacuum of space, far from planets and moons, a small concentration of atoms per cubic meter exists.

If moisture contamination is thought to be minor, changing the filter-drier is an option. In most cases though, you will need to evacuate at least a portion of the system after installing the drier. If a system is known to contain moisture, then consider a thorough evacuation of the entire system to be the primary strategy and best practice. After the refrigerant is recovered, replace the filter-drier and then evacuate the system.

Air is easily extracted from a refrigerant circuit, but water is another matter. Water must be removed by causing it to boil and vaporize. It is then drawn out with a vacuum pump (*Figure 20*) along with air and other vapors. As the vacuum pump reduces the pressure in the circuit, the boiling point of the water is significantly reduced.

Before going further, the units of measure for vacuum need to be reviewed. In HVACR work, vacuum is typically measured in *microns* (μm). When a space reaches a vacuum of 0 μm, it would measure on a manometer as 29.92 in. Hg gauged. However, reaching this level of vacuum is not practical. The best vacuum pumps used in the field can theoretically reach 15 μm, but it is rarely achieved and doing so is unnecessary.

Now consider the previous discussion regarding the vaporization of water. Water will boil and vaporize at 80°F (26.7°C) at 25,400 μm. At a vacuum of 254 μm, water is forced to boil at –24°F (–31.1°C). *Table 4* shows the boiling temperature of water at various pressures. The first row of the table represents atmospheric pressure. Once water has vaporized, it can be swept out of the system by the vacuum pump; it is only designed to pump vapor.

Removing all moisture from a refrigerant circuit can take some time, but it can be done at virtually any temperature you encounter. The keys are a pump that can adequately reduce the pressure and a system free of leaks.

At higher temperatures, it is relatively easy to encourage water to boil. Deeper vacuums are needed for cold temperatures. It is important to remember that encouraging water to vaporize and exit the system is not an event—it is a process. Just as a pot of boiling water on the stove does not suddenly vaporize

TABLE 4 Boiling Point of Water at Various Pressures

Boiling Point (°C)	Boiling Point (°F)	Microns	Inches of Hg Vacuum
100°C	212°F	759,968 μm	0.00" Hg
96°C	205°F	535,000 μm	4.92" Hg
90°C	194°F	425,526 μm	9.23" Hg
80°C	176°F	355,092 μm	15.94" Hg
70°C	158°F	233,680 μm	20.72" Hg
60°C	140°F	149,352 μm	24.04" Hg
50°C	122°F	92,456 μm	26.28" Hg
40°C	104°F	55,118 μm	27.75" Hg
30°C	86°F	31,750 μm	28.67" Hg
26.7°C	80°F	25,400 μm	28.92" Hg
24.4°C	76°F	22,860 μm	29.02" Hg
22.2°C	72°F	20,320 μm	29.12" Hg
20.6°C	69°F	17,780 μm	29.22" Hg
17.8°C	64°F	15,240 μm	29.32" Hg
15°C	59°F	12,700 μm	29.42" Hg
11.7°C	53°F	10,160 μm	29.52" Hg
7.2°C	45°F	7,620 μm	29.62" Hg
0°C	32°F	4,572 μm	29.74" Hg
–6.1°C	21°F	2,540 μm	29.82" Hg
–14.4°C	6°F	1,270 μm	29.87" Hg
–31.1°C	–24°F	254 μm	29.91" Hg

and disappear once it begins to boil, the water in a refrigerant circuit does not vaporize instantaneously. Time with the system in a vacuum is required.

Some of today's synthetic lubricants not only create a strong bond with water, but also attract it. A vacuum pump may not be able to break the bond and remove all moisture, but a filter-drier can usually take care of what remains. One of the most important lessons to learn as a technician is to understand that moisture problems are best avoided by taking the necessary steps to prevent it from ever entering the circuit.

The *deep vacuum method* and the *triple-evacuation method* are the common approaches to evacuation. The deep vacuum method is typically used after a repair that requires the refrigerant charge to be recovered and the circuit opened for a short time. Of course, the work should be done with care to prevent air and moisture from gaining entry whenever possible. The triple-evacuation method is recommended for new systems and when a system is known to be damp. Both methods will be presented in the next section.

On Site

Keeping Moisture Out of Refrigerant Circuits

One experienced technician used bees to make his point about protecting refrigerant circuits and oil from moisture. His approach was to treat the opening in an oil can or refrigerant circuit as if there were bees on the other side. "As soon as the opening appears, block it, plug it, or cap it!"

On Site

Hoses for Evacuation and Recovery

Most hose manufacturers produce heavy-duty $^3/_8$" and even $^1/_2$" hoses with thick walls that are designed to reduce pressure drop and prevent leaks when evacuating a system. Hoses of this type are best dedicated to evacuation and refrigerant recovery to minimize wear and tear. Standard hoses are fine for simply connecting to a system for analysis.

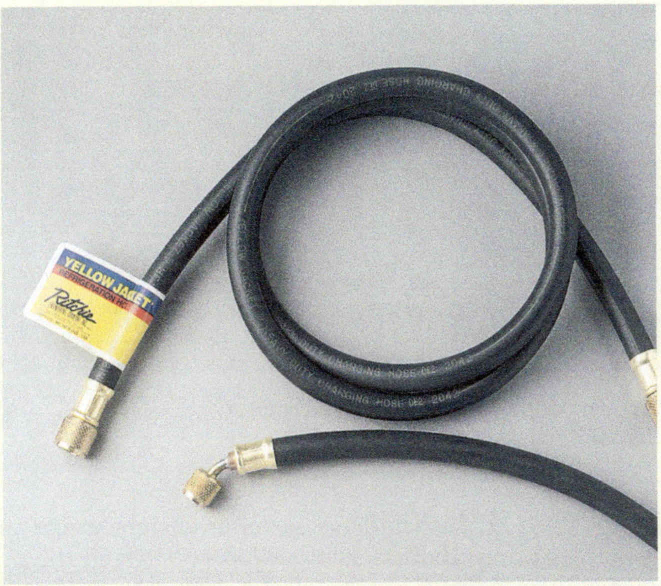

Source: Ritchie Engineering Co., Inc./YELLOW JACKET

3.2.0 Evacuation Equipment and Procedures

Evacuating a system requires a good vacuum pump, an accurate vacuum gauge, a leak-free gauge manifold, and leak-free hoses. When a very deep vacuum is required, copper tubing may even be used in place of the hoses. However, this is rarely necessary in HVACR work, thanks to the quality of today's hoses.

3.2.1 Evacuation Equipment

To properly evacuate a system, several tools are needed. The vacuum pump must be connected to the system, and an accurate means of monitoring progress is essential. The needed equipment is presented in this section.

Vacuum Pumps

The vacuum pump creates a low-pressure condition at its inlet, encouraging the flow of vapors toward it. Once the vacuum pump lowers the pressure in the system enough, any moisture in the system begins to vaporize. Air and vapors pass through the oil-filled pump crankcase and are then discharged to the atmosphere.

A good vacuum pump is essential for the proper evacuation and dehydration of a system. Typically, a good pump can evacuate down to 25 μm (29.90 in. Hg vacuum) under ideal conditions. Note that ideal conditions rarely exist in field work. The pump must also be rugged and easy to handle.

Most vacuum pumps used in the craft are direct-drive, two-stage, rotary vane pumps driven by a close-coupled motor. Belt-driven pumps are also available, although they are much heavier and have lower flow rates. However, belt-driven pumps can also reach very low vacuum levels given the time. For this reason, belt-driven vacuum pumps are often seen in laboratories. The pump shown in *Figure 21* is an advanced model for field use, with a brushless DC motor that is significantly lighter than most pumps of equal capacity. The motor also requires less current and can tolerate lower voltages than pumps with standard AC motors.

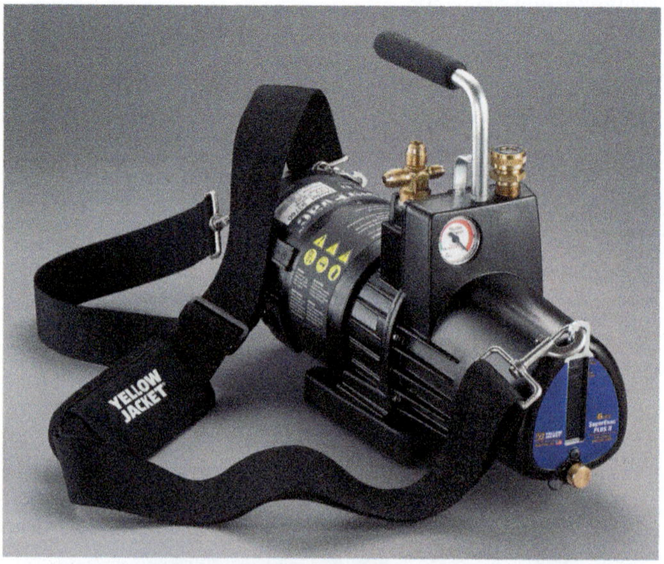

Figure 21 Vacuum pump with a brushless DC motor.
Source: Ritchie Engineering Co., Inc./YELLOW JACKET

Vacuum pumps used in the field have an oil level sight gauge, oil fill port, and oil drain port. A pump that offers a *gas ballast* between the pump's first and second stages is an advantage. This feature permits air from the atmosphere to enter the second stage of the pump, where it combines with any moist vapors passing through the pump from the refrigerant system. This reduces the amount of moisture that condenses in the pump crankcase, which fouls the oil. Oil that has absorbed too much moisture becomes less transparent and turns milky white. When this happens, drain and replace the oil immediately. The quality and condition of the oil has a significant impact on pump performance because it seals all the minute passages between the rotors and the cylinders.

How quickly a pump can evacuate a system is an important consideration. The more vapor the pump can move (measured in cubic feet per minute, or cfm), the faster it can reach the required vacuum level. *Table 5* lists the recommended pump flowrate for various applications.

Remember that hoses and fittings can restrict flow and significantly reduce the flowrate. All too often, technicians purchase vacuum pumps with a high flowrate, and then use standard $^{1}/_{4}$" hoses equipped with valve core depressors to make the connections. The capacity of the pump drops sharply as a result.

TABLE 5 Minimum Recommended Vacuum Pump Capacity

System Capacity	Minimum Recommended Pump Capacity
Up to 10 tons	4 to 6 cfm
Up to 30 tons	6 to 12 cfm
Up to 50 tons	10 to 15 cfm
Up to 70 tons	12 to 20 cfm

All vacuum pumps need periodic maintenance. Most of the maintenance is related to the oil. Because the oil in the pump is contaminated to some degree during an evacuation, the oil should be changed after every 10 hours of operation in the best of conditions. Always drain or change the oil immediately after pumping down a refrigerant circuit known or suspected to contain moisture. Water quickly accumulates in the oil during an evacuation. Leaving a mixture of water and oil in the pump creates sludge and encourages corrosion. In some cases, the oil needs to be changed during the evacuation process, as mentioned previously. Follow the manufacturer's instructions for maintaining the pump.

Note that the oil in the pump can also absorb a great deal of moisture while it is not in use. If oil is left in the pump between uses (common), make sure that all inlets and outlets remain tightly capped. Frequent oil changes cost less than a new pump.

Special oils are used in vacuum pumps (*Figure 22*). Never use oils that are not labeled for vacuum pump service.

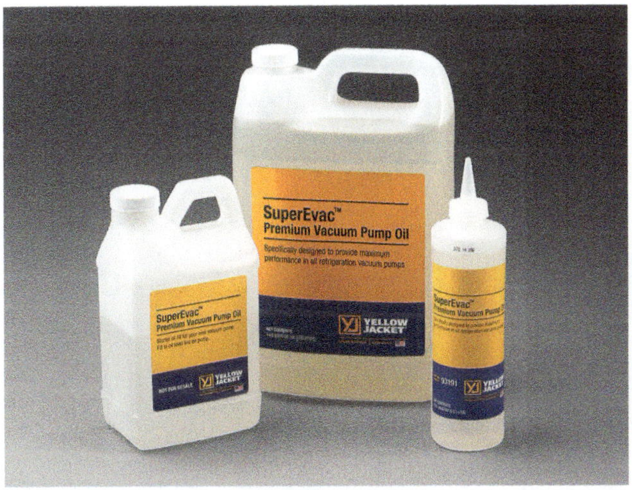

Figure 22 Vacuum pump oil.
Source: Ritchie Engineering Co., Inc./YELLOW JACKET

Vacuum Gauges

An accurate vacuum gauge is essential. Evacuating a system without monitoring the pressure is a wasted effort. Even though the compound gauge on a gauge manifold has vacuum increments, the gauge has very poor accuracy in the vacuum range and cannot provide the information you need.

Some vacuum gauges can read as low as 1 µm. Several vacuum gauges are shown in *Figure 23*. Wireless vacuum gauges allow you to monitor the progress via a smartphone app or a digital display on a compatible digital gauge manifold (*Figure 24*). Some digital manifolds have a built-in vacuum gauge.

The location of the vacuum gauge in the system affects the vacuum reading. The closer it is mounted to the vacuum pump connection, the lower the reading. This can be misleading, as the pressure at the farthest point of the system will be significantly higher. To get a more accurate reading on the system, valve off the vacuum pump and allow the pressure throughout the system to equalize. On a tight system, the pressure will rise somewhat before stabilizing. A small

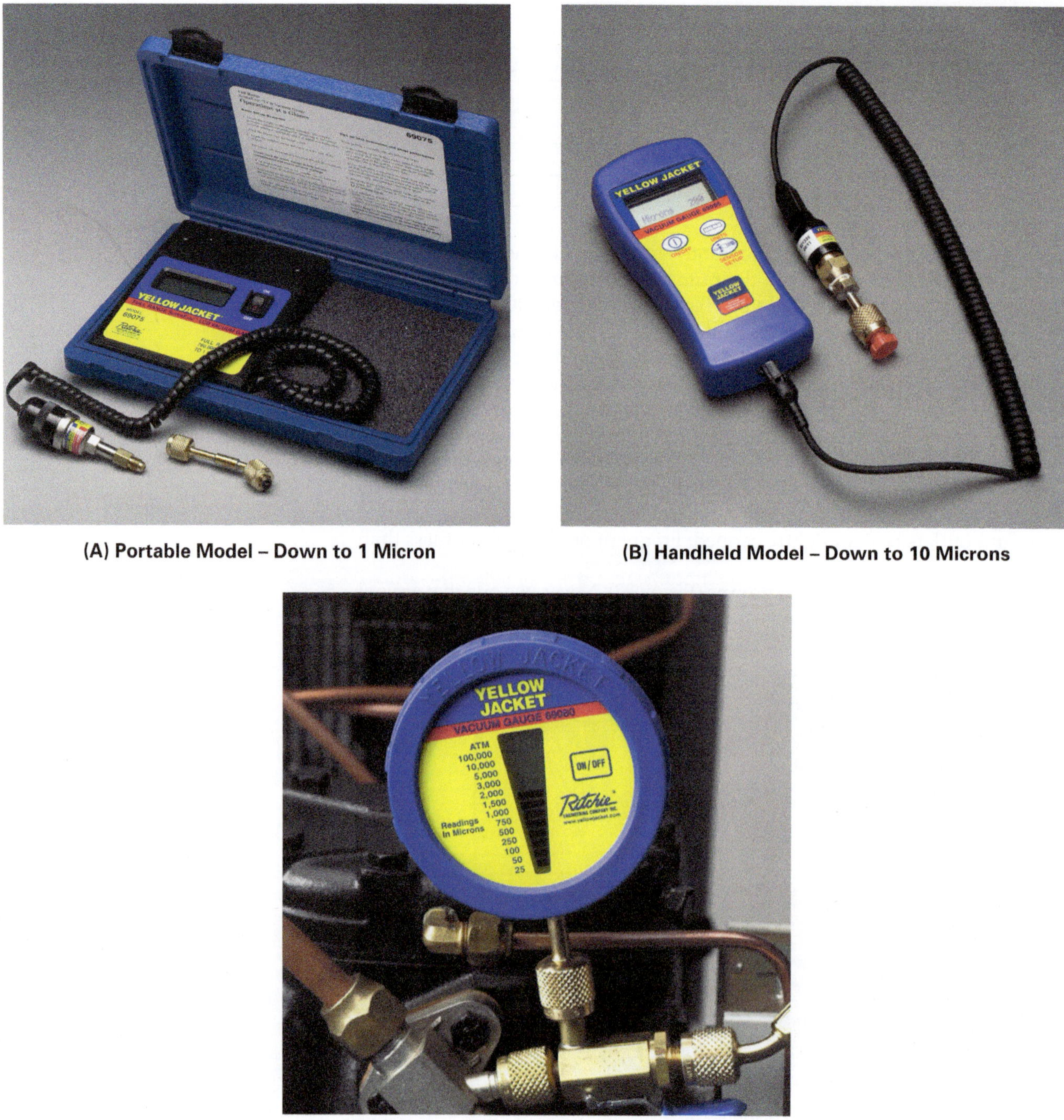

(A) Portable Model – Down to 1 Micron

(B) Handheld Model – Down to 10 Microns

(C) Economy Model – Down to 25 Microns

Figure 23 Examples of vacuum gauges.
Source: Ritchie Engineering Co., Inc., Yellow Jacket Division

increase when the pump is isolated is common and does not necessarily represent a problem.

If the micron level continues to rise after the pump is isolated from the system, there are several possible causes. A leak in the system or in the hose connections could be allowing air to enter. Liquid refrigerant could still be vaporizing from the oil. Even moisture vaporizing in the system will slowly increase the pressure. A thorough system evacuation takes time, so it is important to be patient and monitor the process until vaporizing refrigerant and moisture can be ruled out as the cause of a steady pressure increase.

The vacuum pump and hoses can be quickly tested using a vacuum gauge. Connect the gauge directly to the vacuum pump or use a hose you wish to test,

Figure 24 Wireless vacuum gauge.
Source: Courtesy of Fieldpiece Instruments

then start the pump. A vacuum pump in good shape should quickly pull down to or near its specified rating. A well-used pump might be considered good if it pulls down below 200 μm.

Gauge Manifold

Either a two-valve gauge manifold with evacuation-quality service hoses or a high-capacity evacuation gauge manifold can be used. A good manifold that enhances evacuation processes has four valves, enlarged internal passages, and a ³⁄₈" evacuation hose connection (*Figure 25*). The four valves and ports enable the gauge manifold to be simultaneously connected to the evacuation and charging equipment, without disconnecting and reconnecting the hoses between tasks.

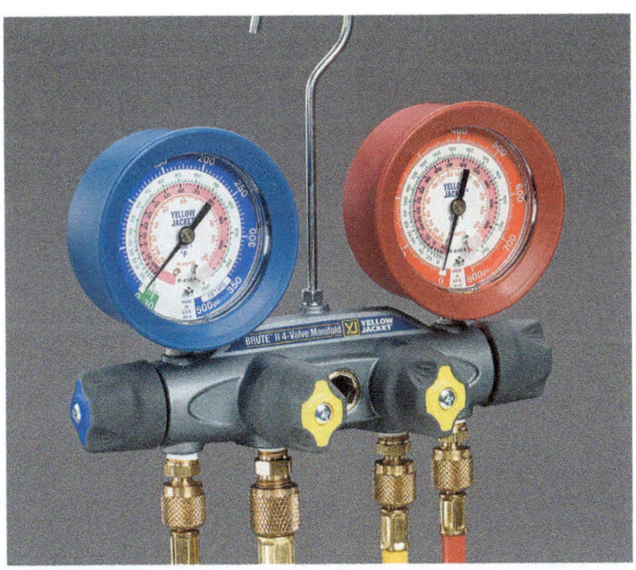

Figure 25 Gauge manifold with four valves and ³⁄₈" evacuation hose port.
Source: Ritchie Engineering Co., Inc./YELLOW JACKET

3.2.2 Evacuation Procedures

The two approaches to evacuation are the deep vacuum method and the triple evacuation method. The triple evacuation method is a more thorough process, designed to enhance dehydration. Both methods are outlined in this section.

> **Did You Know?**
>
> **Refrigerant Gauges**
>
> It is tempting for a technician to choose an inexpensive product when shopping for tools. However, inexpensive gauges may cause problems in HVACR work. Refrigerant gauges are made to various standards; not all are the same.
>
> For example, a *3-2-3 gauge*, also known as a *Class B* gauge, is broken into three different zones of accuracy. At the lower and higher ends of the gauge scale, the accuracy is established as 3% of the full scale. The center of the range is accurate to within 2% of the full scale. So, when working with a gauge that has an 800-psig range, pressure readings at the lower and higher ends of the range can be off as much as 24 psig. The center of the range can be off by as much as 16 psig. Yet the gauge would still be considered calibrated and within its reported accuracy. Note that most gauges aren't that far off, but their specifications allow that much inaccuracy.
>
> A 1%, or *Class 1*, gauge is accurate to within 1% across its entire range. Therefore, an 800-psig, Class 1 gauge is accurate to within 8 psig across the entire range. When trying to determine superheat and subcooling values, the difference in accuracy is very important.

NOTE

The vacuum gauge can be inserted anywhere in the system you wish. When possible, it is a good idea to use one gauge near the pump and another at the far end of the refrigerant circuit. This allows you to monitor the progress throughout the system. Wireless vacuum gauges are especially well suited for this purpose.

Deep Vacuum Evacuation Method

The deep vacuum method relies on the vacuum pump alone to remove moisture from the system. A *deep vacuum* is generally considered to be a vacuum of 500 microns or less. Consider this method to be the standard approach.

When a deep vacuum is established in a closed system, air and moisture in the system are reduced to a negligible level with sufficient time. As long as the ambient temperature is higher than the boiling point of water at a given pressure, the moisture will vaporize and exit the system.

A deep vacuum evacuation is performed as follows:

Step 1 Connect the vacuum pump, vacuum gauge, and gauge manifold to the service valves on the system or component being evacuated, as shown in *Figure 26*.

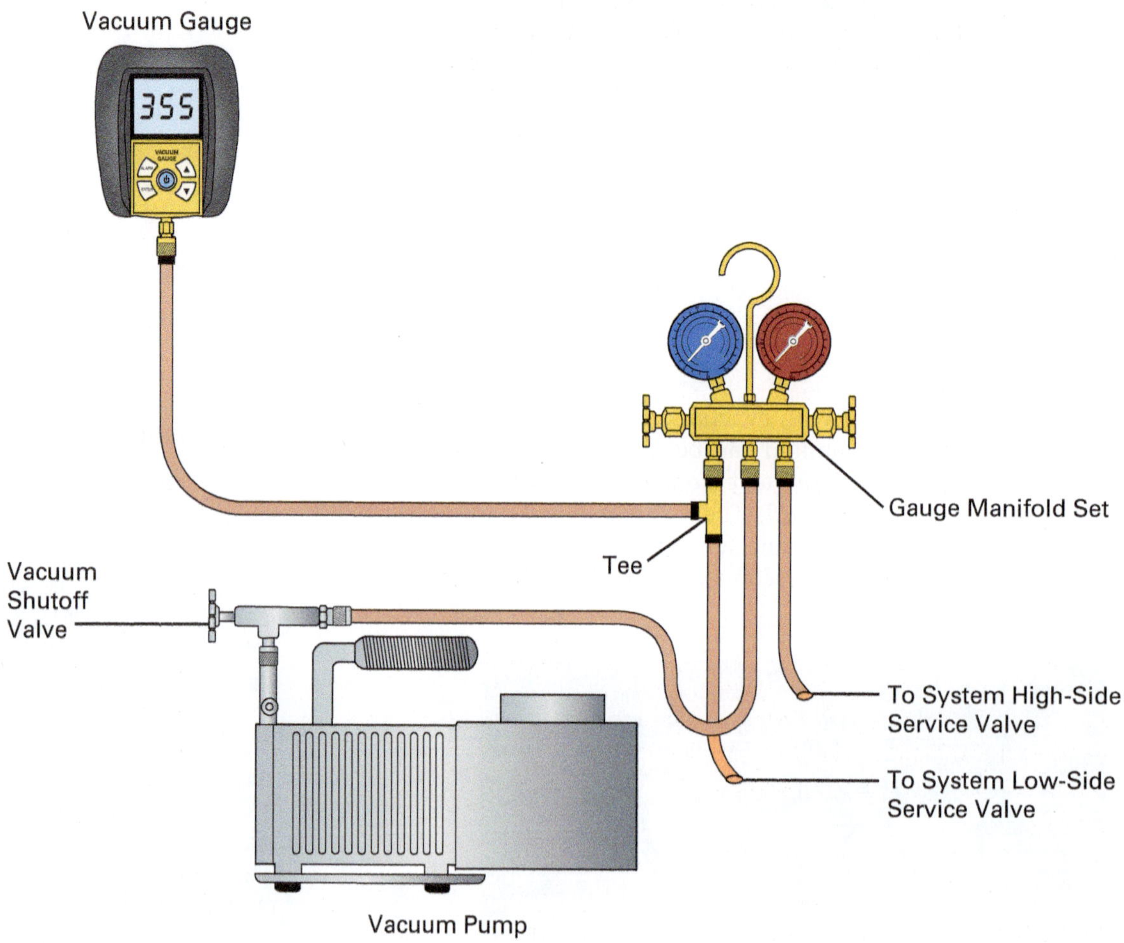

Figure 26 Equipment connections for an evacuation.

Step 2 Turn on the vacuum pump, open the system service valves, and open the gas ballast. Close the gas ballast when the vacuum reaches 2,000–1,000 µm.

Step 3 Pump the system down to 500 µm or less. As you approach 500 µm, close the vacuum shutoff valve, then watch the vacuum gauge to see if a rapid and steady loss of vacuum occurs. If so, check the system, pump, and vacuum hoses for leaks. Repair any leaks found, then proceed with the evacuation. Restart the pump before opening the vacuum shutoff valve. Note that finding a troublesome leak will likely require the system to be pressurized with nitrogen and/or a trace gas.

Step 4 When the vacuum reaches 500 µm or lower, isolate the pump using the shutoff valve and monitor the vacuum gauge. Watch for a steady increase that refuses to stabilize. This either indicates the system is leaking as mentioned in the previous step, or that refrigerant and/or moisture is still vaporizing. When refrigerant is still vaporizing, wet or frosty spots are likely visible on parts of the system, such as the compressor crankcase or the accumulator. Continue the evacuation until there are no more cold spots and the vacuum remains stable at or near 500 µm.

Step 5 Use the vacuum shutoff valve to isolate the vacuum pump from the system. Then turn off the vacuum pump and close the valves to the gauge manifold. Note, however, that the vacuum gauge should remain connected and active.

Step 6 To ensure the system is adequately evacuated, leak test the system under vacuum by monitoring the vacuum gauge. Look for the following indications:

- A constant reading on the vacuum gauge indicates a dry, leak-tight system.
- If the gauge shows a rise in pressure but levels off before reaching 0 psig, the system is likely free of leaks, but liquid refrigerant and/or moisture is still vaporizing. Continue the evacuation process.
- If the gauge shows a rise in pressure to 0 psig (atmospheric pressure) before stopping, there is a leak in the system or in the service connections.

Triple Evacuation Method

The triple evacuation method is recommended by most manufacturers for new or damp systems. This method repeats the deep vacuum method three times. However, after the first and second evacuations, the vacuum is broken with dry nitrogen to help absorb moisture between each evacuation cycle. Dry nitrogen absorbs water vapor well and is readily available.

Did You Know?

All Nitrogen Is Not Dry Nitrogen!

When you pick up a cylinder of dry nitrogen from the supply house, how much water might it contain?

If you visit a nitrogen vendor's website, you might be amazed at how many different grades of nitrogen are on the market. You may see nitrogen products such as *purity*, *vehicle emissions purity*, *high purity*, *ultra-high purity*, *zero*, and *oxygen-free*. Most sites only stock a few different grades for their service area.

Many of the names sound good, but most nitrogen grades focus on specific applications. Some are very dry, while others are not. Nitrogen used for medical purposes and semiconductor work, for example, is usually dried to 0.2–0.5 parts per million (ppm) of water. Other grades may have as much as 3 ppm, while common industrial nitrogen gas may have as much as 100 ppm water. And, as you might expect, purity costs extra. But using wet nitrogen on a system can create a lot of problems, and it may never occur to you that your nitrogen is the source of the problem.

Talk with your preferred gas distributor and identify the correct product for your needs. Know what you are buying. Purchasing medical-grade gas should not be necessary, but you also don't want to use nitrogen that already contains as much or more water as the system you're trying to dry out.

Figure 27 shows the typical equipment connections for a triple evacuation. *Figure 28* shows the sequence of the process. The process can take several hours. The time varies depending on what level of vacuum is reached in each stage, the internal volume of the refrigerant circuit, and the vacuum pump capacity.

The system should be evacuated to at least 1,000 µm during each evacuation; deeper is better. After reaching 1,000 µm, allow the vacuum pump to run for at least 15 minutes longer during each evacuation. This will force any moisture in the system to continue vaporizing, as long as the temperature is at or above 35°F (1.7°C). Remember that the process of vaporizing is not instantaneous; time is required for the water to complete the change of phase.

If some or all the system is below this temperature, deeper vacuums are required to make the water vaporize. If you are evacuating a heat pump in harsh winter conditions, for example, a deeper vacuum is required.

After the first evacuation, the vacuum pump can be turned off and the system vacuum leak tested as previously described in the deep evacuation method. This ensures you aren't wasting your time and resources on a leaking system. However, many agree that it is best to conduct the vacuum leak test after the final evacuation. At this point, the system should be free of all refrigerant and

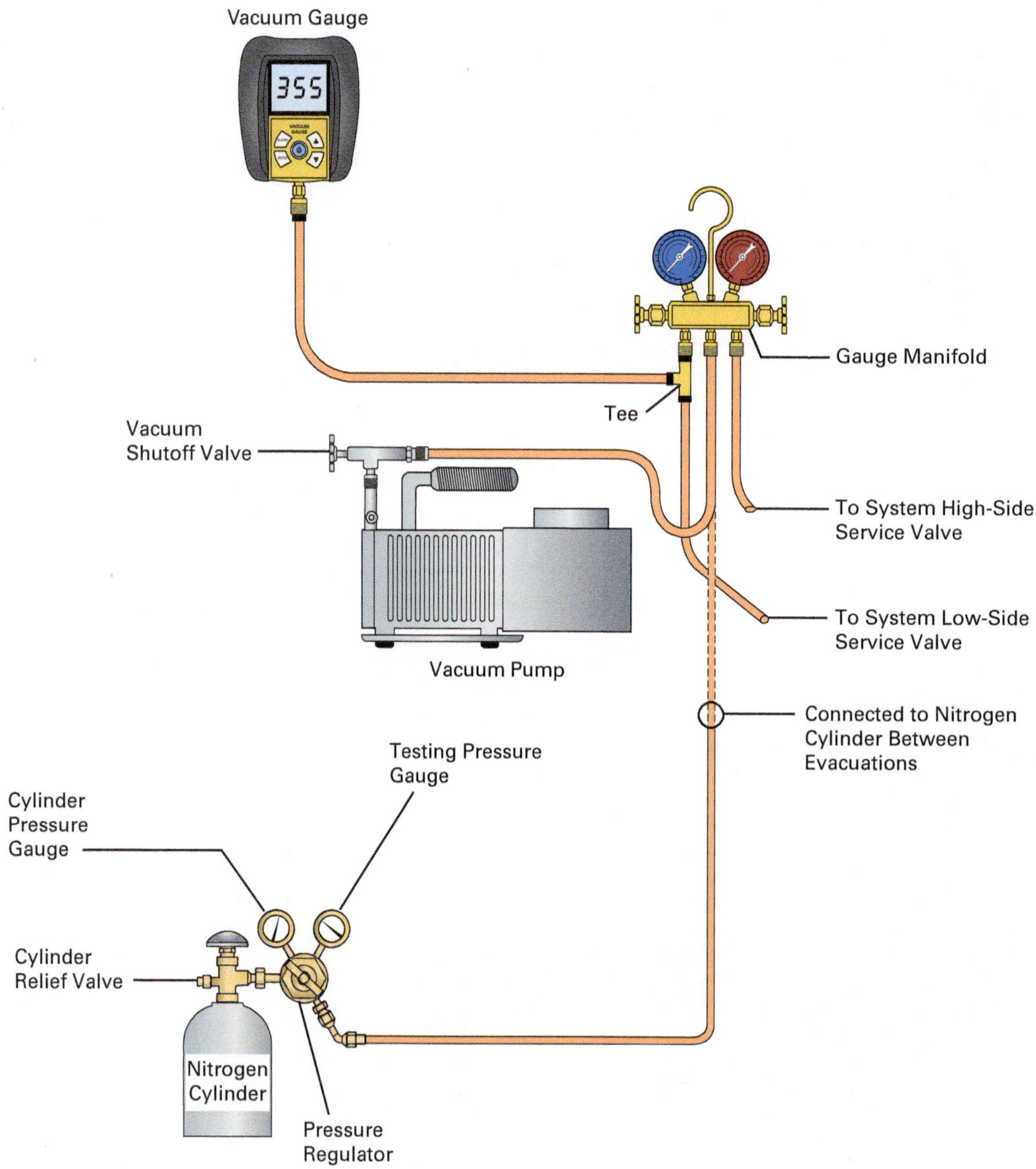

Figure 27 Service equipment connected for a triple evacuation.

moisture that would affect the vacuum leak test. If leaks in the system are a concern, it is much better to leak test the system under pressure before the evacuation process begins.

The initial vacuum is broken with dry nitrogen. Close the vacuum shutoff valve, disconnect the hose from the pump, then connect the hose to the nitrogen regulator. Purge the hose with nitrogen before opening the manifold valves to admit nitrogen into the system.

Pressurize the refrigerant circuit to roughly 10 psig; the nitrogen pressure is not significant to the process, so there is no advantage to raising the pressure further. Allow the system to sit for 30–60 minutes. This allows time for the nitrogen to absorb any water vapor.

After the time has passed, purge the nitrogen from the system. Reconnect the vacuum pump and repeat the evacuation process. Break the vacuum again with

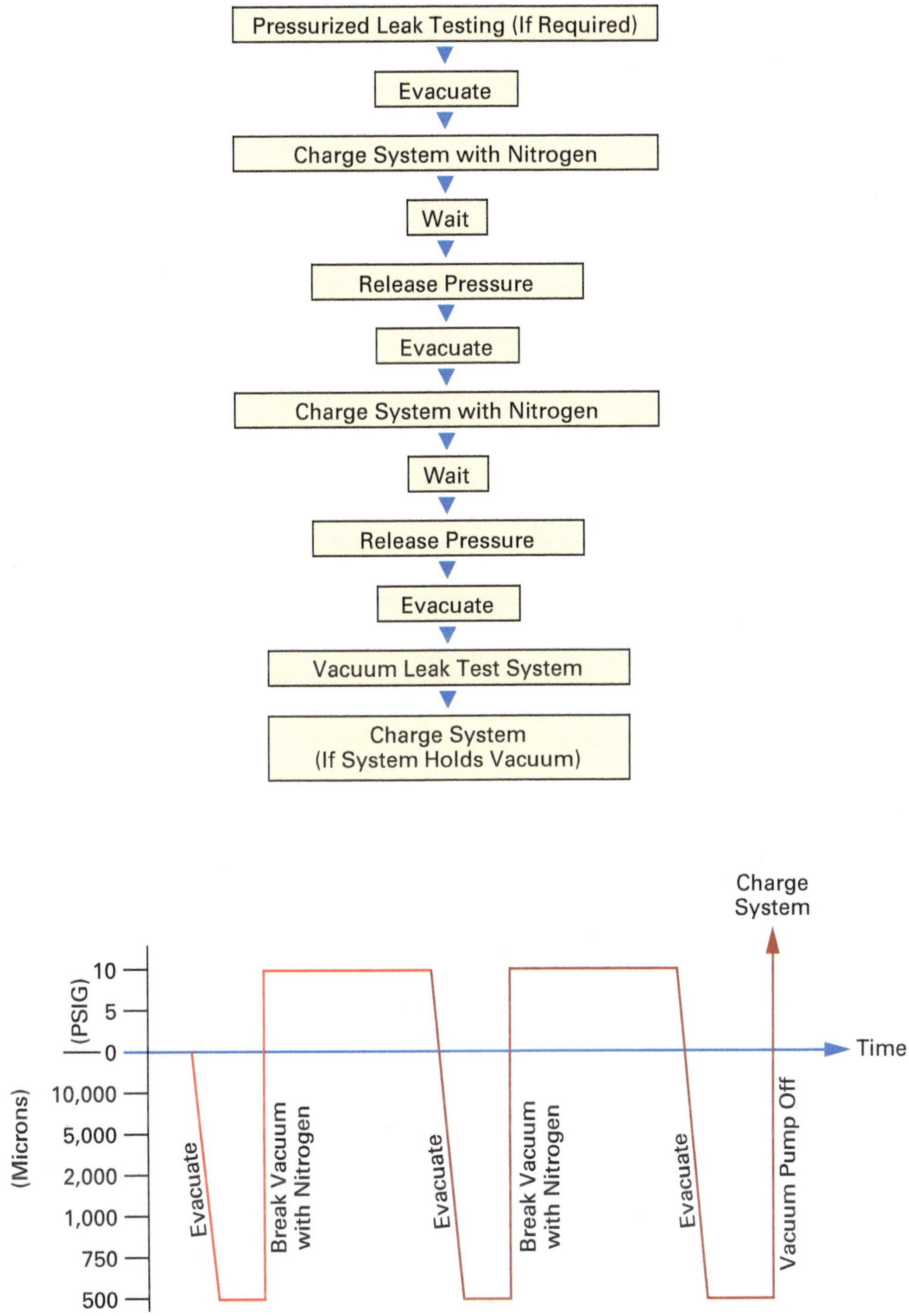

Figure 28 Sequence of events for a triple evacuation.

> **On Site**
>
> **Smoking Vacuum Pumps**
>
> In the early stages of an evacuation, the vacuum pump may be noisy and discharge what appears to be smoke. The "smoke" is actually water vapor. As the vacuum deepens, the pump should become quieter while the visible vapor disappears. If the noise and water vapor does not progressively decrease, a significant leak in the system may be allowing moist air to enter.

nitrogen, allow the system to sit for a while, and then evacuate a third time. This is the final evacuation, and the system should now be free of both air and moisture. Allow the vacuum to stand for some time to ensure the system is tight. The next step is to charge the system with refrigerant.

3.0.0 Section Review

1. Air in a refrigerant circuit will typically migrate to and accumulate in the _____.
   a. compressor
   b. evaporator
   c. accumulator
   d. condenser

2. What feature of a vacuum pump introduces air into the second stage of the pump, where it combines with moist vapors passing through the pump?
   a. Rotary vanes
   b. A condenser
   c. A gas ballast
   d. An air injection pump

4.0.0 Refrigerant Charging

Performance Task

4. Demonstrate how to properly charge a refrigerant circuit using the following methods:
   - By weight
   - By superheat (fixed-orifice metering device)
   - By subcooling (expansion valve)

Objective

Explain how to properly charge refrigerant circuits using various techniques.

a. State important concepts related to refrigerant charging and the related equipment.
b. Explain how to charge a refrigerant circuit by weight.
c. Explain how to charge a refrigerant circuit by superheat.
d. Explain how to charge a refrigerant circuit by subcooling.

A system must be charged with refrigerant to place it in service after the refrigerant circuit has been sealed, leak tested, and evacuated. Some units come factory-charged with refrigerant, but the charge may need to be adjusted in the field at startup. The factory charge is not correct for every installation. This section describes the service equipment and methods commonly used to charge systems, as follows:

- Charging by weight
- Charging by superheat
- Charging by subcooling

NOTE

The information that follows is a review of the content in NCCER Module 03301, *Refrigerants and Oils*. Refer to that module for more information on refrigerants and blends.

To properly charge a refrigerant circuit, you must be familiar with the refrigerant in use, and you must know what type of metering device the system uses. To that end, this section begins with a review of information regarding refrigerant blends and their behavior in the refrigerant circuit.

Azeotropic refrigerants are blends that have a single saturation temperature for a given pressure. Azeotropic refrigerants maintain the proportions of the blend while transitioning from one phase to another. Thus, azeotropic refrigerants behave like compound refrigerants.

Zeotropic refrigerants, however, are blends that have a range of saturation temperatures for a given pressure. Most zeotropes are a blend of three refrigerants and are called *ternary blends*. Some components of a zeotropic refrigerant typically retain their individual characteristics. Thus, zeotropic refrigerants vaporize and condense over a range of temperatures at a given pressure. This causes zeotropic refrigerants to have different liquid and vapor saturation temperatures.

Near-azeotropic refrigerants demonstrate the characteristics of both zeotropes and azeotropes. Most of these blends have only two components and are referred to as *binary blends*. Technically, near-azeotropic refrigerants are zeotropes with a range of saturation temperatures, but the range is so narrow, their behavior in the refrigerant circuit is essentially the same as that of an azeotrope. A near-azeotropic refrigerant, such as HFC-410A, has a PT chart that looks the same as a pure compound—one saturation temperature for a given pressure.

To work with zeotropic refrigerants, you must recall and understand the following terms:

- *Glide* — Glide identifies the range of temperatures at which a zeotropic refrigerant will vaporize and condense for a given pressure. For example, HFC-407C at 80 psig is fully condensed at temperatures below 40°F (4.4°C) and fully vaporized at temperatures above 51°F (10.6°C). Between these temperatures at 80 psig, it is in saturation and is said to have a glide of 11°F (6.1°C). Glide represents the cause of *fractionation*.
- *Fractionation* — Fractionation occurs when one or more refrigerants in a blend are in different phases. When a leak occurs, more of one blend component than another can leak out. The significance of fractionation has a great deal to do with the location of the leak in the system. Where all the refrigerant is in one state or the other—liquid or vapor—fractionation is not an issue. However, if a leak occurs in an area where condensation or vaporization is underway, fractionation can be significant. The result is that the precise proportions of the refrigerants used in the blend are altered. Further, when the system is not running, liquid and vapor can be found together in many parts of the circuit. Thus, a leak in a variety of places is likely to result in some fractionation. The concern is that the refrigerant's performance is compromised.

When there is a concern about fractionation due to a leak, recover the remaining refrigerant, repair the leak, and recharge the system with new refrigerant. Charging the system with virgin refrigerant guarantees that the correct blend proportions are present. The recovered refrigerant is then returned for reclamation.

When charging a system that uses zeotropic refrigerant, always transfer the refrigerant from the cylinder as a liquid. This is necessary to ensure that the proper blend composition is charged into the system. If it is necessary to charge an operating system into the suction line where only vapor is present, draw refrigerant from the cylinder as a liquid and feed it slowly into the low side of the system using your gauges as a metering device. This avoids any possibility of floodback to the compressor.

Zeotropic refrigerants require a slightly different approach to superheat and subcooling calculations. The pressure-temperature (PT) chart for a zeotropic refrigerant has two temperatures for each given pressure: the *saturated vapor temperature* and the *saturated liquid temperature*.

When calculating superheat for zeotropic refrigerants, be sure to use the saturated vapor temperature in the calculation. When calculating subcooling, be sure to use the saturated liquid temperature in the subcooling calculation. Using the wrong value from the chart will cause significant errors. PT charts from manufacturers vary in how they label these columns, however. The saturated vapor temperature column may also be labeled as "Vapor Pressure" or "Dew Point." The saturated liquid temperature column may be labeled as "Liquid Pressure" or "Bubble Point."

4.1.0 Refrigerant Charging Concepts

Regardless of system size, the operating charge of a system is a major factor in its performance. An overcharged system causes high pressures and temperatures that reduce performance and can lead to component failure. An undercharged system leads to insufficient cooling and even evaporator freezing in some conditions. It may also cause the compressor to overheat.

Many of the problems that occur in refrigeration and air conditioning systems are the direct result of overcharging or undercharging. Both are significant

NOTE

Equipment manufactured outside of the US or manufactured for use in other countries may measure refrigerant in units of kilograms and grams. Many digital charging scales can accommodate both systems of measurement.

errors that must be avoided. Regardless of what system you are charging and why, the refrigerant must be added correctly and in the right volume.

At this point in your training, you are aware that refrigerant is measured by weight, in pounds and ounces. A system can be charged exclusively by weight, followed by a check of its performance and the essential values of superheat and/or subcooling. Never assume a weighed charge is accurate—always check the system's performance.

The nameplates of split systems and packaged equipment usually show the weight of the total refrigerant charge. This information is also listed in the manufacturer's service literature for the equipment. Always consult these resources to get the specific charge information. Be sure to account for any additional refrigerant charge needed to compensate for field-installed accessories such as filter-driers and receivers.

The nameplates on split systems provide the charge weight for the system based on a standard refrigerant set length. For newer systems, the line set length used in their calculations may be as little as 15'. Consult the manufacturer's service literature to find out the length of the standard line set for the equipment being serviced. This information changes over time. For example, older systems carried enough refrigerant to support line sets up to 50' in length.

If the system being serviced uses a line set that exceeds the standard length, find the amount of additional refrigerant charge that must be added to compensate for each foot of increased length. If the line set is unusually short, the nameplate refrigerant charge can likely be reduced.

For large, built-up systems for which each major component is purchased separately, the total charge weight can be determined by finding the sum of the charge weight of the individual components and accessories, plus the refrigerant needed to fill the lines with liquid or vapor. The capacity of each component can be found in the manufacturer's literature, or it may be marked on the component nameplate. However, in practice, the charge on these systems is most often determined through performance (superheat, subcooling, and capacity calculations) rather than through math.

NOTE

Small systems that need a very small amount of refrigerant can be challenging to charge accurately. It is suggested that these systems be charged using the simplest gauge manifold with the shortest hoses available. A single refrigerant hose can hold more liquid refrigerant than the entire charge required for some equipment.

When a system has a fixed metering device, such as a capillary tube or a fixed restrictor, it should be charged to a superheat value. Systems that are equipped with any type of modulating metering device are primarily charged by subcooling. This doesn't mean you should completely ignore the other value. For example, if you are charging a system equipped with a thermostatic expansion valve, you must still measure the superheat after charging to ensure it is correct.

Refrigerant may be added to a system in either the vapor or liquid state, depending on the refrigerant. Compound and azeotropic refrigerants can be charged in either state without concern. Zeotropic refrigerants, however, must be charged in the liquid state. These refrigerants do not necessarily need to enter the system in the liquid state, but must exit the refrigerant cylinder as liquid. The gauge manifold valve can be used as a manual metering device to flash some of the liquid to vapor and prevent compressor floodback.

It is very important to remember that most disposable refrigerant cylinders for zeotropic and near-azeotropic refrigerants are made with dip tubes. However, you may encounter some cylinders without dip tubes that contain these refrigerants. Because these refrigerants are always charged in their liquid state, a dip tube ensures that you draw liquid when the cylinder is upright. Recovery cylinders also have dip tubes for liquid extraction, but recovery cylinders have two valves—one for liquid (blue) and one for vapor (red). Refer to *Figure 29*.

It is essential that technicians consistently consider the refrigerants and cylinders they are working with, relative to the presence of a dip tube. If there is any doubt, consult the cylinder label to determine if it has a dip tube. Some near-azeotropic refrigerants, such as HFC-410A, are sold in both types of cylinders. Assumptions lead to errors.

HVACR technicians are also required to use low-loss fittings on gauge manifold hoses (*Figure 30*). Hoses can be purchased with factory-installed low-loss fittings, or the fittings can be purchased as accessories. Low-loss fittings trap

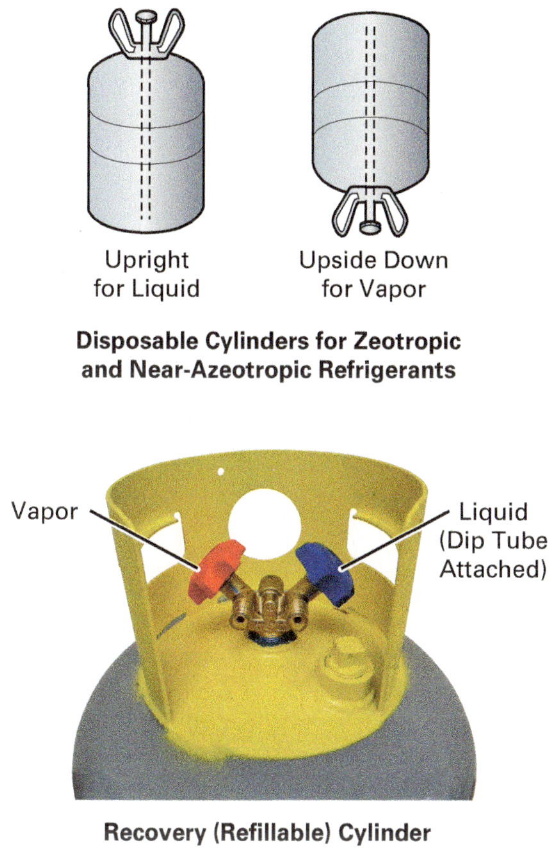

Figure 29 Refrigerant cylinder construction.

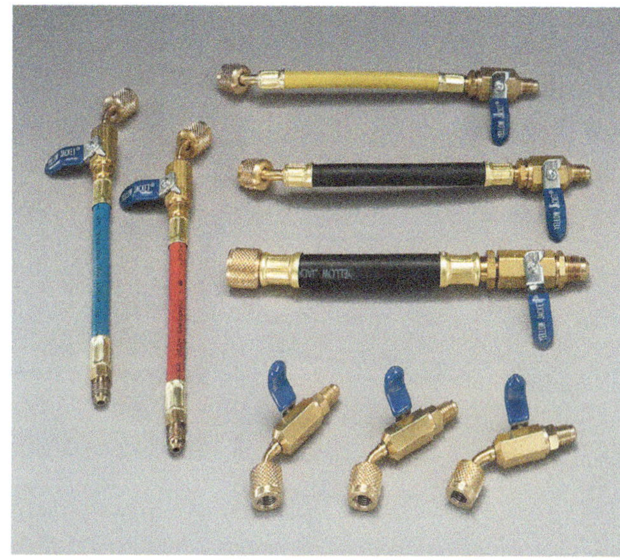

Figure 30 Low-loss adapters for refrigerant hoses.
Source: Ritchie Engineering Co., Inc./YELLOW JACKET

refrigerant in the hose and manifold. When the hose is disconnected, only a small volume of refrigerant is released (a de minimis release). Otherwise, when a hose is disconnected, all the refrigerant in the hose is also released to the atmosphere. On the liquid line especially, several ounces of refrigerant can be lost.

Finally, regardless of how or what you are charging, you must monitor the weight of your refrigerants. Refrigerant is generally invoiced by the pound to the consumer, and all technicians must account for every pound of refrigerant in their care. A charging scale to support this task is essential.

4.1.1 Charging Scales

Charging a system with refrigerant by weight requires an accurate scale. However, regardless of the charging method, technicians need to know exactly how much refrigerant has been used, recovered, or reclaimed. Every technician should be able to account for all the refrigerant they possess or transfer. Always weigh your refrigerant cylinder while you are charging or before and after charging.

Refrigerant charging scales (*Figure 31*) can be used to charge both small and large systems. To prevent damage to the scale, ensure that its platform and range can handle the largest size cylinder you intend to use. Although an accurate bathroom scale can be used to weigh refrigerant, a bathroom scale is not built to withstand the punishment of field work. Today's refrigerant scales, like the model shown in *Figure 31*, also offer wireless technology and other attributes that support HVACR field work.

A scale's most important feature is its accuracy. A rule of thumb is that the accuracy of the scale should be within 1% of the required system charge. For small appliances, a high degree of accuracy is needed because the total refrigerant charge may only be several ounces. When choosing a scale, be sure to note the specified accuracy of each product. Like other precision service instruments, the charging scale should be periodically calibrated as directed by the manufacturer.

Figure 31 Wireless refrigerant charging scale.
Source: Ritchie Engineering Co., Inc./YELLOW JACKET

Electronic scales typically display the cylinder weight using a liquid crystal display (LCD) or light-emitting diodes (LEDs). Wireless models communicate with a smartphone or tablet through a proprietary app. Wireless models may or may not have an integral display on the scale, which is a feature that may affect your choice.

4.1.2 Sight Glasses and Refrigerant Charging

Refrigerant sight glasses are often installed in the liquid line of a cooling system. The purpose of the sight glass is to provide a visual indication of the refrigerant state at that point in the system. However, what a technician sees is often misunderstood, which can lead to unnecessary actions.

When a system is equipped with a sight glass (*Figure 32*), the glass can provide insight regarding the refrigerant charge. However, this is only true if the metering device is an expansion valve rather than a fixed metering device.

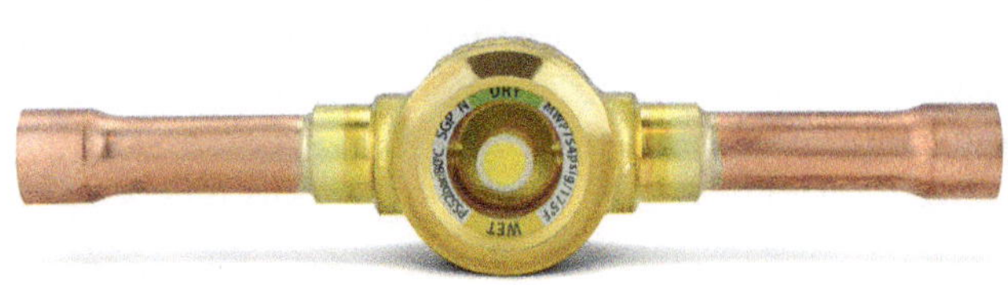

Figure 32 Sight glass with moisture indicator.
Source: Danfoss, Inc.

When the charge is correct and the refrigerant is properly subcooled, only a clear stream of liquid refrigerant passes through the sight glass, even under heavy cooling loads. With the help of refraction provided by the glass, a clear stream of fluid can be distinguished from an empty line through a magnifying effect—a sight glass filled with fluid and without bubbles will appear magnified.

Under an extreme load well outside of normal operating conditions and during system startup, the sight glass may display a fast-moving stream of bubbles, even when the charge is correct. It may take several minutes for the glass to clear following a startup. Under normal operating conditions, a bubbling or flashing sight glass usually indicates a shortage of refrigerant on systems with an expansion valve. Technically, the bubbling and flashing is telling you that the refrigerant is not subcooled at all; it is in saturation.

There are occasions when the sight glass will display a very lazy stream of liquid, clearly moving slowly and without filling the glass to the top. This may be accompanied by bubbles. This can occur when loads are very light and the expansion valve is modulating its position very close to the valve seat. The reduced flowrate slows the movement of refrigerant throughout the system, and a pocket of vapor can form at the top of the glass.

This condition, described as a *low-flow sight glass*, is not an indication of an undercharged system, but merely an indication that the expansion valve is feeding very little refrigerant. When the expansion valve opens further to increase flow, the sight glass will suddenly become clear with a solid stream of liquid. If the condition occurs repeatedly, alternating between normal and a low flowrate, it indicates the expansion valve is **hunting**.

Hunting: Describes the operation of an expansion valve opening and closing excessively and repeatedly, unable to stabilize at a consistent flowrate.

For best performance, only a solid stream of subcooled liquid enters the expansion valve. However, a sight glass is limited in the information it can provide. A sight glass showing a clear stream of liquid does prove that subcooled liquid refrigerant is present. That only means the system has enough refrigerant for the present operating conditions. But the system could be grossly overcharged, or even undercharged if the heat load is minimal at the time. Only a measurement of the subcooling can reveal accurate information.

For this reason, a sight glass alone should never be relied upon to determine if the refrigerant charge is correct. However, the sight glass does provide you with a quick visual indication of the state of the refrigerant at that point in the system.

It is also important to understand that the location of the sight glass in the system is a factor. Technicians often place the sight glass outdoors, adjacent to the condensing unit. The assumption is that, if the sight glass shows a clear stream of liquid there, subcooled liquid is reaching the expansion device. However, consider the possibility of a crimped liquid line, a plugged filter-drier, or a long vertical riser being downstream of the sight glass. All these conditions can cause the liquid to flash, acting as unintentional expansion devices. The result is that the expansion valve may not be receiving subcooled liquid.

For this reason, it is best to install the sight glass near the expansion valve inlet. If you determine there is plenty of subcooling at the condensing unit, but the sight glass at the expansion device is bubbling or flashing, there is likely a restriction or a tall vertical riser between the two points that is causing the problem.

4.2.0 Refrigerant Charging by Weight

Charging by weight is a common approach when a complete charge is to be added to an empty system. It is not practical when a system has some refrigerant in it already since you can't know the weight of the refrigerant already in the system. In these cases, you should charge the system by superheat or subcooling, or recover the refrigerant and then weigh in the total charge. The latter approach is often necessary for heat pump systems being serviced in the winter. Although ductless systems typically have electronic expansion valves and can theoretically be charged by subcooling, their manufacturers generally recommend the charge be weighed in year-round.

As mentioned previously, the total charge weight can be found either on the unit nameplate (*Figure 33*) or in the manufacturer's literature. However, even on packaged equipment, if an accessory has been added to the circuit, such as a large filter-drier to clean up after a burnout, some additional refrigerant is required. The refrigerant holding capacity of accessories can be found in the

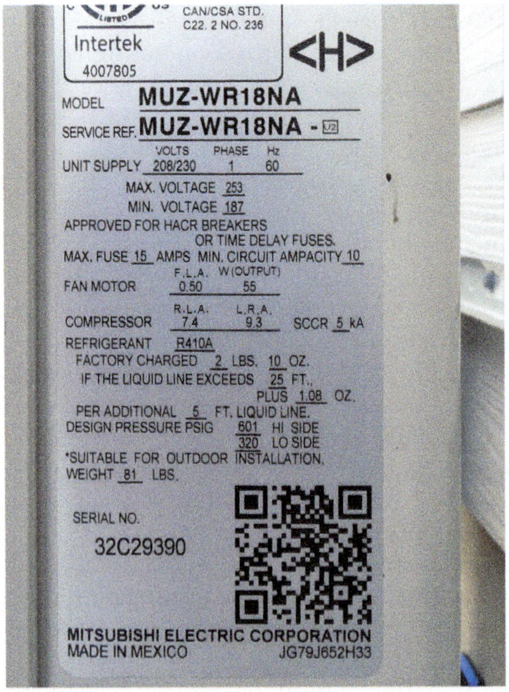

Figure 33 Unit nameplate with refrigerant charge information.

accessory documentation or product catalog. For example, a 16 cubic inch (in^3) filter-drier can hold 7.3 oz of HFO-454B. However, the same filter-drier only holds 3.7 oz of HC-290 refrigerant. The holding capacity is different for every refrigerant due to their differing densities.

For split systems, additional refrigerant may be needed to compensate for long refrigerant lines. In *Figure 33*, the manufacturer has provided the information needed to compensate for the length of the refrigerant lines.

Tables are available to determine the holding capacity of refrigerant lines, based on their size and application. *Table 6* shows the holding capacity of liquid and vapor lines used with HFC-410A. This chart provides the information in ounces per foot of tubing.

The chart also provides technicians with an example of the difference in weight between liquid and vapor. For example, a 5/8" OD liquid line holds 1.64 oz of refrigerant per foot. A vapor line of the same size holds only 0.06 oz of refrigerant per foot. Remember that the weight of refrigerant in a line is different for each refrigerant, due to their differing densities. Always ensure that you are using the correct chart for the refrigerant in use.

NOTE

Some charts indicate the weight per 100' of a given line size.

TABLE 6 Refrigerant Line Holding Capacity for HFC-410A

Liquid Line OD	Ounces/Foot	Vapor Line OD	Ounces/Foot
1/4"	0.19 oz	1/2"	0.04 oz
5/16"	0.33 oz	5/8"	0.06 oz
3/8"	0.51 oz	3/4"	0.09 oz
1/2"	1.01 oz	7/8"	0.12 oz
5/8"	1.64 oz	1 1/8"	0.20 oz
3/4"	2.46 oz	1 3/8"	0.31 oz
7/8"	3.27 oz	1 5/8"	0.43 oz
1 1/8"	5.58 oz	2 1/8"	0.76 oz
		2 5/8"	1.17 oz

Liquid charging is much faster than vapor charging. Liquid charging is the first choice to charge an empty system. When the system is in a deep vacuum, the entire charge can sometimes be introduced into the system while it remains off. At other times, the flow of liquid refrigerant may slow to a trickle or stop as the pressures equalize before the entire charge weight has been added. When this happens, stop the charging process, start the system, and add the remaining charge into the low-pressure side of the system with the compressor running.

Charging by vapor is not only slow, but it also creates another problem. As vapor is drawn from the refrigerant cylinder, the pressure in the cylinder falls and the liquid cools. When the pressure in the cylinder is lower than system pressure, charging cannot continue until the cylinder pressure rises. The cylinder can be warmed by placing it in warm water (around 120°F, or 49°C). This will increase the pressure. However, getting warm water and a tub to the jobsite is rarely practical. Wrap-around heaters for cylinders, like the one shown in *Figure 34*, can be very helpful when it's cold outside.

CAUTION

Never open the high-side valve on the gauge manifold with a refrigerant cylinder also connected while the system is in operation. This will not only place excessive pressure on the refrigerant cylinder, but refrigerant from the system will begin transferring to the cylinder due to the pressure difference.

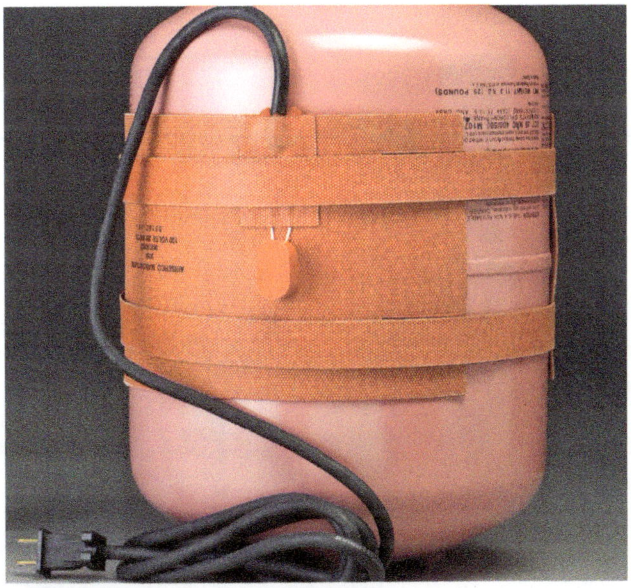

Figure 34 Wrap-around heater for refrigerant cylinders.
Source: Ritchie Engineering Co., Inc./YELLOW JACKET

WARNING!

To prevent the possibility of activating the cylinder relief valve, never heat a refrigerant cylinder above 122°F (50°C) or apply a direct flame to the cylinder body. Electric heaters or hot plates not specifically designed to warm refrigerant cylinders should never be used. When using wrap-around heaters or warming blankets designed for this purpose, monitor the device's operation closely.

Figure 35 shows the equipment connections needed for charging by weight. A cylinder with a dip tube is shown. If the cylinder is not equipped with a dip tube, it must be inverted to dispense liquid.

Liquid charging by weight is initiated by adding the liquid into the high-pressure side of an evacuated system while the system remains off. Running the indoor blower, however, helps vaporize refrigerant as it flows into the evaporator. Do not charge liquid refrigerant into the low-pressure side of a system while the compressor is off.

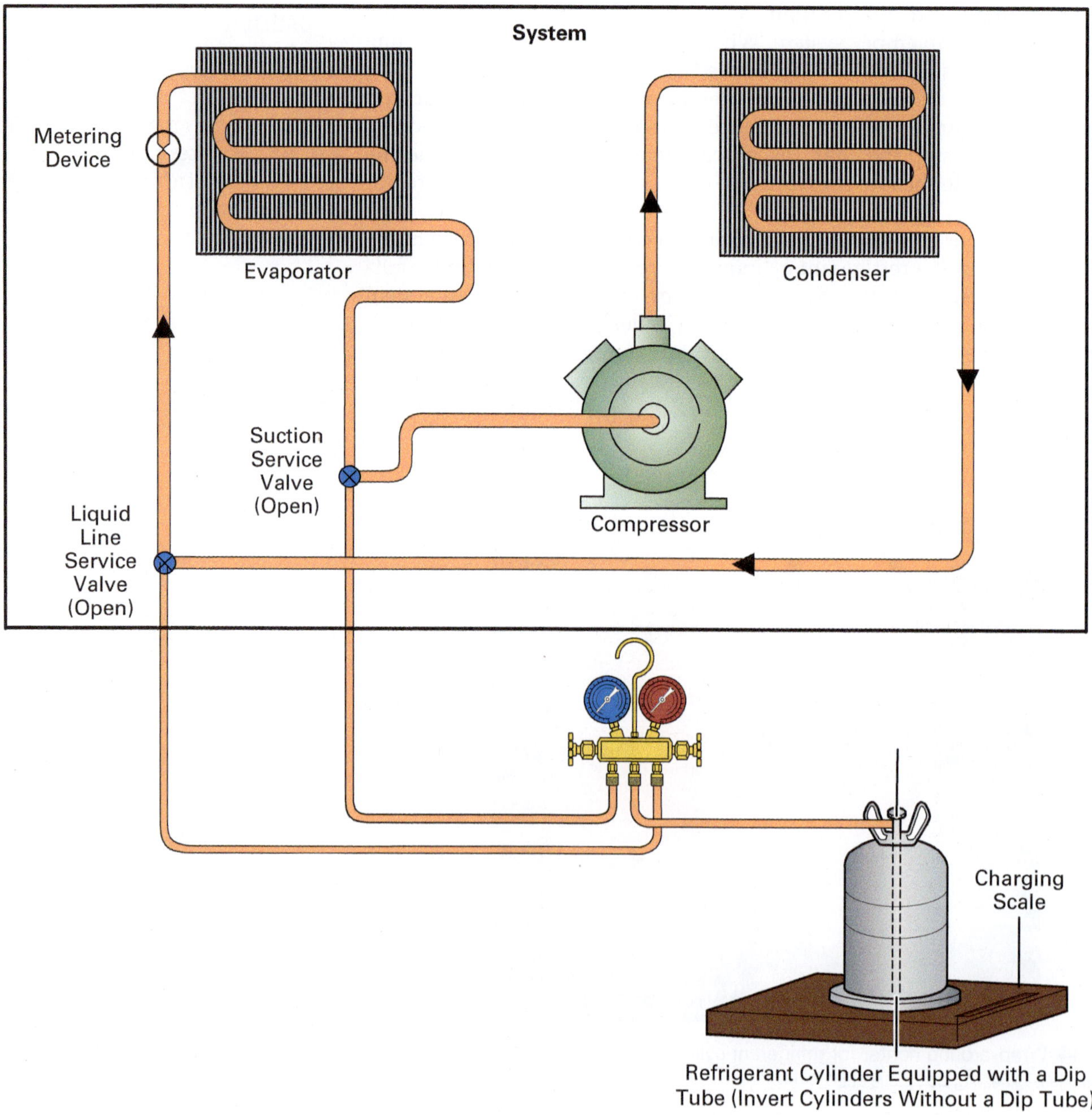

Figure 35 Service equipment connected for liquid charging by weight.

CAUTION

On water-cooled systems and chillers, it is essential to ensure that the condenser water and chilled water pumps are running while liquid refrigerant is charged into the system. Failing to provide consistent water flow may cause the water to freeze and damage the tube bundles.

When the full charge weight refuses to move into a system because the pressures have equalized, the remaining refrigerant must be added with the system running. Use the gauge manifold as a metering device while drawing liquid from the cylinder. Adjust the valve so that the liquid flashes to vapor at the valve seat. A sight glass built into the gauge manifold allows you to observe what is flowing into the charging hose.

Again, do not draw refrigerant vapor directly from a cylinder containing refrigerants other than compounds or azeotropes, regardless of what charging method is being used.

Although charging by weight can be precise, it does not mean that the system subcooling and superheat do not need to be checked. Never charge a system by weight and walk away. Even when charging by weight, it is important to check the system pressures and temperatures, and then calculate and record the superheat and subcooling characteristics to ensure the system is performing properly.

On Site

Hydrocarbon Refrigerant Charging Kit

Systems using hydrocarbon (HC) refrigerants in the US are currently limited to very small refrigerant charges. The refrigerant is often purchased in small cylinders with a net weight of 10 oz to 16 oz.

Charging a very small system by weight, regardless of the refrigerant, can be challenging. A gauge manifold with enlarged passages that works well for recovery and evacuation can hold a lot of refrigerant on its own. Thus, it is difficult to know how much of the refrigerant actually entered the system when using a traditional manifold.

If you anticipate working with this type of equipment, it is best to invest in a charging kit designed for HC refrigerants. The kit shown has a single-valve manifold and relatively short hoses with a reduced inside diameter. The kit also includes a compact charging scale with a resolution of 0.01 oz. Once the hoses have been connected and purged, the weight shown on the scale represents the zero point to begin charging. Adapters for various types of cylinders are also provided. The kit can also be used with other refrigerants.

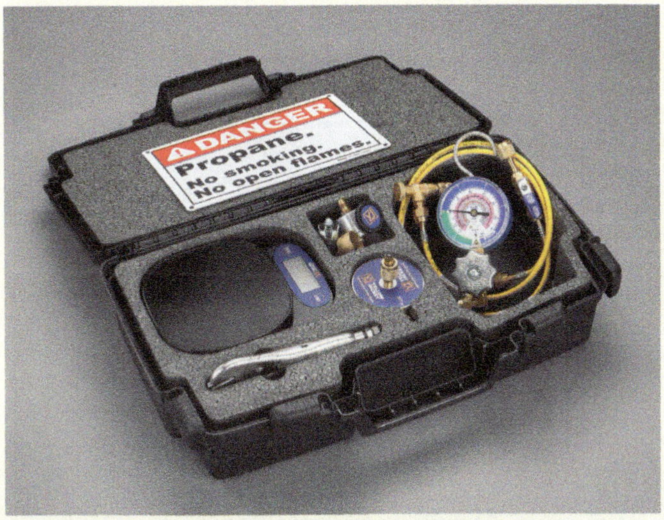

Source: Ritchie Engineering Co., Inc./YELLOW JACKET

Source: Ritchie Engineering Co., Inc./YELLOW JACKET

4.3.0 Refrigerant Charging by Superheat

Superheat represents the sensible heat added to a refrigerant after its change of state to a vapor (a latent heating process) is complete. By determining the superheat, you can tell quite a bit about the charge as well as the performance of the system.

Maintaining the proper superheat is critical to ensure a system delivers its maximum capacity. A lack of superheat means that some amount of liquid is leaving the evaporator, reducing system capacity and exposing the compressor to floodback. Too much superheat indicates that the evaporator is not receiving enough refrigerant, and system capacity is reduced. Because the compressor is cooled by the entering vapor, it runs hotter when the superheat is excessive.

Charging by superheat is performed on systems with fixed metering devices, such as a capillary tube or a fixed orifice. However, the superheat is consistently changing as the system operates. You must determine the correct superheat value for the current operating conditions, then adjust the charge to achieve it.

This method requires equipment-specific charging tables to determine the proper superheat. An example of a charging table for a system with a fixed restrictor is shown in *Figure 36*. The tables are attached to the equipment access door and/or contained in the manufacturer's service literature. Charts like this one are specific to each model.

NOTE

All discussions related to system performance herein assume that the coils are clean and system airflow is as specified. Improper airflow across the evaporator significantly affects the superheat, especially on systems using a fixed metering device.

SUPERHEAT CHARGING CHART
(SUPERHEAT ENTERING SUCTION SERVICE VALVE)

INDOOR AIR CONDITIONS ENTERING EVAPORATOR			OUTDOOR AIR DRY BULB TEMPERATURE (°F) ENTERING CONDENSER COIL									
WB	DB	RH %	65	70	75	80	90	95	100	105	110	110
61	65	80	16	12	8	6						
	70	60	18	14	10	6						
	75	45	20	16	12	8	6					
	80	33	21	17	14	10	6					
	85	23	23	19	16	12	7					
63	70	68	21	17	13	10	8	6				
	75	52	23	19	16	12	9	6				
	80	39	24	20	17	14	10	7				
	85	29	25	21	18	15	12	8	6			
	90	20	26	22	19	16	13	9	7			
65	70	77	24	20	17	13	10	8	6			
	75	59	25	22	19	15	12	10	7			
	80	45	27	24	21	18	14	11	8	6		
	85	33	28	25	22	19	15	12	9	6		
	90	25	29	26	23	20	16	13	10	7		
67	70	86	27	24	21	17	14	11	8	6		
	75	66	28	26	22	18	15	13	10	8	5	
	80	50	30	27	24	21	18	15	12	10	7	
	85	39	31	28	25	22	19	16	13	11	8	
	90	30	32	29	26	23	20	17	14	12	9	6
69	70	95	30	27	25	21	18	15	12	9	7	6
	75	75	31	28	25	22	19	16	13	10	8	6
	80	58	32	29	27	24	21	19	16	14	11	8
	85	45	34	31	28	26	23	20	17	15	12	9
	90	35	35	32	30	27	24	21	19	16	13	11
71	75	82	34	31	29	26	23	21	18	15	12	9
	80	65	36	32	30	28	24	22	20	18	15	13
	85	51	37	34	31	29	26	24	21	19	16	14
	90	39	38	35	32	30	28	25	22	20	17	16
	95	30	39	35	33	30	29	25	23	21	18	17
73	75	92	37	34	32	29	27	24	22	20	17	15
	80	72	37	35	33	30	28	26	24	22	19	17
	85	58	38	36	34	31	29	27	25	23	20	18
	90	45	38	36	34	31	30	28	25	23	21	19
	95	35	38	36	35	32	31	28	26	24	21	19

Charging at these conditions can result in damage to the compressor. Charge to 5 degrees superheat and check superheat when conditions are more favorable.

NOTES: Superheat measurements should be taken at condensing unit service valves. Charge system within 2°F of superheat indicated. Recommended minimum superheat is 5°F.

White area in the chart is the optimum window for charging.

DB = dry-bulb temperature (°F).
WB = wet-bulb temperature (°F).
RH = approximate % of indoor relative humidity.

Find the required superheat at the intersection of the measured outdoor dry-bulb (DB) air temperature and the indoor wet-bulb (WB) and dry-bulb (DB) air temperature.

Example: The intersection of a 95°F DB outdoor temperature and a 71°F WB/90°F DB indoor temperature yields 25°F as the required superheat.

Figure 36 Example charging table for refrigerant charging by superheat.

CAUTION

Do not use the example chart shown in *Figure 36* on the job. Use the manufacturer's resources.

To charge by superheat, proceed as follows. Refer to *Figure 37* for a connection diagram. This example is based on HFC-410A refrigerant.

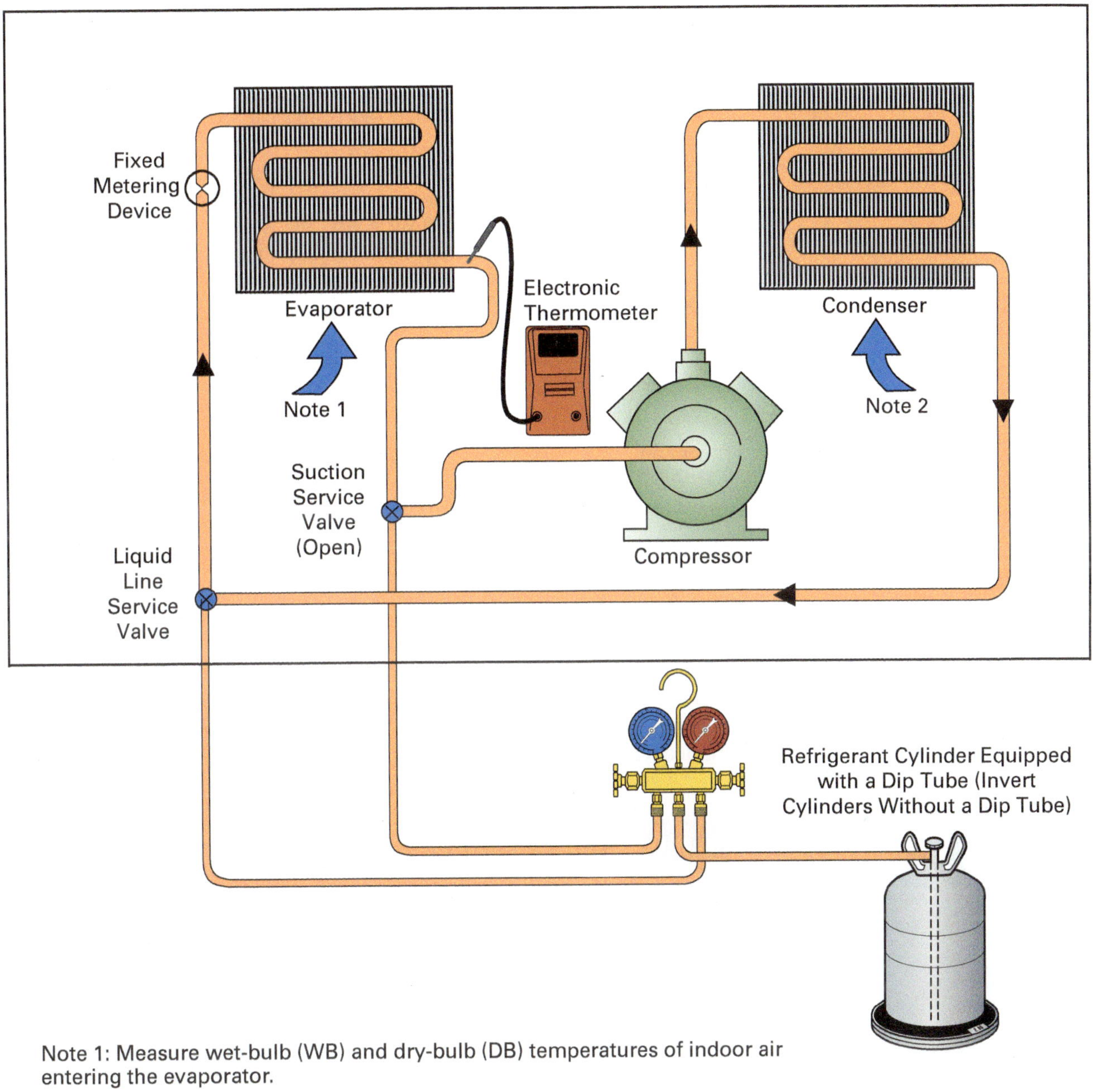

Figure 37 Service equipment connected for charging by superheat or subcooling.

Step 1 Operate the system for 10–15 minutes to stabilize the pressures. The air filter, condenser, and evaporator coil must be clean. Make sure proper indoor airflow has been established. Then measure and record the following operating conditions. The values shown represent those used for this example:

- Suction pressure: 120 psig
- Suction line temperature at the evaporator outlet: 70°F
- Outdoor dry-bulb temperature entering the condenser: 95°F
- Indoor dry-bulb temperature entering the evaporator: 90°F
- Indoor wet-bulb temperature entering the evaporator: 71°F

Step 2 Using the superheat charging chart in *Figure 36*, find the required superheat at the point where the measured outdoor dry-bulb air temperature, indoor dry-bulb temperature, and indoor wet-bulb temperature intersect. As shown in *Figure 36*, the target superheat for our conditions is 25°F.

Step 3 Using a PT chart, find the saturated suction temperature for the measured suction pressure. HFC-410A has a saturated suction temperature of 41°F at 120 psig.

Step 4 Compare the actual suction line temperature measured in the system to the saturated suction temperature of 41°F to determine the present superheat value. Since the measured suction line temperature is 70°F, the current superheat is 29°F. Since the desired superheat in this example is 25°F, the charge must be adjusted slightly. A tolerance of ±2°F is typical before any adjustment is required. This tolerance may vary by manufacturer, but ±2°F typically provides an accurate charge.

Step 5 If the measured superheat is too high, as it is in our example, add refrigerant through the low-side service port with the compressor running until the superheat drops to the correct value. Add refrigerant a little at a time and monitor the system. You can easily overcharge the system by rushing the process and adding too much refrigerant.

Step 6 If the measured superheat is too low, remove refrigerant from the system using a recovery unit until the superheat rises to the correct level. You can also remove refrigerant from the system while it runs using an evacuated recovery cylinder. Use the gauge manifold to route refrigerant from the high-pressure side of the system into the cylinder.

WARNING!

Never connect a recovery unit to an operating system. Both personal injury and recovery equipment damage can result from excessive pressure.

In our example, it is clearly hot in the conditioned space as well as outdoors. In these cases, after adjusting the charge, monitor the system as it cools and reduces the temperature inside. Then record the operating conditions again and recheck the superheat to ensure it is accurate for the new conditions. It is always best to gather new operating data and recheck the superheat and system performance when the conditions are at or near normal.

4.4.0 Refrigerant Charging by Subcooling

Metering devices like TXVs are designed to maintain a consistent superheat over a wide range of load and charge conditions. Thus, the superheat method of charging cannot be used with modulating metering devices.

Subcooling values are used to properly charge these systems. However, the superheat must be checked after charging to ensure the system and the expansion valve are operating properly. Even with the proper charge, the expansion valve may need to be adjusted if the superheat is too high or too low.

CAUTION

Do not use *Table 7* on the job. Use the manufacturer's resources.

Table 7 is an example of a subcooling chart you might encounter for an HFC-410A unit designed to operate with an expansion valve. Examining the table reveals that a subcooling range between 8°F and 14°F is common. The unit nameplate may specify a subcooling value of 10°F, for example. However, adjustments may be needed when the operating conditions are abnormal. For example, consider a system that has been off for several days and it is hot indoors as well as outdoors.

The chart is then used to determine the proper refrigerant charge under abnormal conditions. For example, assume the subcooling should be 12°F and you have a liquid line temperature of 110°F. Per *Table 7*, the high-side pressure should be 430 psig with a proper refrigerant charge. If it is below 430 psig, add refrigerant to increase the pressure. If it is over 430 psig, remove some refrigerant to decrease the pressure. After adjusting the charge, recheck the subcooling, superheat, and other operating characteristics.

This is not the only type of table you will encounter. Manufacturers use different graphs and tables to assist in charging by subcooling. Regardless of the graph or table provided, follow the manufacturer's charging guidance carefully.

TABLE 7 HFC-410A Subcooling Chart

Liquid Line Temp	Design Subcooling						
	8°F	9°F	10°F	11°F	12°F	13°F	14°F
	Liquid Pressure (psig)						
55°F	178	182	185	188	191	195	198
60°F	195	199	202	204	208	211	216
65°F	211	214	218	223	226	230	233
70°F	228	232	236	240	244	247	252
75°F	247	251	256	260	264	267	272
80°F	268	272	275	280	284	288	292
85°F	287	291	296	300	304	309	314
90°F	308	312	317	321	326	330	335
95°F	332	337	342	347	352	356	361
100°F	355	360	365	370	376	381	386
105°F	380	385	391	395	401	406	412
110°F	408	414	419	424	430	436	442
115°F	434	441	446	453	458	464	470
120°F	465	471	477	483	489	496	502
125°F	495	502	506	515	520	526	532

To charge by subcooling, use the following procedure. This example will also be based on HFC-410A:

Step 1 Operate the system for 10–15 minutes to stabilize the pressures. The air filter, condenser, and evaporator coil must be clean. Make sure proper indoor airflow has been established. Then measure and record the following operating conditions. The values shown represent those used for this example:

- Liquid pressure: 439 psig
- Liquid line temperature at the condenser outlet: 114°F

Step 2 Determine the correct subcooling value specified by the equipment manufacturer. This can be found on the unit nameplate or in the manufacturer's service literature for the unit being serviced. For this example, assume that the proper subcooling value is 12°F.

Step 3 Use the measured high-side pressure and a standard PT chart to find the saturation temperature. The saturation temperature for HFC-410A refrigerant at 439 psig is 124°F (*Figure 38*).

Step 4 Calculate the current subcooling value by subtracting the actual temperature measured on the liquid line from the saturation temperature. Since the saturation temperature is 124°F and the measured liquid-line temperature is 114°F, the subcooling value is 10°F. Compare the required subcooling value to the actual value and determine if the charge must be adjusted. If the measured subcooling is within ±2°F of the specified subcooling value, no adjustment in refrigerant charge is typically necessary. The subcooling for this system is acceptable as is, although a small amount of refrigerant could be added to increase the subcooling to 12°F.

Step 5 If the measured subcooling is too high, remove some refrigerant from the system to reduce subcooling. If the measured subcooling is too low, as it is in our example, add refrigerant to the system to increase subcooling.

Allow the system some time to gain control of the indoor temperature, then record the operating conditions again and recheck the subcooling to ensure it is accurate for the new conditions. The superheat must also be checked to ensure the expansion device is operating properly.

°F	R-22	R-134A	R-410A	°F	R-22	R-134A	R-410A
46	77.6	41.1	132.2	100	195.9	124.3	316.4
48	80.7	43.3	137.2	102	201.8	128.5	325.6
50	84.0	45.5	142.2	104	207.7	132.9	334.9
52	87.3	47.7	147.4	106	213.8	137.3	344.4
54	90.8	50.1	152.8	108	220.0	142.8	354.2
56	94.3	52.3	158.2	110	226.4	146.5	364.1
68	97.9	55.0	163.8	112	232.8	151.3	374.2
60	101.6	57.5	169.6	114	239.4	156.1	384.6
62	105.4	60.1	175.4	116	246.1	161.1	395.2
64	109.3	62.7	181.5	118	252.9	166.1	405.9
66	113.2	65.5	187.6	120	259.9	171.3	416.9
68	117.3	68.3	193.9	122	267.0	176.6	428.2
70	121.4	71.2	200.4	124	274.3	182.0	439.6
72	125.7	74.2	207.0	126	281.6	187.5	451.3
74	130.0	77.2	213.7	128	289.1	193.1	463.2
76	134.5	80.3	220.6	130	296.8	198.9	475.4
78	139.0	83.5	227.7	132	304.6	204.7	487.8
80	143.6	86.8	234.9	134	312.5	210.7	500.5
82	148.4	90.2	242.3	136	320.6	216.8	513.4
84	153.2	93.6	249.8	138	328.9	223.0	526.6
86	158.2	97.1	257.5	140	337.3	229.4	540.1
88	163.2	100.7	265.4	142	345.8	235.8	553.9
90	168.4	104.4	273.5	144	354.5	242.4	567.9
92	173.7	108.2	281.7	146	363.3	249.2	582.3
94	179.1	112.1	290.1	148	372.3	256.0	596.9
96	184.6	116.1	298.7	150	381.5	263.0	611.9
98	190.2	120.1	307.5				

Saturation Temperature (From PT Chart)
− Liquid Line Temperature (Measured)

Subcooling Value

Example:
1. Measured Liquid Line Pressure (HFC-410a) = 439 PSIG
2. Saturated Temperature from Chart = 124°F
3. Measured Liquid Line Temperature = 114°F
4. Subcooling = 124°F − 114°F = 10°F

Figure 38 PT chart.

4.0.0 Section Review

1. Fractionation occurs with refrigerants that exhibit which of these characteristics?
   a. Glide
   b. Superheat
   c. Subcooling
   d. Flammability

2. An HFC-410A split system has a line set that is 30' longer than specified for the factory refrigerant charge. The liquid line is ⅜" OD and the suction line is ⅞" OD. How much additional refrigerant needs to be added?
   a. 0.63 oz
   b. 3.6 oz
   c. 15.3 oz
   d. 18.6 oz

3. When the temperature of the suction line leaving the evaporator coil is the same as the saturation temperature of the refrigerant at the current pressure, what does it tell you?
   a. The subcooling is too high.
   b. The system is undercharged.
   c. Some amount of liquid is leaving the evaporator.
   d. The compressor valves have likely been damaged.

4. When charging by subcooling, no adjustment in refrigerant charge is typically necessary if the measured subcooling is within ______.
   a. ±1°F of the specified subcooling
   b. ±2°F of the specified subcooling
   c. ±3°F of the specified subcooling
   d. ±5°F of the specified subcooling

Module 03205 Review Questions

1. In which of the following situations is an ultrasonic leak detector a good choice?
   a. Where stray gases may be encountered
   b. When the refrigerant circuit is at atmospheric pressure
   c. When the leak is expected to be extremely small
   d. When the work area is very noisy or windy

2. True or False: When using a liquid leak detector solution, a very tiny leak will slowly create one very large bubble.
   a. True
   b. False

3. Which of the following gases is commonly added to nitrogen to provide a trace gas for leak detection?
   a. Carbon monoxide
   b. Oxygen
   c. Hydrogen
   d. Argon

4. Which of the following accurately describes refrigerant recovery?
   a. Removing refrigerant from a system and processing it to virgin refrigerant standards.
   b. Circulating refrigerant through filters to remove water and solid contaminants.
   c. Returning refrigerant to a system after it has been previously removed.
   d. Transferring refrigerant from a system into a reusable cylinder.

5. To what pressure must a new, certified recovery unit reach to recover refrigerant from a system containing 212 lb of HCFC-22 refrigerant?
   a. 0" Hg
   b. 4" Hg vacuum
   c. 10" Hg vacuum
   d. 15" Hg vacuum

6. Which of the following events fall into the category of a de minimis release of refrigerant?
   a. A high-pressure relief valve opening and releasing the system refrigerant charge to the atmosphere.
   b. A refrigerant hose being purged with refrigerant before being connected to a recovery unit.
   c. A leak on an industrial refrigeration system that is subsequently reported to the EPA.
   d. A disposable cylinder overheating and rupturing, discharging the entire contents to the atmosphere.

7. Which of the following statements is *true*?
   a. Refrigerant that is recovered from a unit being replaced can be sold to another customer if it is first recycled.
   b. Refrigerant must be contaminated in some way to be sent away for reclamation.
   c. Mixed refrigerants that cannot be reclaimed must be disposed of by the reclaiming facility.
   d. Even systems that have a leak must be recovered down to a 10 in. Hg vacuum.

8. Which of the following recovery methods is the fastest way to recover refrigerant, assuming the proper conditions exist?
   a. Push-pull method
   b. Liquid method
   c. Vapor method
   d. Triple method

9. The primary purpose of evacuating a system is to remove air and lower the system pressure so any moisture will _____.
   a. bond to the refrigerant
   b. condense
   c. vaporize
   d. collect

10. If a system is evacuated to a level of 10,160 μm, the water in the system will boil at or above a temperature of _____.
    a. 32°F (0°C)
    b. 53°F (11.7°C)
    c. 76°F (24.4°C)
    d. 80°F (26.7°C)

11. A deep vacuum is generally considered to be _____.
    a. 25,400 μm and lower
    b. 10,000 μm and lower
    c. 1,240 μm and lower
    d. 500 μm and lower

12. Which of the following types of refrigerants are the most likely to exhibit significant glide and fractionation characteristics?
    a. Azeotropes
    b. Zeotropes
    c. Near-azeotropes
    d. Binary blends

13. When one or more component refrigerants in a blend leak out of a system at different rates, changing the precise proportions of the blend, the process is called ______.
    a. azeotropic reaction
    b. fractionation
    c. superheat
    d. subcooling

14. Which of the following is a correct statement about zeotropic refrigerants?
    a. A zeotropic refrigerant must leave the refrigerant cylinder only as a vapor.
    b. The dew-point column of the PT chart is the saturation temperature for the refrigerant in its liquid form.
    c. The bubble-point column of the PT chart is the saturation temperature for the refrigerant in its liquid form.
    d. Interpolating PT charts for zeotropic refrigerants will result in significant charging errors.

15. A sight glass on a TXV-equipped system that shows refrigerant moving along slowly with some bubbles in it is typically an indication that the ______.
    a. TXV is partially closed and feeding little refrigerant
    b. TXV is wide open and feeding refrigerant rapidly
    c. refrigerant circuit is overcharged
    d. refrigerant circuit is undercharged

Answers to Odd-Numbered Module Review Questions are found in *Appendix A*.

Answers to Section Review Questions

Answer	Section	Objective
Section One		
1. c	1.0.0	1a
2. d	1.2.3	1b
Section Two		
1. c	2.1.1	2a
2. b	2.2.4	2b
Section Three		
1. d	3.1.0	3a
2. c	3.2.1	3b
Section Four		
1. a	4.1.0	4a
2. d	4.2.0; Table 6	4b
3. c	4.3.0	4c
4. b	4.4.0	4d

MODULE 03211

Heat Pumps

Source: Courtesy of Sporlan Division—Parker Hannifin Corporation

Objectives

Successful completion of this module prepares you to do the following:

1. State heat pump operating principles and describe heat pump performance ratings.
   a. Explain how heat pumps transfer heat from air and water.
   b. Identify various heat pump performance ratings and explain how each rating is determined.
2. Describe the operation of heat pump systems.
   a. Describe the refrigeration cycle of heat pumps.
   b. Identify the various types of heat pump systems.
   c. Describe the basic control strategies for heat pumps and defrost cycles.
   d. Identify components that differentiate heat pumps from standard cooling units.
   e. Describe sources of supplemental and/or emergency heat used in heat pump systems.
3. State common heat pump installation practices.
   a. State installation practices associated with split systems.
   b. State installation practices associated with packaged systems.
4. Describe the operation of electric heating equipment commonly used with heat pumps.
   a. Explain how electric heating equipment operates.
   b. Identify the major components of an electric heater.

Performance Tasks

Under supervision, you should be able to do the following:

1. Install a heat pump and complete a proper startup.
2. Test the operation of a heat pump in the heating and cooling modes and record operating data.

Overview

Heat pumps provide both cooling and heating using the mechanical refrigeration cycle. The secret to this capability is that even cold air contains some heat. The heat pump can extract that heat by reversing the refrigeration cycle. Not only are heat pumps capable of transferring heat from what seems to be very cold air, but heat pumps can provide heat using far less energy than conventional electric heating systems.

Digital Resource for HVACR

Scan this code using the camera on your phone or mobile device to view the digital resources related to this craft.

SCAN ME

NCCER Industry-Recognized Credentials

If you are training through an NCCER-accredited sponsor, you may be eligible for credentials from NCCER. The ID number for this module is 03211. Note that this module may have been used in other NCCER curricula and may apply to other level completions. Contact NCCER at 1.888.622.3720 or go to **www.nccer.org** for more information.

You can also show off your industry-recognized credentials online with NCCER's digital credentials. Transform your knowledge, skills, and achievements into credentials that you can share across social media platforms, send to your network, and add to your resume. For more information, visit **www.nccer.org**.

1.0.0 Heat Pump Operating Principles

Performance Tasks

There are no Performance Tasks in this section.

Objective

State heat pump operating principles and describe heat pump performance ratings.

a. Explain how heat pumps transfer heat from air and water.

b. Identify various heat pump performance ratings and explain how each rating is determined.

A heat pump functions by transferring heat from the outdoor air, a water source, or even the ground. That heat is transferred indoors to heat the structure. An air-source heat pump operates on the principle that, no matter how cold it is outdoors, there is still some heat in the air. Water and earth sources typically offer more consistent supplies of heat.

1.1.0 Heat Pump Theory

Air at any temperature above absolute zero, which is 0°K on the Kelvin scale (–460°F), contains some amount of heat. Of course, at warmer temperatures, there is more heat available. As the outdoor air temperature drops, there is less heat available for a heat pump to extract.

Fortunately, we don't need to protect ourselves from temperatures as extreme as absolute zero. It's also important to note that a heat pump can't possibly extract heat, or operate at all, in such extreme temperatures.

Air-source heat pumps do lose capacity as the outdoor temperature falls. Only specific types of heat pumps function effectively at temperatures at or below 5°F (–15°C). Unfortunately, as the temperature falls outdoors, the heating needs of a structure rise while the heating capacity of the heat pump falls. Although heat remains in the air outdoors, heat pumps cannot capture enough heat from extremely cold air. This is due to the limitations of the *vapor-compression cycle*. For that reason, air-source heat pumps are typically installed with some form of auxiliary or supplemental heat, such as an electric heater or a fossil-fuel furnace. These sources can be cycled on automatically when the heat pump alone can no longer meet the demand.

It is important to note, however, that a great deal of research and progress has been made in heat design. Recent developments in heat pumps allow some to operate at extremely low temperatures, down to –22°F (–30°C). These units maintain a significant portion of their heating capacity at these temperatures and can minimize or eliminate the need for supplemental heat in some applications.

The basic refrigerant cycle of an air-source heat pump is a good example of how heat is transferred. It is the same as a conventional air conditioning unit when in the cooling mode (*Figure 1*). Hot refrigerant vapor is sent to the condenser coil to be cooled and condensed, and liquid refrigerant is sent through the metering device to the evaporator coil to provide cooling.

To provide heat, however, the refrigerant must flow in the opposite direction. The indoor coil becomes the condenser, while the outdoor coil becomes the evaporator.

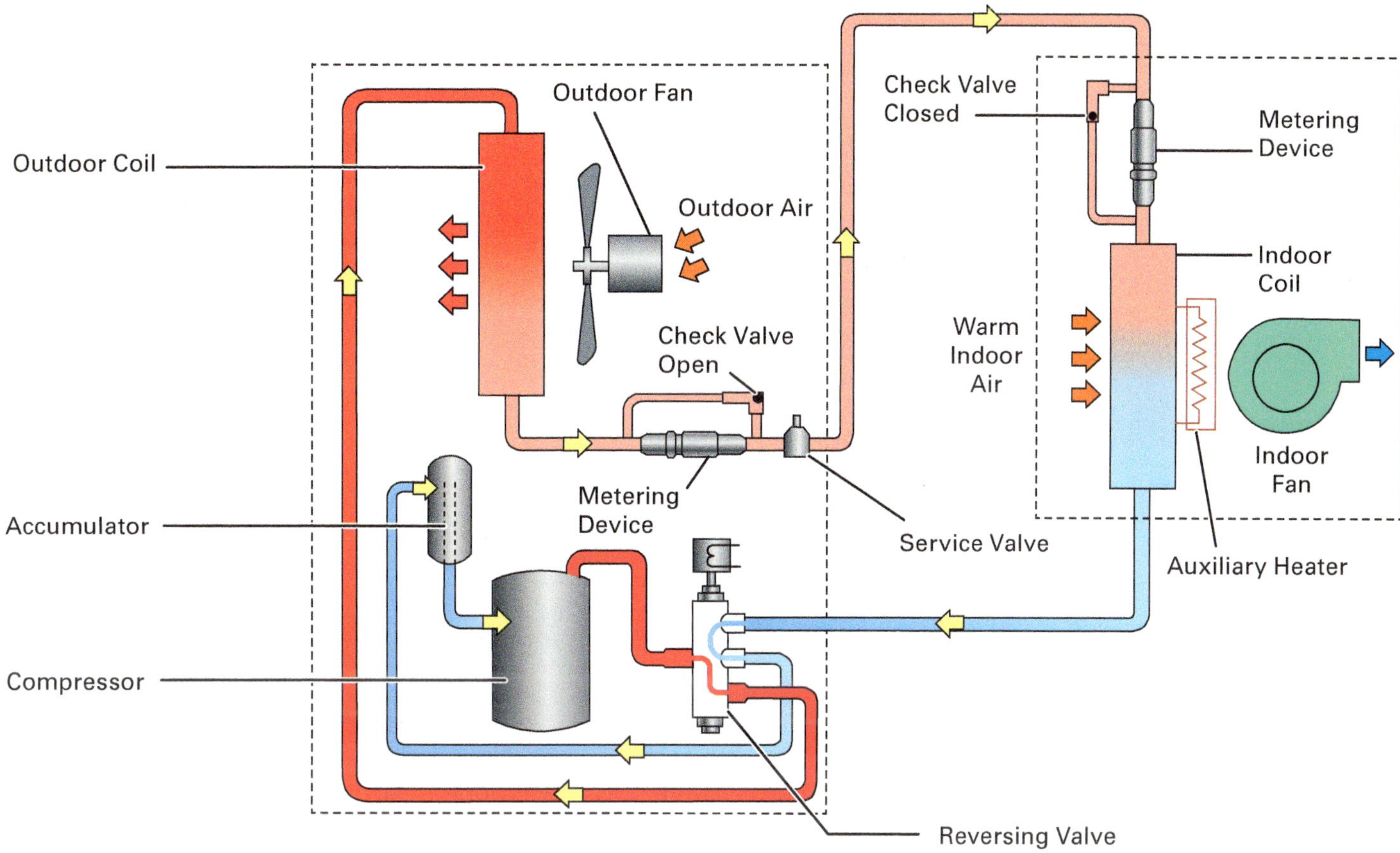

Figure 1 Schematic of an air-to-air heat pump (cooling mode).

As a result, the industry refers to the two coils in a heat pump system as the *indoor coil* and the *outdoor coil* since their roles change as the mode of operation changes.

To transfer heat from the air, we must be able to move refrigerant through the outdoor coil that is colder than the air. Heat can then be added to the refrigerant in the outdoor coil, despite the cold conditions, because the refrigerant entering the outdoor coil is still colder than the surrounding air. However, it is essential to remember that heat in the outdoor coil cannot be captured if the refrigerant in the outdoor coil isn't colder than the air.

For example, if it is 25°F (–4°C) outdoors, the refrigerant sent to the outdoor coil must be below that temperature to gain heat from the outdoor air. As the temperature difference (TD or ΔT) between the air and refrigerant increases, so does the volume and rate of heat transfer. The reverse is also true—a smaller temperature difference reduces the heat that can be captured.

If the temperature difference is large enough, most of the liquid refrigerant in the outdoor coil changes to vapor, just as it does in the evaporator during the cooling cycle. But in the heating mode, it is very common for some liquid to be mixed with the vapor as it returns to the compressor. How that is handled will be explained later in this module.

After compression (adding to the heat content), the hot refrigerant vapor from the outdoor coil is sent to the indoor coil where the heat is transferred to the cooler air circulating inside.

1.2.0 Heat Pump Performance Ratings

An electric heat pump provides more heat to the conditioned space than the electrical power it consumes. This is true even when the outdoor temperature is quite low. However, keep in mind that the heat produced by a heat pump at low outdoor temperatures may not be enough to satisfy the demand, regardless of the efficiency level achieved. Heat pump efficiency is measured using one of several rating methods.

Coefficient of Performance (COP): The ratio of heat produced in relation to the electrical energy used, calculated using the following equation: COP = Btuh ÷ (watts × 3.413).

1.2.1 Coefficient of Performance

The ratio of heat output to the equivalent electrical power consumed is known as the **Coefficient of Performance (COP)**. This ratio compares the heat output of the heat pump with the heat output from electric resistance heat alone. The COP must be determined at specific operating conditions because the performance of the heat pump changes as the outdoor temperature changes.

When electric heat is used, it generates 3,413 Btus of heat for each kilowatt (kW) of power consumed. This can also be stated as 3.413 Btus per watt (W). Because of this, the COP of straight electric resistance heat is always 1—it can never be any higher. The COP for a heat pump provides a way to measure how much more efficiently it produces heat.

A COP of 3.0 at 47°F (8°C), for example, indicates that a given heat pump is three times as effective at producing heat when it is 47°F outdoors than an electric heater consuming the same amount of energy. At 17°F (–8°C), the heat pump will produce less heat, and the COP may be reported as 1.7 at this temperature. At 17°F then, the heat pump is only 1.7 times as efficient as electric heat.

The COP of a heat pump is calculated as follows. Pay close attention to the numbers, as one equation uses watts while the other uses kilowatts:

$$\text{COP} = \frac{\text{Btuh output}}{\text{watts input} \times 3.413}$$

or

$$\text{COP} = \frac{\text{Btuh output}}{\text{kilowatts input} \times 3{,}413}$$

For example, assume that the heating capacity of a heat pump is reported to be 33,480 Btuh (at an unspecified outdoor temperature) with a total power input, including fan motor wattage, of 3.768 kW. Determine the COP as follows:

$$\text{COP} = \frac{\text{Btuh output}}{\text{kW input} \times 3{,}413}$$

$$\text{COP} = \frac{33{,}489}{3.768 \times 3{,}413}$$

$$\text{COP} = \frac{22{,}489}{12{,}860}$$

$$\text{COP} = 2.6 \text{ (rounded)}$$

Study Examples

1. Calculate the COP of a heat pump with a heating capacity of 55,800 Btuh and a total electrical input of 6.28 kW.

$$\text{COP} = \frac{\text{Btuh output}}{\text{kW input} \times 3{,}413}$$

$$\text{COP} = \frac{55{,}800}{6.28 \times 3{,}413}$$

$$\text{COP} = \frac{55{,}800}{21{,}433}$$

$$\text{COP} = 2.6$$

2. Calculate the COP of a heat pump with a heating capacity of 48,000 Btuh while consuming 4,030 watts.

$$\text{COP} = \frac{\text{Btuh output}}{\text{watts input} \times 3.413}$$

$$\text{COP} = \frac{48{,}000}{4{,}030 \times 3.413}$$

$$\text{COP} = \frac{48{,}000}{13{,}754}$$

$$\text{COP} = 3.5 \text{ (rounded)}$$

When the COP is used to compare different heat pump systems or installations, care must be taken. The COPs must be based on the same outdoor and indoor conditions and complete systems. Over the years, common outdoor temperature rating points for comparison have been 47°F (8°C) and 17°F (–8°C). Heat pumps designed for cold climates are also rated at 5°F (–15°C) and potentially at temperatures as low as –15°F (–26°C).

Don't forget that the COP for every unit varies with the operating conditions. Therefore, the COP can't be used to compare systems operating over time. The example problems above do not mention any specific outdoor temperatures. However, the actual heat output of every heat pump changes as the temperature of the heat source (air, water, or ground) changes.

1.2.2 Heating Season Performance Factor

Another means of rating a heat pump's ability to produce heat is known as the **Heating Season Performance Factor (HSPF)**. The HSPF is found by dividing the total heating output for an entire season by the total energy consumed during that season. The advantage of the HSPF is that it can be used to compare systems operating over time, providing a more valuable perspective.

Heating Season Performance Factor (HSPF): A heat pump performance rating that has been adjusted for seasonal operation. It represents the total heating output of a heat pump (in Btus) during its normal annual usage period for heating, divided by the total electric power input in watt-hours during the same period.

For example, consider a heat pump system that produces 115,024 MBtus (115,024,000 Btus) of heat energy while consuming 14,560 kWh (kilowatt/hours) of power. These figures must include the power needed to drive all the electrical components. To calculate the HSPF, use the following formula:

$$\text{HSPF} = \frac{\text{total seasonal heat output (Btu)}}{\text{total power input (watt-hours)}}$$

$$\text{PF} = \frac{\text{115,024,000 Btus}}{\text{14,560,000 watt-hours}}$$

$$\text{HSPF} = \text{7.9 Btus/watt-hour}$$

It is important to note that electric resistance heat has a performance factor of 3.41. Therefore, a heat pump with an HSPF of 7.9 is about 2.3 times as efficient as electric resistance heat over the course of a heating season (7.9 ÷ 3.41 = 2.3). But today's best heat pumps can easily surpass an HSPF of 7.9.

Going Green

Minimum HSPF

The US Department of Energy (DOE) determines the minimum performance requirements for heat pumps to provide consistency and support energy conservation nationwide. Performance requirements are getting more stringent as technology progresses, as shown in the image. Beginning in 2023, manufacturers will be required to produce equipment with a minimum HSPF of 8.8. The most efficient models exceed this value with an HSPF of 10 or more.

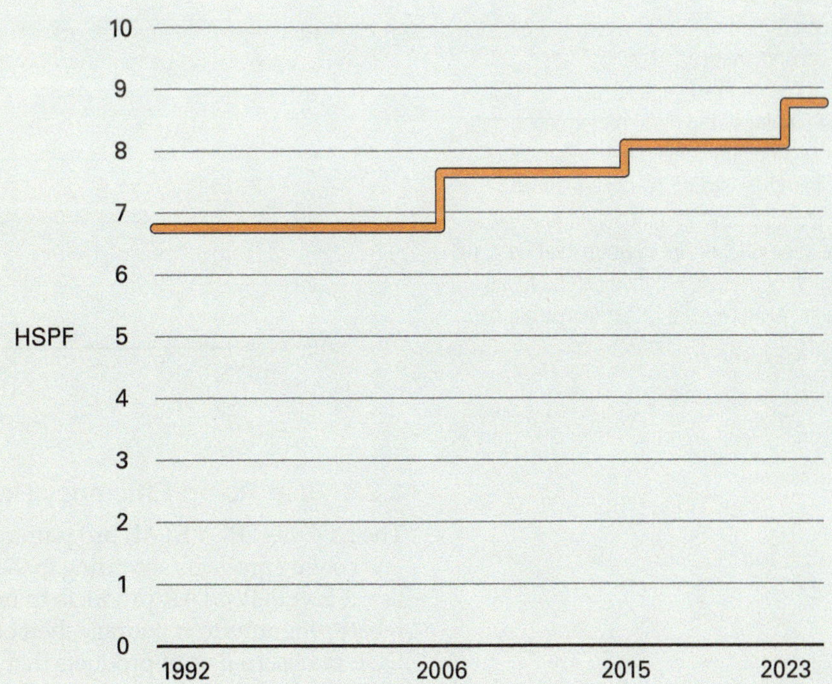

Residential heat pump installations in the southern regions of the US operate with higher HSPFs than similar installations in the northern regions because the average outdoor temperature is higher. Commercial and industrial installations usually operate with higher HSPFs than residential installations due to higher motor and compressor efficiency in large equipment.

1.2.3 Seasonal Energy Efficiency Ratio

Seasonal Energy Efficiency Ratio (SEER): The standard developed by the AHRI for measuring cooling efficiency, representing the total cooling production of a unit (in Btus) during the cooling season divided by the total energy consumed (in watt-hours) during the same period.

The Air Conditioning, Heating, and Refrigeration Institute (AHRI) established the **Seasonal Energy Efficiency Ratio (SEER)** as a standard by which the cooling performance of cooling systems are rated. This includes heat pumps. The efficiency of central air conditioning and heat pump units is governed by US law and regulated by the US Department of Energy (DOE).

Every air conditioning and heat pump unit is assigned a SEER value. It is calculated as the total cooling output (in Btuh) provided by the unit during the cooling season divided by the total energy consumed (in watt-hours) during the same period. In other words, the SEER represents the number of cooling Btuh a unit produces per watt of electricity consumed. Like the HSPF, though, the SEER reflects performance over an entire cooling season, not just under a specific set of operating conditions. When calculating performance under one set of operating conditions, creating a snapshot of performance, it is referred to as the EER.

Higher SEER values are preferred, as it indicates that a unit provides more cooling for the same power consumption as a unit with a lower SEER. SEER values provide a clear picture to consumers of the anticipated energy cost, enabling them to compare various units prior to purchase.

Going Green

Minimum SEER Ratings

The DOE also establishes the minimum SEER for cooling equipment. The graph shown here shows the trend of those requirements over the years. Beginning in 2023, manufacturers must produce cooling equipment with a minimum SEER of 14 for residential systems in the northern part of the US, and 15 in the southern part of the US where more hours of cooling are needed.

The certified capacities, SEER, COP, HSPF ratings, and other performance data for cooling and heating equipment can be found by looking up the equipment by manufacturer and model number in the appropriate AHRI Performance Rating Directory. This information can be found on the AHRI website (*https://www.ahrinet.org/*).

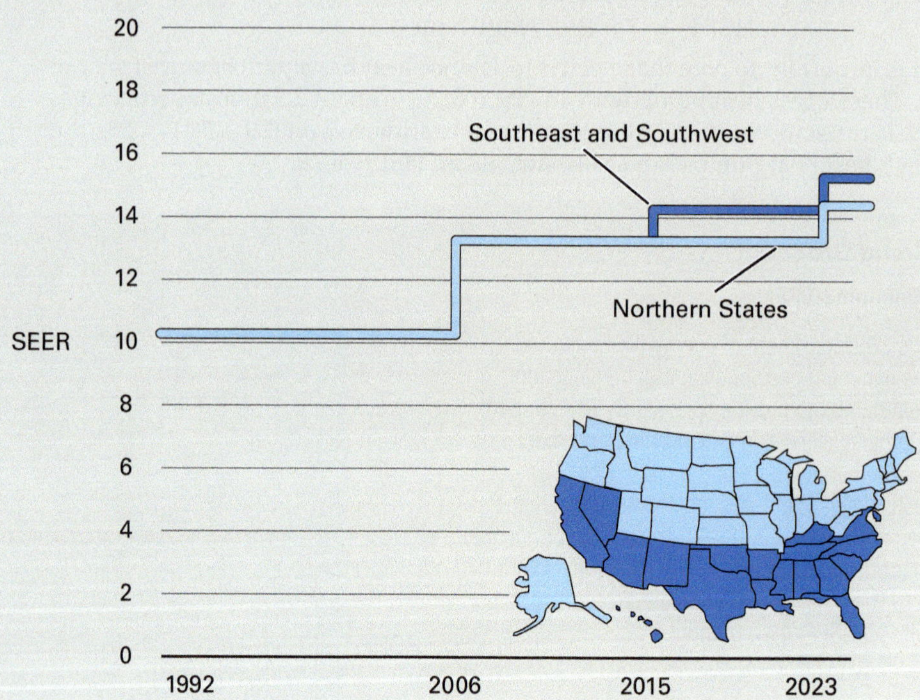

1.2.4 Heat Pump Efficiency Recommendations

The DOE ENERGY STAR program rates various appliances and residential heating and cooling products according to their efficiency. Users are encouraged to purchase listed ENERGY STAR products to help conserve energy. Many local utilities have rebate programs to encourage their customers to buy ENERGY STAR products. The DOE publishes lists of products that qualify for ENERGY STAR ratings. This list is available at *energystar.gov*.

Going Green

The ENERGY STAR Logo

The ENERGY STAR logo appears on all products that meet the required US EPA specifications for energy efficiency, including heat pumps.

Source: Energy Star

1.0.0 Section Review

1. In the heating mode, the outdoor coil of a heat pump ______.
   a. becomes the evaporator
   b. never gets cold
   c. acts as a condenser
   d. remains dry

2. A heat pump's COP is a ratio that compares its efficiency at producing heat to the heat produced by ______.
   a. a gas furnace
   b. an oil furnace
   c. electric heat
   d. all other fuels

2.0.0 Heat Pump System Operation

Objective

Describe the operation of heat pump systems.

a. Describe the refrigeration cycle of heat pumps.
b. Identify the various types of heat pump systems.
c. Describe the basic control strategies for heat pumps and defrost cycles.
d. Identify components that differentiate heat pumps from standard cooling units.
e. Describe sources of supplemental and/or emergency heat used in heat pump systems.

Performance Tasks

1. Install a heat pump and complete a proper startup.
2. Test the operation of a heat pump in the heating and cooling modes and record operating data.

Although the refrigeration cycle of a heat pump operating in the cooling mode is much the same as any other cooling system, the heating mode requires extra components. The role of some common components also differs significantly. The **reversing valve**, also known as a *four-way valve*, is the key to the heat pump's dual function.

Reversing valve: A valve that changes the direction of refrigerant flow in a heat pump refrigerant circuit. Also referred to as a *four-way valve*.

2.1.0 Heat Pump Refrigerant Cycle

The three operating modes for air-to-air heat pumps are cooling, heating, and defrost. *Figure 2* shows a schematic diagram of the heat pump refrigeration cycle in the cooling and heating modes.

In most heat pumps, the reversing valve coil is energized when the thermostat calls for cooling. It is de-energized when it calls for heating. This ensures that heat is available even if the reversing valve or its controls fail. Heat is considered the more critical utility to provide. HVACR technicians might say that the valve "fails to the heating mode" when referring to features like this.

2.1.1 Cooling Mode

When the system is in the cooling mode, the refrigerant flow is the same as that of any cooling unit. However, the refrigerant must still pass through the reversing valve. Notice the path through the valve in *Figure 2A*. The cold, low-pressure refrigerant flowing through the indoor coil absorbs heat from the conditioned space and boils until the liquid has changed to its vapor state. The hot, high-pressure refrigerant vapor leaving the compressor is directed to the outdoor coil, where the heat is rejected. The refrigerant pressures and temperatures shown are typical of those seen in cooling systems using refrigerant HFC-410A.

2.1.2 Heating Mode

In the heating mode, the reversing valve changes its position. The refrigerant leaving the compressor is routed in the opposite direction from that of the cooling cycle. Instead of flowing to the outdoor coil, the hot, high-pressure refrigerant vapor leaving the compressor flows to the indoor coil through the insulated vapor line. As far as the system is concerned, the indoor coil is now acting as a condenser.

Heat is extracted from the refrigerant because the air in the conditioned space is cooler than the refrigerant in the coil. The refrigerant vapor condenses, at least partially, as it passes through. The condensed liquid refrigerant then flows outdoors, through the metering device and into the outdoor coil, where it encounters cold air. Since the refrigerant here, by design, is colder than the air, heat is extracted and transferred to the refrigerant. This causes the refrigerant to begin boiling, returning to its vapor state. However, in many cases, what leaves the outdoor coil is a mixture of liquid and vapor. An *accumulator*, discussed in Section 2.4.3 of this module, is added to collect any liquid before it reaches the compressor.

In the example shown in *Figure 2B*, the saturation temperature of the refrigerant in the outdoor coil is 20°F (–7°C), while the outdoor air temperature is 45°F (7°C). Note the significant differences in the heating mode pressures and temperatures at key points in the system as compared to the cooling mode.

2.1.3 Heat Pump Defrost Operation

As noted in the *Figure 2B* example, the saturated refrigerant temperature in the outdoor coil is 20°F (–7°C). This, of course, is well below the freezing point of water. In addition, the temperature is usually below the dewpoint of the outside air. This means that moisture in the outdoor air will condense on the coil, and then freeze. Note that the outdoor temperature in the example (45°F, or 7°C) is well above freezing, but the moisture will still freeze on the 20°F coil surface. This is a common and expected situation. However, at this outdoor temperature, the frost that develops will usually melt during the off cycle.

At colder temperatures, a layer of frost will build up on the coil, blocking airflow and preventing effective heat transfer (*Figure 3*). If the frost is not removed, the unit will eventually be unable to absorb heat from the air. The longer the unit operates in these conditions, the more frost will develop. Oddly enough, frost formation can be a larger problem at outdoor temperatures between 20°F

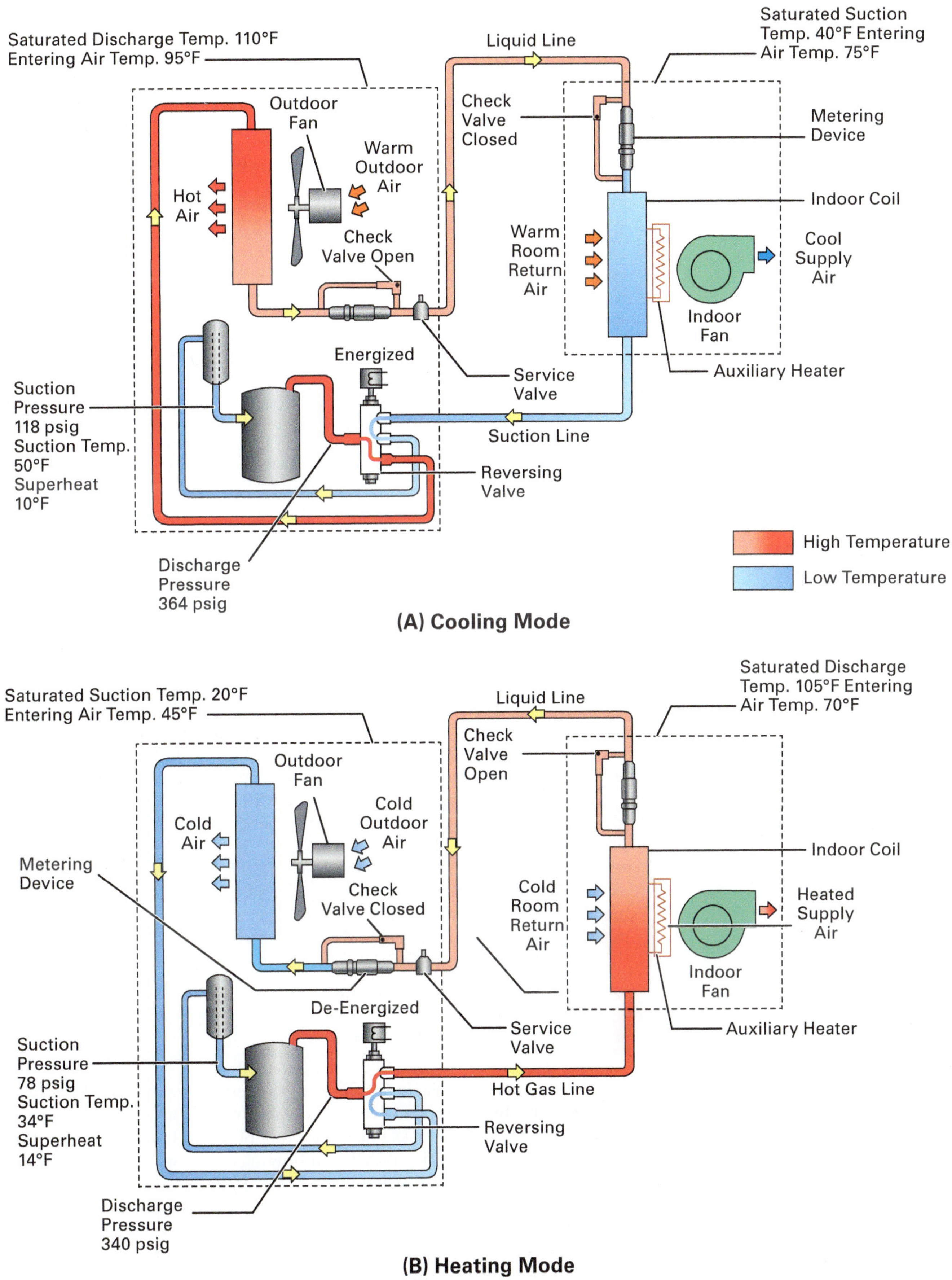

Figure 2 Refrigerant flow in an air-to-air heat pump, using HFC-410A refrigerant.

Figure 3 A heat pump that has frozen over.
Source: HVACGuru

and 35°F (–4°C and 4°C) than it is at lower temperatures. This is because colder air often contains less moisture, and the outdoor coil temperature may not be below the dew point of the air.

To eliminate ice buildup on the outdoor coil, air-source heat pumps have a defrost mode. Technically, the unit is simply switched back to the cooling mode. Hot refrigerant is routed through the outdoor coil to melt the frost buildup. However, while in the defrost mode, the outdoor fan remains off to prevent cold air from being forced across the outdoor coil. The fan would impede the melting process.

The defrost cycle usually lasts from about 2 to 10 minutes, and perhaps longer on units operating in extreme cold. During that time, the heat pump is not providing any heat to the space. In fact, cold air is blowing into the conditioned space since the unit is in the cooling mode. The indoor fan must remain on, as heat from the conditioned space is being transferred to the outdoor coil to melt the ice. To prevent immediate discomfort to building occupants during the defrost cycle, most systems energize an electric heater located downstream of the indoor coil to temper the air.

2.2.0 Heat Pump Types

Heat pumps can be classified according to their heat source and the medium to which the heat is transferred: air, water, or the ground. For example, a water-to-air heat pump in the heating mode picks up heat from water flowing through a coil and transfers it to the air flowing over a second coil. The air is the heating medium while the water is the heat source.

The more common types of heat pumps are air-to-air, air-to-water, water-to-air, and water-to-water. In each case, the heat source is named first. In an air-to-air system, air is the heat source and the heating medium. This is the most common type of heat pump system.

Most heat pump systems are used to heat air. As a result, instead of being called an air-to-air or water-to-air unit, they may be referred to as air-source or water-source units. When the second term is absent, it is assumed that the unit is circulating air to heat the building.

Going Green

Heat Pump Domestic Water Heaters

Heating water with a conventional electric water heater can be expensive. Heat-pump water heaters can heat the same amount of water for significantly less by using heat pump technology.

Some heat-pump water heaters use an electronic control that selects the most efficient operating mode for the conditions. Depending on demand and the ambient temperature around the heat pump, the system can heat water with compression heat alone (most efficient). It also can use both compression heat and electric resistance heat at the same time. Under ideal conditions, power consumption to heat domestic water can be reduced by as much as two-thirds below that of a conventional electric water heater.

Source: Douglas Cliff/Shutterstock

2.2.1 Air-to-Air Heat Pumps

The most common heat-pump heat source is outdoor air (*Figure 4*). Air-to-air models are either self-contained (packaged) or split systems. Packaged heat pumps are available for residential and commercial applications from 2 to 30 tons. Split system air-to-air heat pumps are available in smaller sizes, but they rarely exceed 20 tons. However, today's **variable refrigerant flow (VRF) systems** allow multiple outdoor units to work together with many indoor units to create systems with 60–70 tons of capacity.

Variable refrigerant flow (VRF) systems: Heat pump systems that modulate the flow of refrigerant to match structure heating and cooling load demands simultaneously. VRF systems rely on variable speed compressor technology and can incorporate multiple outdoor units on a single refrigerant circuit.

While air-source units can provide heat in cold climates, these units are more effective in milder climates. Conventional single-speed heat pumps are typically selected based on the cooling load rather than the heating load. Variable-speed heat pumps may be sized to cooling loads, partial heating loads, or full heating loads. Some variable-speed heat pump systems today have more heating capacity than cooling capacity.

Figure 4 Air-to-air split-system heat pumps.
Source: Courtesy of Sporlan Division—Parker Hannifin Corporation

If the heat load exceeds the heating capacity of the selected heat pump unit (common), a form of auxiliary/supplemental heat is needed. Supplemental heat is often required to temper the indoor air during the defrost cycle as well. The most common arrangement in current installations includes supplemental electric heat to both support the heat load and offset the defrost cycle.

Air-to-air heat pumps offer several advantages:

- Air-to-air units do not use water, so the problems related to piping, corrosion, and condenser water disposal are nonexistent.
- Today's air-to-air units are far more energy efficient than electric resistance heat.
- Air-to-air heat pumps are typically easy to access for service and repairs.

2.2.2 Air-to-Water Heat Pumps

An air-to-water system also uses air as the heat source, but it heats and cools water instead of air (*Figure 5*). Heating and cooling for the space is then provided by a water distribution system. The air-to-water system uses fan coil units or similar equipment to transfer the heat to/from the water and the air within the space. Note, though, that the water temperature isn't usually as high as that produced by the average boiler, unless supplemental water heating is added.

Figure 5 Air-to-water heat pump with indoor unit housing the water tank, pump, and supplemental electric water heater.
Source: Enertech Global, LLC

Space heating can also be supplemented by electric resistance heaters in the fan coil units. In another application, the heat pump can pre-heat water for a conventional hydronic heating system with a boiler, thus reducing the use of fossil fuel. Air-to-water systems provide the following benefits:

- Air-to-water units can eliminate the need for large supply and return duct systems because water is piped through the building instead.
- Rooms and areas may be thermostatically controlled more easily in large, central installations than with large supply and return air systems.
- Air-to-water heat pumps can be used as domestic water heaters in addition to providing comfort heating and cooling.

Going Green

Heat Pumps Selected for Extreme Climates

Heat pumps present a unique challenge to the HVAC designer and installer: Should a heat pump be sized to meet the heating needs of a structure, or its cooling needs?

The heating capacity of a traditional single-stage heat pump is related to its cooling capacity: the higher the cooling capacity, the higher the heating capacity. Because of this, it is difficult to match a heat pump's size to both the heating and cooling load of a house. Traditionally, heat pumps have been selected to match the cooling load.

If you select a heat pump with enough capacity to meet the heating load, it will usually have far more capacity than needed for cooling. That means it will cycle on and off frequently and do a poor job of dehumidification. That won't work in the warmer south. Cooling is the priority, and any limited shortfalls in heating capacity are addressed with electric heat or fossil fuel furnaces.

The traditional sizing method doesn't make sense for cold climates, because too much supplemental heating capacity is needed from less efficient sources. So, why bother with the heat pump at all, if it can only provide a small portion of the heat needed? Fortunately, today's advanced heat pumps with variable capacity offer better solutions.

Variable-capacity heat pumps can modulate their output and operate at much lower capacities than their rated maximum. This means that if a system is oversized for the cooling demand, it can operate at a reduced capacity without needing to cycle on and off as frequently. This also improves dehumidification.

Defrost cycles are still needed in the winter, so a supplemental heat source is still a consideration. Heat pumps designed specifically for colder climates have characteristics that help, such as a heater built into the unit base pan, eliminating any buildup of ice in the base. As the owner of this heat pump has learned, elevating and protecting the heat pump from deep snow are essential for good performance.

Source: Fire-n/Shutterstock

2.2.3 Water-to-Air Heat Pumps

Water-to-air heat pumps can use well water or any good, clean water supply as a heat source. These units are similar in size to the air-to-air type and are marketed in the range from 2 to 50 tons. Water-to-air units are even used aboard ships.

Figure 6 is an example of a water-to-air system using two wells. This is considered an *open loop system*. One well supplies the water, while the other provides a path for the water to be returned to the ground.

The water returned to the ground is cooler after heat has been removed during the heating mode, and warmer after heat is added during the cooling mode. Therefore, the two wells must be placed a sufficient distance from each other to prevent discharge water from being drawn back to the system. The two wells can also be drilled to significantly different depths.

Installation of the two wells does increase the installed cost of a system, with the total cost largely dependent upon local geology. In some cases, water treatment may be required to reduce the accumulation of calcium, iron, and other troublesome deposits in the heat exchanger. The water-to-refrigerant heat exchanger coils are generally made of special alloys such as copper-nickel for both good heat transfer and corrosion resistance.

Except in emergency conditions, using a domestic water supply is not cost-effective, due to both the cost of water and its disposal through a sewage treatment system. Indeed, in many areas, the municipal water supply is too precious to use for this purpose.

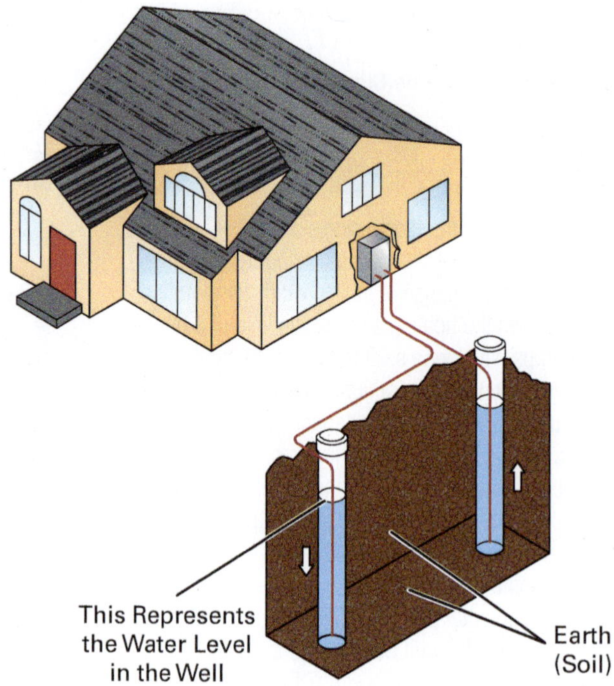

Figure 6 Open loop system using two wells.

Figure 7 shows a commercial-scale water-to-air system using a closed loop. Mechanical systems provide hydronic heating and cooling through a variety of water-source heat pump systems in a commercial setting. The water temperature in the loop is typically maintained between 70°F and 80°F (21°C and 27°C), using a boiler as a heat source and a cooling tower for a heat sink. Mixing valves are used to precisely control water temperature in the loop. The water system can be chemically

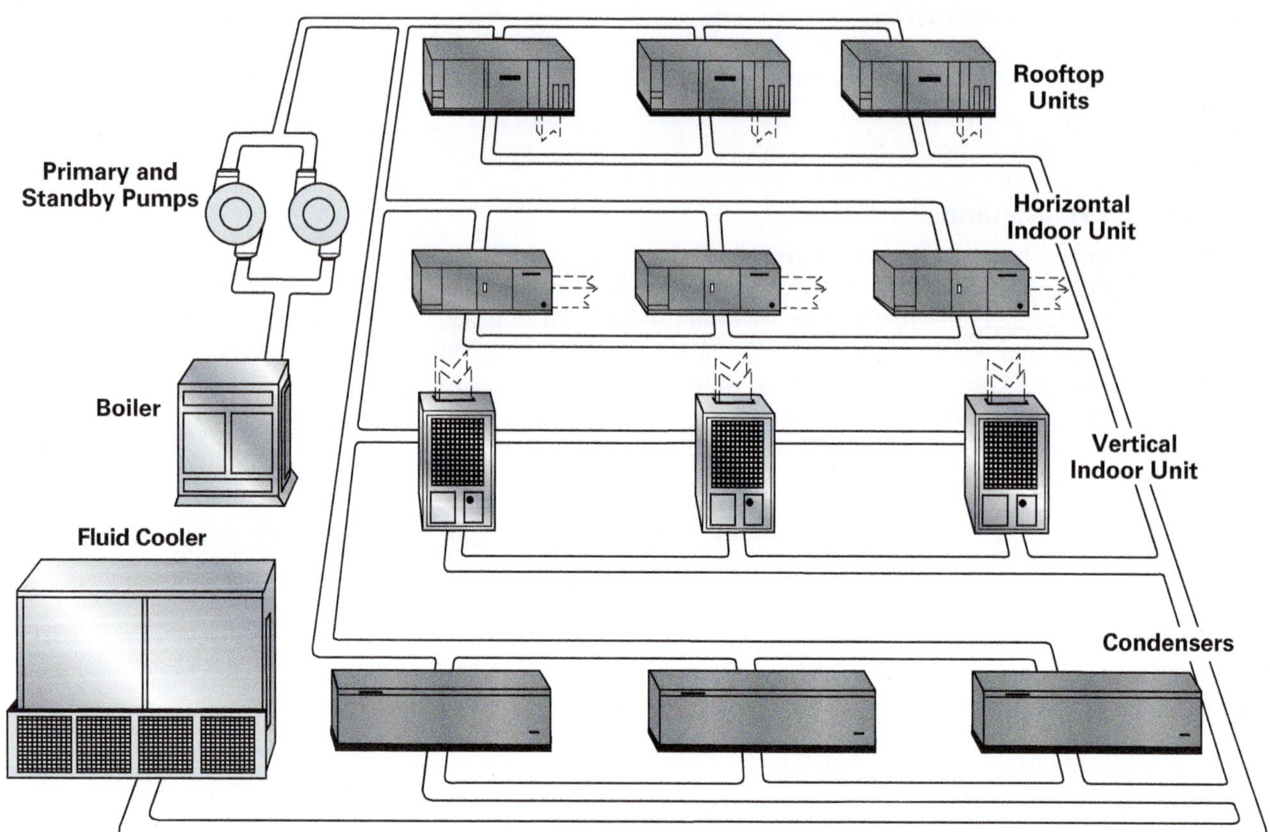

Figure 7 Commercial-scale water-to-air closed loop system.

treated and maintained to avoid some of the issues related to open-loop systems that consistently bring in free oxygen and contaminants along with fresh water.

The following are some of the advantages offered by water-to-air heat pumps served by a water loop maintained at a consistent temperature:

- Water-to-air heat pumps generally offer better energy efficiency than air-source units because the loop water temperature is warmer than winter air or ground water, and the loop temperature is very consistent.
- Self-contained or integral (packaged) water-to-air units have greater application flexibility than integral air-to-air units because these heat pumps are not restricted by the need for good air movement around them. Air-to-water heat pumps can even be installed as free-standing, in-space conditioning units.
- Air-to-water heat pumps do not require a defrost cycle.
- Air-to-water heat pumps rarely require supplemental electric heat as their heating capacity remains stable throughout the winter months and there is no defrost cycle. However, electric or fossil fuel heating equipment may be added to provide an emergency source in the event of a heat pump failure.

2.2.4 Water-to-Water Heat Pumps

Water-to-water heat pumps use water from wells or closed-loop systems to transfer heat to a second circulating loop of water. Because these sources tend to have a more stable temperature than air, the heat output from these systems is stable as well. These systems are more often used to heat domestic water than for comfort heating and cooling, but water-water heat pumps are capable of both.

Like water-to-air systems, supplemental heat is not always needed. But many users feel better having some sort of backup source. However, for comfort heating applications, these systems are often unable to produce water at 160°F to 180°F (71°C to 82°C), which many have used as a design standard for boiler-based hydronic systems. As a result, the heating distribution system must have sufficient heat-transfer surface area to meet the need with cooler water. Although in use since the 1930s, the growth of water-to-water systems has been slow due to the introduction of improved and more cost-effective air-to-air units.

Water quality and water-side maintenance issues are magnified in the water-to-water system, as two separate water loops must be managed. Soluble minerals such as calcium salts, magnesium salts, and iron can form deposits on heat exchanger surfaces and impede heat transfer. Without proper maintenance and treatment techniques, mineral deposits, contaminants, and corrosion can quickly cripple these systems, and the maintenance costs can soar.

Water-to-water heat pumps are sometimes used with closed circuit, flat-plate solar-thermal collectors (*Figure 8*). The solar collectors are used as an additional heat source, and a storage tank can be used to store the heated water or other fluid in these applications.

Going Green

Geothermal Heat Pump Loop Protection

To avoid groundwater contamination, any antifreeze solution used in a buried ground loop must be non-toxic. Typically, propylene glycol or ethanol-based antifreeze is used for this purpose. Both chemicals are also used to winterize potable water systems in homes and recreational vehicles. These chemicals are non-toxic, readily available, and relatively inexpensive. Propylene glycol is preferred in more states for environmental reasons.

Calcium chloride ($CaCl_2$) performs well in geothermal systems. It is non-toxic but extremely corrosive to ferrous metals. On the other hand, if the heat exchanger is constructed of copper-nickel (also called *cupronickel*), and fittings are brass or bronze, then it is a good choice too. But when installing or servicing systems, be aware of local and state codes regarding water additives.

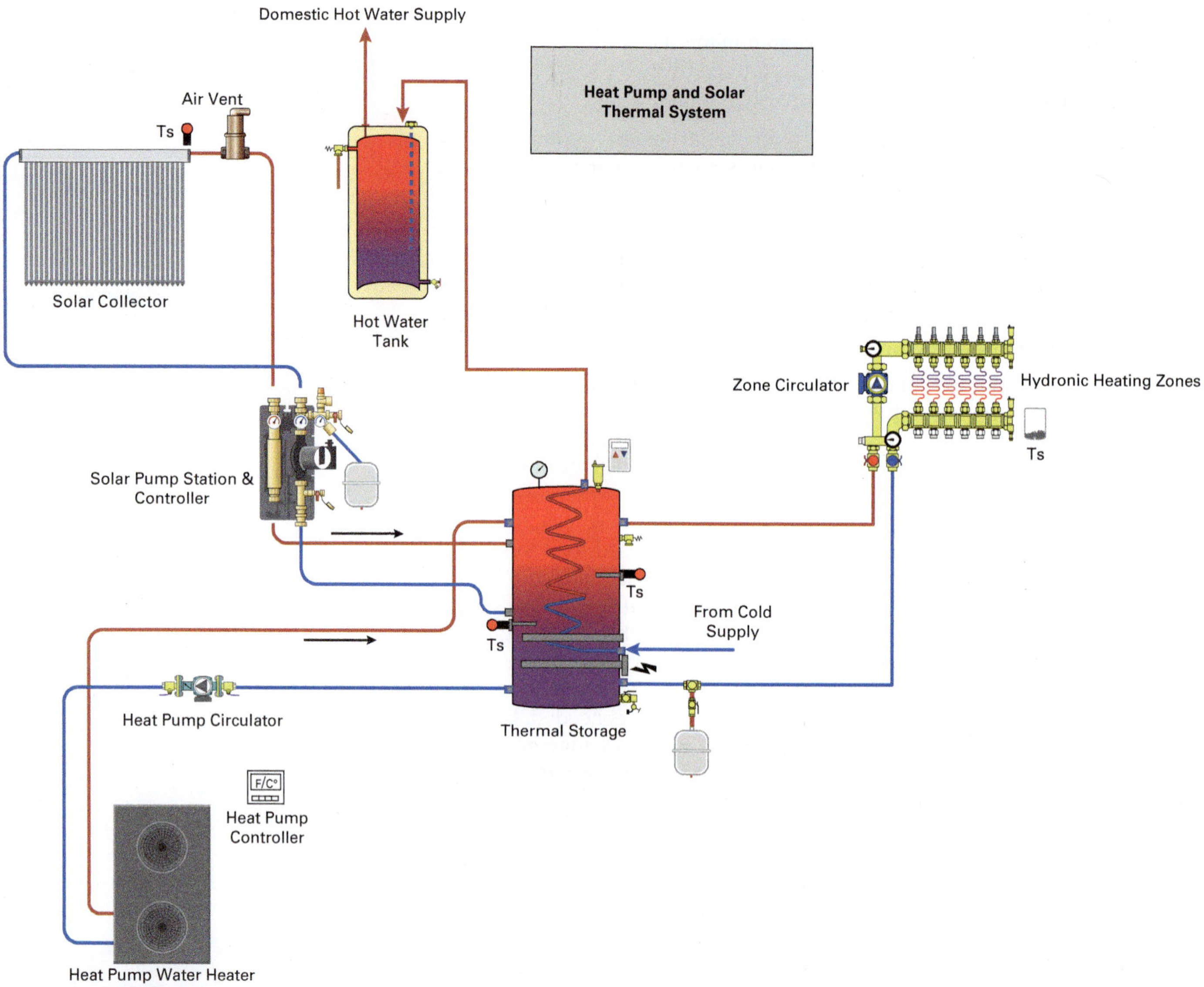

Figure 8 Water-to-water heat pump system with an accessory solar collector.
Source: Arctic Heat Pumps

Geothermal system: A system that uses the earth as a source of heat as well as a heat sink for cooling operation.

2.2.5 Geothermal Heat Pumps

A **geothermal system** relies on coils of tubing buried in the ground to transfer heat to and from the earth. A fluid (water, sometimes with antifreeze products added) is used to transfer the heat (*Figure 9*). Most are water-to-air systems, but water-to-water units are also available.

As is the case with most water-source systems, the consistency of ground temperatures provides a distinct advantage on days of peak demand in both the heating and cooling modes. Since the ground loop is closed, the fluid within the loop can be treated to prevent corrosion and minimize mineral deposits. Antifreeze solutions are often added to prevent any potential for freezing, which can occur wherever the loop is exposed to the weather. These glycol-based solutions can also contain needed corrosion inhibitors.

A variety of piping materials can be used below ground, but plastic dominates the field. Determining the amount of tubing to be buried depends on the heat transfer characteristics of the selected materials, as well as geology. The moisture content of the soil also affects heat transfer; more water in the ground provides better heat transfer.

Although plastic tubing, such as polypropylene or polybutylene, cannot match the heat transfer speed and efficiency of copper tubing, its low cost compensates for the higher volume of tubing needed in the ground. Plastic

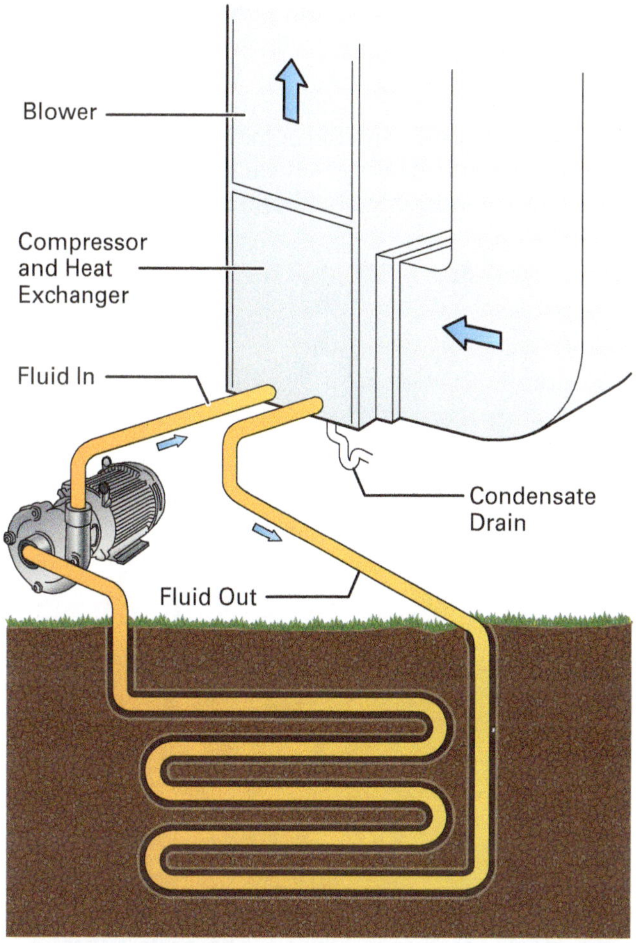

Figure 9 Geothermal heat pump piping arrangement.

tubing also has the advantages of inherent corrosion resistance, flexibility, and ease of connection. Underground connections are often heat fused (*Figure 10*). Geothermal systems enjoy a continued rise in popularity due to both their energy efficiency and low maintenance costs when properly installed.

Figure 10 Heat-fusing an underground connection for a geothermal system loop.
Source: BanksPhotos/Getty Images

Heat transfer tubing is placed into the ground in either a horizontal or a vertical configuration. The choice often depends on the amount of accessible property available. Horizontal loop systems (*Figure 11*) use trenches at an average depth of four to six feet to access stable ground temperature zones. Deeper trenches may be better for extremely cold climates. Horizontal trenching is generally more cost-effective than drilling vertical loops. The tubing layouts can be installed beneath homes, yards, open fields, landscaped areas, sports stadiums, and parking lots.

Figure 11 Arrangement of geothermal tubing in a horizontal trench.
Source: BanksPhotos/Getty Images

The heat transfer tubing can also be placed in trenches with the loops standing vertically (*Figure 12*). Another option, where a suitable body of water exists, is to submerge the loops on the bottom where water temperatures remain more consistent. This system can be an extremely cost-effective installation solution where circumstances allow. However, sediment and other debris naturally deposited on the outer surface of the tubing can impede heat transfer over time.

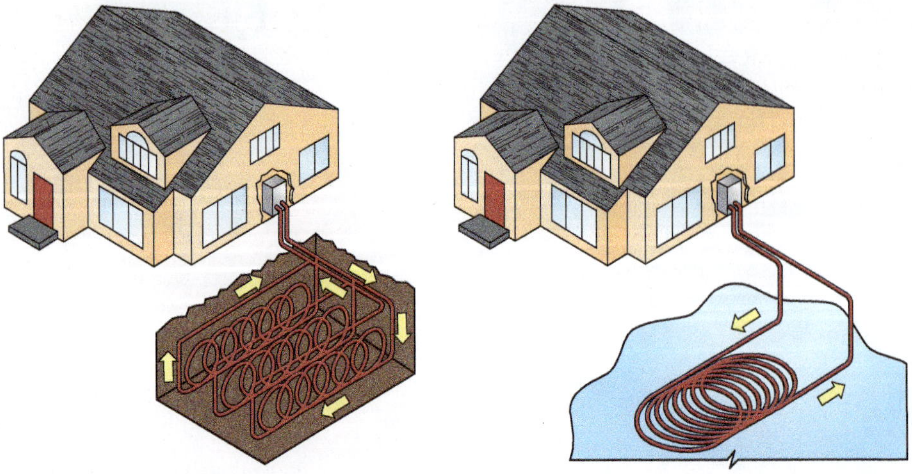

(A) Horizontal Closed Loop System **(B) Pond/Lake System**

Figure 12 Horizontal loop system configurations.

Vertical loop systems (*Figure 13*) use holes drilled into the earth for the loops. A vertical loop system is often used when insufficient land is available for a horizontal loop. The piping is looped down into each hole. As the piping exits the hole, it is joined to the loop in an adjacent hole. After the piping is inserted into the holes and the connections are made, a grout-like material poured into the holes ensures good thermal contact with the surrounding earth (*Figure 14*).

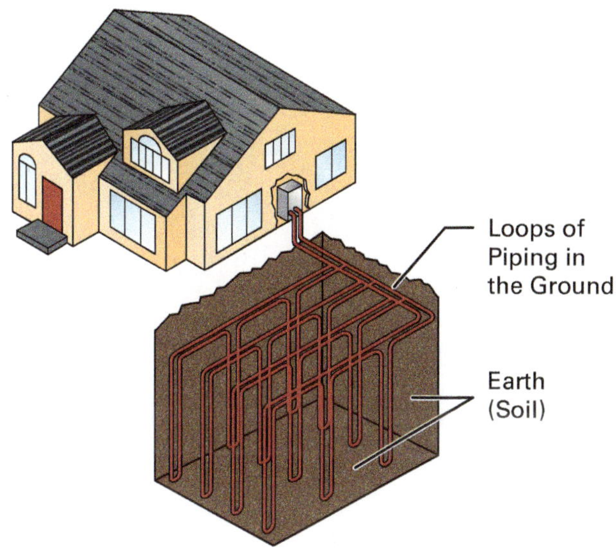

Figure 13 Vertical loop system.

Going Green

Cold Climate Heat Pump Specifications

The Northeast Energy Efficiency Partnership (NEEP) has established specifications for air-source heat pumps that are best suited for use in cold climates. In this case, the term "cold climates" refers to climate zones 4 and higher, as shown here on the map developed by the International Energy Conservation Code (IECC) committee. Ideally, HVAC system designers and practitioners will use the specifications to select equipment that will deliver the best results in cold climates.

The performance requirements include, but are not limited to, the following:

- An HSPF of 10 or higher for non-ducted systems
- An HSPF of 9 or higher for ducted systems
- A COP of at least 1.75 at a 5°F (–15°C) outdoor temperature
- A SEER of 15 or higher

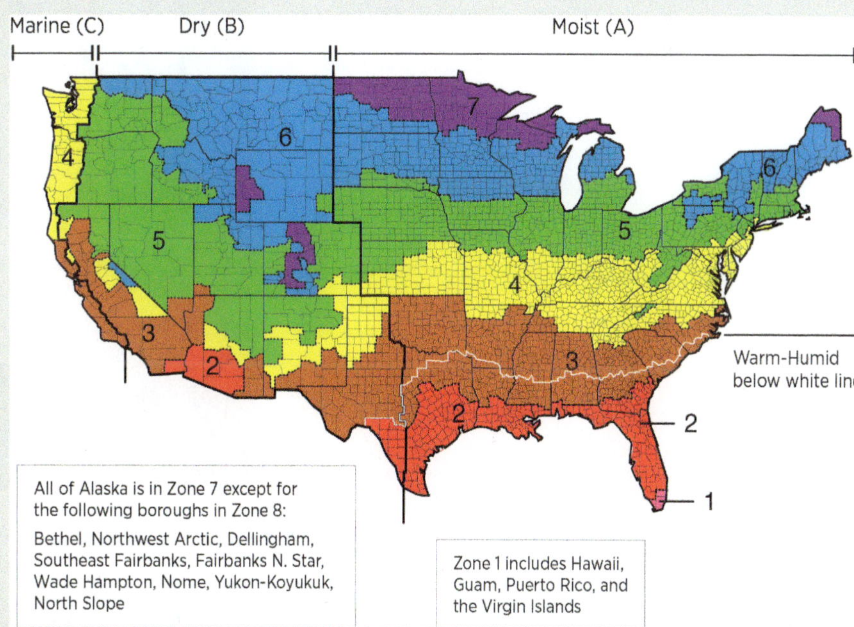

Source: US Department of Energy, Office of Energy Efficiency and Renewable Energy

For more information about energy conservation and high-performance construction, visit the Building America Solution Center, a resource offered by the US Department of Energy's Building Technologies Office and its Office of Energy Efficiency and Renewable Energy (*https://basc.pnnl.gov/*).

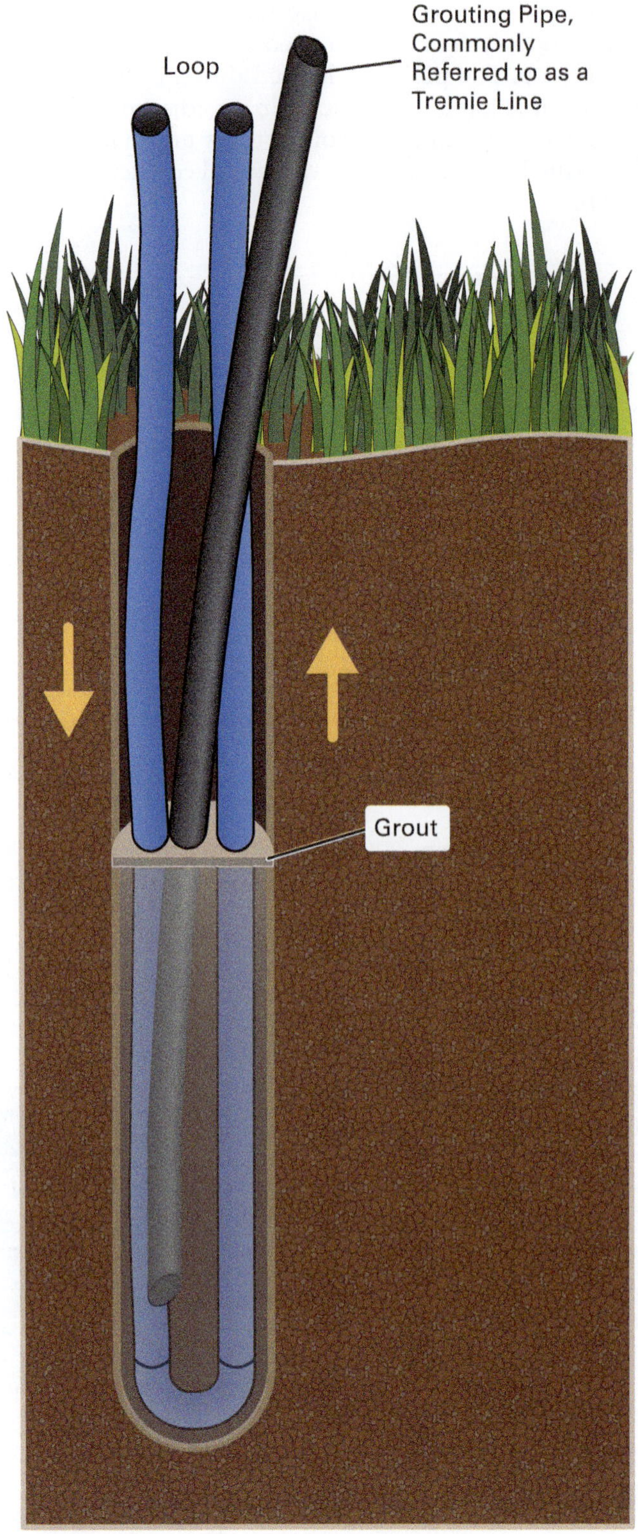

Figure 14 Grouting a geothermal vertical loop.

2.3.0 Heat Pump Controls

The control circuit of a heat pump is a little more complicated than that of a common cooling unit. Refer to *Figure 15* and locate the components or circuits as they are discussed. The numbers in parentheses in the text relate to the circled numbers on the diagram. Try to answer the questions as you progress through the discussion. These questions are included to help you understand the operation of a heat pump. We'll first examine the operation of the defrost cycle.

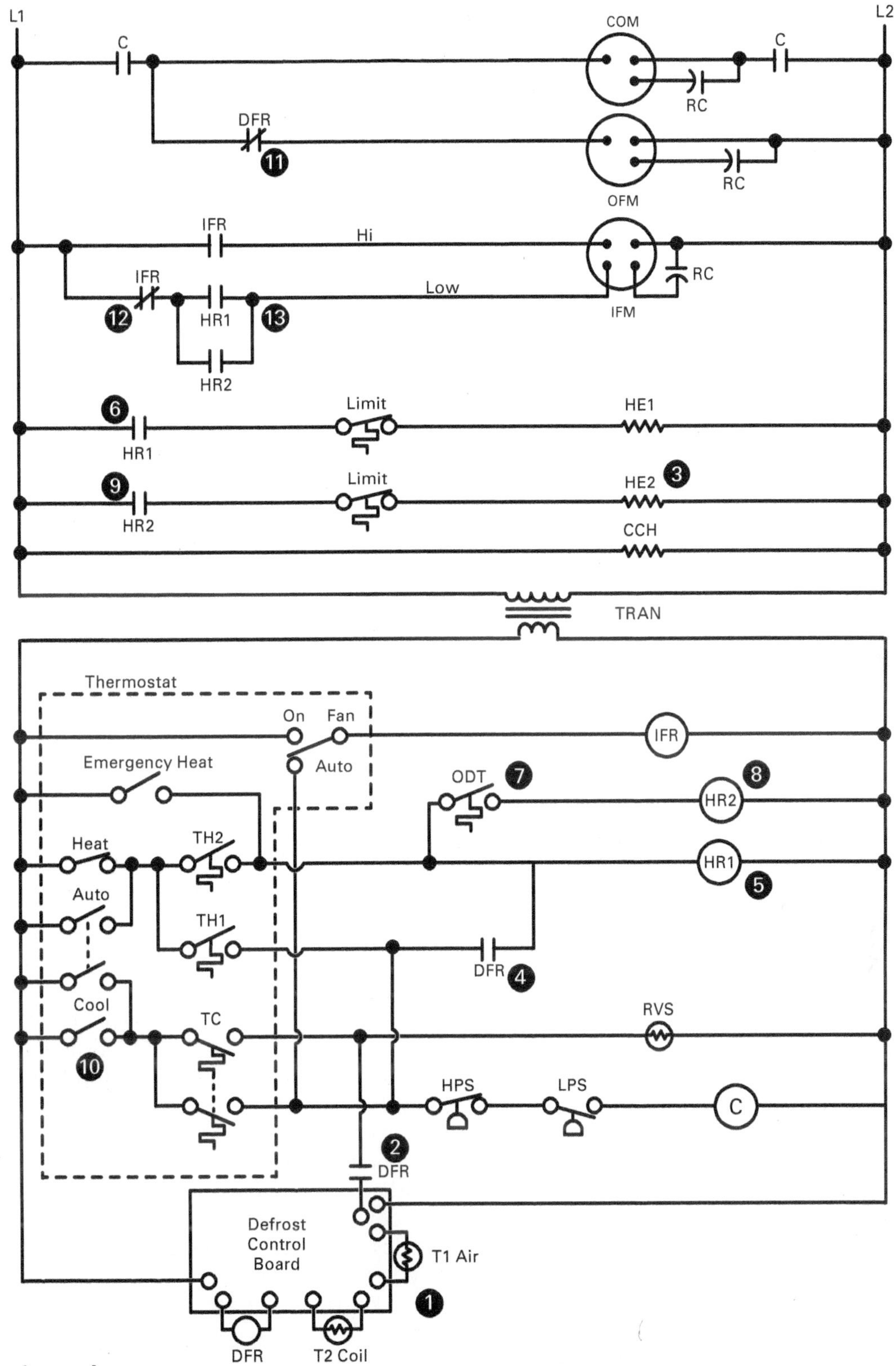

Legend:

C	Compressor Contactor	LPS	Low-Pressure Switch
CCH	Crankcase Heater	ODT	Outdoor Thermostat
COM	Compressor	OFM	Outdoor Fan Motor
DFR	Defrost Relay	RC	Run Capacitor
HE	Heating Element	RVS	Reversing Valve Solenoid
HPS	High-Pressure Switch	T	Thermistor
HR	Heater Relay	TC	Cooling Thermostat
IFM	Indoor Fan Motor	TH	Heating Thermostat
IFR	Indoor Fan Relay	TRAN	Transformer
		- - -	Mechanical Connection

Figure 15 Heat pump ladder diagram.

2.3.1 Defrost Controls Operation

The heat pump requires a defrost control circuit to manage the cycle. The following actions typically need to take place:

- The need for defrost must be monitored to initiate the cycle, either by time or sensing the presence of ice by temperature.
- The reversing valve must change its position, sending the hot refrigerant vapor from the compressor to the outdoor coil.
- The outdoor fan motor must be cycled off to avoid blowing cold air across the coil while the defrost cycle is active.
- The supplemental heat, when present, must be cycled on to temper the cold air in the duct.
- The defrost cycle must be monitored and disabled once the ice has been removed.

In the example shown here, an electronic defrost control board ❶ is used. This board receives the coil temperature and outdoor air temperature from thermistors located on or near the outdoor coil. Most boards also have solid-state timers to further control the cycle.

Question:
Is the reversing valve energized or de-energized in the defrost mode?

Answer:
Energized. Normally open contacts of the defrost relay can control the reversing valve. When the defrost board energizes the defrost relay, its contacts ❷ complete the circuit to the reversing valve. The thermostat is also in control of the reversing valve, energizing it on a call for cooling.

Supplementary electric heaters are usually required on a heat pump. These heaters ❸ perform two primary functions:

Balance point: The outdoor temperature at which the heating capacity of a given heat pump is equal to the heat loss of the structure.

- Provide supplemental heat when the temperature falls below the **balance point**, and the heat pump can't provide enough heat.
- Provide heat indoors while the unit is in the defrost mode.

Question:
How are the supplemental electric heaters activated during the defrost mode?

Answer:
A set of normally open defrost-relay contacts ❹ energizes heater relay HR1 ❺ when the unit goes into the defrost mode. The normally open contacts of HR1 ❻ then close to energize heating element #1. If the outdoor thermostat ❼ is closed, the HR2 relay ❽ is also energized, closing its contacts ❾, and starting the second stage of electric heat as well.

A multi-stage heat-cool thermostat ❿ is needed for a heat pump. The first stage of heating in the thermostat (TH1) controls the heat pump itself; it energizes the compressor contactor when closed. The second stage of heat in the thermostat is the electric resistance heat, which is controlled by TH2.

Two stages of electric heat are available. The second stage, Heating Element #2, is controlled by the outdoor thermostat ❼. The outdoor thermostat keeps the second stage of electric heat off as an energy-saving measure when the outdoor temperature is not so cold. Its setpoint is first determined by calculating the temperature at which the heat load exceeds the capacity of the heat pump combined with the capacity of the Heating Element #1.

Question:
Does this circuit have any means of preventing Heating Element #2 from cycling on without Heating Element #1 being energized first?

Answer:
Yes. Both heating elements are controlled by TH2 of the thermostat. HR2 cannot be energized until both TH2 and the ODT are closed. Therefore, Heating Element #2 will never be energized unless Heating Element #1 is also energized. This is also true in the defrost mode. Note that a failure of the #1 heater

will not stop the #2 heater from energizing, as long as the control circuit remains intact and functional.

Question:
How could you modify this circuit so that the crankcase heater (CCH) is on only when the compressor is off?

Answer:
Add a crankcase heater relay with a set of normally closed contacts in series with the crankcase heater. The relay coil would be wired in parallel with the contactor coil. When the contactor is energized, the relay is also energized, opening the contacts and disabling the crankcase heater.

Question:
Because a defrost cycle is essentially the same as operating in the cooling mode, how is the outdoor fan motor disabled during a defrost cycle?

Answer:
The normally closed contacts of the defrost relay ⓫ are in series with the outdoor fan motor. When the defrost relay energizes, the relay contacts open, and the motor is disabled.

Question:
What feature does this circuit have that makes sure the indoor blower is energized whenever one of the electric resistance heaters is on?

Answer:
Normally closed contacts of the indoor fan relay ⓬ will provide voltage to the low-speed side of the blower whenever either of the heater relays is energized. Those relay contacts are shown at ⓭. This feature is important when the thermostat has an emergency heat control that can be used to turn on the electric heaters if the heat pump fails. Without a fan to disperse the heat created by the heating elements, the unit could overheat.

2.3.2 Defrost Cycle Operation

There are several ways to control the defrost function. One popular method uses a combination of time and temperature. The logic behind this method is that defrost is needed only if the unit has been operating long enough for frost to form on the coil (for example, 90 minutes), and the temperature of the coil is low enough for moisture to freeze on the coil. If both conditions don't exist, defrost is unnecessary. Most timers can be set for 30, 45, 60, or 90 minutes of operation before a defrost cycle will occur. A few new units will allow a two-hour setting as well. A 45-minute cycle may be required in Michigan, while a 90-minute cycle will likely work fine in Florida.

This method requires a timer to keep track of the system's operating time and a means of sensing coil temperature. The latter is easy; a small thermostat or thermistor on the outdoor coil will do the job. Keeping time is easily accomplished with an electronic timer.

Electromechanical Defrost Timers

Electromechanical defrost timers were once widely used and some remain in service. The motor-driven timer runs whenever the compressor runs. *Figure 16* shows such an arrangement. Whenever the compressor contactor is energized, power is supplied to the compressor and the timer. As the defrost timer motor runs, it turns a cam that closes a set of contacts ❺ for 10 seconds every 90 minutes. Because the unit does not always run for 90 minutes at a time, it may take several on-cycles to accumulate 90 minutes.

The 10-second closure of the mechanical contacts is not enough to initiate a defrost. It is a short period provided as an opportunity for the system to evaluate the conditions. The defrost thermostat (DFT) contacts ❻ must also be closed, indicating that the outdoor temperature is below 45°F (7°C) or a similar setting. When both the timer contacts and the DFT are closed, the defrost relay (DFR) can energize to start a defrost cycle.

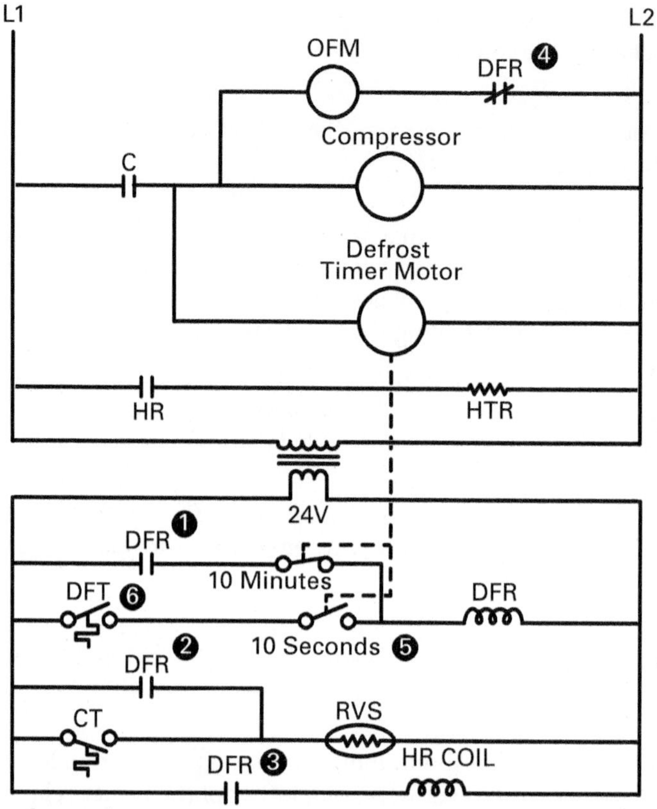

Legend:

C	Compressor Contactor	HR	Heater Relay
OFM	Outdoor Fan Motor	HTR	Heater
CT	Cooling Thermostat	RVS	Reversing Valve
DFR	Defrost Relay	----	Solenoid Mechanical Linkage
DFT	Defrost Thermostat		

Figure 16 Electromechanical defrost control.

Refer again to *Figure 16*. One set of normally open DFR contacts ❶ closes to serve as a holding circuit for the DFR coil. This is needed because the mechanical contacts of the timer ❺ will open again after 10 seconds. Another set of normally open DFR contacts ❷ closes, causing the reversing valve to energize and placing the system in the cooling mode.

After 10 minutes (typical) of defrost operation, the other contacts that are mechanically linked to the timer will open momentarily, stopping the defrost process and restarting the 90-minute timing cycle. If the heating thermostat opens during the defrost cycle, defrost will be suspended. When the thermostat calls for heat again, it will pick up where it left off, as long as the defrost thermostat is still closed.

Notice that a third set of normally open DFR contacts ❸ controls the heater relay (HR). When the defrost cycle begins, HR energizes, and its contacts in the heater circuit close, bringing on the supplemental electric heat to prevent cold air from blowing into the conditioned space. Further, a set of normally closed contacts ❹ opens during the defrost cycle to shut down the outdoor fan motor while defrost is in progress. Electromechanical defrost timers are rarely used in new equipment because they are not as reliable nor as flexible as modern electronic controls.

In addition to mechanical timers, pressure switches that sense airflow through the outdoor coil and thermal controls may also be used to start and stop the defrost cycle. These pressure switches are in the liquid line and open the defrost circuit when the pressure at the outdoor coil reaches 275 psi. This equates to a coil temperature of about 124°F (51°C), which is high enough to indicate that the coil is free of ice. Thus, the pressure switch can terminate the defrost cycle before the total time has elapsed, conserving energy. Defrost cycles represent significant energy losses that must be minimized.

Electronic Defrost Controls

Most modern heat pumps use electronic defrost controls in which all the defrost circuitry, including the relays, is mounted on a printed circuit board (*Figure 17*). The timer is electronic, and therefore less likely to fail. A temperature thermistor sends temperature information from the outdoor coil to the board (*Figure 18*). Instead of keeping the operating time mechanically, an electronic clock in the timing logic receives a signal when the heating thermostat is closed. If defrost

Figure 17 Electronic defrost control board.
Source: HVACGuru

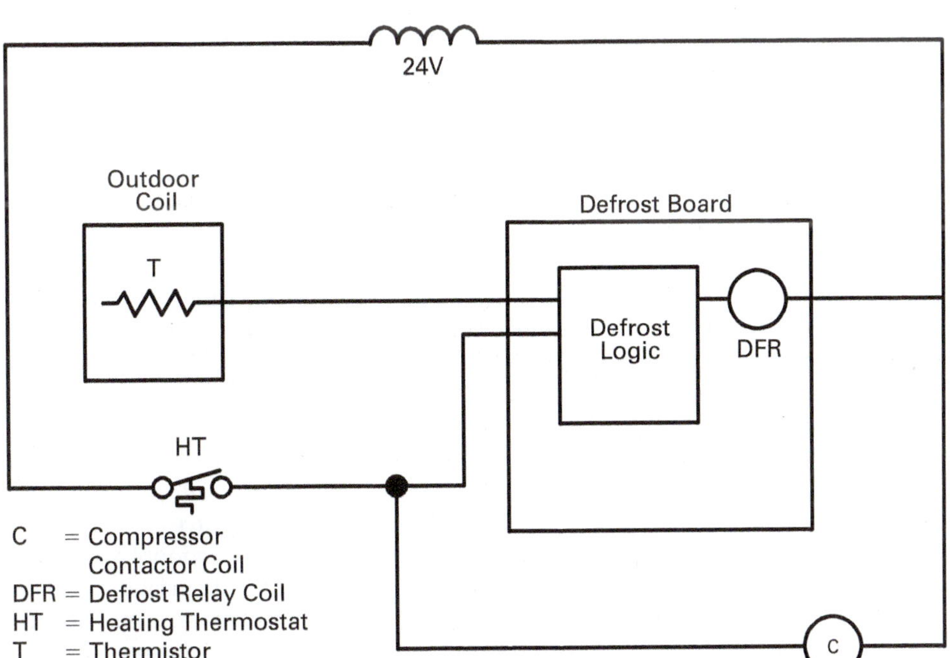

Figure 18 Simplified electronic defrost control schematic.

is interrupted by the opening of the room thermostat, the timing logic will remember where it left off. If the coil temperature still indicates the presence of frost when the unit cycles on again, the unit will automatically go back into the defrost mode before starting a fresh heating cycle.

Thermistors (*Figure 19*) are used as temperature sensors in electronic defrost circuits. A thermistor's resistance changes as the temperature changes. A traditional defrost thermostat or outdoor temperature thermostat responds to a fixed temperature; it is either open or closed. That fixed temperature reduces flexibility when designing a heat pump.

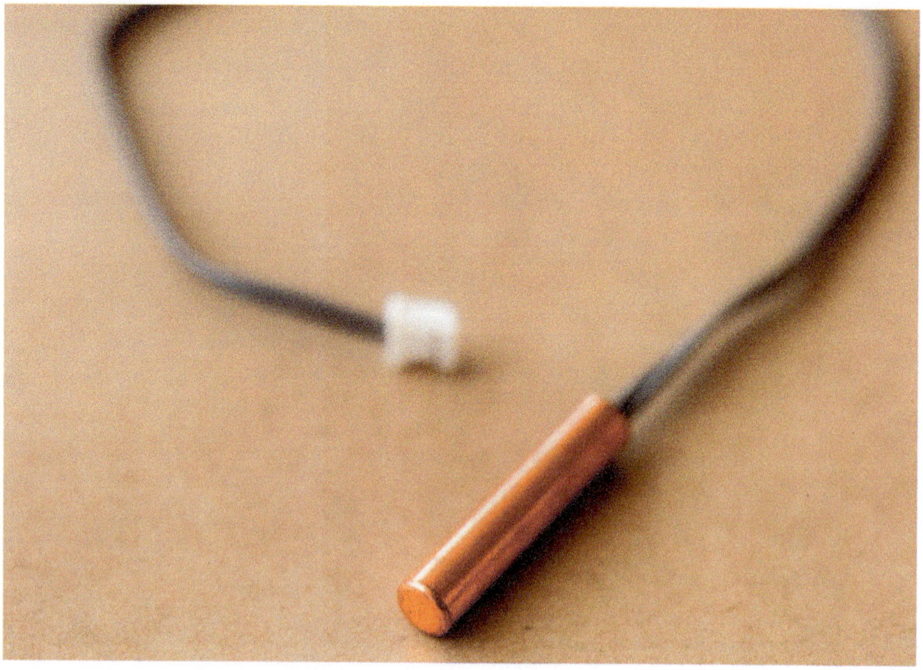

Figure 19 A simple thermistor.
Source: Mr.Guy Zezars/Shutterstock

On the other hand, thermistors respond to temperature change by changing their resistance. The electronic defrost control can use this information and make logical decisions. That allows for greater design flexibility. A single thermistor can also send outdoor temperature information to the room thermostat and provide temperature inputs to enable or disable supplemental heat sources.

A jumper or other selection device on the board allows the installer to select the defrost frequency (30, 45, or 90 minutes, for example) based on local conditions.

An important feature of many of these boards is an integral test function. On most defrost control boards, the technician can short (bridge) two terminals on the board and start an abbreviated defrost cycle. In *Figure 17*, note the posts in the lower left corner of the board that are labeled SPEEDUP. By momentarily bridging these two posts, the defrost cycle timer is overridden and the usual 10-minute defrost cycle can be reduced to just a few seconds.

One major advantage of the electronic defrost control is that it is completely self-contained. If the technician finds that there is a defrost control problem, it is only necessary to isolate the board or a thermistor(s) as the source of the problem.

As previously noted, the heat pump delivers cold air indoors while in a defrost cycle. Electric heat is typically energized during this period to temper the supply air temperature. As a result, the defrost cycle consumes more energy than a typical heating cycle with no real benefit. This must be factored in when considering overall system efficiency. Obviously, minimizing the number of defrost cycles, as well as their length, is a great advantage.

With the advent of newer, more versatile electronic controls, manufacturers have developed demand, or *adaptive*, defrost control strategies. These systems combine information relative to current operating conditions, such as the difference in temperature between the outdoor coil and the ambient temperature. They can precisely determine when a defrost cycle is needed and when it can be terminated to maximize energy efficiency.

On Site

Defrost Timing—Getting It Right

Time/temperature defrost controls are shipped from the factory with a default defrost interval setting (usually 90 minutes). This time interval may be too long for some climatic conditions. For example, if an area is subjected to misting rain or freezing fog in the winter, the coil may ice up so much during the 90-minute interval between defrosts that the ice cannot be cleared during a 10-minute defrost cycle. When this happens, new ice will quickly build on top of the ice that remains.

This creates a snowballing effect in which so much ice builds up, the coil eventually becomes encased in ice. Experience gained while working with heat pumps in your local area will help you determine the interval needed between defrosts to ensure the coil fully defrosts during every cycle.

2.4.0 Heat Pump Components

Although a heat pump and a cooling-only unit are schematically similar, there are certainly differences. Heat pumps not only have more components, but some components they have in common with cooling units are slightly different because they perform a dual role.

In a cooling-only unit, for example, the evaporator can usually be smaller than the condenser because it transfers less heat. The condenser must also reject the heat of compression, which typically represents an additional 3,000 Btuh per ton more than the evaporator will absorb. In a heat pump, the evaporator must also be able to reject the heat gained through the outdoor coil, as well as the heat of compression, which works to the benefit of the user in the heating mode.

The heat pump requires a heavy-duty compressor because it runs year-round. The heat pump compressor must be able to withstand compression ratios of 8:1, while a cooling-only compressor might only reach a ratio of about 4:1.

Some components that are unique to heat pumps will be reviewed here in more detail.

2.4.1 Reversing Valves

All reversing valves (*Figure 20*) operate in basically the same way. Reversing valves have a main valve body containing a sliding piston, plus a three-way pilot valve that is actuated by a solenoid coil.

The reversing valve has four piping connections (*Figure 21*). The flow of refrigerant through two of these connections never changes direction. One is the hot gas discharge connection from the compressor, and the other is the suction line to the compressor. The remaining two ports connect to the indoor and outdoor coils. The direction of refrigerant flow through the two coils is reversed when the valve is energized.

The refrigerant discharge vapor can be directed to either coil depending on the position of the piston in the valve. When the piston shifts, the path of the discharge flow changes from one coil to the other.

The piston moves using the pressure difference between the discharge and suction lines. The position of the main valve is controlled by a pilot valve. When the pilot valve changes position, it directs the flow of high-pressure vapor to one end of the reversing valve. At the same time, a path is created between the opposite end of the valve and the suction line. The pressure difference on each end of the valve causes it to move.

Figure 20 Reversing valve.
Source: Darkdiamond67/Shutterstock

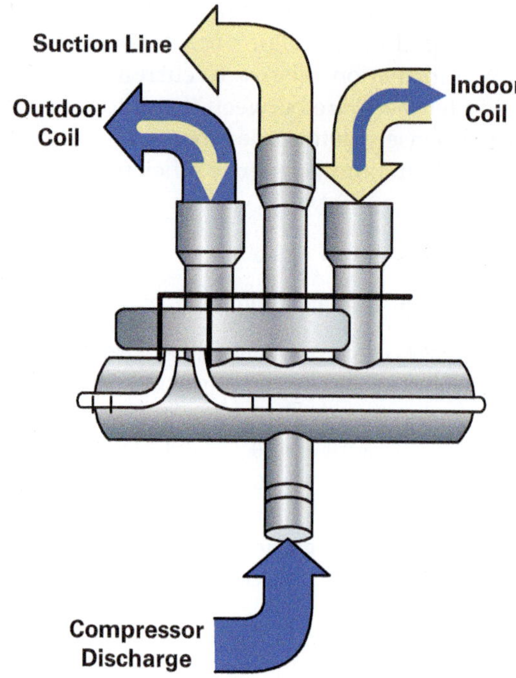

Figure 21 Reversing valve connections.

NOTE

Since a pressure differential must be created before the valve can shift, the compressor may start while the reversing valve is in the incorrect position. The valve coil is energized, but no pressure difference yet exists to make it move. After a few seconds of compressor operation, the valve shifts, often with a clearly audible "whooshing" sound. This is normal and not a concern.

It is important to note that a pressure difference of roughly 100 pounds per square inch gauge (psig) is required to move the piston, per their manufacturers. The valve cannot change position until the compressor has started and developed sufficient differential pressure to reposition the piston. The valve often does move at a lesser pressure differential, but manufacturers consider them to be working properly if they move at a differential of 100 psig or more. The pilot solenoid voltage is usually 24VAC, but it can be 120VAC or even 240VAC in commercial systems.

During the cooling cycle, the valve operates as follows, as shown in *Figure 22*:

- The solenoid is energized, and the pilot-solenoid pin carrier moves to the far left of the pilot solenoid chamber.
- The high-pressure discharge vapor is routed from the compressor discharge line through the capillary tubes to the right end of the piston housing.
- The opposite end of the piston housing is opened to the suction line through the other two capillary tubes. This allows the vapor at the far end of the piston to escape.
- The piston is now free to move to the left end of the housing.
- With the piston in this position, the discharge vapor is directed to the outdoor coil, and the indoor coil is connected to the compressor suction line.

In the heating mode, the solenoid is typically de-energized. The valve, shown in *Figure 23*, operates as follows:

- With the solenoid valve de-energized, a spring pushes the pilot pin carrier to the right.
- The high-pressure discharge vapor is routed from the compressor discharge line through the capillary tubes to the left end of the piston housing.
- The right end of the piston housing is opened to the suction line through the other two capillary tubes. This allows the vapor at the far end of the piston to escape.
- The piston is now free to move to the right end of the housing.
- With the piston in this position, the hot discharge vapor is directed to the indoor coil, and the outdoor coil is connected to the compressor suction line.

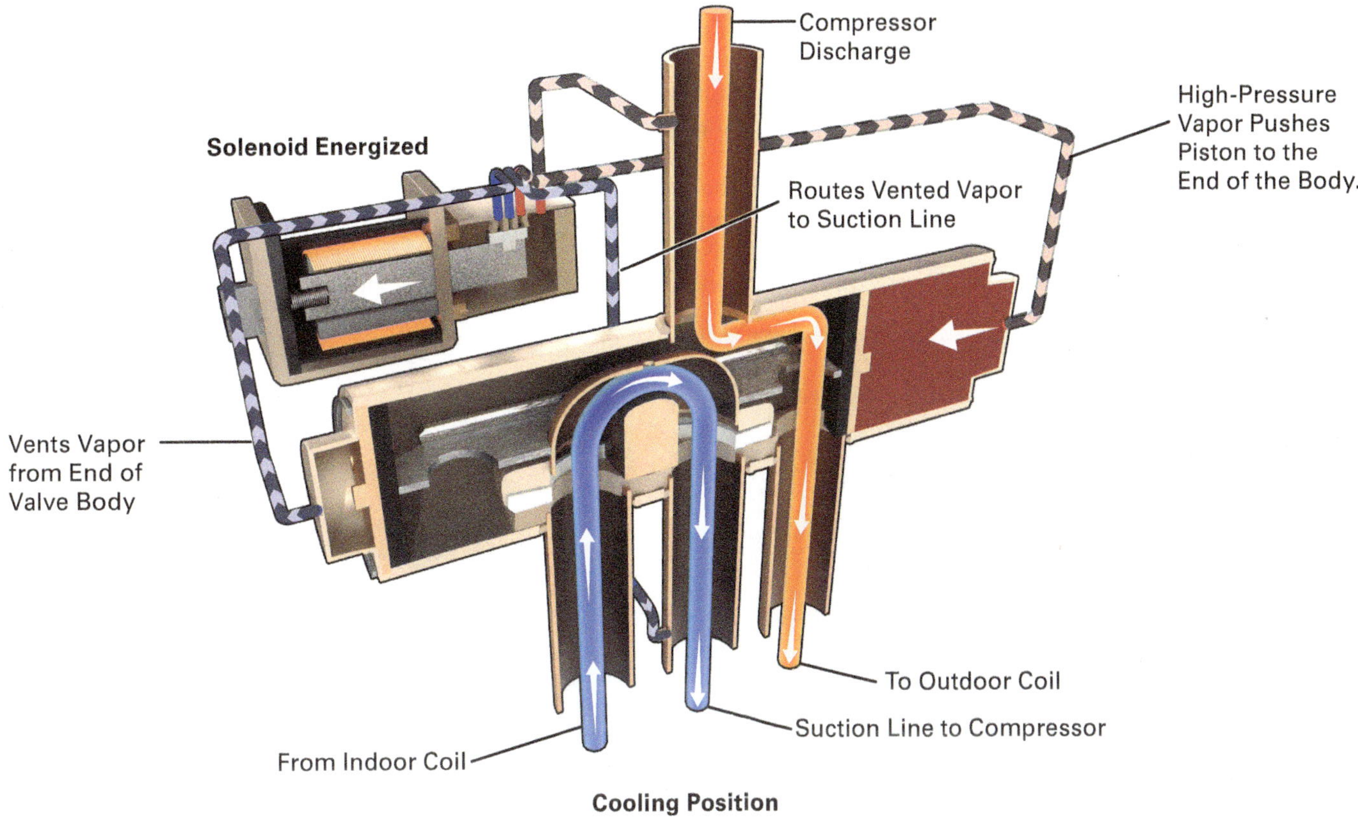

Figure 22 Reversing valve position—cooling cycle.

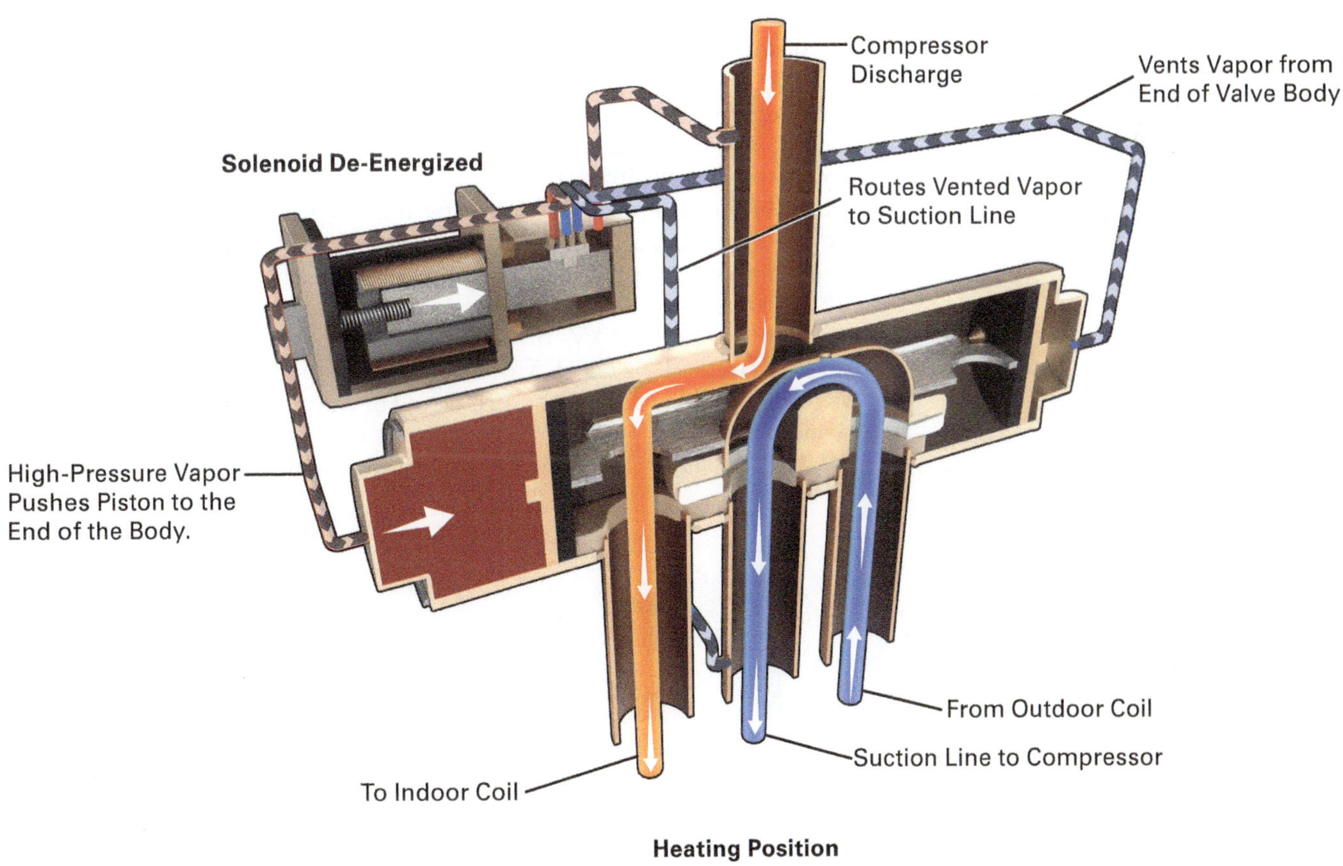

Figure 23 Reversing valve position—heating cycle.

2.4.2 Metering Devices

There are several methods used to meter refrigerant in a heat pump.

In one arrangement, a fixed-orifice metering device is installed at each coil where the liquid line connects (*Figure 24*). This arrangement is unique in that the metering orifice, often called a *piston* (*Figure 25*), moves within the body of the device. In the metering position, pressure in the liquid line forces the piston into a seated position where refrigerant can only flow through the metering orifice. In the bypassed position, pressure in the liquid line unseats the piston and moves it, allowing refrigerant to bypass the metering orifice.

NOTE

Metering devices are presented in detail in NCCER Module 03303, *Metering Devices*.

Figure 24 Fixed-orifice metering device installed on an indoor coil.

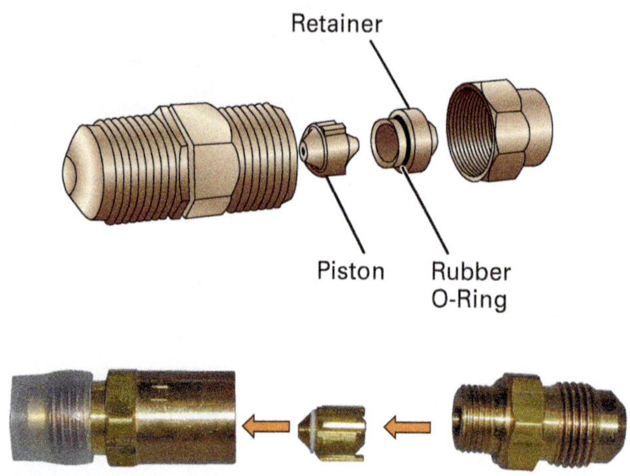

Figure 25 Fixed-orifice assembly.

Since one of these devices is installed at each end of the liquid line (at both coils), one is always metering, and one is always bypassing refrigerant. The body housing the metering piston is designed to accept all pistons of the same shape. The pistons have different orifice sizes based on system capacity. The manufacturer will specify the orifice size needed for the metering device on both coils. In *Figure 24*, the label above the tube connection identifies the default size of the piston within the body.

Which of the two pistons is metering flow depends on the operating mode. In the cooling mode, the outdoor metering device is bypassed while the indoor metering device is metering refrigerant. The opposite is true in the heating mode.

A similar arrangement can be established using thermostatic expansion valves (abbreviated as TXVs or TEVs) and check valves (*Figure 26*). The check valves are simple ball check valves with a spring return. However, today's TXVs have built-in check valves, providing bi-flow capability without an external check valve. Bi-flow TXVs have largely replaced the traditional arrangement, although a few TXV/check valve arrangements remain in service.

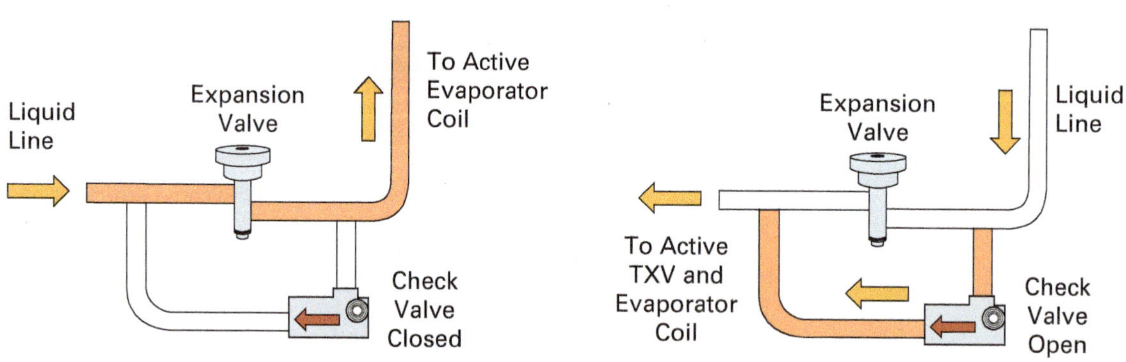

(A) Refrigerant Path Through the Active TXV **(B) Refrigerant Path Around the Inactive TXV**

Figure 26 Refrigerant paths through an arrangement of TXVs and check valves.

Figure 27 and *Figure 28* show a modern bi-flow TXV with a built-in check valve. When the flow enters on the closed side of the check valve, the refrigerant is metered through the pin and seat and a major pressure drop results. When the refrigerant flows in the opposite direction, there is little or no pressure drop through the valve. These valves simplify the arrangement and require far less space than two separate devices.

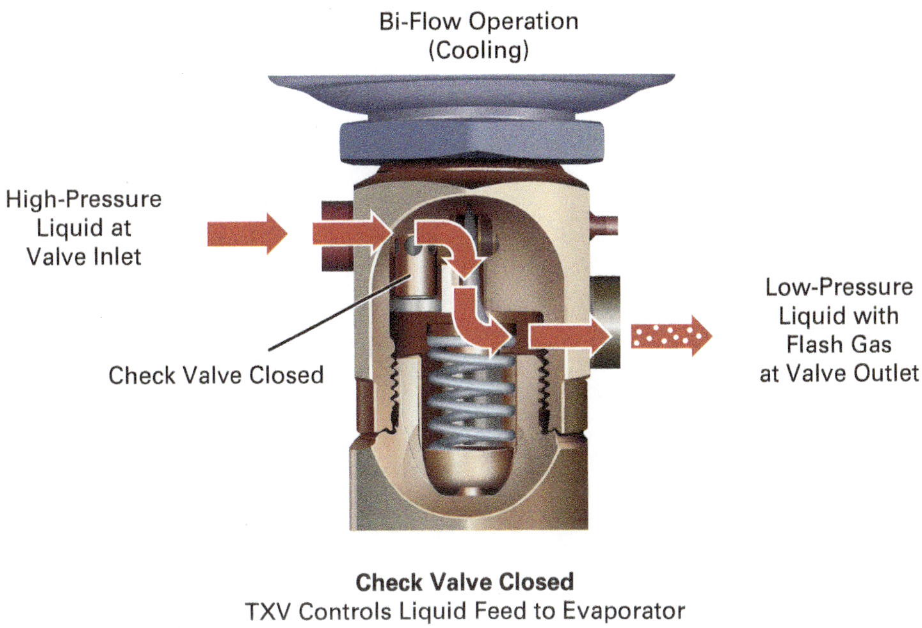

Figure 27 Bi-flow TXV operation as a metering device.
Source: Heil

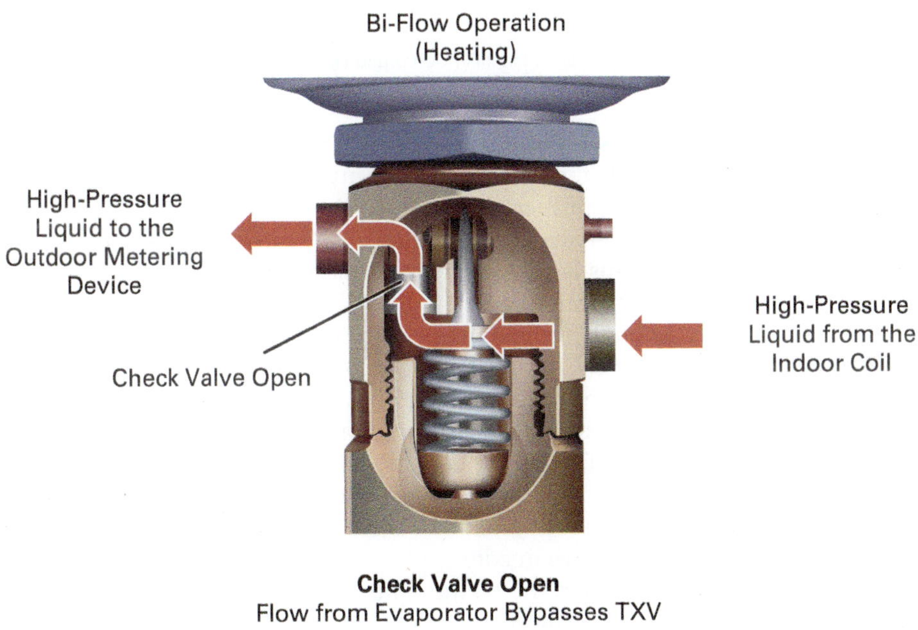

Figure 28 Bi-flow TXV operation when allowing liquid bypass.
Source: Heil

Electronic expansion valves (EEVs) continue to gain popularity due to their ability to precisely control the refrigerant flow. EEVs are also available in a bi-flow configuration for heat pumps (*Figure 29*). Due to the nature of the valve, an internal check valve is not needed. EEVs can easily drive to a fully open position to allow full flow, or they can be positioned as necessary to meter the flow. While some have only several hundred possible positions, many have thousands of them. It is simply a matter of changing the signal sent to the valve.

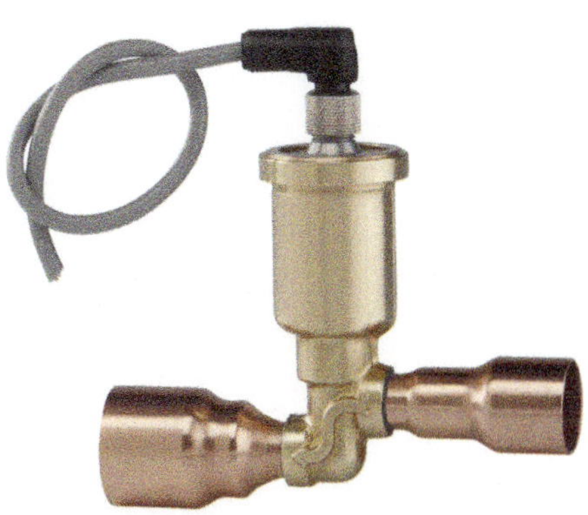

Figure 29 Bi-flow electronic expansion valve (EEV).
Source: Courtesy of Sporlan Division—Parker Hannifin Corporation

2.4.3 Accumulators

An accumulator (*Figure 30*) is a vessel placed in the refrigerant suction line to store excess refrigerant. It is usually located near the compressor inlet to prevent liquid refrigerant from entering the compressor.

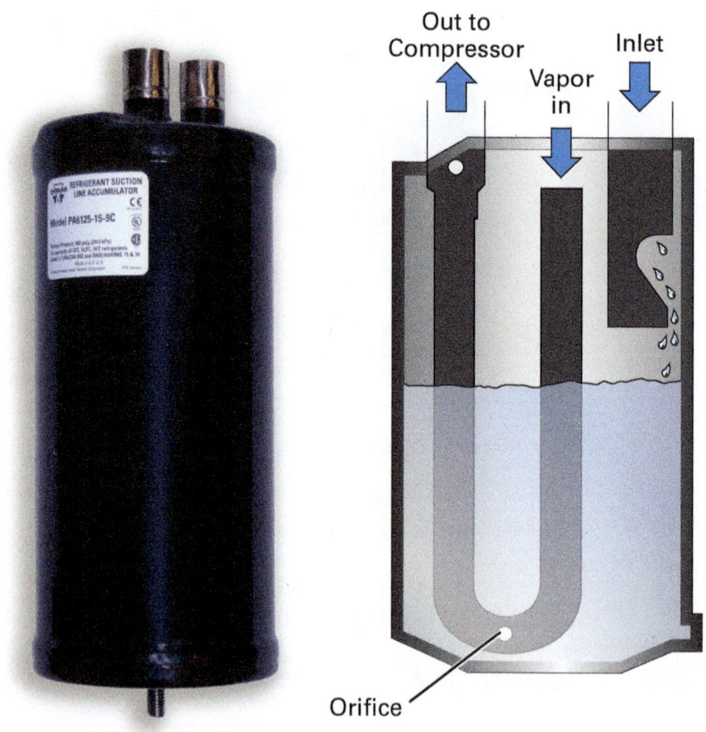

Figure 30 Accumulator.
Source: Courtesy of Sporlan Division—Parker Hannifin Corporation (photo)

Although the two coils used in heat pumps both act as evaporators and condensers, they often have different internal volumes. A sufficient amount of refrigerant is charged in for proper operation in the cooling mode. When the evaporator is used as a condenser (heating mode), the system may contain more refrigerant than it needs. This is because a large volume of the liquid and vapor in the circuit have changed locations.

There is not quite enough room for the refrigerant in this configuration. Without a place to store that excess refrigerant, it could return to the compressor as a liquid, causing damage or failure. The accumulator provides a temporary holding place for this abundance of liquid refrigerant.

When a heat pump is running in the heat mode, it is often evident if there is liquid refrigerant present in the accumulator. When liquid is present and the system is running, a frost line will often be visible. This results from the liquid refrigerant boiling as it vaporizes inside.

The refrigerant that collects in the accumulator must be returned to the compressor slowly. Oil will also be trapped inside with the refrigerant. Both oil and liquid refrigerant are returned to the compressor through a small orifice at the base of the U-shaped tube, as shown in *Figure 30*. The orifice ensures that liquid escapes a little at a time, vaporizing on its way to the compressor.

2.4.4 Thermostats

A two-stage heating, one-stage cooling thermostat is used to control most residential heat pumps. The first heating stage controls compression heat (heat pump), while the second controls the supplemental heat. Many water-source units, those not equipped with supplemental heat, use a simple one-stage heating, one-stage cooling thermostat.

Heat pump thermostats are often equipped with lights or text to indicate the status of the system and display filter change reminders. Most also provide a built-in compressor start delay to prevent **short cycling**. Today's advanced programmable units (*Figure 31*) provide a wide variety of scheduling options and features. For example, most use solid-state logic to lock out or minimize supplemental heat use while recovering from temperature setback. Models with connectivity can be monitored and adjusted remotely using a smart phone, and often provide current weather information from the internet.

Short cycling: Stopping and restarting a compressor too quickly before the refrigerant pressures have equalized.

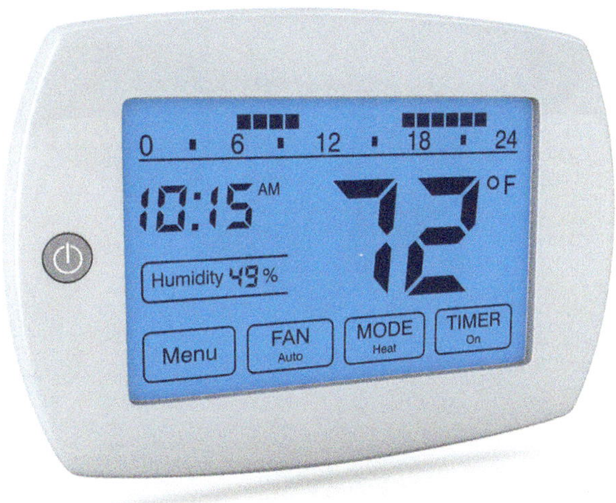

Figure 31 Digital programmable thermostat.
Source: iStock@lucadp

Heating Temperature Setback

For periods when buildings are unoccupied, heating temperature setback is a common means of saving energy. However, setting the temperature back too far on heat pump systems can result in reduced energy savings, especially when done with mechanical thermostats. This is because mechanical thermostats allow the supplemental heat to operate during recovery. The supplemental heat remains on until the space temperature has recovered within 3°F (1.7°C) of the setpoint. Most mechanical thermostats will revert to the heat pump alone within that range. Selecting a more reasonable setback—5°F, or 2.8°C, for example—can significantly reduce the recovery time and energy cost.

Most digital, programmable heat pump thermostats are designed with this problem in mind. Using information gathered in recent days of operation relative to cycling rates and run times, these thermostats determine when to begin the process of recovery to meet the desired occupied setpoint at the appropriate time. In addition, these calculations are often based on the recovery time required without using supplemental heat. This is commonly referred to as *adaptive recovery* or *intelligent recovery*.

Another type of problem can occur when unoccupied temperatures are set too low. It primarily affects split heat pump systems. In cold weather, while the temperature is set back and the compressor is off for a long period, refrigerant begins to migrate to the coldest part of the system (outdoor unit) and condense to a liquid. This leaves very little refrigerant volume remaining in the indoor coil. Once the compressor starts, the suction pressure often drops to an abnormally low value for a short time. If the unit is equipped with a low-pressure safety switch, the pressure may fall below its setpoint, forcing the unit to shut down within seconds of startup when no real problem exists.

One solution is to add a bypass timer to the low-pressure switch circuit (*Figure 32*). This timer, with normally closed contacts in parallel with the low-pressure switch, provides a temporary bypass of the low-pressure switch. This allows the pressures to stabilize before the pressure switch can open the circuit.

Figure 32 Low-pressure switch bypass timer.
Source: ICM Controls

After the timing period has lapsed, the timer contacts open, placing the low-pressure switch back in control. This same strategy is often used in cooling-only systems required to operate in low-ambient conditions.

2.4.5 Pressure Controls

Both high-pressure and low-pressure safety controls are often used on heat pumps. These two pressure controls are in addition to a defrost-termination pressure switch.

Physically, the high-pressure control is installed in the compressor discharge line between the compressor and the reversing valve. The high-pressure control protects the system against excessive pressure in both the heating and cooling modes. Electrically, both pressure switches in series with the coil of the compressor contactor (*Figure 33*).

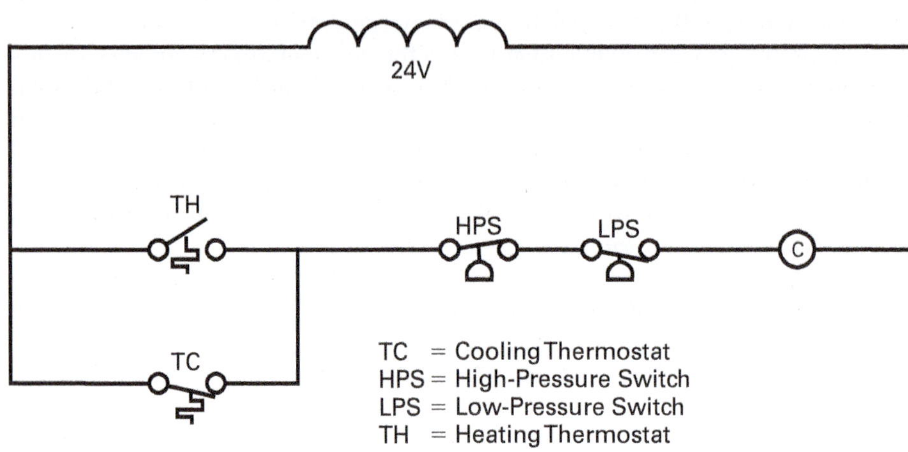

Figure 33 Common pressure switch electrical circuit.

HVACR low-pressure safety switches focus on one of two conditions. One condition is an excessively low pressure in the suction line of the compressor. The other condition is a loss of refrigerant charge. The low-pressure control in many heat pumps is designed to protect the system from a loss of charge. For this reason, you may see the switch labeled as a loss of charge switch, or LOC.

Like the high-pressure switch, it is wired in series with the compressor contactor coil. The switch opens if the pressure falls below the setpoint.

2.4.6 Crankcase Heaters

Heat pump compressors are usually equipped with crankcase heaters. Welded hermetic compressors will usually have an electric crankcase heater wrapped around the base of the compressor (*Figure 34*), or a heater inserted in a well in the bottom of the compressor shell.

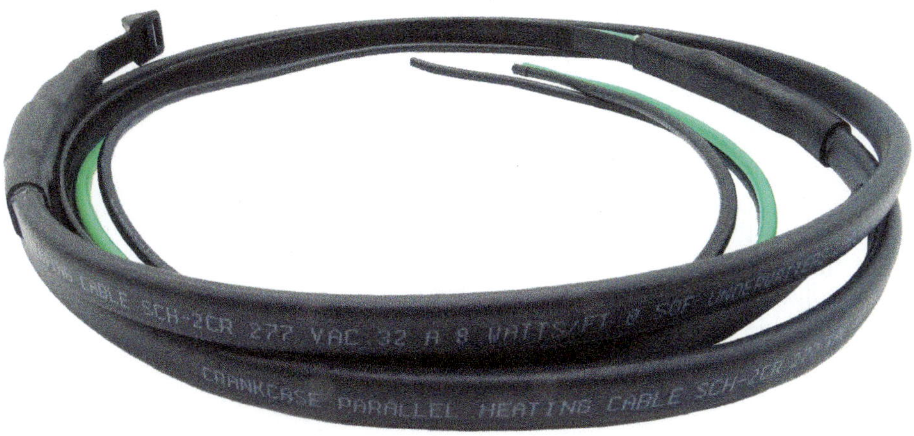

Figure 34 Self-regulating crankcase heater.

The crankcase heater prevents compressor damage by keeping the oil warm. Without it, liquid refrigerant tends to migrate there and condense in the oil during cold weather. Starting a compressor with a crankcase full of condensed refrigerant can cause serious damage as the liquid enters the cylinders. Remember liquids cannot be compressed. The rapidly falling pressure in the crankcase also causes the refrigerant to boil violently, washing the oil out through the cylinders as well.

The crankcase heater is usually powered directly from the line side of the compressor contactor, energized as long as the power source is on. It may also be wired through contacts that open when the compressor starts. Some, such as the one shown in *Figure 34*, are self-regulating. Self-regulating crankcase heaters are always energized but can sense the temperature and vary the power to the heater to maintain a design temperature. Above that temperature, these heaters stop generating heat. The heater in *Figure 34* produces 8 watts of power per foot at 50°F (10°C). The heating cable is bonded to a flexible plastic locking strap similar to nylon cable ties.

If power to the unit is disconnected for more than a few hours, avoid starting the unit for 12–24 hours after power is restored. If it's warm out, that won't be a problem. But if it is cold, a lot of condensed refrigerant could have formed in the crankcase. The heater will vaporize it over time, but not instantly.

A warning label to that effect may be provided with the installation instructions. If so, attach this label to the disconnect switch where the owner can see it.

2.5.0 Supplemental Heat

When an air-source heat pump is used to heat a structure, its heat output drops as the outdoor temperature drops. Remember that the balance point of a heat pump system is the outdoor temperature at which the heating capacity of the heat pump equals the heat loss of the building. The balance point varies from building to building, depending on factors such as construction details and heat pump size.

Figure 35 illustrates the effects of outdoor temperature on the capacity of an air-source heat pump. Line A shows that the heating capacity decreases with

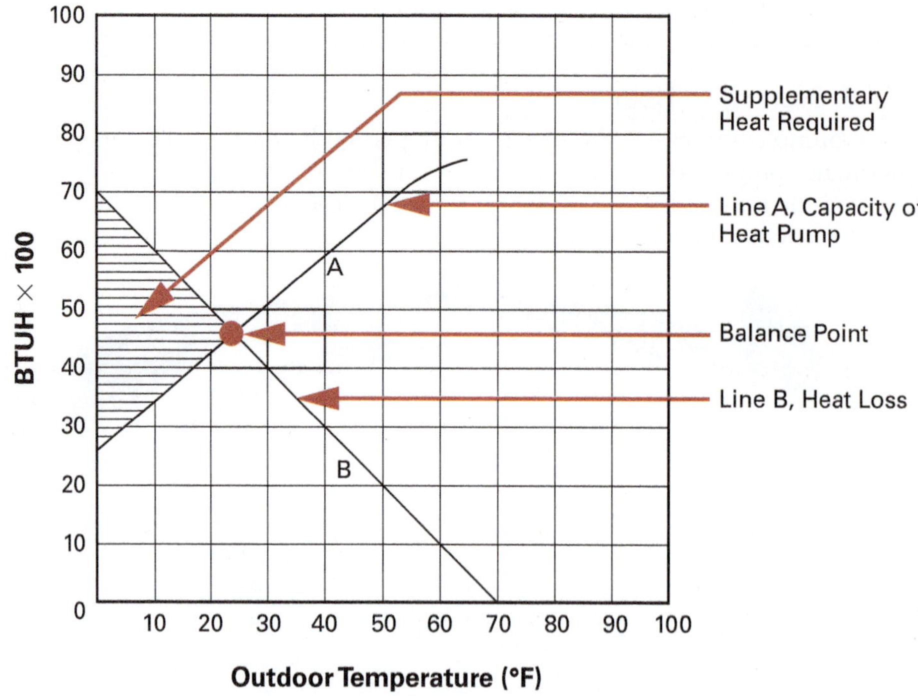

Figure 35 Heat-pump balance point graph.

decreasing outdoor temperatures. Line B shows that the heat loss increases as the outdoor temperature drops. The balance point is found at the intersection of the two lines. In this example, the balance point is about 23°F (–5°C).

In theory, when the outdoor temperature is below 23°F, the heat pump must run continuously. As the temperature outdoors continues to drop, the heat pump can continue to run, but it will be unable to maintain the setpoint. Below the balance point, some additional heat is needed to supplement its output to maintain comfort.

In an all-electric heat pump system, electric resistance heaters are energized as needed to maintain comfort. When two or more heating elements are installed, they can be *staged*. This means that one heating element can operate while the others remain off. If a single heating element along with the heat pump can maintain the setpoint, there is no reason to expend the additional energy to energize the others.

The most common and simplest form of staging control is based on the outdoor temperature. Calculations can be made, beginning at the balance point, to determine the capacity of the heat pump paired with one heating element. Using a similar graph, a secondary balance point can be determined. An outdoor thermostat can then be set to enable the additional heating elements to operate only when the outdoor temperature is below that second point.

The supplemental heat source is often sized to accommodate 100% of the building's heat loss, or nearly so. This ensures that there will be adequate heat in the event of a heat pump failure. Note, however, that building codes may advise that a lesser amount of supplemental heat be installed.

When there is a problem with the heat pump, users can typically switch to the Emergency Heat mode on the thermostat. For most heat pump thermostats, this disables the heat pump, and the electric heat becomes the first stage of heating. In the coldest temperatures, there may not be enough heat available to maintain the setpoint, but it is far better than no heat at all.

NOTE

Although this discussion focuses on supplemental heat, it is important to remember that an air-source heat pump will still need to defrost the outdoor coil occasionally at 23°F (–5°C) outdoors, regardless of its heating capacity. Since the defrost function requires the heat pump to run in the cooling mode, an additional heating source is also needed to temper the cold air supply that is produced.

2.5.1 Dual-Fuel Systems

Another popular use of air-to-air heat pumps is to install them with a gas or oil furnace. This is commonly called a **dual-fuel or hybrid system**. During the heating season, the heat pump supplies heat when outdoor temperature is above the balance point.

Dual-fuel or hybrid system: A system in which a heat pump is combined with a fossil-fuel furnace.

Once the outdoor temperature drops below the balance point, the heat pump is locked out and the gas or oil furnace provides all the heat. The benefit of such a scheme is that the heat pump is likely more economical to operate at the higher outdoor temperatures, while the furnace may be more economical to operate at the lower temperatures. However, as the cost of fossil fuels rises, this may not be the case.

Operating the heat pump and furnace at the same time is a problem. When electric heat is used to supplement a heat pump, the heat pump and the heating elements can operate at the same time, both during a defrost cycle and in the heating mode. In a dual-fuel arrangement, though, the heat pump and furnace can't typically operate at the same time. This is primarily because the hot air produced by the furnace flows over the indoor coil. With that air being 40°F to 70°F (22°C to 39°C) warmer than room air, the heat pump discharge pressures quickly rise beyond an acceptable level, overloading the compressor. The high-pressure safety switch will open as a result, disabling the heat pump.

Figure 36 shows how a simple dual-fuel arrangement works. An outdoor thermostat or thermistor is used to switch between the heat pump and fossil-fuel furnace. When the outdoor temperature falls below its setting, the heat pump is disabled, and the furnace provides 100% of the heating capacity.

NOTE

Advanced gas furnaces with the right controls to modulate furnace heating capacity down to a low value do have the potential to operate while the heat pump is defrosting. Such systems can accommodate heat pump operation below freezing. Modulating furnaces are typically more energy efficient than standard models, but also more expensive.

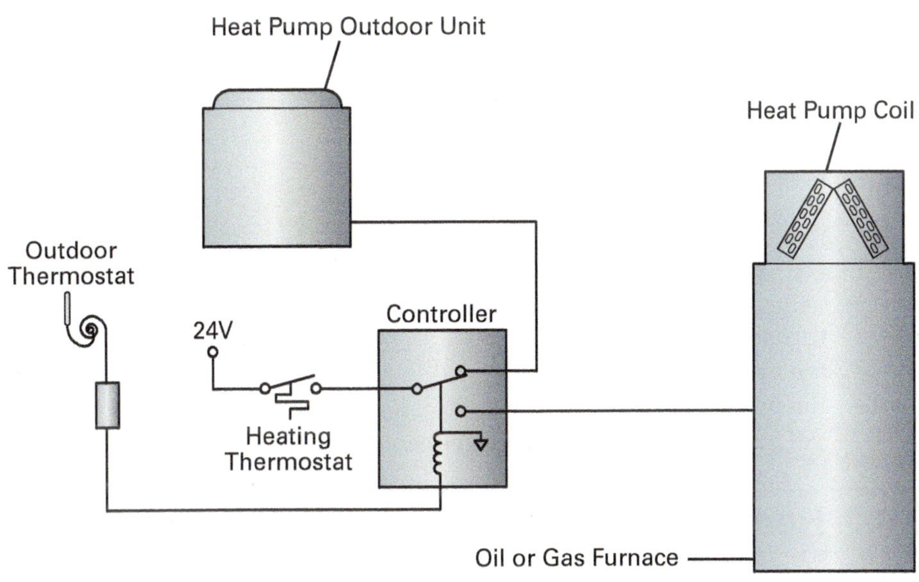

Figure 36 Dual-fuel heat pump controls.

Remember that, in an all-electric system, the electric heat is energized when the heat pump begins a defrost cycle. Since this isn't practical for all furnaces, it may be better if the system switches to the furnace at an outdoor temperature that is above freezing. Any frost that forms on the heat pump when it runs should then melt during the off cycle. However, this approach is not necessarily an energy-conscious choice. Operating costs and excessive energy consumption, especially in the form of fossil fuels, must be factors in the approach to system control.

Cold-climate heat pumps can produce significantly more heat than conventional systems, by design. The performance characteristics of a cold-climate heat pump can shift the balance point significantly lower while keeping the COP high. In the end, the user's preference for energy cost versus comfort should also be considered in the final choice of the changeover point.

On Site

Evolution of the Dual-Fuel Heat Pump

The dual-fuel heat pump is nothing new. It has been around since the 1970s. At that time, gasoline and fuel oil shortages brought on by the oil embargo caused fossil fuel prices to soar. People who heated their homes with oil were especially hard hit.

The dual-fuel heat pump was initially marketed to fuel oil dealers since natural gas supplies and prices were relatively stable. By adding a heat pump to an oil-fired forced-air furnace, customers could maintain comfort and save money at the same time while minimizing their oil consumption. People heating their homes with gas soon realized that they might experience some savings as well, and the idea spread.

2.0.0 Section Review

1. In the heating mode, refrigerant leaves the compressor and enters the indoor coil through the ______.
   a. accumulator
   b. liquid line
   c. vapor line
   d. metering device

2. A geothermal heat pump extracts heat from ______.
   a. air
   b. wells
   c. lakes or ponds
   d. the earth

3. During the defrost cycle, supplemental electric heaters provide heat to ______.
   a. temper the indoor air
   b. the compressor crankcase
   c. the thermostat heat anticipator
   d. melt frost on the outdoor coil

4. Where do you place a fixed-orifice metering device on an air-to-air split-system heat pump?
   a. In series with the accumulator
   b. On both the indoor and outdoor coils
   c. On the indoor coil only
   d. In series with each check valve

5. In a dual-fuel heat pump system, when the outdoor temperature drops *below* the setting of the outdoor thermostat, ______.
   a. furnace operation is locked out
   b. the heat pump is locked out
   c. the furnace can operate at full capacity during a defrost cycle
   d. the furnace and compressor can operate at the same time

3.0.0 Heat Pump Installation Practices

Performance Tasks

1. Install a heat pump and complete a proper startup.
2. Test the operation of a heat pump in the heating and cooling modes and record operating data.

Objective

State common heat pump installation practices.

a. State installation practices associated with split systems.
b. State installation practices associated with packaged systems.

This section covers the basic installation practices for split and packaged heat pump systems.

3.1.0 Split System Installations

A split system (*Figure 37*) is one that has two separate units—an indoor unit and an outdoor unit. The outdoor heat pump unit usually contains the following components:

- Compressor
- Reversing valve

- Outdoor coil
- One of the two metering devices
- Condenser (outdoor fan)
- High- and low-pressure switches, if equipped
- Defrost control board and sensors

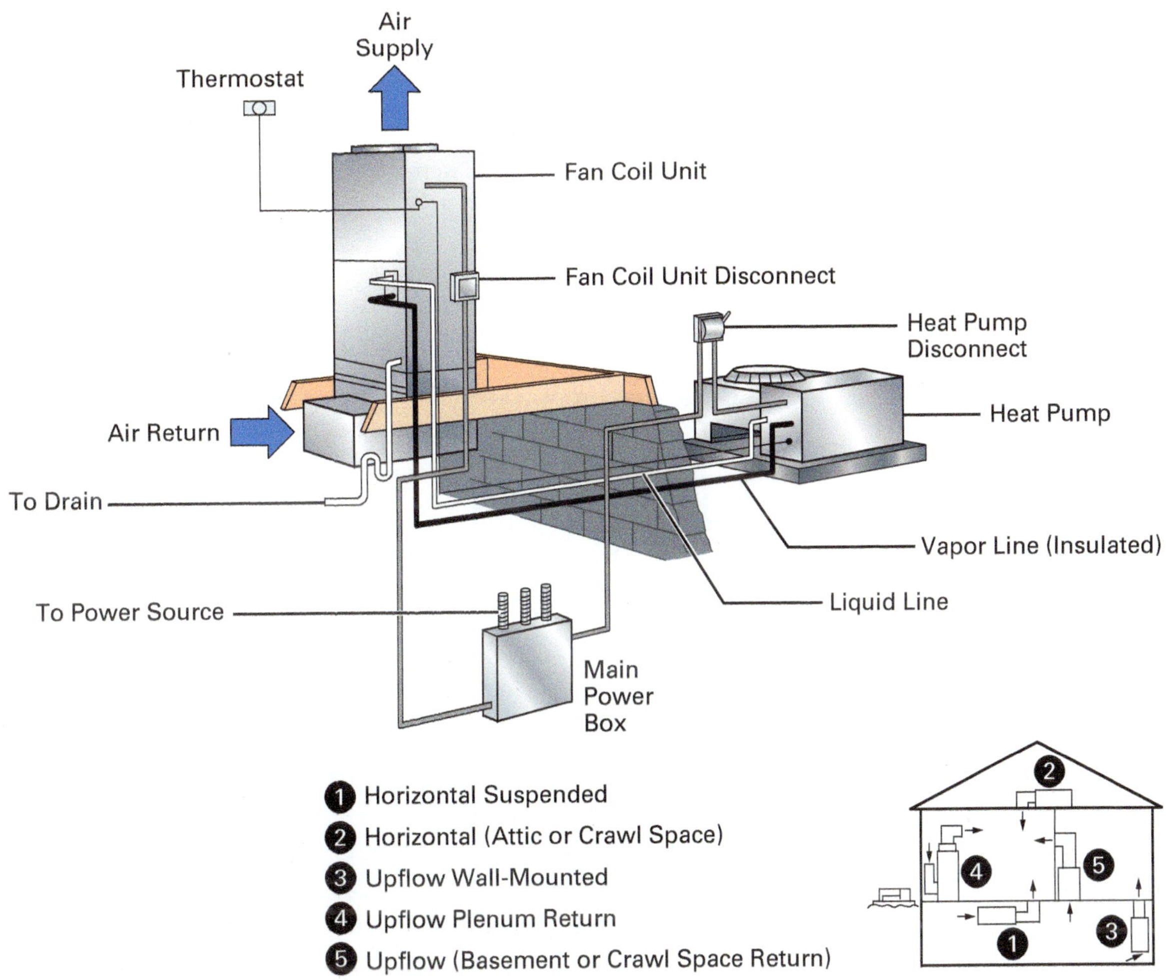

Figure 37 Split system installation.

The indoor unit, which is also known as the *air handler* or *fan coil unit*, contains the following components:

- Indoor fan (blower)
- The second metering device
- Indoor coil
- Primary control and terminal board for connection to the heat pump and wall thermostat
- Control transformer
- Electric heating elements (generally installed on site as an accessory)

As shown in *Figure 37*, indoor units can be found mounted horizontally in attics, basements, and crawl spaces. These units are also found in an upflow or downflow configuration in basements, garages, and equipment closets.

Another type of split system is the *ductless system* (*Figure 38*), which is designed to serve smaller areas. The indoor unit can be mounted high or low on a wall, fitted into a drop ceiling, or installed in a hard ceiling. Some ductless indoor units installed in ceilings can be connected to short runs of duct to serve more than one room, and they may also be connected to an outside air duct for ventilation (*Figure 39*).

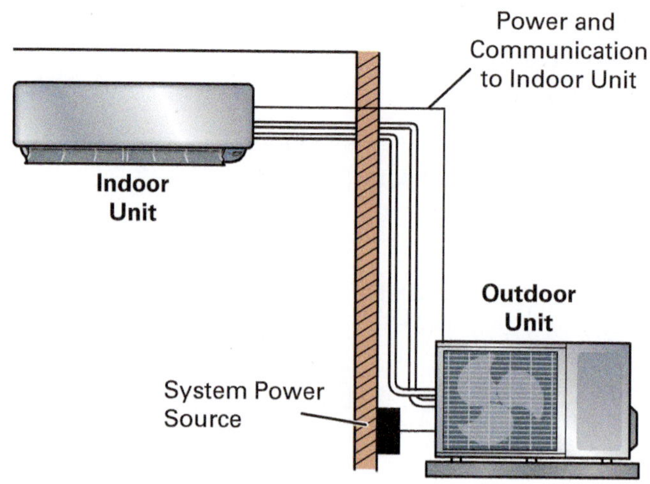

Figure 38 Ductless split system.

Figure 39 Ceiling-installed ductless unit with an outside air duct connection.
Source: Byjeng/Shutterstock

The ductless outdoor unit is typically narrow and rectangular, shaped like a suitcase, making it easy to handle. However, these units are more susceptible to wind than traditional heat pump units, so anchoring them to the slab or other support is needed. Instead of using a traditional room thermostat for temperature control, ductless splits are often controlled with a wireless remote like the remote used to control a television (*Figure 40*).

There are versions of ductless split systems in which a single outdoor unit serves two or more indoor units. These are known as *multi-split* or *multiplexed systems* (*Figure 41*). Each indoor unit has its own remote control, and the compressor is equipped with inverter-based, variable-speed controls, enabling it to economically satisfy the needs of the various indoor zones.

Figure 40 Wireless ductless-system control.
Source: Proxima Studio/Shutterstock

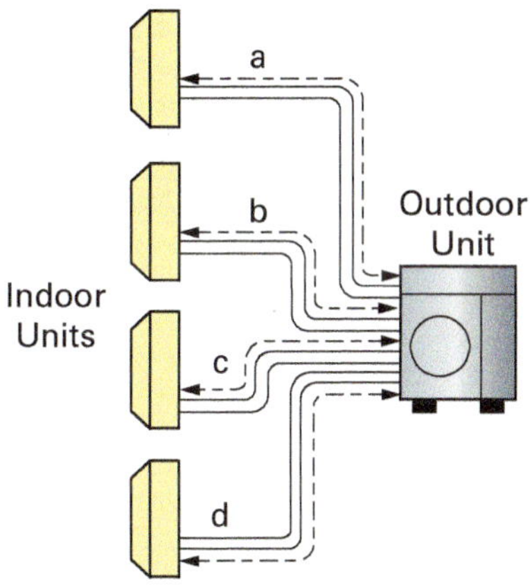

Figure 41 Multi-split ductless system.

3.1.1 Outdoor Unit Location

Several important factors must be considered before selecting the location of the indoor and outdoor units of a split-system heat pump. The placement of the outdoor unit should consider these factors:

- Sound
- Wind direction
- Proximity to the structure
- Proximity to neighboring structures
- Proximity to the indoor unit
- Defrost water drainage

The site is usually adjacent to an outside wall, but away from rooms where the sound might be objectionable. The unit should not be located where its discharge air or sound will disturb neighbors. Local ordinances often impose restrictions on the placement of outdoor units. For that reason, always consult local codes before starting an installation.

If the outdoor unit is to be placed in front of the building, it is best concealed by shrubbery. Ample clearance must be allowed around the unit for air movement and service access. Clearance dimensions are usually specified in the manufacturer's installation instructions. Local codes and even homeowner association bylaws may apply here as well.

The outdoor unit should be located close enough to the indoor unit to eliminate lengthy piping runs, but it should not be placed where water from a roof above will drip onto it. Wind can interfere with defrost too. If the outdoor unit is installed where it is exposed to a prevailing wind, it may be necessary to install a wind baffle around it (*Figure 42*). This shields the coil and prevents wind from blowing across it during defrost.

Wind baffles for ductless systems can serve two functions—they can block the wind and divert discharge air from the outdoor coil upward. Blocking the wind can significantly improve the speed and efficiency of the defrost cycle. This allows multiple units to be installed closer together. Trees or shrubs planted near a heat pump on the ground can also provide a shield against the wind.

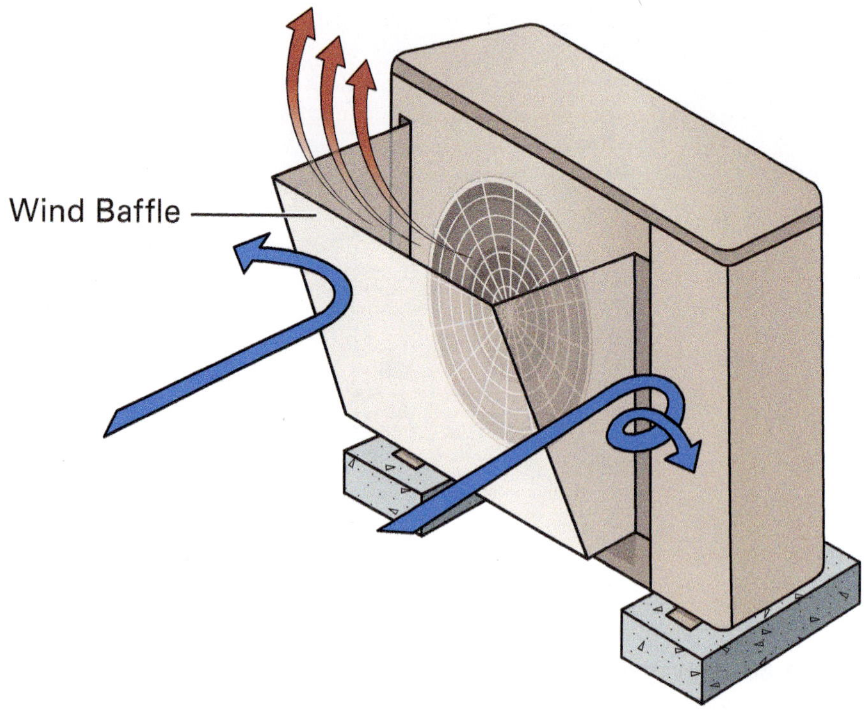

Figure 42 Wind baffles on a ductless unit.

On Site

Locking Caps

The *International Mechanical Code®* now requires locking caps on refrigerant service valves that are located outdoors. This is to prevent unauthorized access into the system by making it difficult to remove the access port caps.

This protects homeowners and vandals from the risks associated with refrigerants, such as frostbite and inhalation hazards. Most locking caps require a special tool or key to remove them. Over time, you will likely collect several different types of cap keys.

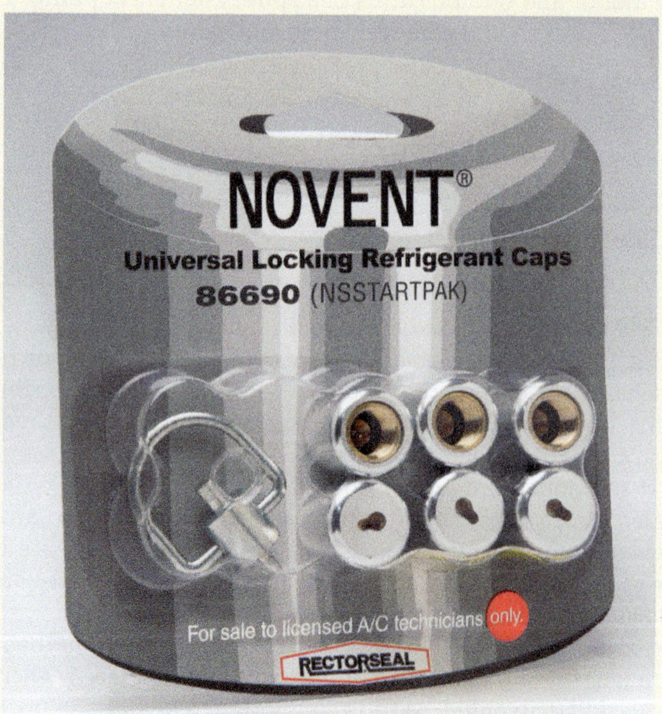

Source: RectorSeal

3.1.2 Outdoor Section Mounting

Once the location of the outdoor unit has been selected, the unit needs a proper foundation beneath it. A concrete pad is recommended, poured or portable, is usually acceptable. Ideally, the bottom of the heat pump is elevated at least 6" (~15 cm) above grade. It should be level and the soil and/or gravel beneath the slab should be firmly packed to keep the unit from sinking.

The base pans of heat pumps have holes or slots beneath the coil. These slots allow defrost water to drain. Keep these slots clear of debris that can block the openings and prevent drainage. If defrost water is trapped in the unit base, it will freeze again after the cycle is complete and create an ice dam.

Where snowfall is heavy, mount the unit on a snow stand that is anchored to a concrete slab. Alternatively, the unit can be wall-mounted (*Figure 43*). The unit should be high enough so that it remains clear of the heaviest anticipated snowfall. It is very important that the stand be firmly anchored to its foundation. If not, high winds might topple the unit. In high snowfall areas, ice can build up on the edge of the roof and heavy accumulations of snow can slide off. Locate the unit so that snow and ice isn't likely to fall on the unit or damage the refrigerant lines.

NOTE

Heat pumps designed for significantly cold climates may offer an accessory electric heater for the base pan to ensure defrost water drains out of the cabinet before it freezes. Ductless heat pumps are often equipped with them.

Figure 43 Wall-mounted heat pump.
Source: HVACGuru

Raising the unit at least 6" (~15 cm) above the ground is recommended, but they can be mounted as high as necessary. Remember to maintain proper clearance around the unit for service access and airflow.

A 12" (~30 cm) wide gravel bed around the foundation is ideal to accommodate defrost water. Care must be taken to make sure that defrost water does not drain directly onto walks or driveways where it can freeze and cause personal injury.

3.1.3 Refrigerant Lines

The refrigerant lines must be ACR-grade copper tubing sized according to the installation instructions. Using the manufacturer's accessory line set is a common approach. In the heating mode, remember that the vapor line becomes the hot-gas discharge line. Pressure pulsations from it can be transmitted to the structure. Isolate the lines from structural members, such as floor joists and wall studs, to reduce noise transmission. Follow the manufacturer's installation instructions.

It is always a good idea to install a filter-drier in the liquid line when installing any split system. If the drier is installed in the refrigerant lines outside the unit, it must be a *bi-flow drier* that allows flow in both directions (*Figure 44*).

On Site

Snow Stands

Heat pump manufacturers offer accessory snow stands engineered to be used with their products. Purchasing an accessory snow stand saves installation time and is probably less costly than the time and materials needed to field-fabricate a snow stand.

Snow "legs" are also available. Snow legs are often made of plastic and come in a set of four. They attach to the bottom of the heat pump and the pad beneath with adhesive tape. Typically, the legs are no more than 8" (~20 cm) tall, so they may not be tall enough for areas with heavy snow fall.

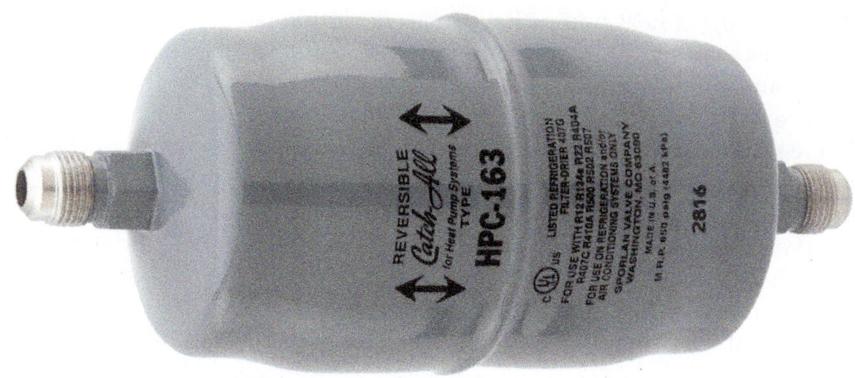

Figure 44 Bi-flow filter-drier.
Source: Courtesy of Sporlan Division—Parker Hannifin Corporation

Ensure that the selected filter-drier is compatible with the refrigerant being used. Additional refrigerant may need to be added to compensate for the internal volume of the filter-drier.

NOTE

Some heat pumps have a factory installed filter-drier inside the unit. Depending on its location, it may not be a bi-flow model. Avoid installing more than one filter-drier in the liquid line. For system cleanup purposes, a second filter-drier may be installed in the suction line. After the system cleanup is complete, the suction line filter-drier should be removed.

3.1.4 Indoor Section Placement

The indoor section of a split system may be placed in a basement, utility room, attic, closet, or attached garage, as shown in *Figure 37*. The indoor location should be selected to allow efficient ductwork design. Wire and conduit lengths from the main service panel to the electric heaters must also be considered.

The supply and return ducts should have flexible connections at the indoor unit. Vibration isolators or sound-dampening pads may also be used, especially when the unit is suspended from, or rests on, structural members.

Provisions must be made for condensate drainage from the indoor unit. In some cases, the condensate must be collected and pumped to a suitable location (*Figure 45*). The condensate pump must be selected to deliver the needed volume and lifting height. If the unit is located above anything that could be damaged by water, safety switches must be installed to disable the cooling equipment if the condensate line backs up or leaks (*Figure 46*). There are many different styles of safety switches from which to choose. When actuated, these switches shut

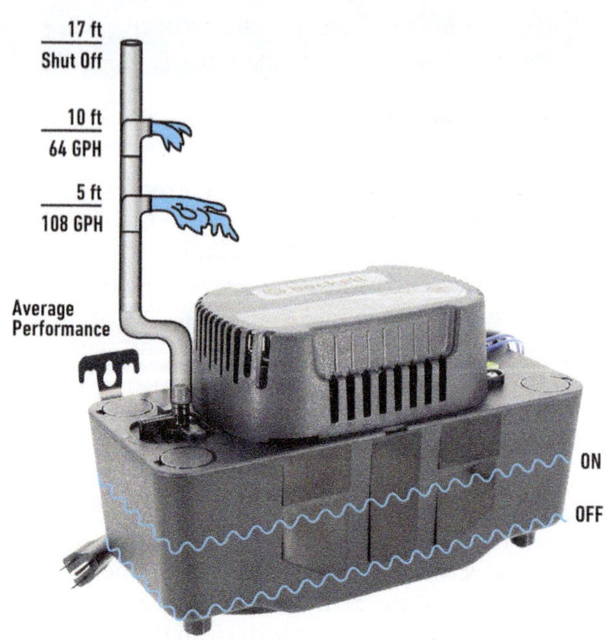

Figure 45 HVAC condensate pump.
Source: Beckett Corporation

Figure 46 Condensate safety switch.
Source: RectorSeal

down the system compressor. The switch shown in *Figure 46* is mounted to the emergency drain pan beneath the unit. A float inside actuates the switch if the pan begins to fill with water.

3.1.5 Dual-Fuel Systems

Due to the cost of fossil fuels, dual-fuel systems offer some advantages. The dual-fuel system shown in *Figure 47* features a coil added to the top of the furnace and connected to a heat pump outdoor unit by refrigerant piping. The heat pump uses the furnace to provide indoor airflow.

An outdoor thermostat or thermistor is generally used to monitor the outdoor temperature and switch between the heat pump and the furnace. Establishing the balance point and thus the changeover point is the responsibility of the installer, but the user should have access to the setting as well. Beyond that feature, the installed system will look much like any furnace and cooling system combination.

It is important to note again that the heat pump and furnace cannot be operated simultaneously unless the furnace is capable of firing at a significantly reduced capacity. With the furnace in operation and hot air flowing across the indoor coil, the head pressure would quickly rise, shutting the compressor off as the high-pressure safety switch opens.

A ducted heat pump system generally requires an indoor airflow of 400 to 450 cfm per ton for proper operation. This is the same nominal airflow rate required by an average ducted cooling system. When a heat pump is added to an existing heating system, the furnace blower must be able to develop the needed static pressure and air volume. All modern furnaces equipped with a multispeed or electronically commutated motor (ECM) can typically support a comparable air conditioning or heat pump unit. When a new installation is considered, it's best to use matching components from the same manufacturer.

The air distribution system must also be capable of handling the required air volume. The air distribution system must be carefully analyzed before adding a cooling or heat pump system to an existing heating-only system. Deficiencies in the air distribution system will make even the best heat pump perform poorly.

On Site

Heat Pumps and Warm Weather

Air-source heat pumps are even more effective at delivering heat at warmer outdoor temperatures. However, they can also be too effective for their own good. Operating an air-source heat pump in the heating mode at high outdoor temperatures can overload the compressor and/or create a high-pressure condition. Check the manufacturer's guidelines for the maximum outdoor operating temperature in the heating mode.

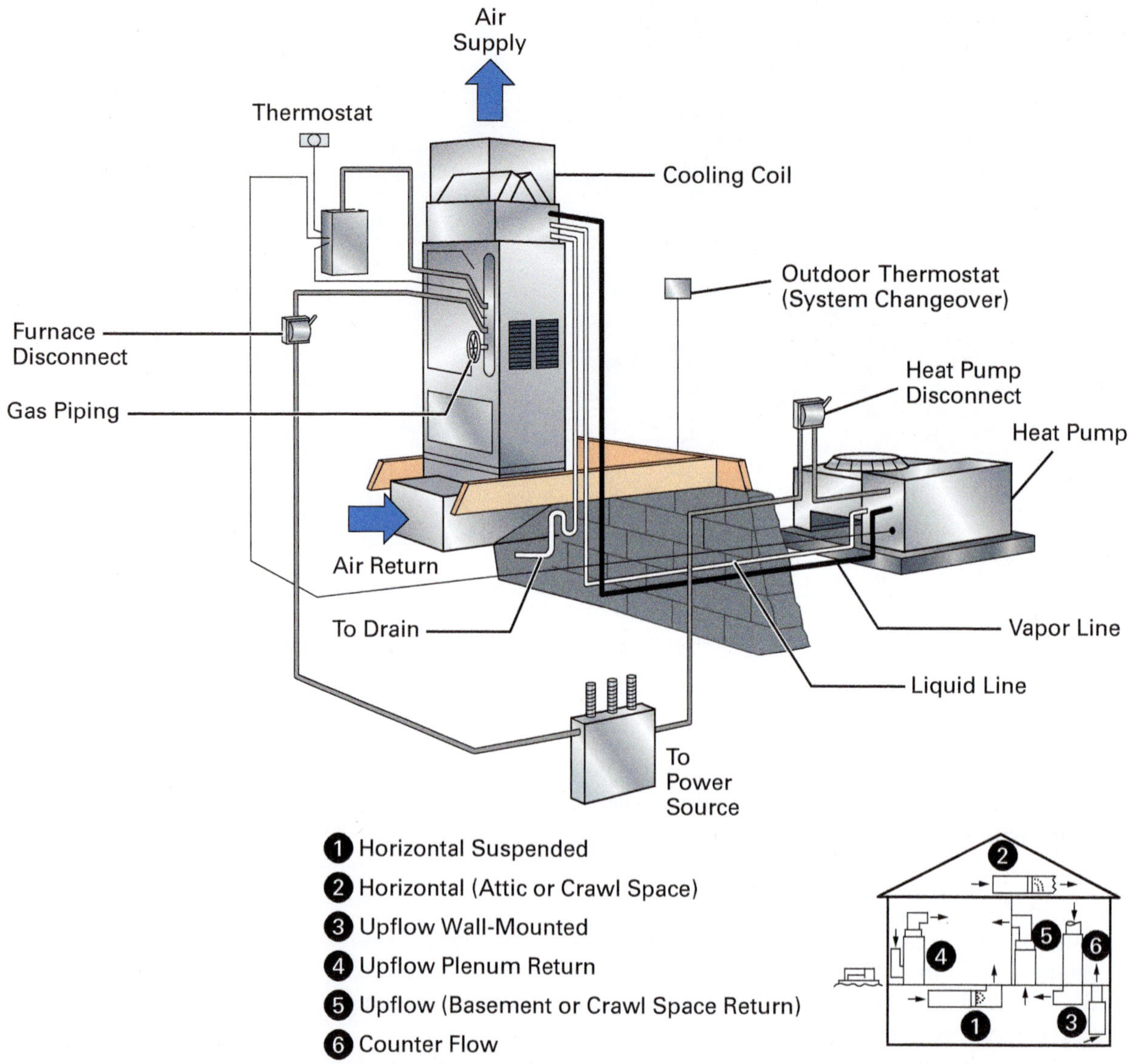

Figure 47 Dual-fuel heat pump installation.

3.2.0 Packaged Unit Installation

A packaged unit (*Figure 48*) contains all the components in a single enclosure, located outdoors. The supply and return ducts of a packaged unit must penetrate the building somewhere. This is significantly different from a split system, where only the refrigerant lines and wiring penetrate the foundation wall or roof. Both standard and dual-fuel packaged heat pumps are available today. The furnaces in dual-fuel packaged units can operate on natural gas or propane. No other secondary fuel is currently supported in packaged equipment.

Packaged heat pump systems require the same installation considerations as the outdoor units of split systems, including elevating the unit for snow and providing for defrost water drainage. Larger and thicker concrete slabs are needed to support packaged units. The slabs are typically poured in place due to their size and weight.

Packaged units are usually found at ground level in residential applications. Homes in the hot and dry southwestern states in the US may have them installed on the roof. For commercial applications, rooftop installation is very common. Ideally, rooftop-mounted packaged units are mounted on a fitted curb with the ducts penetrating the roof beneath the unit (*Figure 49*). In some cases, though, the ducts leave the unit horizontally before turning down through the roof.

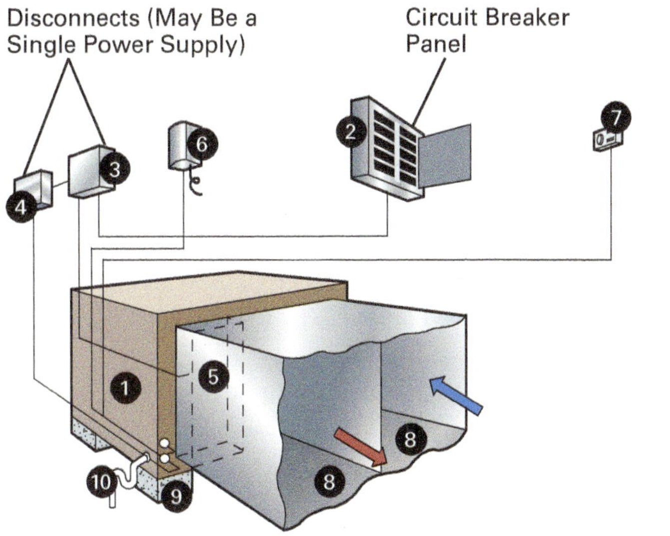

Components of the Packaged Heat Pump System

1. Packaged Heat Pump
2. Main Service Panel
3. Circuit Breaker Panel Electric Heat Disconnect
4. Heat Pump Circuit Breaker Panel Disconnect
5. Supplemental Heaters
6. Outdoor Thermostat
7. Indoor Thermostat
8. Supply and Return Ductwork
9. Mounting Pad or Supports
10. Condensate Piping

Figure 48 Packaged unit installation details.

Figure 49 Packaged unit rooftop installation.
Source: Daikin

The supply and return ductwork must be insulated if it runs through unconditioned spaces. Adequate weatherstripping and sealants must be applied where ducts penetrate the building. The duct penetrations are very vulnerable to water leaks, which ruins the insulation and causes corrosion, biological growth, and water damage to the structure. This is an area that needs extra attention and planning to keep water out.

3.2.1 Startup Procedures

Starting a new heat pump requires attention to detail, regardless of whether it is packaged or split. Always refer to the manufacturer's installation and operation manual (IOM) for the product being installed.

The following tasks are routinely required when starting a new system:

- Ensure the air filters have been installed.
- Ensure that the crankcase heater has been energized for 24 hours before startup. A shorter time is generally OK in warmer weather.
- Check all electrical connections to ensure they are secure.
- Check and record the voltage supplied to the unit.
- Ensure that the system responds to the indoor thermostat and operates in the proper mode.
- Ensure the supplemental heat energizes when called upon by the indoor thermostat.
- Check operation in the Emergency Heat mode. Generally, the heat pump will lock out and all stages of electric heat will be available.
- Initiate a defrost cycle (in test mode, if provided) and ensure it operates as expected.
- Ensure the unit is correctly charged per the manufacturer's guidance.
- Ensure the indoor and outdoor fan motors operate properly.
- Check the supply and return ducts for leakage. Return air leaks are much harder to locate, since air is being drawn in rather than blown out. Leaks must be repaired.
- Check all the safety controls, such as the high- and low-pressure safety switches, for proper operation.

The following data should be documented during a heat pump startup:

- Site address.
- Date of startup (establishes when the warranty begins).
- Make, model, and serial number of all the equipment.
- Outdoor temperature and weather conditions.
- Voltage and current to the equipment. Current to the indoor unit or packaged unit should be measured with all electric heaters energized. Voltage and current imbalance should be calculated on three-phase systems.
- Compressor voltage and current while in operation. Ensure they remain within the specifications.
- Voltage at the electric heat assembly at full load operation.
- Current consumed by each individual heating element.
- System operating pressures in both the heating and cooling mode.
- Calculate and record the superheat and subcooling.
- Duct static pressures.
- Supply and return air temperatures.
- Air temperature rise in heating, both with and without the electric heat operating.
- Air temperature drop in the cooling mode.

Again, consult the manufacturer's literature and follow the startup instructions to ensure that all required tasks are completed, and that the correct information has been recorded.

3.0.0 Section Review

1. Room temperature control using a ductless split-system heat pump is often done with a _____.
   a. line-voltage thermostat
   b. low-voltage thermostat
   c. smart phone
   d. wireless remote

2. At startup, the voltage imbalance and current imbalance should be checked on _____.
   a. three-phase systems
   b. dual-fuel systems
   c. the electric heaters
   d. water-source units only

4.0.0 Electric Heating Equipment

Objective

Describe the operation of electric heating equipment commonly used with heat pumps.

a. Explain how electric heating equipment operates.
b. Identify the major components of an electric heater.

Performance Tasks

1. Install a heat pump and complete a proper startup.
2. Test the operation of a heat pump in the heating and cooling modes and record operating data.

Electric heat is used to supplement most heat pump systems as the weather grows colder. When installed, electric heaters also provide a backup source of heat, and they temper the indoor air during the defrost cycle. While these systems are relatively simple, it is important to understand their construction and operating characteristics.

4.1.0 Electric Heat Operation

Electrical current flows through resistive heating elements, causing the elements to generate heat. The electric heat assembly (*Figure 50*) is positioned in the supply airstream of a fan coil unit (*Figure 51*) or packaged unit where the heat is transferred into the supply air.

Some control schemes disable all or a portion of the heating elements above a predetermined outdoor temperature to reduce unnecessary energy use. The electric heat also provides an alternate heat source if the heat pump fails. When the Emergency Heat mode is selected at the wall thermostat, any controls linking the heating elements to outdoor temperature are generally bypassed and the heat pump itself is locked out.

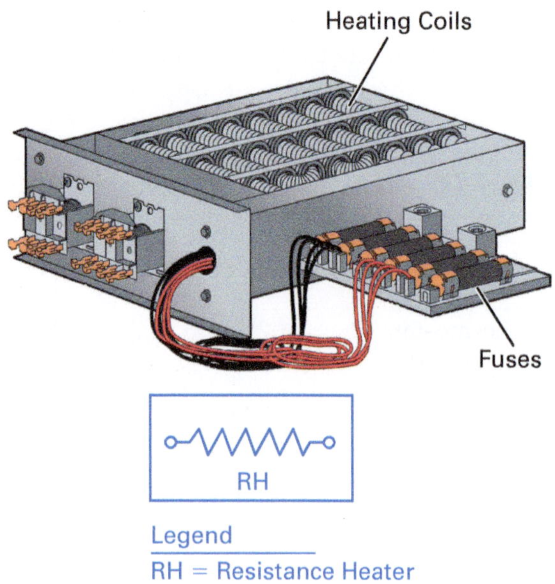

Figure 50 Typical electric heat assembly.

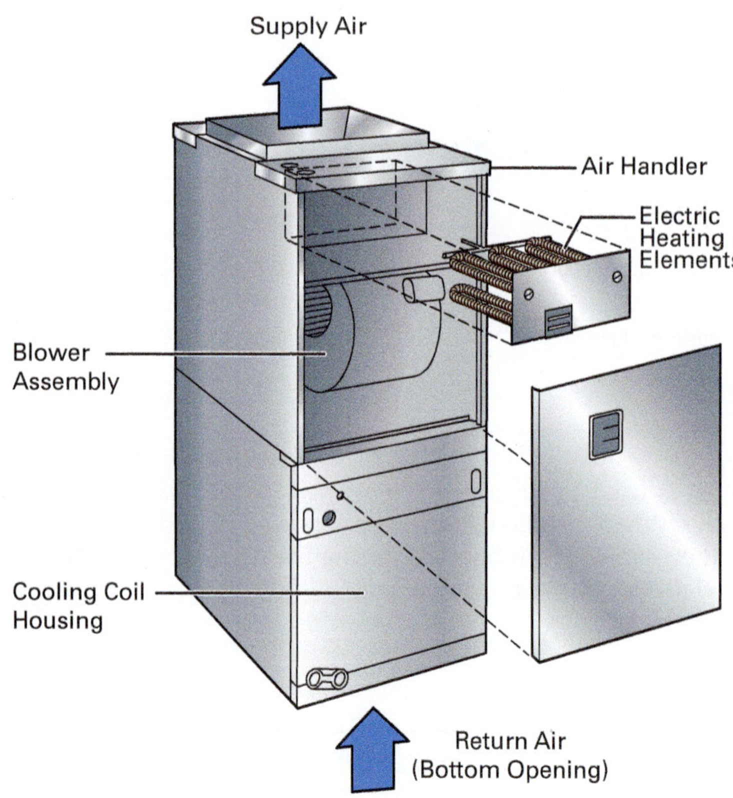

Figure 51 Fan coil unit with electric heat package.

4.1.1 Electrical Power and Performance

Fan coils with electric heaters in residential applications require a 208–240VAC, single-phase, 60Hz power supply. Power is supplied through three wires: two hot leads and a ground wire. The power source serving the fan coil unit contains a disconnect switch. The connection at the air-handling unit is typically done at a single-point terminal block, which distributes power internally to the individual heating elements, blower, and control transformer.

Packaged heat pumps used in commercial applications may require three-phase power. The electric heat packages may also be powered from a three-phase power source. Regardless, since the size of the electric heat assembly can vary, packaged heat pumps often require two separate power supplies.

WARNING!

Packaged air conditioners and heat pumps equipped with electric heaters may be served by more than one electrical disconnect switch. For example, one disconnect switch supplies power to the heat pump itself, while another powers the electric heating elements. Exercise appropriate caution when working around energized electrical circuits and ensure all related power sources are locked out before beginning repairs.

The heating capacity of electric heating elements is rated in thousands of watts or kilowatts (kW). As previously discussed, kilowatts can be translated into Btuh output. Recall that 1 watt provides 3.413 Btuh of heat output. Thus, 1kW = 3,413 Btuh. This is a good number to remember so you can quickly calculate heat output in the field.

The listed capacity of an electric heating element is usually based on a 240VAC input. A typical specification for a heating assembly might be 4.5kW @ 240VAC. As the voltage provided decreases, the kilowatt and resultant heat output also decrease in this resistive circuit. To determine the actual kW of the element for various voltages, use *Table 1* as a guide.

NOTE

The resistance of a heating element changes as it heats up. Use a correction factor of 1.065 when making calculations on a cold heating element, but only if extreme accuracy is required.

TABLE 1 Voltage Multipliers

Voltage	Multiplier
208	Multiply output at 240 volts by 0.715
220	Multiply output at 240 volts by 0.839
230	Multiply output at 240 volts by 0.917

Power formulas can also be used to determine the actual kilowatt output of an element. Use a multimeter to find the resistance (*R*) of an element in ohms (Ω) (*Figure 52*). *Figure 53* shows the power formulas. The following formula can be used to calculate the power output of an element with a known resistance (shown in the upper left quadrant of the wheel):

$$P = E^2 \div R$$

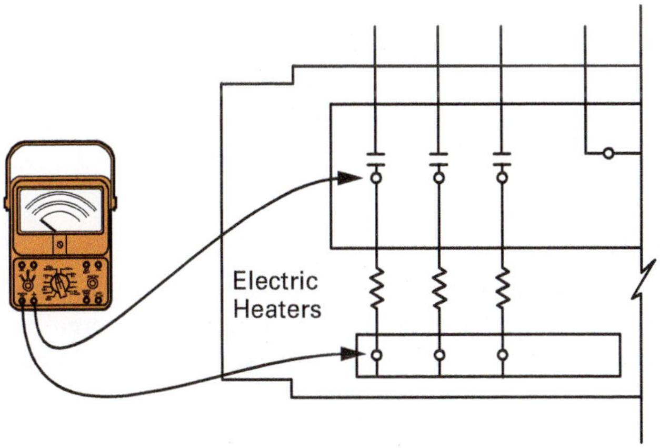

With the power off, and the elements at or near room temperature, measure the resistance of an individual heating element.

Figure 52 Measuring heating element resistance.

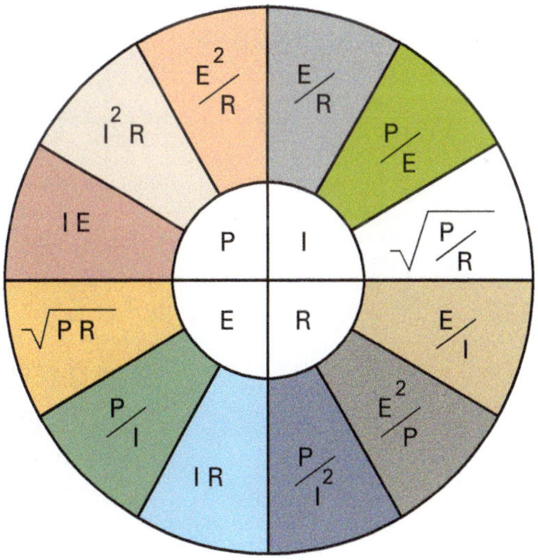

E = Voltage
I = Current
R = Resistance
P = Power

Figure 53 Power formula wheel.

Study Example

Find the kW value of a heating element with a resistance (*R*) of 12.26 ohms. Using the applicable power formula, find *P* (in watts):

$$P = E^2 \div R$$

$$P = 240 \text{ volts}^2 \div 12.26 \text{ ohms}$$

$$P = 4{,}698 \text{ watts, or about } 4.7 \text{ kW}$$

An ammeter is used to measure the current consumed by an electric heating element or assembly in operation. The ammeter is applied to the main power supply to find the current flow on the entire assembly. For individual elements, apply the ammeter to an individual power wire feeding that specific element.

4.2.0 Electric Heater Components

The major components of an electric heater are the heating elements, limit switches, and thermal protective devices. Some electric heat assemblies are equipped with a fuse block, allowing the assembly to be fused for added protection. The control transformer for the system is also commonly mounted on the assembly.

Most of today's air handling units ship without a control power source. The electric heat packages that fit the unit provide the transformer and control power. When electric heat is not required, manufacturers provide a simple transformer and relay package that is installed in the indoor unit. The package is often referred to as a *fan center* since it also provides a relay to control the indoor blower. Packaged units are equipped with control transformers by default.

Did You Know?

The Rise and Fall of Electric Heating

In the 1960s, electricity was plentiful and inexpensive. The all-electric home represented true modern living. Electric baseboard heaters were relatively inexpensive to install in comparison to forced-air systems, which required the installation of ductwork. Electric baseboard heat gave the added advantages of zoned heating and clean operation. In addition, utility companies were offering incentives to people who built homes with electric heat. In some parts of the country, homes were heated with electric furnaces, which are forced-air systems using banks of electric heating elements instead of a fuel-burning heat source.

Over the next decade, the cost of electricity gradually rose, while the cost of fossil fuels remained steady or declined. Heat pumps came into common use in many parts of the country. Eventually, the cost of heating homes with electricity alone far outweighed the cost of other methods, and the owners of all-electric homes began having trouble selling their homes.

Fossil fuels for home heating continued to gain popularity. The entrance of the heat pump as a practical means of heating is credited for having saved electric heating as a viable approach. As fossil fuels continue to rise in cost, and we develop better and cheaper ways to generate electrical power, the heat pump will become even more popular than it is today.

4.2.1 Heating Elements

The heating element provides the heat required for the conditioned space. The element is made from an alloy of nickel and chromium (Nichrome™). The element wire is twisted into a spiral and threaded through a metal holding rack, which has ceramic insulators that prevent the elements from shorting out on the frame of the rack.

A forced-air electric heating system may have one or more banks of heating elements, depending on the required capacity. For example, assume there are four heating elements. If each provides 5kW, the nominal heat capacity would be 20kW. This converts to 68,260 Btuh. At 240VAC, each 5kW element would draw about 21 amps ($I = P \div E$). Because of the high current and related energy cost, individual elements are often controlled independently based on outdoor temperature, mode of operation, or other control input.

Electric heating element assemblies (*Figure 54*) typically range in capacity from about 17,000 Btuh (5kW) with one element to about 120,000 Btuh (35kW) using seven elements. Only one element is shown in *Figure 54*. Large assemblies generally require a three-phase power supply. The equipment manufacturer supplies all the components necessary to install the prewired and preassembled heater packages in a compatible air handling unit. Wiring diagrams and complete instructions are provided with them.

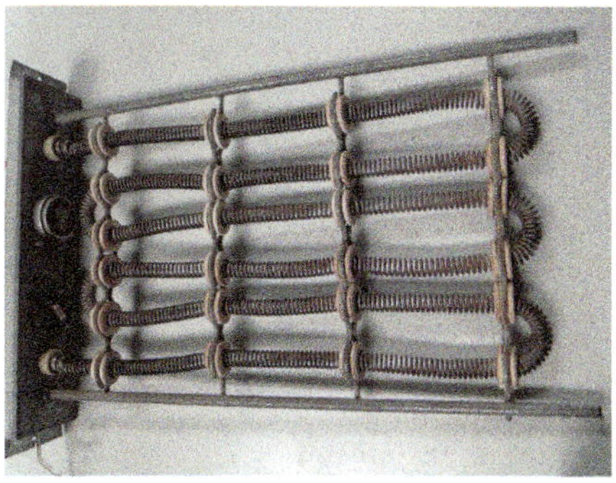

Figure 54 Electric heating elements.

4.2.2 Limit Switches

The function of the limit switch is to prevent the temperature from rising to the point that it causes damage or a fire. When the temperature exceeds the limit switch setting, it opens and de-energizes the circuit. It is important to note that limit switch setpoints can vary. When replacing a limit switch, always ensure that a switch with the correct setpoint is used.

Limit switches are wired in series with the heating elements, and have a disc-type, bimetal sensor placed in a small metal housing (*Figure 55*). The position of the switch is such that the bimetal sensor absorbs radiant heat from the element(s). If the airflow slows or stops altogether, the limit switch overheats and opens the power circuit.

Once the limit switch cools, it automatically closes or resets. This can happen repeatedly. For example, if the airflow is marginal, it may operate for a short period before the limit switch opens. It may open and close several times while the thermostat continues to call for heat. A few limit switches need to be reset manually.

Figure 55 Limit switch and a fusible link.

On Site

Limit Switches

In a correctly installed, properly maintained system, a limit switch may never open during the life of the system. If you encounter electric heating elements cycling on and off due to the limit switch, a system problem is likely. Restricted airflow and dirty air filters are two problems that could cause a limit switch to open. Each of these conditions results in increased temperature due to reduced airflow.

There is a differential separating the temperature at which the contacts open and reclose. The closing setpoint is usually about 35°F (19°C) cooler than the opening setpoint. Again, only a limit switch with the same opening setpoint as the original may be used as a replacement for a defective limit switch.

4.2.3 Fusible Links

Fusible links (*Figure 56*), sometimes called *thermal fuses*, are designed to open the line voltage circuit to the heating element if an abnormally high temperature exists. An installed example is shown in *Figure 55*. It acts as a backup to the limit switch. The fusible links generally have a higher melting point than the setpoint of the limit switch.

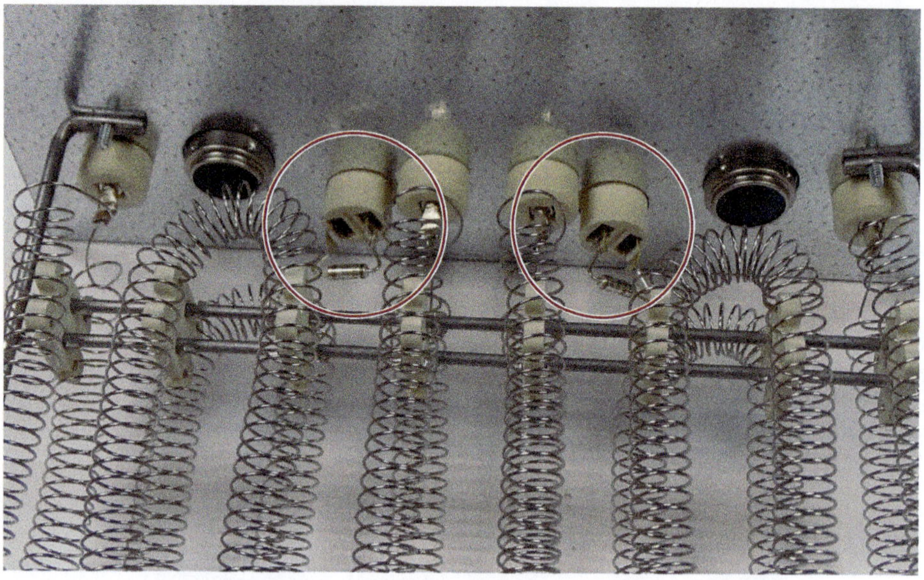

Figure 56 Fusible link for electric heating assemblies.
Source: Warren Technology

The fusible link must be replaced with one of the same rating. If both the limit switch and fusible link open, the cause should be determined and repaired before the unit is returned to service.

Note that fusible links are just like electrical fuses—once they open, they never reclose. Internally, a small piece of wire has melted. The only course of action available is replacement.

WARNING!

Never electrically bypass a failed safety device. Bypassing fusible links and similar safety devices can lead to fires, serious property damage, personal injury, and death. Multiple forms of protection against extreme temperatures are installed on electric heaters to prevent these problems. Bypassing safety devices of any type can result in both personal and corporate liability for the results.

A manual reset limit switch may sometimes be installed in place of a fusible link as backup protection. Since it must be manually reset, the manual reset device is considered to provide the same level of protection as the fusible link. However, this should only be done as a temporary measure, and the opening setpoint must be the same or lower than the fusible link. Before doing so, ensure that the condition that caused the fusible link to open has been resolved. In addition, test the auto-reset limit switch to ensure it functions as it should. Reinstall proper fusible links as soon as possible.

4.0.0 Section Review

1. The resultant heat output of an installed electric resistance heater is largely determined by the ______.
   a. number of heat exchangers installed
   b. voltage provided
   c. the frequency of the voltage
   d. volume of air moved by the blower

2. The heating elements in an electric forced-air heating system are made of ______.
   a. cast iron
   b. stainless steel
   c. a nickel-chromium alloy
   d. a copper-nickel alloy

Module 03211 Review Questions

1. True or False: There is still heat in the air that a heat pump can extract, even at absolute zero.
   a. True
   b. False

2. The COP of an electric resistance heater can never be higher than ______.
   a. 10
   b. 6
   c. 3.413
   d. 1

3. The HSPF provides the ratio of the total heating energy delivered to the energy consumed during a ______.
   a. day
   b. month
   c. period
   d. season

4. True or False: Frost cannot form on the outdoor coil of an operating heat pump unless the outdoor temperature is 32°F (0°C) or less.
   a. True
   b. False

5. Which condition occurs during the defrost cycle of a single-stage air-to-air heat pump?
   a. The indoor fan runs.
   b. The electric heaters remain off.
   c. The outdoor fan runs.
   d. The compressor cycles off.

6. True or False: When a particular heat pump is referred to as a *water-source unit*, it is generally accepted that it is circulating water through the building to provide heat.
   a. True
   b. False

7. Which heat pump type is commonly used to supply domestic hot water?
   a. Air-to-water
   b. Air-to-air
   c. Horizontal loop
   d. Water-to-air

8. Which material is generally used for the buried loop of a geothermal heat pump?
   a. Copper tubing
   b. Galvanized pipe
   c. Plastic tubing
   d. PVC pipe

9. By design, how much of a pressure difference is needed to move the piston inside a reversing valve?
   a. 50 psi or less
   b. 100 psi
   c. 150 psi
   d. 200 psi or more

10. A wind baffle is typically used to ______.
    a. keep debris out of the outdoor coil
    b. direct air across the outdoor coil
    c. prevent air stratification
    d. enhance the defrost cycle

11. Ideally, the bottom of a heat pump is elevated above grade at least ______.
    a. 6" (~15 cm)
    b. 12" (~30 cm)
    c. 18" (~46 cm)
    d. 24" (~61 cm)

12. What happens in an electric heat pump heating system when emergency heat is selected?
    a. Compression heat is activated.
    b. Compression heat is locked out.
    c. Electric heaters are locked out.
    d. Electric heating is de-energized.

13. True or False: An electric forced-air heating unit uses a sealed heat exchanger.
    a. True
    b. False

14. In what increments are electric resistive heater packages usually sized?
    a. 5kW
    b. 1,000 Btuh
    c. 500kW
    d. 3.413kW

15. The purpose of a limit switch in an electric heat fan coil is to ______.
    a. limit system capacity
    b. prevent compressor operation
    c. cycle on the blower motor
    d. prevent overheating

Answers to Odd-Numbered Module Review Questions are found in *Appendix A*.

Answers to Section Review Questions

Answer	Section	Objective
Section One		
1. a	1.1.0	1a
2. c	1.2.1	1b
Section Two		
1. c	2.1.2	2a
2. d	2.2.5	2b
3. a	2.3.1	2c
4. b	2.4.2	2d
5. b	2.5.1	2e
Section Three		
1. d	3.1.0	3a
2. a	3.2.1	3b
Section Four		
1. b	4.1.0	4a
2. c	4.2.1	4b

MODULE 03403

Indoor Air Quality

Source: Girts Ragelis/Shutterstock

Objectives

Successful completion of this module prepares you to do the following:

1. Describe how indoor air quality (IAQ) affects people.
   a. Define IAQ.
   b. Describe good IAQ and how its absence affects people.
2. Identify sources of building air contaminants and how they can be identified.
   a. Describe how building construction, materials, and equipment affect IAQ.
   b. Describe how human occupancy affects IAQ.
   c. Identify and describe external sources that affect IAQ.
   d. Explain how to conduct an IAQ survey.
   e. Describe the air sampling process.
3. Identify various ways to improve and maintain good IAQ.
   a. Explain how building design can affect IAQ.
   b. Explain how ventilation, temperature, and humidity control affect IAQ.
   c. Explain how to control chemical and microbial contaminants.
   d. Describe the effects of duct leakage and explain how to clean air distribution systems.

Performance Tasks

Under supervision, you should be able to do the following:

1. Perform a building indoor air quality evaluation.
2. Make air measurements in a building using a CO or CO_2 detector.

Overview

Indoor air quality is one of the foremost concerns for HVAC engineers, installers, and technicians. Airborne contaminants circulating in a building significantly affect the health and comfort of the occupants. Local and national government agencies require strict compliance with established IAQ codes and standards. To meet and maintain the standards, HVAC technicians must be aware of the problem and its causes to properly service and install the equipment needed to monitor and manage indoor air quality.

Digital Resources for HVACR

SCAN ME

Scan this code using the camera on your phone or mobile device to view the digital resources related to this craft.

NCCER Industry-Recognized Credentials

If you are training through an NCCER-accredited sponsor, you may be eligible for credentials from NCCER. The ID number for this module is 03403. Note that this module may have been used in other NCCER curricula and may apply to other level completions. Contact NCCER at 1.888.622.3720 or go to **www.nccer.org** for more information.

You can also show off your industry-recognized credentials online with NCCER's digital credentials. Transform your knowledge, skills, and achievements into credentials that you can share across social media platforms, send to your network, and add to your resume. For more information, visit **www.nccer.org**.

1.0.0 Indoor Air Quality and Well-Being

Performance Task

1. Perform a building indoor air quality evaluation.

Objective

Describe how indoor air quality (IAQ) affects people.

a. Define IAQ.
b. Describe good IAQ and how its absence affects people.

In the HVAC industry, an awareness of indoor air quality (IAQ) as a health, economic, and environmental issue is very important. HVAC technicians, building owners, and occupants alike need to understand IAQ and their role in managing the indoor environment.

Poor IAQ can cause both long-term and short-term health issues. Economically, poor IAQ can lead to the following problems:

- Direct medical costs for those whose health is affected
- Loss of productivity from absence due to illness
- Decreased work efficiency

The causes of poor IAQ vary and more than one factor is often involved. Consistent attention over the years to energy conservation has led to tighter building construction, HVAC systems with lower capacity, and HVAC system control schemes that minimize noticeable air movement in occupied spaces. These changes have been found to support higher levels of contaminants, although the changes are not the source of the contaminants. Energy conservation and IAQ have often been identified as opposing goals. Poor IAQ also originates from the materials used to construct, furnish, and maintain today's buildings.

Proper HVAC system maintenance is a key factor in the overall health of a building. As a technician, you are largely responsible for the health of the equipment and, in turn, the health of the occupants.

1.1.0 Indoor Environmental Quality

Indoor air quality is a primary part of a larger issue referred to as indoor environmental quality (IEQ). IEQ encompasses all elements of the indoor environment, not just the air. The US Environmental Protection Agency (EPA) estimates that many Americans spend roughly 90% of their time indoors. According to the Rocky Mountain Institute, productivity in the work environment can increase as much as 16% with the correct application of IEQ guidelines.

IEQ factors include thermal comfort and fresh ventilation air, which are IAQ issues. However, IEQ also encompasses seating ergonomics, access to natural light, wall and ceiling colors and textures, and sound levels. Many of these are completely outside the HVACR realm, but are now part of the overall IEQ landscape, of which IAQ is an integral part.

Current building philosophy recognizes that IEQ/IAQ problems are far easier and less expensive to prevent in the construction phase than they are to resolve after the building is finished and occupied. Many issues, such as noise from vibration, can be virtually impossible to cure in existing buildings and the cost can be extraordinary.

The tools used to monitor and improve IAQ include using sensors to monitor undesirable gases, larger volumes of filtered outside air to dilute or flush out those gases, and precise control over air movement and volume. These aspects of HVAC continue to take on a larger role in the design, installation, and maintenance of HVACR systems.

On Site

Indoor Air Quality in Schools

The discussion of indoor air quality generally focuses on problems that occur in homes and commercial office buildings. However, studies show that about one-half of our nation's 130,000 public and private grade schools have problems linked to indoor air quality. These problems can potentially affect 20% of the United States population (about 56 million people) who spend their days in elementary and secondary schools. The concern is that students, especially younger children, are at greater risk from poor air quality because of the hours they spend in school facilities and because they are especially susceptible to pollutants.

Going Green

What Is LEED?

LEED stands for Leadership in Energy and Environmental Design. LEED is an initiative started by the US Green Building Council (USGBC) with the goal of encouraging and accelerating global adoption of sustainable construction standards through a Green Building Rating System™. The rating system addresses six categories. Note that indoor environmental quality (IEQ) is one of the eight categories:

- Location and transportation (LT)
- Sustainable sites (SS)
- Water efficiency (WE)
- Energy and atmosphere (EA)
- Materials and resources (MR)
- Indoor environmental quality (IEQ)
- Innovation in design (ID)
- Regional priority (RP)

LEED is a voluntary program in which the building owner can choose to participate. While most technicians will not have much, if any, say in whether a building is LEED-certified or not, they will have to maintain the systems to LEED standards for the building to retain its LEED rating. In other words, the LEED certification is for the life of the building, not just for the year of commissioning.

The USGBC is not a government entity, but a private, not-for-profit organization that is attempting to motivate and move market forces to a higher level. This initiative is strong enough that ASHRAE created their *Standard 189.1, Standard for the Design of High-Performance Green Buildings*, which many localities adopt as code. Later, the International Code Council (ICC) collaborated with ASHRAE, the USGBC, the Illuminating Engineering Society (IES), and the American Institute of Architects (AIA) to develop a new version of the International Green Construction Code (IGCC), with the most recent version released in 2021.

1.2.0 Good Indoor Air Quality

IAQ is a complex subject. Some debate will always exist as to what good IAQ is and how to best achieve it. For these reasons and like many other professionals, HVACR technicians must remain current on the topic by reading trade publications and journals.

ASHRAE Standard 62.1, *Ventilation and Acceptable Indoor Air Quality* specifies minimum ventilation rates and indoor air quality standards to reduce the potential for adverse health effects. The ASHRAE standard defines acceptable IAQ as, "air in which there are no known contaminants at harmful concentrations as determined by cognizant authorities, and with which a substantial majority (80% or more) of the people exposed do not express dissatisfaction."

To maintain a balance between IAQ and energy consumption, this standard incorporates two approaches to a building's ventilation design. One approach uses calculations related to the ventilation rate, while the other approach is based directly on relevant air quality standards. These approaches provide engineers with some flexibility in their design choices.

To evaluate a building's IAQ and its mechanical systems, the acceptable levels of common contaminants must be known. An acceptable level for each possible contaminant is not always agreed upon across the various standards. For that reason, this module does not provide many specific contaminant-control values. Technicians should obtain copies of the current federal, state, and local codes and standards that affect their local area and use them as a reference.

On Site

Certificate of Occupancy

In most municipalities throughout the United States, local building officials are required to inspect all residential and commercial buildings and issue a Certificate of Occupancy before allowing anyone new to move in. The certificate is proof that a space is suitable for occupancy based on a current inspection. Buildings must typically show compliance with prevailing IAQ requirements as one of the criteria for obtaining the certificate.

1.2.1 Long-Term and Short-Term Effects of Poor IAQ

Health effects resulting from poor IAQ may not show up until years after exposure. Symptoms can develop over long and/or repeated periods of exposure. The long-term effects, which can include respiratory diseases and cancer, can be severely crippling and even fatal. Long-term health effects are associated with potential pollutants such as asbestos, **radon**, and **environmental tobacco smoke (ETS)**. While smoking in buildings is rare today, the list of potential pollutants is long.

Radon is a colorless, odorless, **chemically inert**, and radioactive gas. Radon is a concern in both residences and commercial buildings. It results from the radioactive decay of naturally occurring uranium and thorium in soil and groundwater. Some organizations use the term *thoron* for radon originating from thorium. Radon and thoron are the largest source of radiation exposure (*Figure 1*).

ETS is a combination of smoke from the burning of tobacco products and the exhaled smoke from the smoker. With the introduction of vaping devices (e-cigarettes) and the legalization of marijuana in some states, ASHRAE Standard 62.1 now includes the products of these activities in the ETS category.

Short-term and/or immediate effects of poor IAQ may appear after a single high-dose exposure or because of long-term exposure at lower levels. Symptoms can include irritation of the eyes, nose, and throat, and headaches, dizziness, and fatigue. These conditions are normally treatable, often by simply avoiding exposure to the source of pollution. When a health professional can trace symptoms of a specific illness directly to building contaminants, it is referred to as a **building-related illness (BRI)**.

Radon: A colorless, odorless, radioactive, and chemically inert gas that is formed by the natural decomposition of uranium in soil and groundwater. Radon exposure over an extended period can increase the risk of cancer.

Environmental tobacco smoke (ETS): The combination of smoke from a cigarette, cigar, or pipe and the exhaled smoke from the smoker.

Chemically inert: Describes a substance that does not readily react chemically with other substances, even under high temperatures and pressures, both of which typically accelerate chemical reactions.

Building-related illness (BRI): An illness that can be attributed to building contaminants.

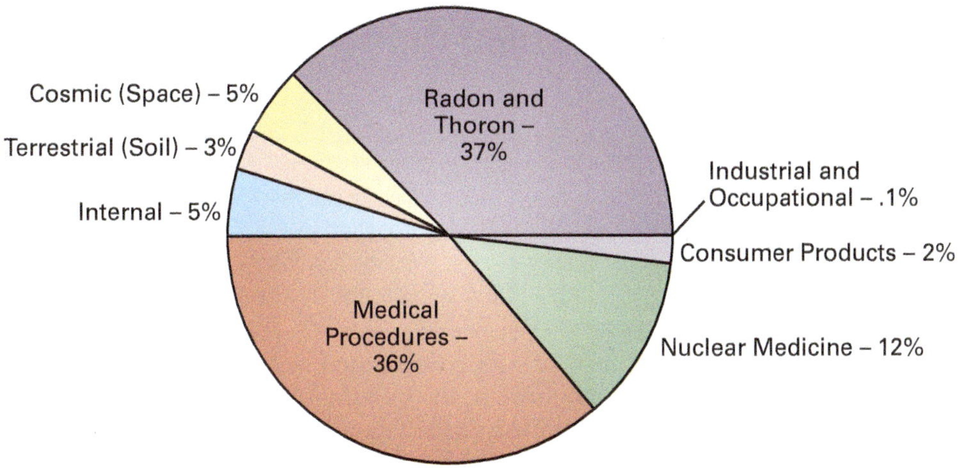

Figure 1 Common radiation sources.

There are also situations in which building occupants experience symptoms that do not fit the pattern of a particular illness. The symptoms are difficult to trace to any one source. These conditions can be temporary, but some buildings have a lengthy history of occupant health issues. IAQ professionals refer to this as **sick building syndrome (SBS)** or **new building syndrome**.

Sick building syndrome is considered to exist when more than 20% of a building's occupants complain during a two-week period of a set of symptoms, such as headaches, fatigue, nausea, and eye irritation, which are alleviated by leaving the building, and which are not known to be caused by any easily identified contaminants. Possible factors include inadequate ventilation, chemical and biological contamination, and other non-pollutant factors such as temperature, humidity, and lighting.

New building syndrome refers to similar problems that affect the occupants of new buildings.

Sick building syndrome (SBS): A physical condition that exists when more than 20% of a building's occupants complain during a two-week period of common symptoms, including headaches, fatigue, nausea, eye irritation, and throat irritation, which are alleviated by leaving the building and are not confirmed to be caused by any specific contaminant.

New building syndrome: A physical condition that is connected to indoor air quality problems in new buildings. The symptoms are generally the same as those for sick building syndrome.

NOTE

The 20% qualifier for SBS is an arbitrary limit based on ASHRAE's early work to define structural air quality. Today, many use it as the minimum limit for initiating remedial action. However, any time there are enough complaints to indicate the possibility of SBS, an investigation must be considered.

1.0.0 Section Review

1. IAQ addresses ______.
   a. seating ergonomics
   b. wall and ceiling colors
   c. thermal comfort and fresh ventilation air
   d. surface colors and textures

2. According to ASHRAE, what percentage of building occupants must express dissatisfaction with the indoor air quality to consider the air quality unacceptable?
   a. 90%
   b. 70%
   c. 50%
   d. 20%

2.0.0 Sources of Building Contaminants

Performance Tasks

1. Perform a building indoor air quality evaluation.
2. Make air measurements in a building using a CO or CO_2 detector.

Objective

Identify sources of building air contaminants and how they can be identified.

a. Describe how building construction, materials, and equipment affect IAQ.
b. Describe how human occupancy affects IAQ.
c. Identify and describe external sources that affect IAQ.
d. Explain how to conduct an IAQ survey.
e. Describe the air sampling process.

Poor IAQ can result from a building's construction or by pollutants released from sources both inside and outside of the building. Potential sources of contaminants include, but are not limited to, the following:

- Building materials and furnishings
- HVAC and other building equipment
- Cleaning compounds and pesticides
- Human occupancy
- Parking garages
- Agricultural operations
- Active construction sites

2.1.0 Indoor Contaminant Sources

Outdoor air can enter a building through infiltration, natural ventilation, and mechanical ventilation. Infiltrating air enters the building uncontrolled through openings such as joints and cracks in walls and unsealed gaps around windows and doors. Natural ventilation occurs through open windows and doors as well as through passive, purpose-built air vents, with a level of manual control. The forces that encourage air to move into a building are created by temperature and pressure differences between the indoor and outdoor environments.

Mechanical ventilation refers to the use of fans and blowers to move fresh air. The rate at which fresh outdoor air replaces indoor air is called the *exchange rate.* When the exchange rate is low, a building is more susceptible to IAQ problems. When the exchange rate is high, IAQ problems are less likely.

Older buildings often allowed more than enough outdoor air to enter by infiltration and natural ventilation to support good IAQ. Newer buildings have much tighter construction (*Figure 2*), resulting from the need to conserve energy. Today's buildings are much tighter than those constructed 50 years ago. The result can be an unhealthy indoor environment if the problem isn't addressed. The absence of proper ventilation allows contaminants to accumulate.

Figure 2 Modern tightly constructed residence.

2.1.1 Building Materials and Furnishings

Building materials and furnishings are the source of many contaminants. These materials can release **volatile organic compounds (VOC)** and may also shed airborne particulates. VOCs include a wide variety of compounds and chemicals found in paints, adhesives, sealants, furniture, carpeting, and wall coverings. VOCs vaporize readily in standard indoor conditions. The vaporization of volatile compounds from materials is known as **off-gassing**.

One common VOC is **formaldehyde**, which manufacturers regularly use in particle board, plywood, and foam insulation. Formaldehyde is a colorless, pungent substance that irritates the eyes and upper respiratory passages. In high concentrations, formaldehyde initiates asthma attacks. Higher temperatures and humidity levels indoors increase the rate at which formaldehyde off-gasses.

Volatile organic compounds (VOC): A group of compounds and chemicals found in such things as solvents, paints, and adhesives, which tend to vaporize at room temperature to become undesirable contaminants.

Off-gassing: The process through which furnishings and building materials release chemicals and other volatile organic compounds into the air.

Formaldehyde: A colorless, pungent byproduct of both synthetic and natural biological processes that causes irritation of the eyes and upper air passages; a known carcinogen.

WARNING!

Structures built before 1978 may contain lead-based paints. While not normally an airborne pollutant, any maintenance or demolition that produces dust from lead-based painted materials could create airborne particulate levels of lead that exceed OSHA standards. Before working with old painted surfaces, workers or occupational health professionals should test the surfaces for lead. When lead is present, occupied areas adjacent to the work should be protected by polyethylene barriers and positive pressure ventilation. Workers in the area must wear respiratory protection approved for use with lead.

New building materials can release a great deal of moisture and VOCs. The volume of VOCs released may remain very high in a new building unless diluted by ventilation during and after construction. New carpet and padding, often major sources of VOCs, are installed near the end of the construction process. When practical, installers should air out carpeting for 24 hours prior to installation, preferably in a warm location to accelerate the rate of VOC emission. The best practice is to delay occupancy for 72 hours following the installation whenever practical.

Going Green

Carpet and Rug Institute Green Label Testing Program

The Carpet and Rug Institute (CRI) is a trade association for the carpet, rug, and flooring industry. In the 2000s, CRI launched the Green Label Plus program for carpets, adhesives, and cushions. The program acknowledges the need to reduce VOCs in relevant products and elevates IAQ standards. The labeling lets buyers know that they are purchasing the very lowest emitting products available on the market. The current Green Label Plus program is the fourth iteration in the industry's efforts to improve and support IAQ. The program is recognized by the U.S. Environmental Protection Agency's (EPA's) *Recommendations of Specifications, Standards, and Ecolabels for Federal Purchasing.*

Source: The Carpet and Rug Institute

Fabrics and soft, porous materials act as incubators for microorganisms. Carpeting, wallpaper, acoustical tiles, and upholstered furniture in newer buildings often replace the tile floors, painted walls, hard ceilings, and metal furniture found in old buildings. However, some of those features, such as hard, cleanable floors have returned. Depending on the temperature and humidity levels in a building, furnishings can provide the ideal environment for biological development.

On Site

Flooring Installation

All flooring adhesives rely on a low moisture content in the subflooring to ensure proper adhesion and product life. Tight schedules associated with new construction often tempt contractors to apply flooring materials to damp subfloors, which can cause long-term IAQ risks such as mold growth and the decomposition of flooring components. Many construction specifications now require testing to ensure the moisture content of subfloors is acceptable prior to installing flooring materials.

Asbestos is another potential contaminant. This naturally occurring mineral was a common component of fireproofing materials, thermal insulation, floor tiles, and coverings for structural members in buildings constructed from 1930 to the mid-1970s. Because asbestos is a known carcinogen, workers must treat asbestos-containing materials with great caution.

Where asbestos is found, it is not normally a problem if it remains encapsulated and undisturbed. However, if the surface of the material is deteriorating or subject to abrasion, asbestos fibers easily become airborne. Brittle, fibrous materials that readily crumble and release airborne particles are considered **friable**.

Friable: Brittle and easily fragmented, releasing particles that can become airborne.

The EPA recommends that building owners leave undamaged asbestos materials alone if the materials are not likely to be disturbed. Qualified contractors must be employed to direct any activities that may disturb asbestos and to perform asbestos removal and cleanup tasks.

On Site

Mold Litigation and Legislation

Health hazards related to mold contamination are always a hot topic of discussion and government regulation. This is the result of a growing number of costly lawsuits pertaining to toxic mold. For example, a Texas homeowner won a $32-million judgment against her insurance company for failing to act properly and promptly to remove toxic mold contamination from her home. She claimed that their inaction caused her family to suffer severe health problems from exposure.

In 2005, Congress passed the Toxic Mold Safety and Protection Act. This law primarily addresses the prevention, inspection, and abatement of toxic mold in public housing and in dwellings with mortgages insured by federal programs. The Act requires the EPA to issue guidelines that define hazardous threshold limits for toxic mold. The Act also mandates that the EPA set mitigation standards and license those who inspect and clean up mold sites.

Source: Smit/Shutterstock

Some types of office equipment, such as photocopiers, can generate pollutants. These devices are sources of irritants such as **ozone**. While ozone is a naturally occurring gas in the atmosphere, when ozone exceeds acceptable levels at ground level, it becomes a pollutant.

Ozone: An unstable form of oxygen with a sharp, pungent odor like chlorine that forms naturally in the presence of electric discharges. Ozone irritates the respiratory tract of animals, damages plant tissues, and represents a hazardous pollutant at ground level while serving in the upper atmosphere to protect us from harmful ultraviolet rays.

Standards assigned by different organizations regarding ozone exposure include the following:

- OSHA requires that people not be exposed to an average concentration exceeding 0.1 part per million (ppm) over the course of an eight-hour shift.

- NIOSH recommends an absolute exposure limit of 0.1 ppm; 5 ppm is immediately dangerous to life or health (IDLH).
- The EPA recommends a maximum average concentration of 0.08 ppm over an eight-hour period.

2.1.2 HVAC and Other Building Equipment

HVAC and refrigeration equipment, combustion (fuel-burning) equipment, and office equipment all have the potential to contribute indoor pollutants. The widespread use of large air distribution systems heightens the concern. Expansive systems are needed to condition large structures, but they can distribute as well as contribute pollutants when poorly maintained. Fiberglass ductboard and duct liner can trap moisture and contaminants that support biological growth.

Cooling coils, condensate collection pans, humidifiers and dehumidifiers, cooling towers, and evaporative coolers and condensers are all places that can become breeding grounds for **biological contaminants**. These contaminants include bacteria, fungi, viruses, algae, insect parts, pollen, and dust. All these can be effectively distributed throughout a home or business through the air distribution system.

Biological contaminants: Living contaminants such as bacteria, fungi, viruses, and algae. Also referred to as *microbial contaminants*.

Legionnaires' Disease and Pontiac Fever

Legionnaires' disease and **Pontiac fever** have been attributed to poorly maintained HVAC systems that supported the incubation and distribution of disease-causing microorganisms. Legionnaires' disease is a serious pneumonia-like lung infection caused by bacteria called *Legionella* (*Figure 3*) Pontiac fever is a similar but milder form of Legionnaires' disease.

Pontiac fever: A mild form of Legionnaires' disease.

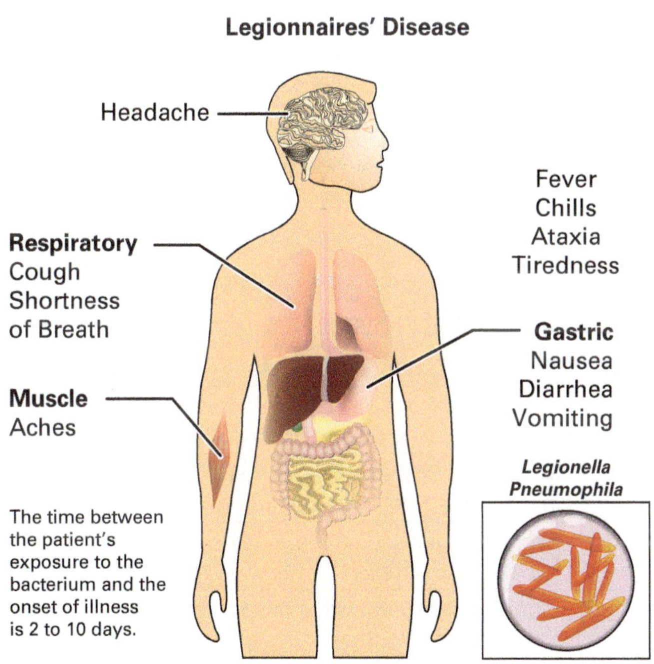

Figure 3 Legionnaires' disease.
Source: ttsz/Getty Images

Legionella bacteria is often found in fresh water but rarely causes illness due to their low concentration. Where water remains consistently warm, though, such as in hot tubs and cooling towers, Legionella can quickly grow out of control. The ideal temperature range for growth is 90°F to 108°F (32°C to 42°C). Outside of this range, the bacterium either lies dormant or dies. When small droplets of water containing the bacteria get into the air, the bacteria are easily inhaled. Cooling towers are designed to expose the water to as much air as possible and thus produce the water droplets that spread the bacteria.

The bacterium was named in 1976, when people who attended an American Legion convention in Philadelphia became ill with similar symptoms. Between 8,000 and 18,000 cases of Legionnaires' disease are diagnosed each year in the United States alone. The disease proves to be deadly in about 15% of the cases. Fortunately, Legionnaires' disease is not generally contagious, but research continues. Proper cooling tower treatment programs are essential to discourage *Legionella* growth as well as other biological activity.

Combustion Byproducts

IAQ problems can also arise in a building due to the byproducts of combustion. Many combustion byproducts can be hazardous and compound pre-existing health problems. Furnaces, boilers, space heaters, wood stoves, gas stoves, and fireplaces can produce pollutants such as carbon monoxide (CO), oxides of nitrogen (NO_X), and sulfur dioxide (SO_2), as well as airborne particles.

CO is a toxic gas that results from incomplete combustion, but it can also be generated during complete combustion. The gas is colorless, tasteless, and nonirritating. High levels of CO are extremely harmful. CO can slowly and quietly build up in the bloodstream where it binds with hemoglobin and displaces oxygen in the blood. Eventually the blood carries so little oxygen that it can no longer support life.

Death from CO poisoning can happen suddenly as a result. Victims are often overcome and helpless before they realize there is a problem. Signs of CO poisoning include headaches, dizziness, fatigue, and nausea. Victims often think they have the flu, COVID, or a common cold because the earliest symptoms are similar. *Table 1* shows symptoms of CO poisoning at different exposure concentrations.

Nitric oxide (NO) and nitrogen dioxide (NO_2) are two common airborne gases, together symbolized as NO_X. All combustion processes can produce NO_X. Oxides of nitrogen also form acids in the earth's lower atmosphere, where they contribute to acid rain. NO_X and hydrocarbons both react with sunlight to produce smog.

NO and NO_2 displace oxygen in the blood. NO_2 can irritate the skin and the mucous membranes in the eyes, nose, and throat. High levels of NO_2 may result in burning and pain in the chest, coughing, and/or shortness of breath. Nitrous oxides released from incomplete combustion can remain in the lungs

TABLE 1 CO Levels and Related Symptoms

Concentration	Exposure Times and Effects
9 ppm	Maximum allowable concentration for short-term exposure in the living area according to ASHRAE
35 ppm	Maximum allowable concentration for continuous exposure in any 8-hour period according to federal law
200 ppm	Slight headaches, fatigue, dizziness, nausea after two to three hours
400 ppm	Frontal headaches within one to two hours, life-threatening after three hours; also, the maximum ppm in flue gas (on a free-air basis) per the EPA
800 ppm	Dizziness, nausea, and convulsions within 45 minutes; unconsciousness within two hours; death within two to three hours
1,600 ppm	Headache, dizziness, and nausea within 20 minutes; death within one to two hours
3,200 ppm	Headache, dizziness, and nausea within 5 to 10 minutes; death within one hour
6,400 ppm	Headache, dizziness, and nausea within one to two minutes; death within 30 minutes
12,800 ppm	Death within one to three minutes

and irritate or damage lung tissue. Inhaling nitrous oxides may also lead to cancer.

Poorly installed and leaking chimneys and flues, as well as furnaces with cracked heat exchangers, are sources of these pollutants. Negative pressures in tight buildings can cause back-drafting of a combustion appliance and the distribution of combustion byproducts throughout the building. In a warm-air furnace, a cracked heat exchanger can cause a consistent buildup of toxic gases in the building. The furnace blower then distributes the gases throughout the conditioned space, causing sickness or death from an unseen source.

Furnace heat exchangers should be inspected for cracks or leaks that could allow combustion byproducts to mix with indoor air. Be aware that a visual inspection may not detect tiny cracks that expand as the furnace heats up. A better inspection method is to test and compare combustion gas readings taken before and after the furnace blower has started. A thorough heat exchanger inspection should be done if there is a change of oxygen (O_2) concentration in the flue gases greater than 0.5%, or a change in the CO level greater than 25 ppm.

Blockages in a chimney or flue vent can cause inefficient combustion and produce dangerous levels of gases. Birds' nests, deterioration, and soot buildup are common sources of blockage.

Routine and proper maintenance of combustion equipment and chimneys is one of the best ways to prevent exposure to CO and NO_X. At a minimum, technicians should inspect, clean, and adjust the equipment annually. Any needed repairs related to the combustion process should be made promptly.

Did You Know?

Carbon Monoxide Detectors

Today, most states require CO detector installation in homes, although some states limit the requirement for CO detectors to homes containing fossil-fuel heating systems or require their installation only when a home is sold.

States are also enacting legislation requiring CO monitoring in schools, hotels, and motels. The National Conference of State Legislatures (NCSL) website, *http://www.ncsl.org/*, includes a current list of all states having CO monitoring laws and regulations.

2.1.3 Cleaning Compounds and Pesticides

Cleaning compounds and pesticides are also sources of building contaminants. Both can contain compounds whose effects range from minor irritation to cancer. Long-term damage to the liver and the central nervous system is also possible.

The safest way to avoid exposure to these materials is to read the product Safety Data Sheet (SDS) and product label, both of which clearly describe any health hazards related to the product. Products should be used only as directed.

Combinations of chemicals and pesticides used or stored in a building may cause some occupants to suffer from what is known as **multiple-chemical sensitivity (MCS)**. MCS is a condition affecting individuals who appear to be vulnerable to a collection of chemicals. However, some debate exists as to whether MCS is a true illness or simply a psychological response to the perceived presence of certain chemicals. Either way, as an HVACR professional, it is important to keep your eyes open for such products that the occupants have not yet recognized as problematic.

Multiple-chemical sensitivity (MCS): A condition found in some individuals who appear to be vulnerable to exposure to certain chemicals and/or combinations of chemicals. Currently, the existence of MCS as a physical condition remains debatable.

2.2.0 Human Occupancy

People spend about 90% of their time indoors. As they breathe, oxygen is consumed, and carbon dioxide (CO_2) is exhaled. A local CO_2 concentration of 400 ppm is generally considered excellent. Air with a CO_2 concentration of 1,000 ppm is minimally acceptable air. In between, there are increments of *good air* and *adequate air.*

Experience has shown that CO_2 concentration levels above 1,000 ppm have a negative impact on the occupants. At concentrations around 1,200 ppm, people tend to get drowsy, have headaches, and/or function at reduced activity levels. For this reason, the monitoring and control of CO_2 concentrations in high-occupancy areas of a building is necessary.

While CO_2 concentrations are widely used as an indicator of IAQ, low CO_2 levels do not mean there aren't other contaminants present. Humans are a source of contamination, emitting bioaerosols from the digestive process, perspiration, and exhalation (*Figure* 4). Our skin sheds, creating airborne particles. When we sneeze or cough, bacteria and viruses are sent flying into the air. Activities such as cooking also add to poor IAQ. There are many possible contaminants and sources.

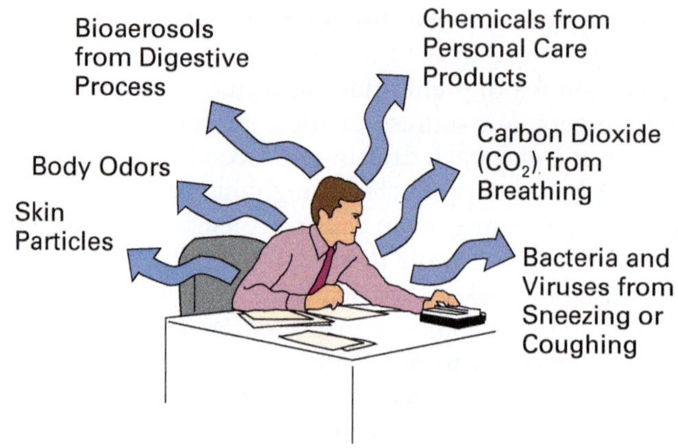

Figure 4 Human sources of air contamination.

On Site

Indoor and Outdoor CO_2 Levels

To dilute CO_2 levels and other indoor air contaminants, mechanical ventilation systems are used to bring in adequate amounts of outdoor air. This approach works well—assuming the outdoor air is less contaminated than the indoor air.

Despite the continuing debate about CO_2 concentrations and control, the industry generally accepts a comparison of the indoor CO_2 level to the local outdoor CO_2 level. Normally, when the CO_2 concentration increases outdoors, there is an associated rise in the indoor CO_2 level. On particularly "smoggy" days, it is possible to bring in air containing more CO_2 than the building air already contains.

For this reason, it is good practice to monitor both the indoor and outdoor CO_2 concentrations to ensure that the outdoor air brought into a building has a lower CO_2 concentration than the indoor air.

2.3.0 External Contaminant Sources

Contaminants from outside a building, especially in urban areas, are a significant cause of poor indoor air quality. Some common sources are shown in *Figure 5*. Contaminants may come from aboveground urban or industrial air

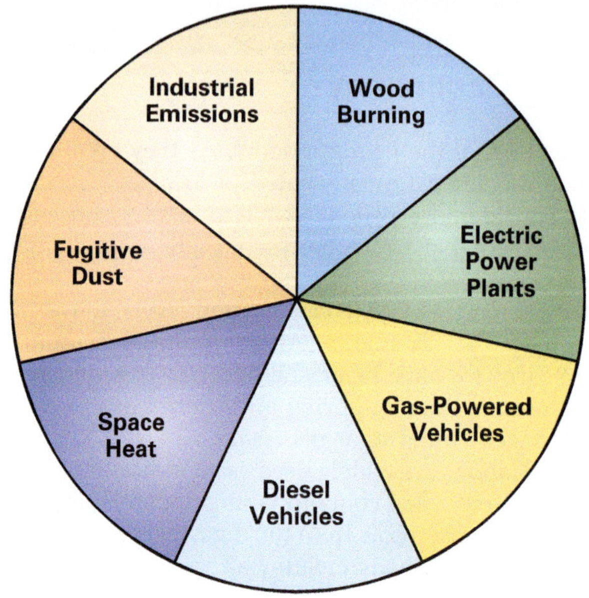

Figure 5 Sources of outdoor air pollutants.

pollution sources or from belowground sources that include pesticides, fertilizers, and radioactive soil gases such as radon.

Outdoor contaminant sources include the following:

- Exhaust from vehicles and loading docks
- Airborne discharges from manufacturing plants
- Urban smog
- Radon
- Pesticides and fertilizers
- Standing water that promotes microbial growth
- Neighboring building construction or demolition
- Garbage collection bins and dumpsters

These contaminants find their way into a building in a variety of ways. Ventilation air intake ducts may be located too close to chimneys, loading docks, or kitchen and bathroom exhaust systems. Perhaps the ducts open into an alley where standing water and dumpsters might be found.

Most of these contaminants cannot be eliminated. Some contaminants can also develop long after a building has been constructed as the surrounding environment changes. However, contaminants can be reduced in volume through the thoughtful location of outside-air intakes in the building structure. Consistent attention to building pressure is also important. Negative pressure in a building encourages unfiltered outside air to enter through every crack and crevice.

2.3.1 Radon Contamination and Testing

Radon and thoron account for 37% of the radiation to which we are all exposed. The most common source of radon is uranium in soil and rocks, but it is also found in well water. As naturally occurring uranium and thorium atoms decay, radon is released. Since radon is inert, it does not react with other components in the atmosphere once released.

Radon enters buildings through cracks in concrete, wall and floor joints, concrete block walls, or openings such as those for sewer pipes and sump pumps (*Figure 6*). Radon from well water is released in a home when water is used for cooking, bathing, and other activities.

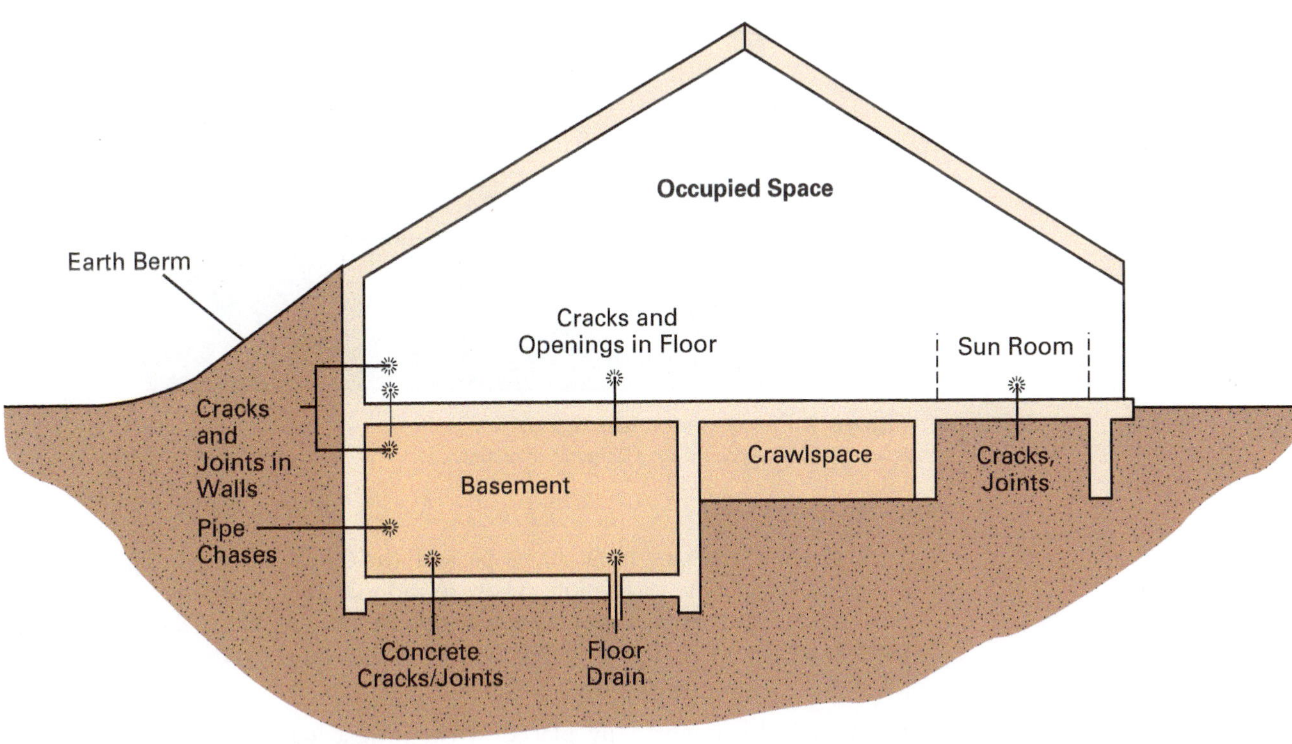

Figure 6 Passage of radon gas into a home.

The level of radon present varies depending on a building's construction features and location. Generally, fewer cracks and openings in a building's foundation reduce radon entry. However, sealing various parts of a building foundation is not a permanent or complete answer. Some openings in the building envelope and foundation may not be accessible, and new paths can develop over time.

Radon produces radioactive decay products called *daughters*. Daughters emit elevated levels of alpha radiation. The daughters are chemically active metals that attach to smoke and dust particles in the air. When inhaled, the particles can lodge in the respiratory system where the lungs are exposed to radiation.

Half-life: The time required for half of radioactive atoms present to experience radioactive decay.

Scintillation detectors: Radiation detectors containing a substance that emits light when it encounters a fast-moving nuclear particle (proton, neutron, or alpha). The sparkling effect caused by these emissions is called *scintillation*. The device includes features that amplify the light flashes and display the rate of particle detection.

Fortunately, each of the radon daughters has a relatively short **half-life**. That means, after being deposited in the lungs, radon daughters will decay in an hour or less. *Thoron*, a similar radioactive gas, is rarely discussed because its half-life is even shorter. It is unlikely that radon-related health issues will occur due to short-term exposure. However, major health organizations agree that extended exposure to radon increases the risk of lung cancer.

Testing for radon is done using electronic radon monitors and/or passive test kits. Inexpensive passive kits (*Figure 7*) are available for consumer use in residences. These passive kits do not require a power source to operate. Charcoal canister and charcoal liquid **scintillation detectors** are the most used passive detectors for short-term testing.

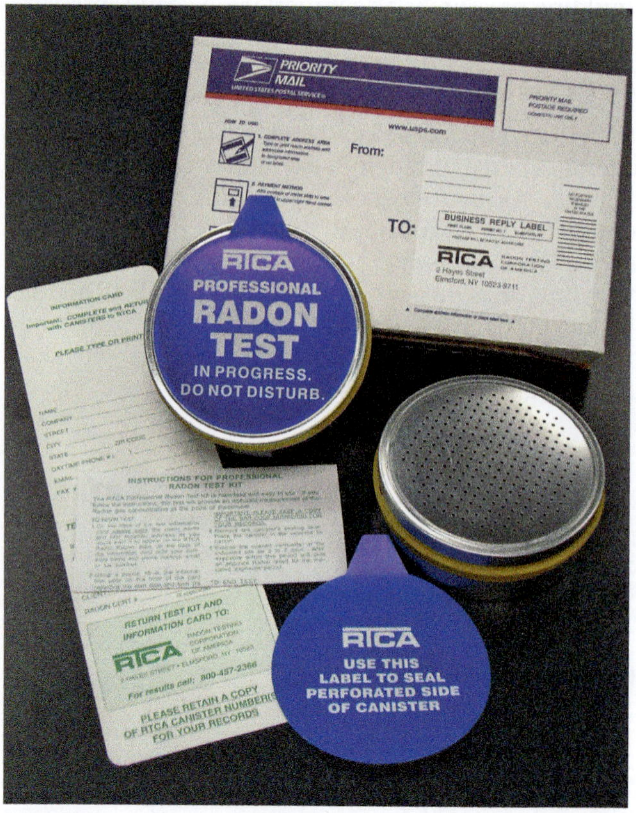

Figure 7 Passive radon test kit.

Source: RTCA®, Radon Testing Corporation of America

Approved test kits must have passed the EPA's testing program or be state-certified. Some of the kits measure radon levels over two to three days; others measure it over one to three months. Follow the manufacturer's instructions carefully to ensure accurate results. The typical procedures for using a passive test kit are as follows:

- The detector is normally placed at the lowest habitable level of the home where the highest concentration of radon is likely to be present. Place it in a location where it will not be disturbed, away from drafts, high temperatures, high humidity, and exterior walls. Placing it at a height nearest the breathing zone is usually recommended.

- If conducting a test for only two or three days, close all windows and outside doors at least 12 hours before beginning the test. Keep the windows and doors closed as much as possible during the entire test period.
- Heating and air conditioning system blowers may be operated, but do not operate heat recovery units or other equipment that bring in outside air during the test.
- Leave the test kit in place for the specified time.
- When the testing is complete, reseal the package and promptly send it for analysis to the lab specified on the package. The results will be shared with the sender soon thereafter.

Long-term testing is performed in much the same way as short-term testing, but it is often extended to 90 days and perhaps beyond. Active detection equipment is generally used for long-term testing.

Active testing devices, which require a power source to operate, are used by professional testers. They are generally considered to be more reliable than passive devices. Continuous radon monitors, such as the one shown in *Figure 8*, are active sampling devices that continuously measure and record radon decay.

Figure 8 Continuous radon monitor.
Source: DURRIDGE Company, Inc.

Active testing devices may be used for both short- and long-term testing. The detector provides a measurement of the radon level in picocuries per liter (pCi/L). The monitor shown in *Figure 8* provides a printout of the collected data. Most active devices also store the data for later analysis.

Some debate continues regarding the acceptable radon level. The Radon Abatement Act of 1988 established a national goal of achieving indoor radon levels no greater than outdoor levels. The average outdoor level is about 0.4 pCi/L. The EPA suggests an indoor level should never exceed 4 pCi/L, but recommends mitigation procedures begin if the level is over 2 pCi/L. To date, the argument remains open as to whether the estimated health benefits justify the costs associated with achieving levels below 2 pCi/L.

The most effective measures for reducing radon are the ones that limit gas entry from the soil into the building. One common method used is *sub-slab depressurization* (*Figure 9*). This method uses an exhaust system to create a negative pressure below the floor. This draws radon gas out of the area and into the atmosphere, preventing entry indoors where a higher pressure exists.

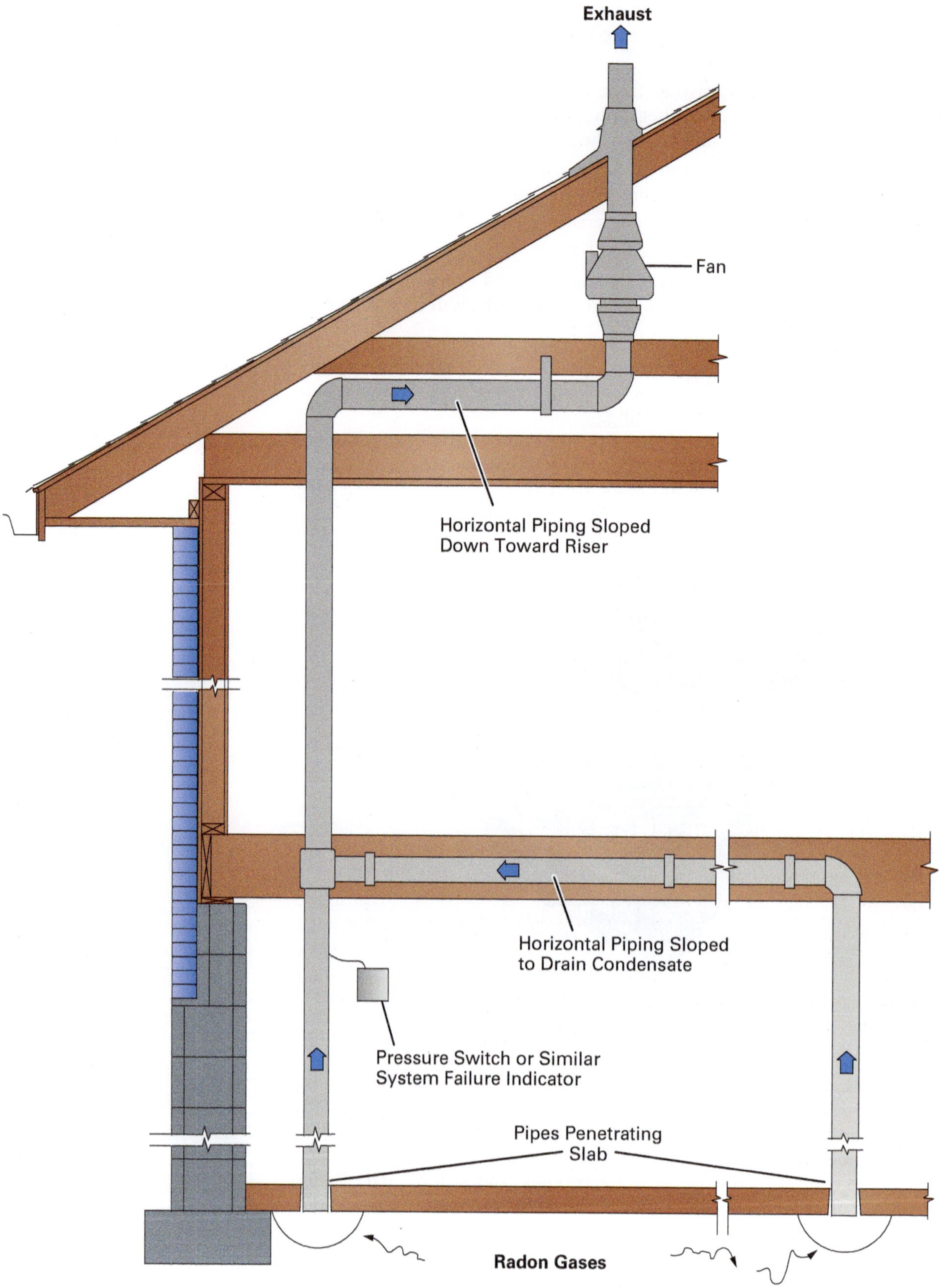

Figure 9 Typical sub-slab depressurization (SSD) system.

Positive pressurization is a similar method used with some success. A positive pressure can be created in a basement, for example, where the walls are also in contact with the soil. This discourages radon gas from entering the building through the floor and walls.

2.4.0 Elements of a Building IAQ Survey

Properly evaluating a building and its systems for IAQ problems is not easy. The problems can be multifaceted because of complex building designs. Some test results may be inconclusive. Ideally, the IAQ inspection team should include a qualified engineer, an industrial hygienist, and an HVAC technician to help them access the equipment and systems.

To successfully quantify and determine the source of IAQ problems, an organized approach is required. The elements of this approach should include the following steps:

- Problem description and interviews
- Site visit and building walk-through
- Building HVAC and ventilation system inspection
- Air sampling and testing for specific contaminants
- Interpretation of test results and proposed corrective actions

2.4.1 Problem Description

To determine the scope of an IAQ problem, the first step is to interview the building staff and the individual(s) who requested an investigation. The occupants' symptoms, as well as the location and duration of those symptoms, can be obtained via interviews or the use of questionnaires. *Appendix 03403A* provides an example of an EPA form that can be used to conduct occupant interviews.

This information helps define and categorize the complaints. The information may also help determine if the problem is localized in the building and identify any other relevant circumstances such as weather, time of day, day of week, and building occupancy levels.

2.4.2 Site Visit and Building Walk-Through

A preliminary evaluation of a building's overall condition and operating systems is the goal of a walk-through. The current set of building plans and specifications is very useful during the visit.

To ensure that potential contamination sources or specific components of the building are not missed during the visit, use a checklist to document the results. This can be a paper or digital form. Also, mobile applications that can aid in documenting an evaluation are available. The EPA publishes a guide entitled *Indoor Air Quality Building Education and Assessment Model (I-BEAM)* that is available on their website.

The I-BEAM materials contain training modules, checklists, references, and links to useful websites. Because this is a US government resource, it can be reproduced without permission. *Appendix 03403B* is a reproduction of the EPA's *HVAC Checklist – Short Form* building IAQ evaluation, also available at the EPA's website.

On Site

IAQ Apps and Monitors

Indoor air quality can be observed using a mobile application and compatible IAQ monitor. An air quality sensor monitors conditions such as temperature, humidity, particle count, VOCs, CO_2, and more. The characteristics detected depend on the specific device and model. A compatible application on a mobile device displays the information for easy viewing.

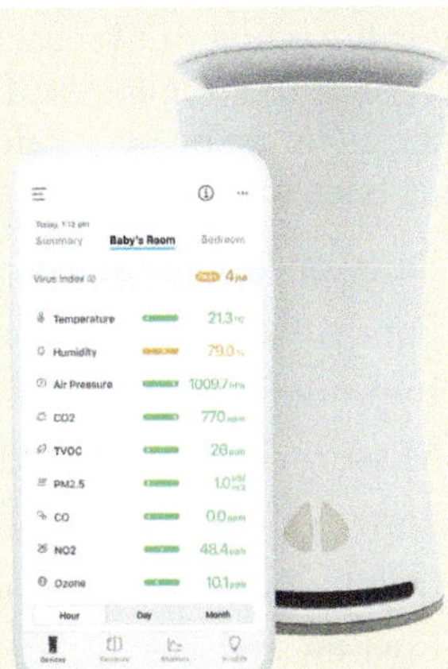

Source: uHoo

The following is a list of some characteristics that should be on the inspector's mind:

- Quality of building construction (loose or tight)
- Adequate ventilation rates
- Rooms properly configured for the intended use
- Adequate separation between activities to control temperature, air movement, odors, liquids, noise, light, or vibration
- Recent addition of partitions or fire walls
- Added windows or doors
- Asbestos, formaldehyde, and/or other harmful substances being released from building materials such as furnishings and carpeting
- New foam insulation added to walls
- Recent use of caulks or sealants
- New openings for ducts, and pipes that allow the transfer of particles, vapors, and odors between occupied spaces

2.4.3 Building HVAC System Inspection

After the structure has been evaluated, the HVAC systems must be checked for potential sources of contamination. Ensure that any building zoning scheme is functional. Pay attention to the number of occupants in each zone and ensure that an adequate volume of outside air is being introduced. If the building uses a variable air volume (VAV) air distribution system, make sure that its layout and control arrangement allow for adequate ventilation. All systems should supply at least the minimum quantity of outside air required by the current edition of ASHRAE Standard 62.1 and local codes.

Even if a building has an adequate ventilation system, the system may not supply enough air to dilute some pollutants. This can happen if the system is shut down too often or for too long, such as during evenings and weekends. Other causes are poor air distribution and poor air mixing. Occupant intervention, such as homemade cardboard diffusers attached to ceiling vents to divert air, boxes or supplies piled on top of floor registers, and office partitions that block airflow also interfere with ventilation.

WARNING!

An HVAC system inspection may expose the inspector to contaminants such as mold, viruses, and other airborne contaminants. Be sure to wear company-prescribed PPE, including gloves and respiratory protection, when inspecting HVAC systems.

Start by examining the ventilation system at the outside air intake. Air intake grilles are sometimes placed too close to building exhausts or sources of pollution, such as cooling towers, loading docks, and garbage bins. Pollutants often find their way to these areas long after construction is complete as the surrounding environment changes. Make sure that air exhausted from the building is not being drawn back into the building.

Ensure that the HVAC filters fit properly and aren't fouled. The replacement schedule must be adequate and consistent. Also check for the following conditions once you make your way to the equipment:

- Dirty or contaminated humidifier reservoirs
- Poorly draining condensate pans
- Improperly pitched drain pans and lines
- Missing or improperly designed traps
- Torn insulation
- Damp internal insulation
- Rusting surfaces
- Mold or other biological growth on internal surfaces
- Improperly adjusted dampers, actuators, and linkages

- Inoperative fans and blowers
- Chemicals stored inside or adjacent to the air handling equipment

2.5.0 Air Sampling and Testing for Specific Contaminants

Air testing can be time consuming and expensive, especially when looking for specific sources of contamination. Air samples are often taken over an extended period (at least a week) to get enough samples to yield reliable results.

2.5.1 Gas Detectors and Analyzers

Gas detectors and analyzers are used to detect the presence of gaseous contaminants. These devices can be either mechanical or electronic. Electronic instruments simplify testing and are more generally accurate. Electronic instruments can also perform automatic sampling and calculations. Some models can produce hard-copy reports of the date, time, and test results. Others can be connected to computers so that the data can be transferred and stored.

Electronic detectors and analyzers are available for use as stationary wall-mounted units or portable test instruments. Some units detect and/or measure only one specific kind of gas. Other units can detect and measure several gases. *Figure 10* shows some common gas detectors and analyzers.

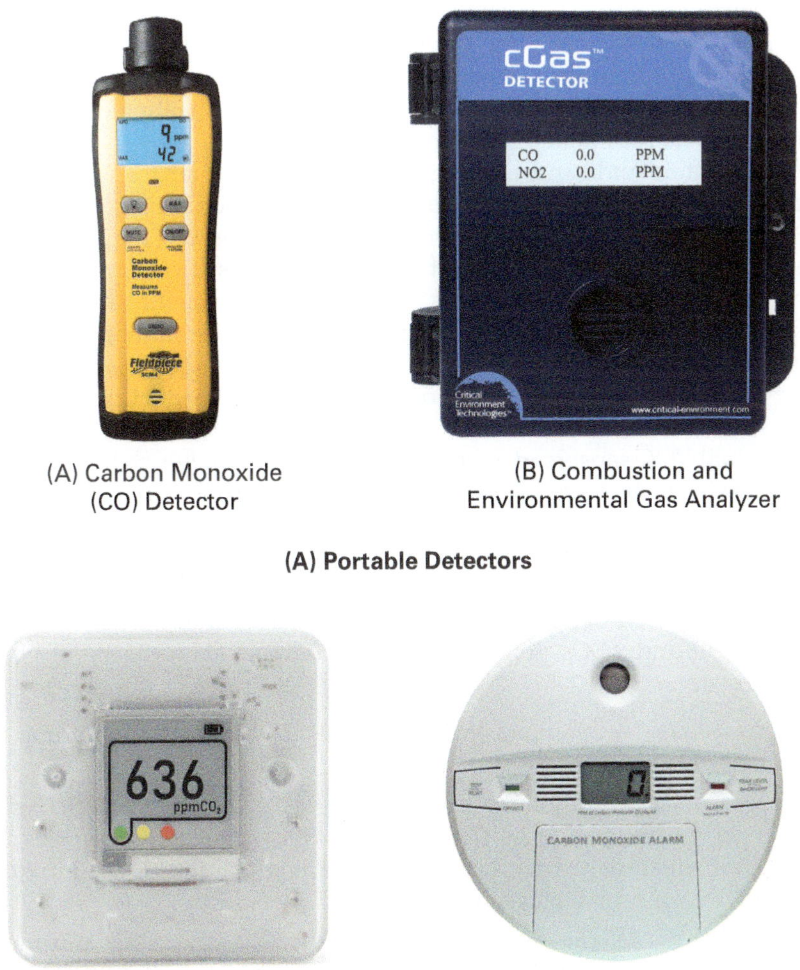

(A) Carbon Monoxide (CO) Detector

(B) Combustion and Environmental Gas Analyzer

(A) Portable Detectors

(C) Carbon Dioxide (CO_2) Monitor

(D) Carbon Monoxide (CO) Monitor

(B) Stationary Detectors

Figure 10 Gas detectors and analyzers.

Sources: (A) Courtesy of Fieldpiece Instruments; (B) Critical Environment Technologies; (C) Zigmar Stein/Shutterstock; (D) Danny E Hooks/Shutterstock

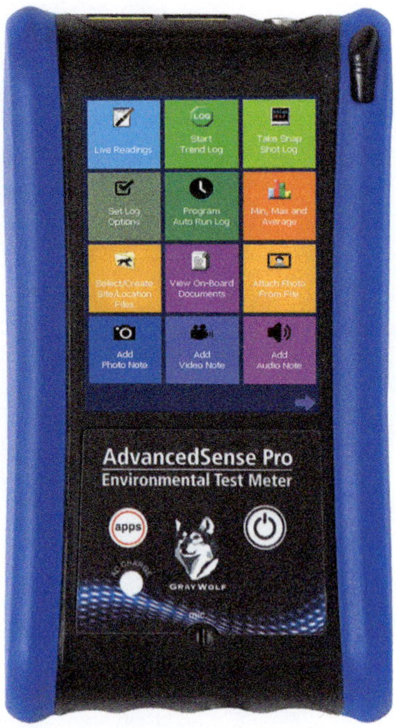

Figure 11 Portable CO_2 detector/IAQ monitor.
Source: Gray Wolf Sensing Solutions

Electronic gas detectors and analyzers should be calibrated periodically according to the manufacturer's recommendations. Individual sensors should also be replaced at the interval specified by the manufacturer.

The level of CO_2 is widely used as an indicator of whether enough fresh air is being provided. Concentrations above 1,000 ppm often indicate ventilation problems. Some CO_2 sensors can be used as ventilation controllers in demand-based ventilation control systems. When used as a ventilation controller, the sensor determines the need for ventilation based on the CO_2 concentration. The controls then modulate the position of the building dampers to reduce the CO_2 concentration. If a space is unoccupied, the CO_2 controller will reduce the ventilation rate to a minimum, saving energy.

The portable CO_2 detector/IAQ monitor shown in *Figure 11* detects real-time temperature and relative humidity (RH) readings as well as CO_2 readings. This detector has an audible alarm that sounds when measured CO_2 concentrations exceed a preset threshold and can also log data for downloading to a computer. Due to the variations in models and brands, it is important to pay attention to the specifications.

A CO detector is a safety device as well as an IAQ device. Stationary CO detectors are typically used as remote sensors in automated systems. The detectors are installed in strategic places throughout a building and will normally activate a contact closure and sound an alarm when a high level of CO is detected.

Portable CO detectors like the one shown in *Figure 10A* are used for testing occupied areas and locating the CO source. The detector shown is very easy to use. To ensure accuracy, it is important to let the detector stabilize at the temperature of the tested environment before taking a reading. A different type of detector is needed to test high-temperature airstreams, such as furnace flue gases. Combustion analyzers are used to test fossil-fuel heating equipment and ensure the burner is operating properly.

VOC detectors and sensors are used to monitor the air for volatile organic compounds. Many VOCs exist, making it difficult to monitor each one individually. Thus, a measurement of *total volatile organic compounds* (TVOCs) was established to monitor the total volume of all possible VOCs. TVOC values represent all the pollutants classified as VOCs. Thus, the environmental analyzer shown in *Figure 10B* reports TVOCs as a single value. The device cannot determine if a specific harmful VOC is present. If the TVOC value is excessive, more detailed testing with compound-specific instruments is needed to identify the substance and its source.

Individual VOC sensors may be tuned to the space in which they are operating. Each sensor can be adjusted so that it provides a low output signal when a low concentration of a gas is present. The sensor increases the signal output as the concentration increases. The sensor output can be used to activate an alarm and control the building ventilation rate.

Many specialized detectors are designed to detect and measure gases other than CO_2, CO, and VOCs. Some of the more common specialized detectors are as follows:

- *Oxygen detector* — Monitors the level of O_2 in the area. The normal atmospheric level is 20.9%.
- *Hydrogen detector* — Monitors the hydrogen (H_2) gas concentration. Hydrogen is dangerous because of its flammability. Hydrogen has a lower explosive limit (LEL) in air of just 4%. This is the level at which the air-gas mixture can explode with a source of ignition.
- *Combustible gas detector* — Monitors the LP (propane, butane) and methane (natural) gas in the area. This detector is also used to check for leaks around fittings.
- *Refrigerant gas detector* — Provides area monitoring and early warning of refrigerant leaks.
- *Ammonia detector* — Provides area monitoring and early warning of ammonia (NH_3). Ammonia is poisonous and is dangerous even at very low levels.

On Site

Integrated Economizer and Demand-Control Ventilation

To save energy and still maintain good indoor air quality, some HVAC systems use demand-control ventilation (DCV) integrated with what is known as an *economizer*. The economizer determines whether outside air can be used for cooling instead of running the system compressor. The DCV CO_2 sensor is in the indoor occupied space. When the sensor detects an increase in occupancy, it commands the system dampers to increase the amount of ventilation to the space.

Some systems also have an air quality sensor located outdoors that is used to determine whether the outdoor air is clean enough to bring indoors. If, according to the sensors, both the indoor air and the outdoor air are of poor quality, an alarm signal is generated to alert the building maintenance personnel. Even if little can be done, it is important for occupants to be aware of the situation.

Economizer operation and control is presented in NCCER Module 03204, *Air Quality Equipment*.

2.5.2 Air Sampling

During an IAQ evaluation and survey, air sampling typically involves measurements outside a building as well as inside. Sampling usually begins in the morning prior to occupancy, with samples taken at multiple locations. The levels of various gases, including CO_2, are recorded in ppm according to the specific location and time of day. The morning measurements are then compared to those taken throughout the day from other sample locations to look for trends in the variations.

The number of sampling periods is determined by the types and duration of activities that occur in the building, including lunch breaks. Samples should also be taken at the end of the workday, just as everyone begins leaving the building, because the increased traffic usually results in higher levels of CO_2. To get a complete picture of the building air pattern, samples should also be taken at night when the building is typically unoccupied.

The building temperature and relative humidity should also be documented as each sample is taken.

2.5.3 Testing for Microbial Contaminants

Microbial contaminants include bacteria, fungi, viruses, algae, insect parts, and pollen. The sources of such contaminants include moist walls, ducts, duct liner, fiberboard, carpet, and furniture. Testing for specific microbial contaminants must be performed by qualified personnel such as microbiologists or industrial hygienists.

Airborne microorganism samples are often inconclusive. Airborne **spores** may only be present in large quantities for short periods of time. Further, the release of spores often depends on the current growing conditions. If ideal growing conditions are intermittent, as is often the case in HVAC systems, then the release of the spores is also intermittent. Significant levels of these contaminants are found only if sampling occurs when spores have recently been released and remain airborne.

Spore: A reproductive cell, typically single-celled, capable of developing into a new individual without fusion with another reproductive cell, often released by lower plants, fungi, and similar microorganisms.

Microbial surface sampling is also used. Periodic sampling provides a historical reference of previous growth and indicates the potential for future growth. Surface sampling relies on sampling strips, each with a pad on one end that contains a growth media for a wide range of organisms. Two strips, each containing a different growth medium, are used at each location.

The strip is activated by removing its sterile protective cover, and one square inch of surface area is wiped with the pad. After taking the sample, the strips are placed in sterile envelopes and returned to the manufacturer's laboratory for incubation and analysis. A report is returned with the results stated in colony counts per square inch. The results may also be ranked on a multistep severity index ranging from very low to severe.

2.5.4 Interpreting Test Results and Taking Corrective Actions

Corrective actions are determined by what has been found during the site walk-through, the HVAC system inspection, and the air sampling and testing phases. Depending on the complexity of the problem, it may be necessary to call in expert help, such as that of an industrial hygienist.

2.0.0 Section Review

1. Off-gassing is associated with which of the following contaminants?
   a. Legionella pneumophilia bacteria
   b. Aspergillus mold spores
   c. Environmental tobacco smoke (ETS)
   d. Volatile organic compounds (VOCs)

2. Occupants of a building affect IAQ by consuming oxygen and exhaling _____.
   a. ammonia (NH_3)
   b. nitric oxide (NO)
   c. carbon dioxide (CO_2)
   d. carbon monoxide (CO)

3. Which of the following is a source of radioactive air pollution?
   a. Ozone from a photocopier
   b. Uranium in soil and rocks
   c. Formaldehyde in particle board
   d. Asbestos in building insulation

4. Ideally, an IAQ inspection team should include, at a minimum, an HVAC engineer and _____.
   a. an industrial hygienist
   b. the building's owner
   c. an occupants' representative
   d. an EPA representative

5. What does the acronym *TVOC* represent?
   a. The most volatile of VOCs
   b. The most dangerous VOCs
   c. All common VOCs
   d. The VOCs most likely to be detected

3.0.0 Achieving Acceptable Indoor Air Quality

Performance Task

1. Perform a building indoor air quality evaluation.

Objective

Identify various ways to improve and maintain good IAQ.

a. Explain how building design can affect IAQ.
b. Explain how ventilation, temperature, and humidity control affect IAQ.
c. Explain how to control chemical and microbial contaminants.
d. Describe the effects of duct leakage and explain how to clean air distribution systems.

Acceptable indoor air quality can generally be achieved through awareness and control of the following areas:

- Initial building design
- Ventilation and thermal comfort control
- Chemical and microbial contaminants
- Scheduled building and equipment maintenance

3.1.0 Building Design

The methods used to conserve energy in a building potentially work against IAQ. Energy conservation without acknowledging IAQ results in discomfort, illness, and lost productivity. Increased energy efficiency can cause buildings to become less inviting environmentally.

Energy conservation is necessary, but it must be done in a manner that also provides for acceptable IAQ. Remember, though, that conditioning large volumes of outside air at varying conditions is expensive. In addition to the energy expended in conditioning outside air, additional equipment must be purchased and maintained throughout the life of the building.

Every occupied space requires a sufficient volume of fresh air. There are two kinds of ventilation: natural and mechanical. In days gone by, natural ventilation in residential dwellings was provided by operable windows and doors. However, this approach was inconsistent and can't be controlled. Energy codes now require tighter construction to reduce infiltration. Therefore, more mechanical ventilation is required to compensate. Applicable codes specify the appropriate ventilation rate.

Codes may also require *blower door testing* on newly constructed homes to determine the infiltration rate and/or to test the air distribution system for leaks.

Owners, architects, and engineers can work toward good and even excellent indoor air quality by implementing the following guidelines:

- Locate buildings on high ground rather than in a valley where pollutants might collect. Sites should avoid major highways and parking lots when possible. Locate buildings upwind from power plants, chemical plants, or other industrial pollution sources.
- Incorporate HVAC equipment and mechanical ventilation systems that best meet the needs of the occupants in terms of comfort and performance.
- Use low-VOC building materials and furnishings that minimize the release of contaminants.
- Consider both the prevailing winds and potential contaminant sources to properly locate and orient outside air intakes.
- Provide adequate air filtration systems and materials.
- Ensure that fossil-fuel equipment and appliances are properly vented and receive enough combustion air.
- Provide proper drainage and seal foundations.
- Use radon-resistant construction techniques.
- Avoid specifying shrubs and trees that produce heavy pollen and can aggravate allergies, such as olive, acacia, oak, pine, and maple trees.

3.2.0 Ventilation and Thermal Comfort Control

Controlling the flow, temperature, and humidity of the air in a building is essential, as it affects the health and comfort of the occupants.

3.2.1 Ventilation Control

The introduction of outside air is the most common approach to maintaining IAQ. Adequate ventilation in commercial buildings is primarily measured by the CO_2 concentration in a building, rather than by the oxygen concentration. As CO_2 or other pollutants accumulate inside a building, ventilation control is used to bring in more outside air to dilute it and ensure plenty of oxygen is available. Even if a specific contaminant, such as formaldehyde, is identified as the cause of a problem, dilution is still the most practical way to resolve it.

Fresh air should be adequately distributed to all areas of a building when it is occupied. In buildings where the heating setpoint is lowered at night, a morning warm-up routine often keeps the outdoor air dampers closed until a specific temperature is reached. Then the system and dampers return to the normal ventilation programming. Morning warm-up routines typically occur before most of the occupants arrive to prepare the building for occupancy.

The ventilation requirements of residential and commercial buildings are constantly changing in response to the latest IAQ standards established by ASHRAE. ASHRAE Standard 62.1 recommends minimum outdoor air ventilation rates for occupied spaces. These rates are not a one-size-fits-all determination. Factors such as space usage, floor area, number of occupants and occupant density, level of human activities, and environmental factors all contribute to the recommended ventilation rates. ASHRAE uses a series of equations to calculate minimum recommended air flows and the introduction of outside air.

Air balancing of the HVAC system may be all that is needed to achieve adequate airflow and improved air quality in buildings. Systems are often designed well enough but are poorly monitored and maintained. Other correctable causes of poor air distribution, including fresh air, include the following:

- Closed or obstructed air grilles, registers, and diffusers.
- Outdoor air dampers and actuators not properly adjusted or programmed. If mechanical ventilation is on, ensure the outdoor air dampers are open. If the building is lightly occupied, make sure the dampers do not close beyond the specified minimum open position. The dampers may have been deliberately set to the closed position by someone to save energy, or with the goal of improving comfort when it is very hot or cold outdoors. It is a choice that works directly against IAQ.
- Supply or exhaust fans that are inoperative.
- Dirty air filters.

On Site

Air Pressure Relationships Within Healthcare Facilities

Most states require hospitals, health clinics, and other healthcare facilities to control and maintain specific air pressure relationships among the activity areas. This is done to protect patients occupying certain areas from exposure to odors and germs emanating from other areas.

For example, laboratories handling infectious materials are maintained at a negative pressure relative to adjacent spaces, ensuring gases only leave the space through dedicated exhaust systems. Operating rooms and patient rooms are often maintained at a positive pressure relative to their surrounding areas, ensuring that air from other areas can't enter.

There are applications where 100% outside air is consistently used for an area. In this case, none of the indoor air is recirculated. One reason is to create a significant positive pressure in a specific area to prevent infiltration from outdoors and adjoining areas. However, in most cases, indoor air is recirculated and mixed with fresh air. In all cases, proper filtration and treatment is essential.

NOTE

Equipment used to enhance indoor air quality, including air filters, humidifiers, and UV lighting systems, is presented in NCCER Module 03204, *Air Quality Equipment*.

Media-based air filters are familiar to most everyone. There are many different types of filters, all of which vary in their ability to capture particles. It is important that the proper type be used. Beyond filtration, recirculated air may also be treated with ultraviolet (UV) light to prevent or reduce the growth of bacteria and other microorganisms. The air may also be humidified to enhance comfort. Because the conditions outdoors vary dramatically, provisions must be made to accommodate those variations as needed to maintain comfort and improve air quality.

3.2.2 Thermal Comfort Control

Occupants are more likely to notice IAQ problems when they are uncomfortable for other reasons. IAQ-related complaints tend to rise when the room temperature is consistently above 75°F (24°C). Thermal comfort plays a significant role in occupant happiness and productivity. When it is too cold or too hot, both happiness and productivity suffer.

Humidity control is also related to comfort. In summer, the human body can tolerate higher temperatures if the humidity level is below 60% RH. In the winter, when there is naturally less moisture in the air, people are generally more comfortable at lower temperatures when the humidity level is above 30% RH.

The HVAC system must maintain the indoor dry-bulb temperature and relative humidity within a comfort zone, as well as provide adequate air circulation. *Figure 12* shows the parameters for a generally accepted comfort zone plotted on a psychrometric chart. The area noted represents the temperature and humidity that will satisfy about 80% of the occupants. The accepted comfort range is 68°F to 78°F (20°C to 26°C) with a relative humidity of 30% to 60%.

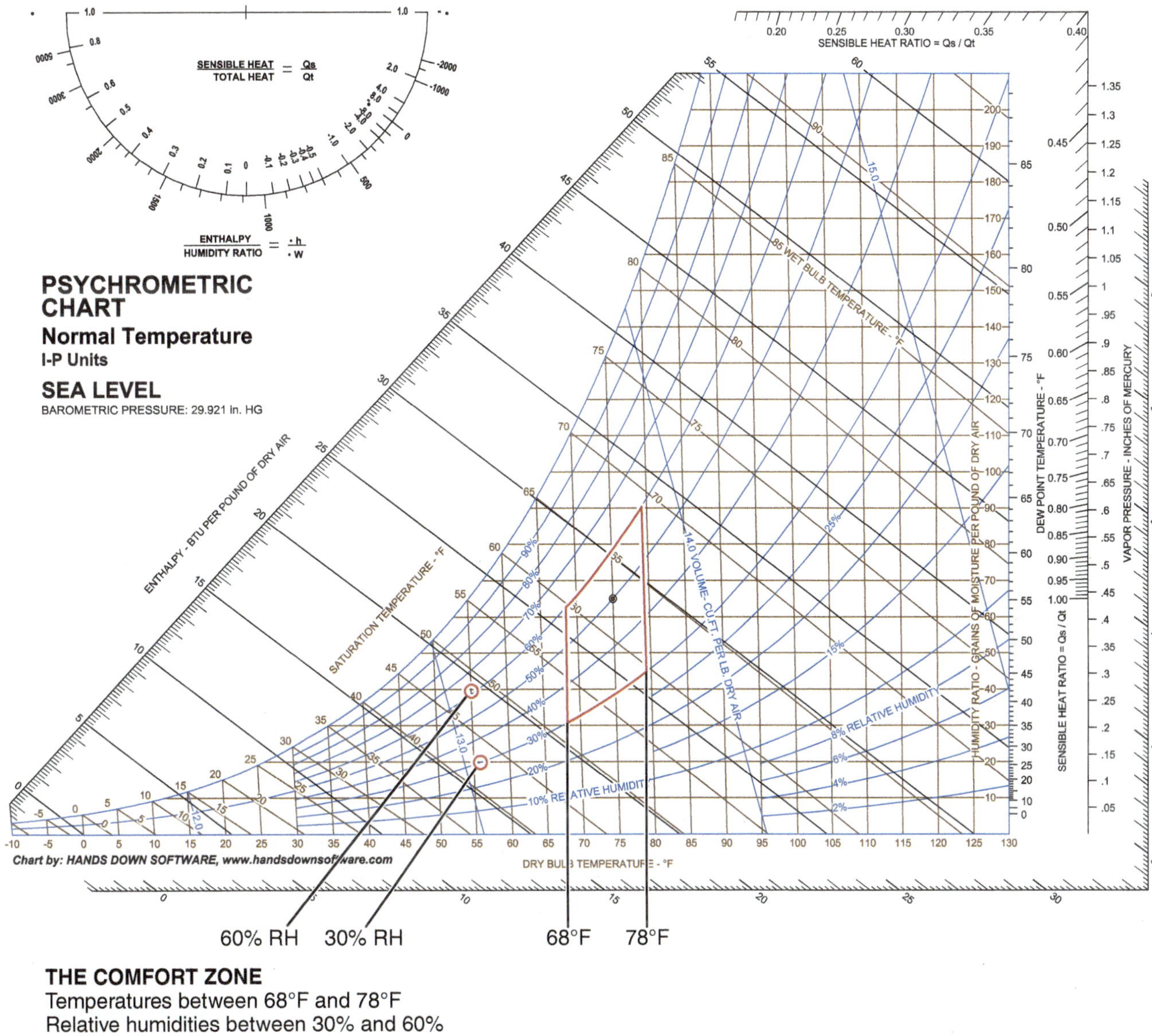

Figure 12 Human comfort zone plotted on a psychrometric chart.
Source: Hands Down Software

3.3.0 Controlling Contaminants

Chemical contaminants can be generated in a building during routine activities such as painting and cleaning. Microbial contaminants can grow in many moist environments, including cooling coils, condensate drain pans, and humidifiers. These contaminants must be controlled.

3.3.1 Controlling Chemical Contaminants

Chemical contaminants can come from VOCs emitted from building materials as well as cleaning and maintenance products. The contaminants can also come from pesticides and similar sources.

Every building must accommodate the storage of various chemicals. The presence of chemicals does not have to affect building IAQ. Exposure to chemical contaminants can be adequately controlled when these guidelines are observed:

- Use local exhaust systems where needed to trap and remove contaminants generated by specific processes or equipment.
- Areas being remodeled, painted, or carpeted should be temporarily isolated from other occupied areas in the building. This includes isolating any related HVAC systems, if possible. As scheduling permits, perform this type of work during evenings and weekends. Provide the maximum amount of ventilation to the affected areas around the clock to help eliminate contaminants.
- Request that pesticides and disinfectants be applied indoors only when a building is unoccupied, then thoroughly ventilate the building before it is reoccupied.
- Request that paints, paint strippers, solvents, wood preservatives, aerosol sprays, and cleaning products be applied only according to the manufacturer's directions. The building should always be thoroughly ventilated before it is reoccupied.
- Make sure that outside air intake grilles are not located close to places where chemical pollutants might collect.

3.3.2 Controlling Microbial Contaminants

Microbial contaminants can usually be controlled by observing the following guidelines:

- Maintain the relative humidity between 30% and 60% in all occupied spaces. Although low humidity negatively affects comfort, higher humidity levels promote microbial growth.
- During the summer, make sure cooling coils are operating below the dewpoint of the indoor air so that dehumidification takes place.
- Regularly check and clean humidifiers and any other equipment continuously exposed to a water source, such as ice machines. Use biocides that are designed for this purpose and ensure they are used carefully in accordance with their labeling.
- Promptly repair all water leaks. Wet ceiling tiles, sheetrock, carpets, and other building materials that absorbed water should be replaced. Ceilings and ceiling tiles are often damaged from leaking condensate drains and condensation from poorly insulated refrigerant lines (*Figure 13*).
- Prevent and/or correct any causes of stagnant water accumulation around cooling coils and air handling units.
- If contamination has occurred in the plenum or ductwork downstream from the coils, downstream filtration may be required. UV light can also be applied to eliminate and prevent microbial growth.

COVID-19

ASHRAE's Epidemic Task Force has addressed the problem of COVID-19. At this writing, the guidelines offered do not extend far beyond what has already been recommended herein to control contaminants. To date, the guidance offered can be summarized as follows:

- *Public Health* — Follow all current regulations and recommendations from heath officials, including vaccinations, wearing effective face masks, social distancing, and maintaining good building and personal hygiene.

NOTE

ASHRAE will continue to update their COVID-19 guidance as the information becomes available. For more information, visit the ASHRAE website at *https://www.ashrae.org*.

Figure 13 Ceiling tile damaged by water.
Source: Inna_liapko/Shutterstock

- *Ventilation* — Provide and maintain the minimum outdoor air requirements specified by applicable codes and standards.
- *Filtration* — Use air filters or a combination of filters with a minimum efficiency reporting value (MERV) of 13 or better for all recirculated air. The MERV rating identifies an air filter's particle capture efficiency. MERV ratings are discussed in detail in NCCER Module 03204, *Air Quality Equipment.*
- *Air Distribution* — Where directed airflow is not required or recommended, promote the mixing of indoor air without creating substantial air currents that may increase transmission of the virus.
- *HVAC System Operation* — Maintain standard temperature and humidity setpoints. To flush spaces with fresh air between occupied periods, operate the system as necessary to complete three air changes. For example, if the volume of the space is 50,000 cubic feet (ft^3), at least 150,000 ft^3 of outside air should pass through the space between occupied periods. Ensure that any contaminated air exhausted from the building does not have a path to return.

3.4.0 Air Distribution System Maintenance and Cleaning

Studies have shown that air distribution systems can lose up to 40% of the conditioned air through duct leaks. Translated into wasted energy, this level of leakage represents an annual fuel usage equivalent to that used by 13 million automobiles! Leaking ducts not only waste energy, but also adversely affect indoor air quality.

Cleanliness is also important. When excess moisture and dirt are present in ducts, biological contaminants can develop and disperse throughout a building. Mold in unlined sheet metal ducts can be successfully treated using an EPA-registered biocide. However, if fiberglass-lined metal ducts or ducts made of fiberglass ductboard become contaminated with mold, cleaning will not prevent regrowth. Unfortunately, no EPA-registered biocides exist for the treatment of porous duct materials. The EPA, National Air Duct Cleaners Association (NADCA), and the North American Insulation Manufacturers Association (NAIMA) all recommend the prompt replacement of wet or contaminated fiberglass liner and ductboard as soon as it is discovered.

3.4.1 Duct Leakage

In an ideal duct system, all ducts are free of leaks. During operation, the system fan causes a pressure differential between the supply duct and the return duct (*Figure 14*). Positive pressure in the supply duct causes conditioned air to flow

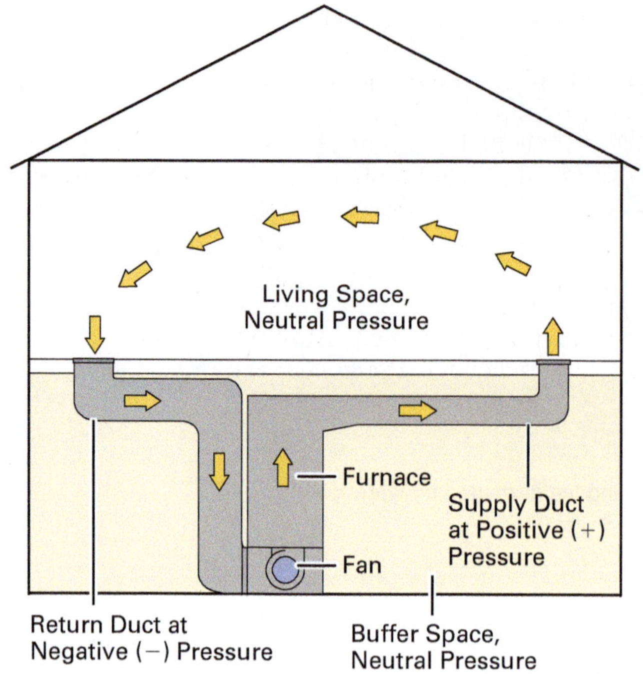

Figure 14 Air distribution system pressure relationships.

into the space. The negative pressure in the return duct draws air into the return grilles. The pressure inside the structure is essentially neutral.

Assume, though, that the return duct is free of leaks, but the supply duct is leaking into an unconditioned space such as an attic or crawl space (*Figure 15*). How does this affect the system and conditioned space?

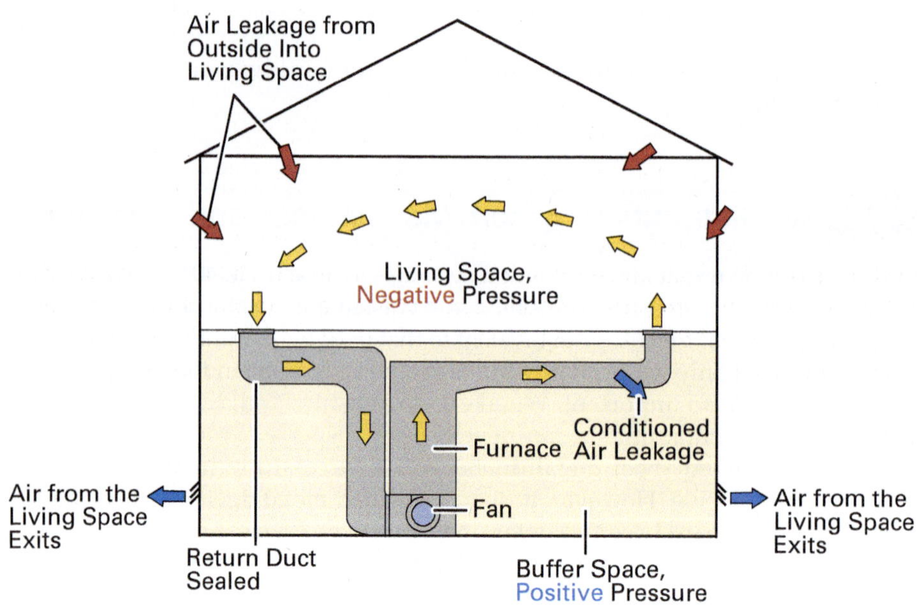

Figure 15 Effects of supply air duct leakage into unconditioned space.

Under these conditions, air lost through supply air leaks comes from the space. This creates a negative pressure in the conditioned space. The reduced pressure in the space encourages outdoor air to enter through cracks around windows and doors. Not only does this waste energy, but it can also cause airborne contaminants, including radon, to be drawn into the structure.

Similarly, assume that the supply duct is free of leaks, but a return duct located outside the conditioned space is drawing air in through leaks (*Figure 16*). Under these conditions, a slightly positive pressure is created indoors because the volume of air delivered is greater than the volume of air drawn from the conditioned space. More importantly, leaks in the return duct allow unconditioned air and contaminants to enter the system from the buffer space. If a fossil-fuel furnace is installed in the same space, byproducts of combustion can also be drawn into the system and distributed throughout the space.

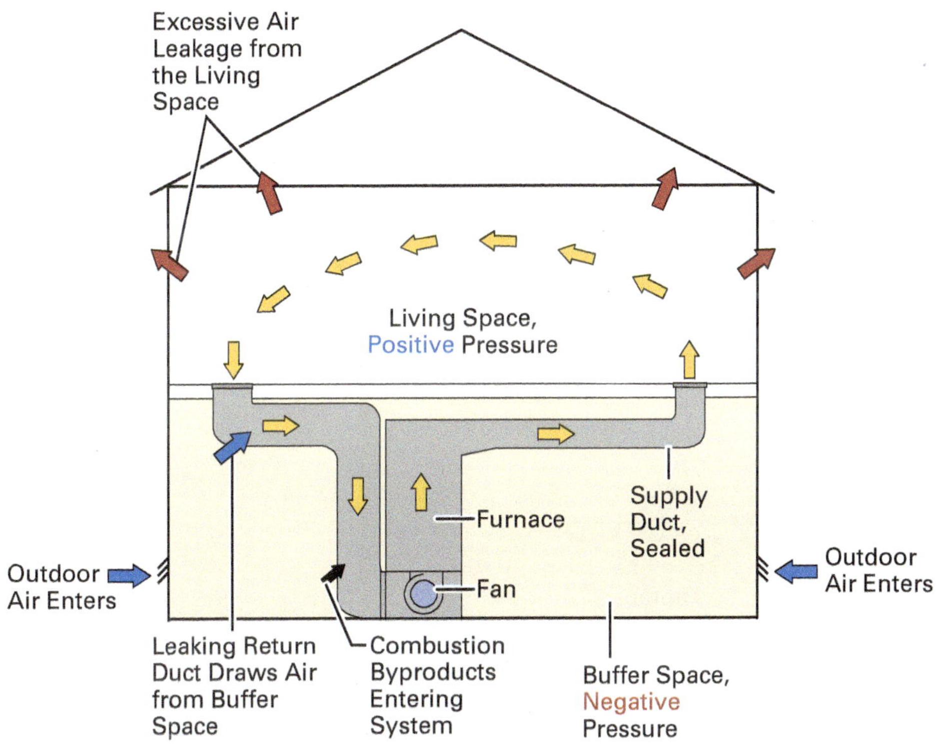

Figure 16 Effects of return air duct leakage in unconditioned space.

The truth is that most systems have leaks on both the supply and return ducts. In some cases, the leaks on each side are relatively equal and cancel each other out, relative to the pressure indoors. But contaminants can still enter through the leaking return duct. In most cases, however, one side of the system leaks more than the other, causing a positive or negative pressure indoors.

3.4.2 Sealing Duct Leaks

The best solution to duct leakage problems is to properly seal them during installation. Unfortunately, it isn't always done properly and may be omitted altogether. Many states and localities, realizing the scope of the problem, have implemented new codes to address the issue. However, this does not resolve the many existing duct leaks, and leaks can still develop over time as structures and duct systems shift. Some leaks are not easily accessed and may be hidden in walls or similar areas.

Duct leaks can be successfully sealed using caulks, **mastic** (*Figure 17*), fiberglass mesh tape, and special foil and film tapes. Large holes, such as an abandoned branch takeoff, can be patched and sealed with mastic and fiberglass mesh tape (*Figure 18*), which serves as a scaffold for the mastic. Oddly enough, common duct tape is not recommended for ductwork. Standard duct tape tends to lose adhesion after a few years and eventually loosens or falls off. Use only tapes approved for duct repairs.

NOTE

Before repairing duct leaks, consult the current Sheet Metal and Air Conditioning Contractors National Association (SMACNA) standards for guidance.

Mastic: A pasty form of sealant that is typically brushed on and hardens to seal joints and repair leaks in ductwork.

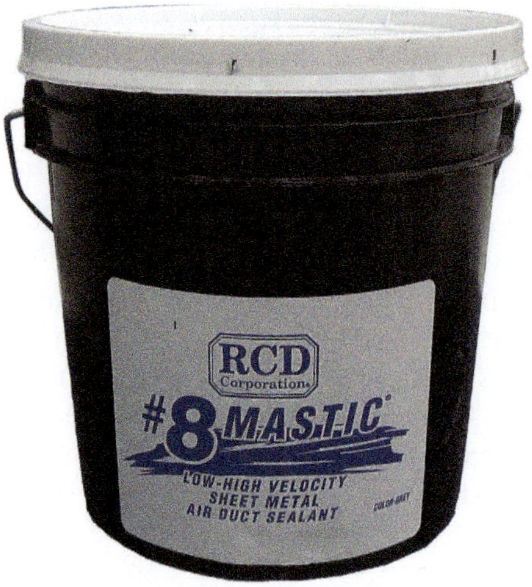

Figure 17 Duct sealing mastic.

Figure 18 Fiberglass mesh tape.

New technologies have recently been developed that focus on sealing ducts from the inside. The process, called *aerosol sealing*, injects small particles of dry adhesive into a pressurized duct system (*Figure 19*). The equipment is isolated from the duct, and all registers or grilles must be carefully blocked. The sealing equipment pressurizes the duct and injects the adhesive particles. The ingredients are non-toxic and similar to ingredients used in chewing gum and baby pacifiers.

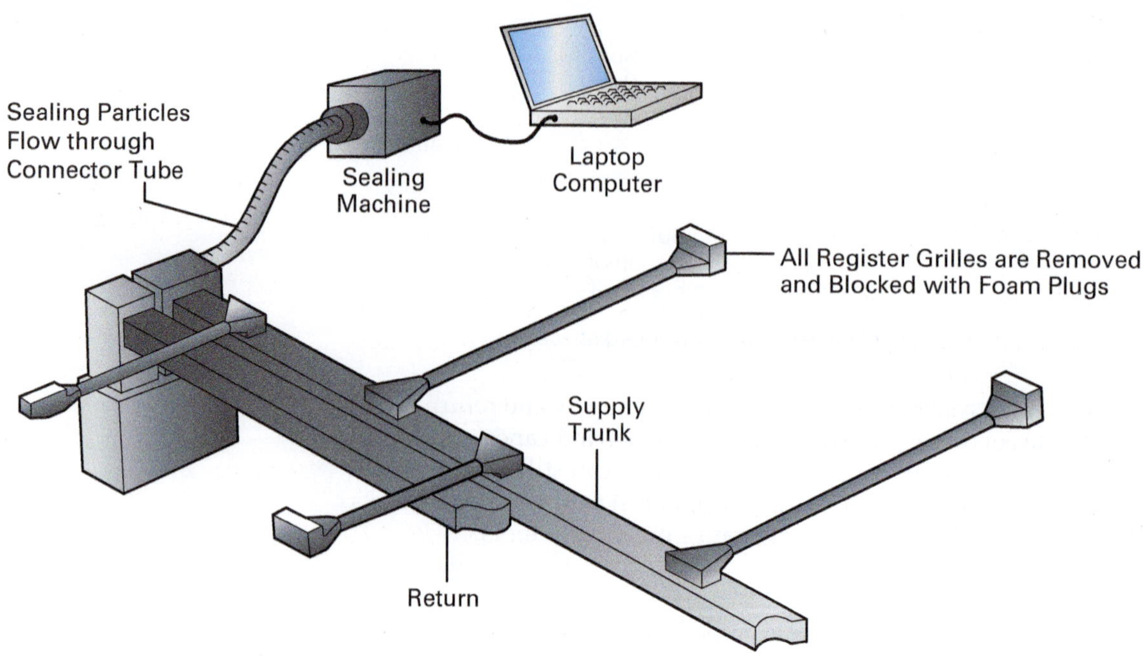

Figure 19 Aerosol duct-sealing system.

With the duct system capped and under pressure, the only place air can theoretically escape is through leaks. The sealant flows toward those locations. Over a period of several hours, enough sealant builds up to block the leaks. The process is monitored by a technician using a laptop computer. The manufacturer for the process claims that leaks up to about $\frac{5}{8}$" (1.6 cm) in diameter can be successfully sealed and that the process can seal up to 90% of the existing leaks in a duct system.

Although aerosol sealing is relatively easy once the preparations are made, it does require a significant amount of work to prepare the system. Most duct leaks are taken care of individually using mastics and other common materials. As a technician, you are far more likely to discover leaks than users and have a responsibility to report them when you do.

3.4.3 Duct Cleaning

Internal duct cleaning generally includes the supply and return ducts, registers, grilles and diffusers, heating and cooling coils, condensate drain pans, and the air handling unit.

Increased emphasis has been placed on the use of duct cleaning as a means of improving indoor air quality. There has been some controversy regarding its effectiveness and the best methods for performing the task. There are also questions such as when duct cleaning should be done and how the quality of the work can be validated. The EPA recommends that air ducts be cleaned only on an as-needed basis, rather than on a periodic schedule.

Duct cleaning alone won't solve IAQ problems. Dirty air distribution systems are most often the result of poor filtration and air treatment. These issues need to be corrected before the ducts are cleaned. If HVAC systems are well maintained, frequent duct cleaning should not be needed. However, when duct cleaning is needed and done properly, it can improve indoor air quality.

WARNING!

Duct cleaning will disturb contaminants that have settled in the ductwork and force them to become airborne. Be sure to wear the appropriate PPE, including respiratory protection, when performing duct cleaning. Any chemical products used must conform to current EPA requirements and local codes.

Members of the duct cleaning industry formed NADCA in 1989. This organization publishes a standard entitled *ACR—The NADCA Standard for Assessment, Cleaning, and Restoration of HVAC Systems*. This standard should be applied to all duct cleaning projects and can be downloaded directly from the NADCA website.

Most HVAC systems should be inspected for cleanliness once or twice a year. Blowers should be cleaned regularly as part of scheduled maintenance, but proper filtration should keep the blower and coil relatively clean. The evaporator coil requires consistent attention since fouling with dust and biological growth affects both its thermal performance as well as system airflow.

Duct cleaning equipment includes portable and/or truck-mounted vacuum systems equipped with **high-efficiency particulate air (HEPA) filters**. Mechanical tools, such as rotary brushes and air whips to dislodge dirt and debris are also needed (*Figure 20*). *Figure 21* shows typical video equipment used to inspect and document the conditions inside the ductwork before and after cleaning. Fiber optic scopes provide excellent images from inaccessible areas.

NOTE

ACR—The NADCA Standard for Assessment, Cleaning, and Restoration of HVAC Systems is a comprehensive standard that provides guidance in many areas, including duct inspection, work plans, cleaning procedures, and cleanliness verification. The standard should be reviewed, and its guidance applied, to all duct cleaning projects to ensure effective practices and legitimate results.

High-efficiency particulate air (HEPA) filters: Extended media dry type filters mounted in a rigid frame. They have a minimum efficiency of 99.7% for 0.3-micron and larger particles when clean filters are tested at their rated airflow capacity.

On Site

As-Built Drawings

Before cleaning expansive duct systems in commercial buildings, it is good practice to obtain a set of *as-built drawings*. As-built drawings are prepared after an installation is complete to show any changes that were made as the project progressed.

The drawings can be used to identify obstructions such as turning vanes and dampers in the duct system. They are also used to plan the location of access points in the ductwork and to divide the system into workable sections for cleaning. Ideally, individual sections designated for cleaning should be no more than about 25' (8 m) in length.

(A) HEPA Vacuum Collector

(B) Powered Brushes

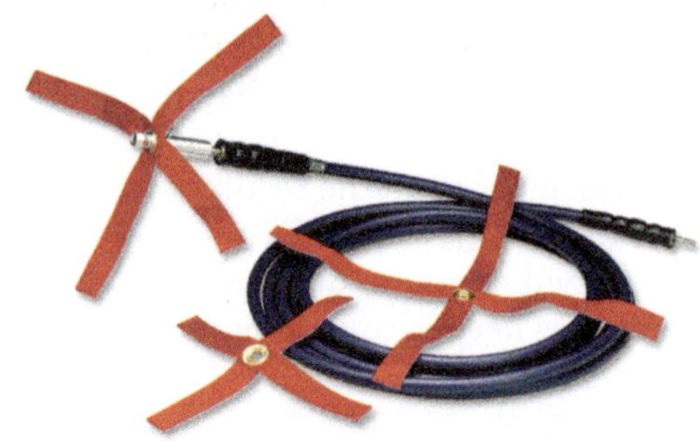

(C) Powered Whips

Figure 20 Duct cleaning equipment.
Source: Abatement Technologies

(A) Video Inspection System with Color Monitor

(B) Camera

Figure 21 Duct inspection equipment.
Source: Abatement Technologies

Before cleaning begins, the HVAC system must be turned off and locked out using approved lockout/tagout procedures. Each ductwork section being cleaned must be isolated from adjacent sections to prevent debris from returning to a clean section. This is often accomplished using inflatable bladders (*Figure 22*) inserted into the ductwork on each end of the section. When the bladders, often referred to as *zone bags*, are inflated, they conform to the interior shape of the ductwork, sealing off the section being cleaned.

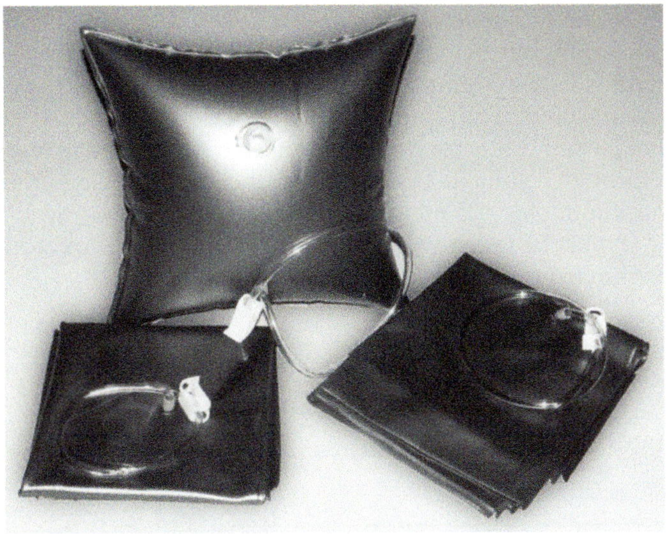

Figure 22 Inflatable zone bags.
Source: Abatement Technologies

Cleaning Methods

The system size and construction characteristics, along with the condition of the duct, determine the best method to clean the duct. There are three approaches, and all three may be used on the same system. *Contact vacuuming* refers to using an attachment on the end of a hose to clean accessible surfaces (*Figure 23*). HEPA filters must be used on the vacuum equipment. Starting at the return side of the system, vacuuming begins at the opening furthest away from the air handling unit. The work proceeds toward the air handling unit.

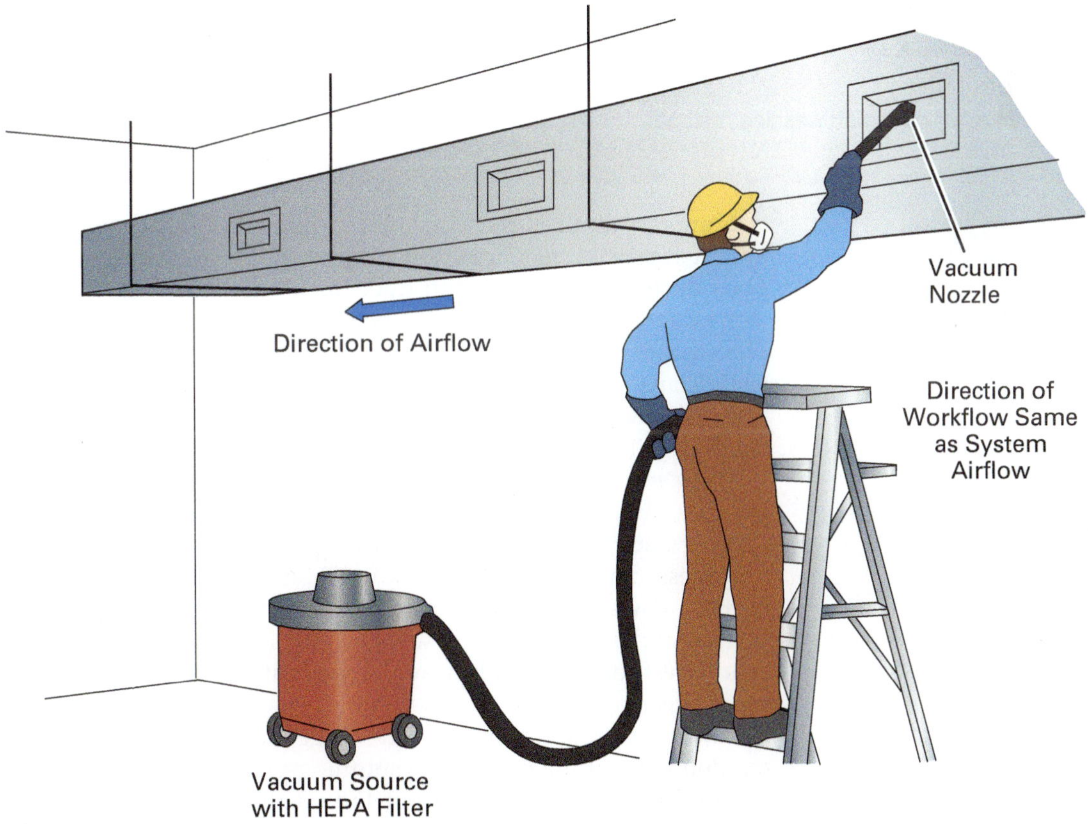

Figure 23 Contact vacuuming.

Each duct section is inspected for cleanliness before moving to the next section and repositioning the bladders. This method is generally used for cleaning simple, straightforward systems with easy access.

In the *air washing* method (*Figure 24*), a HEPA vacuum collection unit with an 8" to 12" (20 to 30 cm) diameter hose is connected to the downstream end of a duct section through a suitable opening. The duct section being cleaned is subjected to a minimum of 1" (2.5 mbar) negative air pressure to draw out loosened materials. Care must be taken with a powerful vacuum to avoid collapsing the duct from excessive negative pressure.

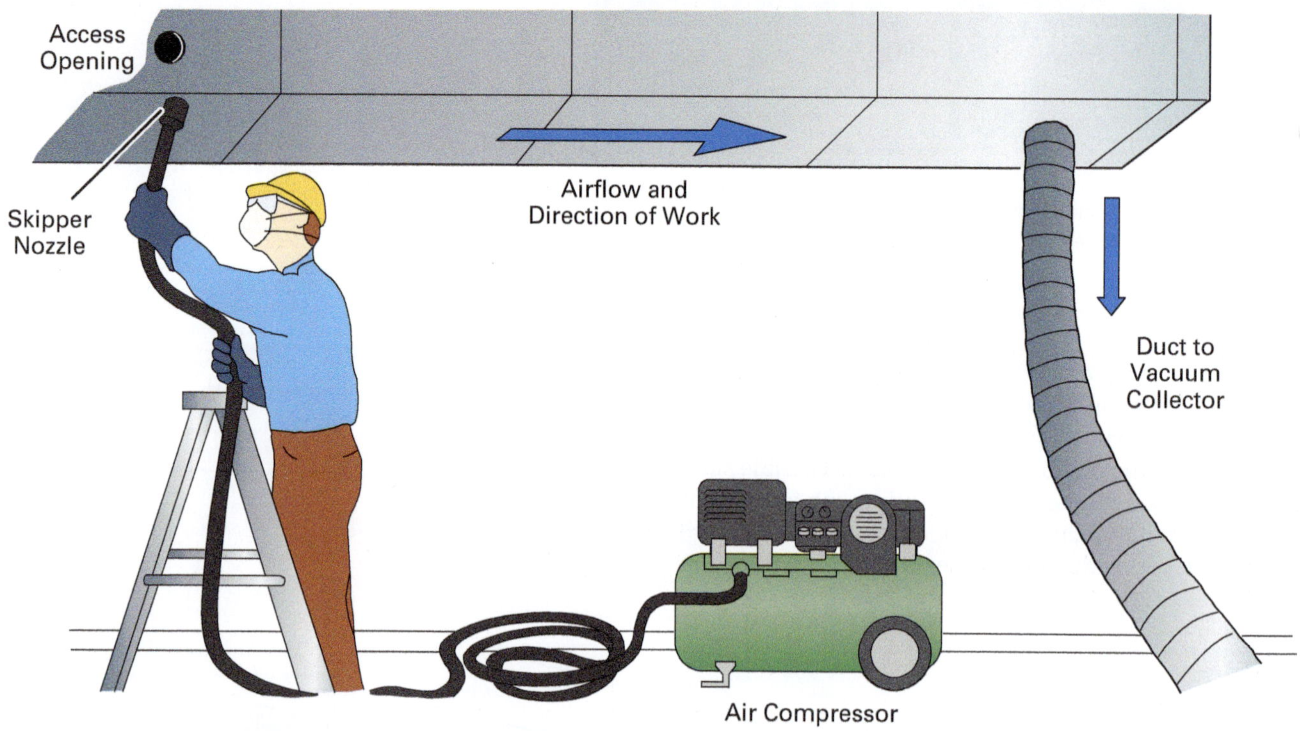

Figure 24 Duct cleaning using the air washing method.

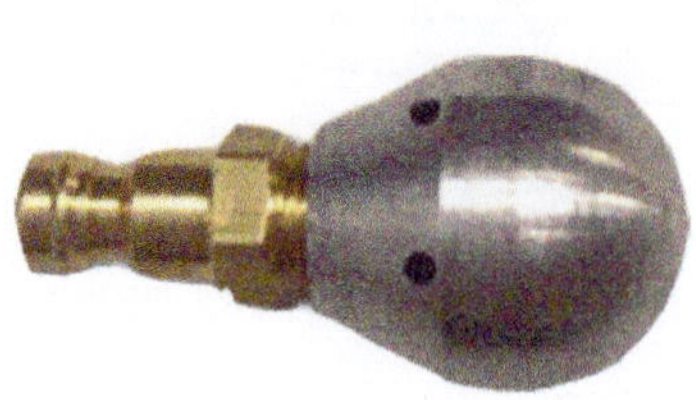

Figure 25 Skipper nozzle.
Source: Nikro Industries, Inc.

Compressed air is then introduced into the duct through a hose equipped with a *skipper nozzle* (*Figure 25*). Skipper nozzles are primarily used on fiberglass ducts, metal ducts with duct liner, and round ducts. The smooth sides won't damage duct liner or ductboard as it uses jets of air to dislodge debris. The smooth nozzle is propelled through the duct by the air jets and/or by the technician pushing it forward using the air hose. For the air washing method to be effective, the compressed air source should be able to produce and sustain between 160 and 200 psi (1,100–1,380 kPa) air pressure.

The *power brushing* method (*Figure 26*) provides the best agitation and debris removal, but it is too aggressive for fiberglass and lined metal ducts. Pneumatic or electric spinning brushes and air whips are used to dislodge debris (*Figure 27*). Compressed air is used to push any remaining debris toward the vacuum connection. Power brushing can be used with all types of ducts and fiberglass surfaces if the brush isn't too stiff and isn't held in one place too long. This process is best for ducts with heavy accumulations of debris that developed over years of neglect.

Power brushing usually requires larger access openings in the duct to allow manipulation of the brush. The spinning brush is inserted into the duct at the opening farthest upstream from the vacuum. The brush is moved steadily downstream to dislodge dirt and dust particles. After each section of duct is cleaned, the brush is removed and inserted through the next opening where the process continues.

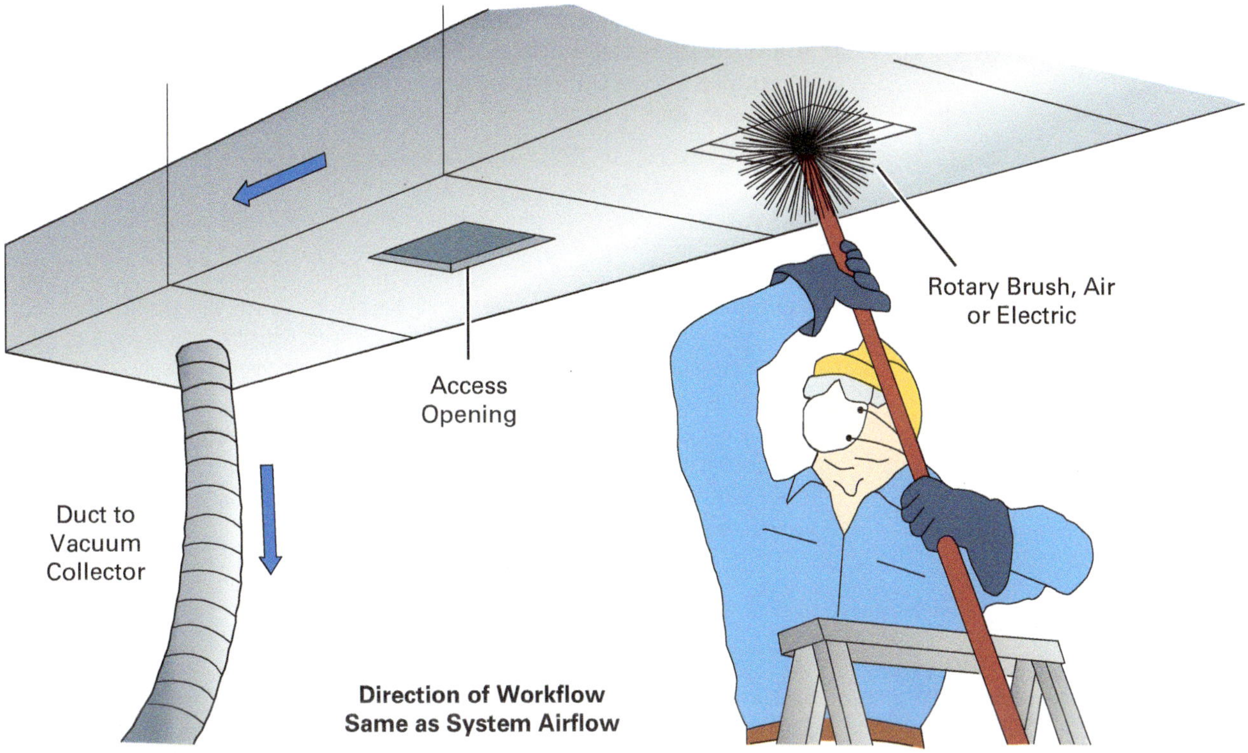

Figure 26 Power brushing ductwork.

Figure 27 Mechanical agitation during duct cleaning.
Source: Abatement Technologies

On Site

Fiberglass Duct and Liner

Care must be taken to prevent damaging fiberglass ductboard and duct liner. Stiff brushes can destroy the surface and literally tear the material apart. Fiberglass duct is also more susceptible to collapse than metal duct when exposed to the negative pressure produced by a vacuum collector.

NAIMA has published a manual that includes guidelines for cleaning insulated ductwork, *Cleaning Fibrous Glass Insulated Duct (Ductboard/Duct Board) Systems—Recommended Practices*. This and many other valuable resources can be found at the NAIMA website at *https://insulationinstitute.org/tools-resources/resource-library/*.

Regardless of the cleaning method used, the duct must be inspected to verify that it is clean as the work progresses and after it is complete. In some cases, photographs and specific tests may be contractually required upon completion to prove the duct is clean.

Did You Know?

Responsiveness

Heightened awareness of IAQ issues among the public requires an urgent level of response to building occupants when a potential IAQ issue is raised. Although statistically less than 10% of all IAQ complaints are related to an actual contaminant, there is no way of knowing if the complaint is valid until the problem has been acknowledged and researched.

If the occupants feel their complaints are being ignored, the issue is more likely to escalate. It may also manifest itself in other management issues due to staff frustration. Building managers and contractors must assume all IAQ complaints are real and handle IAQ complaints urgently to avoid further escalation.

3.0.0 Section Review

1. In the design of a building, energy conservation must be achieved in a way that also considers ______.
   a. environmental improvements
   b. indoor air quality
   c. a low intake/exhaust air ratio
   d. maximum return on investment

2. The primary method used to promote IAQ is ______.
   a. the introduction of outside air
   b. interior oxygen supplementation
   c. forcing CO_2 to bond with other gases
   d. ozone purging and regeneration

3. What should one do when applying pesticides in a building?
   a. Cover all the exhaust grilles for the area being treated.
   b. Ensure the HVAC system is not isolated from the area being treated.
   c. Thoroughly ventilate the building before the occupants are allowed to return.
   d. Maintain the humidity level between 60% and 80%.

4. If a fossil-fuel furnace is installed in an unconditioned space with a leaking return air duct, ______.
   a. a negative pressure will develop in the conditioned space
   b. it benefits indoor air quality
   c. the system's heating capacity will be significantly reduced
   d. byproducts of combustion can be drawn into the system

Module 03403 Review Questions

1. Minimum ventilation rates and indoor air quality standards acceptable for human occupants are specified in ______.
   a. ASHRAE Standard 52.2
   b. ASHRAE Standard 90.1
   c. ASHRAE Standard 62.1
   d. ASHRAE Standard 189.1

2. Good IAQ requires that cognizant authorities have determined that the air contains ______.
   a. no more than 19.5% pure oxygen
   b. no known contaminants at harmful concentrations
   c. no known carcinogens at any level
   d. at least 75% relative humidity

3. Buildings that have high air exchange rates ______.
   a. require less attention to air filtration
   b. are less susceptible to poor indoor air quality
   c. have low levels of dissolved oxygen
   d. have high levels of carbon dioxide (CO_2)

4. The process through which furnishings and building materials release chemicals and other volatile organic compounds into the air is called ______.
   a. venting
   b. evaporation
   c. sublimation
   d. off-gassing

5. Human occupants are directly responsible for which of the following contaminants?
   a. Bacteria and viruses
   b. Carbon monoxide (CO)
   c. Ozone and radon
   d. Carbon tetrachloride

6. Radon is released when ______.
   a. ground water levels decrease
   b. the ground warms in the summer months
   c. uranium is exposed to sunlight
   d. uranium and thorium atoms decay

7. The most common passive radon detectors for short-term testing are called ______.
   a. scintillation detectors
   b. radiometers
   c. electron traps
   d. UV emitters

8. The first step in conducting an IAQ survey is usually ______.
   a. making a site visit and conducting a building walk-through
   b. taking air samples and test for specific contaminants
   c. conducting interviews to determine the scope of the problem
   d. inspecting the building's HVAC equipment and ventilation system

9. Basic air sampling typically involves measuring the CO_2 levels ______.
   a. of air being exhausted from the building
   b. at the air handler
   c. at all air intake points around the building
   d. both inside and outside the building

10. Air sampling for IAQ purposes usually begins ______.
    a. in the morning
    b. at midday
    c. in the afternoon
    d. in the evening

11. Which of the following building design choices are practical and will likely improve its IAQ?
    a. Locating the building in the lowest point of a valley.
    b. Installing every possible IAQ enhancement and air treatment device, regardless of expense.
    c. Using low-VOC building materials and furnishings that minimize contaminant release.
    d. Encouraging the use of acacias, oaks, pines, and similar plants in the landscape.

12. Which of the following is a reason that a space might be consistently provided with 100% outside air?
    a. To prevent any contaminants in the space from escaping.
    b. To maintain a positive pressure in the space.
    c. To save energy.
    d. To avoid the need for CO_2 sensors.

13. In the winter, when there is naturally less moisture in the air, people are generally more comfortable at lower indoor temperatures when ______.
    a. there is significant air movement
    b. CO_2 levels are held below 500 ppm
    c. the exchange rate is increased
    d. the humidity indoors is above 30% RH

14. The best way to minimize IAQ problems caused by microbial growth is to ______.
    a. maintain the building humidity between 30% and 60%
    b. increase the exchange rate
    c. close all system outside air dampers
    d. relocate the equipment that is the source of the contaminants

15. Common duct cleaning methods include contact vacuuming, power brushing, and ______.
    a. aerosol steaming
    b. air washing
    c. dry scrubbing
    d. power washing

Answers to Odd-Numbered Module Review Questions are found in *Appendix A*.

Answers to Section Review Questions

Answer	Section	Objective
Section One		
1. c	1.1.0	1a
2. d	1.2.0	1b
Section Two		
1. d	2.1.1	2a
2. c	2.2.0	2b
3. b	2.3.1	2c
4. a	2.4.0	2d
5. c	2.5.1	2e
Section Three		
1. b	3.1.0	3a
2. a	3.2.1	3b
3. c	3.3.1	3c
4. d	3.4.1	3d

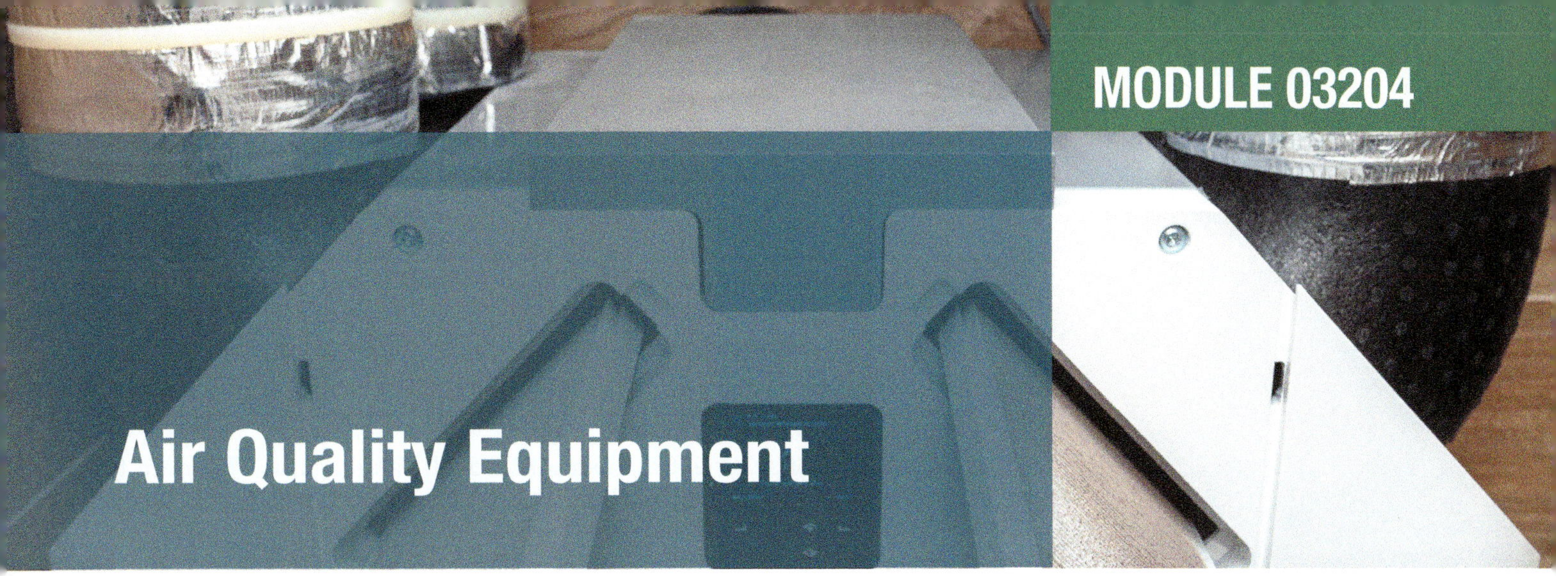

MODULE 03204

Air Quality Equipment

Source: Kinek00/Shutterstock

Objectives

Successful completion of this module prepares you to do the following:

1. Identify and describe the operation of equipment used to provide fresh air in buildings.
   a. Explain how dampers and economizers control the introduction of fresh air.
   b. Identify and describe the function and operation of energy and heat recovery ventilation equipment.
2. Describe the processes and equipment used to control indoor humidity.
   a. Explain the relationship between air and moisture content.
   b. Describe the equipment used to humidify and dehumidify air.
3. Describe the equipment and devices used to control air cleanliness.
   a. Identify various types of media-based air filters.
   b. Describe the operation of nonmedia-based air filtration and treatment equipment.

Performance Task

Under supervision, you should be able to do the following:

1. Measure the pressure drop across a bank of air filters using a manometer or differential pressure gauge.

Overview

Many accessories can be added to an HVAC system to improve indoor air quality. The quality of the air in buildings is affected by dust, mold, viruses, and other contaminants. While some accessories are designed to dilute or eliminate these contaminants, others, such as humidifiers, modify the characteristics of the air to improve comfort. Because these accessories are very common and the demand is growing, it is essential that HVAC technicians understand these systems and how they are applied.

NCCER Industry-Recognized Credentials

If you are training through an NCCER-accredited sponsor, you may be eligible for credentials from NCCER. The ID number for this module is 03204. Note that this module may have been used in other NCCER curricula and may apply to other level completions. Contact NCCER at 1.888.622.3720 or go to **www.nccer.org** for more information.

You can also show off your industry-recognized credentials online with NCCER's digital credentials. Transform your knowledge, skills, and achievements into credentials that you can share across social media platforms, send to your network, and add to your resume. For more information, visit **www.nccer.org**.

Digital Resources for HVACR

Scan this code using the camera on your phone or mobile device to view the digital resources related to this craft.

SCAN ME

1.0.0 Fresh Air Introduction and Control

Performance Tasks

There are no Performance Tasks in this section.

Objective

Identify and describe the operation of equipment used to provide fresh air in buildings.

a. Explain how dampers and economizers control the introduction of fresh air.

b. Identify and describe the function and operation of energy and heat recovery ventilation equipment.

Figure 1 Carbon monoxide (CO) detector.
Source: Danny E Hooks/Shutterstock

The introduction of sufficient outdoor air through ventilation is important to the health of building occupants. The dilution of various airborne contaminants in a building is one reason fresh air is needed. Some contaminants are common in an occupied environment, but contaminants can't be allowed to increase to an unsafe level. Additional oxygen is also needed indoors, especially if a building is heavily occupied.

Carbon monoxide (CO) is one gaseous contaminant that can't be allowed to collect in an occupied environment. CO monitors (*Figure 1*) are widely used in both homes and businesses. Because CO is lighter than air, the detector should be mounted high on the wall or on the ceiling. A detector with a digital readout may provide additional peace of mind to occupants and allow them to see trends in the concentration before the alarm level is reached.

Many states require CO detectors in homes with a fossil-fuel appliance through statutes. States that have adopted the *International Residential Code®* (*IRC®*) require CO detectors to be installed where relevant. The deadly effects of CO are well documented.

> **Did You Know?**
>
> **Carbon Monoxide**
>
> According to the US Center for Disease Control and Prevention (CDC), CO is the second most common cause of non-medicinal deaths by poisoning. CO exposure forces more than 10,000 people in the US to seek medical treatment each year, and more than 400 die annually as the result of significant exposure. As an HVACR technician, you are positioned to help reduce these numbers, although many CO poisonings are not related to HVAC systems.

Figure 2 Handheld CO detector.
Source: Courtesy of Fieldpiece Instruments

CO results from the incomplete combustion of carbon-based fuel. The gas is colorless, odorless, and tasteless. Early detection of CO is nearly impossible without a CO detector. CO detectors and monitors are often combined with smoke detectors. Surface-mounted units are common in homes, while duct-mounted models are common in the commercial setting. Many models can be connected to the HVAC equipment control system and/or the building management system (BMS) to execute an equipment shutdown when high levels are detected. Handheld CO detectors are used by technicians for spot checks (*Figure 2*).

Carbon dioxide (CO_2), on the other hand, is constantly present indoors since exhaling CO_2 is part of normal respiration. CO_2 is a colorless, odorless, tasteless gas released from both humans and combustion processes.

The concentration of CO_2 in a commercial building is commonly used as an indicator of indoor air quality (IAQ) and oxygen levels. Because it is not yet practical to scrub CO_2 from the air in buildings, the common approach to managing it is through dilution—mixing fresh outdoor air with indoor air. Concentrations below 1,000 parts per million (ppm) generally indicate that the building's ventilation system (or natural infiltration) is adequately diluting the CO_2 and increasing the oxygen content. CO_2 levels above 1,000 ppm generally indicate the need for more fresh air.

When there is excess CO_2 indoors, it is accepted that the concentration of oxygen is lower than desired. Studies show that a low concentration of oxygen causes sleepiness and increases one's sense of anxiety. The loss of an occupant's mental awareness is therefore likely. A CO_2 concentration of 1,400 ppm, for example, typically indicates the oxygen levels are low enough to hamper our ability to make decisions and think strategically. Unfortunately, we are generally not aware of the loss, or we fail to associate minor symptoms with a lack of oxygen.

CO_2 sensors are used to monitor CO_2 levels in a building. In commercial buildings, most are wall-mounted to monitor the air directly in an occupied zone (*Figure 3*). Some models are used as a ventilation system controller, determining the volume of fresh air introduced into the building. During unoccupied periods, ventilation rates are often reduced to a minimum as an energy conservation measure.

NOTE

It is important to understand that a concentration of 1,000 ppm of CO_2 indoors is neither hazardous nor concerning. The concentration of CO_2 indoors is predominantly used to predict the amount of oxygen present. CO_2 is an easy gas to detect, and the sensors are less expensive than oxygen sensors.

Concentrations greater than 20,000 ppm can cause panting or a heightened sense of oxygen deprivation. Extended exposure to CO_2 concentrations above 250,000 ppm—25% of the air in a sample—can lead to death. Fresh air typically contains about 400 ppm of CO_2.

Figure 3 Wall-mounted CO_2 sensor with display.
Source: Zigmar Stein/Shutterstock

Maximizing oxygen levels while minimizing air pollutants indoors requires the introduction of fresh outdoor air. However, fresh air must be added in a controlled manner. Two problems associated with the introduction of fresh air are comfort and energy consumption. Both problems result from the temperature and humidity conditions of the fresh air supply. Outside air very rarely has the same characteristics as the air in the building. Thus, the incoming air must be heated, cooled, dried, or moistened. Taking these steps does maintain comfort, but at a significant energy cost.

To manage and control the volume of fresh air introduced into a building, certain equipment and controls are needed. This equipment includes **economizers**, **heat recovery ventilators (HRVs)**, and **energy recovery ventilators (ERVs)**.

Economizers: Mechanical devices, largely consisting of dampers and their controls, that are integrated into an HVAC unit or system to draw in outdoor air and mix it with the indoor air.

Heat recovery ventilators (HRVs): Specialized HVAC equipment that saves energy in the winter using a heat exchanger to transfer heat from an exhausted airstream to cold ventilation air entering a building.

Energy recovery ventilators (ERVs): Specialized HVAC equipment that saves energy year-round using a heat exchanger to transfer heat both to and from an exhausted airstream to ventilation air entering a building.

1.1.0 Economizers

An economizer is an accessory often found on commercial rooftop heating/cooling units (*Figure 4*). One of its functions is to reduce operating costs by allowing the use of outdoor air for cooling when the temperature and humidity conditions are suitable. The term *economizer* comes from this function because the intent is to reduce energy consumption. However, the same apparatus can also be used to admit outside air in response to IAQ problems.

Outside air intakes like the one shown in *Figure 4*, are provided on most commercial rooftop units, as an established minimum volume of outside air is required. The additional dampers, controls, and other devices needed to create an economizer are optional accessories for most systems, allowing engineers to configure the equipment to suit their needs.

Regardless of whether a manual volume damper or a fully automated economizer is used, the outside air must be filtered. Bird and bug screens are often needed upstream of the filters. These filters are very important and must be well fitted to the opening. The filters must also be maintained along with the rest of the system. Some outside air filters are metal and washable because of their exposure to the weather. Others are enclosed behind the outside air damper, protected from the weather, like those in the unit shown in *Figure 4*.

Figure 4 Packaged rooftop unit with outside air intake.
Source: aquariagirl1970/Shutterstock

The economizer automatically positions the outside air dampers according to the system settings. It admits outdoor air and mixes it with conditioned air in a proportion that depends on the outdoor air conditions and the indoor demand. For example, if the space needs cooling and the outdoor air temperature is 58°F (14°C) at a low humidity, it represents an opportunity for "free" cooling. Note that this means that no energy-consuming mechanical process is needed to condition the air. In this scenario, the outside air dampers open all the way to bring in the maximum amount of outside air. At the same time, a set of dampers in the return air opening close. This allows the system to draw 100% of its air from outdoors, and indoor air is no longer being recirculated.

There are many possible scenarios like this one. The outside air and return air dampers in some economizers can position themselves in a virtually infinite number of positions between open and closed. In all cases, the programming and setpoints are established to conserve energy while maintaining comfort.

1.1.1 Building Pressure

When economizer dampers move to the fully open position, the return air dampers often close all the way. Only a small volume of indoor air may sometimes be recirculated. These damper positions take full advantage of the free cooling available.

Consider, though, what it's like to blow up a small paper bag. Eventually, the bag fills and you're exerting all the pressure on it that you can muster from your lungs. However, air stops going in unless you happen to have strong enough lungs to tear the bag!

Buildings react much the same way. When the system fills the building with air from outside, eventually the pressure in the building rises to the capacity of the fan. The air stops going in, and the building remains at a positive pressure while the economizer is activated.

All buildings, though, have some air leaks. In some buildings, there are enough places where air can escape that the pressure change doesn't represent a serious problem. If air can escape somewhere, outside air will continue to move into the building, replacing what leaks out. However, when the incoming air volume is very large and the building is tightly constructed, a path out of the building must be provided.

In *Figure 5*, note the lower set of dampers, which are *relief air dampers*. Behind the two damper assemblies is a centrifugal blower. It is referred to as the *relief blower* or *exhaust blower*. When the economizer opens fully, this blower starts and exhausts air from the building via the system return duct, at a volume

Figure 5 Relief air dampers.

comparable to that flowing in through the economizer. The indoor blower operates as well, forcing fresh air into the building. This maintains building pressure and ensures that fresh air continues to enter the building.

In some cases, only relief air dampers are provided. These dampers are *barometric*, meaning that they simply swing open when the building pressure increases. A relief blower is not required in these cases.

1.1.2 Economizer Operation

The primary factors in economizer operation are the temperature and humidity of both the indoor and outdoor air, and the demand for cooling and/or ventilation in the conditioned space. Remember that the dampers must be arranged so that the minimum required volume of outside air is always provided when the building is occupied. This means the outside air damper should always open to some degree during occupied hours. When a manual outside air damper alone is provided, it is set to a specified position at startup and remains fixed in that position.

NOTE

This introduction to economizer operation is based on simple rooftop equipment and systems that do not offer variable air volume or modulated airflow. Economizers serving variable air volume systems with a high level of automation have more complex and varied operating schemes.

The economizer is only valid for cooling operation. When building systems are in the heating mode and occupied, the economizer dampers will remain in the position necessary to provide the minimum outside air necessary. *Table 1* shows the most common damper responses during the various modes of operation.

TABLE 1 Typical Economizer Operation

Mode of Operation	Outdoor Air Damper (OAD)	Return Air Damper (RAD)
Off	100% closed	100% open
Fan Only and Mechanical Cooling	Open to minimum position for ventilation	Modulates slightly toward closed to complement the OAD
Free (Economizer) Cooling	100% open or modulated to provide the programmed mixed air temperature	Modulates as needed to complement OAD
Heating	Open to minimum open position for ventilation	Modulates slightly toward closed to complement the OAD

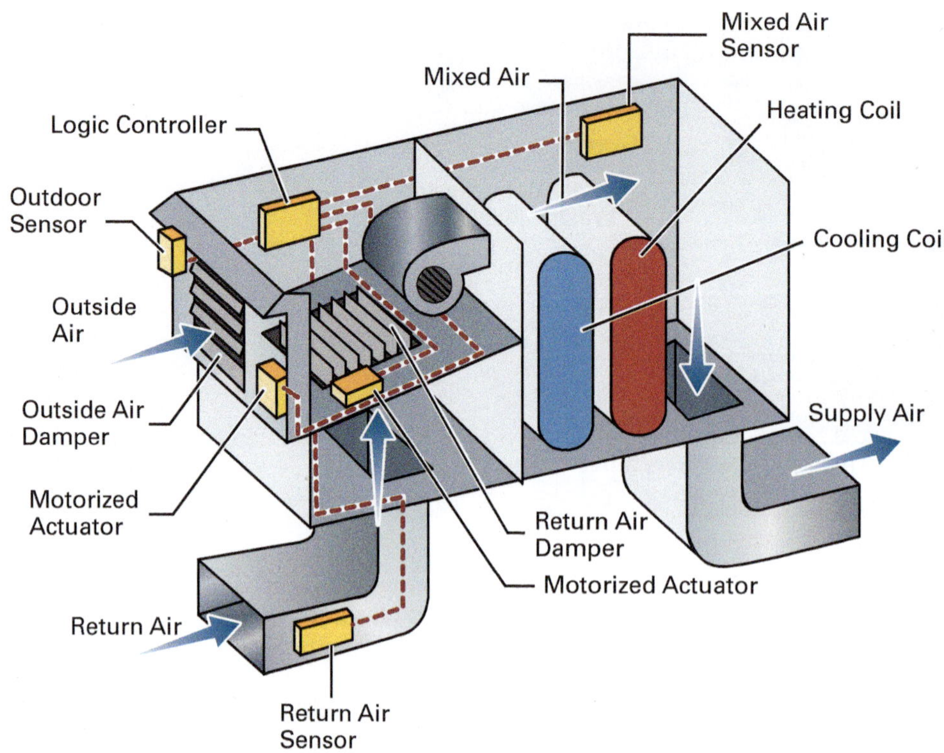

Figure 6 Basic economizer setup.

Refer to *Figure 6*. Four factors are generally used to control economizer operation:

- Outside air conditions — temperature and, sometimes, the humidity
- Return air temperature
- Mixed air temperature
- Design outside air volume required to maintain indoor air quality

Enthalpy sensor: A sensor that reports the total heat contained in air, accounting for both the dry-bulb temperature and the volume of water content in an air sample. Wet air, for example, contains more heat than dry air at the same temperature.

Outdoor air conditions are sensed by either a dry-bulb temperature sensor or an **enthalpy sensor** that can measure both the temperature and humidity. Enthalpy-based controls are typically used in areas where high humidity persists. These controls are considered better because they respond to the total amount of heat contained in the outdoor air, rather than just the dry-bulb temperature.

Three common approaches to airside economizer control are as follows:

- *Dry Bulb* — The simplest form with the lowest cost to install and maintain. The outdoor sensor only monitors the dry-bulb temperature of the outside air.
- *Single Enthalpy* — This type monitors the total heat contained in the outside air by evaluating both the dry-bulb and wet-bulb temperature of the outside air.
- *Differential Enthalpy* — The most complex arrangement but also the most effective, especially in areas where high humidity persists. The enthalpy of both the outside air and return air streams is monitored and compared to determine which airstream represents the best option.

When the outdoor air sensor detects that the outdoor air conditions are above its setpoint, cooling for the building is provided in the conventional way through mechanical cooling. When the outdoor air conditions fall below the setpoint, the economizer controls can cool the building using outside air. As described earlier, the outside air and return air dampers modulate to the needed position. The outdoor air setpoint is normally determined by the system designer.

The mixed air temperature (MAT) sensor monitors the air temperature flowing into the building. One of its primary functions is to ensure that the air circulating in the building isn't too cold. For example, the economizer may open to cool the building when the outside air temperature is 40°F (4°C). To avoid discomfort, the system will respond to the MAT sensor by repositioning the outside air and return air dampers somewhat, mixing the two airstreams to provide a supply air temperature of 55°F (13°C).

Equipment manufacturers may provide their own economizer control packages and devices, but they often use components from other manufacturers for at least a part of the system. *Figure 7* shows a popular controller that handles the inputs from the various sensors and provides the necessary outputs to the dampers and relief air fan (when installed). Complete kits include compatible sensors and damper actuators. The display and keypad allow the controller to be programmed without the need for a laptop PC or similar device to interface with it.

NOTE

Residential zoning systems are presented in NCCER Module 03315, *Zoning, Ductless, and Variable Refrigerant Flow Systems*.

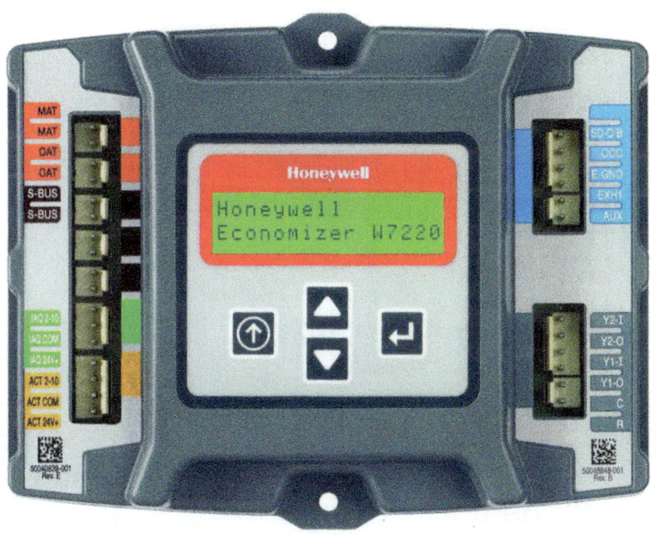

Figure 7 Economizer system controller.
Source: Courtesy of Kele, Inc.

Damper actuators (*Figure 8*) rotate linkage shafts to control the damper position, usually through 90 degrees of movement from open to close (or vice versa). While most operate on 24VAC power, other models are designed for line voltage. The simplest models are powered to move the damper to one fixed position, then a spring pulls it back to the original position when the actuator is de-energized. These models are referred to as *spring-return dampers*. You'll encounter this type often when servicing or installing residential zoning systems. The installer determines whether the damper is installed for powered-open or powered-close operation.

Economizer dampers, though, must be able to turn in both directions and stop in many possible positions. These dampers are referred to as *modulating dampers*. Traditional actuator motor designs have two separate motor windings. One winding causes clockwise rotation while the other causes counterclockwise rotation. Only one winding can be energized at a time. Modulating dampers also require a signal of some sort to determine how far to move and in which direction. The economizer controller provides the information to the actuator through either a variable DC voltage or a variable current. Common signal ranges include 1–5VDC, 2–10VDC, and 4–20mA. The actuator accepts the signal and moves as instructed.

Damper actuators can be mechanically connected to the damper in many ways. *Figure 9* shows a geared drive design that saves precious space along the sides of the damper.

Figure 8 Typical economizer damper actuator.
Source: Belimo, 2022, Fail-Safe Damper Actuator, www.belimo.us

Figure 9 Geared economizer drive with actuator.
Source: Cambridgeport

1.2.0 Energy and Heat Recovery Ventilation Equipment

The primary function of ERV and HRV systems is to provide fresh air to a structure to improve indoor air quality. Introducing outside air into a structure is a significant expense, as that air must eventually be conditioned in some way. Although the introduction of fresh air is required by various codes and standards, it is best done in the most energy-efficient way possible. ERVs (*Figure 10*) and HRVs reduce the energy cost associated with ventilation by transferring thermal energy from indoor air leaving the structure to the fresh air entering to replace it.

On Site

Commercial ERVs

Not all ERVs and HRVs are compact units designed for residential use. The ERV shown here is a commercial rooftop model capable of providing up to 11,000 cfm of air. Air can move in and out of the unit through the sides or the bottom. Much larger units that are built without integral blowers (engineered and installed separately) can handle over 70,000 cfm of airflow.

Source: Image courtesy of RenewAire LLC

Figure 10 Energy recovery ventilator (ERV) with cover removed.
Source: Image courtesy of RenewAire LLC

The introduction of fresh air in commercial and public buildings has been required for many years. Multiple factors must be considered when determining the proper volume of fresh air required for a given building. However, there are also standards related to ventilation for residences that are easier to understand and apply.

The American Society of Heating, Refrigerating and Air-Conditioning Engineers (ASHRAE) standards recommend that a residence's indoor air be exchanged for fresh outdoor air at a minimum rate of 0.35 air changes per hour (ACH), but not less than 15 cubic feet per minute (cfm) per occupant. A rate 0.35 ACH means that the entire volume of air in the home is replaced every three hours. An alternate method recommended by ASHRAE calls for an exchange rate of 15 cfm per person considered in the home's design, 20 cfm per bathroom, and 25 cfm per kitchen.

To understand the term *air changes*, consider the inside volume of home. A simple single-story home, for example, that is 40′ × 50′ with an 8′ ceiling has an internal volume of 16,000 cubic feet (ft^3). One air change per hour would mean that all the air inside be replaced with fresh air each hour. To replace all 16,000 ft^3 of air each hour, you would need to introduce about 267 cfm of outside air into the home around the clock.

$$16{,}000\ \text{ft}^3/\text{hour} \div 60\ \text{minutes/hour} = 267\,\text{cfm}$$

Since the ASHRAE standard requires 0.35 air changes per hour, the required outside air volume is significantly less than that.

$$(16{,}000\ \text{ft}^3/\text{hour} \div 60\ \text{minutes/hour}) \times 0.35 = 93\ \text{cfm required outside air volume}$$

The alternate method offered by ASHRAE may be more favorable, but not always. It depends on the number of occupants and bathrooms. Assuming the example home has four occupants, two bathrooms, and a single kitchen, the calculation would look like this:

$$4 \text{ occupants} \times 15 \text{ cfm} = 60 \text{ cfm}$$
$$2 \text{ bathrooms} \times 20 \text{ cfm} = 40 \text{ cfm}$$
$$1 \text{ kitchen} \times 25 \text{ cfm} = 25 \text{ cfm}$$
$$60 \text{ cfm} + 40 \text{ cfm} + 25 \text{ cfm} = 125 \text{ cfm required outside air volume}$$

In this case, the primary method of calculating the outside air volume is the most advantageous, at least from the perspective of energy consumption.

The required outside air can be introduced through a central HVAC system. An outside air duct can be attached to the return air duct, and the airflow balanced to allow the required 93 cfm of outside air to enter whenever the indoor blower is running. However, remember that the requirement is constant. That means the blower would need to run constantly, and unless the heating or cooling equipment is running as well, the air would not be conditioned or tempered in any way.

This is where ERVs and HRVs can be applied. ERVs and HRVs can operate independently of the HVAC system to provide outside air. In addition, these units exhaust a volume of indoor air equal to what is being added from outside, rather than relying on stale air leaving the home randomly through structural leaks. While doing so, ERVs and HRVs can also transfer some of the heat between the two airstreams to conserve energy. ERVs and HRVs are typically capable of recovering about 60% to 80% of the thermal energy from the exiting indoor air.

Know the Code

ERVs/HRVs

The 2021 *International Residential Code*® requires an ERV or HRV to be installed in new homes built in US Climate Zones 7 and 8. These areas represent the coldest portion of the United States.

1.2.1 ERV/HRV Operation

Before discussing how these units work, it's important to understand the difference between an ERV and an HRV.

HRVs are generally installed in homes and buildings in colder climates that have long heating seasons, such as in the northern part of the United States and Canada. HRVs are designed to transfer only sensible heat from one airstream to the other. ERVs, on the other hand, can transfer both sensible and latent heat between the two airstreams. The difference is in the construction of the heat exchanger. ERVs are much more effective in hot, humid climates. Generally, an ERV is unnecessary in the coldest climates due to the short or non-existent cooling season. Beyond the design of the heat exchanger media, though, there is little difference between the two units.

Refer to *Figure 11*. Fresh outside air is drawn in and filtered before passing through the heat exchanger. At the same time, indoor air is drawn into the opposite end of the unit. That air is filtered as well, to protect the heat exchanger from dust and debris. As the two airstreams pass through the same heat exchanger, the airstreams do not mix. In a unit such as that shown in *Figure 10*, each airstream has its own small, variable-speed blower. Setting the air volume needed is a simple matter of adjusting the fan speed.

Figure 12 shows one example of how a static-plate heat exchanger core is constructed. Although the structure of the heat exchanger is roughly the same in ERVs and HRVs, the heat transfer materials used are different. In an ERV, water

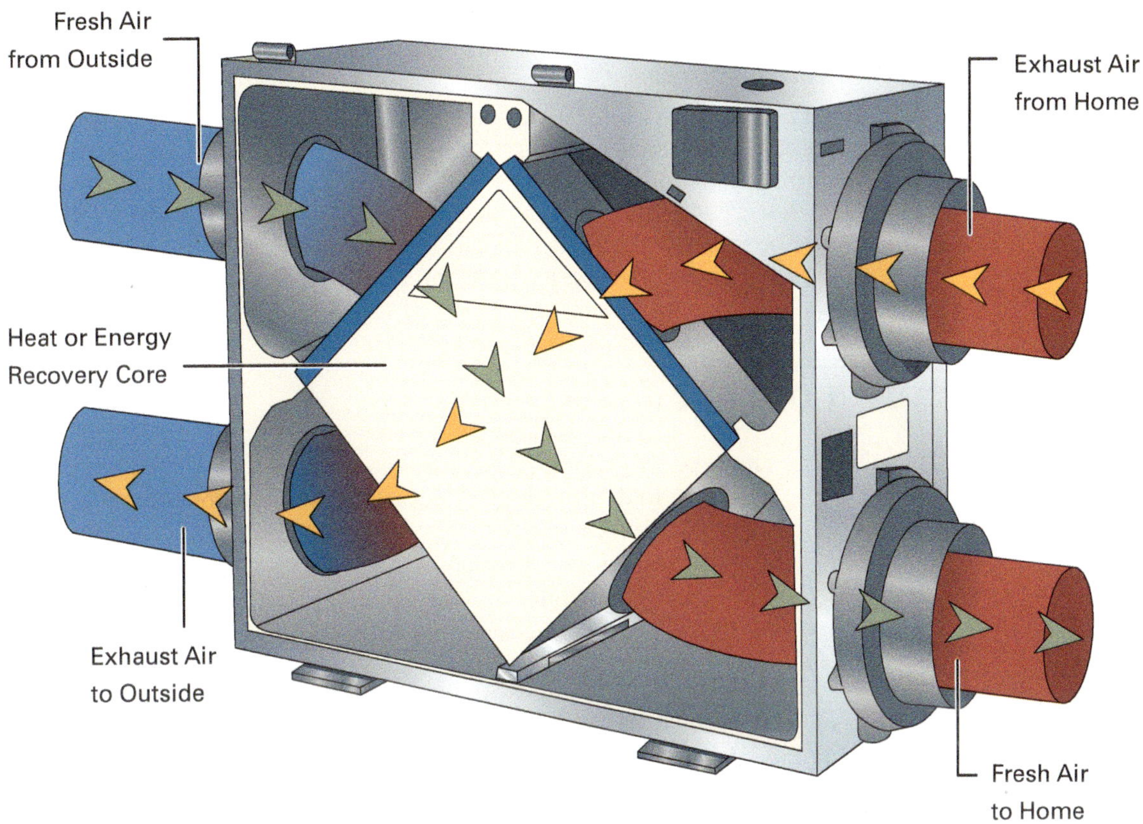

Figure 11 ERV/HRV operation.

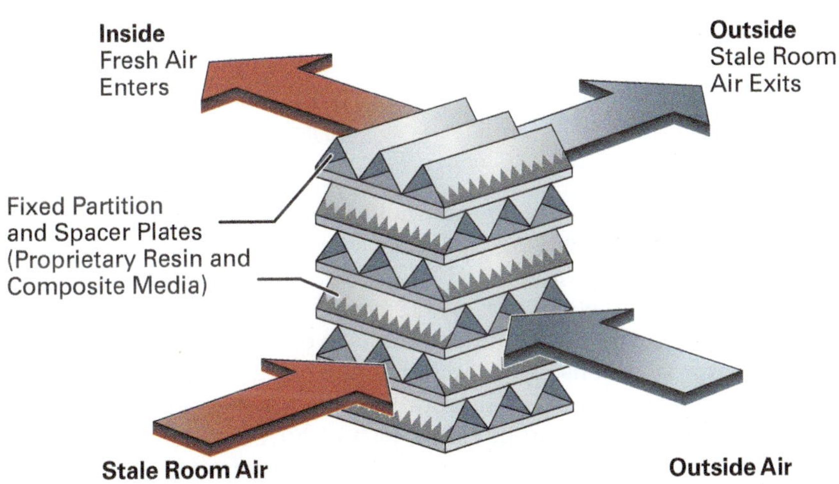

Figure 12 Static-plate heat exchanger core construction.
Source: Image courtesy of RenewAire LLC

vapor must be able to pass through the material as well as thermal energy. A portion of the water vapor in the two airstreams will naturally move from the wetter airstream to the dryer one through the unique *hygroscopic* material used to construct the core.

Think About It

ERV/HRV Operation

Air enters and exits a home in many ways and at many different times. Bathroom and kitchen exhaust fans, for example, operate intermittently and can exhaust more than 100 cfm of air per fan when in operation. Clothes dryers also send a significant amount of indoor air out of the home.

Now consider that an ERV/HRV consistently exhausts indoor air and replaces it with fresh air from outside. How might the various points of exhaust affect each other? How is the air pressure in the home affected? If you see a problem, how might you solve it in the most energy efficient way?

1.0.0 Section Review

1. The type of economizer controller that evaluates both temperature and humidity to determine if outdoor air is suitable for cooling is the _____.
   a. dry-bulb controller
   b. psychrometric controller
   c. enthalpy controller
   d. free cooling controller

2. Per ASHRAE standards, the air in a residence should be completely exchanged for fresh air approximately every _____.
   a. hour
   b. two hours
   c. three hours
   d. four hours

2.0.0 Humidity Control Equipment

Performance Tasks

There are no Performance Tasks in this section.

Objective

Describe the processes and equipment used to control indoor humidity.

a. Explain the relationship between air and moisture content.
b. Describe the equipment used to humidify and dehumidify air.

Human comfort relies heavily on the amount of moisture in the air around us. At a given temperature, we typically feel a bit cooler when there is less moisture in the air than when it is damp. A 95°F (32°C) day in Arizona has a very different feel than a similar day on the Florida coast. The dryer air of Arizona feels far more comfortable at the same temperature. However, when the desert gets cold at night, the dry conditions make it more uncomfortable.

Although we can't affect the humidity of the air outdoors, controlling the humidity of the indoor environment is part of air conditioning. This section provides an important discussion of humidity in general and introduces some of the equipment used to control it.

2.1.0 Air and Water Vapor

The moisture level in the air has a great deal to do with human comfort. The human body partially cools itself through sweating. As moisture evaporates from the skin, a cooling effect takes place. The water in perspiration requires heat to change from its liquid state to its vapor state (evaporation). A significant amount of that heat comes from the body itself. Evaporation is a key factor in keeping the body cool.

The evaporation of sweat from our skin is a constant process, whether perspiration is visible or not. When perspiration is clearly visible on the skin, the body is producing more heat than it can reject at a normal pace. Two primary factors affect the rate at which perspiration evaporates: air motion and the *relative humidity*. The faster air moves across the skin, the faster perspiration evaporates. This is because air movement keeps fresh, relatively dry air adjacent to the skin. When there is little or no air movement, the air close to the skin absorbs all the moisture it can, and the rate of evaporation slows dramatically.

Relative humidity (RH) is a measure of how much moisture is in the air compared to the maximum amount of moisture it can hold at a given temperature. In other words, it is the amount of moisture in air relative to the maximum it can hold before the water condenses and falls out as dew. It is expressed as a percentage of the maximum. The lower the percentage, the drier the air. Drier air absorbs moisture from the skin at a much faster rate. Therefore, the evaporation rate decreases as the relative humidity increases.

A comfort air conditioning system is ideally controlled to maintain indoor conditions within a range called the *comfort zone*. *Figure 13* shows the comfort zone plotted on a psychrometric chart and highlighted in red.

The recommended indoor RH for comfort ranges from about 30% in winter to 60% in summer. A higher percentage in winter is more comfortable, but too

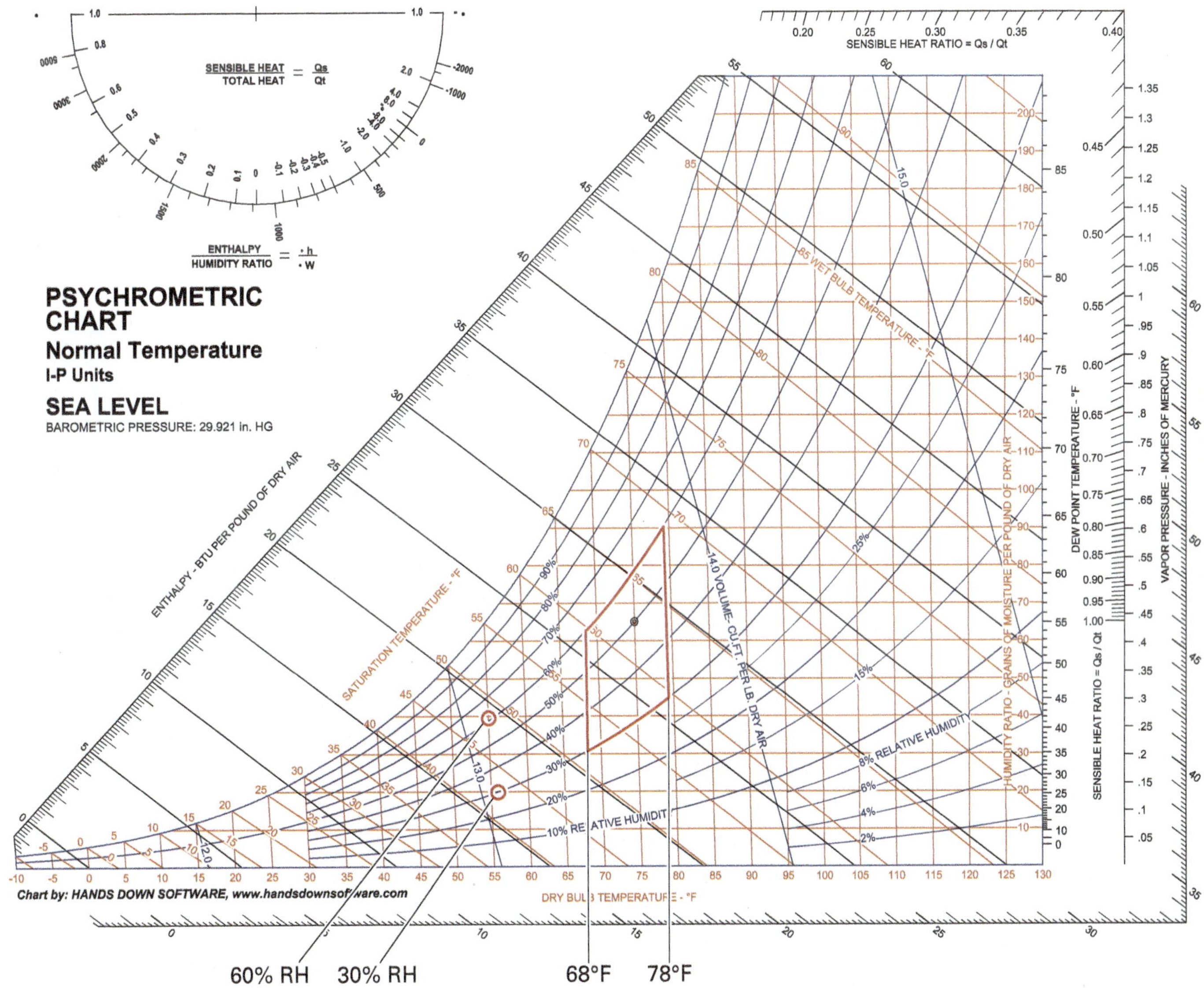

THE COMFORT ZONE
Temperatures between 68°F and 78°F
Relative humidities between 30% and 60%

Figure 13 The comfort zone plotted on a psychrometric chart.
Source: Hands Down Software

much airborne moisture leads to problems with condensation in or around the structure. RH levels below 30% and above 60% at any time of year can cause problems for both the occupants and the structure.

Low humidity, typically a problem in the winter, can cause the following problems:

- Dry, itchy skin and chapped lips
- Static electricity shocks
- Clothing static cling
- Sinus problems
- Consistent chilly feeling
- Sickly pets and plants
- Loose furniture and flooring joints
- Reduction in overall personal comfort

Excessive humidity indoors can cause the following problems:

- Condensation on the inside surface of windows
- Condensation inside walls or even on the surface
- Accelerated development of bacteria, viruses, fungi, and mite infestations
- Warping of wood furniture and floors
- Reduction in personal comfort

2.1.1 Winter Conditions

Some moisture enters a building along with outside air through infiltration and mechanical ventilation. Cold winter air, though, can be very dry compared to indoor air. For an example, refer to *Figure 14*.

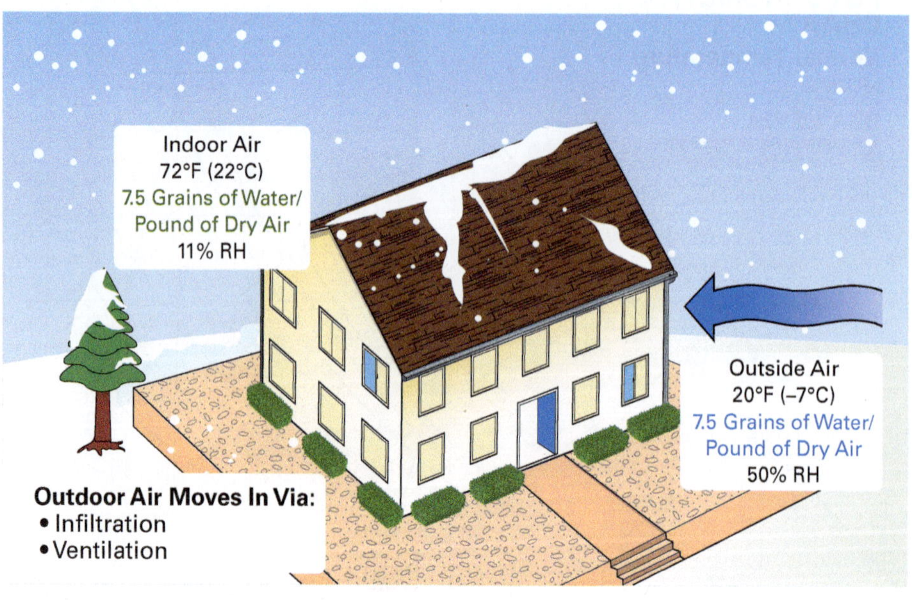

Figure 14 Comparison of outdoor and indoor air with the same specific humidity—winter conditions.

NOTE

This scenario is a good example of the value of an ERV that can transfer both heat and moisture from one airstream to the other. Cooling coils are nearly always exposed to both sensible and latent loads.

If the outdoor temperature is 20°F (–7°C) and at 50% RH, the volume of water in the air is about 7.5 grains of moisture per pound of dry air (GPP). This ratio is referred to as the *humidity ratio* or *specific humidity*. When this same air enters a building and is heated to 72°F (22°C), its moisture content remains at 7.5 GPP, but its RH drops significantly because the warmer air can hold far more water vapor. The resulting relative humidity is 11%. This is well outside of the comfort zone, and it also has a negative effect on the structure. Air at this condition is very "thirsty." Human skin and woodwork are readily available sources of moisture.

In the winter, though, we must also avoid making the indoor air too humid. It may enhance winter comfort somewhat, but it can result in water vapor condensing on windows and even inside walls (*Figure 15*). In the example shown in *Figure 14*, the indoor temperature of the home is 72°F (22°C). If we maintain an indoor humidity of 50%, will that create a condensation problem?

To determine the answer, we must determine the dewpoint of that air. Per the psychrometric chart, at 72°F and 50% RH, air has a dewpoint of 60°F (16°C). That means water vapor in the air will begin to condense and collect on any surface that is at or below 60°F. When it is very cold outdoors, it is not hard to imagine the inside surface of a window dropping to or below 60°F. You can tell when the conditions are less than favorable for the structure—condensate forms on windows and a pool of water often collects beneath them.

If the indoor humidity is maintained at 40%—still well within the comfort zone—the dewpoint of the air falls to about 57°F (14°C). That may not seem like a substantial difference, but it can easily be the difference between wet and dry windows in the winter.

Figure 15 Condensation on a window, encouraging mold growth.
Source: Daisy Daisy/Shutterstock

2.1.2 Summer Conditions

In the summer, when the indoor RH is above 60%, people complain about feeling sticky and uncomfortable, regardless of the room temperature. Indeed, people are uncomfortable outdoors in the summer when the humidity is high. This is because moisture-laden air slows down the body's evaporative cooling process.

But hot summer weather also brings cooling system operation into the equation. Mechanical cooling systems dehumidify naturally because the average indoor coil operates at a temperature well below the dewpoint of the indoor air. As air passes over the indoor coil, water vapor condenses on the fins, collects in the pan beneath it, and is drained away through the condensate piping.

Did You Know?

Humidity and Air Conditioning

A humidity-related problem really initiated the development of comfort air conditioning. Dr. Willis Carrier was not trying to invent a comfort cooling system when he discovered what we generally refer to as "air conditioning." He was trying to develop a system to remove moisture from the air in a printing plant. Excess humidity in the plant caused the paper to curl up, resulting in jammed presses. As it turned out, the process he developed to remove water vapor in the plant also removed sensible heat.

The user can't really stop this process; it is unavoidable. The efficiency of the process, however, can be affected by adjustments to the blower speed. For example, higher air velocities across the coil reduce the dehumidification rate, while lower velocities increase the dehumidification rate. In this case, however, the natural tendency of the cooling system to dehumidify is to the occupants' benefit.

NOTE

The cooling of the air as it passes through a cooling coil is a sensible process since a measurable temperature change occurs. The condensing of water vapor on the coil surface, however, represents a change of state and is therefore a latent process.

Figure 16 helps demonstrate why dehumidification by the cooling system is important. When the temperature outdoors is 90°F (32°C) with a relative humidity of 50%, it contains about 106 GPP. Indoor air at a temperature of 72°F (22°C) containing the same volume of water vapor has a relative humidity of 90%. All the air entering from outdoors migrates to these conditions. Some form of dehumidification is clearly necessary, and it also presents a compelling case for using ERVs that can reduce the impact on both the cooling system and personal comfort.

Only in unique situations does a home's indoor air need additional moisture in the summertime. In western states where the humidity is very low, some additional moisture may enhance comfort. Thanks to the dehumidification process offered by cooling systems, additional dehumidification equipment is also rarely needed. Laboratories and computer rooms, for example, are places where the humidity might need to be tightly controlled year-round. But the average home or business does not need to address indoor humidity during the summer months.

In some cases, dehumidification is needed in the winter months, when the cooling system is not usually operating. Basements beneath homes, for example,

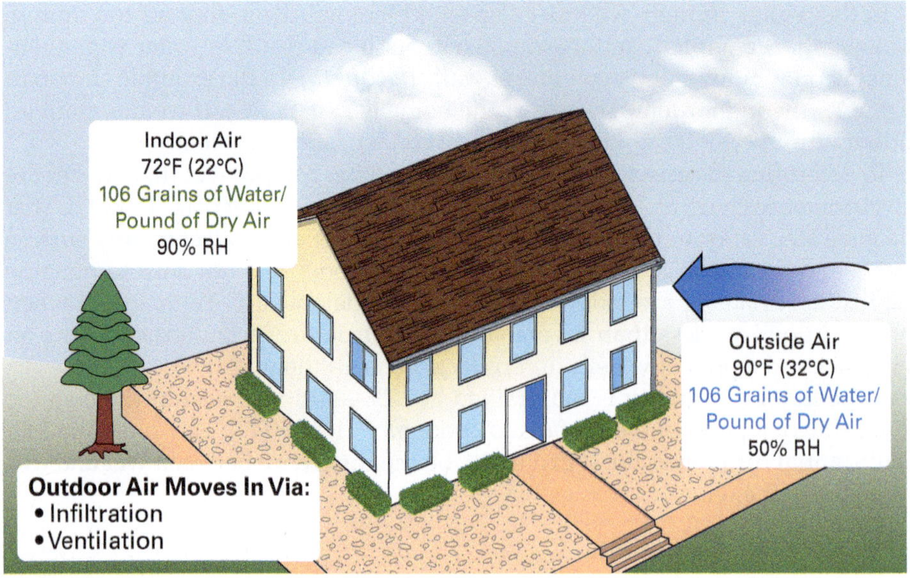

Figure 16 Comparison of outdoor and indoor air with the same specific humidity—summer conditions.

may require dehumidification during the winter. Since basements tend to be more humid because of water migrating through the walls and floor, dehumidification may even be needed year-round. This helps reduce mold and mildew growth.

2.2.0 Humidifiers and Dehumidifiers

You can now see that there are cases when the indoor humidity must be controlled to ensure comfort and protect the structure and its contents. To do that, we use humidifiers and dehumidifiers. This section will introduce some of the equipment used to manage indoor humidity.

2.2.1 Humidifiers

Humidifiers are designed to add water vapor to the indoor environment. A water supply at the installation location is required, and very often a power source as well. A drain connection may also be required. But long before a humidifier is installed, a unit with the needed capacity must be chosen.

Humidifier capacity is typically rated in gallons of water per day. The required capacity depends on several factors:

- The air volume of the building, in cubic feet. Some charts may use a building's square footage instead, with an 8′ ceiling height assumed.
- The airtightness of the structure. Leaky buildings require more moisture to be added since more air is coming and going.
- The volume of outside air entering through mechanical ventilation.

Humidifier sizing is not always an exact science. *Figure 17* shows an example graph used for the selection of residential humidifiers. Similar graphs and charts are available from most humidifier manufacturers. However, many manufacturers do not appear to consider air entering the home through mechanical ventilation.

To size a humidifier accurately, a structure should be studied carefully and calculations made to determine how much outside air is coming and going through all possible entry and exit points. However, in-depth studies like this are rarely conducted for homes. Generally, selecting the humidifier capacity based on average, or even loose, construction is a reasonably safe choice when mechanical ventilation is involved.

Because the humidifier can be automatically cycled on and off as needed, oversizing is not a common problem. However, it is best to either follow the

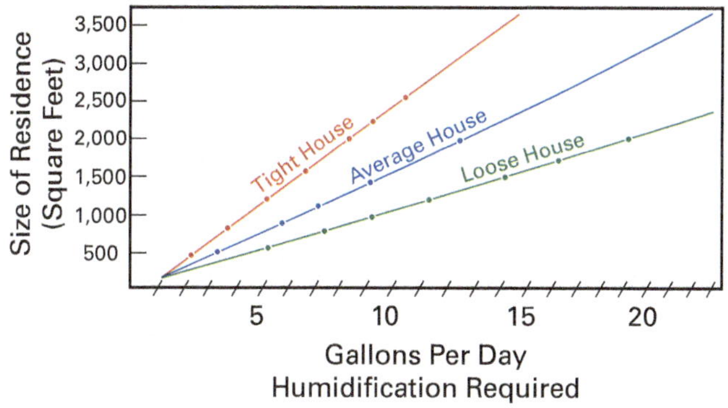

Figure 17 Example of a humidifier sizing chart (not for field use).

sizing recommendations offered by the equipment manufacturer or seek guidance from a competent engineer for accurate calculations.

Loose, average, and *tight* homes are typically defined as follows:

- *Loose* — The home has little insulation or vapor barriers, and no storm doors or storm windows. The home may also have a fireplace with a loose-fitting damper.
- *Average* — The home is insulated, has vapor barriers, and has reasonable storm doors or storm windows. The fireplace typically has a manual damper.
- *Tight* — The home is well insulated, continuous vapor barriers, and tight storm doors or storm windows. Tighter homes, which are a significant priority in today's construction standards, generally require mechanical ventilation to ensure acceptable indoor air quality.

There are four basic humidifier classes in common use:

- Wetted element
- Atomizing
- Infrared
- Steam

Remember that heat plays a critical role in the absorption of water vapor into an airstream. Warmer air picks up water vapor much faster than cool air. Keep this fact in mind as you consider the operation of these four types of units. Always follow the manufacturer's installation instructions carefully to avoid problems.

Wetted-Element Humidifiers

Wetted-element humidifiers are evaporative units commonly found in residential systems. These units have a porous media called a *water panel* that is kept moist with water while it is operating (*Figure 18*). Air passes through the wet media, picking up moisture and carrying it into the duct system. When the controlling humidistat closes, a solenoid valve in the small water line opens and fills a trough across the top of the panel. The water trickles down through the media and is collected below. Some models recycle the water while others route the excess water to a drain.

Two common types of wetted-element humidifier include the *bypass humidifier* and the *fan-powered humidifier* (*Figure 19*). A bypass humidifier (*Figure 20*) can be mounted on the supply or return duct. When a bypass humidifier is mounted on the return duct, the bypass duct is connected to the supply duct. Since there is a negative pressure in the return duct and a positive pressure in the supply duct, warm air from the furnace or air handling unit travels through the bypass duct, through the wetted media, and back through the furnace. When a bypass humidifier is mounted on the supply duct, the air travels in the same direction, but it passes through the wetted media before entering the bypass duct.

In *Figure 19*, the controlling humidistats are shown mounted on the return air ducts. This is one option, but humidistats can also be wall mounted in the conditioned space. In addition, many of today's digital thermostats offer humidity control as well as temperature control.

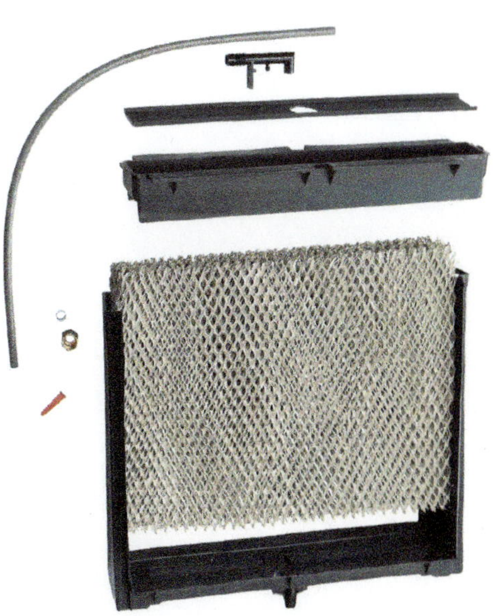

Figure 18 Typical humidifier water panel assembly.
Source: Aprilaire

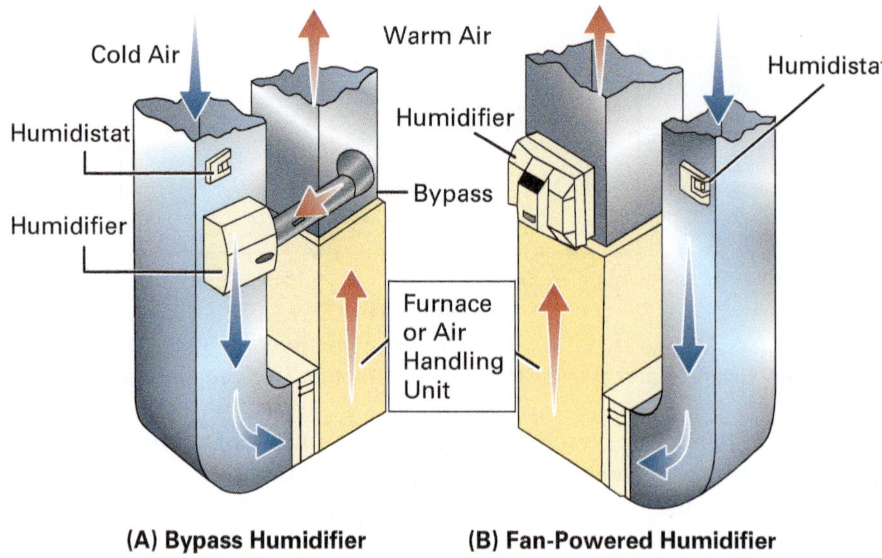

Figure 19 Wetted-element humidifier installations.

Figure 20 Bypass humidifier.
Source: oasisamuel/Shutterstock

When a bypass humidifier is mounted on the supply duct, warm air from the heating unit passes immediately through the wetted media, then travels through the bypass duct to the return duct and back through the HVAC unit. Note that manufacturers typically recommend that hot water, up to 140°F (60°C), be supplied to a bypass humidifier to increase the evaporation rate.

A variation on the bypass humidifier with a water panel is the rotating drum humidifier (*Figure 21*). As the name implies, a rotating drum partially submerged in a water sump absorbs water as it turns. A bypass duct is still required to provide airflow. As shown in *Figure 21*, some can also be installed on the bottom of a horizontal duct, where the warm air blows across the rotating media. This under-duct style must be installed on the supply duct to be most effective.

A fan-powered humidifier (*Figure 22*) relies on a small integral fan to move air across the water panel. Openings in the humidifier allow the fan to draw warm air directly from the plenum to which it is mounted, force it through the water panel, and back into the plenum or duct. Fan-powered humdifiers can be installed on a return duct, but they are far more effective when installed on the supply duct where heated air can be drawn across the media. Like bypass humidifiers, a hot water supply (up to 140°F) is recommended for best performance.

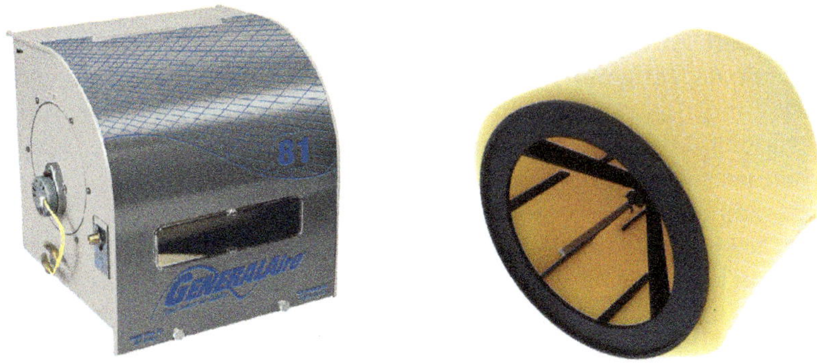

A) Rotating-Drum Bypass Humidifier **B) Rotating Drum Media**

Figure 21 Rotating drum humidifier and media wheel.
Source: General Aire Parts

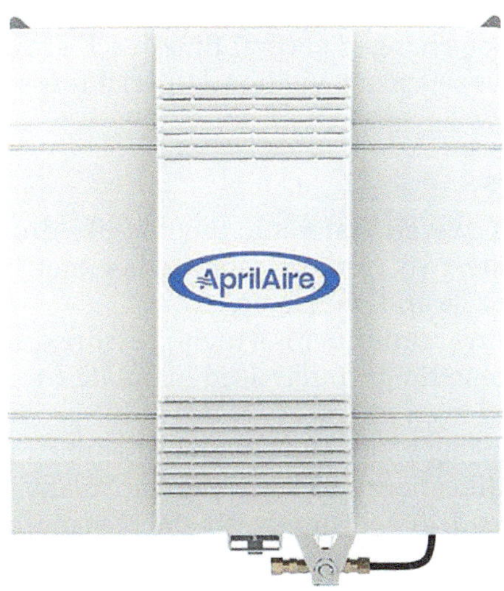

Figure 22 Fan-powered humidifier.
Source: Courtesy of Aprilaire

On Site

Self-Contained Central Humidifier

Some wetted-element humidifiers designed for residential and light commercial use are self-contained and are installed independent of the HVAC system. A return and supply duct are routed to the installation location, which is typically in the crawl space. Since only room-temperature air passes through the water panel, a hot water supply is needed for these humidifiers to be most effective.

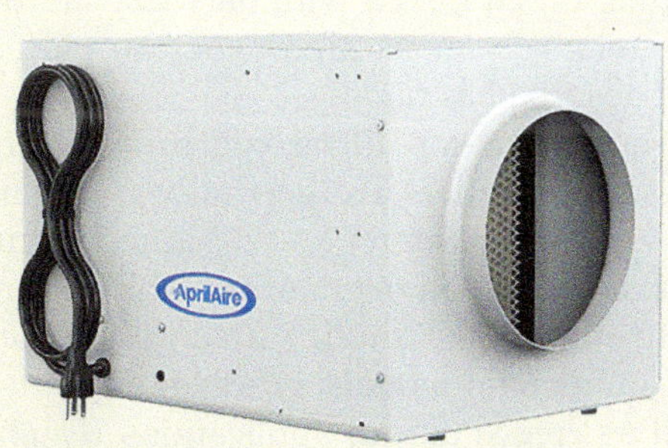

Source: Courtesy of Aprilaire

The wetted-element humidifiers shown in this section have a maximum capacity of about 18 gallons per day. However, this assumes ideal conditions that include a hot water supply, a high supply air temperature, and constant operation.

Wetted-element models are highly dependent on the temperature of the air passing through to maximize evaporation. If room temperature air is blown across the media, little moisture is picked up and distributed. Although these models can and do operate with only the HVAC system blower running, the rate of evaporation is much slower. The efficiency may also be reduced somewhat when used with a heat pump rather than a fossil-fuel furnace, since the supply air temperature is cooler.

To operate properly, a control interlock between the humidifier and the HVAC system blower is required. When there is a demand for increased humidity, but the thermostat is not calling for heat, the humidifier must be able to start the blower. This interlock is automated when a single digital thermostat is used to control both room temperature and humidity.

The level of maintenance required for a wetted-element humidifier largely depends on the quality of the water supply. As water is evaporated from the element, minerals remain as deposits. As these deposits grow, the effectiveness of the wetted element to absorb and transfer water to the airstream is reduced. The wetted element must be replaced to restore performance. In areas where humidification needs are high and the water contains lots of minerals, the water panel or drum may need to be changed several times each season to ensure proper performance. Whenever servicing any humidifier, it is essential to remove any signs of mold or mildew.

Atomizing Humidifiers

Atomizing humidifiers convert water into tiny fog-like droplets for release into the airstream. There are two common approaches that rely on atomization: high-pressure spray nozzle and ultrasonic.

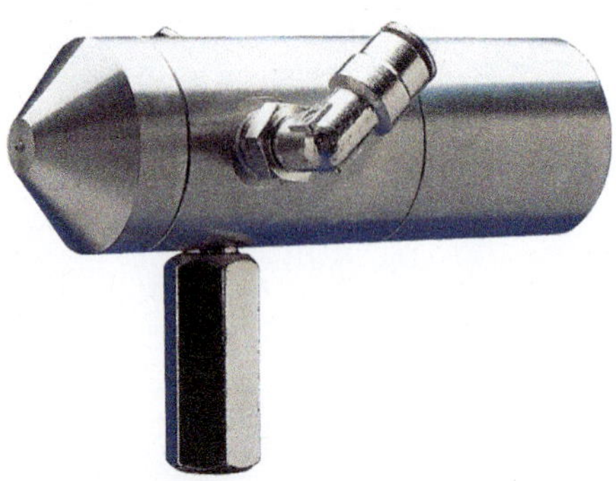

Figure 23 High-pressure humidifier spray nozzle.
Source: Condair Ltd.

Piezoelectric crystals: Crystalline materials, such as quartz, that generate an electrical potential across opposing faces when exposed to mechanical stress.

The simplest atomizing humidifier is a high-pressure spray nozzle (*Figure 23*). You have likely seen something similar used in public areas to help people cool off, in greenhouses, and even in grocery stores in the produce department. The nozzles can create fog using water and a high-pressure air source.

Spray-nozzle humidification was once used in many residential systems, with the nozzle mounted in the supply air duct. Many problems ensued, as some tended to drip water into the duct or spray more water into the duct than the air could absorb. Both conditions were common and resulted in significant biological growth and corrosion. However, spray-nozzle humidifiers are still used today in open spaces, and in ducts where proper engineering prevents the problems that plagued residential systems (*Figure 24*).

Ultrasonic humidifiers (*Figure 25*) use what are known as **piezoelectric crystals**. The crystals vibrate at a very high frequency when an electric current is applied to them. The crystals, often referred to as *transducers*, are submerged in water where the high-frequency vibrations atomize it. A plume of water vapor results. When mixed with very dry air, the plume is scarcely visible. However, in a relatively moist airstream, the plume is clearly visible since the moisture isn't absorbed as rapidly.

Many table-top ultrasonic humidifiers are sold for home and office use. These humidifiers are generally quiet and offer good capacity for their size. However, there aren't any designed for use in residential air distribution systems currently on the market. Table-top models don't have the necessary capacity to effectively humidify an entire home.

Commercial models, however, are often used in air distribution systems as well as free-standing applications (*Figure 26*). Commercial/industrial models have a great deal of capacity. For example, the type shown in *Figure 26* is available with a maximum capacity of 80 pounds of water per hour (lb/hr). Because water weighs about 8.35 pounds per gallon (1 kg per liter), this equates to 9.6 gallons per hour (gph), or 230 gallons per day (871 liters per day). That is a lot of water vapor from a single unit!

Ultrasonic humidifiers are also very energy efficient. Studies conducted by the US Department of Energy (DOE) indicate that retrofitting steam humidifiers with ultrasonic models can reduce energy consumption by more than 90%.

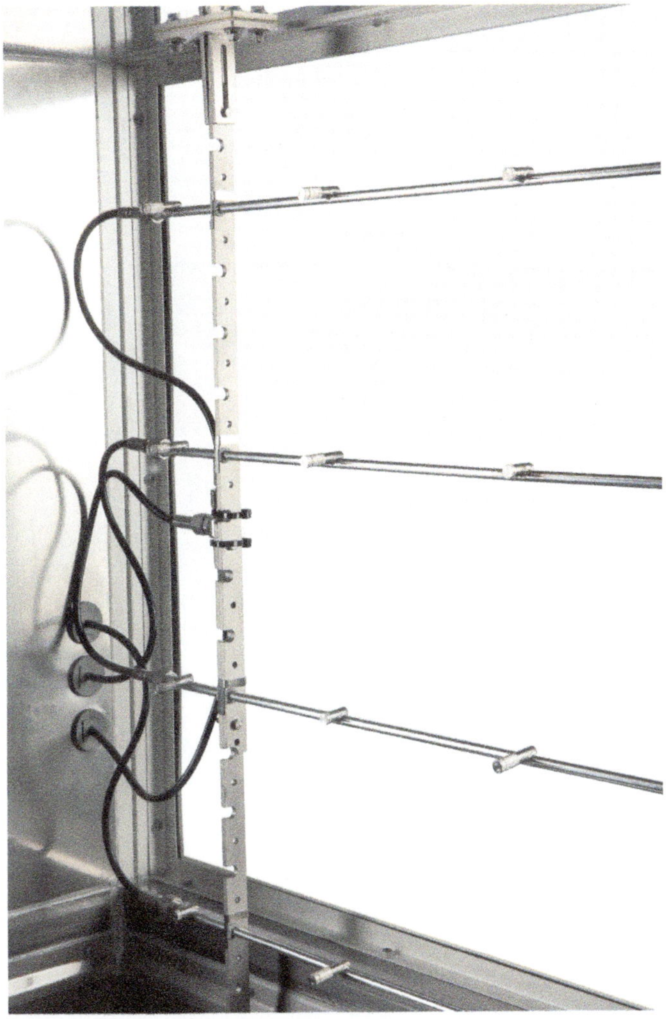

Figure 24 Commercial/industrial atomizing humidifier array.
Source: Condair Ltd.

Figure 25 Self-contained ultrasonic humidifier in operation.
Source: Courtesy of Humidifirst Co.

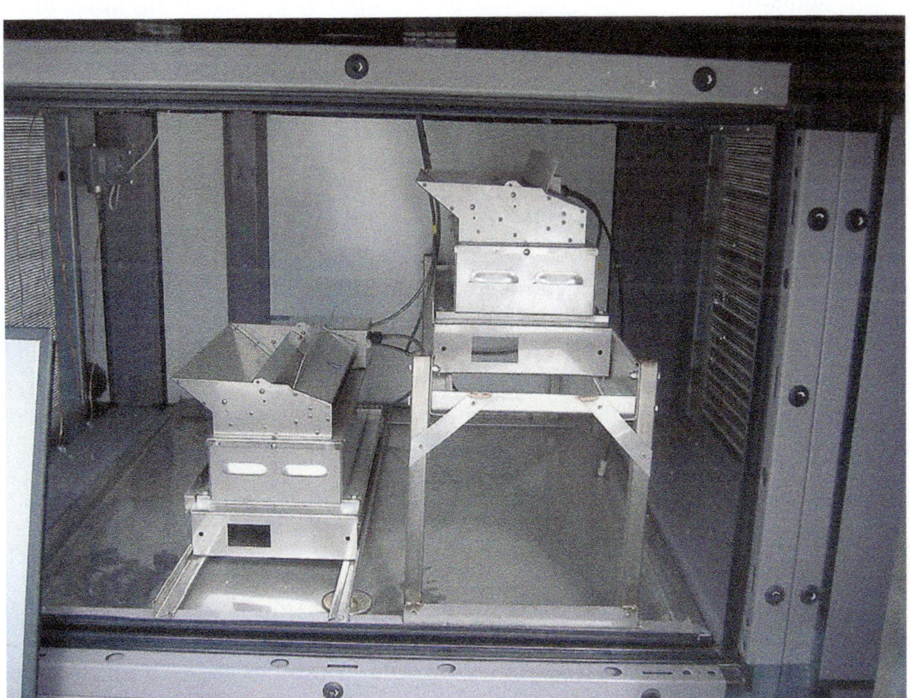

Figure 26 Commercial/industrial ultrasonic humidifiers inside an air handling unit.
Source: Courtesy of Humidifirst Co.

This is because no heat is required to create the vapor, while generating steam requires a great deal of heat. In addition, in a cooling application, the heat contained in the steam must be removed by the cooling system.

Case History

A Humidification Retrofit for NASA's Goddard Space Flight Center

NASA's Goddard Space Flight Center in Maryland required humidification in 12 labs. The calculated humidification load was 95 lb/hr of water vapor, or about 11.3 gph. Before proceeding, a study was conducted to determine the overall energy cost associated with two selected humidification units from different manufacturers. One was an ultrasonic unit while the other produced steam.

The study revealed that the overall energy cost associated with the steam equipment was $6.09 per hour. This included the energy cost associated with removing heat added by the introduction of steam into the space, as well as the energy required to generate the steam.

However, the projected energy cost associated with the ultrasonic units was only $0.31 per hour. Although the overall operating cost would be affected by the need to use deionized water in the ultrasonic units, the ultrasonic units offered a 95% savings in energy cost. The additional cost of the deionized water was easily offset by the cost of maintenance. Ultrasonic units require less maintenance than steam humidifiers that rely on electrode cannisters to generate steam.

Source: NASA Image Collection/Alamy

Atomizing humidifiers rely on a sufficiently warm airstream to pick up and absorb moisture. These units can add water vapor to the airstream at a consistent and controlled rate. In many commercial/industrial applications, the airstream is not heated. Any water vapor generated that is not absorbed into the air simply falls to the bottom of the duct or on room surfaces. This creates problems such as corrosion and possibly hazardous biological growth. Therefore, atomizing humidifiers must be operated and controlled very carefully to avoid these problems. The volume of moisture introduced must not exceed the ability of the airstream to absorb it and carry it into the conditioned space.

It is also important to note that minerals contained in the water supply remain in the water vapor produced by atomizing units. In wetted-element units, the minerals remain behind and collect on the water panel. As the water vapor is absorbed into the airstream, the minerals become dry airborne dust. The dust eventually falls and collects, and the volume of dust is often surprising. For this reason, manufacturers either require or recommend that **deionized (DI) water** or water cleansed through **reverse osmosis** be used.

Deionized (DI) water: Water from which ions, or charged particles, have been effectively removed, resulting in water that is largely free of salts and other minerals.

Reverse osmosis: A process used to demineralize or deionize water by passing it through a special filtration material under pressure.

Infrared Humidifiers

An infrared humidifier generally consists of a horizontal water reservoir and infrared lamps with reflectors above it. Infrared humidifiers do not require a warm air supply to vaporize water. Instead, energy from the infrared lamps is reflected toward the water where radiant heat evaporates the water.

Infrared humidifiers are most often found in computer rooms and data processing facilities where precise control of the humidity is required. Infrared humidifiers are usually integrated directly into the HVAC equipment serving the facility.

Since water is evaporated directly from the reservoir, the minerals are left behind. The deposits can quickly become rock-hard if not addressed. Most units use a timed flushing cycle to reduce buildup, but eliminating the deposits altogether is unlikely. Using deionized water or water treated by reverse osmosis is an option that can reduce or virtually eliminate the problem.

Steam Humidifiers

Steam humidifiers offer the advantage of high capacity and multiple approaches to steam generation. The additional heat contained in the steam, as well as the minute size of the vapor particle, help the vapor to be readily absorbed into an airstream. For residential applications in dry climates, steam humidifiers are typically the most effective approach.

The electrode steam humidifier shown in *Figure 27* is a popular option for residences. Fresh water enters the bottom to fill the steam canister. A tube at the top of the unit directs the steam into the supply air plenum of the furnace or air handling unit.

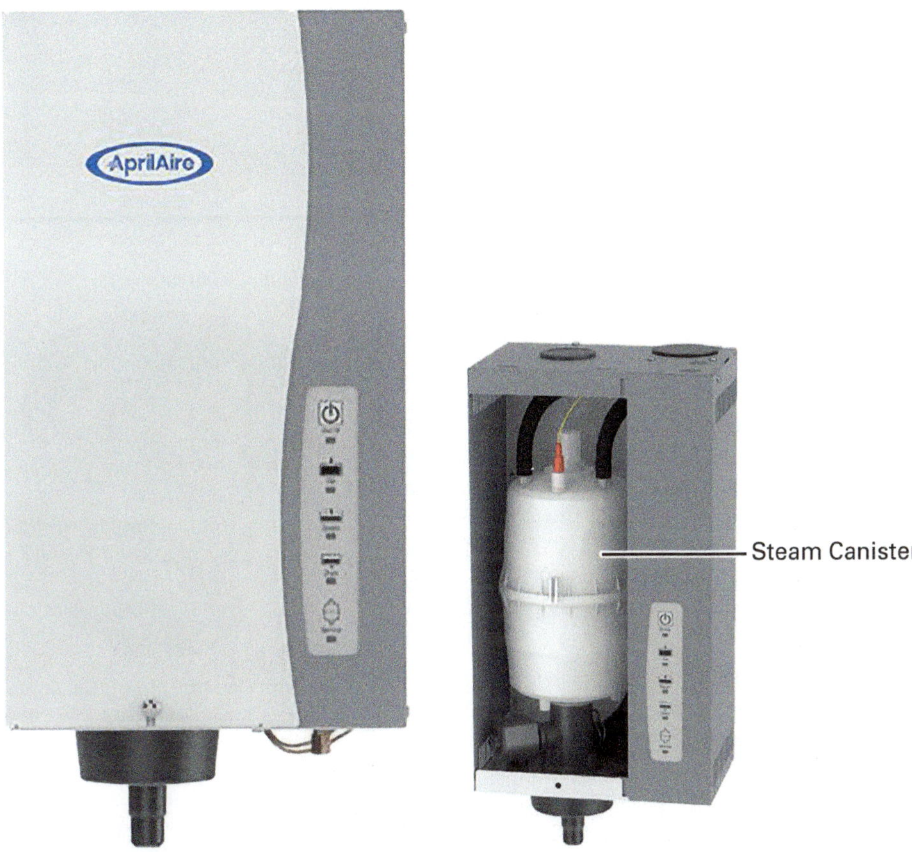

Figure 27 Duct-mounted steam humidifier.
Source: Courtesy of Aprilaire

Steam is generated inside the steam canister. As shown in *Figure 28*, the canister contains electrodes that are submerged in water as it fills. The unit relies on the conductivity of the water to establish an electrical current between the electrodes and quickly heat it to boiling. Electrode steam humidifiers have the capability to modulate steam production to match the demand, which is a significant advantage. This allows them to maintain a precise humidity setpoint.

Although some humidifiers can use deionized water to reduce mineral deposits, an electrode humidifier depends on the mineral content of the water to support conductivity. Thus, the water quality plays a major role in the humidifier's performance as well as in the life of the canister. Over time, the minerals

NOTE

Pure water, such as distilled and deionized water, does not conduct electricity and is actually an excellent insulator. However, a few ions can remain in water purified by these processes.

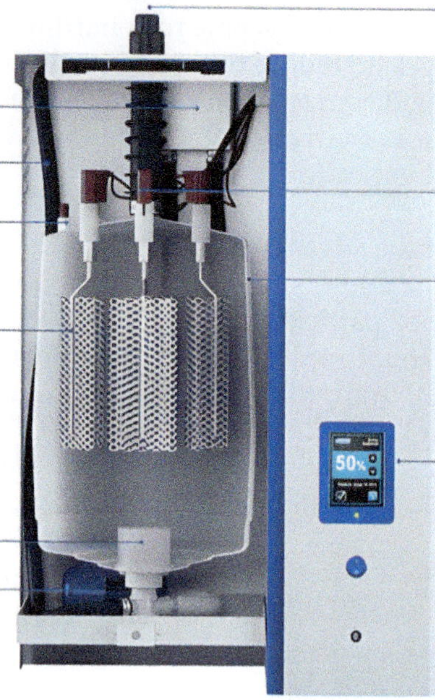

Figure 28 Electrode steam humidifier components.
Source: Condair Ltd.

Know the Code

Humidifier Water Sources

Humidifiers can contain contaminated water that should not be allowed to backflow into the potable water system. Recent changes to international plumbing codes require the water supply to a humidifier to be protected by a backflow preventer or a similar means of preventing backflow. Very few models on the market are currently equipped with a backflow preventer.

One alternative is to provide an air gap in the water supply line. An air gap is simply an open vertical space between the device sump and the water line that allows the water to freefall into the sump. This may not be practical in some cases.

Another alternative is the atmospheric vacuum breaker, shown here. A simple form of backflow prevention, it is designed to respond to a negative pressure that occurs when water is being siphoned back toward the source. When that occurs, a valve opens and allows air to enter the supply line rather than water from the sump.

Source: Courtesy of www.wattswater.com

do collect on the electrodes, reducing the effectiveness. This is an unavoidable characteristic of electrode humidifiers.

Once the electrodes become fouled with deposits, the canister is usually replaced as a unit. Most replacement canisters are sealed and considered disposable. Some, though, are serviceable. Canister replacement or refurbishment is considered standard periodic maintenance for these units. Beyond that and annual cleaning of components, little maintenance is generally required. Commercial units can be controlled locally or by a building management system (BMS), making them flexible as well as reliable.

The humidifier can be connected to an air distribution system or operate as a stand-alone system. When connected to ducts, some type of distribution manifold is needed inside to ensure optimal air/steam mixing. For smaller systems,

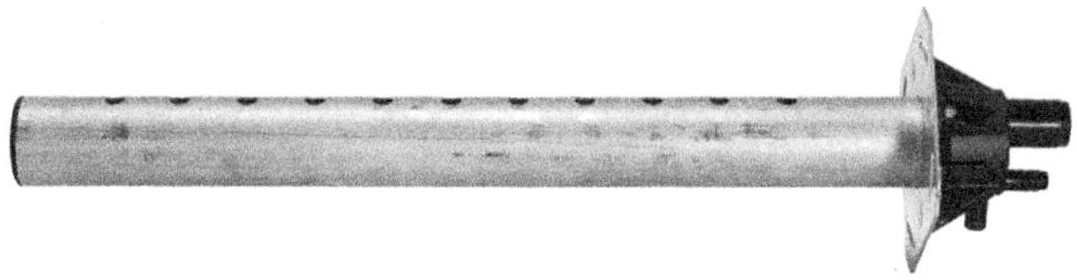

Figure 29 Steam distribution tube for duct insertion.
Source: Condair Ltd.

a manifold that inserts through the duct wall is sufficient (*Figure 29*). For larger systems and ducts, a much larger and more elaborate manifold is necessary (*Figure 30*). Note that, in all cases, the manifold design and installation must allow for any condensate that forms inside the manifold to drain back to the humidifier. The humidifier itself is usually mounted on a nearby wall or equipment cabinet and is connected to the manifold with flexible hoses.

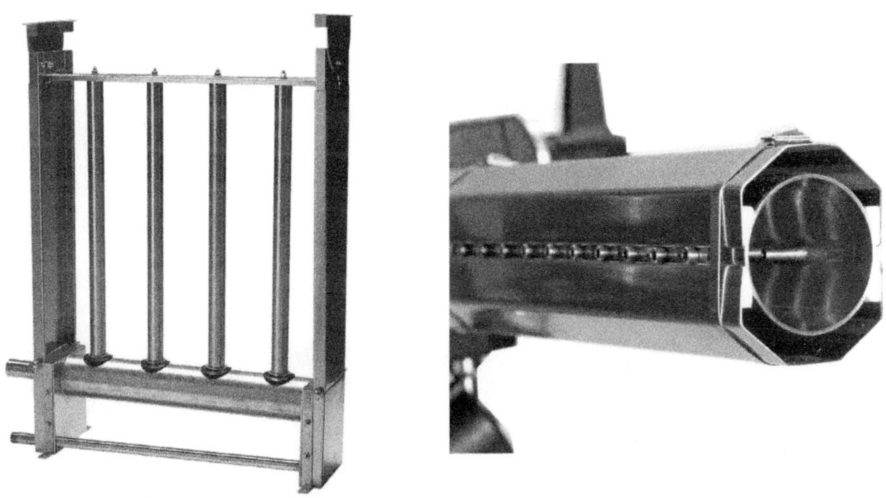

(A) Duct-Mounted Distribution Manifold **(B) Distribution Tube Construction**

Figure 30 Steam distribution manifold and tube cross-section.
Source: Condair Ltd.

When used as a stand-alone unit, these humidifiers are mounted on a wall. An accessory blower and distribution package is needed to provide the airflow (*Figure 31*).

Steam humidifiers that use traditional heating elements submerged in water are also used. A commercial model, referred to as an *electric humidifier*, is shown in *Figure 32*. Steam humidifiers can operate with potable water, deionized water, or reverse osmosis water. However, the minerals in potable water will slowly collect on the heating element and harden, just as in a coffee maker or similar appliance. The blue vessel on the bottom of the unit is a scale collection tank. Periodically, the unit completes a patented scale-control cycle that encourages the scale to break off and fall into the collection tank. This significantly extends the life of the heating elements.

A simpler version is available for residential use. Shown in *Figure 33*, this type is a simple stainless-steel reservoir with a submerged heating element. It is designed to mount on the underside of an air duct. The element heats the water, converting it to steam. Units like this can produce as much as 23 gallons (87 liters) per day. Again, potable water is most often used, and the minerals do eventually foul both the pan and heating element. The expansion and contraction of the heating element helps prevent minerals from bonding to it. Timed flushing cycles also help reduce mineral build-up in the pan.

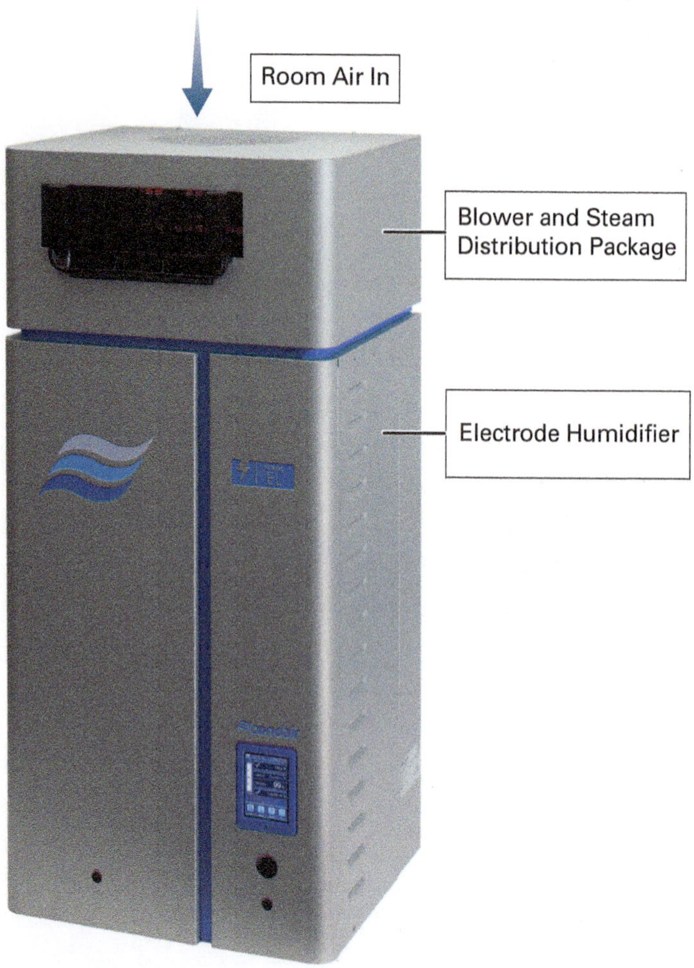

Figure 31 Electrode humidifier with accessory blower and steam distribution package.
Source: Condair Ltd.

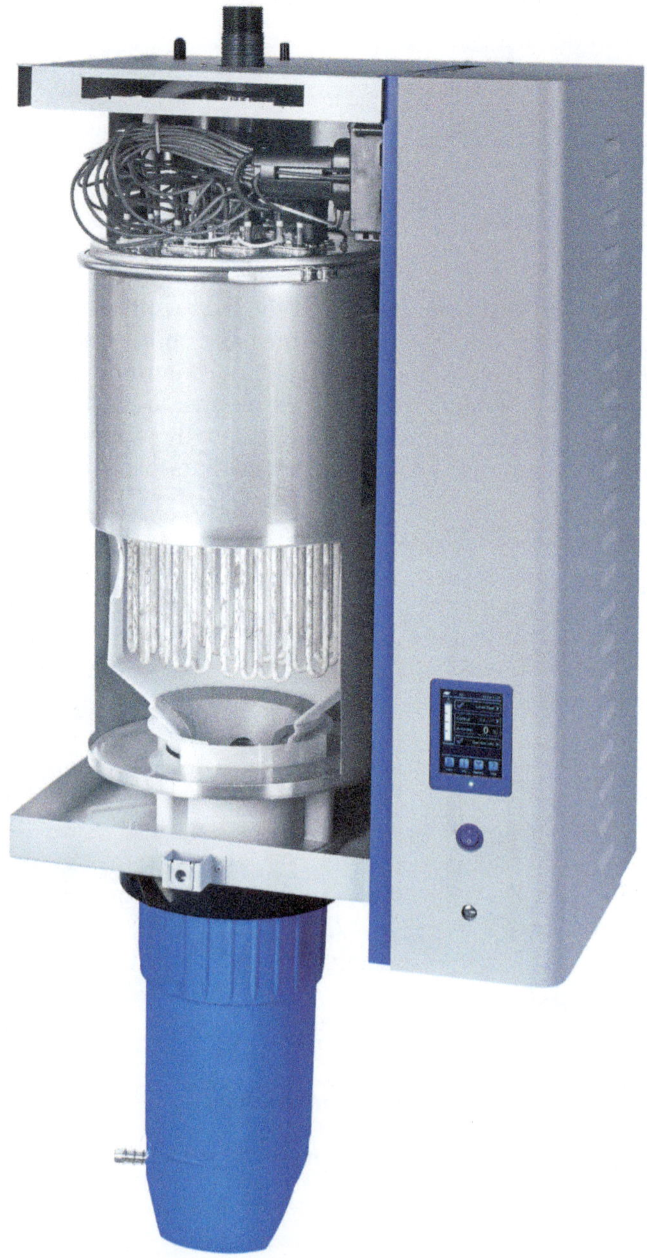

Figure 32 Electric humidifier.
Source: Condair Ltd.

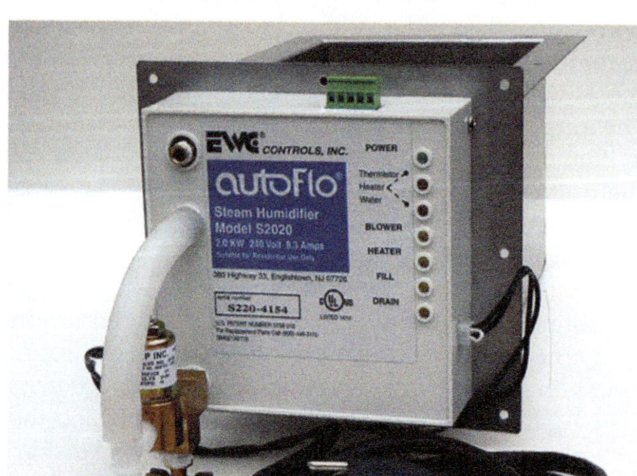

Figure 33 Residential electric humidifier.

Despite the precise level of control offered by today's steam humidifiers, it is important to note that a sufficiently dry airstream is required to prevent the fallout of unabsorbed vapor. Excessive amounts of steam injected into an airstream can cause water damage, rapid corrosion, and mold growth. When inspecting duct-mounted humidifiers, always look for evidence of excess moisture being released.

2.2.2 Dehumidification Equipment

Dehumidification can be applied for human comfort, structural health, industrial and lab processes, or to preserve sensitive materials. Dehumidification is not often necessary for comfort alone, however. As discussed earlier, dehumidification is ongoing when air conditioning systems are in operation. Because evaporator coils for comfort cooling usually operate at around 40°F (4°C), water vapor naturally condenses on the cold surface.

As a result, indoor humidity levels in homes during the cooling season tend to remain in the comfort zone without any additional effort. Basements, though, may need dedicated dehumidification to minimize biological growth and improve comfort. The exposure of basement walls to soil often allows moisture to slowly migrate indoors. In the commercial/industrial world, dehumidification may be required for a variety of reasons.

It's worth noting that technicians can influence how well an air conditioning system dehumidifies. Lowering the indoor blower speed enhances dehumidification in two ways. First, reducing the blower speed may reduce the coil temperature slightly. This causes a bit more condensation to occur, while also reducing the supply air temperature. A reduced blower speed allows the air to remain in contact with the coil surface longer, increasing the volume of condensation. Today's sophisticated thermostats and blowers driven by electronically commutated motors (ECMs) are designed to reduce the blower speed when the thermostat calls for the dehumidification mode. This minimizes sensible cooling and maximizes latent cooling.

Note that the thickness of an evaporator coil also affects dehumidification. Like a reduced blower speed, a thicker coil allows the air to remain in contact with the coil surface longer. The longer the air is in contact with the coil, the greater the condensation volume. Some evaporator coils are thin and designed to maximize sensible cooling. When dehumidification is the primary goal, the evaporator coil will be thicker. Most evaporators in residential and commercial unitary equipment represent a happy medium between sensible and latent cooling.

On Site

Computer Room Environmental Systems

Computer rooms and data processing facilities are always looking for ways to reduce static electricity around the equipment, due to the damage it can cause. Maintaining higher humidity levels goes a long way toward the prevention of static electricity.

These facilities very rarely need dehumidification since any moisture removed from the air will likely need to be replaced to maintain the humidity setpoint. For that reason, the evaporator coils are usually relatively thin compared to other equipment with equal cooling capacity. This decreases the amount of condensate collected and provides more sensible cooling.

There are two main types of equipment used to dehumidify air. A refrigerated dehumidifier is little more than an air conditioning unit modified to maximize latent cooling and dehumidification. The evaporator coil is thick, and the blower moves less air than a conventional air conditioning unit.

There are also units that attract and capture water vapor using a **desiccant**. Desiccants have a significant thirst for water. There is no refrigerant circuit required for a desiccant humidifier to function. Both types of units are introduced in this section.

Desiccant: A substance such as silica gel or calcium chloride (CaCl) that attracts water and water vapor, creating and maintaining a state of dryness around it, limited by its water-holding capacity.

Refrigerated Dehumidifiers

A simple refrigerated dehumidifier (*Figure 34*) relies on the familiar refrigeration cycle to condense and collect water vapor. Refrigerated dehumidifiers are generally portable and commonly found in residences and office areas. Integral controls are common.

Figure 34 Portable refrigerated dehumidifier.
Source: Daria Kulkova/Shutterstock

When the humidity in the surrounding air exceeds the humidistat setpoint, the unit starts. Water vapor will condense on the cooling coil as it does on any air conditioning unit evaporator.

What sets these units apart from similar cooling systems is that the condenser coil is in series with the evaporator coil in the airstream. Because the unit is designed to remove moisture without significantly changing the room temperature, the air leaving the evaporator coil must be reheated. As moist air flows across the evaporator, the air becomes cooler as well as drier. The air then flows across the condenser where it is reheated.

Since the heat of compression is added to the refrigerant in the condenser, the leaving air temperature is usually a little higher than the room temperature. After the unit operates for a while, the room temperature tends to rise. The additional heat must be removed from the room by the air conditioning system.

The condensate is often collected and stored onboard in a removeable tank. The tank is equipped with a shutoff switch to stop the unit when the tank is full. The tank must be emptied manually. Most units offer a hose connection so the condensate can flow to a drain instead. Condensate pumps are also an option. Both options allow the unit to operate continuously without constant attention to the tank.

Units like the one shown in *Figure 34* are considered appliances. When problems arise, the units are usually returned to the manufacturer or simply disposed of. Although it is unlikely that you will service one of these units, it is important to understand them. However, you may be called upon to service or install larger units with the same principle of operation.

Refrigerated dehumidifiers have a few weaknesses. One is their inability to work well with cooler air temperatures. The units are far more effective when the air is warm. When the process air temperature is below 60°F (16°C), their performance suffers, and the coil tends to develop frost. Large refrigerated units also tend to be loud and require significant maintenance. On the other hand, they use less energy and are less expensive than desiccant dehumidifiers.

Desiccant Dehumidifiers

Desiccant dehumidifiers remove water vapor through a completely different process. As the name implies, desiccants are used to capture moisture from an airstream. The major components are a rotating wheel that contains the desiccant; a blower that moves the air to be dried; a blower to move *reactivation air*; and a heat source.

Desiccants can only absorb so much moisture. Once the desiccant material is saturated, the units become ineffective. Therefore, the desiccant must be reactivated by driving the moisture out. To operate continuously, a heat source is needed for consistent reactivation. Large units typically use a gas-fired burner.

Refer to *Figure 35*. The desiccant-filled wheel rotates very slowly. The wheel is exposed to both the process airstream and the reactivation airstream at the same time. As the desiccant rotates into the reactivation airstream, the moisture is driven out by the heated airstream. The desiccant is long-lasting, but it does eventually reach a point where it is less effective and must be replaced. There are several desiccant types used, but silica gel is a common choice (*Figure 36*).

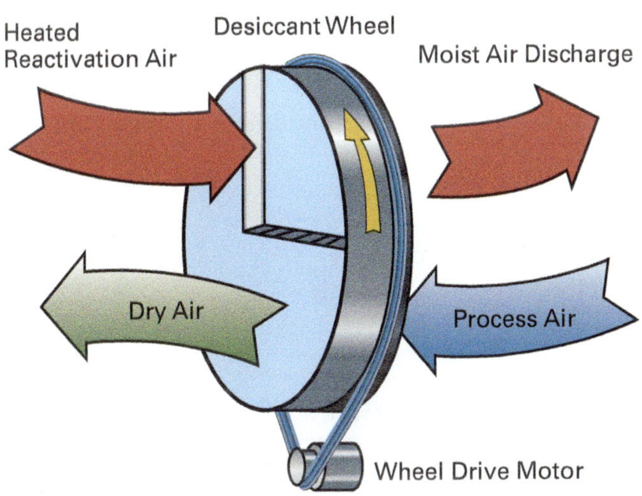

Figure 35 Desiccant dehumidifier operation.

Figure 36 Silica gel.
Source: Sendo Serra/Shutterstock

Large dehumidification units are often provided by equipment rental companies that also provide transportation and setup services. These units can help dry out large buildings where a substantial amount of water may be involved. Large desiccant dehumidifiers can remove a stunning 969 gallons (3,785 L) of water vapor per day under ideal conditions. Large units are also needed in a variety of industrial applications, such as pharmaceuticals and printing.

Surprisingly, desiccant dehumidification usually consumes more energy than refrigerated dehumidification. Desiccant models require a very warm reactivation airstream to operate effectively. The desiccant retains a great deal of this heat, which is transferred to the process airstream. Thus, these systems add a significant amount of heat to the process airstream that must be removed by other equipment. However, desiccant units have an extended life cycle and can dehumidify a cold airstream with an extremely low dewpoint. Thus, they can dehumidify environments that are well outside the range of a refrigerated unit.

On Site

Indoor Pools

Maintaining a comfortable and healthy environment around indoor pools represents a significant challenge to the HVAC industry. The area must be kept warm enough for swimmers, without being uncomfortably warm. Warm air is thirsty air, and having a massive, exposed water surface provides plenty of moisture for thirsty air to absorb. Indoor pool water is also heated, further increasing the evaporation rate. High humidity levels create comfort problems, as well as increased concerns for corrosion and biological growth. Pool chemicals that are corrosive to some metals add to the challenge. The pool environment is, simply put, very difficult to maintain.

Highly sophisticated units are designed specifically for these environments. These units must incorporate the ability to cool, heat, dehumidify, and heat pool water into a single package. The units must also operate efficiently and save energy in every possible way. For example, hot refrigerant vapor from the compressor can be used to heat pool water; reheat the air during a dehumidification cycle; and provide capacity control in the cooling mode. An integrated economizer is also part of the package. With so many different modes of operation, many of which overlap, troubleshooting these sophisticated systems can challenge even the best technician.

2.0.0 Section Review

1. True or False: When a system is operating in the cooling mode, increasing the velocity of the air passing over the indoor coil also increases the dehumidification rate.
   a. True
   b. False

2. To maximize performance, wetted-element humidifiers are highly dependent on what factor?
   a. The humidifier voltage
   b. The outdoor temperature
   c. The velocity of the air passing through
   d. The temperature of the air passing through

3.0.0 Air Cleanliness Equipment

Performance Task

1. Measure the pressure drop across a bank of air filters using a manometer or differential pressure gauge.

Objective

Describe the equipment and devices used to control air cleanliness.

a. Identify various types of media-based air filters.
b. Describe the operation of nonmedia-based air filtration and treatment equipment.

Micron (μm): One-millionth of a meter, or about 0.00003937" (3.937" × 10^{-5}). In the metric system, the micron is referred to as a *micrometer*.

Clean air is an essential requirement for good indoor air quality. *Figure 37* shows some of the common particles that contaminate indoor air. These particles have diameters that range in size from less than 0.01 **micron (μm)** to more than 100 μm. The size of most airborne particles, though, is less than 1 μm. The size of SARS-CoV-2 (COVID) virus, for example, ranges between 0.1 μm and 0.5 μm. The remainder consists of larger, heavier particles such as dust, lint, and pollen.

Most indoor air contaminants originate inside the building. For example, adhesives, carpeting, upholstery, manufactured wood products, copy machines, pesticides, and cleaning agents can release volatile organic compounds (VOCs), including formaldehyde. Research shows that some VOCs can cause chronic and acute health issues at high concentrations, and some are carcinogens. Low to moderate levels of some VOCs can produce acute reactions. Combustion byproducts such as CO and nitrogen dioxide (NO_2) come from unvented kerosene and gas space heaters, as well as woodstoves, fireplaces, and gas stoves.

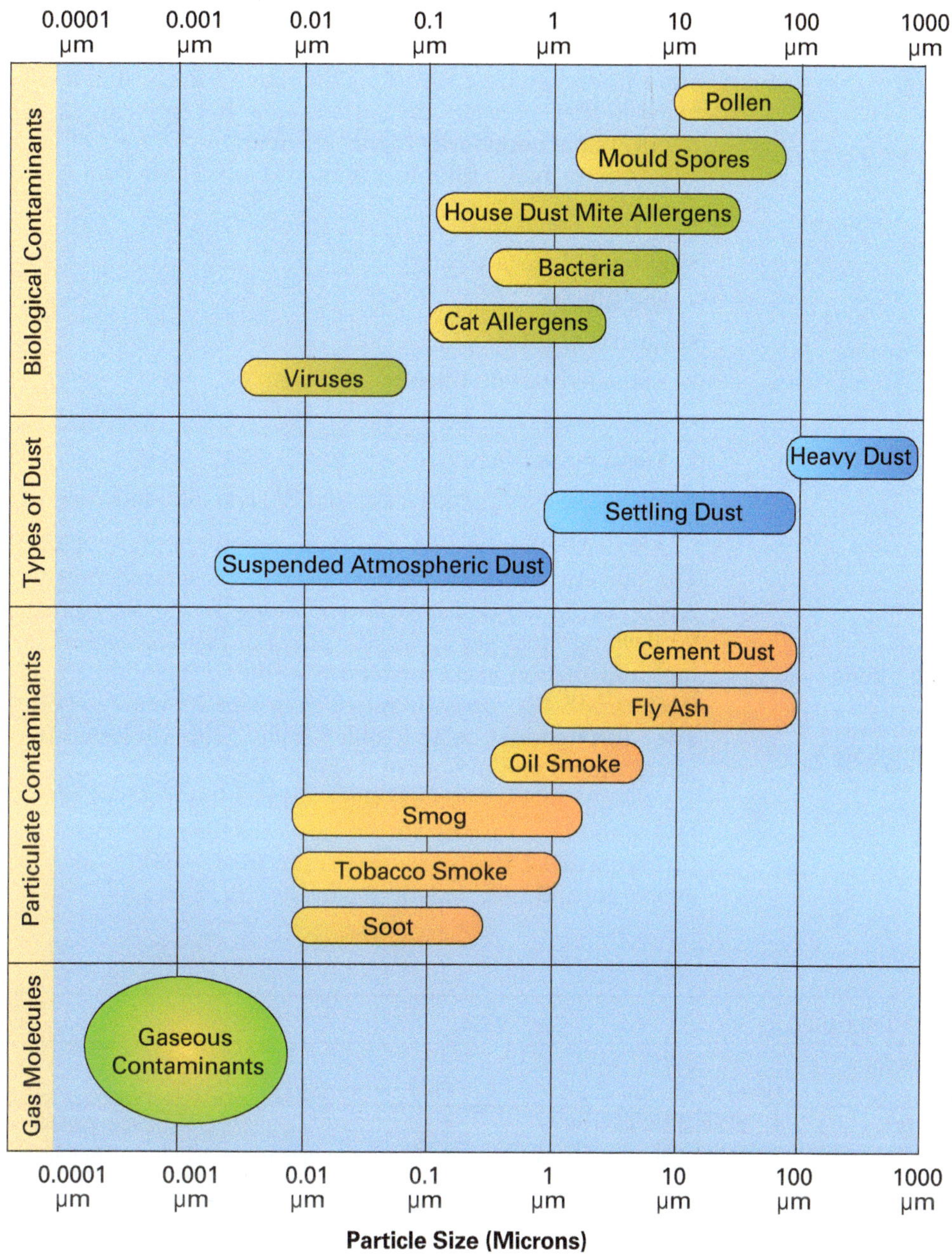

Figure 37 Sizes of common airborne particles.
Source: Jisaac9 via Wikimedia

The ventilation air entering a building is also a source of contaminants. For example, pollutants from motor vehicle exhausts, plumbing vents, and building exhaust airstreams can enter the building through air intakes, windows, doors, and even floor drains.

While ventilation can reduce the concentration of many contaminants, particles in the air need to be captured before entering the building or while remaining airborne inside. This section will present some of the common methods and equipment used to capture particles and mitigate bacteria and viruses.

3.1.0 Media-Based Air Filtration

Media-based filters are those that rely on a fabric to physically capture particles. Most of these filters are designed to be replaced on a periodic basis. Some filters, such as those that use a metal fabric, can be cleaned and reused.

The efficiency of air filters varies widely. The efficiency is reflected in the price and, often, in the usable life. However, the term *efficiency* has two meanings in the filter market. Air filters are rated by ASHRAE and other organizations by the size of the particle the filters can capture. More efficient filters capture smaller particles. However, manufacturers and marketing materials may use the term *high efficiency* to describe filters that have a longer life cycle thanks to a pleated material that increases the surface area. But in terms of particle capture, these filters are not necessarily highly efficient.

Common media filter forms and types include the following:

- Conventional filters
- Extended-surface filters
- Bag filters
- Cartridge filters
- Activated carbon filters
- Electrostatic filters
- Metal mesh filters
- High-efficiency particulate air (HEPA) and ultra-low penetration air (ULPA) filters

ASHRAE Standard 52 was the original standard related to filter efficiency. The current standard used to define the testing process and rate filters is ASHRAE Standard 52.2, *Method of Testing General Ventilation Air-Cleaning Devices for Removal Efficiency by Particle Size.* For HVACR technicians, the most important information to understand is a filter's **minimum efficiency reporting value (MERV).** MERVs currently range from 1 to 20, with 20 being the most efficient. The MERV should be accessible for any properly tested filter product, and it is often printed on the filter or its packaging.

Minimum efficiency reporting value (MERV): A rating of air filter particle-capture efficiency based on testing outlined by ASHRAE Standard 52.2.

Table 2 provides general filter application guidelines based on their MERV. HEPA and ULPA filters are the most effective, approaching 100% efficiency capturing particles that measure 0.3 μm in size. HEPA and ULPA filters are very expensive and used only in the most critical areas.

TABLE 2 Filter Application Guide

MERV Rating	Particle Size (μm)	Uses	Filter Material/Form
1–4	>10.0 μm	• Minimal filtration for residential applications • Portable equipment	• Disposable fiberglass or synthetic panels • Aluminum mesh, latex-coated animal fibers, or foam rubber panels • Electrostatic filters with self-charging woven polycarbonate panels
5–8	3.0 to 10.0 μm	• Commercial applications • Better residential performance • Industrial • Paint booth inlet air	• Extended-surface disposable filters with cotton-polyester media in a cardboard frame • Disposable synthetic media panels • Cartridge filters with graded-density cube or pocket filters and synthetic media
9–12	1.0 to 3.0 μm	• Superior residential performance • Better commercial performance • Some labs	• Bag filters 12" to 36" (300 to 900 mm) deep and 6 to 12 pockets with flexible microfine fiberglass or synthetic media • Rigid box filters 6" to 12" (150 to 300 mm) deep with lofted or paper media
13–16	0.30 to 1.0 μm	• Super commercial performance • Hospital rooms • General surgery • Smoking lounges	• Bag filters 12" to 36" (300 to 900 mm) deep and 6 to 12 pockets with flexible microfine fiberglass or synthetic media • Rigid box filters 6" to 12" (150 mm to 300 mm) deep with lofted or paper media
17–20	<0.30 μm	• Radioactive applications • Pharmaceutical manufacturing • Clean rooms	• HEPA- and ULPA-class filters

Air filters should generally be located as directed by the equipment manufacturer. *Figure 38* shows typical mounting locations for smaller equipment. Filtration must always be provided on or near the return air inlet of the air handling unit, or where the air enters the return duct. In critical applications, the supply air may also be filtered. HEPA or ULPA filters, for example, may be placed at each supply air outlet of a system serving a surgical area or clean room.

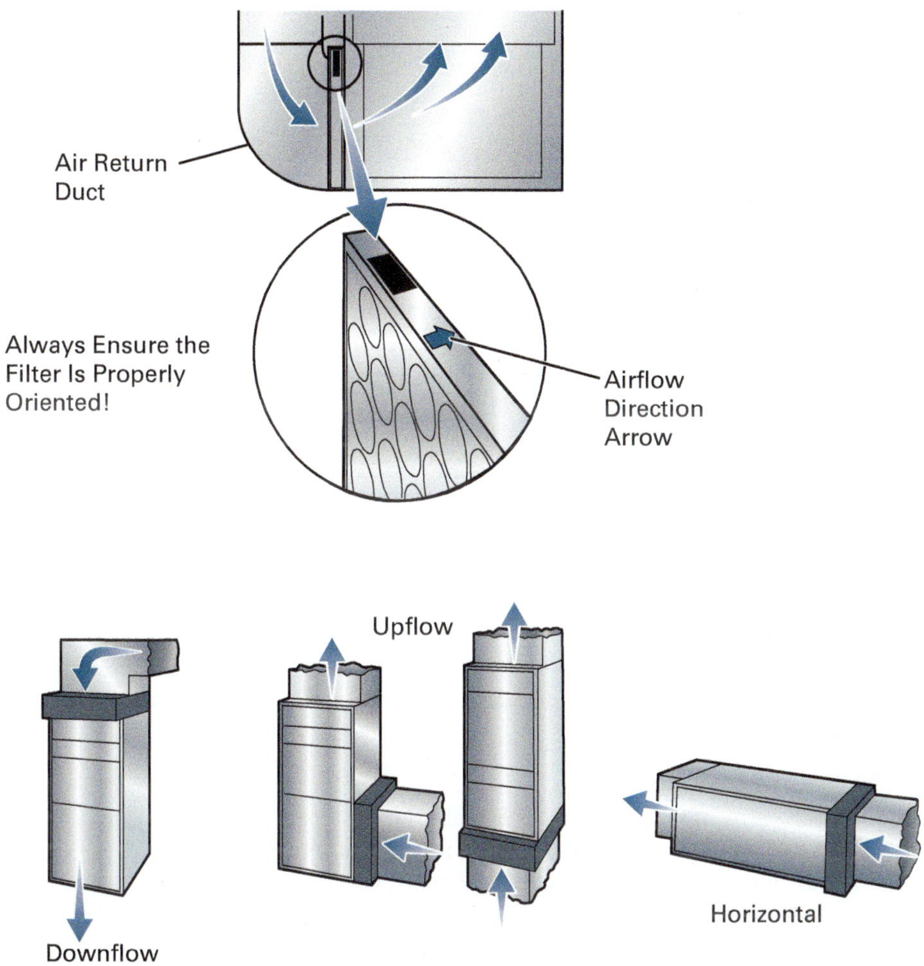

Figure 38 Common filter mounting locations.

Although most of the discussion about air filters is related to air quality, never forget that air filters also protect the equipment. Poorly fitted and mismanaged air filters can quickly result in dirty ducts, clogged coils, and dirty blower wheels (*Figure 39*). This encourages biological growth and reduces the efficiency of the equipment significantly. Indoor coils and blowers are often very difficult to access and clean, especially in ducted systems.

It is extremely important that air filters be replaced in a timely manner. Although it is easy to simply replace air filters based on elapsed time, such as monthly or quarterly, only experience with a system can truly determine how long the filter should remain in service. Filter life is affected by many factors, including the cleanliness of the environment and the filter surface area provided per cubic foot of airflow. Homeowners often care for their own air filters, but they may need your input regarding timing. Clogged filters reduce airflow and lead to inefficient system operation.

Figure 39 Fouled blower wheel and coil.
Source: spyarm/Shutterstock

Before introducing some of the most common filter types, it is also important to understand that all media air filters affect airflow. While media filters rated as MERV 4 and lower have a limited affect on airflow, those with high MERV ratings can significantly impede airflow even when clean. Although it may seem like a good idea to install a HEPA filter in a residential system to capture viruses and other tiny particles, a residential air handling unit does not have the capacity to move air through a HEPA filter. A lot of horsepower is required to move air through high-efficiency filters. All media filters must be selected to match the capabilities of the system in which they are installed.

It is also essential to ensure that air filters fit properly. Air will seek the path of least resistance. Any openings around the perimeter will allow air and debris to bypass the filter, defeating the purpose.

Figure 40 Fiberglass-media conventional filters.
Source: Parker HVAC Filtration Division

Know the Code

Air Filtration

The *International Mechanical Code*® requires heating and air conditioning systems to be fitted with air filters. Filters must be provided in a manner that ensures return, outdoor air, and make-up airstreams are filtered before encountering any heat exchanger or coil. Ducts must be constructed to ensure the even distribution of air across the face of the filter. Any adhesive coatings applied are required to have a flash point that is equal to or higher than 325°F (163°C).

3.1.1 Conventional Filters

Conventional disposable filters include those made from fiberglass strands, latex-coated natural fibers, and open-cell foam material (*Figure 40*). The filtration media may be coated with an adhesive that captures airborne particles. The media is also available in bulk rolls from which panels can be cut and placed in permanent metal frames. Conventional filters are often provided with a new residential furnace or air handling unit for installation.

Some conventional filter media gets progressively denser as the air passes through it. For this reason, and because the filter may be structurally enhanced on one side to keep the media from being drawn out by the blower, media filters must be placed in the airstream correctly. Most media filters are marked with an arrow that indicates the correct direction of airflow.

Conventional filters are offered in many different sizes, and custom sizes can be fabricated by manufacturers or local vendors. Most have a nominal thickness of 1" (2.5 cm), but 2" (5 cm) filters are also common.

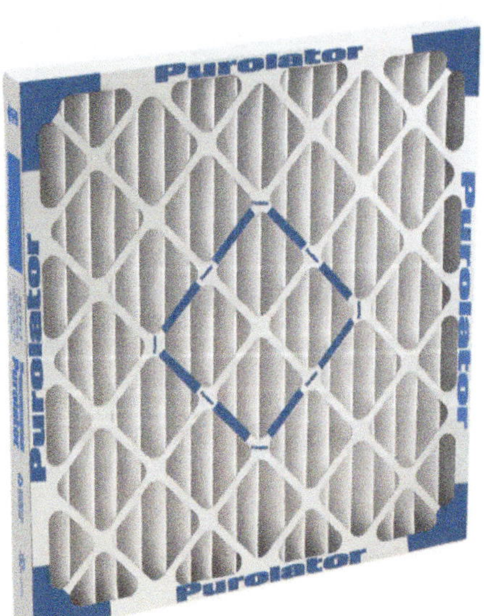

Figure 41 MERV 8 pleated filter.
Source: Parker HVAC Filtration Division

3.1.2 Extended-Surface Filters

The term *extended-surface filters* generally refers to *pleated* filters (*Figure 41*), which is the term most technicians are familiar with. By adding pleats, the surface area of the filter is substantially increased. This allows the filter to capture a larger volume of debris before replacement is required. Pleated filters potentially provide two to four times longer life than conventional filters. The nonwoven, cotton-synthetic material used also offers a higher MERV rating than conventional filter media. The efficiency rating of pleated filters generally ranges from MERV 4 to MERV 12.

In the residential and light commercial markets, pleated filters dominate. Like conventional filters, pleated filters are offered in many different sizes. The nominal thickness of pleated filters ranges from 1" to as much as 6". However, note that the thicker filters are not more efficient in terms of particle capture, as these filters offer the same MERV value as the thinner versions. The extra depth merely increases the volume of debris that can be captured before replacement is needed. In the average residential or commercial environment, a 6" pleated filter can remain in service a very long time.

3.1.3 Bag Filters

Disposable bag filters (*Figure 42*) are used where high-performance, high-volume filtration is required in commercial and industrial environments. MERV ratings of 10 to 16 are common. The extended holding capacity can be attributed to their depth. The depth of the pockets ranges from 12" to 36" (30 cm to 91 cm). Each filter assembly typically has 4 to 10 pockets attached to a frame at the open end. The opening dimensions of bag filters are typically divisible by 6, such as 12" × 24" or 24" × 24".

Bag filters are made from several different types of media, depending primarily on the capture efficiency needed. Both finely woven fiberglass and synthetic media are common. Each pocket may be reinforced internally or on the leaving-air side of the media.

Figure 42 24" × 24" bag filter with eight pockets.
Source: Parker HVAC Filtration Division

3.1.4 Cartridge Filters

Cartridge filters offer a different filter form. Cylindrical cartridge filters (*Figure 43*) are rarely used in comfort systems. Cartridge filters are most often found in industrial dust collectors and air compressors. These filters come in a variety of unique sizes to meet the needs of equipment manufacturers.

Another cartridge style that is used in comfort systems is most accurately called a *box filter* (*Figure 44*). Box filters generally have MERV ratings above 10 and are used for primary filtration or as prefilters for HEPA filters. They are generally constructed in a metal casing that keeps them rigid and resists distortion. This type of filter, as well as the cylindrical version, relies on gaskets around the perimeter to ensure that particles cannot bypass the filter.

Figure 43 Cylindrical cartridge filters.
Source: Parker HVAC Filtration Division

Figure 44 Box filter.
Source: CamFil USA

The media used in cylindrical cartridge filters is most often pleated and not notably different from the media used in pleated panels. Some have a metal cage around them to ensure stability and prevent blowouts. Box filters use one of several different media types. Cartridge and box filters are often subjected to air moving through at a significant pressure compared to the average comfort system. This is another reason for the use of gaskets—to prevent air leakage at higher pressures.

3.1.5 Activated Carbon Filters

Adsorption: A process through which a solid material captures and holds gas or liquid atoms, ions, or molecules on its surface as a thin film.

Activated-carbon pleated filters (*Figure 45*) provide both particle filtration and a measure of odor removal. Activated carbon acts like a porous sponge that neutralizes odors through **adsorption**. However, there is a limit to how much gas the carbon can adsorb.

Figure 45 Activated-carbon pleated filter.
Source: Parker HVAC Filtration Division

Standard pleated filters have the carbon impregnated into the media, so the volume of carbon present is relatively low. Activated-carbon pleated filters have the same MERV ratings as their standard pleated counterparts. Activated-carbon pleated filters are easily identified by their blackened or dark gray media.

Industrial applications usually require a more aggressive approach. When odor or gas adsorption is the priority and the air is being cleansed of particles another way, activated carbon in granular form is loaded into frames or trays and placed in the airstream (*Figure 46*). This approach puts a large volume of carbon to work. Panel filters can also be constructed by filling the void between two layers of filter media sealed around the perimeter.

3.1.6 Electrostatic Filters

Electrostatic filters (*Figure 47*) clean the air using an inherent electrostatic charge stored in layered pairs of woven synthetic (plastic) filter material. Electrostatic action attracts small, naturally charged particles, and causes them to collect and settle on the filter material.

Electrostatic filters are washable rather than disposable, which reduces landfill waste. These filters typically have a MERV rating between 8 and 10, limiting the value. Note that electrostatic filters, despite the name, do not require a power source.

Figure 46 Deep-bed gas scrubber module with activated carbon-filled filter cells.
Source: CamFil USA

Figure 47 Electrostatic air filter.
Source: Permatron Corporation

3.1.7 Metal Mesh Filters

Metal mesh filters are most often used in outside air intakes as a prefilter and a means to keep insects and birds out of the equipment plenum. Metal mesh filters are also used extensively as grease filters in both residential and commercial kitchens. Although washable and durable, these filters do not capture small particles like the others discussed.

3.1.8 HEPA and ULPA Filters

HEPA filters are made from a randomly arranged pattern of fibers, usually fiberglass. To be classified as a HEPA filter, the filter must be able to trap particles as small as 0.3 μm. This translates to 99.97% of all airborne particles that can be captured. HEPA filters are generally found in box form, the form of which is shown in *Figure 44*, since they must remain rigid and undistorted against a gasketed surface. However, HEPA panel filters are also available.

ULPA filters take filtration to another level. An ULPA filter can remove 99.99% of particles of 0.12 μm and larger.

Both HEPA and ULPA filters are used in applications where the utmost cleanliness is required. This includes hospital surgical suites, clean rooms, and facilities that process nuclear material. HEPA and ULPA filters are expensive, but when proper prefilters are used and the application is already exceptionally clean, they can last up to 10 years, according to manufacturers. Prefilters are often used with these filters to avoid fouling the filters with relatively large debris. Only the tiniest of particles should remain in the airstream as it encounters HEPA and ULPA filters. When prefilters are used, the HEPA or ULPA filters are referred to as *final filters*.

Both types do have a significant resistance to airflow. Blowers used in systems equipped with HEPA or ULPA filters require a lot of horsepower to overcome the resistance, which substantially increases the system operating cost.

3.2.0 Non-Media Filtration Systems

Not all filtration and air cleaning equipment rely on some form of media to capture particles. The most common of these are electronic air cleaners, ultraviolet (UV) light systems, and needlepoint bipolar ionization systems. All three approaches require a power source.

3.2.1 Electronic Air Cleaners

Electronic air cleaners (*Figure 48*), also known as *electrostatic precipitators*, capture airborne particles in a unique way. The filtration section consists of a prefilter, ionizer section, and a collector section. Some models also provide an activated carbon postfilter to assist with odor removal. A high-voltage, solid-state power supply, ranging from 6,000 VDC to 10,000 VDC, is used to electrically charge (ionize) particles as they enter the filter housing. This section consists of a series of wires pulled taut across the opening. In the collector section, the charged particles are captured on metal plates that carry an opposing charge. Electronic air cleaners are typically assigned a rating of MERV 11.

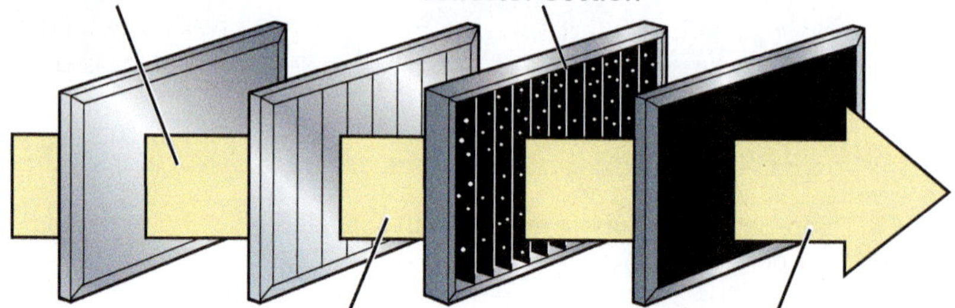

Figure 48 Electronic air cleaner and operating principle.

While ducted models are designed for installation in the return air duct adjacent to the furnace, standalone units are also available. The prefilters are generally washable, and the collector section of most residential products can be placed in a dishwasher for cleaning.

Electronic air cleaner operation is sensitive to the velocity of airflow. When undersized and exposed to excessive airflow, these air cleaners become less effective. When the airflow is reduced below the manufacturer's recommended minimum, an excessive amount of ozone can be generated.

Ozone is a needed component in the upper atmosphere since a layer of ozone helps protect us from the UV rays of the sun. However, at ground level, it is considered a significant pollutant. Per the EPA, ozone can irritate the lungs and cause adverse health effects. These facts should be considered before installing an electronic air cleaner or any other equipment that may generate ozone. Ensure that any electronic air cleaner installed conforms to United Laboratories (UL) Solutions Standard 2998, *Environmental Claim Validation Procedure (ECVP) for Zero Ozone Emissions from Air Cleaners*.

Note that electrostatic precipitation has other applications far beyond the home. This technology has been applied in power generation for many years to capture smoke and harmful pollutants from the combustion byproducts generated at fossil fuel power plants. Electronic air cleaners' capability to capture smoke particles also led to their use in commercial settings where smoking indoors is allowed. Thus, electronic air cleaners have often been referred to as *smoke eaters*.

3.2.2 UV Light

Ultraviolet (UV) light is used to help prevent the growth of bacteria and other microorganisms. There are many manufacturers and designs of UV air treatment equipment, but they all share the same basic operating principles.

C-band UV light (UVC) energy in a narrow range of wavelengths destroys viruses and similar microorganisms by penetrating the cell walls, damaging the protein structure of the cells and/or chemically altering the cell's DNA. Once this occurs, the organisms die or fail to reproduce. The germicidal effectiveness (killing power) of UV light is directly related to the dose applied, which is a function of both time and light intensity. Since UV light can be applied in several different ways, using it in this manner is referred to as *ultraviolet germicidal irradiation* (UVGI).

HVAC systems apply UVGI in two ways: treatment of fixed objects and treatment of a moving airstream. In fixed-object treatment, it is usually applied to the indoor coil and condensate drain pan as well as the surrounding area (*Figure 49*). These are the most likely locations for biological growth to occur, but certainly not the only places. The time required to destroy microorganisms depends on several factors, including the distance at which the UVC emitter is mounted from a surface; the size and intensity of the UVC light source; and the temperature of the air.

In a moving airstream, the process is generally less effective because of the short time the air is exposed to the light. Therefore, purification of an airstream normally requires multiple UV light sources and reflectors to lengthen the time of exposure.

Figure 50 shows two examples of UVC light fixtures. For smaller systems, single lamps can be mounted on the duct wall, with the lamp extending through the duct and the power supply on the outside. Multiple lamps can be used as needed to gain the necessary coverage. Larger fixtures are installed entirely within an air handling unit, again using multiple lamps to ensure all surfaces are exposed to the light.

Figure 49 UV light system in a commercial air handling unit.
Source: Fresh-Aire UV/Triatomic Environmental, Inc.

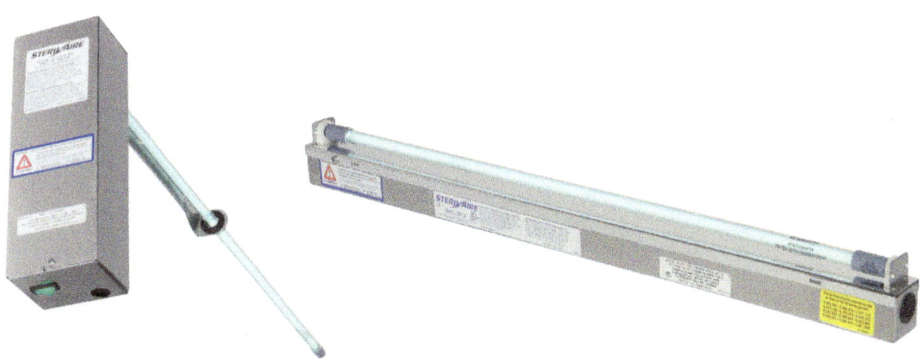

(A) Single-Ended UVC Fixture **(B) Double-Ended UVC Fixture**

Figure 50 Examples of UVC lamp fixtures.
Source: Steril-Aire, Inc.

There are four important points for technicians to remember about UVGI systems:

- Human exposure to UV light must be minimized. UV light, and especially UVC light, can cause serious sunburn, skin darkening, and skin cancer. In addition, it can result in retinal burns and cause *photokeratitis*. This condition causes a sensation of sand in the eyes. Unfortunately, there are no immediate warnings of overexposure. Symptoms often appear hours after exposure has occurred.

WARNING!

Overexposure to UV light can cause serious skin and eye damage. Avoid exposure whenever possible. If short-term exposure is required, cover all exposed areas of the skin with suitable clothing. Wear eye protection, such as a face shield, that is ANSI-certified for UV protection. Note that safety glasses offering UV protection are also available, but do not protect the face.

NOTE

NBPI can cause a rise in CO_2 levels indoors. While it is unlikely that CO_2 levels will increase to a point of concern, many commercial ventilation systems rely on CO_2 readings to control the volume of outside air introduced into the building. Consequently, systems may respond with increased ventilation rates that also increase system energy consumption.

- UV lamps lose their ability to disinfect over time. Although the lamps may remain functional, the ability to kill bacteria is significantly diminished after roughly 9,000 hours of operation. For most applications, annual replacement of the lamps is necessary.
- To properly select and install UVGI systems, follow the product manufacturer's guidance carefully. Failure to do so often results in an ineffective installation that leaves users unprotected.
- UV light outside of the narrow spectrum that is ideal for disinfection can produce ozone, so any model applied should conform to United Laboratories (UL) Solutions Standard 2998, *Environmental Claim Validation Procedure (ECVP) for Zero Ozone Emissions from Air Cleaners.*

3.2.3 Needlepoint Bipolar Ionization

Needlepoint bipolar ionization (NPBI) is another method used today to help cleanse air. Ions occur naturally and are simply atoms that have gained or lost electrons, giving them a positive or negative potential. The opposite charges of ions are attracted to each other and combine to form something new. Common salt (sodium chloride, or NaCl), for example, is a common ionic compound.

NBPI systems generate both positive and negative ions. The ionic charges cause small particles to come together, creating larger particles. This makes them easier to capture in filter media unless the ions become heavy enough to fall out of the airstream altogether. The ions produced are also capable of deactivating some pathogens, including viruses, by disrupting proteins on their surfaces. In addition, harmful VOCs can be broken down into compounds such as nitrogen, oxygen, CO_2, and water vapor.

NBPI devices come in several forms. A duct-mounted model like the one shown in *Figure 51* has a self-cleaning cycle and is virtually maintenance free. Many can accept a variety of voltages, from 24VAC to 240VAC.

Figure 51 Duct-mounted ionization system.
Source: GPS Air

NBPI is still considered an emerging technology by the EPA. Since NBPI equipment can produce harmful ozone under some conditions, any model applied should conform to United Laboratories (UL) Solutions Standard 2998, *Environmental Claim Validation Procedure (ECVP) for Zero Ozone Emissions from Air Cleaners.*

3.0.0 Section Review

1. The thickest extended-surface air filters ______.
   a. cannot be used in residences
   b. have efficiency ratings above MERV 16
   c. often have media made from metal
   d. can hold far more debris than their thinner counterparts

2. True or False: UV light fixtures are typically more effective at treating a moving airstream than a fixed surface.
   a. True
   b. False

Module 03204 Review Questions

1. What is the typical concentration of CO_2 in outdoor air?
   a. 100 ppm
   b. 400 ppm
   c. 1,000 ppm
   d. 2,500 ppm

2. In the heating mode for an occupied building, an economizer outdoor air damper is ______.
   a. fully open
   b. fully closed
   c. set to the minimum open position for ventilation
   d. modulating as needed to raise the supply air temperature

3. True or False: Modulating dampers are generally required for economizers to operate properly.
   a. True
   b. False

4. For a home that has an internal volume of 22,000 ft^3, how much outside air, in cubic feet per minute, must be provided to reach 0.35 air changes per hour? Round your answer to the nearest whole number.
   a. 128 cfm
   b. 462 cfm
   c. 847 cfm
   d. 7,700 cfm

5. The *humidity ratio* of air refers to the ______.
   a. volume of water in one cubic foot of air
   b. volume of water per pound of dry air
   c. volume of wet air compared to the volume of dry air
   d. difference in relative humidity between the indoor and outdoor air

6. Infrared humidifiers are most often found in ______.
   a. homes
   b. greenhouses
   c. office buildings
   d. computer rooms

Figure RQ01
Source: CamFil USA

7. Air filters with an exceptionally high MERV rating _____.
   a. are the least efficient
   b. are the least expensive
   c. can significantly impede airflow
   d. are required for residential applications

8. The filter shown in *Figure RQ01* is a(n) _____.
   a. box filter
   b. bag filter
   c. conventional filter
   d. activated carbon filter

9. Which of these types of filters is *not* considered disposable?
   a. Bag filter
   b. HEPA filter
   c. Extended-surface filter
   d. Electrostatic filter

10. When the airflow through an electronic air cleaner is well below the manufacturer's recommendation, _____.
    a. it will overheat
    b. an excessive amount of ozone can be generated
    c. particles cannot be captured
    d. the internal power supply will not energize

Answers to Odd-Numbered Module Review Questions are found in *Appendix A.*

Answers to Section Review Questions

Answer	Section	Objective
Section One		
1. c	1.1.2	1a
2. c	1.2.0	1b

$(22{,}000\ \text{ft}^3/\text{hour} \div 60\ \text{minutes/hour}) \times 0.35\ \text{ACH} = 128.3$, rounded to 128 cfm

Answer	Section	Objective
Section Two		
1. b	2.1.2	2a
2. d	2.2.1	2b
Section Three		
1. d	3.1.2	3a
2. b	3.2.2	3b

MODULE 03202

Chimneys, Vents, and Flues

Source: Alexander Kirch/Shutterstock

Objectives

Successful completion of this module prepares you to do the following:

1. State the principles of fossil-fuel combustion and explain the need to vent combustion byproducts.
   a. Define combustion and explain how flame characteristics reflect the state of combustion.
   b. Distinguish between complete and incomplete combustion.
   c. Describe the composition of flue gases and state the importance of managing condensation within a venting system.
2. Identify the basic requirements and components of a furnace venting system.
   a. State the basic principles of combustion ventilation.
   b. Identify vented appliance categories and common venting materials.
3. State the basic venting considerations for common gas-fired heating equipment.
   a. State the venting considerations for induced-draft furnaces.
   b. State the venting considerations for condensing furnaces.

Performance Task

1. Use manufacturer's tables to determine the proper vent size for an induced-draft or condensing gas furnace according to an instructor-outlined installation scenario.

Overview

To operate safely and efficiently, furnaces and other fuel-burning appliances must have an adequate supply of air to support combustion. The proper venting of the combustion byproducts is equally important to both safety and performance.

There are well-defined standards covering the materials and methods used to vent furnaces. These standards are reflected in the furnace manufacturer's installation instructions along with various codes and standards. These codes and standards must be applied to avoid the creation of a hazardous environment for building occupants that also affects the operation of fossil-fuel equipment.

Digital Resources for HVACR

Scan this code using the camera on your phone or mobile device to view the digital resources related to this craft.

SCAN ME

NCCER Industry-Recognized Credentials

If you are training through an NCCER-accredited sponsor, you may be eligible for credentials from NCCER. The ID number for this module is 03202. Note that this module may have been used in other NCCER curricula and may apply to other level completions. Contact NCCER at 1.888.622.3720 or go to **www.nccer.org** for more information.

You can also show off your industry-recognized credentials online with NCCER's digital credentials. Transform your knowledge, skills, and achievements into credentials that you can share across social media platforms, send to your network, and add to your resume. For more information, visit **www.nccer.org**.

1.0.0 Principles of Combustion

Performance Tasks

There are no Performance Tasks in this section.

Objectives

State the principles of fossil-fuel combustion and explain the need to vent combustion byproducts.

a. Define combustion and explain how flame characteristics reflect the state of combustion.
b. Distinguish between complete and incomplete combustion.
c. Describe the composition of flue gases and state the importance of managing condensation within a venting system.

All fossil-fuel furnaces produce combustion byproducts comprised of various gases and substances. Combustion byproducts often contain substances that are hazardous to human health. The byproducts also include moisture, soot, and various acids that can damage equipment.

Combustion byproducts are also referred to as *flue gases*. Flue gases must be vented to the outdoors where the gases are diluted by the air, protecting the building occupants. The design of the flue vent, or *vent system*, depends on the building construction, the type of furnace and fuel, and the temperature of the flue gases. *Figure 1* shows examples of furnace venting using a masonry chimney with a metal liner and using a dedicated metal chimney.

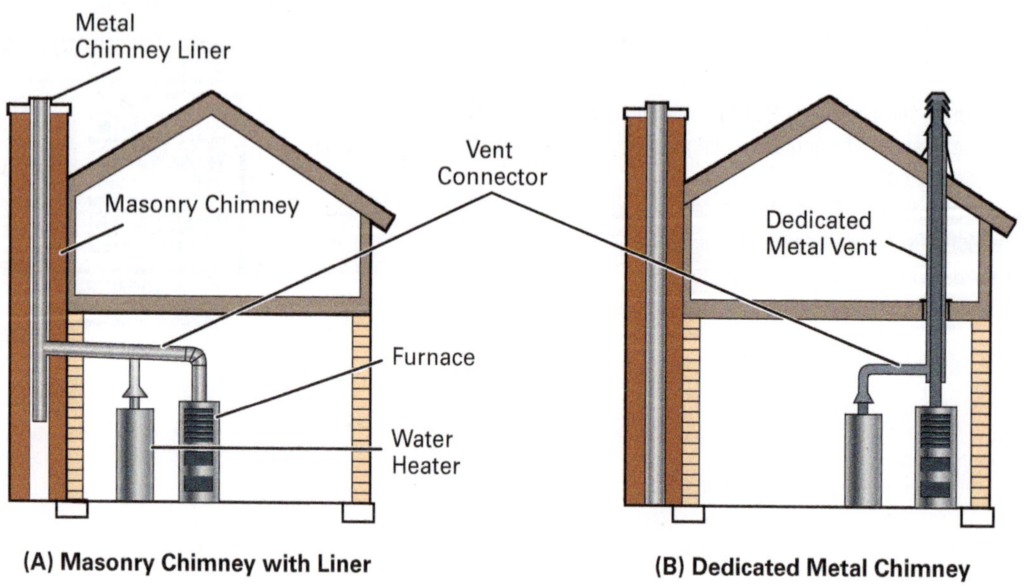

Figure 1 Examples of furnace venting.

Proper venting is especially important in induced-draft furnaces. Induced-draft furnaces have an Annual Fuel Utilization Efficiency (AFUE) rating of 78% to 85%. Because of their cooler flue gas temperatures when compared to natural-draft models, condensation inside the vent system is more likely to occur. Condensation can damage both the furnace and the vent materials.

High-efficiency condensing furnaces—those with an AFUE of 90% and higher—are vented using different materials and procedures than both natural-draft and induced-draft furnaces. A secondary or condensing heat exchanger removes additional heat before the flue gases reach the vent. As a result, the flue gases are cool enough to vent through plastic pipe. Since condensation forms in the furnace by design, condensing furnaces are designed to collect and dispose of the resulting water.

1.1.0 Combustion

During a combustion process, oxygen combines with fuel to release stored energy in the form of heat. Three conditions are necessary for combustion to take place:

- A form of fuel must be present. The fuel can be gaseous, such as natural gas; liquid, such as fuel oil; or solid, such as coal. Two elements that all fuels have in common are carbon (C) and hydrogen (H).
- Oxygen must be present to combine with carbon and hydrogen, releasing heat as it forms water and carbon dioxide (CO_2).
- A source of ignition is required to initiate the process. An existing flame, an electronic spark, and a sufficiently heated surface are common approaches to providing ignition.

The three conditions for combustion are often shown in a graphic called the *combustion* or *fire triangle* (*Figure* 2). If any of the three legs of the triangle are missing, combustion cannot take place.

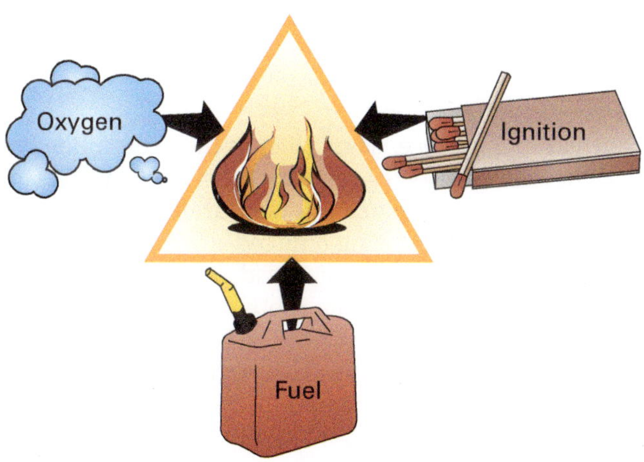

Figure 2 The combustion triangle.

1.1.1 Flame Characteristics

The characteristics of the flame and the intensity with which it burns are characteristic of the combustion process and its quality. Pressure-type oil burners, for example, produce a yellow flame. Natural gas burners typically produce a blue flame. The difference is mainly due to the way the air (oxygen) is mixed with the fuel.

The color of a gas flame is often indicative of the volume of air being supplied for combustion. A yellow flame is produced when gas is ignited as it flows out of the open end of a pipe. A blue flame, though, is produced when about 50% of the required combustion air is mixed with the gas before it is ignited. Combustion

NOTE

Since air is the typical source of oxygen for combustion, *air* is often used in place of the word *oxygen* throughout this module.

air introduced this way is called *primary air* (*Figure 3*). The balance of the needed combustion air, called *secondary air*, is supplied where combustion takes place.

The primary air is drawn, or *induced*, into the gas stream as it enters the burner. The flow of gas through the orifice creates a slight negative pressure in the surrounding area, pulling in air as shown in *Figure 3*. The secondary air is drawn into the flame at the burner assembly. The primary air volume entering an in-shot burner (as opposed to an atmospheric burner on a natural-draft furnace) is rarely adjustable. The secondary air volume also cannot be adjusted. Most in-shot burners in today's furnaces are carefully designed to ensure the correct amount of primary and secondary air is available to the burner, assuming the building pressure is reasonably neutral.

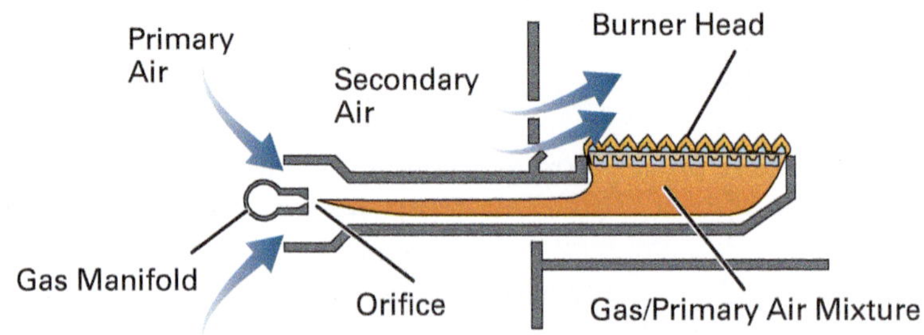

Figure 3 Primary and secondary combustion air paths.

CAUTION

Remember that furnaces must be provided with a sufficient supply of combustion air for the room or enclosure in which they are installed. In some cases, outside air grilles or ducts are required to ensure an adequate supply of combustion air is available to the equipment. Also note that significant negative pressures in a structure will also interfere with proper burner operation and venting performance.

The characteristics of a gas flame reflect the balance between fuel and combustion air. Certain flame characteristics indicate inferior combustion that can be caused by too much or too little primary or secondary air.

Know the Code

The *International Mechanical Code®* (*IMC®*) is often adopted by state and local authorities. Note, however, that the IMC addresses the installation and venting of fuel-fired appliances other than gas. The *International Fuel Gas Code®* addresses the installation of gas piping and the venting of gas-fired systems and appliances.

1.1.2 Flue Gas Temperature

When fuel is burned in a furnace, heat is lost in the vented flue gases. Some of the heat loss was necessary in older systems to establish a draft in the venting system. Because today's furnaces use a forced-draft approach, more heat can be extracted from the flue gases. Using a forced-draft approach allows the path through the heat exchanger to be longer and more complex. These changes increase the surface area and result in a more effective exchange of heat. Lost heat should be minimized for the furnace to operate at peak efficiency. For example, if the heat loss is 20% of the input capacity, the furnace is 80% efficient.

Air entering the furnace at room temperature or lower is heated to flue gas temperatures that range from 100°F to 600°F (38°C to 316°C), depending on the furnace design. The flue gas temperature is an important factor in selecting and installing venting systems. The range of flue gas temperatures in various furnaces is as follows:

- Natural-gas furnaces — 350°F to 600°F (177°C to 316°C)
- Induced-draft furnaces — 275°F to 400°F (135°C to 204°C)
- Condensing furnaces — 100°F to 125°F (38°C to 52°C)

1.2.0 Complete and Incomplete Combustion

The combustion process is not always perfect. The ratio of air to fuel determines whether the combustion process is considered *complete* or *incomplete*.

Complete combustion takes place when carbon combines with oxygen to form carbon dioxide (CO_2). CO_2 is nontoxic in small quantities and can be exhausted to the outdoors. Hydrogen combines with oxygen to form water vapor (H_2O). The water vapor is vented along with the other combustion byproducts.

Incomplete combustion results when too little oxygen is available to support efficient combustion. The result is the formation of undesirable byproducts, such as carbon monoxide (CO), pure carbon (soot), and **aldehydes**. Both CO and aldehydes can reach toxic levels. When present, aldehydes give off a sharp and pungent odor. CO has no odor at all. Soot coats the heating surfaces of the furnace and reduces heat transfer. Thus, providing sufficient air to achieve complete combustion is essential to avoid these negative results.

When incomplete combustion is indicated by the flame characteristics, a lack of primary air is often the cause. When the primary air volume is not adjustable, look for obstructions in the burner that could impede the natural flow of air into the gas stream.

Too much primary air, however, can reduce the combustion efficiency of the furnace. Although you have no way to accurately measure the primary air volume in the average furnace, 15% to 30% excess primary air has been found to provide satisfactory combustion without significantly reducing burner efficiency. Operating burners with less than 15% of excess primary air is not worth the risk, as the small improvement in efficiency does not offset the hazards created by incomplete combustion. Furnace design engineers consider these facts during the design phase.

Complete combustion: Combustion supported by enough oxygen to prevent the significant formation of carbon monoxide (CO).

Incomplete combustion: Combustion without enough oxygen to prevent the formation of carbon monoxide (CO).

Aldehydes: Highly reactive organic compounds that are characterized by a carbonyl group (CHO) attached to a hydrogen atom and often a distinctive odor. A commonly recognized form of aldehyde is *formaldehyde*.

1.3.0 Flue Gases and Condensation

The primary components of flue gases include CO_2, water vapor, excess air (with much of its oxygen consumed), and small amounts of other elements. If incomplete combustion occurs, the gases also include CO, aldehydes, and soot, all of which are potential hazards.

Proper venting refers to the removal of all combustion byproducts, as well as excess air and any **dilution air**, from the structure. Many furnaces are vented through a chimney or metal vent routed from the furnace area up and through the roof. Some furnaces are also vented horizontally through an outside wall. Different furnaces and boilers vary significantly in their venting requirements and options.

A near-horizontal metal pipe called a **vent connector** is often used to connect induced-draft furnaces to a chimney or **vent riser**, as shown in *Figure 1*. Condensing furnaces can be vented using PVC plastic pipe passing through the roof or an outside wall. This is possible because the flue gases from these furnaces are much cooler than that of other furnaces.

Sizing the vent system and its various components is a critical step. The volume of the flue gases is easily underestimated. The total flue gas volume is many times greater than the volume of fuel burned. The water vapor contained in the flue gases can also be troublesome if it is allowed to condense inside the venting system.

Condensation not only initiates corrosion inside the vent, but it can also trickle down into the furnace heat exchangers and leak from joints in the venting system. The water acts as a carrier for acids that accelerate corrosion. Looking for problems related to condensation and corrosion in the vent system is an important part of any furnace inspection.

In burning 100 cubic feet (ft^3) of natural gas, a furnace can produce 200 ft^3 of water vapor. In its condensed (liquid) form, this represents about one gallon of water. If the surface temperature of the vent system is at or below the dewpoint of the flue gases, the water vapor will begin to condense.

Consider that, when any furnace makes a cold start with some or all of the venting system at a low temperature, some condensation will occur until the

Dilution air: Air added to the flue gases in a natural-draft furnace to aid in flue gas removal.

Vent connector: A near-horizontal metal pipe section of the vent system that connects the appliance(s) to the vent riser or chimney.

Vent riser: The vertical section of a metal vent pipe.

vent surface is above the dewpoint from beginning to end. The water that condenses inside the venting system soon after the burner starts may not remain long, though, as the rising hot air will re-evaporate the moisture as the operating cycle continues and the venting system warms up.

In condensing furnaces that extract more heat from the flue gases, the secondary heat exchanger condenses a great deal of the water vapor the gases contain. A system to collect and remove the condensate is included in the design. Although the flue gases still contain some moisture as they exit, the dewpoint is much lower. A small amount of water vapor may still condense inside the PVC vent piping, but a proper installation allows for the condensate to drain out of the piping.

1.3.1 Carbon Monoxide Hazards

Incomplete combustion produces CO instead of CO_2 in the flue gas. However, even if a burner is properly adjusted and combustion is complete, it can still produce small amounts of CO.

Various agencies have different maximum CO exposure levels that they consider safe. The Occupational Safety and Health Administration (OSHA), for example, prohibits workers from being exposed to more than 50 ppm of CO as averaged over eight hours.

The US Consumer Product Safety Commission (CPSC) reports that most people will not experience significant symptoms of CO poisoning during prolonged exposure at or below 70 ppm. Above 70 ppm though, most people acknowledge symptoms. Concentrations between 150 and 200 ppm cause disorientation and often render people unconscious. If a person falls unconscious within the contaminated area, death is likely. A higher exposure over a very short time, however, may not produce any symptoms of CO poisoning.

According to the CPSC, about 170 people die each year from CO poisoning produced by sources other than automobiles. The US Center for Disease Control (CDC) reports that more than 100,000 people seek emergency treatment for CO poisoning each year. Many of the deaths result from improper installation, a lack of periodic maintenance, and venting system failures. A qualified technician should inspect all fossil-fuel appliances annually to ensure safe operation.

Many states and local codes require CO detectors (*Figure 4*) to be installed in newly constructed single- and multi-family dwellings in which a fuel-burning appliance is present. The installation of CO detectors in older structures is also highly recommended. The CPSC recommends CO detectors be placed near every separate sleeping area and on each level of a home. The CPSC also recommends that detectors not be placed in kitchens or directly above a fuel-burning appliance.

Figure 4 Carbon monoxide (CO) detector.
Source: Danny E Hooks/Shutterstock

Did You Know?

CO Poisoning

CO poisoning causes a variety of symptoms. Low to moderate exposure can cause the following symptoms:

- Headache
- Fatigue
- Shortness of breath
- Nausea
- Dizziness

Exposure to higher levels of CO can cause the following:

- Confusion
- Vomiting
- Loss of muscle control
- Unconsciousness
- Death

1.0.0 Section Review

1. How is the primary air for a gas burner mixed into the gas stream?
   a. A small vacuum pump draws the air in.
   b. A negative pressure from gas flow induces air to mix with the gas.
   c. A positive pressure from gas flow pushes air into the gas stream.
   d. The indoor blower provides the pressure needed to mix air with the gas.

2. Which substance forms during incomplete combustion, has a sharp odor, and is hazardous?
   a. CO_2
   b. Water vapor
   c. CO
   d. Aldehydes

3. According to the CPSC, most people will not experience any symptoms of CO poisoning as long as the concentration remains below what level?
   a. 200 ppm
   b. 150 ppm
   c. 100 ppm
   d. 70 ppm

2.0.0 Furnace Venting Systems

Objectives

Identify the basic requirements and components of a furnace venting system.

a. State the basic principles of combustion ventilation.
b. Identify vented appliance categories and common venting materials.

Performance Task

1. Use manufacturer's tables to determine the proper vent size for an induced-draft or condensing gas furnace according to an instructor-outlined installation scenario.

The nature of flue gases makes the proper venting of gas furnaces a serious matter. This section explains the principles of venting combustion appliances, identifies the categories of vented gas appliances, and describes the construction of common gas-appliance venting systems. Due to their popularity, this module primarily focuses on gas-fired systems.

2.1.0 Basic Venting Concepts

Furnace and boiler vent systems should include the following basic characteristics:

- Low resistance to flue gas flow
- Limited mass to allow the system to warm-up quickly
- Reasonable insulating properties to maintain the flue gas temperature and protect nearby flammable materials

Installing a venting system requires the same technical understanding and planning as the rest of the HVAC system. It is vital to understand the basic principles of venting and the factors that interfere with system performance.

The volume of flue gases produced by a gas furnace is roughly 30 times the volume of gas burned. For example, a 100,000-Btuh gas furnace consumes about 100 ft^3 of natural gas during each hour of operation. This is based on the natural gas supply having a heat content of 1,000 Btus per ft^3, which is the average

On Site

Gas Furnace Venting

The *National Fuel Gas Code*, which is published jointly by the American Gas Association (AGA) and the National Fire Protection Association (NFPA), allows induced-draft furnaces to use indoor air for combustion in certain circumstances. However, it is important to keep in mind that the environment in which the furnace is installed can change over time. What once worked fine can suddenly become a serious hazard. Consider the following examples:

- A homeowner may remodel a basement where a furnace is installed, adding walls that isolate the furnace from its previous source of combustion air.
- Additional gas appliances such as stoves, water heaters, and clothes dryers might be added, increasing the demand for combustion air within the structure.
- A homeowner may replace a ventilated closet door where a furnace is installed with a solid door that restricts combustion airflow, not realizing that the ventilated door served an important purpose.
- Users may cover the combustion air duct and grilles in an equipment closet, to keep cold air from migrating in, unaware of the need for combustion air.

When servicing and inspecting vented fossil-fuel systems, always be on the lookout for situations like these.

heat-content value assigned to natural gas. The 30:1 air/gas ratio suggests that 3,000 ft^3 of air per hour is also consumed when the burner runs.

Assume a furnace is installed indoors and in a home with an area of 1,250 square feet (ft^2) and a volume of about 10,000 ft^3. Thus, to satisfy the combustion needs of the furnace alone, 30% of the home's air volume must be replaced for each hour of operation. This volume of outside air passing through the home represents a significant heat load on its own. This fact alone is why you'll find many of today's furnaces installed in crawl spaces and attics, where plenty of unconditioned air is available for combustion.

In days gone by, normal air infiltration around doors and windows was enough to satisfy the needs of a gas furnace, although it wasn't an energy-efficient approach. Homes were not as tightly constructed. Modern building construction standards require a much tighter structure. In addition, more home appliances and various exhaust fans require a source of air.

Essentially, the demand for air inside a home has risen over the years, while the paths for air to enter naturally have been reduced. This is one of the reasons why heat or energy recovery ventilators are needed in many of today's homes. Combustion air that passes through the structure must be heated too, to keep the structure comfortable, so it pays to reduce the energy required.

Better system performance and greater energy efficiency can be achieved by providing a furnace with a dedicated supply of combustion air from outdoors. Condensing furnaces typically use outdoor air for combustion by design, but

On Site

The Heat Content of Fuels

The heat content of natural gas varies across the country. A value of 1,000 Btus/ft^3 is widely used in the HVAC industry as an average value. However, the heat content of natural gas locally is often higher or lower than the accepted average. Utilities can provide a local heat content value for the fuel gas supplied to their customers. For example, Piedmont Natural Gas in South Carolina reports that their natural gas heating content averages 1,031 Btu/ft^3.

Fuel oil grade #2, which is typically used for oil heating systems, has a heat content of approximately 138,000 Btus per gallon with some variance. Propane has a heat content of roughly 90,000 Btus per gallon.

all furnaces can benefit from a dedicated air supply. Remember that a lack of combustion air not only affects combustion, but it also affects the ability of the venting system to properly eliminate combustion byproducts.

Several installation characteristics and furnace features affect the performance of the venting system. This section explores characteristics and features related to venting system performance and the issue of condensation that technicians should consider when installing or servicing furnaces.

2.1.1 Venting Code Requirements

Be aware of the local codes and regulations governing vent systems and review them periodically. You should also refer to the latest edition of the *National Fuel Gas Code* (NFPA 54/ANSI Z223.1) published by the American Gas Association (AGA) and the National Fire Protection Association (NFPA). Remember, though, that local codes can be more stringent than national or international codes and standards.

2.1.2 Clearances

Local codes and manufacturers' installation instructions usually specify the required minimum distance between the furnace and venting components and nearby combustible materials.

WARNING!

Flammable materials must always be kept away from the furnace heat exchangers, burners, and any other hot surfaces. Adhere to the minimum clearances documented by the equipment manufacturer.

Accessibility clearance requirements are usually much larger than the minimum fire protection clearances. Always consult the manufacturer's installation instructions for the appropriate clearance to be provided around the furnace itself. Listed venting components have specific clearances that must be maintained as well. For example, single-wall metal vent pipe must typically remain 6" from combustible materials, while double-walled materials generally require only 1" of clearance.

2.1.3 Combustion Air Supply

All return air ducts for the furnace must extend outside the room or closet in which the furnace is installed. In other words, the furnace must not draw its return air from within the same space. Exceptions can be made when all the supply air is flowing into the same space, maintaining a balance in room pressure.

Furnaces located in a confined area must be provided with sufficient air for combustion. Generally, two permanent openings must be provided for combustion air to enter a furnace in a confined area. One opening must be within 12" of the ceiling, while the other must be within 12" of the floor. Each of the two openings must have a *free area* of at least 1 square inch (in^2) per 1,000 Btuh of input rating for all the gas-fired equipment in the enclosure (*Figure 5*). Gas-fired water heaters are often installed near a gas furnace in a confined area. Thus, its input must also be included in the calculations.

It is important that the combustion air supply be free of substances such as cleaning solvent vapors, aerosol sprays, strong detergents, bleaches, commercial air fresheners, and similar substances. The vapors from these materials can damage the furnace and venting system as they are incinerated.

2.1.4 Furnace Sizing and Venting

Furnaces must not be severely oversized. However, it is better to slightly oversize than it is to undersize. Generally, the furnace output rating should be from 95% to 120% of the calculated heating load. The load is established by an analysis that considers building construction, type and amount of insulation, amount of glass, and other factors. *Manual J Residential Load Calculation*, published by

On Site

High-Altitude Installations

Refer to the manufacturer's instructions for furnaces to be installed at altitudes above 2,000'. The *National Fuel Gas Code* also provides guidelines for these installations.

Generally, the furnace capacity must be derated by 4% for every 1,000' a furnace is located above sea level. For example, at 5,000' above sea level, a furnace with an input rating of 100,000 Btu will have an actual input rating of only 80,000 Btus—a 20% reduction.

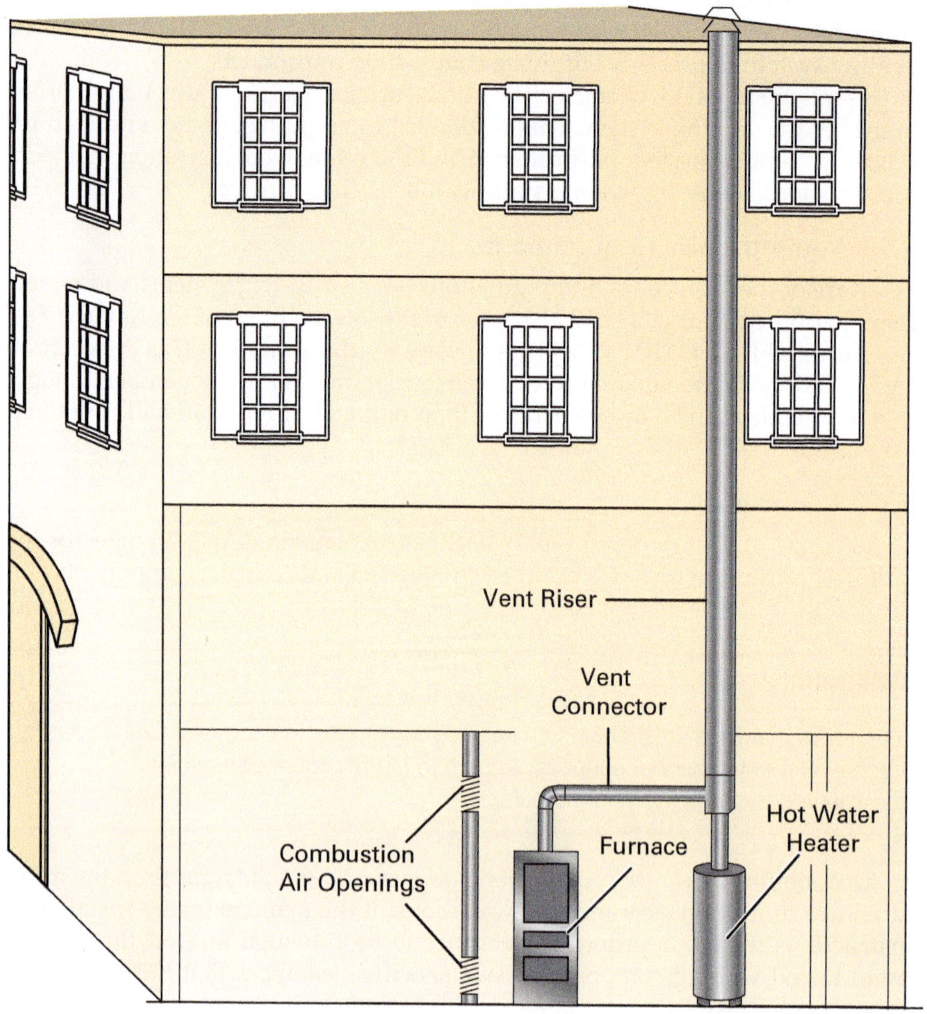

Figure 5 Combustion air for an isolated furnace room.

the Air Conditioning Contractors of America (ACCA), provides guidance in residential heating and cooling load calculation.

If a furnace is grossly oversized, the most effective way to correct the problem is to replace it. Do not attempt to de-rate the furnace by reducing the gas orifice size, or by reducing the gas manifold pressure, without the manufacturer's model-specific guidance. De-rating a furnace can cause condensation to form in the furnace and/or vent system and may create a safety hazard. Always follow the installation instructions supplied with the unit.

A correctly sized furnace will have a longer on-cycle and a shorter off-cycle, particularly when the outdoor temperature nears the heat-loss design temperature. This keeps the furnace and vent system from cooling off substantially between cycles and helps maintain a draft in the vent system for the next heating cycle. A significantly oversized furnace will deliver a large amount of heat during a short on-cycle. With short run times, the furnace and vent system will cool quickly, allowing condensation to form when the warm flue gas enters the vent during the next startup.

2.1.5 Temperature Rise Adjustment

Temperature rise refers to the temperature difference between the furnace supply airstream and the return airstream. The amount of room air flowing over the heat exchangers determines the temperature rise, assuming the burner capacity is fixed. If there is too much airflow, the heat exchangers will not be able to reach normal operating temperature. Moisture can condense on the heat exchangers and in the vent system as a result. If there is too little airflow, the heat exchanger may overheat on each cycle, also shortening its life. The furnace nameplate (*Figure 6*) will specify the correct temperature rise range. A range of 40°F to 70°F (22°C to 39°C) is a common example.

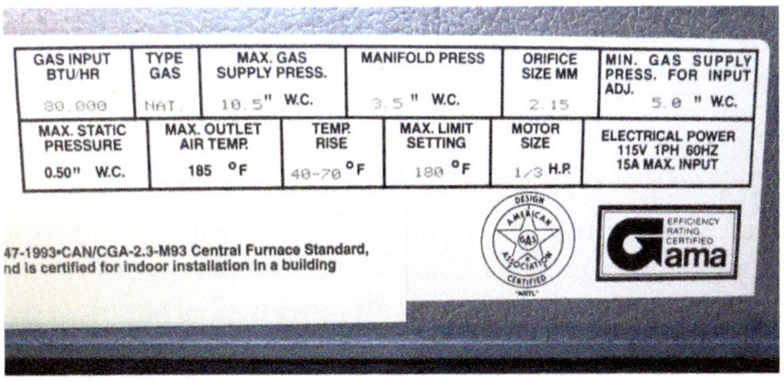

Figure 6 Furnace nameplate showing temperature rise range.

The temperature rise is measured by drilling holes in the supply and return ducts and measuring the temperatures through those openings (*Figure 7*). Note that the hole should not be placed where the temperature measuring probe has a clear line-of-sight to the heat exchanger. For example, do not place the hole in the supply plenum directly above the heat exchanger of an upflow furnace. The reading will be affected by radiant heat from the heat exchanger surface.

The difference between the two temperatures must be within the specified range. Ideally, it should be around the midpoint of the range. If it is not, the blower speed can be changed to adjust the temperature rise.

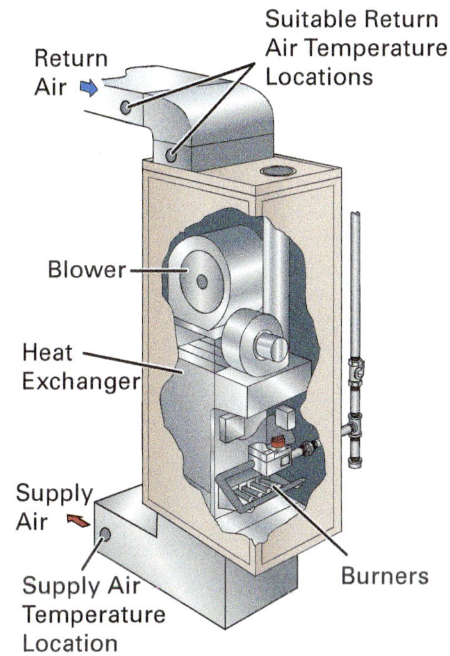

Figure 7 Suitable points for measuring supply and return air temperature.

2.1.6 Heating Cycle Rate

Many electronic thermostats (*Figure 8*) have what is known as a *cycle rate adjustment*. The default cycle rate setting may or may not match the ideal cycle rate for each furnace and application. A cycle rate that is too high will cause the furnace to operate for shorter periods. This can prevent the venting system from reaching its normal operating temperature, increasing the potential for condensation to occur.

Check the thermostat literature for the default cycle rate and how to access the thermostat service menu for changes. Then adjust the cycle rate to match the requirements of the installation. Determining the correct cycle rate for each installation is largely based on observation over time, so technicians often need to rely on the user to provide some input. A good rule of thumb is to set the cycle rate for three cycles per hour, then allow time for observation to determine if a change is needed.

Figure 8 Electronic room thermostat.
Source: lucdp/iStock/Getty Images

2.2.0 Vent System Categories and Common Materials

The code-required vent system characteristics are largely based on the furnace type. Furnaces are categorized based on the pressure applied to the vent system and the likelihood of condensate forming inside the vent piping. The *National Fuel Gas Code* identifies vented appliance categories and describes them as follows:

- *Category I* — An appliance that operates with a non-positive vent static pressure and with a vent gas temperature that avoids excessive condensate production in the vent.
- *Category II* — An appliance that operates with a non-positive vent static pressure and with a vent gas temperature that may cause excessive condensate production in the vent.
- *Category III* — An appliance that operates with a positive vent static pressure and with a vent gas temperature that avoids excessive condensate production in the vent.

- *Category IV* — An appliance that operates with a positive vent static pressure and with a vent gas temperature that may cause excessive condensate production in the vent.

This module focuses primarily on Category I and Category IV appliances. Induced-draft furnaces fall into Category I while condensing furnaces fall into Category IV. Category III furnaces include *sidewall-vented* 80% AFUE induced-draft furnaces. Neither Category II nor Category III furnaces are currently common.

2.2.1 Common Vent System Materials

Masonry and factory-built chimneys, along with metal and plastic venting materials, comprise the basic venting system components for coal, gas, oil, and wood-burning appliances. Factory-built chimneys with an inner wall of stainless steel that comply with UL Solutions (UL) Standard 959 are suitable for all these fuels. UL-listed factory-built chimneys are sometimes referred to as *all-fuel chimneys.*

Figure 9 shows the components of a typical factory-built chimney. The temperature-rise limit for these applications is set at 1,730°F (~940°C). Factory-built chimneys are widely used with oil furnaces, appliances that burn solid fuel such as wood and coal, and fireplaces. They are not always necessary for gas-fired appliances that produce lower flue gas temperatures.

Metal venting materials referred to as *Type B* are double walled. The inner wall is aluminum, while the outer wall is galvanized steel (*Figure 10*). Type B products are commonly used to vent gas furnaces. The vent piping is round and available in a variety of sizes.

The double-wall construction of Type B vent pipe supports a quick warm-up while reducing heat loss. The outer wall remains much cooler than the inner wall, allowing the material to be installed closer to combustible materials. Only 1" of clearance from combustible materials is typically required.

Due to their temperature rating, Type B vent products can only be used with gas-fired appliances. Type B products are often used in new construction because they are far less expensive than building a lined masonry chimney and are equally effective, if not better.

WARNING!

Do not vent oil-fired or solid-fuel appliances using Type B venting components. Type B products are not rated for the flue gas temperatures produced by these fuels.

Type B-W venting materials have the same double-wall construction as Type B, but have an oval shape. Type B-W venting products are designed exclusively for venting gas-fired wall furnaces. Because the typical wall cavity is only $3\frac{1}{2}$" deep, an oval shape is needed to accommodate the volume of vented byproducts in a narrow space.

Type B and Type B-W products are usually used together. Type B-W is connected to the flue collar of the furnace and installed in the wall cavity where round Type B vent would not fit. An oval-to-round adapter is then used where the oval vent exits the stud cavity. From there and through the roof, standard Type B materials are used. Roof caps and similar components are not available in Type B-W form.

When installing Type B or Type B-W vent components, all the components (transitions, supports, vent caps, etc.) must be from the same manufacturer. Never mix vent system components from different manufacturers. Doing so will void any listing agency rating for the product as well as the manufacturer's warranty, as it may create an unsafe condition.

Type L venting products are also constructed with a double wall. Similar to Type B products, Type L products have a stainless-steel inner wall with a galvanized steel outer wall. In general, Type L can be used in any application where Type B is suitable. The reverse is not true, however. Type L vents can be used

On Site

Chimney Fires

One cause of chimney fires is the failure to maintain the required clearance or air space between the chimney and adjacent combustible materials. For this reason, a chimney must be installed in strict accordance with local codes as well as the manufacturer's instructions.

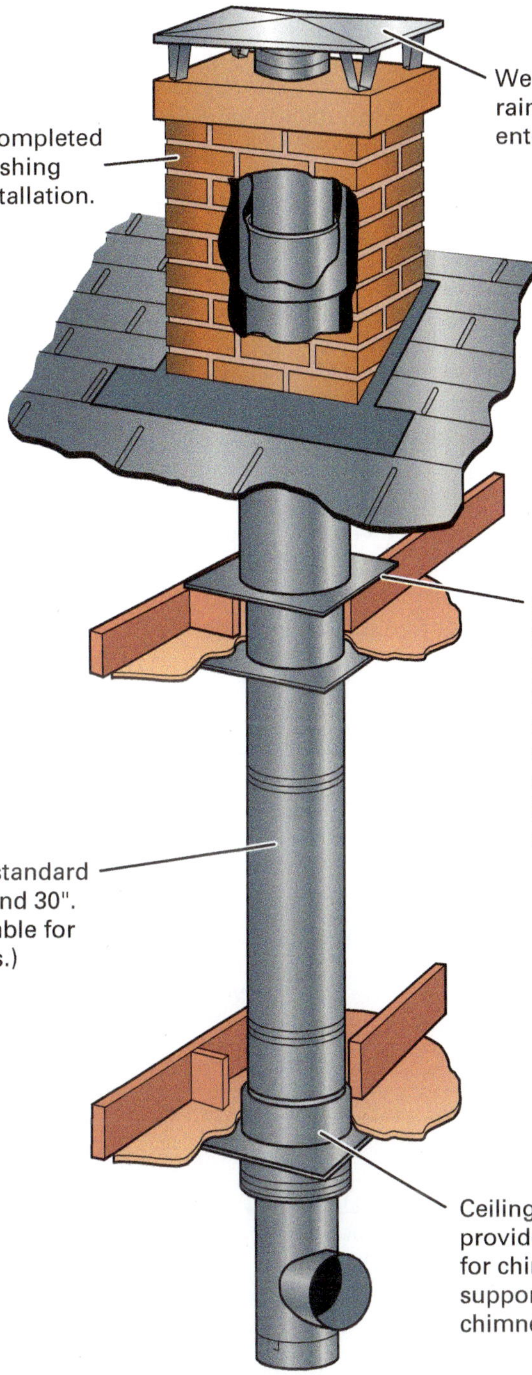

Figure 9 Components of a factory-built chimney.

in venting gas or oil appliances, as well as residential incinerators and certain appliances equipped with draft hoods.

Category III equipment requires venting materials rated to withstand a positive pressure. A few pieces of gas-fired heating equipment, such as sidewall-vented induced-draft unit heaters, fall into this category. Category III equipment was once vented using special high-temperature plastic products designed to withstand the acidic condensate that might be produced, as well as the heat. Due to problems associated with these products, plastic products are no longer used. Instead, most Category III products now require single-wall stainless steel vents. Stainless steel can handle the heat and condensate produced in the flue, although at a significantly higher cost than plastic products. Because the venting system operates at a positive pressure, most products use proprietary clamping and gasket systems to ensure a strong, leak-free installation.

Figure 10 Type B double-walled 45-degree ell.
Source: Duravent Group

Figure 11 Polypropylene vent products for condensing furnaces.
Source: Duravent Group

Common Schedule 40 PVC pipe is used to vent Category IV (condensing) furnaces. Because the vent pipes operate at a positive pressure while the combustion air piping operates at a negative pressure, the PVC joints must be sealed carefully to avoid leaks. Manufacturers also offer polypropylene venting products for use with condensing furnaces (*Figure 11*).

2.2.2 General Guidelines for Metal Vents and Vent Connectors

The *National Fuel Gas Code* contains tables and instructions for selecting the diameter and type of metal vents and vent connectors for induced-draft furnaces and other gas appliances. Chapter 13 of the 2021 *National Fuel Gas Code* provides the information and various tables related to sizing Category I venting systems. Relevant portions of this data are provided in the installation instructions for furnaces. The Code often refers users to the manufacturer's installation instructions and emphasizes that following those instructions is a code requirement. This section summarizes some of the furnace venting requirements found in the *National Fuel Gas Code*.

Gas appliances may be vented through a metal pipe or a metal-lined chimney. The *vent riser* is the vertical section of the vent. The *vent connector* is the horizontal section that connects the appliance(s) to the vent riser (*Figure 12*). Note in the figure that the horizontal vent connector is not perfectly horizontal. Some upward pitch is required to help establish and maintain a draft.

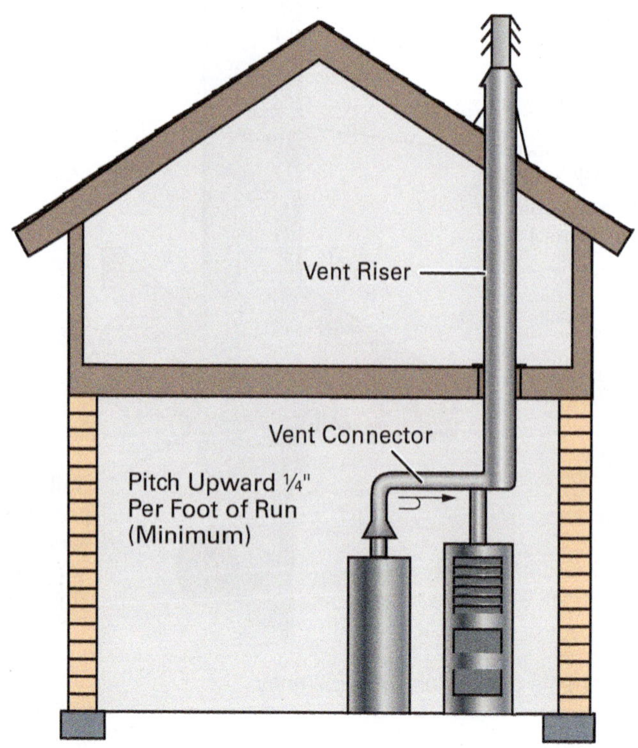

Figure 12 Furnace vent riser and vent connector.

Because of the potential for condensation, and to allow the vent to be relatively close to combustible materials, gas furnace metal vents must be of double-wall (Type B) construction. Vent connectors for induced-draft furnaces should also be constructed of Type B materials when practical. However, a single-wall vent connector can be used with induced-draft furnaces under the following conditions:

- The furnace must be vented with another gas appliance, such as a gas water heater, to the same riser.
- The length of the vent connector manifold, in feet, cannot exceed 1.5 times the connector's diameter in inches. For example, a 6" vent connector cannot

exceed 9' in length. The selection tables are used to determine the diameter of the vent connector. However, the capacity of the vent size shown in the tables must be reduced by 10% to determine the size of the connector.

Single-wall pipe loses heat quickly, which allows the combustion byproducts inside to cool rapidly as well. Using a double-wall Type B vent connector does slightly increase the cost of the installation, but that is far outweighed by the long-term benefits. Using a single-wall vent connector may hamper installation flexibility because of the limited length allowed. In addition, single-wall vent pipe must remain 6" from combustible materials, while Type B pipe only requires a 1" clearance. Type B pipe is also safer for occupants, as vent connectors are often in the open and can easily be touched or bumped into. The exterior of Type B pipe will remain cooler.

Vent connectors should always be pitched up toward the vent riser at a slope of no less than $\frac{1}{4}$" per foot of run. Vent connectors should also be as short in length as possible. Avoid elbows in a vent connector whenever possible because they significantly increase the resistance to flow. If a vent connector must be used, avoid sharp turns. Two 45-degree fittings to create an offset are much better than two 90-degree fittings.

If a single-appliance vent system requires more than two 90-degree turns, a 10% reduction in the maximum capacity shown in the sizing tables must be made for each additional 90-degree elbow. A 5% reduction in capacity must be taken for each additional elbow that is 45 degrees or less.

When a tile-lined masonry chimney is available, it can be used to vent gas appliances (*Figure 13*). In this case, *tile* refers to a product made from clay that may also be referred to as *terracotta*. However, an induced-draft furnace cannot be vented through a tile-lined masonry chimney unless it is connected through a vent connector with another gas appliance. The second appliance helps to keep the chimney warmer and reduces the risk of condensation. If only the gas furnace is to be vented, the chimney must be equipped with a metal liner.

An unlined masonry chimney must never be used for gas venting. Every masonry chimney must have a protective liner, or corrosive substances in the flue gases and condensate will attack the mortar and cause the chimney to deteriorate. At best, an unlined masonry chimney will be seriously damaged over time.

NOTE

Section 13.2.3 of the *National Fuel Gas Code* offers several special cases that allow the length of the vent connector to be increased.

Figure 13 Chimney with two tile-lined flue passages.
Source: Richard Nantais/Shutterstock

On Site

Flexible Chimney Liners

Flexible chimney liner kits have made lining masonry chimneys relatively easy. The most difficult part of the job is working on high and/or steeply pitched roofs. In some cases, using scaffolding or a motorized lift is the safest option. Proper fall prevention and arrest techniques must be applied in all cases.

To install the liner, a weighted rope is dropped down the chimney. The flexible liner is attached to the rope at one end and pulled up or down through the chimney. Since dragging the liner through a home is not always practical, it is most often pulled down through the chimney. Then the liner is attached to the apparatus below and to the top plate and vent cap on the roof.

A furnace and/or similar gas appliance cannot be vented through a chimney that serves a fireplace or other equipment that burns solid fuel such as wood or coal. Downdrafts can carry the byproducts of combustion directly into the living space. It is not unusual to find masonry chimneys with two flue passages beneath the masonry—one dedicated to a furnace and the other to a fireplace or wood stove. *Figure 13* shows an example of a chimney with two tile-lined flue passages.

Ideally, a masonry chimney used for furnace venting should be contained inside the conditioned space (*Figure 14*). If the chimney is exposed to outdoor conditions below the roofline, it will probably need a metal liner for all uses. In some cases, the metal liner must be insulated to further protect it from getting too cold.

Figure 14 Chimneys with a lot of exterior exposure cool very quickly.
Source: Suzi Nelson/Shutterstock

Note that outdoor exposure is a very important consideration for metal venting as well. Metal cools very quickly to the ambient temperature, making it difficult to establish a draft when the appliance starts. Cold air in a chimney or metal vent is dense and heavy, making the air difficult to displace and lift when the furnace starts.

The vent selection tables will identify the chimney area (in square inches) required to match the furnace capacity. Masonry chimneys are often too large (in area) for furnace applications, making it difficult to establish a proper draft and increasing condensation. Reducing the size of the chimney is another reason why liners are installed in masonry chimneys.

Type B metal vent products can also be used as a liner for a masonry chimney. However, rigid Type B vent pipe may be difficult or impossible to install inside a chimney. Thus, flexible chimney liners are more often used (*Figure 15*). Flexible metal liners are expandable; a 30' long vent might be shipped in a 3' long box.

Flexible metal liner sizes are not interchangeable with solid metal vent sizes. This is due to the added resistance to flow created by the corrugations that make liners flexible and compressible. The series of ridges inhibits the flow of flue gas more than smooth pipe of an equal size. Per the *National Fuel Gas Code*, the same table used for Type B vents is used to size corrugated chimney liners, but the maximum capacity shown must be reduced by 20%.

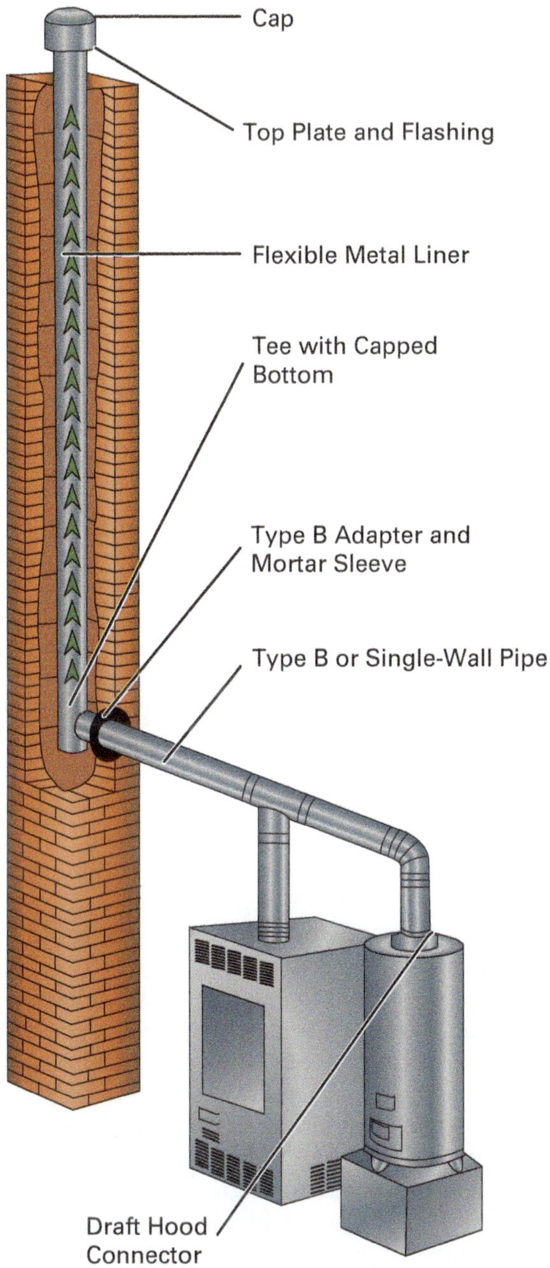

Figure 15 Flexible chimney liner installation.

2.0.0 Section Review

1. Based on the average heat content of natural gas, about how much gas would an 80,000 Btuh input furnace consume during each hour of operation?
   a. 80 ft^3
   b. 94 ft^3
   c. 1,000 ft^3
   d. 1,116 ft^3

2. Which *National Fuel Gas Code* category includes condensing furnaces?
   a. Category I
   b. Category II
   c. Category III
   d. Category IV

3.0.0 Venting Gas-Fired Furnaces

Performance Task

1. Use manufacturer's tables to determine the proper vent size for an induced-draft or condensing gas furnace according to an instructor-outlined installation scenario.

Objectives

State the basic venting considerations for common gas-fired heating equipment.

a. State the venting considerations for induced-draft furnaces.

b. State the venting considerations for condensing furnaces.

Induced-draft and condensing furnaces have different venting requirements, primarily due to the significant difference in flue gas temperatures. The following sections focus on the installation requirements for vent systems serving these two furnace types.

Because natural-draft furnaces are no longer manufactured, it is unlikely that you will be involved in the installation of a vent system for one of these furnaces. However, you may encounter existing installations in which something has been done to reduce the amount of combustion and dilution air available to the furnace. Perhaps something has been done to cause a negative pressure indoors, such as the installation of an exhaust fan. This can result in a *backdraft*, where air flows down through the vent instead of up and out. Thus, some or all combustion byproducts that would normally be vented outdoors exit through the draft hood indoors.

Changes that can cause a backdraft are the addition of building exhaust fans; the addition of vented combustion appliances such as gas fireplaces or wood stoves; and renovations that reduce air infiltration around the home such as new windows, insulation, or siding. A negative pressure in the building can also affect induced-draft and condensing furnaces, but these forced-draft furnaces can overcome the condition if it is minor.

While natural-draft furnaces are no longer available, natural-draft gas-fired water heaters are still on the market. It has been very common to install an induced-draft furnace adjacent to a natural-draft water heater. When any furnace is connected to a common vent connector with a water heater, you must follow the manufacturer's venting guidelines for both pieces of equipment. This is true even if you are only installing the furnace. However, the installation of advanced gas water-heating equipment is growing rapidly, especially in new construction. Thus, fewer furnaces and water heating systems that share a venting system are being installed.

3.1.0 Induced-Draft Gas Furnace Venting

In induced-draft furnaces (*Figure 16*), condensation can be a problem because the flue gases are cooler than those of a natural-draft furnace. Natural-draft furnaces have flue gas temperatures of 350°F to 600°F (177°C to 316°C), making the vent system less likely to have condensation problems unless the furnace was significantly oversized, or the vent system was improperly designed or installed. The flue gas temperature of induced-draft furnaces is in the range of 275°F to 400°F (135°C to 204°C). Thus, less heat must be removed to initiate condensation in the venting system.

To avoid condensation, five key issues must be considered when installing an induced-draft furnace:

- The furnace should be properly sized. An oversized furnace will short cycle, preventing the vent from warming and establishing a good draft.
- The burners must be firing as close as possible to 100% of their rated input so that the maximum amount of heat is available to warm the vent.

Figure 16 Induced-draft furnace.

- The correct amount of air must be flowing over the furnace heat exchangers to maintain the temperature rise at the correct level. A low temperature rise (indicating excessive airflow through the furnace) reduces the vent temperature, potentially allowing condensation to occur in the furnace and/or vent.
- The thermostat cycle rate or heat anticipator must be correctly adjusted. A setting that is too low causes short cycling.
- The furnace must be vented in accordance with the manufacturer's instructions as well as local and national codes.

Vent sizing tables are used to size vent components for Category I induced-draft furnaces. Sizing tables are available in the latest edition of the *National Fuel Gas Code.* No single table covers all installations. To use the vent sizing tables, first determine the installation configuration, and then choose the appropriate table. For example, the vent sizing table used when a single induced-draft furnace is vented through a Type B vent is different from the venting table used when an induced-draft furnace and a natural-draft water heater share a vent connector.

NOTE

Vent sizing tables are also part of the manufacturer's installation instructions and may be used as well.

Induced-draft furnaces can actually benefit from being vented with a natural-draft water heater. When the furnace is running intermittently, heat from the water heater keeps the vent system warmer, reducing the possibility of condensation developing when the furnace starts.

Factors that affect vent sizing include the following:

- Total burner input
- Height of the vent
- Horizontal distance from the appliance(s) to the vent riser
- Vent connector material (single-wall, Type B, etc.)
- The number of elbows in the vent connector and riser

Type B pipe is recommended for the vent connector in all induced-draft furnace installations. The double-wall construction allows the vent connector to retain heat longer. Single-wall pipe can be used for the vent connector if it is vented with another appliance (*Figure 17*). The vent sizing table will show the maximum permitted length of a single-wall vent connector.

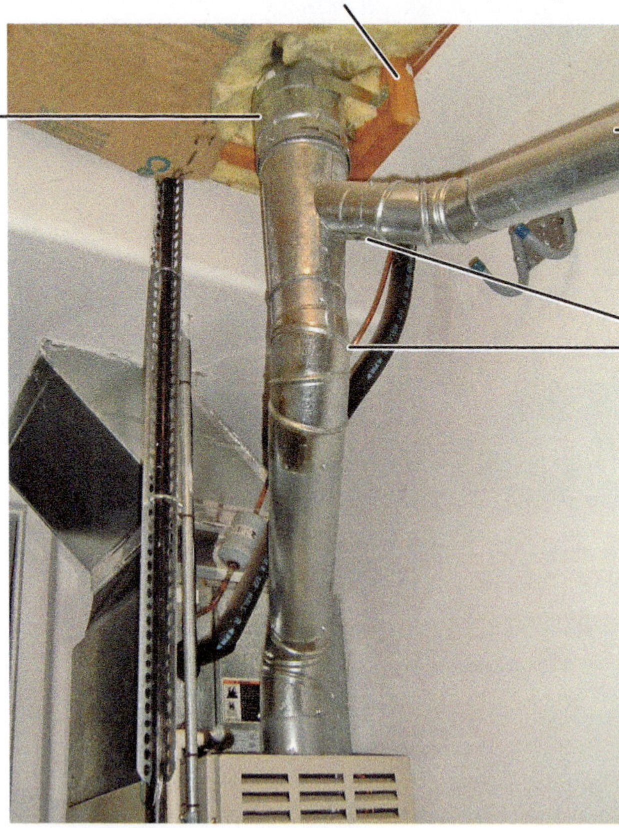

Figure 17 Induced-draft furnace with short vent connector and a common-vented gas-fired water heater. Note the increase in vent diameter where the water heater vent attaches.

The ambient temperature in which the furnace is installed also affects the choice of material for the vent connector. If the area remains cold in the winter, such as in an unheated basement or garage, Type B is best used, even if single-wall pipe is permitted.

In very close quarters, the construction features of Type B vent components make them challenging to work with, as connections cannot be made to random lengths of the material. A length of Type B cannot be cut and assembled, as the connection on each end cannot be fabricated in the field. But when space is available, Type B venting materials are always the best and safest choice for venting induced-draft furnaces.

Figure 18 shows a typical installation and the Type B fittings required. An adjustable length of pipe is shown on the left side. These sections can be extended or reduced several inches and can be used as necessary to make up odd lengths.

3.1.1 Induced-Draft Furnace Vent Sizing Example

Before sizing a venting system for a furnace, select a diagram in the *National Fuel Gas Code* that matches the installation and use the venting tables related to that diagram. In this example, the vent will be sized based on a single-wall vent connector connected to a Type B double-wall vent riser.

NOTE

Diagrams of various approaches of venting are found in Annex F of the *National Fuel Gas Code*, along with sizing calculation examples.

To size the vent for this arrangement, the total height of the vent and the lateral length of the near-horizontal vent connector must be known. Assume a new induced-draft furnace with an 80,000 Btuh input needs to be vented. The furnace is in the basement of a single-story home with a total vent height of 20'. The lateral length of the single-wall vent connector will be 5'. *Figure 19*

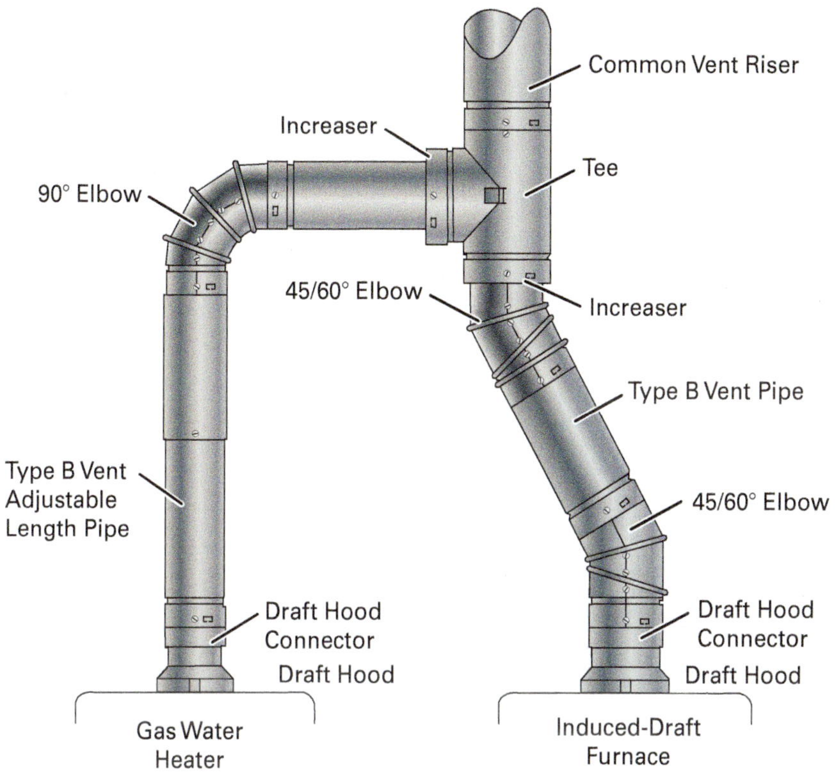

Figure 18 Typical Type B vent assembly.

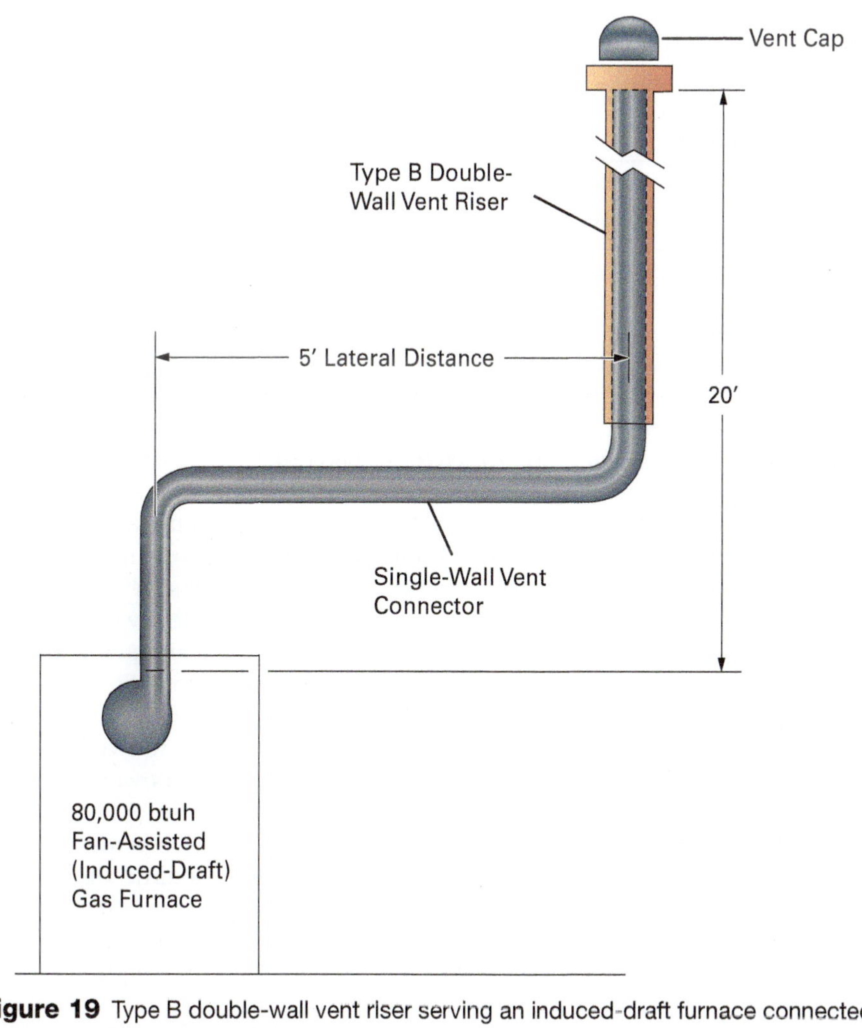

Figure 19 Type B double-wall vent riser serving an induced-draft furnace connected to the riser with single-wall pipe.

CAUTION

When sizing vents, always use the input rating of the appliances, rather than the output rating.

shows the arrangement and the key dimensions. Note that there are two 90-degree elbows in the vent connector. The venting system is not exposed to outdoor conditions below the roof line, which can be an additional factor in the calculations.

Refer to *Table 1*. This table is similar to the tables found in the *National Fuel Gas Code*. Note the Fan Min and Fan Max columns. These columns represent the minimum and maximum induced-draft furnace capacity that a given vent size can manage at the listed vent connector length. The Nat Max column refers to natural-draft equipment, which is not a factor in this example.

TABLE 1 Vent Sizing for Category I Appliances Using Type B Vent and a Single-Wall Vent Connector

		Vent Diameter (D)*											
		3"			4"			5"			6"		
		Appliance Input Rating (Btuh × 1,000)											
Height (feet)	Lateral (feet)	Fan Min	Fan Max	Nat Max	Fan Min	Fan Max	Nat Max	Fan Min	Fan Max	Nat Max	Fan Min	Fan Max	Nat Max
15'	0	36	93	57	56	190	111	80	325	186	116	499	283
	2'	38	69	47	57	136	93	80	225	149	115	337	224
	5'	51	63	44	75	128	86	102	216	140	144	326	217
	10'	NA	NA	39	95	116	79	128	201	131	182	308	203
	15'	NA	NA	NA	NA	NA	72	158	186	124	220	290	192
20'	0	35	96	60	54	200	118	78	346	201	114	537	306
	2'	37	74	50	56	148	99	78	248	165	113	375	248
	5'	50	68	47	73	140	94	100	239	158	141	363	239
	10'	NA	NA	41	93	129	86	125	223	146	177	344	224
	15'	NA	NA	NA	NA	NA	80	155	208	136	216	325	210
	20'	NA	NA	NA	NA	NA	NA	186	192	126	254	306	196

* Not for field use; refer to the current edition of local and national codes.

NOTE

When the exact vent height is not shown in the tables, mathematical interpolation is allowed. Extrapolation outside of the table values is not allowed.

At a height of 20' with a 5' lateral single-wall vent connector, a 3" vent can handle 50,000 to 68,000 Btuh. This will not satisfy the needs for our example. A 4" vent, however, will handle 73,000 to 140,000 Btuh.

Note that, when choosing a 4" vent, the vent connector could not be as long as 10' long, as our 80,000 Btuh furnace falls outside the minimum and maximum range of 93,000 to 129,000 Btuh. It is important to note that the minimum and maximum capacity allowed when the vent connector is constructed of Type B vent is much broader. Under the same conditions with a Type B vent connector, the related *National Fuel Gas Code* table indicates a 4" vent can handle 29,000 to 143,000 Btuh. Even a vent connector up to 20' in length is possible when using Type B pipe for the vent connector.

Remember that the tables for single appliances allow up to two 90-degree elbows in the installation. For each additional 45-degree elbow, the capacity shown in the tables must be reduced by 5%. For each additional 90-degree elbow, the capacity shown must be reduced by 10%. Also remember that additional changes in direction add to the overall length of the vent connector.

The *National Fuel Gas Code* prohibits the size of vent connectors being more than two pipe sizes larger than the connection on the equipment. For example, if a furnace has a 4" vent collar, the maximum size of the vent connector is 6".

On Site

Power Venting

When venting a furnace or boiler through a conventional chimney or vertical vent isn't practical, a manufacturer-approved power vent can often be installed, like the one shown here. This allows the byproducts of combustion to be vented through a sidewall, using a fan to establish the needed draft.

However, power venting is not always permitted. Always check local codes and seek the guidance of the equipment manufacturer before planning an installation like this.

Note that the combustion air for this oil burner is also ducted outdoors.

Source: Field Controls LLC

3.1.2 Vent Dampers

Vent dampers (*Figure 20*) are energy-saving devices that can be added to some furnaces and boilers. In some cases, vent dampers are incorporated into the equipment by the manufacturer. Vent dampers are automated devices installed in the vent that remain closed when the burner is not operating to prevent air from escaping the conditioned space unnecessarily.

Vent dampers are typically powered open and close by a small motor. When the burner shuts off, the motor is energized and closes the damper. This also opens redundant switches wired in series with the burner control circuit. The two open switches ensure the burner cannot start while the damper is closed.

The damper opens on a call for heat from the system control. Once it opens, the safety switches close and the burner starts. The damper can also be locked in the open position for service and troubleshooting.

Vent dampers were more prevalent when natural-draft furnaces were popular, primarily when the furnace was installed inside the conditioned space. A vent damper increases the AFUE rating of natural-draft furnaces. The dampers have little or no value in systems that use outdoor air for combustion and venting. Vent dampers are relatively easy to install, but can be a health and fire hazard if the burner runs while it is closed.

Consult local codes and the equipment manufacturer's documentation before installing a vent damper. Dampers are more likely to be found on natural-draft gas boilers today, as few natural-draft warm-air furnaces remain in service. Their installation on induced-draft furnaces is not beneficial. Due to the typically S-shaped construction of the heat exchanger, a natural draft through the furnace rarely develops.

Figure 20 Vent damper.
Source: Field Controls LLC

WARNING!

Vent dampers are rarely allowed or necessary on induced-draft furnaces. Never install a vent damper on any appliance without specific consent from the equipment manufacturer.

If you are assigned the task of installing equipment that is fitted with a vent damper, note that the *National Fuel Gas Code* outlines some changes in the table values. For example, the maximum capacity of the vent system is determined using the Nat Max column of the table. The minimum capacity is determined using the Fan Min column. Be sure to review the code requirements carefully for all installations to ensure compliance and the safety of the building occupants.

3.2.0 Condensing Gas Furnace Venting

Because condensing furnaces produce low-temperature flue gases, these furnaces are typically vented with Schedule 40 PVC pipe (*Figure 21*). New materials have entered the market as well. Vent pipe and fittings made from polypropylene are available and accepted by many equipment manufacturers and codes. These new products are designed specifically for venting, while PVC pipe was not specifically designed for the application. Despite this fact, PVC performs very well and is very widely accepted.

Figure 21 Condensing furnace PVC combustion air and vent lines.

Other common plastic piping products commonly approved for venting are Schedule 40 ABS-DWV and Schedule 40 PVC-DWV. The DWV refers to their designation as drain, waste, and vent products that are not usable for pressurized applications.

Condensing furnaces may draw their combustion air from outdoors or from inside the space. These two approaches are referred to as *direct venting* and *non-direct venting*. Direct venting requires a dedicated combustion air and venting system routed to an outside wall or roof. Non-direct venting means that the furnace draws its combustion air directly from the space, with only the byproducts of combustion vented outdoors.

When a condensing furnace is configured for non-direct venting, you must make all the necessary calculations called for in the *National Fuel Gas Code* to ensure an adequate volume of combustion air is provided. Refer to *Figure 5* and its relevant text.

The diameter of the combustion air and vent pipes in a direct venting application depends on the furnace input, the length of the pipe run, and the number and type of elbows used. Changes of direction create additional friction. Pipe fittings are therefore assigned a length, in feet, that is equivalent to the same size of straight pipe. This is referred to as a fitting's *equivalent length*.

For example, installing one standard 90-degree elbow in the vent line adds 10' to the measured length of the vent line. This is because it adds a similar amount of friction to a passing fluid as 10' of straight pipe. Choosing a long-radius, or *long-sweep*, elbow instead of a standard elbow adds only 5' to the length.

Thus, before consulting the installation instructions to size the combustion air and vent pipes, you must determine the measured (or scaled from a drawing) length of the lines and the equivalent length of each elbow needed. Summing the measured and equivalent lengths provides a total length needed to consult the manufacturer's charts. The sum of the straight pipe run length and the equivalent lengths of the fittings is referred to as the total equivalent length of the piping run.

Table 2 is an example of a pipe sizing chart you may encounter in the manufacturer's installation instructions. Other table styles and approaches to pipe sizing are available as well, but all are based on the same principles. The total equivalent length of the piping run and the capacity of the furnace largely determine the pipe size.

TABLE 2 Example Direct-Venting Pipe Sizing Chart

Maximum Equivalent Length versus Pipe Size*				
Furnace Model	**Input BTUH**	**Pipe Size**	**Max Equivalent Length**	**Max Number of Elbows****
CF60	58,000	2"	65'	6
		3"	100'	6
CF70	70,000	2"	65'	6
		3"	100'	6
CF85	83,000	2"	30'	2
		3"	100'	6
CF100	98,000	2"	20'	2
		3"	85'	6

* Not for field use.
** Long-radius 90-degree elbows; subtract an additional 5' per elbow for short-radius elbows.

Table 3 shows another example of a pipe selection chart you might see in the installation instructions. Note that the table values will differ for different furnace heating capacities; this table represents data for a single size of furnace. In this particular table, since the pipe size is adjusted for the number of elbows, the pipe length in the left column is the measured length rather than the equivalent length.

TABLE 3 Pipe Selection Chart

SCH 40 PVC or ABS Pipe Size – 080 Furnace Size*					
Pipe Length (Feet)	**Number of 90° Elbows (Long Radius)**				
	0	**2**	**4**	**6**	**8**
5'	2"	2"	2"	2"	2"
10'	2"	2"	2"	2"	2"
20'	2"	2"	2"	2"	2½"
30'	2"	2"	2"	2½"	2½"
40'	2"	2"	2½"	2½"	2½"
50'	2"	2½"	2½"	2½"	2½"
60'	2½"	2½"	2½"	2½"	3"
70'	2½"	2½"	2½"	3"	3"
80'	2½"	2½"	3"	3"	3"
90'	2½"	3"	3"	3"	3"

* Not for field use.

Sizing tables provided by the furnace manufacturer are used to size the combustion air and vent pipe of a condensing furnace. The tables are often altitude-sensitive, so be sure to apply any correction factors that are provided. Furnace heating capacity is reduced at higher altitudes due to differences in oxygen levels and air density. The length of each pipe run and the number of 90-degree elbows needed must also be known.

Here is an example of a vent sizing scenario. Assume a condensing furnace with an input of 83,000 Btuh requires 30' of pipe and two elbows to construct the combustion air intake, and 30' of pipe and three elbows to construct the vent pipe. Refer to *Table 2*.

Table 2 shows that up to 30' of 2" pipe can be used with two elbows for the combustion air intake. However, the table also shows that 2" pipe cannot be used if there are more than two elbows. As a result, 3" pipe must be used for the vent.

Because the 2" pipe is also borderline in size for the combustion air intake, the best practice here would be to use 3" pipe for both the combustion air intake and the vent pipe. Assuming the job is being properly planned in advance, using 3" pipe for both lines also allows for any field changes that may require additional elbows to be added. For example, an obstruction may be present that was not present during the planning stages.

Going Green

Condensing Furnace Efficiency

Condensing furnaces are very efficient because they capture a great deal of heat that would otherwise be lost through the vent. Much of that heat is contained in the water vapor that is a byproduct of combustion.

Changing one pound of water to vapor (vaporization) requires 970 Btus of heat. In the secondary heat exchanger of a condensing furnace, air moving over the coil absorbs heat from the vapor, causing it to condense back to its liquid state. As it changes back to water, it releases 970 Btus for every pound of water vapor that condenses.

When 100 ft^3 of natural gas burns, it produces 200 ft^3 of water vapor, which is roughly equal to a gallon of liquid water. One gallon of water weighs about 8.3 lb. For each gallon of condensed water (which is drained from the furnace during operation), over 8,000 Btus of heat is released (970 Btus $\times$ 8.3 lb/gallon $=$ 8,051 Btuh). This recovered heat increases the AFUE of the furnace significantly.

Also note that condensate pumps used with condensing furnaces must be acid resistant. If not, the acid must be neutralized before it enters the pump. Otherwise, the acid can be neutralized in the pump discharge line. Some furnace condensate pumps on the market today incorporate acid neutralization.

Did You Know?

Condensing Furnace Condensate

The condensate produced by a condensing furnace is mildly acidic. Many local plumbing codes do not allow this condensate to be discharged directly into sewer systems without treatment. To avoid this problem, an acid-neutralizing accessory can be installed to neutralize the condensate before it is discharged.

Source: Axiom Industries Limited

Installers must also take precautions to ensure the condensate line cannot freeze, which will cause the condensate pump to overfill and shut off the heating system.

Here are some additional tips on installing plastic vent piping for non-direct systems:

- Horizontal piping should be sloped back toward the furnace at least ¼" per foot. The slope allows any condensate that may form in the vent to drain back to the furnace where it can be collected and expelled. *Figure 22* shows the result of the vent pipe sloping away from the furnace. As condensate forms in the pipe, it drains outdoors to create "ice sculptures" like the one shown.
- The piping must be properly supported. Standard plastic piping practices are generally applied when fabricating pipe joints, including the use of a PVC/ABS cleaner prior to the application of the cement. However, it is important to review the installation instructions to ensure the installation meets the manufacturer's requirements.
- No portion of the vent or combustion air intake should be placed underground.
- Portions of the vent piping may need to be insulated to prevent moisture from freezing inside.
- Some manufacturers specify a minimum vent pipe length of 5', although this minimum length is not applied to the combustion air inlet.

Figure 22 Vent pipe from a condensing furnace pitched in the wrong direction.
Source: oasisamuel/Shutterstock

3.2.1 Direct-Vented Piping Terminations

In a direct vent system, combustion air is drawn in through one pipe while the byproducts of combustion are vented through the second pipe. Note that two or more furnaces cannot be connected to a single combustion air or vent pipe; each condensing furnace or appliance must have dedicated combustion air and vent piping.

A direct-vented condensing furnace is shown in *Figure 23*. In horizontal piping configurations like the one shown, a drain trap is often required in the combustion air intake pipe. This prevents water that may condense and collect in the intake pipe from entering the combustion chamber. A *pigtail trap* prevents air from entering through the drain line, since the combustion air intake operates at a slight negative pressure. The flexible drain tubing drains into the furnace condensate drain line, but the connection is not a secure connection. An air gap is generally required. In other words, the soft plastic tubing is simply dropped inside an open standpipe on the furnace condensate drain line.

The termination of the piping outdoors is critical to proper furnace operation. Pressure safety switches typically protect the furnace from problems in the piping system, and errors in the piping and terminations can cause these switches to open intermittently, shutting the furnace down. For example, both pipes must terminate outdoors in the same pressure zone, meaning the air intake and vent

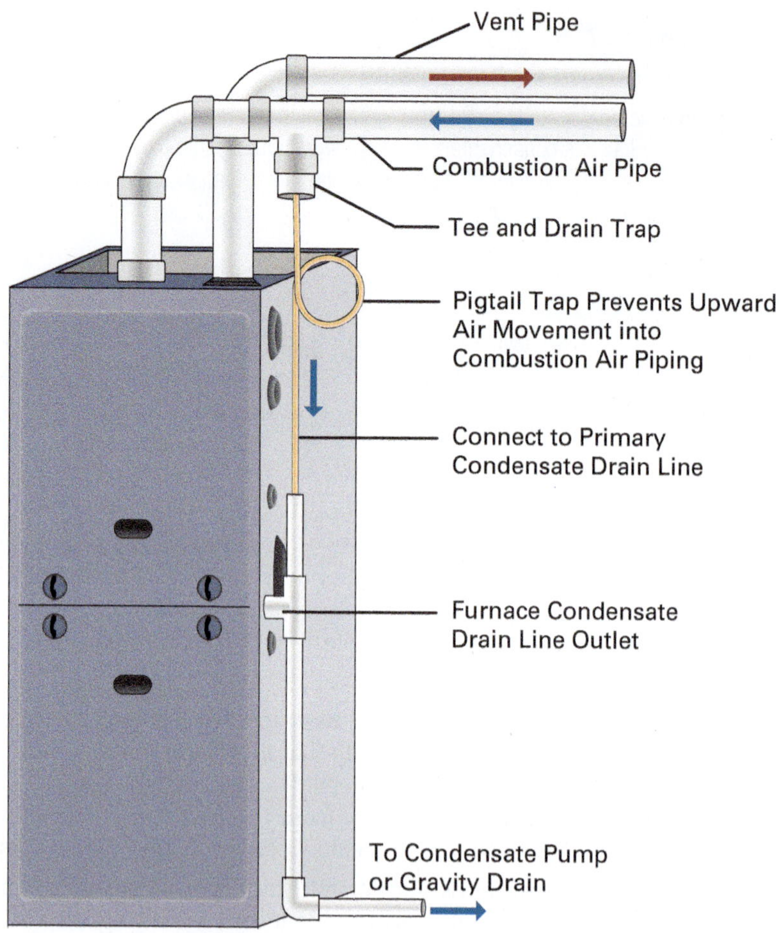

Figure 23 Condensing furnace direct-vent arrangement.

pipes must be close together. If the terminations are placed around a corner from each other, then the prevailing winds can create a pressure differential between the two pipes.

Technicians must follow the manufacturer's installation instructions for each furnace on every job. Never assume the requirements are the same for all condensing furnaces. The following are common installation practices for vents and intake piping, although some variations and special situations will always exist:

- The vent and intake pipes should always terminate in the same atmospheric pressure zone, whether they run through the roof or through an exterior wall. The two pipe openings should also be close together; separations between the two pipes of 3" on a roof and 6" on a sidewall are standard requirements.
- For vertical terminations through a roof, the vent pipe must typically be 12" higher than the combustion air intake. This prevents the combustion byproducts from being drawn back into the combustion airstream. To further separate the vent and combustion air intake, and to prevent rain from entering the combustion chamber, the combustion air intake is goosenecked down toward the roof.
- The combustion air intake should be a minimum of 3" above the roof on a vertical termination (*Figure 24*). This places the minimum height of the vent pipe at 15" above the roof. However, if snow is common, the minimum distance must be 3" above the average snow depth on the roof.
- Also keep the termination well above expected snow levels when venting through a wall. In cold climates, the outdoor portion and any part of the vent pipe that runs through an unconditioned space should be insulated.

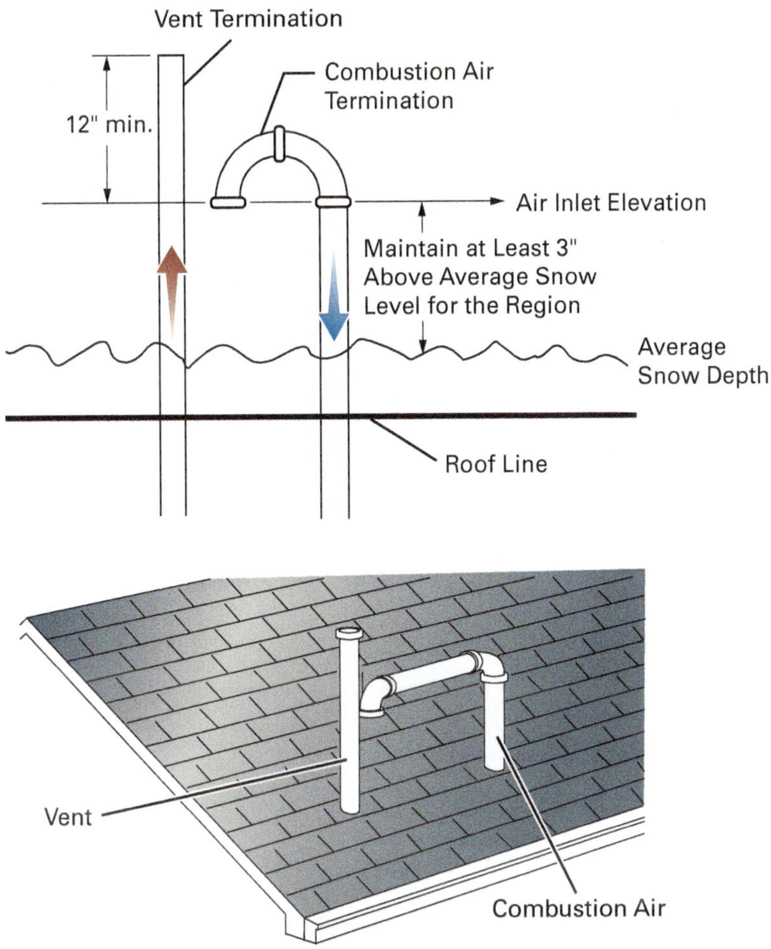

Roof Installation

Figure 24 Combustion air and vent termination details.

This helps keep the vent warmer, reducing the volume of condensate forming in the pipe. Most manufacturers require the vent to be either 12" above grade or the same distance above the anticipated snow level.

- *Figure 25* shows the common relationship between the vent and intake pipes outdoors when venting through a wall. As shown, the vent pipe should be higher than the intake pipe by 4" to 12". The two pipes should only be separated by 3" to 5" horizontally. However, this only applies to the final position of the openings. The two pipes may penetrate the wall outside of these dimensions. The intake pipe is typically turned toward the ground as shown in *Figure 26*. Pay close attention to the manufacturer's requirements in this area.

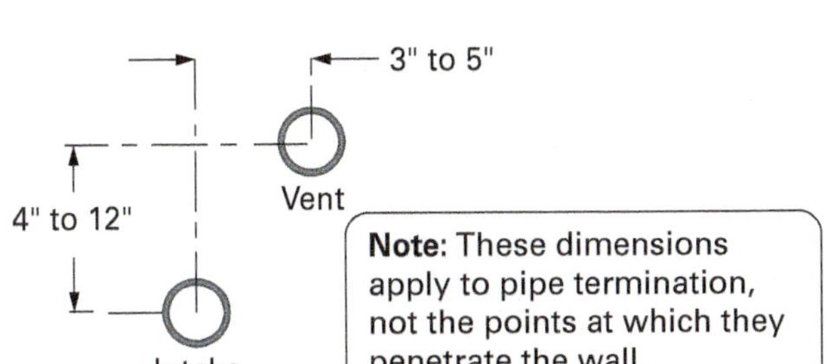

Figure 25 Common sidewall furnace vent and intake pipe termination dimensions.

Figure 26 Condensing furnace vent piping through an outside wall.

- When venting through a sidewall, avoid terminating the pipes at an inside corner, under a deck, or in dense shrubbery to prevent the recirculation of vent gases. Also avoid terminating pipes beside doors and beneath windows. Doing so can allow combustion byproducts to enter the building.
- When the pipe size must be increased to 2½" or 3" from the common 2" connection size at the furnace, manufacturers often require the pipe size to be reduced back to 2" before penetrating the roof or wall. This increases the velocity of the air and helps ensure the two airstreams do not mix.

Special concentric termination packages are available from most manufacturers (*Figure 27*). These devices allow combustion and exhaust gases to be carried through the same opening in the roof or exterior wall. Concentric packages also generally require a 2" pipe connection, meaning that larger pipe sizes must be reduced to 2" before being connected.

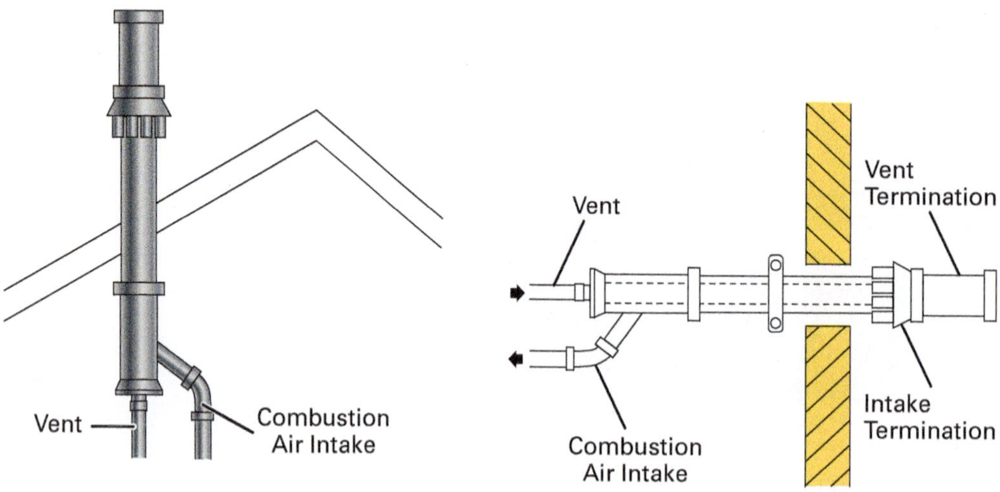

Figure 27 Concentric venting terminations.

It is not unusual for the installation instructions for a condensing furnace to have 20 or more pages devoted to venting and combustion air requirements. This speaks to the importance of the task as well as the many possible variations that are encountered in fieldwork. Be patient and review the requirements carefully to ensure that the installation is safe and allows the furnace to operate properly in all weather conditions.

3.0.0 Section Review

1. A low temperature rise in an induced-draft furnace _____.
   a. reduces the vent temperature
   b. can cause the furnace to overheat
   c. causes short cycling
   d. interferes with furnace ignition

2. A direct-vented condensing furnace requires _____.
   a. Type B vent pipe
   b. a condensate drain in the vent pipe
   c. a combustion air intake routed outside of the space
   d. stainless steel vent piping

Module 03202 Review Questions

1. A blue gas flame is typically produced when how much of the required combustion air is mixed with the gas before it is ignited?
   a. 20%
   b. 35%
   c. 50%
   d. 75%

2. What is the temperature range of the flue gases produced by an induced-draft furnace?
   a. 100°F to 125°F (38°C to 52°C)
   b. 160°F to 275°F (71°C to 135°C)
   c. 275°F to 400°F (135°C to 204°C)
   d. 350°F to 600°F (177°C to 316°C)

3. If there are indications that incomplete combustion is occurring in a natural-draft furnace, the problem may be solved by ______.
   a. increasing the fuel pressure
   b. increasing the amount of primary air
   c. increasing the size of the flue vent
   d. using a standing pilot

4. What is the term for the near-horizontal metal pipe that connects a furnace vent to a masonry chimney?
   a. Furnace connector
   b. Vent riser
   c. Chimney pipe
   d. Vent connector

5. What will happen if a person is exposed to CO levels above 200 ppm for an extended period?
   a. No symptoms
   b. Labored breathing
   c. Mild headache
   d. Loss of consciousness and eventual death

6. What is the average heat-content value assigned to natural gas?
   a. 1,000 Btus per cubic foot
   b. 2,200 Btus per cubic foot
   c. 5,300 Btus per cubic foot
   d. 10,000 Btus per cubic foot

7. To what does the term *temperature rise* refer?
   a. A natural draft that propels combustion byproducts through the vent.
   b. The volume of heat contained in the flue gases.
   c. The temperature difference between the supply and return airstreams.
   d. The heat remaining in the heat exchanger when the burner cycles off.

8. Regarding venting, induced-draft furnaces fall into which category?
   a. Category I
   b. Category III
   c. Category IV
   d. Category VI

9. In what location are Type B-W vent products used?
   a. outside the structure
   b. inside walls
   c. where the vent passes through an attic
   d. where a water heater is vented with a furnace

10. The vent connector for an induced-draft gas furnace ______.
    a. can be single-wall metal pipe in some cases
    b. can be constructed of plastic such as PVC
    c. must be constructed of Type B material
    d. must be constructed of Type B-W material

11. Under what circumstances can you vent a gas furnace through a lined masonry chimney serving a wood fireplace?
    a. Natural-draft furnaces can be vented with the fireplace.
    b. In mild climates, any furnace can be vented with the fireplace.
    c. The furnace can be vented with the fireplace if Type B vent is used for the vent connector.
    d. Under no circumstances should a furnace be vented with a fireplace through the same chimney.

12. How much clearance to combustibles is typically required for Type B vent pipe?
    a. No clearance is required.
    b. 1"
    c. 2"
    d. 6"

13. Which condition can cause a backdraft in a natural-draft furnace?
    a. Incorrect heat anticipator adjustment
    b. The installation of a vent damper
    c. A new wood-burning fireplace in a home
    d. High return air temperatures

14. True or False: The vent pipe of a condensing furnace must be sloped downward in the direction of the furnace.
    a. True
    b. False

15. Which of the following statements regarding condensing furnaces is *correct*?
    a. Short sections of vent piping can be placed underground as necessary.
    b. When venting vertically through a roof, the combustion air intake should always be lower than the vent opening.
    c. To ensure that vented byproducts and combustion air cannot mix, the two pipes can be routed through opposite walls of the building.
    d. All concentric discharge packages are designed to accept 2" or 3" pipe sizes, but not $2\frac{1}{2}$".

Answers to Odd-Numbered Module Review Questions are found in *Appendix A*.

Answers to Section Review Questions

Answer	Section	Objective
Section One		
1. b	1.1.1	1a
2. d	1.2.0	1b
3. d	1.3.1	1c
Section Two		
1. a	2.1.0	2a
2. d	2.2.0	2b
Section Three		
1. a	3.1.0	3a
2. c	3.2.0	3b

APPENDIX A Answers to Odd-Numbered Module Review Questions

MODULE 01 (03316)

Answer	Section
1. c	1.1.0
3. d	1.2.1
5. c	2.1.1
7. a	2.1.2
9. d	2.2.4

MODULE 02 (03215)

Answer	Section
1. a	1.1.1
3. a	1.2.1
5. b	1.3.2
7. d	1.3.3
9. b	2.1.1
11. b	2.2.3
13. d	3.1.3
15. b	3.2.2
17. d	4.2.2
19. a	4.4.1

MODULE 03 (03313)

Answer	Section
1. d	1.1.2
3. b	1.1.5
5. b	1.2.2
7. c	1.3.1
9. a	1.4.2
11. b	2.1.1
13. d	2.1.3
15. c	2.2.1

MODULE 04 (03206)

Answer	Section
1. b	1.1.1
3. b	1.2.1
5. d	2.2.1
7. b	3.1.0
9. c	3.2.1
11. b	3.2.6
13. d	4.2.1
15. b	4.3.1

MODULE 05 (03302)

Answer	Section
1. a	1.0.0; *Table 1*
3. c	1.2.0
5. c	1.6.0
7. a	2.1.5; *Figure 21*
9. a	3.1.1
11. a	3.2.2
13. c	3.2.4
15. c	4.2.1
17. c	5.1.1
19. c	5.1.1

MODULE 06 (03303)

Answer	Section
1. b	1.2.0
3. d	1.2.0
5. d	2.1.1
7. c	2.3.0
9. a	3.1.0
11. b	3.3.1
13. d	3.3.2
15. a	3.3.4
17. a	3.4.3
19. b	4.1.4

MODULE 07 (03301)

Answer	Section
1. b	1.0.0
3. d	1.2.3
5. b	1.6.0; *Table 4*
7. a	2.1.2
9. d	3.1.3
11. d	3.2.2
13. c	3.4.0
15. b	3.5.1

MODULE 08 (03205)

Answer	Section
1. a	1.1.1
3. c	1.2.3
5. c	2.1.3; *Table 2*
7. c	2.1.4
9. c	3.1.0
11. d	3.2.2
13. b	4.0.0
15. a	4.1.2

MODULE 09 (03211)

Answer	Section
1. b	1.1.0
3. d	1.2.2
5. a	2.1.3
7. a	2.2.2
9. b	2.4.1
11. a	3.1.2
13. b	4.1.0
15. d	4.2.2

MODULE 10 (03403)

Answer	Section
1. c	1.2.0
3. b	2.1.0
5. a	2.2.0
7. a	2.3.1
9. d	2.5.2
11. c	3.1.0
13. d	3.2.2
15. b	3.4.3

MODULE 11 (03204)

Answer	Section
1. b	1.0.0
3. a	1.1.2
5. b	2.1.1
7. c	3.1.0
9. d	3.1.6

MODULE 12 (03202)

Answer	Section
1. c	1.1.1
3. b	1.2.0
5. d	1.3.1
7. c	2.1.5
9. b	2.2.1
11. d	2.2.2
13. c	3.0.0
15. b	3.2.1

APPENDIX B Gas Furnace Inspection Form

Gas Furnace Inspection Form

NCCER National Center for Construction Education and Research

General:	**Condensing Furnaces:**	**Furnace Data:**
☐ Check thermostat operation	☐ Inspect/clean furnace condensate drain and pump	Make ____________
☐ Replace batteries annually		Model ____________
☐ Reset filter reminders	**Combustion Analysis:**	Serial # ____________
☐ Record voltage and indoor blower current (PSC only)	O2 _____%	Type
Voltage _____ VAC	CO _____ ppm	☐ Induced Draft
Current _____ A	CO2 _____ ppm	☐ Condensing
☐ Inspect and clean indoor blower as needed	Draft _____ in w.c.	
☐ Inspect furnace vestibule and all wiring connections	Efficiency _____%	
☐ Inspect igniter; set gap or test resistance (HSI)		
☐ Inspect and clean flame sensor	**Notes:**	
☐ Remove and clean burners as needed		
☐ Inspect heat exchanger		
☐ Inspect draft inducer assembly		
☐ Inspect flue piping or PVC air inlet/outlet piping		
System Performance:		
☐ Start furnace and observe startup sequence		
☐ Observe ignition and burner flame		
☐ Measure manifold gas pressure _____ in w.c.		
☐ Ensure indoor blower starts after delay		
☐ Measure temperature rise _____°F		
Suppy air temp _____ °F		
Return air temp _____ °F		
☐ Disconnect draft pressure switch and ensure burner shuts off		
☐ Test ignition lockout circuit (close gas supply valve to disable burner)		
☐ Test high-temperature limit switch		

APPENDIX C Cooling Inspection Form

Cooling Inspection Form

NCCER National Center for Construction Education and Research

General:	**System Performance:**	**Indoor/Outdoor Conditions:**
☐ Check thermostat operation	☐ Record superheat and/or subcooling	Return air temp ____ °F
☐ Replace batteries annually	Metering device:	Return air wet bulb ____ °F
☐ Reset filter reminders	☐ TXV or EXV	Return air enthalpy ____ Btu/lb
☐ Record voltage; record current for PSC motors only	☐ Fixed	Supply air temp ____ °F
Voltage ____ VAC	Superheat ____ °F	Supply air wet bulb ____ °F
Indoor blower current ____ A	Subcooling ____ °F	Supply air enthalpy ____ Btu/lb
☐ Inspect and clean indoor blower as needed	☐ Indoor coil ΔT and/or ΔH	System air volume ____ cfm
☐ Inspect and clean indoor coil as needed	ΔT ____ °F	Outdoor temp ____ °F
☐ Inspect and clean condensate drain system	ΔH ____ Btuh	
☐ Clear drain line	Indoor coil pressure drop ____ in w.c.	
☐ Inspect/clean pump (if present)		
	Indoor Unit Data:	**Notes:**
Outdoor Unit:	Make ____________	
☐ Check all wiring terminations	Model ____________	
☐ Inspect contactor contacts	Serial # ____________	
☐ Evidence of refrigerant leaks	Type	
☐ Inspect condenser fan blade and motor	☐ Induced Draft	
☐ Clean condenser coil as needed	☐ Condensing	
☐ Record voltage and operating current	☐ Cased Coil (if present):	
Voltage ____ VAC	☐ Make ____________	
Compressor current ____ A	☐ Model ____________	
Outdoor fan current ____ A	☐ Serial # ____________	
☐ Record operating pressures		
High pressure ____ psig	**Outdoor Unit Data:**	
Low pressure ____ psig	Make ____________	
	Model ____________	
	Serial # ____________	
	Refrigerant ____________	

APPENDIX D Typical SDS for Refrigerant Oil

Nu-Calgon

SAFETY DATA SHEET

Date Prepared: 05/28/2015
SDS No: NC 0113

ZEROL® 150 (4310-07, 4310-05, 4310-01)

1. PRODUCT AND COMPANY IDENTIFICATION

PRODUCT NAME: ZEROL® 150 (4310-07, 4310-05, 4310-01)
PRODUCT DESCRIPTION: Synthetic Alkylate

DISTRIBUTOR
Nu-Calgon
2008 Altom Court
St. Louis, MO 63146
Emergency Phone: 800-554-5499

24 HR. EMERGENCY TELEPHONE NUMBERS
CHEMTREC: 800-424-9300

2. HAZARDS IDENTIFICATION

GHS LABEL

This product is not classified as hazardous according to 29CFR 1910.1200.

3. COMPOSITION / INFORMATION ON INGREDIENTS

Chemical Name	Wt.%	CAS
Synthetic Alkylate(s)	*	Proprietary

COMMENTS: Ingredients and composition are withheld as trade secret in accordance with 29CFR1910.1200(i).

4. FIRST AID MEASURES

EYES: Immediately flush eyes with plenty of water. Get medical attention, if irritation persists.

SKIN: Immediately wash skin with soap and plenty of water. Remove contaminated clothing. Get medical attention if symptoms occur. Wash clothing before reuse.

INGESTION: Rinse mouth with water. Drink plenty of water. Do not induce vomiting. Get immediate medical attention if you feel unwell. Never give anything by mouth to an unconscious person.

INHALATION: Remove to fresh air. If breathing is difficult, give oxygen. If not breathing, give artificial respiration. Get immediate medical attention.

SIGNS AND SYMPTOMS OF OVEREXPOSURE

EYES: May cause eye irritation.

SKIN: May cause skin irritation.

5. FIRE FIGHTING MEASURES

FLAMMABLE CLASS: Non-flammable.

GENERAL HAZARD: Do NOT use water stream or water jet.

EXTINGUISHING MEDIA: Dry Chemical, CO2, Foam, Fog or Water Spray.

HAZARDOUS COMBUSTION PRODUCTS: Oxides of carbon.

OTHER CONSIDERATIONS: Keep adjacent containers cool with water spray.

FIRE FIGHTING PROCEDURES: As in any fire, wear self-contained breathing apparatus pressure-demand, (MSHA/NIOSH approved or equivalent) and full protective gear.

FIRE FIGHTING EQUIPMENT: In case of fire wear a full-face positive pressure self-contained breathing apparatus and protective suit.

6. ACCIDENTAL RELEASE MEASURES

SMALL SPILL: Contain and soak up spill with non-combustible absorbent material. Dampen and scoop spilled material into suitable containers, seal and label. Avoid mixing spilled material with other chemicals or debris. Avoid runoff into storm sewers and ditches which lead to waterways.

Source: Nu-Calgon

LARGE SPILL: Shut off leak if safe to do so. Remove all possible sources of ignition. Contain spill with inert, absorbent material. Reclaim into recovery or salvage drums or vacuum truck for proper disposal. Treat residues as small spills.

ENVIRONMENTAL PRECAUTIONS

WATER SPILL: Prevent material from entering drains or watercourses using sand, earth or other appropriate barriers

LAND SPILL: Prevent contamination of the soil.

GENERAL PROCEDURES: Avoid contact with skin or eyes. Avoid contact with spilled materials. Use appropriate personal protective equipment. Ensure adequate ventilation.

SPECIAL PROTECTIVE EQUIPMENT: Use protective equipment recommended in Section 8.

7. HANDLING AND STORAGE

GENERAL PROCEDURES: Handle in accordance with good industrial hygiene and safety practices.

HANDLING: Avoid breathing (dust, vapor, mist, gas). Avoid contact with skin and eyes. Ground and bond containers when transferring material. Wash thoroughly after handling.

STORAGE: Keep container tightly closed in a dry, well-ventilated place away from heat, open flame, or sources of ignition. Store away from incompatible materials (see section 10).

STORAGE TEMPERATURE: Ambient.

LOADING TEMPERATURE: Ambient.

ELECTROSTATIC ACCUMULATION HAZARD: Bond and ground during loading, unloading and transferring material to prevent electrostatic accumulation.

8. EXPOSURE CONTROLS / PERSONAL PROTECTION

ENGINEERING CONTROLS: Provide adequate ventilation. Eye washes and showers for emergency use.

PERSONAL PROTECTIVE EQUIPMENT

EYES AND FACE:

SKIN: Use protective clothing, sleeves, boots, face protection and chemical resistant gloves. Use extra precautions when working with heated materials.

RESPIRATORY: Use appropriate protection if there is risk of breathing the mists or vapors or heated materials. Use suitable respiratory equipment if risk of insufficient ventilation.

PROTECTIVE CLOTHING: Wear protective clothing appropriate to the risk of exposure.

WORK HYGIENIC PRACTICES: Wash hands, forearms and face before eating, drinking, smoking and using the restroom. Launder contaminated clothing before re-use.

9. PHYSICAL AND CHEMICAL PROPERTIES

PHYSICAL STATE: Liquid

APPEARANCE: Clear colorless liquid

PHYSICAL STATE COMMENTS: Clear, bright, viscous.

pH: Not Available

FLASH POINT AND METHOD: 184°C COC-ASTM D-92

AUTOIGNITION TEMPERATURE: Not Available

VAPOR PRESSURE: Not Available

BOILING POINT: Not Available

POUR POINT: -36°C

THERMAL DECOMPOSITION: Not Available

SPECIFIC GRAVITY: 0.87 at 15°C

VISCOSITY #1: 30 cSt at 40°C

VISCOSITY #2: 4.3 cSt at 100°C

10. STABILITY AND REACTIVITY

HAZARDOUS POLYMERIZATION: None

Source: Nu-Calgon

STABILITY: Stable under normal conditions.

CONDITIONS TO AVOID: Excessive heat, sparks or open flames.

POSSIBILITY OF HAZARDOUS REACTIONS: None Expected.

HAZARDOUS DECOMPOSITION PRODUCTS: Under normal conditions, no hazardous decomposition products should be produced.

INCOMPATIBLE MATERIALS: Strong oxidizers.

11. TOXICOLOGICAL INFORMATION

ACUTE

DERMAL LD_{50}: > 2000 mg/kg

ORAL LD_{50}: > 2000 mg/kg

EYE EFFECTS: Minimal irritation.

SKIN EFFECTS: Minimal irritation.

CARCINOGENICITY

Notes: Contains no ingredient listed as a carcinogen.

REPEATED DOSE EFFECTS: None Expected.

IRRITATION: Not a primary irritant.

SENSITIZATION: None Expected.

REPRODUCTIVE EFFECTS: None Expected.

MUTAGENICITY: None Expected.

12. ECOLOGICAL INFORMATION

ENVIRONMENTAL DATA: Not classified as an environmental hazard.

ECOTOXICOLOGICAL INFORMATION: This product is not expected to be toxic to aquatic life.

BIOACCUMULATION/ACCUMULATION: Expected to have a very slow rate of bioaccumulation.

DISTRIBUTION: Insoluble in water. Adsorbs to soil and has low mobility.

CHEMICAL FATE INFORMATION: Expected to be partially or slowly biodegradable.

13. DISPOSAL CONSIDERATIONS

DISPOSAL METHOD: Dispose of contents/container to a licensed contractor or facility in accordance with local, state and federal regulations.

PRODUCT DISPOSAL: Materials are suitable for burning or disposal at a supervised incineration facility.

EMPTY CONTAINER: Clean container thoroughly and dispose to licensed disposal contractor. Do not weld, cut, or braze empty containers, or allow product residue to come in contact with other sources of ignition as they may contain residue which could ignite.

14. TRANSPORT INFORMATION

DOT (DEPARTMENT OF TRANSPORTATION)

PROPER SHIPPING NAME: Not regulated for Transport

ROAD AND RAIL (ADR/RID)

PROPER SHIPPING NAME: Not regulated for Transport

AIR (ICAO/IATA)

SHIPPING NAME: Not regulated for Transport

VESSEL (IMO/IMDG)

SHIPPING NAME: Not regulated for Transport

15. REGULATORY INFORMATION

UNITED STATES

SARA TITLE III (SUPERFUND AMENDMENTS AND REAUTHORIZATION ACT)

Source: Nu-Calgon

311/312 HAZARD CATEGORIES: Not hazardous.

313 REPORTABLE INGREDIENTS: No ingredients at detectable levels.

TSCA (TOXIC SUBSTANCE CONTROL ACT)

TSCA STATUS: All components listed or exempt.

CANADA

DOMESTIC SUBSTANCE LIST (INVENTORY): All components are listed or exempt.

16. OTHER INFORMATION

APPROVED BY: Audris King **TITLE:** HS&E Manager

PREPARED BY: A. King **Date Prepared:** 05/28/2015

HMIS RATING

HEALTH	1
FLAMMABILITY	1
PHYSICAL HAZARD	0
PERSONAL PROTECTION	

NFPA CODES

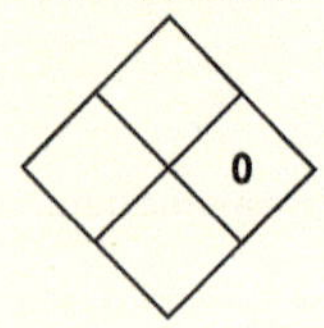

MANUFACTURER DISCLAIMER: The information is based on the data of which we are aware and is believed to be correct as of the data hereof. Since the information contained herein may be applied under conditions beyond our and with which we may be unfamiliar and since data made available subsequent to the date hereof may suggest modification of the information, we do not assume any responsibility for the result of its use. This information is furnished upon condition that the person receiving it shall make his own determination of the suitability of the material for his particular purpose.

Source: Nu-Calgon

APPENDIX E Indoor Air Quality Interview Form

IBEAM: OCCUPANT COMPLAINT RECORDS page 1 of 2

IAQ Interview Form

D3: Indoor Air Quality Interview Form

This form is used in conjunction with the Occupant Complaint Form. If the occupant called in the complaint and has not filled out the Occupant Complaint Form, ask him/her to do so as needed. Assure the occupant that all information provided will be kept confidential.

Building Name ____________Address____________________File #__________

Occupant Name______________________________________

Room Number/Location________________________________Phone________

Interviewed by ________________ Date ______

Symptom Patterns

What kind of symptoms or discomfort are you experiencing?

Are you aware of other people with similar symptoms or concerns? Yes____No____. If yes, what are their names and locations?

Do you have any health conditions that may make you particularly susceptible to environmental problems?

• contact lenses • allergies	• chronic cardiovascular disease • chronic respiratory disease • chronic neurological problems	• undergoing chemotherapy or radiation therapy • immune system suppressed by disease or other causes

Spatial Patterns

Where are you when you experience symptoms or discomfort? ______________________

Where do you spend most of your time in the building? ________________________

Where else in the building do you frequent? ______________________________

Timing Patterns

When did the symptoms/problems start? _______________________________

When are symptoms/problems generally worse?

Beginning of week	Morning	Afternoon	Spring	Summer
End of week			Fall	Winter
Particular days of week	Particular times of day		Particular months	
______ ______ ______	______ ______ ______		______ ______ ______	

Source: US Environmental Protection Agency

IBEAM: OCCUPANT COMPLAINT RECORDS
IAQ Interview Form

page 2 of 2

Other pattern or no pattern (explain)

____

____

When do symptoms/problems go away?

After you leave the building ____yes____no: If yes, how long does it take to go away? ______

After you leave the space _____yes____no: If yes, how long does it take to go away? ______

Have you noticed any particular events, activities that you or others engage in, weather conditions, temperature or humidity conditions, odors, or other things that tend to occur around the same time or before your symptoms?

Additional Information

Do you have any observations about the building conditions that might need attention or might help explain your symptoms?

Do you have other comments?

Source: US Environmental Protection Agency

APPENDIX F HVAC Checklist – Short Form

HVAC Checklist - Short Form

Page 1 of 4

Building Name: ________________ Address: ________________

Completed by: ________________ Date: ________ File Number: ________

Sections 2, 4 and 6 and Appendix B discuss the relationships between the HVAC system and indoor air quality.

MECHANICAL ROOM

- Clean and dry?________________ Stored refuse or chemicals?________________
- Describe items in need of attention ________________

MAJOR MECHANICAL EQUIPMENT

- Preventive maintenance (PM) plan in use?________________

Control System

- Type________________
- System operation ________________
- Date of last calibration ________________

Boilers

- Rated Btu input ________ Condition ________________
- Combustion air: is there at least one square inch free area per 2,000 Btu input? ________________
- Fuel or combustion odors ________________

Cooling Tower

- Clean? no leaks or overflow?________________ Slime or algae growth?________________
- Eliminator performance ________________
- Biocide treatment working? (list type of biocide) ________________
- Spill containment plan implemented? ________________ Dirt separator working?________________

Chillers

- Refrigerant leaks? ________________
- Evidence of condensation problems? ________________
- Waste oil and refrigerant properly stored and disposed of? ________________

Source: US Environmental Protection Agency

HVAC Checklist - Short Form

Page 2 of 4

Building Name: ______________________ Address: ______________________

Completed by: ______________________ Date: __________ File Number: __________

AIR HANDLING UNIT

- Unit identification ______________________ Area served ______________________

Outdoor Air Intake, Mixing Plenum, and Damper

- Outdoor air intake location ______________________
- Nearby contaminant sources? (describe) ______________________
- Bird screen in place and unobstructed? ______________________
- Design total cfm ________ outdoor air (O.A.) cfm ________ date last tested and balanced ________
- Minimum % O.A. (damper setting) ________ Minimum cfm O.A. $\frac{\text{(total cfm x minimum \% O.A.)}}{100}$ = ________
- Current O.A. damper setting (date, time, and HVAC operating mode) ______________________
- Damper control sequence (describe) ______________________
- Condition of dampers and controls (note date) ______________________

Fans

- Control sequence ______________________
- Condition (note date) ______________________
- Indicated temperatures supply air______ mixed air______ return air______ outdoor air______
- Actual temperatures supply air______ mixed air______ return air______ outdoor air______

Coils

- Heating fluid discharge temperature________ ΔT______ cooling fluid discharge temperature______ ΔT______
- Controls (describe) ______________________
- Condition (note date) ______________________

Humidifier

- Type______________________ if biocide is used, note type ______________________
- Condition (no overflow, drains trapped, all nozzles working?) ______________________
- No slime, visible growth, or mineral deposits? ______________________

Source: US Environmental Protection Agency

HVAC Checklist - Short Form

Page 3 of 4

Building Name: ______________________ Address: ______________________

Completed by: ______________________ Date: __________ File Number: ______________

DISTRIBUTION SYSTEM

Zone/ Room	System Type	Supply Air		Return Air		Power Exhaust		
		ducted/ unducted	cfm*	ducted/ unducted	cfm*	cfm*	control	serves (e.g. toilet)

Condition of distribution system and terminal equipment (note locations of problems)

- Adequate access for maintenance?______________________
- Ducts and coils clean and obstructed?______________________
- Air paths unobstructed? supply______ return______ transfer______ exhaust______ make-up______
- Note locations of blocked air paths, diffusers, or grilles______________________
- Any unintentional openings into plenums?______________________
- Controls operating properly?______________________
- Air volume correct?______________________
- Drain pans clean? Any visible growth or odors?______________________

Filters

Location	Type/Rating	Size	Date Last Changed	Condition (give date)

Source: US Environmental Protection Agency

HVAC Checklist - Short Form

Page 4 of 4

Building Name: ______________________ Address: ______________________

Completed by: ______________________ Date: __________ File Number: __________

OCCUPIED SPACE

Thermostat types______________________

Zone/ Room	Thermostat Location	What Does Thermostat Control? (e.g., radiator, AHU-3)	Setpoints		Measured Temperature	Day/ Time
			Summer	Winter		

Humidistats/Dehumidistats type______________________

Zone/ Room	Humidistat/ Dehumidistat Location	What Does It Control?	Setpoints (%RH)	Measured Temperature	Day/ Time

- Potential problems (note location)______________________
- Thermal comfort or air circulation (drafts, obstructed airflow, stagnant air, overcrowding, poor thermostat location)

____

____

- Malfunctioning equipment______________________
- Major sources of odors or contaminants (e.g., poor sanitation, incompatible uses of space)

____

Source: US Environmental Protection Agency

GLOSSARY

Active listening: Listening attentively while using both verbal and nonverbal indications to demonstrate your interest and engagement.

Active recovery: Recovery process that relies on a self-contained refrigerant recovery unit to draw or push the refrigerant out of a refrigerant circuit.

Adsorption: A process through which a solid material captures and holds gas or liquid atoms, ions, or molecules on its surface as a thin film.

Aldehydes: Highly reactive organic compounds that are characterized by a carbonyl group (CHO) attached to a hydrogen atom and often a distinctive odor. A commonly recognized form of aldehyde is *formaldehyde*.

Alternator: A device that generates alternating current through conductors rotating in a magnetic field.

Amplitude: The maximum height and depth reached by the sine wave in a cycle, and the point on the wave at which the highest voltage is being generated.

Armature: The rotating component of a generator.

Axial load: An external force acting lengthwise, parallel to the shaft centerline.

Azeotropes: Blended refrigerants that are not true compounds but behave much like a compound. The refrigerant evaporates and condenses at specific pressure and temperature combinations. From the technician's perspective, working with an azeotropic blend is like working with a compound refrigerant.

Balance point: The outdoor temperature at which the heating capacity of a given heat pump is equal to the heat loss of the structure.

Belt whip: A condition where excess slack in a drive belt causes it to begin riding up and out of the sheave grooves, primarily on the slack side, causing a whipping action. The slack side is the side exiting the drive pulley as it rotates.

Binary blends: Blended refrigerants consisting of two components.

Biological contaminants: Living contaminants such as bacteria, fungi, viruses, and algae. Also referred to as *microbial contaminants*.

Borescope: An optical device that typically uses light-transmitting fiber to display an image through a narrow tube, allowing for the inspection of otherwise difficult or impossible areas to see.

Break-away torque: The torque required to loosen a fastener.

Bubble point: The point at which a zeotropic refrigerant begins to vaporize as heat is added. It is also the point at which it completes the condensing process as heat is removed. See *dew point*.

Building-related illness (BRI): An illness that can be attributed to building contaminants.

Burnout: A condition in which the failure or degradation of the motor winding insulation causes the windings to electrically short to ground.

Capacitor: A passive electrical energy storage device containing two metal plates separated by an insulating (dielectric) material.

Capacity control: The manipulation of compressor operating characteristics to reduce the volume of vapor compressed to better match the load conditions. Other methods of capacity control can also be applied to a system without affecting compressor operation or the volume of vapor moving through it.

Caustics: Strong alkaline substances that destroy or damage organic tissue and cause significant metal corrosion.

Centistoke (cSt): A metric unit of measure for viscosity, representing centimeters of movement per gram per second.

Chemically inert: Describes a substance that does not readily react chemically with other substances, even under high temperatures and pressures, both of which typically accelerate chemical reactions.

Clearance volume: The amount of clearance between the cylinder head and the piston when it is at the top dead-center position of its travel.

Coefficient of Performance (COP): The ratio of heat produced in relation to the electrical energy used, calculated using the following equation: COP = Btuh ÷ (watts × 3.413).

Commutator: The movable contact surface on an electric generator or motor.

Complete combustion: Combustion supported by enough oxygen to prevent the significant formation of carbon monoxide (CO).

Compliant Scroll™ compressor: A version of the scroll compressor designed to allow the orbiting scroll to temporarily shift from its normal operating position, allowing liquid refrigerant that may enter to pass through without causing significant damage.

Compounds: Elemental combinations in a fixed ratio that exhibit their own unique characteristics. Refrigerants are typically compounds or blends of compounds.

Compression ratio: For refrigerant compressors, the ratio of discharge pressure to suction pressure. Note that *compression ratio* has a different meaning when applied to internal combustion engines.

Compression: The reduction in volume of a vapor by mechanical means.

Contactor: A control device consisting of a magnetic coil and one or more sets of contacts used as switching devices in high-voltage, high-current circuits, such as those serving compressors.

Crimped: Formed through compression, such as a connector on the end of an electrical conductor.

Current imbalance: For a three-phase power supply, the maximum deviation of the current in any phase from the average current in the three phases, divided by the average current. It is generally expressed as a percentage.

De minimis release: *De minimis* is a Latin expression referring to small or insignificant things. In the HVACR craft, the term refers to a minute release of refrigerant that is unavoidable. The very brief, low-pressure purging of a refrigerant hose prior to connection is an example of a de minimis release.

Deionized (DI) water: Water from which ions, or charged particles, have been effectively removed, resulting in water that is largely free of salts and other minerals.

Demulsifiers: Used to separate emulsions, such as oil and water; the opposite of an emulsifying agent.

Desiccant: A substance such as silica gel or calcium chloride (CaCl) that attracts water and water vapor, creating and maintaining a state of dryness around it, limited by its water-holding capacity.

Dew point: The point at which a zeotropic refrigerant stops vaporizing as heat is added. It is also the point that the refrigerant starts condensing when heat is removed. See *bubble point*.

Dielectric strength: The ability of an electrical insulating material to resist breakdown in the presence of voltage.

Dielectric: A material that strongly resists the passage of current.

Digital linear actuator (DLA): A device that changes the rotational force of an EEV stepper motor to a push/pull (linear) motion needed to open and close the valve. The geared portion of the device also serves to increase the torque of the tiny motor.

Dilution air: Air added to the flue gases in a natural-draft furnace to aid in flue gas removal.

Direct-expansion (DX) cooling system: A system dependent upon a refrigerant circuit in which the cooling effect is achieved by the expansion of a refrigerant as it changes from a liquid to a vapor state.

Dispersants: An additive used to disperse, or separate, groups of particles to prevent them from coagulating or joining together.

Dropping point: The temperature at which grease is soft enough for a drop of oil to fall away or flow from the bulk of the grease.

Dual-fuel or hybrid system: A system in which a heat pump is combined with a fossil-fuel furnace.

Economizers: Mechanical devices, largely consisting of dampers and their controls, that are integrated into an HVAC unit or system to draw in outdoor air and mix it with the indoor air.

Effective voltage: The value of AC voltage that will produce as much power when connected across a load as an equivalent amount of DC voltage. Also known as root-mean-square (RMS) voltage.

Empathy: The ability to identify with another person's situation or feelings.

Emulsifying agents: Chemical compounds that support the mixing of two or more normally immiscible liquids, such as oil and water. Emulsifying agents also help prevent the liquids from separating.

Energy recovery ventilators (ERVs): Specialized HVAC equipment that saves energy year-round using a heat exchanger to transfer heat both to and from an exhausted airstream to ventilation air entering a building.

Enthalpy sensor: A sensor that reports the total heat contained in air, accounting for both the dry-bulb temperature and the volume of water content in an air sample. Wet air, for example, contains more heat than dry air at the same temperature.

Environmental tobacco smoke (ETS): The combination of smoke from a cigarette, cigar, or pipe and the exhaled smoke from the smoker.

Externally equalized: Describes an expansion valve that has a pressure connection to the suction line downstream of the evaporator coil to serve as a closing force. External equalizer lines are used when there is a significant pressure drop through the evaporator and/or distributor.

Fire point: The temperature at which a lubricant will burn if an ignition source is present.

Flash gas: Refrigerant vapor that forms spontaneously due to a rapid change in pressure that forces the refrigerant to boil.

Flash point: The temperature at which lubricant vapors mixed with air can ignite but are unable to sustain combustion.

Floc point: The temperature at which a 90%/10% mixture of oil and refrigerant forms a cloudy precipitate of wax in the mixture.

Floodback: The return of liquid refrigerant or a liquid refrigerant/oil mixture to the crankcase of an operating compressor.

Flooded starts: Compressor startups during which slugging, foaming, and inadequate lubrication occur due to oil in the compressor having absorbed condensed refrigerant during the off cycle, or due to the presence of liquid refrigerant at the compressor suction inlet. When the compressor crankcase pressure drops quickly at startup, any liquid refrigerant present boils rapidly, causing foaming and potentially washing the oil out of the crankcase.

Foot-pound (ft-lb): A torque unit equal to the force of one pound applied to a lever one foot long. One foot-pound equals 12 inch-pounds. Also called the *pound-foot (lb-ft)*.

Formaldehyde: A colorless, pungent byproduct of both synthetic and natural biological processes that causes irritation of the eyes and upper air passages; a known carcinogen.

Fractionation: A process in which one or more components of a blended refrigerant vaporize while its other components are not changing state.

Frequency: The number of complete cycles of an alternating current, sound wave, or vibrating object that occur over a predetermined time.

Friable: Brittle and easily fragmented, releasing particles that can become airborne.

Geothermal system: A system that uses the earth as a source of heat as well as a heat sink for cooling operation.

Glide: The temperature range across which a zeotropic refrigerant blend evaporates and condenses.

Global warming potential (GWP): A measure of a substance's contribution to global warming, expressed as a number. For practical comparison, carbon dioxide (CO_2), a known greenhouse gas, is assigned a GWP of 1. The higher the number, the greater the warming potential.

Global warming: A measurable increase in the temperature of earth's atmosphere, primarily attributed to the greenhouse effect from carbon dioxide (CO_2) and other materials released into the atmosphere that impede the movement of heat away from the planet.

Greenhouse effect: The result of atmospheric gases such as CO_2 trapping heat in the atmosphere in the same way that a greenhouse captures and holds solar heat.

Half-life: The time required for half of a number of atoms to undergo radioactive decay.

Halogens: The elements chlorine, fluorine, bromine, iodine, and astatine. Halogens are reactive nonmetallic elements that form strongly acidic compounds and form simple salts when combined with metals such as calcium.

Heat anticipator: A resistive heating element in a thermostat that warms the thermostat, affecting the cycle rate. The heat anticipator helps tune the operating cycle and prevents the room temperature from overshooting the setpoint.

Heat recovery ventilators (HRVs): Specialized HVAC equipment that saves energy in the winter using a heat exchanger to transfer heat from an exhausted airstream to cold ventilation air entering a building.

Heating Season Performance Factor (HSPF): A heat pump performance rating that has been adjusted for seasonal operation. It represents the total heating output of a heat pump (in Btus) during its normal annual usage period for heating, divided by the total electric power input in watt-hours during the same period.

Hertz (Hz): The unit of measure for the frequency of alternating current. One Hertz equals one cycle per second.

High-efficiency particulate air (HEPA) filter: An extended media, dry-type filter mounted in a rigid frame. It has a minimum efficiency of 99.97% for 0.3-micron and larger particles when a clean filter is tested at its rated airflow capacity.

Horsepower: A unit of measure for power equal to 550 foot-pounds per second, or 746 watts.

Hunting: Describes the operation of an expansion valve opening and closing excessively and repeatedly, unable to stabilize at a consistent flowrate.

Hydrocarbons (HCs): Compounds comprised of only carbon and hydrogen atoms. When some or all the hydrogen atoms in a hydrocarbon are combined with a halogen, the result is a compound known as a *halocarbon*. Hydrocarbon refrigerants are typically flammable.

Hydrostatic testing: A method of testing pressure vessels, piping systems, and similar components for leaks by filling them with a liquid (typically water) and then applying pressure using an inert gas to test the component's structural integrity.

Hygroscopic: Having a tendency to absorb water and water vapor readily and aggressively, much like a desiccant.

Impedance: The effective resistance of a circuit or component to the flow of current.

Inch-pound (in-lb): A small torque unit equal to the force of one pound applied to a lever one inch long. One inch-pound equals $\frac{1}{12}$ foot-pound. Also called the *pound-inch (lb-in)*.

Incomplete combustion: Combustion without enough oxygen to prevent the formation of carbon monoxide (CO).

Induction motors: Electric motors that operate through induction.

Induction: The generation of a current in a conductor by placing it in a moving magnetic field.

Inertia: The tendency of a body in motion to remain in motion and a body at rest to remain at rest.

Inrush current: The maximal, short-lived current drawn by an inductive electrical device when first energized.

Internally equalized: Describes an expansion valve that uses the pressure value at its own outlet to serve as a closing force. This effectively disregards the pressure drop through the evaporator.

Ionizing wires: Thin wires in an electronic air cleaner that impart a positive charge to passing particles. The charged particles are then attracted to the negatively charged collector plates.

Isolation transformer: A transformer with a one-to-one turn ratio, used to isolate electrical circuits or prevent electrical interference.

Journal: The part of a shaft, axle, or spindle that is supported by and revolves inside a bearing.

Lantern ring: A perforated ring positioned near the center of a pump stuffing box, providing a point of access and a distribution point for fluid entering to cool and lubricate the packing and shaft.

Leak reactant solution: Noncorrosive liquid solution that readily foams or forms bubbles in the presence of a leak.

Line-duty device: An electrical component designed to handle the voltage and current of load devices such as motors.

Lugs: Wiring terminals used on wire sizes No. 8 AWG and larger; also, a term used to describe terminals that apply pressure to a stripped conductor with a screw.

Mandrel: A cylindrical or tapered rod or shaft around which other materials can be shaped or forged.

Mastic: A pasty form of sealant that is typically brushed on and hardens to seal joints and repair leaks in ductwork.

Megohmmeter (megger): A test instrument used to test high-resistance circuits using a high-voltage test signal.

Microfarad (mfd): A unit of measure for capacitance, equaling one-millionth of a farad.

Micron (µm): One-millionth of a meter, or about 0.00003937" (3.937" × 10^{-5}). In the metric system, the micron is referred to as a *micrometer*.

Minimum efficiency reporting value (MERV): A rating of air filter particle-capture efficiency based on testing outlined by ASHRAE Standard 52.2.

Miscible: Having the necessary properties to mix with or dissolve into another material. Many refrigerants, for example, are miscible in refrigeration oil.

Multiple-chemical sensitivity (MCS): A condition found in some individuals who appear to be vulnerable to exposure to certain chemicals and/or combinations of chemicals. Currently, the existence of MCS as a physical condition remains debatable.

Near-azeotropes: Zeotropes with azeotropic properties. Refrigerant HFC-410A is an example of a near-azeotrope. Like zeotropes, near-azeotropes do exhibit glide characteristics, but they are slight and not a factor technicians need to consider during handling and charging.

New building syndrome: A physical condition that is connected to indoor air quality problems in new buildings. The symptoms are generally the same as those for sick building syndrome.

Newton-meter (Nm): A metric torque unit equal to the force of one newton applied to a lever one meter long.

Nominal size: A means of expressing the size of a bolt or screw using its approximate diameter.

Off-gassing: The process through which furnishings and building materials release chemicals and other volatile organic compounds into the air.

Oil slugging: The entrance of oil in circulation into a compressor cylinder during operation in an unmanageable volume, leading to the potential for serious mechanical damage if it enters the compressor cylinder.

Oscillation: The repetitive and/or periodic variation of some measure in relation to a central point.

Oxidation resistance: The ability to resist the degradation process that is caused by elements freely combining with oxygen. All petroleum products react with oxygen to some degree, and the process accelerates as temperature increases.

Ozone depletion potential (ODP): A measure of a substance's ability to deplete atmospheric ozone, expressed as a number. The baseline ODP of 1 is assigned to refrigerant CFC-11 for practical comparison.

Ozone: An unstable, highly reactive form of oxygen with a sharp, pungent odor like chlorine that forms naturally in the presence of electric discharges. Ozone is composed of three oxygen atoms (O_3) and is slightly soluble in water. Ozone represents a hazardous pollutant at ground level while serving in the upper atmosphere to protect us from harmful ultraviolet rays.

Paraphrasing: Expressing what you believe to be the meaning of a speaker's message in different words to ensure you understand the message as intended.

Passive recovery: Refrigerant recovery process that makes use of system pressure and/or its compressor to push the refrigerant out; no dedicated recovery unit is used. The refrigerant is captured in a non-pressurized container, such as an evacuated refrigerant cylinder.

Piezoelectric crystals: Crystalline materials, such as quartz, that generate an electrical potential across opposing faces when exposed to mechanical stress.

Pilot-duty device: An electrical component designed only to handle small current values such as those found in a control circuit.

Piston: In this context, the component of a fixed-orifice metering device that contains the orifice. The piston is free to move back and forth inside the housing. At one end of its travel, the piston meters refrigerant flow through the orifice; at the other end, it allows refrigerant to flow both through the orifice and around its outside perimeter.

Pontiac fever: A mild form of Legionnaires' disease.

Positive displacement compressors: Compressors that capture vapors in a chamber, then the size of the chamber is mechanically reduced to increase the pressure.

Potential relay: A relay energized by back EMF voltage generated during startup by the start winding in a capacitor-start motor.

Pour point: An oil quality related to viscosity, defined as the temperature at which oil first starts to flow. See *viscosity*.

Pressure (force-fed) lubrication system: A compressor lubrication system that uses a crankshaft-driven oil pump to draw oil from the crankcase and move it to the internal parts through the lubrication circuit.

Pump-down: Operating a compressor to clear most or all refrigerant from the low-pressure side of the system or a portion thereof. A pump-down cycle may be applied at the end of an operating cycle, or as a means of transferring refrigerant into the condenser for service work.

Qualified person: One who has the required skills and knowledge related to the construction and operation of electrical equipment and installations and has received relevant safety training to recognize and avoid the hazards involved.

Radial load: A side, or radial, force acting perpendicular to a shaft.

Radon: A colorless, odorless, radioactive, and chemically inert gas that is formed by the natural decomposition of uranium in soil and groundwater. Radon exposure over an extended period can increase the risk of cancer.

Rapport: A close and agreeable relationship between people that understand and accept each other's thoughts and connect well.

Recovery: The act of transferring refrigerant from a system or other vessel into a reusable recovery cylinder for future use, recycling, or reclamation.

Reverse osmosis: A process used to demineralize or deionize water by passing it through a special filtration material under pressure.

Reversing valve: A valve that changes the direction of refrigerant flow in a heat pump refrigerant circuit. Also referred to as a *four-way valve*.

Rotor: The rotating component of an induction motor.

Run winding: The stator winding of a motor that draws current during the entire running cycle of the motor.

Run-down resistance: The torque required to overcome the resistance of associated hardware, such as locknuts and lock washers, when tightening a fastener.

Saturation: The condition of a refrigerant when it exists in both the liquid and vapor states. Saturation is the point at which a refrigerant can be in transition from one state to the other or at a stable ratio at a fixed temperature and pressure. The ratio of liquid to vapor is not a consideration; as long as both liquid and vapor are present in some amount, the refrigerant is considered saturated.

Scintillation detectors: Radiation detectors containing a substance that emits light when it encounters a fast-moving nuclear particle (proton, neutron, or alpha). The sparkling effect caused by these emissions is called *scintillation*. The device includes features that amplify the light flashes and display the rate of particle detection.

Seasonal Energy Efficiency Ratio (SEER): The standard developed by the AHRI for measuring cooling efficiency, representing the total cooling production of a unit (in Btus) during the cooling season divided by the total energy consumed (in watt-hours) during the same period.

Seismic activity: Movement of the earth's crust caused by the release of volcanic or other geologic stress.

Seismic restraint: A device or cable arrangement used to control the movement of equipment when the release of geologic stress causes seismic events.

Set or seizure: An apparent locking effect felt during the last stages of rotation as you approach the final torque value for a fastener, often accompanied by a noticeable popping effect.

Sheave: A wheel with a groove in which a drive belt rides.

Short cycling: A condition in which a compressor is restarted very soon after it has shut down, before the refrigerant pressures have equalized. Short cycling may be accompanied by very short run times as well.

Sick building syndrome (SBS): A physical condition that exists when more than 20% of a building's occupants complain during a two-week period of common symptoms, including headaches, fatigue, nausea, eye irritation, and throat irritation, which are alleviated by leaving the building and are not confirmed to be caused by any specific contaminant.

Sine wave: A waveform that displays periodic oscillations in the generation of AC power.

Single-phasing: A condition in which a three-phase load receives power from only two of the three phases (phase loss).

Slugging: The entrance of excess liquid refrigerant and/or oil into a compressor cylinder during operation.

Splash lubrication system: A compressor lubrication system that relies on oil splashed up from the crankcase during each revolution of the crankshaft to provide lubrication to the internal parts.

Spore: A reproductive cell, typically single-celled, capable of developing into a new individual without fusion with another reproductive cell, often released by lower plants, fungi, and similar microorganisms.

Start winding: The stator winding of a motor that is used to provide starting torque.

Stator: The stationary portion of a motor, primarily consisting of the windings.

Stepper motor: A type of motor used to open and close an electronically controlled expansion valve. The motor rotates in small increments called *steps*, based on signals received from an external controller.

Synchronous belts: Belts with teeth that correspond to grooves in matching sheaves, eliminating slippage and maintaining synchronized rotation of the shafts; also referred to as *timing belts*.

Synchronous speed: The speed at which the magnetic field in a motor rotates, determined by the number of stator windings and the frequency of the applied voltage.

Tensile strength: The maximum stress a material can endure before being pulled apart.

Ternary blend: A blended refrigerant consisting of three components.

Thermostatic element: An assembly filled with refrigerant in saturation with an attached sensing bulb that responds to the temperature of the suction line. The pressure from the element acts as an opening force in a thermostatic expansion valve. It is also referred to as a *power element*.

Thread class: A classification related to the amount of tolerance the threads provide between a bolt and nut, designated by a number and letter combination. Higher numbers indicate a tighter fit.

Thread series: The combination of diameter and thread pitch, used to identify the coarseness or fineness of a thread.

Thread standards: An established set of standards for machining threads.

Thrust: A force acting along the axis of a shaft, in either direction.

Torque: A measure of force that causes an object, such as a motor shaft, to rotate.

Torquing: Applying a specific amount of twisting force to a threaded fastener.

Turns ratio: The ratio between the number of turns in a transformer's primary winding and a secondary winding.

Variable frequency drive (VFD): An approach to motor speed control that manipulates the voltage and frequency of power delivered to an AC motor to fully modulate its speed.

Variable refrigerant flow (VRF) systems: Heat pump systems that modulate the flow of refrigerant to match structure heating and cooling load demands simultaneously. VRF systems rely on variable speed compressor technology and can incorporate multiple outdoor units on a single refrigerant circuit.

Vent connector: A near-horizontal metal pipe section of the vent system that connects the appliance(s) to the vent riser or chimney.

Vent riser: The vertical section of a metal vent pipe.

Vibration isolators: Devices that help prevent or reduce the transmission of vibrations from machinery to the surrounding structure.

Viscosity index (VI): A measure of the rate at which an oil's viscosity changes with temperature.

Viscosity: The resistance of a fluid to flow, often related to its thickness. The viscosity of a fluid typically varies, depending on its temperature. For refrigeration oils, the chosen viscosity must allow the oil to flow and provide proper lubrication over the system's range of operating temperatures.

Volatile organic compounds (VOC): A group of compounds and chemicals found in such things as solvents, paints, and adhesives, which tend to vaporize at room temperature to become undesirable contaminants.

Voltage drop: A decrease of electrical potential along the path of a current flowing in an electrical circuit, often caused by insufficient wire sizing for circuit length and poor wiring connections.

Voltage imbalance: For a three-phase power supply, the maximum deviation of the voltage of any phase from the average voltage of the three phases, divided by the average voltage. It is generally expressed as a percentage.

Zeotropes: Blended refrigerants that are not compounds. As a result, one or more components vaporize and condense over a range of temperatures instead of at a single temperature/pressure combination.

REFERENCES

Module 1 03316 Customer Relations

Applied Communication Skills for the Construction Trades. Upper Saddle River, NJ: Prentice Hall.

HVAC Customer Service Handbook. Fourth Edition. Steve Coscia. Havertown, PA: Coscia Communications Inc.

Tools for Success: Critical Skills for the Construction Industry. Upper Saddle River, NJ: Prentice Hall.

Module 2 03215 Basic Maintenance

Center for Disease Control and Prevention (CDC). *www.cdc.gov*.

Encyclopedia of Lubricants and Lubrication. 2014. Berlin, Germany: Springer-Verlag Berlin Heidelberg.

Lubricants and Lubrication. 2017. Theo Mang and Wilfried Dresel. Hoboken, NJ: Wiley-Blackwell.

Refrigeration and Air Conditioning: An Introduction to HVAC/R. Current edition. Larry Jeffus. Upper Saddle River, NJ: Pearson Education.

Module 3 03313 Fasteners and Hardware

Fundamentals of HVACR. Latest Edition. Carter Stanfield and David Skaves. Upper Saddle River, NJ: Prentice Hall.

VISCMA 412, *Installing Seismic Restraints for Mechanical Equipment*. Latest Edition. Wayne, PA: Vibration Isolation and Seismic Control Manufacturers Association.

Module 4 03206 Alternating Current

Electric Power System Basics for the Nonelectrical Professional (IEEE Press Series on Power and Energy Systems). Steven W. Blume. 2nd Edition. New York, NY: Wiley-IEEE Press.

Fieldpiece Instruments, Inc. *https://www.fieldpiece.com/*.

measureQuick® Multibrand Wireless Tool Platform. *https://measurequick.com/*.

Ugly's Electric Motors and Controls. Latest Edition. Burlington, MA: Jones & Bartlett Learning.

Module 5 03302 Compressors

Service Guide for 06D, 06E, and 06CC Semi-Hermetic Reciprocating Compressors, 2023. Syracuse, NY: Carlyle Compressor Company.

Copeland (formerly Emerson Climate Technologies), www.copeland.com

Module 6 03303 Metering Devices

TEV & AEV Theory and Application. Catalog E-1A. Sporlan Division of Parker Hannifin Corporation.

Thermostatic Expansion Valves—Theory of Operation, Application, and Selection. Bulletin 10–9. Sporlan Division of Parker Hannifin Corporation.

Thermostatic Expansion Valves with Selective Thermostatic Charges. Bulletin 10–10. Sporlan Division of Parker Hannifin Corporation.

Thermostatic Expansion Valves—Installation, Field Service, and Assembly. Bulletin 10–11. Sporlan Division of Parker Hannifin Corporation.

Electric Valves for Refrigerant Control. Bulletin 100–9. Sporlan Division of Parker Hannifin Corporation.

Parker Hannifin Corporation, www.parker.com/

Module 7 03301 Refrigerants and Oils

American Society of Heating, Refrigeration, and Air Conditioning Engineers (ASHRAE), *https://www.ashrae.org/*

Clean Air Act—Section 608 Technician Certification (US Environmental Protection Agency), *https://www.epa.gov/section608/section-608-technician-certification-0*

Freon™ Refrigerants. Chemours, *https://www.freon.com/en*

Honeywell Refrigerants. Honeywell International, Inc., *https://sustainability.honeywell.com/us/en/products/refrigerants*

Significant New Alternatives Policy (SNAP) Program (US Environmental Protection Agency), *https://www.epa.gov/snap*

Stationary Refrigeration and Air Conditioning Resources—Section 608 of the Clean Air Act (US Environmental Protection Agency), *https://www.epa.gov/section608*

The Air Conditioning, Heating, and Refrigeration Institute (AHRI), *https://www.ahrinet.org/*

Module 8 03205 Leak Detection, Evacuation, Recovery, and Charging

US Environmental Protection Agency, *https://www.epa.gov/section608*

40 *CFR* Part 82, Subpart F, *www.ecfr.gov*

NCCER Module 03301, *Refrigerants and Oils*.

Module 9 03211 Heat Pumps

Building America Solution Center. Access expert information on hundreds of high-performance construction topics, including air sealing and insulation, HVAC components, windows, indoor air quality, and much more. A service of the US Department of Energy, Office of Energy Efficiency and Renewable Energy. *https://basc.pnnl.gov/*.

Building Science Education Solution Center. Relevant training modules from the US Department of Energy, Office of Energy Efficiency and Renewable Energy. *https://bsesc.energy.gov/training-modules*.

Heat Pumps—Fundamentals and Applications. Walter Grassi. 2018. New York, NY: Springer Publishing.

Regional Energy Efficiency Organizations. Independent non-profit organizations partially funded by utilities and the US Department of Energy, serving almost every state with a mix of policy and programs to help advance energy efficiency. *https://eepartnership.org/about/regional-energy-efficiency-organizations-network/*.

Solar Assisted Ground Source Heat Pump Solutions. Francesco Reda. 2017. New York, NY: Springer Publishing.

US Department of Energy, Office of Energy Efficiency and Renewable Energy. https://www.energy.gov/eere/office-energy-efficiency-renewable-energy.

Module 10 03403 Indoor Air Quality

ACR—The NADCA Standard for Assessment, Cleaning, and Restoration of HVAC Systems. Latest edition. Washington, DC: National Air Duct Cleaners Association.

ASHRAE Standard 62.1, *Ventilation for Acceptable Indoor Air Quality*. Latest edition. Atlanta, GA: American Society of Heating, Refrigerating and Air-Conditioning Engineers (ASHRAE).

Building Air Quality: A Guide for Building Owners and Facility Managers. Latest edition. Washington, DC: US Environmental Protection Agency.

Cleaning Fibrous Glass Insulated Air Duct Systems—Recommended Practices. Latest edition. Alexandria, VA: North American Insulation Manufacturers Association (NAIMA).

Indoor Air Quality: The Latest Sampling and Analytical Methods. Kathleen Hess-Kosa. 2020. Boca Raton, FL: CRC Press.

Module 11 03204 Air Quality Equipment

Handbook of Indoor Air Quality. Yinping Zhang, Philip K. Hopke, and Corinne Mandin. Singapore: Springer Nature Singapore Pte Ltd.

Managing Indoor Air Quality. H.E. Burroughs and Shirley J. Hansen. 5th Edition. Gistrup, Denmark: River Publishers.

Module 12 03202 Chimneys, Vents, and Flues

National Fuel Gas Code (NFPA 54/ANSI Z223.1), Latest Edition. Quincy, MA: National Fire Protection Association.